Nutrient Use Efficiency: from Basics to Advances

Amitava Rakshit • Harikesh Bahadur Singh •
Avijit Sen
Editors

Nutrient Use Efficiency: from Basics to Advances

Springer

Editors
Amitava Rakshit
Soil Science & Agricultural Chemistry
Institute of Agricultural Sciences
Varanasi,
Uttar Pradesh
India

Harikesh Bahadur Singh
Mycology and Plant Pathology
Institute of Agricultural Sciences
Varanasi
Uttar Pradesh
India

Avijit Sen
Department of Agronomy
Institute of Agricultural Sciences
Varanasi
Uttar Pradesh
India

ISBN 978-81-322-2168-5 ISBN 978-81-322-2169-2 (eBook)
DOI 10.1007/978-81-322-2169-2
Springer New Delhi Heidelberg New York Dordrecht London

Library of Congress Control Number: 2014959273

Printed on acid-free paper

Springer (India) Pvt. Ltd. is part of Springer Science+Business Media (www.springer.com)

Dedicated to our wives

प्रोफेसर पंजाब सिंह
Prof. Panjab Singh
FNAAS, FIAS, FNIE, FBRSI, FSEE
(EX-SECRETARY, DARE, GOI & DG, ICAR)

Foreword

I am very happy that this comprehensive edited book on "*Nutrient Use Efficiency: From Basics to Advances*" has been prepared by Dr. Amitava Rakshit along with his colleagues, Prof. H.B. Singh and Prof. A. Sen. Optimum use of nutrients by crops is essential for sustainable agricultural production. With increasing world demands for food and energy, this is set to become a priority agenda in times to come. Fertilisers are costly to produce and need scientific application matching on environmental needs. There is an absolute requirement to maximise efficiency using both agronomic and plant breeding approaches. Nutrient use efficiency is one mode through which this issue can be addressed holistically. Improving yield alone may be at the expense of energy intensive nutrient content and hence quality, therefore yield improvements must be matched by appropriate increases in nutrient dynamics in plant system. Improving nutrient efficiency is a worthy goal and fundamental challenge facing the input industry, and agriculture in general. The opportunities are there and tools are available to accomplish the task of improving the efficiency of applied nutrients. Multiple complex processes contribute to the overall nutrient use efficiency as well as multidisciplinary approaches involving agronomy, soil science; traditional breeding and biotechnology will contribute to future improvements in this direction.

This book contains chapters with applied aspects of nutrient use efficiency authored by leading experts in the field. This book is meant for academics, researcher and extension workers who would like to learn more about the principles and practices of integrated resource management. The book gives good discussions of topics on current interest in increasing efficiency of nutrient use by plants in greater details. I congratulate the editors for getting such a valuable book published.

(Panjab Singh)

Dated the 18th June, 2014 Varanasi

डा. एस. अय्यप्पन
सचिव एवं महानिदेशक
Dr. S. AYYAPPAN
SECRETARY & DIRECTOR GENERAL

भारत सरकार
कृषि अनुसंधान और शिक्षा विभाग एवं
भारतीय कृषि अनुसंधान परिषद
कृषि मंत्रालय, कृषि भवन, नई दिल्ली 110 001

GOVERNMENT OF INDIA
DEPARTMENT OF AGRICULTURAL RESEARCH & EDUCATION
AND
INDIAN COUNCIL OF AGRICULTURAL RESEARCH
MINISTRY OF AGRICULTURE, KRISHI BHAVAN, NEW DELHI 110 001
Tel.: 23382629; 23386711 Fax: 91-11-23384773
E-mail: dg.icar@nic.in

FOREWORD

Global population growth is increasing demand for food, fiber, fuel and energy, with limited land and water resources available for crop production and higher costs of inorganic fertilizer, environmental concerns are relevant for the development of more nutrient-efficient crops, which are likely to play a major role in maintaining or increasing crop yields for sustainable agriculture production in the future, especially after changing climate scenario. We have to accept the need for increased efficiency, sustainability and resilience of our agricultural production, giving due importance on the responsibility of feeding the world, justifying food versus fuel debate without impairing natural resource base biodiversity and climate change. Fertilizer nutrients have helped spare millions of acres of land while sustaining crop production increases to meet the demand, and its management holds the key for quantitative and qualitative improvement of various crops. Escalating fertilizer prices, environmental concerns and stagnant crop prices have growers looking for ways to increase the nutrient use efficiency of their crops.

Addition of fertilizers and/or amendments are essential for a proper nutrient supply and maximum yields. Estimates of overall efficiency of applied fertilizer have been reported to be about or lower than 50% for N, less than 10% for P, about 40% for K and about 2-5% for micronutrients. Plants that are efficient in absorption and utilization of nutrients greatly enhance the efficiency of applied fertilizers, reducing cost of inputs, and preventing losses of nutrients to ecosystems. Improvement of nutrient use efficiency (NUE) is an essential pre-requisite for expansion of crop production into marginal lands with low nutrient availability. Efficiency improvement is key, therefore, for cropping systems as a whole, and for nutrient use within them and are known to be under genetic and physiological control and are modified by plant interactions with environmental modulations. Effective nutrient use coupled with best management practices and vibrant breeding programs involving traits like nutrient absorption, transport, utilization, and mobilization on developing cultivars with high NUE protect, improve and promote soil, water and air quality.

I am very happy that this comprehensive edited book on "*Nutrient Use Efficiency: From Basics to Advances*" has been prepared by Dr. Amitava Rakshit along with his colleagues, Prof. H.B. Singh and Prof. A. Sen. This book contains chapters written by leading experts in the field. It provides a comprehensive, inter-disciplinary description of the problems of nutrient use efficiency in the overall context of biogeochemical cycle of nutrients, its environmental and health implications, as well as various approaches to nutrient use efficiency. The topics vary from physiology and adaptive mechanisms, molecular biology, and applied aspects in agronomy and soil science. This book will be useful for all interested in strategies for rational and effective nutrient use under peer biotic and abiotic stresses. I congratulate the editors for this meaningful exercise.

(S. Ayyappan)

Dated the 19th June, 2014
New Delhi

Acknowledgement

An edited book of this expanse does not become possible without the contribution of several willing souls. Many who are engaged in resource use efficiency in agriculture all over the world have contributed to this book. It is impossible to refer to all the correspondents who have furnished informations and have so freely reported their findings. We are indebted to a number of Societies for permission to reproduce information and illustrations which have already been published. We have been lucky that several colleagues and students helped us in our endeavor to bring this book to light. Special mention must be made of Prof. Norbert Claassen and Dr. V. C. Baligar who saw value in this effort at the outset!

Finally, the production team members Surabhi, Raman, Kiruthika and Pandian deserve special appreciation for guiding us through the process of publishing a new work. Last but not least, we would thank our family, immediate and extended, for their unconditional support, unflagging love in putting everything together and inspiration.

Amitava Rakshit
Avijit Sen
Harikesh Bahadur Singh

I can no other answer make, but, thanks, and thanks (Shakespeare (Twelth N, Act iii, Sc.3))

Contents

About the Authors

Amitava Rakshit, an IIT Kharagpur alumnus, is presently the faculty member in the Department of Soil Science and Agricultural Chemistry at the Institute of Agricultural Sciences, Banaras Hindu University (IAS, BHU). Dr. Rakshit worked in the Department of Agriculture, Government of West Bengal, in administration and extension roles. He has visited Norway, Finland, Denmark, France, Austria, Russia, Thailand, Egypt and Turkey pertaining to his research work and presentation. He was awarded with TWAS Nxt Fellow (Italy), Biovision Nxt Fellow (France), Darwin Now Bursary (British Council), Young Achiever Award and Best Teacher's Award at UG and PG levels (BHU & ICAR, New Delhi). He is serving as review college member of British Ecological Society, London, since 2011. He has published 70 research papers, 15 book chapters, 28 popular articles and one manual, and co-authored five books.

Avijit Sen is presently Professor and Head in the Department of Agronomy, IAS, BHU. He has presented more than 40 research papers and delivered lectures in conferences/seminars/refresher courses within and outside the country besides handling five ad hoc projects. Currently he is the Deputy Coordinator of the prestigious Special Assistance Programme awarded to the Department by UGC. He is recipient of CWSS Gold Medal 2013. With 33 years of experience, he has published more than 60 research papers and articles, and written 3 chapters of books till date.

Harikesh Bahadur Singh is presently Professor of Mycology and Plant Pathology at IAS, BHU. He served the State Agriculture University, Central University and CSIR Institute in teaching, research and extension roles. In recognition of Prof. Singh's scientific contributions and leadership in the field of plant pathology, he was honored with several prestigious awards, notable being CSIR Prize for Biological Sciences, Vigyan Bharti Award, Prof. V.P. Bhide Memorial Award, BRSI Industrial Medal Award, Bioved Fellowship Award, Prof. Panchanan Maheshwari, IPS Plant Pathology Leader Award, CSIR–CAIRD Team award, Environment Conservation Award, CST Vigyan Ratna Award and many more. Dr. Singh has been the Fellow of National Academy of Agricultural Sciences. Professor Singh has written two books, several training modules and manuals and more than 150 research publications and has more than 18 US patents and 3 PCTs to his credit.

Contributors

Tapan Adhikari Indian Institute of Soil Science, Bhopal, MP, India

Richa Agnihotri Directorate of Soybean Research, ICAR, DARE, Ministry of Agriculture, Indore, MP, India

V.C. Baligar USDA-ARS Beltsville Agricultural Research Center, Beltsville, MD, USA

Kartikay Bisen Department of Mycology and Plant Pathology, Institute of Agricultural Sciences, Banaras Hindu University, Varanasi, Uttar Pradesh, India

M.C.S. Carvalho National Rice and Bean Research Center of EMBRAPA, Santo Antônio de Goiás Brazil

Girish Chander Resilient Dryland Systems, International Crops Research Institute for the Semi-Arid Tropics (ICRISAT), Patancheru, Andhra Pradesh, India

Anchal Dass Division of Agronomy, Indian Agricultural Research Institute, New Delhi India

Avishek Datta Agricultural Systems and Engineering Program, School of Environment, Resources and Development, Asian Institute of Technology, Pathumthani Thailand

N.K. Fageria National Rice and Bean Research Center of EMBRAPA, Santo Antônio de Goiás, GO, Brazil

Zannatul Ferdous Agricultural Systems and Engineering Program, School of Environment, Resources and Development, Asian Institute of Technology, Pathumthani Thailand

B.N. Ghosh Central Soil and Water Conservation Research and Training Institute, Dehradun, Uttrakhand, India

D.C. Ghosh Institute of Agriculture, Birbhum, West Bengal, India

H.S. Gupta Indian Agricultural Research Institute, New Delhi India

A.B. Heinemann National Rice and Bean Research Center of EMBRAPA, Santo Antônio de Goiás Brazil

F. Hossain Indian Agricultural Research Institute, New Delhi India

Shankar Lal Jat DMR, IARI Campus, New Delhi India

B.K. Kandpal Directorate of Rapeseed-Mustard Research, Sewar, Bharatpur, Rajasthan, India

Nanjappan Karthikeyan National Bureau of Agriculturally Important Microorganisms (NBAIM), Mau, Uttar Pradesh, India

Chetan Keswani Department of Mycology and Plant Pathology, Institute of Agricultural Sciences, Banaras Hindu University, Varanasi, Uttar Pradesh, India

Kalyani S. Kulkarni Department of Agricultural Biotechnology, Anand Agricultural University, Anand, Gujrat, India

Manish Kumar National Bureau of Agriculturally Important Microorganisms, Kushmaur, Maunath Bhanjan, Uttar Pradesh, India

Manoj Kumar Central Potato Research Station, Patna, Bihar, India

Rajesh Kumar Division of Agricultural Chemicals, Indian Agricultural Research Institute, New Delhi India

Sunil Kumar Indian Grassland and Fodder Research Institute, Jhansi, UP, India

Sushil Kumar Department of Agricultural Biotechnology, Anand Agricultural University, Anand, Gujarat, India

M.G. Mallikarjuna Indian Agricultural Research Institute, New Delhi India

P.K. Mandal National Research Centre on Plant Biotechnology, New Delhi India

Sayaji T. Mehetre Nuclear Agriculture and Biotechnology Division, Bhabha Atomic Research Centre, Mumbai, Maharashtra, India

J.S. Mishra Department of Agronomy, Directorate of Sorghum Research, Hyderabad India

P.K. Mishra Central Soil and Water Conservation Research and Training Institute, Dehradun, Uttrakhand, India

Sandhya Mishra Department of Mycology and Plant Pathology, Institute of Agricultural Sciences, Banaras Hindu University, Varanasi, Uttar Pradesh, India

Prasun K. Mukherjee Nuclear Agriculture and Biotechnology Division, Bhabha Atomic Research Centre, Mumbai, Maharashtra, India

Ravi Naidu CRC CARE-Cooperative Research Centre for Contamination Assessment and Remediation of the Environment, Salisbury, SA, Australia

T. Nepolean Division of Genetics, Indian Agricultural Research Institute, New Delhi India

S.P. Pachauri Department of Soil Science, G.B. Pant University of Agriculture & Technology, Pantnagar, Uttarakhand, India

Sumita Pal Department of Mycology and Plant Pathology, Institute of Agricultural Science, BHU, Varanasi, UP, India

D.R. Palsaniya Indian Grassland and Fodder Research Institute, Jhansi, UP, India

Balraj S. Parmar Division of Agricultural Chemicals, Indian Agricultural Research Institute, New Delhi India

J.V. Patil Department of Agronomy, Directorate of Sorghum Research, Hyderabad India

Ratna Prabha National Bureau of Agriculturally Important Microorganisms, Kushmaur, Maunath Bhanjan, Uttar Pradesh, India

Om Prakash Microbial Culture Collection, National Centre for Cell Science, Pune, Maharashtra, India

O.P. Premi Directorate of Rapeseed-Mustard Research, Sewar, Bharatpur, Rajasthan, India

Praveen Rahi Microbial Culture Collection, National Centre for Cell Science, Pune, Maharashtra, India

A.K. Rai Indian Grassland and Fodder Research Institute, Jhansi, UP, India

Amitava Rakshit Department of Soil Science and Agricultural Chemistry, Institute of Agricultural Science, BHU, Varanasi, UP, India

K.S. Rana Division of Agronomy, Indian Agricultural Research Institute, New Delhi India

S.S. Rathore Directorate of Rapeseed-Mustard Research, Sewar, Bharatpur, Rajasthan, India

Deepa Rawat Department of Soil Science, G.B. Pant University of Agriculture & Technology, Pantnagar, Uttarakhand, India

Vangimalla R. Reddy Crop Systems and Global Change Laboratory, USDA-ARS, Beltsville, MD, USA

Supradip Saha Division of Agricultural Chemicals, Indian Agricultural Research Institute, New Delhi India

Binoy Sarkar CERAR-Centre for Environmental Risk Assessment and Remediation, Building X, University of South Australia, Mawson Lakes, SA, Australia

Amrita Saxena Department of Mycology and Plant Pathology, Institute of Agricultural Sciences, Banaras Hindu University, Varanasi, Uttar Pradesh, India

Arun K. Sharma National Bureau of Agriculturally Important Microorganisms, Kushmaur, Maunath Bhanjan, Uttar Pradesh, India

Mahaveer P. Sharma Directorate of Soybean Research, ICAR, DARE, Ministry of Agriculture, Indore, MP, India

Ramavtar Sharma Central Arid Zone Research Institute, Jodhpur India

Rohit Sharma Microbial Culture Collection, National Centre for Cell Science, Pune, Maharashtra, India

Kapila Shekhawat Directorate of Rapeseed-Mustard Research, Sewar, Bharatpur, Rajasthan, India

Sangam Shrestha Water Engineering and Management Program, School of Engineering and Technology, Asian Institute of Technology, Pathumthani Thailand

Manoj Shrivastava Centre for Environment Science and Climate Resilient Agriculture, Indian Agricultural Research Institute, New Delhi India

Akanksha Singh Department of Mycology and Plant Pathology, Institute of Agricultural Science, BHU, Varanasi, UP, India

D.P. Singh National Bureau of Agriculturally Important Microorganisms, Kushmaur, Maunath Bhanjan, Uttar Pradesh, India

Govind Singh Plant Biotechnology Centre, S.K. Rajasthan Agricultural University, Bikaner, Rajasthan, India

Harikesh Bahadur Singh Department of Mycology and Plant Pathology, Institute of Agricultural Science, Banaras Hindu University, Varanasi, Uttar Pradesh, India

Raman Jeet Singh Central Soil and Water Conservation Research and Training Institute, Dehradun, Uttrakhand, India

Shardendu K. Singh Crop Systems and Global Change Laboratory, USDA-ARS, Beltsville, MD, USA

Wye Research and Education Center, University of Maryland, Queenstown, MD, USA

S.K. Sinha National Research Centre on Plant Biotechnology, New Delhi India

R. Srinivasan National Research Centre on Plant Biotechnology, New Delhi India

P.C. Srivastava Department of Soil Science, G.B. Pant University of Agriculture & Technology, Pantnagar, Uttarakhand, India

Kumai Sunita Department of Soil Science and Agricultural Chemistry, Institute of Agricultural Science, BHU, Varanasi, UP, India

S.P. Trehan Central Potato Research Station, Jalandhar, Punjab, India

Rajneet K. Uppal Resilient Dryland Systems, International Crops Research Institute for the Semi-Arid Tropics (ICRISAT), Patancheru, Andhra Pradesh, India

Céline Vaneeckhaute Laboratory of Analytical and Applied Ecochemistry, Faculty of Bioscience Engineering, Ghent University, Ghent Belgium

M. Vignesh Indian Agricultural Research Institute, New Delhi India

Suresh Walia Division of Agricultural Chemicals, Indian Agricultural Research Institute, New Delhi India

Suhas P. Wani Resilient Dryland Systems, International Crops Research Institute for the Semi-Arid Tropics (ICRISAT), Patancheru, Andhra Pradesh, India

Cho Cho Win Agricultural Systems and Engineering Program, School of Environment, Resources and Development, Asian Institute of Technology, Pathumthani Thailand

Nutrient Use Efficiency in Plants: An Overview

V.C. Baligar and N.K. Fageria

Abstract

In modern agriculture use of essential plant nutrients in crop production is very important to increase productivity and maintain sustainability of the cropping system. Use of nutrients in crop production is influenced by climatic, soil, plant, and social-economical condition of the farmers. Overall, nutrient use efficiency by crop plants is lower than 50 % under all agro-ecological conditions. Hence, large part of the applied nutrients is lost in the soil-plant system. The lower nutrient use efficiency is related to loss and/or unavailability due to many environmental factors. The low nutrient use efficiency is not only increase cost of crop production but also responsible for environmental pollution. Nutrient use efficiency in the literature is defined in several ways. The most common nutrient use efficiency is designated as nutrient efficiency ratio, agronomic efficiency, physiological efficiency, agrophysiological efficiency, apparent recovery efficiency, and utilization efficiency. Definition and methods of calculation of these deficiencies are presented. Improving nutrient use efficiency is essential from economic and environmental point of view. The most important strategies to improve nutrient use efficiency are the use of adequate rate, effective source, timing, and methods of application. In addition, decreasing abiotic and biotic stresses and use of nutrient efficient crop species and genotypes within species are also important in increasing nutrient use efficiency.

Keywords

Nutrient use efficiency • Physiology • Fertilizers • Abiotic stress agronomy • Management • Soils

V.C. Baligar (✉)
USDA-ARS Beltsville Agricultural Research Center, Beltsville, MD 20705-2350, USA
e-mail: VC.Baligar@ars.usda.gov

N.K. Fageria
National Rice and Bean Research Center of EMBRAPA, Caixa Postal 179, Santo Antonio de Goias, GO CEP 75375000, Brazil
e-mail: nand.fageria@embrapa.br

1 Introduction

World agriculture is faced with serious challenge of feeding adequate and healthy food for over 7 billion (B) people and by 2050 world population expected to reach over 9 B (FAO 2013).

A. Rakshit et al. (eds.), *Nutrient Use Efficiency: from Basics to Advances*,
DOI 10.1007/978-81-322-2169-2_1, © Springer India 2015

Table 1 Potential element deficiencies and toxicities associated with major soil orders[a]

Soil order/US taxonomy	Soil group, FAO	Element Deficiency	Toxicity
Alfisols/Ultisols (Albic) (poorly drained)	Planosol	Most nutrients	Al
Alfisols/Aridisols/Mollisols (Natric) (high alkali)	Solonetz	K, N, P, Zn, Cu, Mn, Fe	Na
Andisols (Andepts)	Andosol	P, Ca, Mg, B, Mo	Al
Aridisols	Xerosol	Mg, K, P, Fe, Zn	Na
Aridisols/Arid Entisols	Yermosol	Mg, K, P, Fe, Zn, Co, I	Na, Se
Aridisols (high salt)	Solonchak		B, Na, Cl
Entisols (Psamments)	Arenosol	K, Zn, Fe, Cu, Mn	
Entisols (Fluvents)	Fluvisol		Al, Mn, Fe
Histosols	Histosol	Si, Cu	
Mollisols (Aqu), Inceptisols, Entisols, etc. (poorly drained)	Gleysol	Mn	Fe, Mo
Mollisols (Borolls)	Chernozem	Zn, Mn, Fe	
Mollisols (Ustolls)	Kastanozem	K, P, Mn, Cu, Zn	Na
Mollisols (Aridis) (Udolls)	Phaeozem		Mo
Mollisols (Rendolls) (shallow)	Rendzina	P, Zn, Fe, Mn	
Oxisols	Ferralsol	P, Ca, Mg, Mo	Al, Mn, Fe
Spodosols (Podsols)	Podzol	N, P, K, Ca, and micronutrients	Al
Ultisols	Acrisol	N, P, Ca, and most other	Al, Mn, Fe
Ultisols/Alfisols	Nitosol	P	Mn
Vertisols	Vertisol	N, P, Fe	S

[a]Modified from Baligar et al. (2001), Clark (1982), Dudal (1976) and personal communications, Buol SW North Carolina State University, Raleigh, NC and Eswaran H USDA, NRCS, Washington, DC

Such an increase in population growth will intensify the pressure on world's resource base (land, water, air) to achieve higher food production. Expanding the land under crops and by increasing yield per unit area could help in increasing food production. According to estimate of FAO (2013), about 1.54 billion ha of worlds land is arable and is under permanent crops. Most of the land that could be brought under cropping has been utilized with exception of some land areas of sub-Saharan Africa and South America which are very fragile to degradation to bring under any forms of cultivation.

Soil degradation due to inappropriate management and intensive cultivations and increased abiotic and biotic stresses have posed serious challenges to achieve reasonable good yields of annual and perennial crops worldwide. Adequate water and soil nutrients (fertilizers) along with superior genetic materials (cultivars, genotypes) are vital to achieve higher yields and high-quality food materials. Many of the world soils are deficient in many of the essential nutrients and contain often toxic elements to achieve higher crop yields (Table 1) (Dudal 1976; Clark 1982; Baligar et al. 2001). Salinity, alkalinity, acidity, anthropogenic processes, nature of farming, and erosion can lead to soil degradation and lowering of soil fertility. In the world, close to four billion ha of the ice-free land area is considered having soil acidity and about 950 million ha of land area is salt affected and to bring some of these large areas of land under cultivation require enormous costly inputs such as irrigations, soil amendments, and fertilizers. On degraded and infertile soils addition of fertilizers and/or amendments are essential for proper nutrient supply and to achieve higher yields.

Currently, world uses about 105 million tons of N, 20 million tons of P, and 23 million tons of K for crop production (FAO 2013). Recovery of applied fertilizer efficiency is low, and overall efficiency of applied fertilizer have been about or lower than 50 % for N, less than 10 % for P, and about 40 % for K (Baligar and Bennett 1986a, b). Lower efficiency of applied fertilizer is attributed to leaching and run-off, gaseous losses, fixation by soil, and use of inefficient nutrient absorbing/utilizing plant species or cultivars. Nutrient losses can potentially contribute to degradation of soil and water and degradation of environment.

In this overview, the plant, soil, fertilizer, agronomic, abiotic, and biotic factors how they affect nutrient use efficiency (NUE) and remedial measures adaptable to improve NUE in plant is being presented. No attempt has been made to cover the extensive literature published in these areas and readers can refer or consult to some of the excellent publications (Alam 1999; Baligar and Duncan 1990; Baligar and Fageria 1997; Baligar et al. 2001; Blair 1993; Fageria and Baligar 2005; Fageria et al. 2002, 2008; Gerloff and Gabelman 1983; Marshner 1995; Mengel and Kirkby 2001; Vose 1987). Fageria et al. (2006) have covered extensively on the physiological basis of macro-micro nutrient nutrition in crop plants. In this review we only present the overview of the issue.

2   Estimation of NUE in Plants

The NUE in plants is profoundly influenced by genetic and physiological components and their influence on plants ability to absorb and utilize nutrients under various environmental and ecological conditions. Determination of NUE is useful to differentiate plant species genotypes and cultivars for their ability to absorb (uptake) and utilize (assimilation) nutrients for maximum production of dry mater/yields. The NUE is based on (a) uptake efficiency (acquire from soil, influx rate into roots, influx kinetics, radial transport in roots are based on root parameters per weight or length and uptake is also related to the amounts

of the particular nutrient applied or present in soil), (b) incorporation efficiency (transports to shoot and leaves are based on shoot parameters), and (c) utilization efficiency (based on remobilization, whole plant, i.e., root and shoot parameters). Generally, NUE in plant can be defined as the maximum economic yield or dry matter produced per unit of any nutrient that is applied or unit of that particular nutrient taken up. Graham (1984) defined nutrient use efficiency of a genotype (for each element separately) as the ability to produce a high yield in a soil that is limited in that element for a standard genotype. Clark (1990) defined nutrient efficient crop species or genotypes within species as those that produce more dry matter or have a greater increase in harvested portion per unit time, area, or applied nutrient, have fewer deficiency symptoms, or have greater incremental increases and higher concentrations of mineral nutrients than other plants grown under similar conditions or compared to a slandered genotype. Blair (1993) defined nutrient efficiency as the ability of a genotype/cultivar to acquire nutrients from growth medium and/or to incorporate or utilize them in the production of shoot and root biomass or utilizable plant material (seed, grain, forage). Gourley et al. (1994) defined nutrient efficient plants as germplasm that requires fewer nutrients than an insufficient one for normal metabolic process. Fageria et al. (2008) defined efficient plant as that produces higher economic yield with a determined quantity of applied or absorbed nutrient compared to other or a standard plant under similar growing conditions.

Commonly used NUE definitions are given below, and for extensive coverage of this area, readers are referred to Baligar and Duncan (1990); Baligar et al. (2001); Blair (1993); Fageria et al. (2008); and Gerloff and Gabelman (1983).

Nutrient Efficiency Ratio (NER) The NER was suggested by Gerloff and Gabelman (1983) to differentiate genotypes into efficient and inefficient nutrient utilizers. It can be defined as the yield in kg per unit of nutrient in kg. The equation for calculating NER is:

$$\text{NER } \left(\text{kg kg}^{-1}\right) = \frac{\text{Yield in kg}}{\text{Nutrient in plant tissue in kg}}$$

Agronomic Efficiency (AE) Is defined as the economic production obtained per unit of nutrient applied. It can be calculated by using the following equation:

$$\text{AE } \left(\text{kg kg}^{-1}\right) = \frac{\text{Yield of fertilized plot in kg} - \text{Yield of unfertilized plot in kg}}{\text{Quantity of nutrient applied in kg}}$$

Physiological Efficiency (PE) Is defined as the biological yield obtained per unit of nutrient uptake. It can be calculated by using the following equation:

$$\text{PE } \left(\text{kg kg}^{-1}\right) = \frac{\text{BY}_f \text{ in kg} - \text{BY}_{uf} \text{ in kg}}{\text{NU}_f \text{ in kg} - \text{NU}_{uf} \text{ in kg}}$$

where BY_f = biological yield (grain plus straw) of fertilized plot, BY_{uf} = biological yield of unfertilized plot, NU_f = nutrient uptake in grain and straw of fertilized plot, and NU_{uf} = nutrient uptake in grain and straw of unfertilized plot.

Agrophysiological Efficiency (APE) Is defined as the economic production obtained per unit of nutrient uptake. It can be calculated by using the following equation:

$$\text{APE } \left(\text{kg kg}^{-1}\right) = \frac{\text{GY}_f \text{ in kg} - \text{GY}_{uf} \text{ in kg}}{\text{NU}_f \text{ in kg} - \text{NU}_{uf} \text{ in kg}}$$

where GY_f = grain yield of fertilized plot, GY_{uf} = grain yield of unfertilized plot, NU_f = nutrient uptake in grain plus straw of fertilized plot, and NU_{uf} = nutrient uptake in grain plus straw of unfertilized plot.

Apparent Recovery Efficiency (ARE) Is defined as the quantity of nutrient uptake per unit of nutrient applied. It can be calculated by using the following equation:

$$\text{ARE } (\%) = \frac{\text{NU}_f - \text{NU}_{uf}}{\text{Quantity of nutrient applied}} \times 100$$

where NU_f = nutrient uptake in grain plus straw of fertilized plot and NU_{uf} = nutrient uptake in grain plus straw of unfertilized plot.

Utilization Efficiency (EU) Is product of physiological efficiency and apparent recovery efficiency.

$$\text{UE } \left(\text{kg kg}^{-1}\right) = \text{PE} \times \text{ARE}$$

where PE = physiological efficiency and ARE is apparent recovery efficiency as defined above.

3 Factors That Influence the NUE and Ways to Manipulate Them to Improve NUE

The efficiency of nutrient acquisition, transport, and utilization by plants grown in soil is controlled by (a) the capacity of the soil to supply the nutrients and (b) the ability of the plants to absorb, utilize, and remobilize the nutrients and this is referred to as NUE in plants. The NUE is partitioned into uptake efficiency (nutrient uptake/capture by roots, transport through roots, and shoot) and utilization efficiency (nutrient conversion to dry matter and grain).

3.1 Plant Factors

Selection of improved genotypes/cultivars adaptable to wide range of soils and climatic changes has been a major contributor to the overall gain in crop productivity. Crop genetic improvement through breeding and with improved management practices are the major contributors for achieving higher yields of crops during second half of the twentieth century. However average yields of many important crops at farm level are still two to four times lower than the recorded maximum yield potentials (Baligar et al. 2001).

Table 2 Total population, NPK use, total arable, and cropped land and fertilizer use/ha in 2010[a]

Region	Population (10^6)	N-P-K use (10^6 Mg)	Arable and cropped land (10^6 ha)	Fertilizer use (kg/ha)
Africa	1,022	3.83	256.4	14.93
Eastern	324	0.80	69.4	11.51
Middle	127	0.03	27.3	1.04
Northern	209	2.01	47.4	42.31
Southern	58	0.55	14.5	38.02
Western	304	0.43	97.8	4.39
America	935	33.03	395.4	83.54
Northern	344	19.24	210.8	91.26
Central	156	1.95	36.2	53.93
Southern	392	11.60	141.2	82.13
Caribbean	42	0.24	6.9	34.07
Asia	4,164	90.54	553.4	163.60
Central	61	0.91	32.6	27.97
Eastern	1,574	48.19	135.4	335.88
Southern	1,704	28.81	231.5	124.47
Western	232	2.44	43.7	55.79
South Eastern	593	10.19	110.2	92.50
Europe	738	18.55	290.7	63.83
Northern	99	2.97	19.3	153.64
Southern	155	3.06	39.6	77.31
Eastern	295	7.20	196.4	36.67
Western	189	5.31	35.3	150.35
Oceania	37	2.29	45.2	50.67
World	6,896	148.25	1,541.1	96.20

[a]Source: FAO FAOSTAT, 2013. Available at http://faostat.fao.org/Rome

Borlaug and Doswell (1994) stated that soil fertility is the single most important factor that limits crop yields and as much as 50 % of the increase in crop yields worldwide during the twentieth century is due to the use of chemical fertilizers. High or low crop yield in different parts of the world could be correlated to level of fertilizer use per unit of land (Table 2).

Genetic variability within plant species and cultivar within species is responsible for the differences in NUE and such differences in NUE in plants could be related to differences in absorption, translocation, shoot demand, dry matter production per unit nutrient absorbed, and plant environmental interactions. Genotypic differences for uptake and utilization efficiency of nutrients are governed by different soil and plant mechanisms and processes (Table 3) (Baligar et al. 2001). Adkinson (1990) and Fageria et al. (2002) state that plant factors such as root and root hair morphology (length, density, surface area); root-induced changes (secretion of H^+, OH^-, HCO_3^-); root exudation of organic acids (citric, malic, tartaric, oxalic, phenolic), sugars, and non-proteinogenic amino acids (phytosiderophores); secretion of enzymes (phosphatases); plant demand; plant species/cultivars; and microbial associations (enhanced CO_2 production, rhizobia, mycorrhizae, rhizobacteria) have profound influences on plant's ability to absorb and utilize nutrients from soil.

Fageria et al. (2008) states that capacity of some plant species or cultivar/genotypes within species to absorb nutrients at higher rate at low nutrient concentration of the growth medium is one of the mechanisms responsible for efficient nutrient use by plants. Plant induces several changes in rhizosphere as follows: (a) modification of rhizosphere pH; (b) oxidation potentials; (c) exudation of

Table 3 Components of NUE and processes that influence genotypic differences in NUE in plants[a]

A. Nutrient acquisition
1. Diffusion and mass flow in soil: buffer capacity, ionic concentration, ionic properties, tortuosity, soil moisture, bulk density, temperature
2. Root morphological factors: number, length, root hair density, root extension, root density
3. Rhizosphere modification
4. Physiological: root: shoot ratio, root microorganisms (rhizobia, azotobacter, mycorrhizal fungi), nutrient status, water uptake, nutrient influx and efflux, nutrient transport in roots and shoots, affinity to uptake (Km), threshold concentration (Cmin)
5. Biochemical: enzyme secretion (phosphatase), chelating compounds, phytosiderophores, proton exudates, organic acid exudates (citric, trans-aconitic, malic acid)
B. Nutrient movement in root
1. Transfer across endodermis and transport within root
2. Compartmentalization/binding within roots
3. Rate of nutrient release to xylem
C. Nutrient accumulation and remobilization in shoot
1. Demand at cellular level and storage in vacuoles
2. Retransport from older to younger leaves and from vegetative to reproductive tissues
3. Rate of chelates in xylem transport
D. Nutrient utilization and growth
1. Nutrient metabolism at reduced tissue concentration
2. Lower element concentration in supporting structure, particularly stem
3. Elemental substitution (Na for K, Fe for Mn, Mo for P, Co for Ni)
4. Biochemical (nitrate reductase for N-use efficiency, glutamate dehydrogenase for N metabolism, peroxidase for Fe efficiency, pyruvate kinase for K deficiency, metallothionein for metal toxicities, ascorbic acid oxidase for Cu, carnonic anhydrase for Zn)

[a]Modified from Baligar et al. (2001), Duncan and Baligar (1990), Fageria and Baligar (2005)

organic acids, chelators, reductants, and oxidants; (d) extracellular enzymes to turn over organically bound nutrients; and (e) providing substrate for microbial biomass (Fageria et al. 2008; Sauerbeck and Helal 1990).

Breeding programs should consider plant characteristics such as the ability to produce near maximum yields at low nutrient levels, and extensive root systems efficient in exploring large soil volumes to develop cultivars with high NUE. Breeding cultivars for high tolerance to low levels of nutrient supply and abiotic and biotic constraints will have a better chance of improving NUE (Baligar et al. 2001). Breeding plant cultivars with superior NUE depends upon: (a) the genetic variability present for particular trait (s) that governs NUE and (b) development of methodology to accurately quantify the physiological parameters that reflect the efficient NUE (Baligar et al. 2001; Fageria et al. 2008; Duncan and Baligar 1990; Duncan and Carrow 1999;

Fageria and Baligar 1994; Gerloff 1987; Gerloff and Gabelman 1983; Vose 1984, 1987).

Recent reviews have summarized broader aspects of identified genes responsible for NUE and plant environment on the expression of these genes (Agrama 2006; Masclaux-Daubresse et al. 2010; Pathak et al. 2011; Xu et al. 2012). Vinod and Heuer (2012) state that molecular breeding now provides a real opportunity to develop varieties with multiple tolerance traits—provided that large-effect QTLs/genes are available. Extensive discussion on molecular aspects of NUE is beyond the scope of this review.

It has been well documented the presence of most efficient (E) and most in efficient (I) NUE plants in different species and cultivars/genotypes within species (Baligar and Duncan 1990; Clark and Duncan 1991; Baligar et al. 2001; Fageria and Baligar 2005; Fageria et al. 2008; Gerloff and Gabelman 1983). Table 4 lists the E and I for nutrient efficiency ratio

Table 4 Variations in Nutrient Efficiency Ratio (NER) values for P, K, Ca, and Mg of most efficient (E) and inefficient (I) entries of selected crop species[a]

Species	Efficiency	NER (kg kg^{-1}) P	K	Ca	Mg
Rice	E	1,125	–	–	–
	I	563	–	–	–
Maize	E	625	46	256	476
	I	171	18	115	333
Wheat	E	188	–	–	–
	I	125	–	–	–
Sorghum	E	1,000	44	208	417
	I	476	23	123	278
Bean	E	671	294	–	–
	I	562	154	–	–
Red clover	E	1,012	104	91	670
	I	470	61	53	476
Alfalfa	E	1,369	113	57	689
	I	25	16	37	428
Tomato	E	–	357	434	–
	I	–	173	381	–

[a]Modified from Baligar et al. (2001)

Table 5 Various nutrient use efficiency parameters for N, P, and K of irrigated rice genotypes at medium soil fertility (MFL) and high soil fertility (HFL) levels in a low land acid soil of central Brazil[a]

Genotypes	N MFL	N HFL	P MFL	P HFL	K MFL	K HFL
Nutrient efficiency ratio (NER) (kg kg^{-1})						
Alianca	104	77	183	167	72	85
Metica I	81	108	140	236	56	120
Physiological efficiency (PE) (kg kg^{-1})						
Alianca	158	121	360	324	89	97
Metica I	130	113	491	458	89	121
Agronomic efficiency (AE) (kg kg^{-1})						
Alianca	48	43	93	99	76	87
Metica I	50	51	95	117	78	102
Agrophysiological efficiency (APE) (kg kg^{-1})						
Alianca	69	61	158	162	39	49
Metica I	73	51	252	207	37	54
Apparent nutrient recovery efficiency (ANR) (%)						
Alianca	71	63	51	52	81	92
Metica I	61	99	30	52	61	99

[a]Modified from Baligar and Fageria (1997)

(NER) in different crop species for P, K, Ca, and Mg. In all these species efficient entries for N, P, and N for NUE were far superior in utilization of absorbed nutrients than the inefficient entries.

NUE parameters are influenced by levels of soil applied nutrients (Baligar and Fageria 1997; Fageria and Baligar 2005; Fageria et al. 2008) and genotypes and level of soil fertility for N, P, and K (Table 5). Based on plant response to nutrient levels, Gerloff (1987) and Blair (1993) grouped genotype/cultivars as follows (Fig. 1): (a) efficient responders, plants that produce high yields at low levels of nutrients and that respond to higher levels of nutrient additions; (b) inefficient responders, plants with low yields at low levels of nutrition and that have a high response to high levels of added nutrients; (c) efficient nonresponders, plants that produce high yields at low levels nutrition but do not respond to nutrient additions; and (d) inefficient nonresponders, plants that produce low yields at low levels of nutrition and do not respond to nutrient addition. Such type of classifications will help in identification of genotypes/cultivars that are efficient in nutrient use under stresses or non-stressed systems. Superior genotypes with

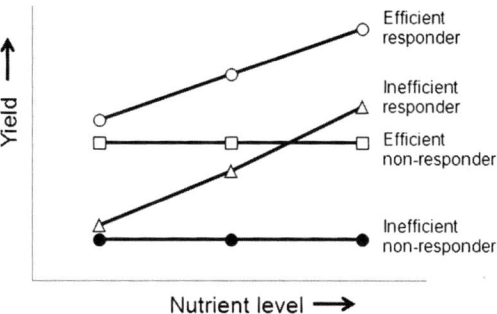

Fig. 1 Classes of plant relative to yield responses to nutrient level in the growth medium (Gerloff 1987; Blair 1993)

high NUE are useful in breeding of nutrient use efficient cultivars that are adaptable to stress ecosystems and to produce reasonable higher crop yields.

3.2 Soil Factors

Production potentials of many soils in the world are decreased by low supplies of essential nutrients from adverse soil physical (bulk density, hardpan layers, structure and texture, surface sealing and crusting, water holding capacity, water-logging, drying, aeration, temperature) and chemical (salinity, acidity, elemental deficiencies/toxicities, low SOM) constraints. These soil constraints affect transformation (mineralization, immobilization), fixation (adsorption, precipitation), and leaching or surface runoff of indigenous and added nutrients (Barber 1995; Baligar and Fageria 1997; Fageria et al. 2002, 2008; Dudal 1976) and reduce nutrient bioavailability.

Mineral nutrient deficiencies and toxicities due to extreme soil pH affect growth (dry mass, root: shoot ratio) and morphology (length, thickness, surface area, density) of roots and root hairs (Baligar et al. 1998). Such changes in root growth and morphology affect plant ability to absorb and translocate minerals from soil to meet plant demands (Fageria 2013).

Acid soil occupy close to 3.95 billion ha of area world's ice-free land area (von Uexkull and Mutert 1995). Poor plant productivity on acid soils is due to combination of toxicities of Al, Mn, and H and deficiencies of N, P, K, Ca, Mg, and some micronutrients (e.g., Fe, Zn). Additions of lime and fertilizers are an effective way to correct soil acidity constraints. Plant species and genotypes/cultivars within species differ widely in tolerance to soil acidity constraints (Foy 1983, 1984; Baligar and Fageria 1997, 1999). Baligar and Fageria (1997) reported that Al-tolerant genotypes had lower NER for essential nutrients in sorghum (except Ca) and red clover (except P) and had higher NER in maize (except K) and alfalfa (except Ca and MG) than Al-sensitive genotypes. However efforts in selection of acid soil-tolerant plant cultivars with high dry matter accumulations and NUE is very much unknown.

The total area of salt-affected soil has been estimated at about 952 million ha (Szabolcs 1979). Saline soils contain predominantly chlorides and sulfates of Na, Ca, and Mg, and alkaline soils contain excess of $NaHCO_3$ and exchangeable Na and these ions are commonly occurring in toxic concentrations and are very toxic to plants (Gupta and Abrol 1990). In saline soils, plants are affected by water deficit, ion

toxicity (Cl, Na), and nutrient imbalances due to depression in uptake and transport. In alkaline soils, Fe deficiency, B toxicity, and salinity are the most obvious problems for successful crop production. Large differences in salt tolerance have been reported for plant species and cultivars within species (Maas 1986; Marshner 1995). During growth cycle in both acidic and salt-affected soils short- or long-duration drought adversely affects physiological processes, shoot and root morphology, and NUE. Use of salt-tolerant species and cultivars could help in overcoming moderate soil salt toxicity and improving crop yields and NUE.

Soils in many areas of the world have toxic levels of heavy metals (Cd, Cr, Ni, Hg, Pb, Cu, Zn). Heavy metals that alter root growth, morphology, and roots' ability to absorb nutrients from soil and interfere in these changes have significant effects on nutrition and NUE in plants (Baligar et al. 1998; Mengel and Kirkby 2001; Marshner 1995). By adapting soil amendments addition (lime, organic matter, fertilizers) and use of hyper metal accumulator by adapting phytoremediation techniques, toxic levels of heavy metals in agricultural soils could be remediated (Fageria et al. 2011).

Use of improved and high-yielding crop species and genotypes/cultivars within species, in combination with improved best management practices for acid soils (lime, fertilizers, organic matter) and saline/alkali soils (irrigation, organic matter, macro-micronutrient fertilizers), is an important strategy to reduce cost of production and improve crop yields and NUE in various types of soils. Detail discussions of and reviews of plant and soil factors that affect nutrient uptake and NUE are available (Baligar et al. 2001; Barber 1995; Clark and Baligar 2000; Fageria et al. 2002; Marshner 1995; Mengel and Kirkby 2001).

3.3   Fertilizer Factors

Barber (1976) defined fertilizer efficiency as the amount of increase in yield of the harvested portion of the crop per unit of fertilizer nutrient applied where high yields are obtained. Fertilizer recovery efficiency could be improved by selecting right type of fertilizer formulation, correcting soil adverse chemical constraints by applications of soil amendments, and adjusting the method, dose, and time of application based on soil, plant, and climatic factors to reduce losses (leaching and runoff, denitrification, ammonia volatilization, fixation). Single- or multi-nutrient slow release fertilizers (SRF) and controlled release fertilizers (CRF) have added advantages in improving the recovery efficiency by plants, by lowering rate of release, thereby limiting emissions/volatilization (N_2O, NH_3) and leaching losses (NO_3-N, K), reducing P fixation, and providing constant availability of nutrients during the entire plant growing season. CRF are designed to synchronize nutrient availability (rate of release) to the plant nutrient uptake. However there is a need for research to understand the efficiency of these fertilizer formulations and their interaction.

Over the year several nitrification inhibitors have been developed and widely used: N-serve/nitropyrin [2-chloro-6 (trichloromethyl) pyridine], AM [2-amino-4-chloro 6 methyl pyrimidine], DCD [dicyandiamide cyanoguanidine], and DMPP (3,4-dimethylpyrazole phosphate) (Hauck 1985; Peoples et al. 1995; Prasad and Power 1995; Subbarao et al. 2006; Trenkel 2010) to reduce N loss. Nitrification inhibitors delay bacterial oxidation of ammonia ion thereby control the loss of nitrate by leaching or the production of nitrous oxides by denitrification (Trenkel 2010). Urease inhibitors such as NBPT [N-(n-Butyl) thiophosphoric triamide], PPD/PPDA [phenylphosphorodiamidate], and hydroquinone have been widely used (Hendrickson 1991, 1992; Kiss and Simihaian 2002; Trenkel 2010) and have improve applied urea efficiency by preventing or suppressing the transformation of amide-N in urea to ammonium hydroxide and ammonium through the hydrolytic action of the enzyme urease (Trenkel 2010). Application of NBPT with urea is successful in reducing the rate of urea hydrolysis and improves its efficiency (Hendrickson 1991, 1992). Site-specific (precision) technology along with sound management systems could lead to reduced fertilizer inputs, thereby improving costs of fertilizer input and the degradation of the environment.

Under certain circumstances of adverse soil (extreme pH, low SOM) and climatic (extreme rainfall and temperatures) conditions, plants develop nutrient disorders (deficiencies) that can affect plants' growth, development, and NUE. Timely application of nutrients through foliar is the most effective methods to correct specific nutrient disorder and to improve plant growth and NUE. Extensive discussions on foliar fertilization and their impact on NUE are given by Fageria et al. (2009) and Kannan (1990).

3.4 Agronomic/Management

Soil quality parameters (physical, chemical, biological), SOM, and nutrient distribution throughout different soil horizons are influenced by nature of tillage operations such as traditional tillage, no tillage, minimum tillage, and conservation tillage. Tillage methods have impact on rooting pattern, water holding capacity, water penetration, aeration, soil compaction, soil temperature, and soil microbial activities. Such changes greatly affect nutrient bioavailability and subsequently affect NUE in plants. Inclusion of cover crops either in crop rotation or along with main crops could reduce soil erosion, improve soil fertility and SOM, and suppress weeds. Beneficial effects of cover crops include increased soil organic matter content, improved soil quality parameters and availability nutrients, increased concentration of nutrients at surface layers, reduced leaching losses of nutrients, better erosion control, improved soil structure and texture, decreased soil acidity and compaction, cut fertilizer input costs, improved water holding capacity, increased biological activities, weed suppression, decreased disease, and reduced pest problems (Fageria et al. 2011). Improved nutrient cycling and soil quality parameters by the use of cover crops will improve crop growth and NUE.

3.5   Abiotic Stresses

Invariably plants are subjected to various degrees of abiotic (soil acidity, mineral deficiencies and/or toxicities, drought/floods, light quality/ shade, temperatures extremes) and biotic stresses (pest, diseases, weeds). These stresses have tremendous effect on plant growth and development resulting in lower absorption and utilization of absorbed nutrients and consequently to reduced NUE (Fageria et al. 2008; Lyda 1981). Cultivars/ genotypes of plants that have high macro-micro- nutrient use efficiency ratio under abiotic and biotic stresses may have advantage in adapting mineral stress ecosystems and in overcoming the prevailing abiotic and biotic stresses.

Cultivar selection, crop improvement, and management practices must be tailored to prevailing climatic conditions. In crop improve- ment programs (selection and breeding), it is vital to incorporate physiological traits that improve the plants' ability to tolerate multiple climatic variables and their extremes. Precipita- tion, solar radiation, and temperature have major impact on nutrient transformation and availabil- ity in soil and plants' ability to take up and utilize the nutrients (Baligar and Fageria 1997; Baligar et al. 2001; Barber 1995; Fageria et al. 2006, 2011; Marshner 1995). Drought stress is a major constraint to crop production and yield stability. Impact of water stress is a function of duration crop growth stage, type of crop species or cultivar, soil type, and management practices. Water deficit has adverse effects on plant growth (dry matter accumulation), morphology (reduc- tion of cell growth and enlargement, leaf expan- sion, increased leaf thickness, root growth, epicuticular wax), physiology (photosynthesis, stomatal regulation, protein metabolism, synthe- sis of amino acids, nitrate reductase), NUE, and yield (flowering, anthesis, grain fill/pod fill). Fageria et al. (2006) have covered extensively on the nature of short- and long-duration drought and their impact on crop plant growth, develop- ment, and yields. The amount of solar radiation has a direct effect on photosynthesis which in turn influences demand for nutrient uptake and

utilization of absorbed nutrients. Adequate solar radiation maximizes the NUE in plants. Rate of nutrient release in soil from organic and inor- ganic sources, uptake of nutrients by roots, and subsequent translocation and utilizations are influenced by soil and aerial temperatures. Avail- ability of most nutrients tend to decrease at low soil temperature and low soil moisture contents because of reduced root activity and low rates of dissolution and diffusion of nutrients.

3.6   Biotic Stresses

Infection of diseases and insects reduce crop growth and yield and consequently NUE. Diseases and insects mostly affect leaves, stem, and roots. Infection of leaves reduces photosyn- thetic activity leading to reduced utilization of absorbed nutrients. Soil-borne pathogens such as actinomycetes, bacteria, fungi, nematodes, and viruses present in the soil around roots lead to pathogenic stress and bring changes in morphol- ogy and physiology of roots and shoots, thereby reducing roots' ability to absorb nutrients from the soil and using absorbed nutrients more effec- tively (Baligar et al. 2001; Fageria 1992; Lyda 1981). Deficiency/toxicity of nutrients in soil greatly that influences extent and severity of plant diseases and balance nutrition has an important affect in determining plants' resistance or susceptibility to diseases (Huber 1980). Defi- ciency of Ca, Mg, Cu, Fe, B, Mn, Mo, Ni, and Si is known to induce various diseases in plants (Engelhard 1989; Fageria et al. 2011; Graham and Webb 1991; Huber 1980). Adequate levels of nutrients in plant tissue increase resistance by maximizing the inherent resistance of plants, facilitating disease escape through increased nutrient availability or stimulated plant growth, and altering external environments to influence survival, germination, and penetration of pathogens (Huber 1980; Engelhard 1989; Fageria et al. 2011). Micronutrient concentrations in plants are important in host ability to resist or tolerate infectious pathogens (Fageria et al. 2002). Insects, mites, aphids, and nematodes are harmful to the health of the plant. One of the major roles of nematodes played in ecosystem is the release of nutrient in soil for plant uptake (Coleman et al. 1984). Proper biological, chemical, physical, and cultural management practices can be used to alleviate the pathogenic stress.

The soil fauna consist of microfauna, mesofauna, and macrofauna. Soil microfauna play an important role in decomposition of crop residues, nutrient cycling, and plant nutrient availability (Paul and Clark 1996). Therefore meso- and microfaunal activities should be included in the assessment of NUE in plants as these organisms greatly affect the nutrient cycling, soil biochemical processes, and physical characteristics of soil and eventually affecting plant growth and development. Microorganisms in soil play a key role in biochemical cycling process that tend to regulate nutrient cycling through decomposition (release) and sequestration/binding (retention) process thereby sustaining natural ecosystems and agricultural production. Rhizobia, diazotrophic bacteria, and mycorrhizae in the rhizosphere have improved root growth by fixing atmospheric N_2, suppressing pathogens, producing phytohormones, enhancing root surface area to facilitate uptake of less mobile nutrients such as P and micronutrients and mobilization, and solubilizing of unavailable organic/inorganic nutrients (Baligar et al. 2001). Microorganisms in soil and rhizosphere have profound influence on plant growth through their phyto-effective metabolic activities (Curl and Truelove 1986). Rhizosphere microorganisms produce plant growth regulators (PGR) that results in modifications in plant growth and development and NUE.

Mycorrhizal fungi are involved in ecosystem processes such as litter decomposition, ammonification, of organic N and nitrification, weathering of soil minerals, and influence on soil structure (Sturmer and Siqueira 2006). Arabuscular mycorrhizal fungi (AMF) live in symbiotic association with plant roots thereby increasing root surface area which assist roots in exploring larger soil volumes thereby bringing more ions closure to roots and contributing to higher nutrient in floe (Sanders et al. 1977; Smith et al. 1993). Weeds compete with crop plants for water, nutrients, and sunlight, thereby reducing crop yields and consequently NUE. Appropriate weed control methods (chemical, manual, crop rotations) should be followed to alleviate weeds.

Conclusions

Nutrient deficiency is one of the most yield constraints in crop production in most of the agro-ecological regions of the world. It is estimated that overall contribution of essential plant nutrients in increasing crop yields is about 40 % in low fertility soils when other productions factors (suitable cultivar, availability of adequate water, control of diseases, insects and weeds) are at an adequate levels. Overall, cost of fertilizer inputs in modern agriculture is about 30 %. Furthermore, the recovery efficiency of applied fertilizers is very low. It is estimated that in cereals N recovery efficiency is about 33 % worldwide. Similarly, the recovery efficiency of P is less than 20 % and K less than 40 %. In addition the recover efficiency of micronutrients varied from 10 to 15 %. Hence, most of the nutrients applied as inorganic fertilizers are lost in soil-plant system or unavailable to crops. Under these situations, improving nutrient use efficiency is important not only to increase crop yields but also to reduce cost of crop production and environmental pollution. Important management strategies to improve nutrient use efficiency are judicious use of fertilizers (adequate rate, effective source, methods, and time of application), supply of adequate water, and control of diseases, insects, and weeds. In addition, planting nutrient-efficient crop species and/or genotypes/cultivar within species is an important strategy in improving nutrient use efficiency in crop plants.

References

Adkinson D (1990) Influence of root system morphology and development on the need for fertilizer and efficiency of use. In: Baligar VC, Duncan RR (eds) Crops as enhancers of nutrient use. Academic, San Diego, pp 411–451

Agrama H (2006) Application of molecular markers in breeding for nitrogen use efficiency. J Crop Improv 15:175–211

Alam SM (1999) Nutrient uptake by plants under stress conditions. In: Pessarakli M (ed) Handbook of plant and crop stress, 2nd edn. Marcel Dekker Inc, New York, pp 285–313

Baligar VC, Bennett OL (1986a) Outlook on fertilizer use efficiency in the tropics. Fertil Res 10:83–96

Baligar VC, Bennett OL (1986b) NPK-fertilizer efficiency. A situation analysis for the tropics. Fertil Res 10:147–164

Baligar VC, Duncan RR (eds) (1990) Crops as enhancers of nutrient use. Academic, San Diego

Baligar VC, Fageria NK (1997) Nutrient use efficiency in acid soils: nutrient management and plant use efficiency. In: Monitz AC, Furlani AMC, Fageria NK, Rosolem CA, Cantarells H (eds) Plant-soil interactions at low pH: sustainable agriculture and forestry production. Brazilian Soil Science Society, Campinas, pp 75–93

Baligar VC, Fageria NK (1999) Plant nutrient efficiency: towards the second paradigm. In: Siqueira JO, Moreira FMS, Lopes AS, Guilherme V, Faquin AE, Neto F, Carvalho JG (eds) Soil fertility, soil biology and plant nutrition interrelationships. Brazilian Soil Science Society-Federal University of Lavras Soil Science Department (SBCS/UFLA/DCS), Lavras, pp 183–204

Baligar VC, Fageria NK, Elrashidi MA (1998) Toxicity and nutrient constraints in root growth. Hort Sci 36:960–965

Baligar VC, Fageria NK, He Z (2001) Nutrient use efficiency in plants. Commun Soil Sci Plant Anal 31:921–950

Barber SA (1976) Efficient fertilizer use. In: Patterson FL (ed) Agronomic research for food, Special publication 26. American Society of Agronomy, Madison, pp 13–29

Barber SA (1995) Soil nutrient bioavailability: a mechanistic approach, 2nd edn. Wiley, New York

Blair G (1993) Nutrient efficiency—what do we really mean. In: Randall PJ, Delhaize E, Richards RA, Munns R (eds) Genetic aspects of plant mineral nutrition. Kluwer Academic Publishers, Dordrecht, pp 205–213

Borlaug NE, Doswell CR (1994) Feeding a human population that increasingly crowds a fragile planet 15th world congress of soil science. Acapulco, Mexico, Supplement to Trans, 10 p

Clark RB (1982) Plant response to mineral element toxicity and deficiency. In: Christiansen MN, Lewis CF (eds) Breeding plants for less favorable environments. Wiley, New York, pp 71–73

Clark RB (1990) Physiology of cereals for mineral nutrient uptake, use, and efficiency. In: Baligar VC, Duncan RR (eds) Crops as enhancers of nutrient use. Academic, San Diego, pp 131–209

Clark RB, Baligar VC (2000) Acidic and alkaline soil constraints on plant mineral nutrition. In: Wilkinson RW (ed) Plant-environment interactions. Marcel Dekker Inc, New York, pp 113–177

Clark RB, Duncan RR (1991) Improvement of plant mineral nutrition through breeding. Field Crops Res 27:219–240

Coleman DC, Cole CV, Elliott ET (1984) Decomposition, organic matter turnover and nutrient dynamics in agroecosystems. In: Lowrance R, Stinner BR, House GJ (eds) Agricultural ecosystems: unifying concepts. Wiley-Interscience, New York, pp 83–104

Curl EA, Truelove B (1986) The rhizosphere. Springer, New York

Dudal R (1976) Inventory of the major soils of the world with special reference to mineral stress hazards. In: Wright MJ (ed) Plant adaptation to mineral stress in problem soils. Cornell University Press, Ithaca, pp 3–13

Duncan RR, Baligar VC (1990) Genetics, breeding and physiological mechanisms of nutrient uptake and use efficiency: an overview. In: Baligar VC, Duncan RR (eds) Crops as enhancers of nutrient use. Academic, San Diego, pp 3–35

Duncan RR, Carrow RN (1999) Turfgrass—molecular genetic improvement for abiotic/edaphic stress environment. Adv Agron 67:233–306

Engelhard AW (ed) (1989) Soil borne plant pathogens management of diseases with macro-and microelements. The American Phytopathological Society Press, St. Paul

Fageria NK (1992) Maximizing crop yields. Marcel Dekker Inc, New York

Fageria NK (2013) The role of plant roots in crop production. CRC Press/Taylor & Francis Group, Florida

Fageria NK, Baligar VC (1994) Screening crop genotypes for mineral stresses. In: Maranville JW, Baligar VC, Duncan RR, Yohe JM (eds) Adaptation of plants to soil stress. University of Nebraska, INTSORMIL-USAID, Lincoln, pp 152–159

Fageria NK, Baligar VC (2005) Enhancing nitrogen use efficiency in crop plants. Adv Agron 88:97–185

Fageria NK, Baligar VC, Clark RB (2002) Micronutrients in crop production. Adv Agron 77:185–268

Fageria NK, Baligar VC, Clark RB (2006) Physiology of crop production. Food Product Press/The Haworth Press Inc, New York

Fageria NK, Baligar VC, Li YC (2008) The role of nutrient efficient plants in improving crop yields in twenty first century. J Plant Nutr 31:1121–1157

Fageria NK, Barbosa Filho MP, Moreira A, Guimaraes CM (2009) Foliar fertilization of crop plants. J Plant Nutr 32:1044–1063

Fageria NK, Baligar VC, Jones CA (2011) Growth and mineral nutrition of field crops, 3rd edn. CRC Press/Taylor & Francis Group, Boca Raton

FAO (2013) Food and Agricultural Organization. FAOSTAT. http://faostat.fao.org/Rome

Foy CD (1983) The physiology of plant adaptation to mineral stress. Iowa State J Res 54:355–391

Foy CD (1984) Physiological effects of hydrogen, aluminum and manganese toxicities in acid soils. In: Adams

F (ed) Soil acidity and liming, 2nd edn, Agronomy monograph. American Society of Agronomy, Madison, pp 57–97

Gerloff GC (1987) Intact-plant screening for tolerance of nutrient-deficiency stress. In: Gabelman WH, Loughman BC (eds) Genetic aspects of plant mineral nutrition. Martinus Nijhoff Publisher, The Hague, pp 55–68

Gerloff GC, Gabelman WH (1983) Genetic basis of inorganic plant nutrition. In: Lauchli A, Bieleski RL (eds) Inorganic plant nutrition, vol 15B, Encyclopedia and plant physiology, new series. Springer, New York, pp 453–480

Gourley CJP, Allan DL, Russell MP (1994) Plant nutrient efficiency: a comparison of definitions and suggested improvement. Plant Soil 158:29–37

Graham RD (1984) Breeding for nutritional characteristics in cereals. In: Tinker PB, Lauchli A (eds) Advances in plant nutrition, vol 1. Praeger Publisher, New York, pp 57–102

Graham RD, Webb MJ (1991) Micronutrients and disease resistance and tolerance in plants. In: Mortvadt JJ, Cox FR, Shuman LM, Welch RM (eds) Micronutrients in agriculture, 2nd edn, Soil science society American book series no. 4. Soil Science Society of America, Madison, pp 329–370

Gupta RK, Abrol IP (1990) Salt affected soils: their reclamation and management for crop production. In: Lal R, Stewart BA (eds) Soil degradation, vol 11, Advances in soil science. Springer, New York, pp 223–288

Hauck R (1985) Slow-release and bioinhibitor-amended nitrogen fertilizers. In: Engelstad OP (ed) Fertilizer technology and use. Soil Science Society of America, Madison, pp 293–322

Hendrickson L (1991) Exploring urease inhibitor. Solution (Manchester) 35:36–50

Hendrickson LL (1992) Corn yield response to the urease inhibitor NBPT: five year summary. J Prod Agric 5:131–137

Huber DM (1980) The role of mineral nutrition in defense. In: Horsfall JG, Cowling EB (eds) Plant pathology an advanced treaty. Academic, New York, pp 381–406

Kannan S (1990) Role of foliar fertilization on plant nutrition. In: Baligar VC, Duncan RR (eds) Crops as enhancers of nutrient use. Academic, San Diego, pp 313–348

Kiss S, Simihaian M (2002) Improving efficiency of urea fertilizers by inhibition of soil urease activity. Kluwer Academic Publishers, Dordrecht

Lyda SD (1981) Alleviating pathogen stress. In: Arkin GF, Taylor HM (eds) Modifying the root environment to reduce crop stress, ASAE monograph no. 4. American Society of Agricultural Engineers, St. Joseph, pp 195–214

Maas EV (1986) Salt tolerance of plants. Appl Agric Res 1:12–26

Marshner H (1995) Mineral nutrition of higher plants, 2nd edn. Academic, New York

Masclaux-Daubresse C, Daniel-Vedele F, Dechorgnat J, Chardon F, Gaufichon L, Suzuki A (2010) Nitrogen uptake, assimilation and remobilization in plants: challenges for sustainable and productive agriculture. Ann Bot 105:1141–1157

Mengel K, Kirkby EA (2001) Principles of plant nutrition, 3rd edn. International Potash Institute, Berne

Pathak RR, Lochab S, Raghuram N (2011) Plant systems: improving plant nitrogen-use efficiency. In: Moo-Young M (ed) Comprehensive biotechnology, vol 4, 2nd edn. Elsevier, New York, pp 209–218

Paul EA, Clark LE (1996) Soil microbiology and biochemistry. Academic, San Diego

Peoples MB, Freney JR, Mosier AR (1995) Minimizing gaseous losses of nitrogen. In: Bacon PE (ed) Nitrogen fertilization in the environment. Marcel Dekker Inc, New York, pp 565–602

Prasad R, Power JE (1995) Nitrification inhibitors for agriculture, health and environment. Adv Agron 54:233–281

Sanders FE, Tinker PB, Black RL, Palmerley SM (1977) The development of endomycorrhizal root system. I. Spread of infection and growth promoting effects with four species of vesicular-arbuscular mycorrhizae. New Phytol 78:257–268

Sauerbeck DR, Helal HM (1990) Factors affecting the nutrient efficiency in plants. In: El Balsam N, Dambroth M, Loughman BC (eds) Genetic aspects of plant mineral nutrition. Kluwer Academic Publishers, Dordrecht, pp 11–17

Smith SE, Robson AD, Abbott LK (1993) The involvement of mycorrhizas in assessment of genetically dependent efficiency of nutrient uptake and use. In: Randall PJ, Delhaize E, Richards RA, Munns R (eds) Genetic aspects of plant mineral nutrition. Kluwer Academic Publishers, Dordrecht, pp 221–231

Sturmer SL, Siqueira JO (2006) Diversity of arbuscular mycorrhizal fungi in Brazilian ecosystems. In: Moreira FMS, Siqueira JO, Brussaard L (eds) Soil biodiversity in Amazonian and other Brazilian ecosystems. CABI Publishing, Wallingford, pp 206–235

Subbarao GV, Sahrawat KL, Berry WL, Nakahara K, Ishikawa T, Watanabe T, Suenaga K, Rondon M, Rao IM (2006) Scope and strategies for regulation of nitrification in agricultural systems- challenges and opportunities. Crit Rev Plant Sci 25:302–335

Szabolcs I (1979) Review of research on salt-affected soils, Nature resource research 15. UNESCO Press, Paris, p 137

Trenkel ME (2010) Slow- and controlled-release and stabilized fertilizers: an option for enhancing nutrient use efficiency in agriculture. International Fertilizer Industry Association (IFA), Paris

Vinod KK, Heuer S (2012) Approaches towards nitrogen- and phosphorus-efficient rice. AoB PLANTS 2012: pls028. doi:10.1093/aobpla/pls028 von Uexkull HR, Mutert E (1995) Global extent, development and economic impact of acid soils. In: Date RA, Grundon NJ, Raymont GE, Probert ME (eds) Plant–soil interactions at low pH: principles and management. Kluwer Academic Publishers, Dordrecht, pp 5–19

Vose PB (1984) Effect of genetic factors on nutritional requirement of plants. In: Vose PB, Blixt SG (eds) Crop breeding—a contemporary basis. Pergamon Press, Oxford, pp 67–114

Vose PB (1987) Genetic aspects of mineral nutrition-progress to date. In: Gabelman WH, Loughman BC (eds) Genetic aspects of plant mineral nutrition. Martinus Nijhoff Publisher, Boston, pp 3–13

Xu C, Fisher R, Wullschleger SD, Wilson CJ, Cai M (2012) Toward a mechanistic modeling of nitrogen limitation on vegetation dynamics. PLoS ONE 7(5): e37914. doi:10.1371/journal.pone.0037914

Part I

Nutrients as a Key Driver of Nutrient Use Efficiency

Soil and Input Management Options for Increasing Nutrient Use Efficiency

B.N. Ghosh, Raman Jeet Singh, and P.K. Mishra

Abstract

Public interest and awareness of the need for improving nutrient use efficiency is great, but nutrient use efficiency is easily misunderstood. Four indices of nutrient use efficiency are reviewed, and an example of different applications of the terminology shows that the same data set might be used to calculate a fertilizer N efficiency of 21 or 100 %. Fertilizer N recovery efficiencies from researcher-managed experiments for major grain crops range from 46 to 65 %, compared to on-farm N recovery efficiencies of 20–40 %. Fertilizer use efficiency can be optimized by fertilizer best management practices that apply nutrients at the right rate, time, and place and accompanied by the right agronomic practices. The highest nutrient use efficiency always occurs at the lower parts of the yield response curve, where fertilizer inputs are the lowest, but effectiveness of fertilizers in increasing crop yields and optimizing farmer profitability should not be sacrificed for the sake of efficiency alone. There must be a balance between optimal nutrient use efficiency and optimal crop productivity.

Keywords

Balanced fertilization • Fertilizer best management practices • Nitrogen efficiency • Right rate • Right time • Right place

1 Introduction

Awareness of an interest in improved nutrient use efficiency has never been greater. Driven by a growing public belief that crop nutrients are excessive in the environment and farmer concerns about rising fertilizer prices and stagnant crop prices, the fertilizer industry is under increasing pressure to improve nutrient use efficiency (Dibb 2000). However, efficiency can be defined in many ways and is easily misunderstood and misrepresented. Definitions differ, depending on the perspective. Environmental nutrient use efficiency can be quite different than agronomic or economic efficiency and maximizing efficiency may not always be

B.N. Ghosh (✉) • R.J. Singh • P.K. Mishra
ICAR-Central Soil and Water Conservation Research and Training Institute, 218, Kaulagarh Road, Dehradun, 248 195, India
e-mail: bnghosh62@rediffmail.com; rdxsingh@gmail.com; pkmbellary@rediffmail.com

A. Rakshit et al. (eds.), *Nutrient Use Efficiency: from Basics to Advances*,
DOI 10.1007/978-81-322-2169-2_2, © Springer India 2015

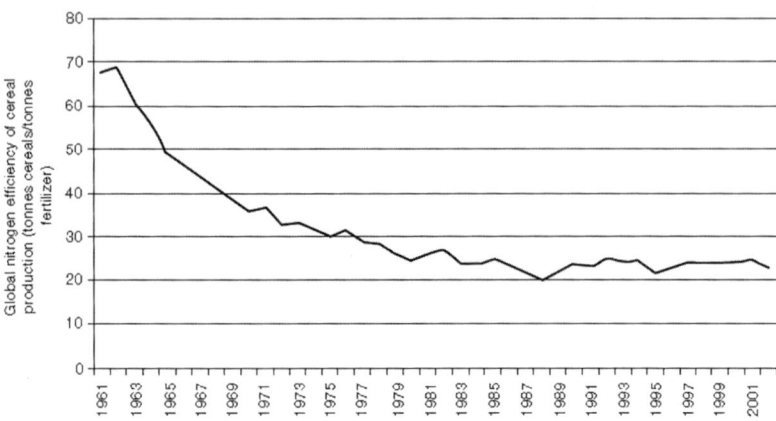

Fig. 1 Global nitrogen fertilizer efficiency of cereal production (annual global cereal production in tonnes divided by annual global nitrogen fertilizer production in tonnes for domestic use in agriculture) (*Source*: Tilman et al. 2002)

advisable or effective. Agronomic efficiency may be defined as the nutrients accumulated in the above-ground part of the plant or the nutrients recovered within the entire soil-crop-root system (Fageria et al. 2008). Economic efficiency occurs when farm income is maximized from proper use of nutrient inputs, but it is not easily predicted or always achieved because future yield increases, nutrient costs, and crop prices are not known in advance of the growing season (Tilman 2000). Environmental efficiency is site-specific and can only be determined by studying local targets vulnerable to nutrient impact. Nutrients not used by the crop are at risk of loss to the environment, but the susceptibility of loss varies with the nutrient, soil and climatic conditions, and landscape. In general, nutrient loss to the environment is only a concern when fertilizers or manures are applied at rates above agronomic need. Though perspectives vary, agronomic nutrient use efficiency is the basis for economic and environmental efficiency. As agronomic efficiency improves, economic and environmental efficiency will also benefit.

In the past decades, an increase in the consumption of nitrogen and phosphorus fertilizers has been observed globally. By 2050, nitrogen fertilization is expected to increase by 2.7 times and phosphorus by 2.4 times on a global scale (Tilman 2001). However, increased fertilizer application rates exhibit diminishing marginal returns such that further increases in fertilizer are unlikely to be as effective in increasing cereal

yield as in the past. A declining trend in global nitrogen efficiency of crop production (annual global cereal production divided by annual global nitrogen application) is shown in Fig. 1 (Tilman et al. 2002). It is estimated that today only 30–50 % of applied nitrogen fertilizers (Smil 2002; Ladha et al. 2005) and 45 % of phosphorus fertilizers (Smil 2000) are used for crops. For example, only 20–60 % of nitrogen fertilizers applied in intensive wheat production is taken up by the crop, 20–60 % remains in the soil, and approximately 20 % is lost to the environment (Pilbeam 1996). The phosphorus-use efficiency can be as high as 90 % for well-managed agroecosystems (Syers et al. 2008) or as low as 10–20 % in highly phosphorus-fixing soils (Bolland and Gilkes 1998).

2 Nutrient Use Efficiency Terminology

Nutrient use efficiency can be expressed in several ways. Mosier et al. (2004) described four agronomic indices commonly used to describe nutrient use efficiency: partial factor productivity (PFP, kg crop yield per kg nutrient applied); agronomic efficiency (AE, kg crop yield increase per kg nutrient applied); apparent recovery efficiency (RE, kg nutrient taken up per kg nutrient applied); and physiological efficiency (PE, kg yield increase per kg nutrient taken up). Crop

Table 1 Fertilizer N efficiency of maize from 56 on-farm studies in north central USA

Average optimum N fertilizer rate, kg ha^{-1}	103
Fertilizer N recovered in the crop, kg ha^{-1}	38
Total N taken up by crop, kg ha^{-1}	184
N removed in the harvested grain, kg ha^{-1}	103
N returned to field in crop residue, kg ha^{-1}	81
Crop recovery efficiency (38 kg N recovered/ 103 kg N applied), %	37
Crop removal efficiency (103 kg N applied/103 kg N in grain), %	100

Cassman et al. (2002), source of data, Bruulsema et al. (2004), source of calculations

removal efficiency (removal of nutrient in harvested crop as percent of nutrient applied) is also commonly used to explain nutrient efficiency. Available data and objectives determine which term best describes nutrient use efficiency. Fixen (2005) provides a good overview of these different terms with examples of how they might be applied.

Understanding the terminology and the context in which it is used is critical to prevent misinterpretation and misunderstanding. For example, Table 1 shows the same maize data from the north central USA can be used to estimate crop recovery efficiency of nitrogen (N) at 37 % (i.e., crop recovered 37 % of added N) or crop removal efficiency at 100 % (N removed in the grain was 100 % of applied N) (Bruulsema et al. 2004). Which estimate of nutrient use efficiency is correct? Recovery of 37 % in the aboveground biomass of applied N is disturbingly low and suggests that N may pose an environmental risk. Assuming the grain contains 56 % of the above-ground N, a typical N harvest index; only 21 % of the fertilizer N applied is removed in the grain. Such low recovery efficiency prompts the question – where is the rest of the fertilizer going and what does a recovery efficiency of 37 % really mean?

In the above data, application of N at the optimum rate of 103 kg ha^{-1} increased aboveground N uptake by 38 kg ha^{-1} (37 % of 103). Total N uptake by the fertilized maize was 184 kg ha^{-1}; 146 from the soil and 38 from the fertilizer. The N in the grain would be 56 % of

184, or 103 kg ha^{-1}: equal to the amount of N applied. Which is correct – a recovery of 21 % as estimated from a single-year response recovery in the grain or 100 % as estimated from the total uptake (soil N + fertilizer N) of N, assuming the soil can continue to supply N in long term? The answer cannot be known unless the long-term dynamics of N cycling are understood.

Fertilizer nutrients applied, but not taken up by the crop, are vulnerable to losses from leaching, erosion, and denitrification or volatilization in the case of N, or they could be temporarily immobilized in soil organic matter to be released at a later time, all of which impact apparent use efficiency. Dobermann et al. (2005) introduced the term system level efficiency to account for contributions of added nutrients to both crop uptake and soil nutrient supply.

3 Current Status of Nutrient Use Efficiency

A recent review of worldwide data on N use efficiency for cereal crops from researcher-managed experimental plots reported that single-year fertilizer N recovery efficiencies averaged 65 % for corn, 57 % for wheat, and 46 % for rice (Ladha et al. 2005). However, experimental plots do not accurately reflect the efficiencies obtainable on-farm. Differences in the scale of farming operations and management practices (i.e., tillage, seeding, weed and pest control, irrigation, harvesting) usually result in lower nutrient use efficiency. Nitrogen recovery in crops grown by farmers rarely exceeds 50 % and is often much lower. A review of best available information suggests average N recovery efficiency for fields managed by farmers ranging from about 20 to 30 % under rainfed conditions and 30 to 40 % under irrigated conditions.

Cassman et al. (2002) looked at N fertilizer recovery under different cropping systems and reported 37 % recovery for corn grown in the north central USA (Table 2). They found N recovery averaged 31 % for irrigated rice grown by Asian farmers and 40 % for rice under field-specific management. In India, N recovery

Table 2 Nitrogen fertilizer recovery efficiency by maize, rice, and wheat from on-farm measurements

Crop	Region	Number of farms	Average N rate, kg ha^{-1}	N recovery, %
Maize	North Central USA	56	103	37
Rice	Asia – farmer practice	179	117	31
	Asia – field-specific management	179	112	40
Wheat	India – unfavorable weather	23	145	18
	India – favorable weather	21	123	49

Cassman et al. (2002)

averaged 18 % for wheat grown under poor weather conditions, but 49 % when grown under good weather conditions. Fertilizer recovery is impacted by management, which can be controlled, but also by weather, which cannot be controlled. The above data illustrate that there is room to improve nutrient use efficiency at the farm level, especially for N.

While most of the focus on nutrient efficiency is on N, phosphorus (P) efficiency is also of interest because it is one of the least available and least mobile mineral nutrients. First-year recovery of applied fertilizer P ranges from less than 10 % to as high as 30 %. However, because fertilizer P is considered immobile in the soil and reaction (fixation and/or precipitation) with other soil minerals is relatively slow, long-term recovery of P by subsequent crops can be much higher. There is little information available about potassium (K) use efficiency. However, it is generally considered to have a higher use efficiency than N and P because it is immobile in most soils and is not subject to the gaseous losses that N is or the fixation reactions that affect P. First-year recovery of applied K can range from 20 to 60 %.

4 Optimizing Nutrient Use Efficiency

The fertilizer industry supports applying nutrients at the right rate, right time, and in the right place as a best management practice (BMP) for achieving optimum nutrient efficiency.

Right Rate Most crops are location and season specific depending on cultivar, management

practices, climate, etc., and so it is critical that realistic yield goals are established and that nutrients are applied to meet the target yield. Over- or under-application will result in reduced nutrient use efficiency or losses in yield and crop quality. Soil testing remains one of the most powerful tools available for determining the nutrient supplying capacity of the soil, but to be useful for making appropriate fertilizer recommendations, good calibration data is also necessary. Unfortunately, soil testing is not available in all regions of the world because reliable laboratories using methodology appropriate to local soils and crops are inaccessible or calibration data relevant to current cropping systems and yields are lacking.

Other techniques, such as omission plots, are proving useful in determining the amount of fertilizer required for attaining a yield target (Witt and Dobermann 2002). In this method, N, P, and K are applied at sufficiently high rates to ensure that yield is not limited by an insufficient supply of the added nutrients. Target yield can be determined from plots with unlimited NPK. One nutrient is omitted from the plots to determine a nutrient-limited yield. For example, an N omission plot receives no N, but sufficient P and K fertilizer to ensure that those nutrients are not limiting yield. The difference in grain yield between a fully fertilized plot and an N omission plot is the deficit between the crop demand for N and indigenous supply of N, which must be met by fertilizers.

Nutrients removed in crops are also an important consideration. Unless nutrients removed in harvested grain and crop residues are replaced, soil fertility will be depleted.

Right Time Greater synchrony between crop demand and nutrient supply is necessary to improve nutrient use efficiency, especially for N (Johnson et al. 1997). Split applications of N during the growing season, rather than a single, large application prior to planting, are known to be effective in increasing N use efficiency (Cassman et al. 2002). Tissue testing is a well-known method used to assess N status of growing crops, but other diagnostic tools are also available. Chlorophyll meters have proven useful in fine-tuning in-season N management (Francis and Piekielek 1999), and leaf color charts have been highly successful in guiding split N applications in rice and now maize production in Asia (Witt et al. 2005). Precision farming technologies have introduced, and now commercialized, on-the-go N sensors that can be coupled with variable rate fertilizer applicators to automatically correct crop N deficiencies on a site-specific basis.

Another approach to synchronize release of N from fertilizers with crop need is the use of N stabilizers and controlled-release fertilizers. Nitrogen stabilizers (e.g., nitrapyrin, DCD [dicyandiamide], NBPT [n-butyl-thiophosphor-ictriamide]) inhibit nitrification or urease activity, thereby slowing the conversion of the fertilizer to nitrate (Havlin et al. 2005). When soil and environmental conditions are favorable for nitrate losses, treatment with a stabilizer will often increase fertilizer N efficiency. Controlled-release fertilizers can be grouped into compounds of low solubility and coated water-soluble fertilizers.

Most slow-release fertilizers are more expensive than water-soluble N fertilizers and have traditionally been used for high-value horticulture crops and turf grass. However, technology improvements have reduced manufacturing costs where controlled-release fertilizers are available for use in corn, wheat, and other commodity grains (Blaylock et al. 2005). The most promising for widespread agricultural use are polymer-coated products, which can be designed to release nutrients in a controlled manner. Nutrient release rates are controlled by manipulating the properties of the polymer coating and are generally predictable when average temperature and moisture conditions can be estimated.

Right Place Application method has always been critical in ensuring fertilizer nutrients are used efficiently. Determining the right placement is as important as determining the right application rate. Numerous placements are available, but most generally involve surface or subsurface applications before or after planting. Prior to planting, nutrients can be broadcast (i.e., applied uniformly on the soil surface and may or may not be incorporated), applied as a band on the surface, or applied as a subsurface band, usually 5–20 cm deep. Applied at planting, nutrients can be banded with the seed, below the seed, or below and to the side of the seed. After planting, application is usually restricted to N and placement can be as a topdress or a subsurface sidedress. In general, nutrient recovery efficiency tends to be higher with banded applications because less contact with the soil lessens the opportunity for nutrient loss due to leaching or fixation reactions. Placement decisions depend on the crop and soil conditions, which interact to influence nutrient uptake and availability.

Plant nutrients rarely work in isolation. Interactions among nutrients are important because a deficiency of one restricts the uptake and use of another. Numerous studies have demonstrated those interactions between N and other nutrients, primarily P and K, impact crop yields, and N efficiency. For example, data from a large number of multi-location on-farm field experiments conducted in India show the importance of balanced fertilization in increasing crop yield and improving N efficiency (Table 3).

Adequate and balanced application of fertilizer nutrients is one of the most common practices for improving the efficiency of N fertilizer and is equally effective in both developing and developed countries. In a recent review based on 241 site-years of experiments in China, India, and North America, balanced fertilization with N, P, and K increased first-year recoveries an average of 54 % compared to

Table 3 Effect of balanced fertilization on yield and N agronomic efficiency

| Crop | Yield, t ha^{-1} | | | Agronomic efficiency, kg grain kg N^{-1} | | |
	Control	N alone	+PK	N alone	+PK	Increase
Rice (wet season)	2.74	3.28	3.82	13.5	27.0	13.5
Rice (summer)	3.03	3.45	6.27	10.5	81.0	69.5
Wheat	1.45	1.88	2.25	10.8	20.0	9.2
Pearl millet	1.05	1.24	1.65	4.7	15.0	10.3
Maize	1.67	2.45	3.23	19.5	39.0	19.5
Sorghum	1.27	1.48	1.75	5.3	12.0	6.7
Sugarcane	47.2	59.0	81.4	78.7	227.7	150.0

Assumes a typical N harvest index of 56 %

Table 4 Effect of cropping system and fertility level on agronomic N use efficiency, physiological N use efficiency, apparent N recovery, N efficiency ratio, physiological efficiency index of N, and N harvest index in Bt cotton (mean data of 2 years)

Treatment	ANUE (kg seed cotton kg N^{-1})[a]	PNUE (kg seed cotton kg N^{-1})[b]	ANR (%)[c]	NER (kg DM kg N uptake^{-1})[d]	PEIN (kg seed cotton kg N uptake^{-1})[e]	NHI (%)[f]
Cropping system						
Sole cotton	–	–	–	46.25	17.0	38.0
Cotton + groundnut (1:3)	–	–	–	44.1	13.8	33.5
Fertility level (recommended dose of N: 150 kg ha^{-1})						
Control (0N)	–	–	–	55.8	15.1	35.3
100 % urea	8.2	11.8	69.3	41.7	13.7	32.9
75 % urea + 25 % FYM	9.5	11.44	83.3	40.0	13.4	37.6
50 % urea + 50 % FYM	5.27	13.17	40.0	50.0	14.5	42.6

[a](Yield in treatment plot-yield in control)/kg N applied
[b](Yield in treatment plot-yield in control)/(N uptake in treatment plot−N uptake in control)
[c](N uptake in treatment plot−N uptake in control)/kg N applied
[d](Dry matter yield/N accumulated at harvest)
[e](Seed cotton yield/N absorbed by biomass)
[f](N uptake by seed cotton/N uptake by whole plant)*100

recoveries of only 21 % where N was applied alone (Fixen et al. 2005).

A variety of practices and improvements are suggested in the scientific literature to increase nutrient use efficiency in agriculture, such as the adoption of multiple cropping systems, improved crop rotations, or intercropping. Because of escalating costs of chemical fertilizers, the nutrient uptake and utilization in field crops should be most efficient to cause reductions in the cost of production and achieve greater profit for resource-poor farmers. To arrive at these objectives, it is important to understand and enhance nutrient use efficiency. Singh and Ahlawat (2012) concluded that substitution of 25 % recommended dose of N (RDN) through FYM recorded the greatest Agronomic Use Efficiency (ANUE) and Apparent Nitrogen Recovery (ANR) followed by 100 % RDN through urea, whereas 50 % RDN substitution recorded the least ANUE and ANR (Table 4). Substitution of 50 % RDN followed by 25 % RDN substitution recorded the greatest Physiological Nitrogen Use Efficiency (PNUE), whereas 100 % RDN through urea recorded the least PNUE. Sole cotton maintained the greatest Nitrogen Efficiency Ratio (NER), Physiological Efficiency Index of Nitrogen (PEIN), and Nitrogen Harvest Index

(NHI) over cotton + groundnut. The greatest NER, PEIN, and NHI were recorded in the unfertilized control treatment followed by 50 % RDN substitution through FYM. The least NER, PEIN, and NHI were recorded with 25 % RDN substitution. The greatest ANUE and ANR by application of 25 % RDN substitution through FYM could be attributed to increase in seed cotton yield with combined application of inorganic and organic sources of N (Bandyopadhyay et al. 2009; Rao et al. 1991). Another reason might be that it improved N uptake of crop because of the increased humus content of soil, which would have slowed down release of ammoniacal N and its conversion to nitrates, thereby reducing the leaching loss of N (Silvertooth et al. 2001; Fritschi et al. 2004). High N availability in 25 % RDN substitution through FYM stimulated the development of larger plants and a more extensive root system capable of supplying the increased water and nutrients demanded by the larger plants. The cotton crop, therefore, drew from a larger pool of both added and indigenous N, which influenced the efficiency of fertilizer N (recovery vs. applied) as well overall N efficiency (Boquet and Breitenbeck 2000). The greatest NER, PEIN, and NHI were attributed to the better physical, chemical, and biological properties of soil that would have caused greater nutrient uptake and yield, leading to better fertilizer use efficiencies.

Mohanty et al. (1998) observed relatively higher NUE of rice with urea as compared with combined use of GM and urea up to 80 kg N ha^{-1} (Table 5). However, the trend was reverse at 120 kg N ha^{-1}.

Agroforestry, which includes trees in a cropping system, may improve pest control and increase nutrient- and water-use efficiency. Also, cover crops or reduced tillage can reduce nutrient leaching. Nutrient use efficiency is increased by appropriately applying fertilizers and by better matching temporal and spatial nutrient supply with plant uptake (Tilman et al. 2002). Applying fertilizers during periods of highest crop uptake, at or near the point of uptake (roots and leaves), as well as in smaller and more frequent applications have the potential to reduce losses while maintaining or improving crop

Table 5 Nitrogen use efficiency in rice through integrated nutrient management

Treatment	ANR(%)		AE	PE
	1st rice	2nd rice	1st rice	1st rice
		N0		
GM-N40 + N0	24.8	28.0	18.0	72.7
N40	43.3	44.9	23.5	54.5
GM-N40 + N40	35.6	35.7	15.5	43.5
N80	46.3	43.9	17.1	37.0
GM-N40 + N80	44.3	45.6	14.4	32.5
N120	31.8	30.8	10.3	31.4
GM-N40 + N120	34.4	38.7	9.7	28.2

yield quantity and quality (Cassman et al. 2002). However, controlled release of nitrogen (e.g., via using nitrogen inhibitors) or technologically advanced systems such as precision farming appear to be too expensive for many farmers in developing countries (Singh 2005).

Many of the aforementioned management practices can be supported by targeted research (e.g., on improving efficiency and minimizing losses from both inorganic and organic nutrient sources; on improvements in timing, placing, and splitting of fertilizer applications, as well as by judicious investments, for example, in soil testing).

4.1 Efficient Does Not Necessarily Mean Effective

Improving nutrient efficiency is an appropriate goal for all involved in agriculture, and the fertilizer industry, with the help of scientists and agronomists, is helping farmers work toward that end. However, effectiveness cannot be sacrificed for the sake of efficiency. Much higher nutrient efficiencies could be achieved simply by sacrificing yield, but that would not be economically effective or viable for the farmer, or the environment. This relationship between yield, nutrient efficiency, and the environment was ably described by Dibb (2000) using a theoretical example. For a typical yield response curve, the lower part of the curve is characterized by very low yields, because few nutrients are available or applied, but very high efficiency. Nutrient use efficiency is high at a low yield level, because any small amount of nutrient applied could give a large yield response. If nutrient use efficiency were the only goal, it would be achieved here in the lower part of the yield curve. However, environmental concerns would be significant because poor crop growth means less surface residues to protect the land from wind and water erosion and less root growth to build soil organic matter. As you move up the response curve, yields continue to increase, albeit at a slower rate, and nutrient use efficiency typically declines. However, the extent of the decline will be dictated by the BMPs employed (i.e., right rate, right time, right place, improved balance in nutrient inputs, etc.) as well as soil and climatic conditions.

The relationship between efficiency and effective was further explained when Fixen (2006) suggested that the value of improving nutrient use efficiency is dependent on the effectiveness in meeting the objectives of nutrient use, objectives such as providing economical optimum nourishment to the crop, minimizing nutrient losses from the field, and contributions to system sustainability through soil fertility or other soil quality components. He cited two examples. Saskatchewan data from a long-term wheat study where 3 initial soil test levels were established with initial P applications followed by annual additions of seed-placed P. Fertilizer P recovery efficiency, at the lowest P rate and at the lowest soil test level, was 30 %, an extremely high single-year efficiency. However, this practice would be ineffective because wheat yield was sacrificed.

The second example is from a maize study in Ohio that included a range of soil test K levels and N fertilizer rates. N recovery efficiency can be increased by reducing N rates below optimum yield that is sacrificed. Alternatively, yield and efficiency can be improved by applying an optimum N rate at an optimum soil test K level. Nitrogen efficiency was improved with both approaches but the latter option was most effective in meeting the yield objectives.

5 Different Computation Methods

5.1 Nitrogen Fertilizer Use Efficiency

In isotopic-aided fertilizer experiments, a labeled fertilizer is added to the soil and the amount of fertilizer nutrient that a plant has taken up is determined. In this way different fertilizer practices (placement, timing, sources, etc.) can be studied.

1. *Percent nitrogen derived from fertilizer (Ndff):*

The first parameter to be determined when studying the fertilizer uptake by a crop by means of the isotope techniques is the fraction of the nutrient in the plant derived from the (labeled) fertilizer, i.e., fdff (fraction derived from fertilizer).

$$Y = S/F \times 100,$$

where Y = Amount of labeled fertilizer N in sample (%Ndff)
S = Atom % ^{15}N excess in sample
F = Atom % ^{15}N excess in the labeled fertilizer

2. *Uptake of nitrogen by plants:*

The grain and straw uptake of nitrogen is calculated as follows:

$$\text{Uptake by grain or straw (kg/ha)} = \frac{\%\text{N content in grain or straw} \times \text{grain or straw yield (kg/ha)}}{100}$$

3. *N use efficiency (NUE):*

$$= \frac{\text{Total N uptake (kg/ha)} \times \% \text{ Ndff}}{\text{Rate of fertilizer N applied (kg/ha)}}$$

4. *Residual fertilizer N in soil (kg ha^{-1}):*

$$= \frac{\text{Total N in soil (kg/ha)} \times \% \text{ Ndff}}{100}$$

5. *Unaccounted fertilizer N (%):*

$$= 100 - [\text{fertilizer} - \text{N recovery (\%)} + \text{residual fertilizer} - \text{N in soil}]$$

^{15}N as tracer studies have yielded valuable information on the aspects of:

- Availability of native soil N to crops
- Influence of N carriers associated with the plant recovery studies.
- Impact of immobilization in soil on plant uptake
- Studies of biological interchange in which mineralization and immobilization proceed simultaneously in the same system
- Denitrification loss in or from soil
- Influence of added available N on mineralization
- The relative uptake of NH_4^+ and NO_3^- ions by crop plants and microorganisms
- Placement position in root zone on availability of N fertilizer to crops
- Balance studies as influenced by time and method of N application

5.2 Phosphorus Fertilizer Use Efficiency

Generally phosphorus losses are largely from erosion and surface runoff (Shepherd and Withers 2001). However, P leaching can occur where soil P sorption is low as in sandy soils and with repeated P fertilizer application. The problem of P leaching is accelerated under high input P, and with frequent and heavy rainfall events (Sims et al. 1998). In a sandy loam soil with low P

sorption saturation, P leaching was higher than from a clay (Djodjic et al. 2004). Phosphorus from inorganic fertilizer can be leached to beneath 1.1 m soil depth (Eghball et al. 1996).

6 Nutrient Efficient Plants

Soil Science Society of America (1997) defined nutrient efficient plant as a plant that absorbs, translocates, or utilizes more of a specific nutrient than another plant under conditions of relatively low nutrient availability in the soil or growth media. In the twenty-first century, nutrient efficient plants will play a major role in increasing crop yields compared to the twentieth century, mainly due to limited land and water resources available for crop production, higher cost of inorganic fertilizer inputs, declining trends in crop yields globally, and increasing environmental concerns. Nutrient efficient plants are defined as those plants, which produce higher yields per unit of nutrient, applied or absorbed than other plants (standards) under similar agro-ecological conditions (Fageria et al. 2008). During the last three decades, much research has been conducted to identify and/or breed nutrient efficient plant species or genotypes/cultivars within species and to further understand the mechanisms of nutrient efficiency in crop plants. However, success in releasing nutrient efficient cultivars has been limited. The main reasons for limited success are that the genetics of plant responses to nutrients and plant interactions with environmental variables are not well understood. Complexity of genes involved in nutrient use efficiency for macro- and micronutrients and limited collaborative efforts between breeders, soil scientists, physiologists, and agronomists to evaluate nutrient efficiency issues on a holistic basis have hampered progress in this area. Hence, during the twenty-first century agricultural scientists have tremendous challenges, as well as opportunities, to develop nutrient efficient crop plants and to develop best management practices that increase the plant efficiency for utilization of applied fertilizers. During the twentieth century, breeding for nutritional traits has been proposed as a strategy to improve the efficiency of fertilizer use or to obtain higher yields in low-input agricultural systems. This strategy should continue to receive top priority during the twenty-first century for developing nutrient efficient crop genotypes (Fageria et al. 2008).

Conclusion

Improving nutrient efficiency is a worthy goal and fundamental challenge facing the fertilizer industry and agriculture in general. The opportunities are there and tools are available to accomplish the task of improving the efficiency of applied nutrients. However, we must be cautious that improvements in efficiency do not come at the expense of the farmers' economic viability or the environment. Judicious application of fertilizer BMPs, right rate, right time, right place, and right agronomic practice targeting both high yields and nutrient efficiency will benefit farmers, society, and the environment alike.

References

Bandyopadhyay KK, Prakash AH, Sankranarayanan K, Dharajothi B, Gopalkrishnan N (2009) Effect of irrigation and nitrogen on soil water dynamics, productivity and input use efficiency of Bt cotton in a Vertic Ustropept. Indian J Agric Sci 79(6):448–453

Blaylock AD, Kaufmann J, Dowbenko RD (2005) Nitrogen fertilizer technologies. In: Proceedings of the western nutrient management conference, vol 6, Salt Lake City, Utah, 3–4 March 2005, pp 8–13

Bolland MDA, Gilkes RJ (1998) The chemistry and agronomic effectiveness of phosphate fertilizers. In: Rengel Z (ed) Nutrient use in crop production. Haworth Press, New York, pp 139–163

Bruulsema TW, Fixen PE, Snyder CS (2004) Fertilizer nutrient recovery in sustainable cropping systems. Better Crops 88:1517

Bouquet DJ, Breitenbeck GA (2000) Nitrogen rate effect on partitioning and dry matter of cotton. Crop Sci 40:1685–1693

Cassman KG, Dobermann A, Walters D (2002) Agroecosystems, nitrogen-use efficiency, and nitrogen management. AMBIO 31:132–140

Dibb DW (2000) The mysteries (myths) of nutrient use efficiency. Better Crops 84:3–5

Djodjic F, Börling K, Bergström L (2004) Phosphorus leaching in relation to soil type and soil phosphorus content. J Environ Qual 33:678–684

Dobermann A, Cassman KG, Waters DT, Witt C (2005) Balancing short- and long-term goals in nutrient management. In: Proceedings of the XV international plant nutrient colloquium, 14–16 September 2005, Beijing, China

Eghball B, Binford GD, Baltensperger DD (1996) Phosphorus movement and adsorption in a soil receiving long-term manure and fertilizer application. J Environ Qual 25:1339–1343

Fageria NK, Baligar VC, Li YC (2008) The role of nutrient efficient plants in improving crop yields in the twenty first century. J Plant Nutr 31(6):1121–1157. doi:10.1080/01904160802116068

Fixen PE (2005) Understanding and improving nutrient use efficiency as an application of information technology. In: Proceedings of the symposium on information technology in soil fertility and fertilizer management, a satellite symposium at the XV international plant nutrient colloquium, 14–16 September 2005, Beijing, China

Fixen PE (2006) Turning challenges into opportunities. In: Proceedings of the fluid forum, fluids: balancing fertility and economics. Fluid Fertilizer Foundation, 12–14 February 2006, Scottsdale, Arizona

Fixen PE, Jin J, Tiwari KN, Stauffer MD (2005) Capitalizing on multi-element interactions through balanced nutrition—a pathway to improve nitrogen use efficiency in China, India and North America. Sci China Ser C Life Sci 48:1–11

Francis DD, Piekielek WP (1999) Assessing crop nitrogen needs with chlorophyll meters. Site-specific management guidelines. Potash & Phosphate Institute. SSMG-12. Reference 99082/Item#10-1012

Fritschi FB, Roberts BA, Rains DW, Travis RL, Hutmacher RB (2004) Fate of nitrogen-15 applied to irrigated Acala and Pima cotton. Agron J 96:646–655

Havlin JL, Beaton JD, Tisdale SL, Nelson WL (2005) Soil fertility and fertilizers. An introduction to nutrient management. Pearson Education, Inc, Upper Saddle River

Johnson JW, Murrell TS, Reetz HF (1997) Balanced fertility management: a key to nutrient use efficiency. Better Crops 81:3–5

Ladha JK, Pathak H, Krupnik TJ, Six J, Kessel CV (2005) Efficiency of fertilizer nitrogen in cereal production: retrospects and prospects. Adv Agron 87:85–156

Mohanty SK, Panda MM, Mosier AR, Mahapatra PK, Reddy MD (1998) 15^N balance studies in a rice-green gram cropping system. J Indian Soc Soil Sci 46:232–238

Mosier AR, Syers JK, Freney JR (2004) Agriculture and the nitrogen cycle. Assessing the impacts of fertilizer use on food production and the environment, Scope-65. Island Press, London

Pilbeam CJ (1996) Effect of climate on the recovery in crop and soil of ^{15}N-labelled fertilizer applied to wheat. Fertil Res 45:209–215

Rao ACS, Smith JL, Papendick RI, Parr JF (1991) Influence of added nitrogen interactions in estimating recovery efficiency of labeled nitrogen. Soil Sci Soc Am J 55:1616–1621

Shepherd MA, Withers PJ (2001) Phosphorus leaching from liquid digested sewage sludge applied to sandy soil. J Agric Sci (Camb) 136:433–441

Silvertooth IC, Navarro JC, Nortan ER, Gladima A (2001) Soil and plant recovery of labeled fertilizer nitrogen in irrigated cotton. Arizona Cotton Report, University of Arizona, College of Agriculture and Life Sciences, index at http://ag.arizona.edu/Pubs/Crops/az1224/

Sims JT, Sinnard RR, Joern BC (1998) Phosphorus loss in agricultural drainage: historical perspective and current research. J Environ Qual 27:277–293

Singh U (2005) Integrated nitrogen fertilization for intensive and sustainable agriculture. J Crop Improv 15:259–288

Singh RJ, Ahlawat IPS (2012) Dry matter, nitrogen, phosphorous, and potassium partitioning, accumulation and use efficiency in transgenic cotton based cropping systems. Commun Soil Sci Plant Anal 43 (20):2633–2650. doi:10.1080/00103624.2012.716125

Smil V (2000) Phosphorus in the environment: natural flows and human interferences. Annu Rev Energy Environ 25(1):53–88

Smil V (2002) Nitrogen and food production: proteins for human diets. AMBIO 31(2):126–131

Soil Science Society of America (1997) Glossary of soil science terms. Soil Science Society of America, Madison

Syers JK, Johnston AE, Curtin D (2008) Efficiency of soil and fertilizer phosphorus use, FAO fertilizer and plant nutrition bulletin 18. Food and Agriculture Organization of the United Nations, Rome

Tilman D (2000) Causes, consequences and ethics of biodiversity. Nature 405:208–211

Tilman D (2001) Forecasting agriculturally driven global environmental change. Science 292:281–284

Tilman D, Cassman K, Matson P (2002) Agricultural sustainability and intensive production practices. Nature 418:671–677

Witt C, Dobermann A (2002) A site-specific nutrient management approach for irrigated, lowland rice in Asia. Better Crops Int 16:20–24

Witt C, Fairhurst TH, Griffiths W (2005) Proceedings of 5th national ISP seminar, Johor, Bahru, Malaysia, 27–28 June 2005. Incorporated Society of Planters, pp 1–22

Nutrient and Water Use Efficiency in Soil: The Influence of Geological Mineral Amendments

Binoy Sarkar and Ravi Naidu

Abstract

Mineral amendments are known to improve the physical, chemical and biological properties of soil, which in turn can enhance the efficiency of nutrient and water use by plants. This chapter discusses the current state of the knowledge regarding the application of geological mineral amendments in soil which either helps to retain nutrients in soils or prevents losses of nutrients from soil and directly or indirectly contributes to improve the overall nutrient use efficiency (NUE). A critical analysis of the currently available research information recommends a site-specific (precision) management approach in order to explore the most beneficial effects of the mineral materials for increasing plants' nutrient and water use efficiency. The management practices should include an integrated plant nutrition system (IPNS) for the best utilisation of resources including mineral materials, fertilisers and organic inputs. This holds the potential for leading to a reduced fertiliser input in modern agriculture and therefore may lower the cost of agricultural production without impacting the crop yield.

Keywords

Nutrient use efficiency • Mineral amendments • Integrated plant nutrition system • Agricultural production

B. Sarkar (✉)
CERAR-Centre for Environmental Risk Assessment and Remediation, Building X, University of South Australia, Mawson Lakes, SA 5095, Australia
e-mail: binoy.sarkar@unisa.edu.au

R. Naidu
CRC CARE-Cooperative Research Centre for Contamination Assessment and Remediation of the Environment, P.O. Box 486 Salisbury, SA 5106, Australia
e-mail: ravi.naidu@crccare.com

1 Introduction

Numerous agricultural soils in the world are inherently deficient in one or more essential nutrients which are required as part of a sustainable crop production system. In the current era where scientific developments are being made almost weekly, the most intensive farming systems ever known are being introduced to

A. Rakshit et al. (eds.), *Nutrient Use Efficiency: from Basics to Advances*,
DOI 10.1007/978-81-322-2169-2_3, © Springer India 2015

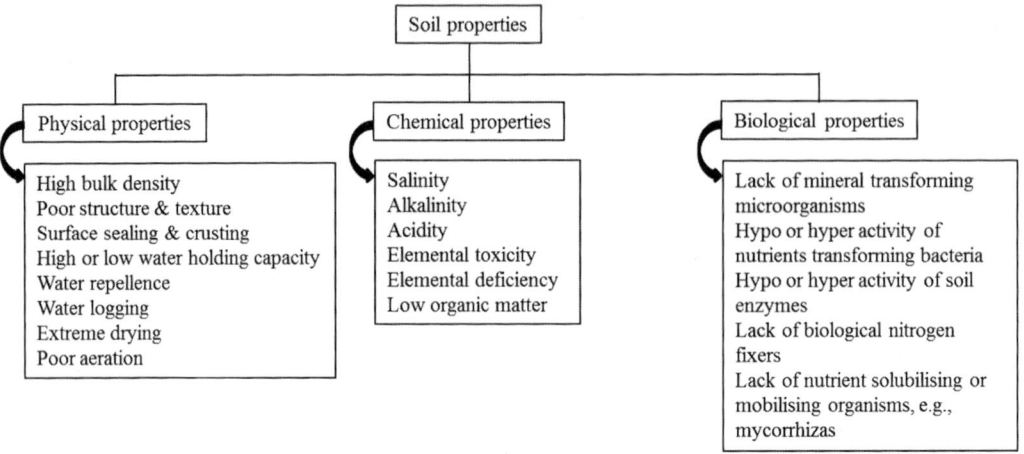

Fig. 1 Soil properties which directly or indirectly reduce NUE by plants

meet the food requirements of the growing global population, which is projected to reach between 8.3 and 10.9 billion by 2050. However, intensive cultivation systems invariably lead to numerous soil degradation issues (e.g. salinity, acidity, alkalinity, erosion, pollution and water scarcity). Numerous natural and synthetic inputs including chemical fertilisers have been successfully applied as part of best practice for improving soil fertility and productivity. Amongst the numerous inputs, chemical fertilisers are one of the most expensive items commonly used by farmers in order to increase crop yields. However, the recovery and utilisation of applied nutrients by plants in many soils is generally low. It has been estimated that the overall efficiency of applied fertilisers is approximately 50 % or lower for N, less than 10 % for P and 20–40 % for K (Baligar and Bennett 1986a, 1986b). Losses by different processes such as leaching, run-off, gaseous emission and fixation by soil are the significant factors that lower the use efficiencies of the applied nutrients (Baligar et al. 2001). These losses not only lower the crop yields, but they also contribute to potential soil degradation and can impair the local water quality (Baligar et al. 2001). Therefore, this chapter discusses the current state of the knowledge regarding the application of geological mineral amendments in soil which either helps to retain

nutrients in soils or prevents losses of nutrients from soil and directly or indirectly contribute to improve the overall nutrient use efficiency (NUE).

2 NUE and Soil Properties Affecting It

Several authors have discussed the topic of plant NUE (Epstein 1972; Vitousek 1982; Blair 1993; Baligar et al. 2001; Hawkesford 2011). The NUE is defined as the ability of a genotype/cultivar to acquire nutrients from growth medium and/or to incorporate or utilise them in the production of shoot and root biomass or utilisable plant material (e.g. seed, grain, fruit and forage) (Blair 1993). Under heterogeneous environmental and ecological conditions, the genetic and physiological traits of a crop plant primarily control its NUE. Furthermore, numerous soil properties directly or indirectly affect the NUE by plants (Baligar et al. 2001). Soil properties encompass a range of physical, chemical and biological factors (Fig. 1). By applying geological mineral amendments to the soil, many of these properties can be enhanced or repaired, which results in improved NUE by plants.

Table 1 Use of mineral amendments for retaining nutrients in compost materials

Mineral	Compost type	Nutrients form retained	Retention mechanism	References
Pyrite	Phosphocompost	NH_4^+-N and P	pH reduction reduces NH_3 volatilisation	Bangar et al. (1988)
Alum	Swine manure compost	NH_4^+-N	pH reduction reduces NH_3 volatilisation	Bautista et al. (2011)
Clinoptilolite	Swine manure compost	NH_4^+-N	Zeolite exchange sites adsorb NH_4^+-N	Bautista et al. (2011)
Clinoptilolite and sepiolite	Pig slurry and wheat straw compost	NH_4^+-N	Zeolite exchange sites adsorb NH_4^+-N	Bernal et al. (1993)
Clinoptilolite	Municipal solid waste compost	NH_4^+-N	Zeolite exchange sites adsorb NH_4^+-N	Gamze Turan and Nuri Ergun (2007)
Clinoptilolite	Cattle manure applied to soil	NH_4^+-N and NO_3^--N	Zeolite exchange sites adsorb NH_4^+-N and render it unavailable to nitrifying bacteria; NO_3^--N leaching is therefore reduced	Gholamhoseini et al. (2013)
Goethite, gibbsite and allophane	Poultry manure compost	Potentially mineralisable N	Clay materials stabilise C in the composts	Bolan et al. (2012)

3 Geological Mineral Amendments

The amendment of soils with geological minerals can directly or indirectly influence the nutrient transformation, nutrient retention, nutrient losses, water retention and their use by plants. Table 1 summarises a number of examples of geological mineral amendments which are used to retain nutrients in composted materials, which improve crop NUE post application. The most commonly used mineral amendments to soils include clay minerals, zeolites, calcite and dolomite, gypsum, phosphate rock, pyrite, alum, waste mica and mineral mixtures in some other industrial by-products (e.g. fly ash and red mud).

3.1 Clay Minerals

Water holding capacity (WHC) is an important factor which affects nutrient chemistry and, hence, availability to plants. Light-textured soils usually have poor WHC. In some parts of the world, especially in arid or semiarid regions, soils can become water repellent due to capping of the soil particles by some hydrophobic organic compounds. In those soils, the application of clays or clay-rich subsoils, commonly known as 'claying' or 'clay spreading', has been reported to be very effective in adequately preserving the soil moisture for crop production. The clay is spread over the organic coated sand grains, masking the hydrophobic sand surface and exposing a hydrophilic clay surface (Ward and Oades 1993; Cann 2000). Approximately 43 years ago, Clem Obst, a South Australian farmer, accidentally discovered the ability of clays to counteract the effects of water-repellent sand (Cann 2000). More than 37,000 ha of land in South Australia is now clay spread, of which 32,000 ha is in the south-east of South Australia (Cann 2000). The application of clay to those soils has improved the nutrient and moisture retention in the topsoil, the germination, the establishment and yield of pasture plants and crops and increased the effectiveness of preemergent herbicides (Cann 2000).

Claying (the addition of clay-rich subsurface soils) and deep ripping (breaking up the compacted soil layers using tines down to a depth of 35–50 cm to loosen hard layers of soil) in water-repellent sand plain soils improve water and nutrient retention and are therefore an effective long-term management technology for increasing crop NUE. The low organic carbon and clay content of sand plain topsoils results in poor nutrient-holding capacity (CEC < 3) (Hall et al. 2010). Research has shown that the effects of claying include increased concentrations of organic carbon by 0.2 %, potassium (K) by 47 mg kg^{-1} and cation exchange capacity (CEC) by 1.3 cmol (p$^+$) kg^{-1} in the topsoil (Hall et al. 2010), respectively. The authors reported that claying improved the yield of canola, lupin and barley by as much as 102 % in soils; however, it reduced the rainfall-limited yield potential of these soils to 30–50 % (Hall et al. 2010). The increase in yield was due to a combination of effects including higher plant emergence, improved plant nutrition (in particular K) and near surface water infiltration and distribution (Hall et al. 2010). Full yield potential could not be achieved due to higher soil strength as a result of the clay being applied. Deep ripping may increase the yield by 11–20 % (Hall et al. 2010). A time period of up to 6 years may be required before the combination of claying and deep ripping technologies fully overcomes the water-repellent nature of soils (Hall et al. 2010). The addition of beneficiated bentonite (bentonite saturated with Ca^{2+}, Mg^{2+} and K^+ in a ratio of 8:4:1) at a rate up to 40 t ha^{-1} to degraded Oxisol and Ultisol in tropical Australia permanently improved the basic surface charge which concomitantly caused a significant and sustained increase in forage sorghum yields on both the soil types (Noble et al. 2001). Similarly, acid waste bentonite (a by-product from vegetable oil bleaching) co-composted with rice husk, rice husk ash and chicken litter showed a highly significant increase in maize yields over two consecutive cropping cycles grown on a degraded soil in northern Thailand (Soda et al. 2006).

The increase in organic carbon contents of soil due to the application of clays occurs by the physical binding and protection of organic materials from microbial decomposition by the added minerals (Hall et al. 2010; Bolan et al. 2012; Churchman et al. 2013). The addition of goethite, gibbsite and allophane could potentially increase the half-life of poultry manure compost from 139 days to 620, 806 and 474 days, respectively (Bolan et al. 2012). The stabilisation of carbon in compost by clays was not reported to impair the quality of composts in terms of their ability to improve post-application soil quality parameters (e.g. potentially mineralisable nitrogen and microbial biomass carbon) (Bolan et al. 2012). Following application to soils, these clay-rich composts improve the organic carbon content of soil and reduce carbon loss as CO_2.

The water use efficiency of plants is closely related to the NUE and hence the crop yield. The application of clay to light-textured soils is known to improve the water use efficiency, growth and yield of crop plants (Al-Omran et al. 2005, 2007, 2010; Ismail and Ozawa 2007). Desirable crop performance can be obtained by applying clays to sandy soils which have poor irrigation utilisation efficiency. Remarkable improvement was achieved in cucumber and maize yields (2.5 times as compared to control) through improved water retention and water use efficiency when sandy soils were amended with a clay-rich soil which contained 21 % clay (Ismail and Ozawa 2007). Either by overlaying or incorporating methods (in top 20 cm depth), the clay application reduced water usage by approximately 45–64 % in areas under cucumber and maize cultivation (Ismail and Ozawa 2007). The water content distributions in the root zone area of squash (*Cucurbita pepo*) grown in sandy calcareous soils under surface, and subsurface drip irrigation was significantly improved by the amendment with clay deposits (Fig. 2), which provided a yield increment up to 13 % (Al-Omran et al. 2005).

Microorganisms play crucial roles in the cycling of nutrients in soils. Some specific microorganisms take part in nutrient transformation and make them available to the growing plants, and therefore, the NUE of plants is improved. The most widely used organisms are *Rhizobium* and *Bradyrhizobium* spp., which have the capacity to fix atmospheric N_2 into soil in a symbiotic relationship with leguminous plants (Elsas and Heijnen 1990). In addition to N

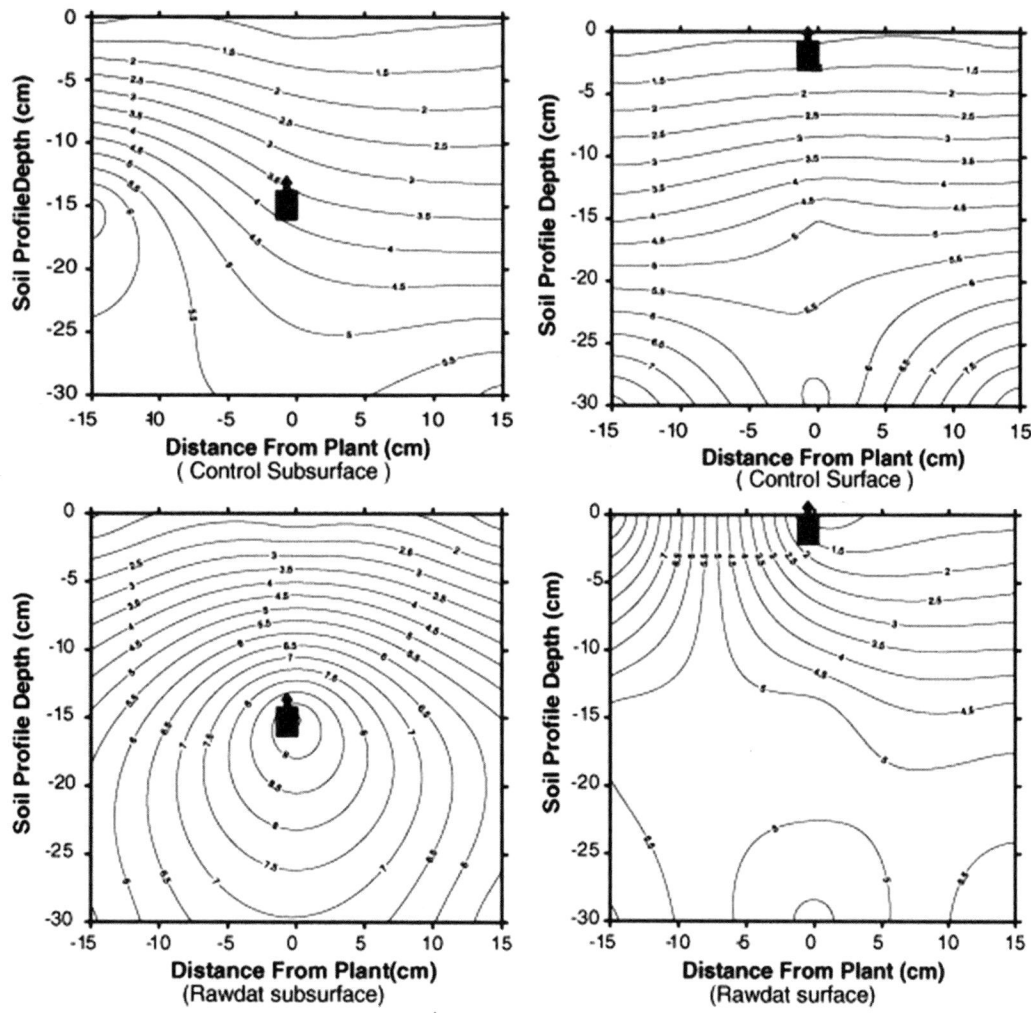

Fig. 2 Water content distributions in the root zone area of squash (*Cucurbita pepo*) grown in sandy calcareous soils amended with Rawdat clay deposit (59 % clay; high smectite content) under surface and subsurface drip irrigation. The control treatment contains no clay amendment. The *black rectangle* represents the position of the irrigation emitter (Al-Omran et al. 2005)

nutrition, the available P status of soils can also be improved by certain species of bacteria, for example, *Bacillus polymyxa*, *Bacillus megaterium* and *Pseudomonas fluorescens*. To obtain the optimum effect in desired nutrient transformation, an application method that facilitates the survival of a sufficiently high number of bacteria in the soils for a longer period of time is required (Elsas and Heijnen 1990; Heijnen and van Veen 1991; Heijnen et al. 1992). At an application rate of 5 %, bentonite clay was successful in improving the

survival of *Rhizobium leguminosarum* biovar *trifolii* which was introduced into a loamy sand soil through creation of large amount of microhabitats which protected the bacterium from protozoan predation (Heynen et al. 1988; Heijnen and van Veen 1991). Provided the rhizobial culture and the clay were mixed thoroughly before introduction into soil, concentrations as low as 0.5 % bentonite improved rhizobial survival compared to a control without bentonite amendments (Heijnen et al. 1992). With the use of less clay, some deleterious effects of clay addition including excessive swelling of the soil could be avoided (Heijnen et al. 1992). The protective effects of clay on the survival of introduced bacteria in soil were the result of an increase in the number of pores (almost doubled) with an equivalent neck diameter <6 μm (Heijnen et al. 1993). The growth and survival of bacterial species including *Pseudomonas fluorescens* and *Bacillus subtilis,* which are known to improve soil available P, was also better supported in clay-rich fine-textured soil as compared to silt loam and coarser loamy sand soils (van Elsas et al. 1986).

3.2 Zeolites

The addition of zeolite in combination with pure organic amendments (e.g. cellulose) improved the nitrogen use efficiency in calcareous sandy soils under citrus cultivation (He et al. 2002). With the application of clinoptilolite and cellulose (both 15 g kg^{-1}), ammonia volatilisation from NH_4NO_3-, $(NH_4)_2SO_4$- and urea-treated soils (applied at the rate 200 mg N kg^{-1} soil) decreased by 4.4-, 2.9- and 3.0-fold, respectively, compared to soils without amendments (He et al. 2002). The organic input in the form of cellulose provided favourable conditions for microbial growth and increased the microbial biomass, which consequently caused N retention by microbial immobilisation. The mineral zeolite improved the N retention further by adsorption of NH_4^+-N in the ion-exchange sites. The interactive effect of cellulose and zeolite amendment was additive on soil microbial biomass which improved nutrient retention and availability of nutrients to microorganisms and citrus plants.

The application of zeolite along with cattle manure was shown to improve the N use efficiency directly in sunflower grown in sandy soils under semiarid conditions (Gholamhoseini et al. 2013). The treatment which combined urea, cattle manure and clinoptilolite (14–21 %) was considerably more effective than urea alone or urea with cattle manure with respect to improving the most quantitative and qualitative traits of sunflower (Gholamhoseini et al. 2013).

Zeolites are usually reported to adsorb the NH_4^+-N contained in composts in their pores. The nitrifying bacteria, which use NH_4^+-N as the precursor for NO_3^- production, cannot access the zeolite pores (Gholamhoseini et al. 2013). Thus, zeolites render NH_4^+-N unavailable to the nitrifying bacteria and decrease the transformation of NH_4^+ to NO_3^-, hence preventing NO_3^- leaching. Since zeolite reduces N leaching in such conditions, it increases the plant-available N and consequently the N use efficiency. In addition, the application of clinoptilolite reduced P leaching; however, the effect was more prominent in reducing the N leaching (Gholamhoseini et al. 2013).

As a naturally occurring mineral, zeolite has the potential to stabilise nutrients in various compost materials (Bernal et al. 1993; Lefcourt and Meisinger 2001; He et al. 2002; Gholamhoseini et al. 2013). The degree of retention of NH_4^+-N by zeolites may vary depending on the pore diameter of these materials. For example, clinoptilolite has a higher capacity for water adsorption than phillipsite due to its larger pore diameter (Hayhurst 1978), and NH_4^+-N adsorption by zeolites is inversely related to water adsorption capacity (Bernal et al. 1993). In these materials, the adsorbed water blocks the internal channels against NH_4^+-N adsorption. Therefore, the effectiveness of zeolites in retaining NH_4^+-N, and consequently their optimum rate of application, is largely dependent on the water loss characteristics during the composting process.

3.3 Calcite and Dolomite

The application of calcite ($CaCO_3$) and dolomite ($CaMg(CO_3)_2$) as liming materials is well known for the management of acid soils (Haynes and Naidu 1998; Naidu et al. 1990a, b; Naidu and Syers 1992; Fageria and Baligar 2008). The application of lime increases the soil pH at which a negative charge on the surface clay colloids develops and a repulsive force between soils particles dominates (Naidu et al. 1990b; Haynes and Naidu 1998). It also causes an increase in both the Ca^{2+} concentration and the ionic strength in the soil solution. As a result, a compression of the electrical double layer occurs which promotes flocculation of the soil particles at a higher lime rate (Haynes and Naidu 1998). This improves the quantity and quality of organic matter, the soil structure and the congenial nutrient transformation in soils subsequently.

In addition, weathering of silicate minerals is a significant source of plant nutrients in soils. Bacteria primarily take part in this type of mineral weathering. Amendment with calco-magnesium mineral (liming) in mountainous forest soils (which typically have a high organic matter content and acidic pH) could improve the weathering rate of phyllosilicate minerals by bacteria compared to that in the unamended soils (Balland-Bolou-Bi and Poszwa 2012). The calco-magnesium mineral amendment increased the availability of some existing nutrients in the soils, which subsequently enhanced bacterial silicate mineral weathering potential (Balland-Bolou-Bi and Poszwa 2012).

The decline of forest vegetation due to anthropogenic acidification is a common problem in mountainous soils in Europe and elsewhere. The application of dolomite in those soils raises the pH and the concentrations of the base cations (Ca and Mg), which concomitantly decreases heavy-metal toxicity to plants (Ingerslev 1997). The diversity of the acidobacterial and gram-positive groups declines, and the diversity of the proteobacterial community improves, due to dolomite application to soils (Clivot et al. 2012). The ratio between *Proteobacteria* and *Acidobacteria*, which increases as a result of dolomite application, serves as a microbial indicator of soil quality improvement (Hartman et al. 2008; Clivot et al. 2012).

The application of calcite ($CaCO_3$) to soil as a liming material can result in long-term changes in the humus structure of soils. Soil humus can be classified into three types, namely, mor, moder and mull (Green et al. 1993). Mor is the type of humus which arise under conditions of low-biological activity in soil and contains a C:N ratio of more than 20 and sometimes 30–40. Moder is a transitional form of humus between mull and mor; it has a C:N ratio of 15–25. Mull is a well-humified organic matter, which is produced in biologically very active habitat and contains a C:N ratio of 10. After approximately 22 years of application, one study reported that the humus structure in calcite-treated forest soils receiving NPK fertilisers evolved from the 'moder' to the 'mull-moder' type as compared to the moder-type humus in control soils (Deleporte and Tillier 1999). This subsequently altered the soil faunal communities; the lumbricid population (epigeic species) in those calcite-treated soils increased (Deleporte and Tillier 1999).

3.4 Gypsum

Gypsum ($CaSO_4·2H_2O$) is a key mineral material for maintaining agricultural production in soils affected by sodicity. The altered electron and proton activities (pE and pH) in sodic soils, which are produced as a result of a degraded soil structural environment, create nutrient constraints in such soils (Naidu and Rengasamy 1993). Upon application to sodic soils, gypsum increases the stability of soil organic matter, leads to the formation of more stable soil aggregates, improves water penetration into the soil and facilitates more rapid seed emergence (Wallace 1994). The application of gypsum is known to increase the growth and yield of numerous crops (e.g. wheat (Whitfield et al. 1989; Thomas et al. 1995), sorghum (Thomas et al. 1995), maize (Toma et al. 1999) and alfalfa (Toma et al. 1999)). Gypsum helps to improve soil structure by better flocculation and aggregation and rupture soil strength (Rengasamy and Olsson 1991; Rengasamy et al. 1993). The long-term effects of gypsum application to soils are the alteration of soil pH and increased amounts of exchangeable Ca and SO_4^{2-} (Toma et al. 1999). Soil pH plays a crucial role in the transformation of fertilisers in soils; the loss of NH_3 from nitrogenous fertiliser as a result of volatilisation is accelerated in an alkaline soil, and thus, the N use efficiency is reduced (Bolan et al. 2004). In such cases, the correction of soil pH which results from applying amendments can improve NUE by plants.

3.5   Phosphate Rock

The mineral constituents in phosphate rocks are generally apatites, crandallites, millisites, silica and calcite. Amongst these hydroxyapatite, $(Ca_{10}(PO_4)_6(OH)_2)$ is considered as the most important P-containing mineral which can supply P to plants following direct application to soils. However, the agronomic effectiveness of phosphate rocks depends on many factors: (a) the chemical nature and physical form of the product, (b) soil properties, (c) type of crop species grown, (d) climatic conditions, and (e) method of measuring reactive P in phosphate rock and soil (Bolan et al. 1990; Chien and Menon 1995). The agronomic effectiveness is assessed against a standard P-supplying fertiliser (e.g. SSP) as shown below (Eq. 1) and referred to as relative agronomic effectiveness (RAE) (Bolan et al. 1990).

$$RAE = 100$$
$$\times \frac{[(\text{Yield with phosphate rock}) - (\text{Yield in control})]}{[(\text{Yield with SSP}) - (\text{Yield in control})]}$$
$$(1)$$

Aluminium (Al) toxicity in acid soils and associated low P availability can be effectively addressed by direct application of phosphate rocks to soils (Easterwood et al. 1989; Rajan et al. 1996). This mineral amendment is ideally suited for long-term crops such as permanent pastures and plantation crops, but can show immediate seasonal effects in plants grown under acidic soil conditions (Rajan et al. 1996). Since depending on their sources phosphate rocks might contain significant amount of heavy trace elements, care should be taken for their direct application to soils for avoiding a potential heavy-metal build-up (Raven and Loeppert 1997).

The low reactivity of phosphate rocks (less plant-available P fraction) in certain soil types (neutral and alkaline) can be overcome by simple partial acidulation, co-composting and inoculation with microorganisms (Begum et al. 2004; Biswas and Narayanasamy 2006; Biswas et al. 2009; Biswas 2011). Microorganisms such as fungi including arbuscular mycorrhiza and P-solubilising bacteria have been proved to be efficient inoculant for increasing P availability from phosphate rocks both in soils and compost materials (Vassilev et al. 1995; Toro et al. 1997; Biswas and Narayanasamy 2006; Park et al. 2010).

3.6   Pyrite and Alum

Some mineral materials (e.g. pyrite and alum) have the ability to correct high pH and reduce nutrient loss from compost materials during their production and after application to soils. Approximately 33–62 % of the initial total N of manure may be lost during composting if some critical parameters (e.g. pH, moisture content and temperature) are not properly controlled (Kithome et al. 1999). The loss of N from phosphocompost enriched with nitrogenous compounds is significantly reduced by amending the compost with Fe-bearing mineral (e.g. pyrite (FeS_2)) (Bangar et al. 1988). Since a high pH value promotes N loss through NH_3 volatilisation, the role of pyrite in retaining N in compost is attributed to a pH reduction effect. The addition of other amendments like alum and zeolite to swine manure also offers a high potential for reducing NH_3 loss and increases the stability of the final compost (Bautista et al. 2011). A systematic application of these two amendments can reduce NH_3 emissions by 85–92 %, with the final compost retaining three fold more NH_4^+-N than the unamended control (Bautista et al. 2011). Furthermore, by sequestering 44 % of the retained NH_4^+-N at exchange sites, zeolite remarkably improves the quality of the compost, which acts like a slow-release fertiliser upon application to agricultural soils (Bautista et al. 2011). Thus, mineral-amended composts after application to soils improve the NUE by plants.

The application of pyrite to some problem soils (e.g. calcareous soils) can increase the availability of certain trace elements to crops, increase the nutritive parameters of forages and increase overall dry matter production. This is a very effective strategy to revegetate abandoned mine site soils. An example of the successful application of pyrite occurs in calcareous Cambisol soils in western Portugal (Castelo-Branco et al. 1999). The pyrite did not appear to pollute the surface waters or promote toxicological problems in grazing animals (Castelo-Branco et al. 1999). Thus, pyrite obtained from the ore milling industry was effective both as a soil amendment and a fertiliser for agricultural crops.

3.7 Waste Mica

Waste mica is generated during the processing of raw micas. Low-grade waste mica contains about 8–10 % K which is not readily available to plants. Most of the K in waste mica exists as structural and non-exchangeable forms. However, waste mica can effectively supply K nutrition to plants following suitable chemical and/or biological modifications. Composting has recently evolved as an efficient technology for bringing the unavailable K in waste mica into plant-available forms (water soluble and exchangeable) (Nishanth and Biswas 2008; Biswas et al. 2009; Basak and Biswas 2009, 2010; Biswas 2011). The acidic environment which prevails during composting facilitates the process (Nishanth and Biswas 2008; Biswas et al. 2009). In addition, bio-intervention of waste mica with K-solubilising bacteria (*Bacillus mucilaginosus*), N-fixing bacteria (*Azotobacter chroococcum*) and fungi (*Aspergillus awamori*) in the presence or absence of phosphate rock was effective in providing K, N and P nutrition to various crops (sudan grass, wheat and maize) (Nishanth and Biswas 2008; Basak and Biswas 2009, 2010; Singh et al. 2010).

3.8 Mineral Mixtures in Other Industrial By-Products

Several mineral-rich industrial by-products have been applied to agricultural soils in the recent years. These mainly include fly ash and red mud; the former is a by-product of coal-fired thermal power plants, whereas the latter is generated in bauxite refining factory.

3.8.1 Fly Ash

The mineralogical composition of fly ash is highly heterogeneous, depending on the raw materials (lignite, bituminous coal, etc.) and the composition and source of the coal used. Fly-ash components may include feldspars, calcite, anhydrite, quartz, calcium silicates, silica, alumina, iron oxides and high amounts of amorphous phases (Koukouzas et al. 2007; Kostakis 2009; Mishra and Das 2010). Fly ash typically contains both available and fixed forms of nutrient elements. It can be used as a more efficient source of plant nutrients than chemical fertiliser, because of the availability of nutrient elements in the former over a longer period of time (Ramesh et al. 2007; Pandey et al. 2009; Pandey and Singh 2010; Ukwattage et al. 2013). However, excessive application may cause deleterious effects including an increase in heavy-metal concentrations and the immobilisation of nutrients (Ramesh et al. 2007; Singh et al. 2008). If applied judiciously to agricultural soils, fly ash has been proven to be beneficial to crops in terms of nutrient availability and improved water retention. The beneficial effects of fly-ash application on crop productivity and nutrient uptake are improvements to soil texture and water holding capacity (WHC), reduced soil crusting and increased availability of nutrients (Srivastva and Chhonkar 2000; Gaind and Gaur 2002; Seshadri et al. 2013). A judicious application of fluidised bed combustion (FBC) ash could increase P nutrition to Indian mustard (*Brassica hirta* L.) by mineralising organic P into available P forms and immobilising inorganic P and later mobilising the bound P into available P for the second crop (Seshadri et al. 2013). A list of agricultural crops grown with fly-ash amendment in soils is given in Table 2.

Fly-ash application has distinctive effects on the soil physical properties that promote crop growth and yield. Its application to texturally variable soils (e.g. sandy clay loam, sandy and sandy loam soils) increased moisture retention at

Table 2 Application of fly ash for growing agricultural crops

Crop	Amendment	Reason for increased production	Toxicity consideration	References
Rice (*Oryza sativa* L.)	Fly ash applied to soil	Fly ash supplies plant-available silicate	Boron toxicity was avoided at a fly-ash application rate of 120 t ha^{-1}	Lee et al. (2008)
	Fly ash applied to soil	Fly ash improved physical properties of the soil, improved the germination percentage of rice seeds and improved uptake of K, P, Mn, Zn and Cu	Rice grain accumulated heavy metals below allowable limits at a fly-ash application rate of 10 t ha^{-1}	Mishra et al. (2007)
Wheat (*Triticum aestivum* L.)	Fly ash applied as foliar spray	Fly ash reduced the infestation of *Alternaria triticina* and improved the uptake of S, P, K and Ca	No adverse effect was observed at a fly-ash application rate of 5.0 g plant^{-1} day^{-1}	Singh and Siddiqui (2003)
	Fly ash applied to soil	Fly ash reduced the hydraulic conductivity and improved moisture retention at field capacity and wilting point and marginally increased the uptake of micronutrients	Wheat yield increased without any adverse effects at an application rate of up to 20 t ha^{-1}	Kalra et al. (1998)
Maize (*Zea mays* L.)	Fly ash applied to soil	Fly ash reduced the hydraulic conductivity and improved moisture retention at field capacity and wilting point and marginally increased the uptake of micronutrients	Maize yield increased up to an application rate of 10 t ha^{-1} without any adverse effects	Kalra et al. (1998)
	Fly ash applied to soil	Fly ash raised the pH of acidic soils and helped to reduce the metal solubility and availability to plants	Maize yield increased without any adverse effects up to an application rate of 5 t ha^{-1}	Shende et al. (1994)
Soybean (*Glycine max* L.)	Fly ash applied to soil along with *Pseudomonas striata*	*P. striata* solubilised P from fly ash and improved P uptake by grain	A does rate of 40 t ha^{-1} did not adversely affect the inoculated bacteria and improved the yield	Gaind and Gaur (2002)
Mustard (*Brassica juncea* L.)	Fly ash applied to soil	Fly ash reduced the hydraulic conductivity and improved moisture retention at field capacity and wilting point; however, it marginally increased uptake of micronutrients	Mustard yield increased without any adverse effects up to an application rate of 10 t ha^{-1}	Kalra et al. (1998)
Palak (*Beta vulgaris* L.)	Fly ash applied to soil	Negatively impacted the palak growth and yield	An application rate of 5 % was unsuitable for leafy vegetable production; plants accumulated toxic level of heavy metals	Singh et al. (2008)
Tomato (*Lycopersicon esculentum*)	Fly ash applied to soil	Fly ash increased porosity, water holding capacity, pH, conductivity, CEC, sulphate, carbonate, bicarbonate, chloride, P, K, Ca, Mg, Mn, Cu, Zn and B uptake	An application rate of 60 % gave optimum growth without any toxic effect	Khan and Khan (1996)

field capacity, whereas the reverse trend was noted for clayey soil (Kalra et al. 2000). The moisture retention at wilting point, however, improved in all the soils with fly-ash application (Kalra et al. 2000). The incorporation of the ash in soils created significant modification in the macro- and microparticles in the soils and their pore sizes, which consequently improved the moisture retention constants (Kalra et al. 1997, 2000; Yunusa et al. 2011).

The dose at which the application of fly ash to soils is not harmful in the soil-plant systems depends largely on the soil type. For example, an application rate of up to $100 \, t \, ha^{-1}$ fly ash was believed to be safe for the microbial communities living in a tropical red lateritic soil (Roy and Joy 2011). The grain yield of maize and mustard increased in fly-ash-amended soils with a maximum dose of $10 \, t \, ha^{-1}$ (Kalra et al. 1998). The P- and S-mineralising microbial functions (e.g. phosphatase and aryl sulphatise enzyme activities) were unaffected by comparatively higher application rates of fly ash in soils (Roy and Joy 2011; Seshadri et al. 2013).

Fly ash may promote heavy-metal accumulation in soils and their increased uptake by plants (Singh et al. 2008). A number of studies have reported the immobilisation or stabilisation of heavy metals by fly ash in contaminated soils and reduced plant uptake (Dermatas and Meng 2003; Bertocchi et al. 2006). The risk of entrenching heavy metals in the food chain through the application of fly ash could be avoided if the material is used judiciously at the optimum dose, or it is used to grow nonedible economic plants (e.g. trees in the forest). A considerably greater dose of fly ash (66 % by volume) mixed with compost was reported to be a better alternative source of nutrients and a good amendment for the creation of more favourable soil conditions in dry land forests (Ramesh et al. 2007). The growth of teak (*Tectona grandis*) and leucaena (*Leucaena leucocephala*) was enhanced by an increased availability of major nutrients (e.g., P, K, Ca and Na) as supplied by fly ash which was mixed with compost (Ramesh et al. 2007). The enhanced growth of a biodiesel plant *Jatropha curcas* (measured in terms of chlorophyll content in the leaves) was reported as a result of soil amendment with fly ash at a dosage of up to 20 % (Mohan 2011).

There is evidence that fly ash promotes biological activities in soils and thereby improves plant nutrient uptake. At an application rate of up to $40 \, t \, ha^{-1}$, fly ash was compatible with P-solubilising bacteria (PSB) (e.g., *Pseudomonas striata*) in a sandy loam soil and significantly improved soybean productivity by increasing P

supply (Gaind and Gaur 2002). The nitrogen uptake by willow plants (*Salix* spp.) grown in a fly-ash dump was greatly improved by inoculation with *Sphingomonas* sp. which stimulated the formation of ectomycorrhizae with an autochthonous *Geopora* sp. strain (Hrynkiewicz et al. 2009). This significantly increased the shoot growth of two *Salix viminalis* clones and the root growth of a *S. viminalis* × *caprea* hybrid clone grown on fly-ash-amended soil (Hrynkiewicz et al. 2009). A greater yield of maize was achieved in soil layers overlying coal fly ash which was colonised by two arbuscular mycorrhizal fungi (*Glomus mosseae* and *Glomus versiforme*) (Bi et al. 2003). The results were attributed to the greater absorption of nutrients by the mycorrhizal plants than the non-mycorrhizal controls grown in fly-ash-amended soil (Bi et al. 2003).

3.8.2 Red Mud

India is amongst the major producers of alumina in the world and also produces approximately 4 million tons of red mud as a by-product annually (Samal et al. 2013). The mineral constituents in red mud include boehmite (AlOOH), kaolinite ($Al_2Si_2O_5(OH)_4$), quartz (SiO_2), anatase/rutile (TiO_2), diaspore (AlO(OH)), haematite (α-Fe_2O_3), calcite ($CaCO_3$), goethite (FeO(OH)), muscovite ($KAl_2(AlSi_3O_{10})(F,OH)_2$) and tricalcium aluminate ($Ca_3Al_2O_6$) (Liu et al. 2011). In the recent years, red mud has received significant research attention in order to promote its use as an amendment for pollutants in solid (soils) and liquid (wastewater) phases (Bhatnagar et al. 2011; Liu et al. 2011; Feigl et al. 2012; Samal et al. 2013). It can also play a crucial role in reducing the eutrophication of rivers and waterways by retaining nutrients on infertile sandy soils (Ward and Summers 1993; McPharlin et al. 1994; Summers et al. 1996b; Snars et al. 2004). Red mud can also contribute to improved water retention in light-textured soils and can neutralise acidic soils (Ward and Summers 1993).

Like fly ash, red mud also poses a pollution risk to plants due to an extremely high pH and dispersion of soil particles due to excessive sodium. The potential harm to plants caused by a high pH value is often addressed by the incorporation of gypsum into the red mud (Summers et al. 1996a, b). A judicious application rate is required to harness the optimum effects. Red mud improved pasture production in a coarse acidic sandy soil in Western Australia when applied at rates less than 80 t ha^{-1} (Summers et al. 1996a). The improvement in production was attributed to the liming effect of the remnant alkali in red mud (present as Na_2CO_3 which is more soluble than traditional lime $CaCO_3$) (Summers et al. 1996a). The high pH value of red mud-amended soil can also be managed by biological intervention which can subsequently improve the crop yield. A phosphate-solubilising fungi *Aspergillus tubingensis* was effective in reducing the alkalinity of red mud (pH values dropped by 2–3 units) after its application to soil and improved growth and yield of maize (Krishna et al. 2005). The production of edible crops in red mud-amended soils has not obtained much interest, possibly due to the potential adverse effects of red mud. However, some success has been achieved in revegetating land that was believed to be uncultivable and barren (Chauhan and Ganguly 2011). A combination of 55 % red mud, 25 % farm yard manure (FYM), 15 % gypsum and 5 % vegetative dry dust, inoculated with bacteria and mycorrhizae, resulted in good growth of tree species (e.g. kikar (*Acacia nilotica*), karanj (*Pongamia pinnata*) and babul (*Prosopis juliflora*)) (Chauhan and Silori 2010; Chauhan and Ganguly 2011).

3.9 Other Commercial Materials

If applied to soil, a number of other mineral materials can act as a direct source of nutrients to plants. For example, granite meal is an organic fertiliser which is rich in K and contains a high concentration of silica. It greatly enhances soil structure and promotes healthier plants. Upon application, it does not alter the soil pH. Similarly, aragonite (94 % $CaCO_3$) is a rich source of calcium which is a secondary nutrient for plants. Few mineral materials can act as a source of micronutrients in soils (e.g. Azomite contains more than 67 elements beneficial to plant growth). Greensand is another organic source of K and approximately 30 trace elements. It acts as a slow-release K fertiliser. Granite powder (<70 μm) could also act as a slow-release K fertiliser and improved the yields of clover and ryegrass grown on acidic sandy soils over control treatment (Coroneos et al. 1995). Few siliceous volcanic rocks (e.g. perlite) and basaltic or andesitic rock (e.g. scoria) have also recently found their limited applications in gardening and landscaping activities. However, many of these minerals are expensive and not commonly used by the farmers in routine cultivation practices.

Conclusions

Mineral amendments to soils can improve the efficiency of nutrient use by plants by directly or indirectly influencing soil physical, chemical and biological parameters, which in turn control the nutrient transformation processes in soils. This amendment provides additional advantages in light-textured soils than in clayey soils, as many of the beneficial effects are due to an improvement in the physical characteristics of soils. Another direct influence can be observed where problem soils are reclaimed, by using mineral amendments. However, site-specific (precision) management technology is required to explore the beneficial effects of the mineral materials for increasing plants' NUE. The suitable application rate of various mineral amendments under heterogeneous soil and climatic conditions is also an important topic for further research. This will lead to reduced fertiliser inputs and therefore lower the cost of agricultural production. Management practices should include an integrated plant nutrition system (IPNS) for the improved utilisation of resources including mineral materials, fertilisers and organic inputs.

Acknowledgements Authors thank the editors for the invitation to contribute the chapter. The authors acknowledge the financial support from the Cooperative Research Centre for Contamination Assessment and Remediation of the Environment (CRC CARE). The authors also thank Prof GJ Churchman for his valuable comments to improve the quality of the manuscript.

References

Al-Omran AM, Sheta AS, Falatah AM, Al-Harbi AR (2005) Effect of drip irrigation on squash (*Cucurbita pepo*) yield and water-use efficiency in sandy calcareous soils amended with clay deposits. Agric Water Manag 73:43–55

Al-Omran AM, Sheta AS, Falatah AM, Al-Harbi AR (2007) Effect of subsurface amendments and drip irrigation on tomato growth. WIT Trans Ecol Environ 103:593–601

Al-Omran AM, Al-Harbi AR, Wahb-Allah MA, Mahmoud N, Al-Eter A (2010) Impact of irrigation water quality, irrigation systems, irrigation rates and soil amendments on tomato production in sandy calcareous soil. Turk J Agric For 34:59–73

Baligar VC, Bennett OL (1986a) NPK-fertilizer efficiency- a situation analysis for the tropics. Fertil Res 10:147–164

Baligar VC, Bennett OL (1986b) Outlook on fertilizer use efficiency in the tropics. Fertil Res 10:83–96

Baligar VC, Fageria NK, He ZL (2001) Nutrient use efficiency in plants. Commun Soil Sci Plant Anal 32:921–950

Balland-Bolou-Bi C, Poszwa A (2012) Effect of calcomagnesian amendment on the mineral weathering abilities of bacterial communities in acidic and silicate-rich soils. Soil Biol Biochem 50:108–117

Bangar KC, Kapoor KK, Mishra MM (1988) Effect of pyrite on conservation of nitrogen during composting. Biol Wastes 25:227–231

Basak BB, Biswas DR (2009) Influence of potassium solubilizing microorganism (*Bacillus mucilaginosus*) and waste mica on potassium uptake dynamics by sudan grass (*Sorghum vulgare* Pers.) grown under two Alfisols. Plant Soil 317:235–255

Basak B, Biswas D (2010) Co-inoculation of potassium solubilizing and nitrogen fixing bacteria on solubilization of waste mica and their effect on growth promotion and nutrient acquisition by a forage crop. Biol Fertil Soils 46:641–648

Bautista J, Kim H, Ahn D-H, Zhang R, Oh Y-S (2011) Changes in physicochemical properties and gaseous emissions of composting swine manure amended with alum and zeolite. Korean J Chem Eng 28:189–194

Begum M, Narayanasamy G, Biswas DR (2004) Phosphorus supplying capacity of phosphate rocks as influenced by compaction with water-soluble P fertilizers. Nutr Cycl Agroecosyst 68:73–84

Bernal MP, Lopez-Real JM, Scott KM (1993) Application of natural zeolites for the reduction of ammonia emissions during the composting of organic wastes in a laboratory composting simulator. Bioresour Technol 43:35–39

Bertocchi AF, Ghiani M, Peretti R, Zucca A (2006) Red mud and fly ash for remediation of mine sites contaminated with As, Cd, Cu, Pb and Zn. J Hazard Mater 134:112–119

Bhatnagar A, Vilar VJP, Botelho CMS, Boaventura RAR (2011) A review of the use of red mud as adsorbent for the removal of toxic pollutants from water and wastewater. Environ Technol 32:231–249

Bi YL, Li XL, Christie P, Hu ZQ, Wong MH (2003) Growth and nutrient uptake of arbuscular mycorrhizal maize in different depths of soil overlying coal fly ash. Chemosphere 50:863–869

Biswas DR (2011) Nutrient recycling potential of rock phosphate and waste mica enriched compost on crop productivity and changes in soil fertility under potato–soybean cropping sequence in an Inceptisol of Indo-Gangetic Plains of India. Nutr Cycl Agroecosyst 89:15–30

Biswas DR, Narayanasamy G (2006) Rock phosphate enriched compost: an approach to improve low-grade Indian rock phosphate. Bioresour Technol 97:2243–2251

Biswas DR, Narayanasamy G, Datta SC, Singh G, Begum M, Maiti D, Mishra A, Basak BB (2009) Changes in nutrient status during preparation of enriched organomineral fertilizers using rice straw, low-grade rock phosphate, waste mica, and phosphate solubilizing microorganism. Commun Soil Sci Plant Anal 40:2285–2307

Blair G (1993) Nutrient efficiency – what do we really mean? In: Randall PJ, Delhaize E, Richards RA, Munns R (eds) Genetic aspects of plant mineral nutrition, vol 50. Springer, Dordrecht, pp 205–213

Bolan N, White R, Hedley M (1990) A review of the use of phosphate rocks as fertilizers for direct application in Australia and New Zealand. Aust J Exp Agric 30:297–313

Bolan NS, Saggar S, Luo J, Bhandral R, Singh J (2004) Gaseous emissions of nitrogen from grazed pastures: processes, measurements and modelling, environmental implications, and mitigation. Adv Agron 84:37–120

Bolan NS, Kunhikrishnan A, Choppala GK, Thangarajan R, Chung JW (2012) Stabilization of carbon in composts and biochars in relation to carbon sequestration and soil fertility. Sci Total Environ 424:264–270

Cann MA (2000) Clay spreading on water repellent sands in the south east of South Australia- promoting sustainable agriculture. J Hydrol 231/232:333–341

Castelo-Branco MA, Santos J, Moreira O, Oliveira A, Pereira Pires F, Magalhães I, Dias S, Fernandes LM, Gama J, Vieira e Silvaa JM, Ramalho Ribeiro J (1999) Potential use of pyrite as an amendment for calcareous soil. J Geochem Explor 66:363–367

Chauhan S, Ganguly A (2011) Standardizing rehabilitation protocol using vegetation cover for bauxite waste (red mud) in eastern India. Ecol Eng 37:504–510

Chauhan S, Silori CS (2010) Rehabilitation of red mud bauxite wasteland in India (Belgaum, Karnataka). Ecol Restor 28:12–14

Chien SH, Menon RG (1995) Factors affecting the agronomic effectiveness of phosphate rock for direct application. Fertil Res 41:227–234

Churchman J, Noble A, Bailey G, Chittleborough D, Harper R (2013) Clay addition and redistribution to enhance carbon sequestration in soils. IUSS global soil carbon workshop. Madison, Wisconsin, USA

Clivot H, Pagnout C, Aran D, Devin S, Bauda P, Poupin P, Guérold F (2012) Changes in soil bacterial communities following liming of acidified forests. Appl Soil Ecol 59:116–123

Coroneos C, Hinsinger P, Gilkes RJ (1995) Granite powder as a source of potassium for plants: a glasshouse bioassay comparing two pasture species. Fertil Res 45:143–152

Deleporte S, Tillier P (1999) Long-term effects of mineral amendments on soil fauna and humus in an acid beech forest floor. For Ecol Manag 118:245–252

Dermatas D, Meng X (2003) Utilization of fly ash for stabilization/solidification of heavy metal contaminated soils. Eng Geol 70:377–394

Easterwood GW, Sartain JB, Street JJ (1989) Fertilizer effectiveness of three carbonate apatites on an acid ultisol. Commun Soil Sci Plant Anal 20:789–800

Elsas JD, Heijnen CE (1990) Methods for the introduction of bacteria in soil: a review. Biol Fertil Soils 10:127–133

Epstein E (1972) Mineral nutrition of plants: principles and perspective. Wiley Publisher, New York

Fageria NK, Baligar VC (2008) Chapter 7: Ameliorating soil acidity of tropical oxisols by liming for sustainable crop production. Adv Agron 99:345–399

Feigl V, Anton A, Uzigner N, Gruiz K (2012) Red mud as a chemical stabilizer for soil contaminated with toxic metals. Water Air Soil Pollut 223:1237–1247

Gaind S, Gaur AC (2002) Impact of fly ash and phosphate solubilising bacteria on soybean productivity. Bioresour Technol 85:313–315

Gamze Turan N, Nuri Ergun O (2007) Ammonia uptake by natural zeolite in municipal solid waste compost. Environ Prog 26:149–156

Gholamhoseini M, Ghalavand A, Khodaei-Joghan A, Dolatabadian A, Zakikhani H, Farmanbar E (2013) Zeolite-amended cattle manure effects on sunflower yield, seed quality, water use efficiency and nutrient leaching. Soil Tillage Res 126:193–202

Green RN, Trowbridge RL, Klinka K (1993) Towards a taxonomic classification of humus forms. For Sci 39: a0001–z0002

Hall DJM, Jones HR, Crabtree WL, Daniels TL (2010) Claying and deep ripping can increase crop yields and profits on water repellent sands with marginal fertility in southern Western Australia. Soil Res 48:178–187

Hartman WH, Richardson CJ, Vilgalys R, Bruland GL (2008) Environmental and anthropogenic controls over bacterial communities in wetland soils. Proc Natl Acad Sci 105:17842–17847

Hawkesford MJ (2011) An overview of nutrient use efficiency and strategies for crop improvement. In: Hawkesford MJ, Barraclough P (eds) The molecular and physiological basis of nutrient use efficiency in crops. Wiley-Blackwell, Oxford, pp 3–19

Hayhurst DT (1978) The potential use of natural zeolites for ammonia removal during coal-gasification. In: Sand LB, Mumpton FA (eds) Natural zeolites. Occurrence, properties and use. Pergamon Press, Oxford, pp 503–508

Haynes RJ, Naidu R (1998) Influence of lime, fertilizer and manure applications on soil organic matter content and soil physical conditions: a review. Nutr Cycl Agroecosyst 51:123–137

He ZL, Calvert DV, Alva AK, Li YC, Banks DJ (2002) Clinoptilolite zeolite and cellulose amendments to reduce ammonia volatilization in a calcareous sandy soil. Plant Soil 247:253–260

Heijnen CE, van Veen JA (1991) A determination of protective microhabitats for bacteria introduced into soil. FEMS Microbiol Ecol 8:73–80

Heijnen CE, Hok-A-Hin CH, Van Veen JA (1992) Improvements to the use of bentonite clay as a protective agent, increasing survival levels of bacteria introduced into soil. Soil Biol Biochem 24:533–538

Heijnen CE, Chenu C, Robert M (1993) Micromorphological studies on clay-amended and unamended loamy sand, relating survival of introduced bacteria and soil structure. Geoderma 56:195–207

Heynen CE, Van Elsas JD, Kuikman PJ, van Veen JA (1988) Dynamics of *Rhizobium leguminosarum* biovar *trifolii* introduced into soil; the effect of bentonite clay on predation by protozoa. Soil Biol Biochem 20:483–488

Hrynkiewicz K, Baum C, Niedojadło J, Dahm H (2009) Promotion of mycorrhiza formation and growth of willows by the bacterial strain *Sphingomonas sp.* 23L on fly ash. Biol Fertil Soils 45:385–394

Ingerslev M (1997) Effects of liming and fertilization on growth, soil chemistry and soil water chemistry in a Norway spruce plantation on a nutrient-poor soil in Denmark. For Ecol Manag 92:55–66

Ismail SM, Ozawa K (2007) Improvement of crop yield, soil moisture distribution and water use efficiency in sandy soils by clay application. Appl Clay Sci 37:81–89

Kalra N, Joshi HC, Chaudhary A, Choudhary R, Sharma SK (1997) Impact of flyash incorporation in soil on germination of crops. Bioresour Technol 61:39–41

Kalra N, Jain MC, Joshi HC, Choudhary R, Harit RC, Vatsa BK, Sharma SK, Kumar V (1998) Flyash as a soil conditioner and fertilizer. Bioresour Technol 64:163–167

Kalra N, Harit RC, Sharma SK (2000) Effect of flyash incorporation on soil properties of texturally variant soils. Bioresour Technol 75:91–93

Khan MR, Khan MW (1996) The effect of fly ash on plant growth and yield of tomato. Environ Pollut 92:105–111

Kithome M, Paul JW, Bomke AA (1999) Reducing nitrogen losses during simulated composting of poultry manure using adsorbents or chemical amendments. J Environ Qual 28:194–201

Kostakis G (2009) Characterization of the fly ashes from the lignite burning power plants of northern Greece based on their quantitative mineralogical composition. J Hazard Mater 166:972–977

Koukouzas N, Hämäläinen J, Papanikolaou D, Tourunen A, Jäntti T (2007) Mineralogical and elemental composition of fly ash from pilot scale fluidised bed combustion of lignite, bituminous coal, wood chips and their blends. Fuel 86:2186–2193

Krishna P, Reddy MS, Patnaik SK (2005) *Aspergillus tubingensis* reduces the pH of the bauxite residue (red mud) amended soils. Water Air Soil Pollut 167:201–209

Lee SB, Lee YB, Lee CH, Hong CO, Kim PJ, Yu C (2008) Characteristics of boron accumulation by fly ash application in paddy soil. Bioresour Technol 99:5928–5932

Lefcourt AM, Meisinger JJ (2001) Effect of adding alum or zeolite to dairy slurry on ammonia volatilization and chemical composition. J Dairy Sci 84:1814–1821

Liu Y, Naidu R, Ming H (2011) Red mud as an amendment for pollutants in solid and liquid phases. Geoderma 163:1–12

McPharlin IR, Jeffery RC, Toussaint LF, Cooper M (1994) Phosphorus, nitrogen, and radionuclide retention and leaching from a Joel sand amended with red mud/gypsum. Commun Soil Sci Plant Anal 25:2925–2944

Mishra DP, Das SK (2010) A study of physico-chemical and mineralogical properties of Talcher coal fly ash for stowing in underground coal mines. Mater Charact 61(11):1252–1259

Mishra M, Sahu R, Padhy R (2007) Growth, yield and elemental status of rice (*Oryza sativa*) grown in fly ash amended soils. Ecotoxicology 16:271–278

Mohan S (2011) Growth of biodiesel plant in flyash: a sustainable approach response of *Jatropha curcus*, a biodiesel plant in fly ash amended soil with respect to pigment content and photosynthetic rate. Procedia Environ Sci 8:421–425

Naidu R, Rengasamy P (1993) Ion interactions and constraints to plant nutrition in Australian sodic soils. Soil Res 31:801–819

Naidu R, Syers JK (1992) Influence of sugarcane millmud, lime, and phosphorus, on soil chemical properties and the growth of *Leucaena leucocephala* in an oxisol from Fiji. Bioresour Technol 41:65–70

Naidu R, Syers JK, Tillman RW, Kirkman JH (1990a) Effect of liming on phosphate sorption by acid soils. J Soil Sci 41:157–164

Naidu R, Tillman RW, Syers JK, Kirkman JH (1990b) Effect of liming and added phosphate on charge characteristics of acid soils. J Soil Sci 41:165–175

Nishanth D, Biswas DR (2008) Kinetics of phosphorus and potassium release from rock phosphate and waste mica enriched compost and their effect on yield and nutrient uptake by wheat (*Triticum aestivum*). Bioresour Technol 99:3342–3353

Noble AD, Gillman GP, Nath S, Srivastava RJ (2001) Changes in the surface charge characteristics of degraded soils in the wet tropics through the addition of beneficiated bentonite. Soil Res 39:991–1001

Pandey VC, Singh N (2010) Impact of fly ash incorporation in soil systems. Agric Ecosyst Environ 136:16–27

Pandey VC, Abhilash PC, Singh N (2009) The Indian perspective of utilizing fly ash in phytoremediation, phytomanagement and biomass production. J Environ Manag 90:2943–2958

Park J, Bolan N, Mallavarapu M, Naidu R (2010) Enhancing the solubility of insoluble phosphorus compounds by phosphate solubilizing bacteria. In: Proceedings of the 19th world congress soil science, Brisbane, Australia, pp 65–68

Rajan SSS, Watkinson JH, Sinclair AG (1996) Phosphate rocks for direct application to soils. Adv Agron 57:77–159

Ramesh V, Korwar GR, Mandal UK, Sharma KL, Venkanna K (2007) Optimizing fly ash dose for better tree growth and nutrient supply in an agroforestry system in semi-arid tropical India. Commun Soil Sci Plant Anal 38:2747–2766

Raven KP, Loeppert RH (1997) Trace element composition of fertilizers and soil amendments. J Environ Qual 26:551–557

Rengasamy P, Olsson K (1991) Sodicity and soil structure. Soil Res 29:935–952

Rengasamy P, Naidu R, Beech TA, Chan KY, Chartres C (1993) Rupture strength as related to dispersive potential in Australian soils. Catena Suppl 24:65–75

Roy G, Joy VC (2011) Dose-related effect of fly ash on edaphic properties in laterite cropland soil. Ecotoxicol Environ Saf 74:769–775

Samal S, Ray AK, Bandopadhyay A (2013) Proposal for resources, utilization and processes of red mud in India – a review. Int J Miner Process 118:43–55

Seshadri B, Bolan N, Choppala G, Naidu R (2013) Differential effect of coal combustion products on the bioavailability of phosphorus between inorganic and organic nutrient sources. J Hazard Mater 261:817–825.

Shende A, Juwarkar AS, Dara SS (1994) Use of fly ash in reducing heavy metal toxicity to plants. Resour Conserv Recycl 12:221–228

Singh LP, Siddiqui ZA (2003) Effects of *Alternaria triticina* and foliar fly ash deposition on growth, yield, photosynthetic pigments, protein and lysine contents of three cultivars of wheat. Bioresour Technol 86:189–192

Singh A, Sharma RK, Agrawal SB (2008) Effects of fly ash incorporation on heavy metal accumulation, growth and yield responses of *Beta vulgaris* plants. Bioresour Technol 99:7200–7207

Singh G, Biswas DR, Marwaha TS (2010) Mobilization of potassium from waste mica by plant growth promoting Rhizobacteria and its assimilation by maize (*Zea mays*) and wheat (*Triticum aestivum* L.): a hydroponics study under phytotron growth chamber. J Plant Nutr 33:1236–1251

Snars K, Hughes JC, Gilkes RJ (2004) The effects of addition of bauxite red mud to soil on P uptake by plants. Aust J Agric Res 55:25–31

Soda W, Noble AD, Suzuki S, Simmons R, Sindhusen L, Bhuthorndharaj S (2006) Co-composting of acid waste bentonites and their effects on soil properties and crop biomass. J Environ Qual 35:2293–2301

Srivastva A, Chhonkar PK (2000) Effect of fly ash on uptake of P, K and S by Sudan grass and oat grown in acid soils. J Indian Soc Soil Sci 48:850–853

Summers R, Guise N, Smirk D, Summers K (1996a) Bauxite residue (red mud) improves pasture growth on sandy soils in Western Australia. Soil Res 34:569–581

Summers R, Smirk D, Karafilis D (1996b) Phosphorus retention and leachates from sandy soil amended with bauxite residue (red mud). Soil Res 34:555–567

Thomas G, Gibson G, Nielsen R, Martin W, Radford B (1995) Effects of tillage, stubble, gypsum, and nitrogen fertiliser on cereal cropping on a red-brown earth in south-west Queensland. Aust J Exp Agric 35:997–1008

Toma M, Sumner ME, Weeks G, Saigusa M (1999) Long-term effects of gypsum on crop yield and subsoil chemical properties. Soil Sci Soc Am J 63:891–895

Toro M, Azcon R, Barea J (1997) Improvement of arbuscular mycorrhiza development by inoculation of soil with phosphate-solubilizing Rhizobacteria to improve rock phosphate bioavailability (^{32}P) and nutrient cycling. Appl Environ Microbiol 63:4408–4412

Ukwattage NL, Ranjith PG, Bouazza M (2013) The use of coal combustion fly ash as a soil amendment in agricultural lands (with comments on its potential to improve food security and sequester carbon). Fuel 109:400–408

van Elsas JD, Dijkstra AF, Govaert JM, van Veen JA (1986) Survival of *Pseudomonas fluorescens* and *Bacillus subtilis* introduced into two soils of different texture in field microplots. FEMS Microbiol Lett 38:151–160

Vassilev N, Baca MT, Vassileva M, Franco I, Azcon R (1995) Rock phosphate solubilization by *Aspergillus niger* grown on sugar-beet waste medium. Appl Microbiol Biotechnol 44:546–549

Vitousek P (1982) Nutrient cycling and nutrient use efficiency. Am Nat 119:553–572

Wallace A (1994) Use of gypsum on soil where needed can make agriculture more sustainable. Commun Soil Sci Plant Anal 25:109–116

Ward P, Oades J (1993) Effect of clay mineralogy and exchangeable cations on water repellency in clay-amended sandy soils. Soil Res 31:351–364

Ward SC, Summers RN (1993) Modifying sandy soils with the fine residue from bauxite refining to retain phosphorus and increase plant yield. Fertil Res 36:151–156

Whitfield DM, Smith CJ, Gyles OA, Wright GC (1989) Effects of irrigation, nitrogen and gypsum on yield, nitrogen accumulation and water use by wheat. Field Crops Res 20:261–277

Yunusa IM, Manoharan V, Odeh IA, Shrestha S, Skilbeck CG, Eamus D (2011) Structural and hydrological alterations of soil due to addition of coal fly ash. J Soils Sediments 11:423–431

Resource Conserving Techniques for Improving Nitrogen-Use Efficiency

Anchal Dass, Shankar Lal Jat, and K.S. Rana

Abstract

The use of nitrogen fertilisers has played an instrumental role in enhancing agricultural productions the world over including India. Currently, about 83 million tons N is used in agriculture globally. A large portion of applied N is lost through leaching, volatilisation and runoff, and only 50 % of applied N is assimilated by the crop plant. Recently, there have been serious concerns about environmental footprints of N fertilisers, particularly greenhouse gas emissions from the rice fields and escalating costs of fertilisers beyond farmers' reach. To meet the growing need for N fertilisers due to the rise in food requirement for ever multiplying population on the one hand and an increasing environmental and atmospheric pollution on the other, improving nitrogen-use efficiency (NUE) appears to be a viable solution. Certain resource conserving techniques, such as laser land levelling, zero or minimum tillage (save fuel), direct seeding, permanent or semi-permanent residue cover, new varieties that use plant nutrients more efficiently, furrow irrigated raised bed (FIRB) technology, system of rice intensification (SRI), direct seeded rice (DSR), precision farming techniques, use of leaf colour chart (LCC), chlorophyll meter, GreenSeeker, etc. have been shown to increase crop yields and NUE. For example, the use of optical sensors like GreenSeeker, chlorophyll meter and FIRB saved 25–50 % N. Even laser levelling has been reported to increase NUE by 6–7 % in India. Hence the use of such resource conserving technologies should be facilitated and supported for the sustainability of agricultural production and the natural resource base (land and water).

A. Dass (✉) • K.S. Rana
Division of Agronomy, Indian Agricultural Research Institute, New Delhi 110012, India
e-mail: anchal_d@rediffmail.com; ksrana@iari.res.in

S.L. Jat
DMR, IARI Campus, New Delhi 110012, India
e-mail: sljat@icar.org.in

A. Rakshit et al. (eds.), *Nutrient Use Efficiency: from Basics to Advances*,
DOI 10.1007/978-81-322-2169-2_4, © Springer India 2015

Keywords

Laser levelling • FIRB • Optical sensors • Zero tillage • Crop rotation •
Nitrogen-use efficiency

1   Introduction

Nitrogen (N) is the most important plant nutrient
determining the crop production. The doubling
of agricultural food production worldwide over
the past four decades has been associated with a
seven-fold increase in the use of N fertilisers.
Over the years, wide spread deficiency of N in
soil has become a serious concern the world over
as 50 % of the human population relies on N
fertiliser for food production. This has led to
greater use of N fertilisers. The present use of N
in the world is 83 m t which is a 100-fold increase
over the last 100 years. Although all crops
(except leguminous crops) require large amounts
of N for producing high yields, 60 % of global N
fertiliser is used for producing the world's three
major food crops, such as rice, wheat, and maize.
As the world population is multiplying at a very
fast rate and it will reach 9.3 billion by 2050,
50–70 % more cereal grain will be required by
2050 to ensure food security. This population and
food requirement scenario of the world points to
the fact that the N requirement for crop production is going to increase sharply. This can be met
by either enhancing the nitrogen-use efficiency
and/or by pouring in more amounts of N fertiliser
and manures. Numerous studies have shown that
N fertiliser-recovery efficiency by the first crop is
30–50 %. Some quantity of the unutilised N gets
deposited in soil. The recovery of this N in the
succeeding crops is very limited (<7 % of
applied N up to six consecutive crops). Some
portion of the unutilised chunk of applied N
gets dissociated from the soil-plant system and
causes environmental problems, like atmospheric pollution, groundwater pollution, global
warming, etc. It has been observed that generally, 50 % of the N applied is not assimilated by
the plant and is a potential source of environmental pollution. Many N^{15} recovery experiments

have reported losses of N fertiliser in cereal production from 20 to 50 %. These losses were
ascribed to joint effect of denitrification,
volatilisation, and/or leaching.

Both the recent and future intensification of
the use of N fertilisers in agriculture already has
and will continue to have major detrimental
impacts on the diversity and functioning of the
nonagricultural neighbouring bacterial, animal,
and plant ecosystems. The most typical examples
of such an impact are the eutrophication of freshwater (London 2005) and marine ecosystems
(Beman et al. 2005) as a result of leaching
when high rates of N fertilisers are applied to
agricultural fields (Tilman 1999). In addition,
there are also gaseous emission of N oxides
reacting with the stratospheric ozone and the
emission of toxic ammonia into the atmosphere
(Ramos 1996; Stulen et al. 1998). Despite the
detrimental impact on the biosphere, the use of
fertilisers (N in particular) in agriculture,
together with an improvement in cropping
systems, mainly in developed countries, has
provided a food supply sufficient for both animal
and human consumption (Cassman 1999). However, declining N fertiliser-use efficiency (NUE)
continuously in cereal production is a serious
concern in twenty-first century. Therefore, the
challenge now and for the next decades will be
to accommodate the needs of the expanding
world population by developing a highly productive agriculture, while at the same time preserving the quality of the environment. Furthermore,
farmers are facing increasing economic pressures
with the rising fossil fuel costs required for the
production of N fertilisers. All these factual
issues compel the agricultural researchers as
well as practitioners of agriculture to find out
the feasible means that would enhance fertiliser
NUE, especially for the main cereals of the
world, that is, rice, wheat, and maize. Much

research has been conducted during the past decades to improve N-use efficiency by developing fertiliser management strategies based on better synchronisation between the supply and requirement of N by the crop. Importantly, some of these techniques are being adopted on a large scale by the farmers. For further adoption of the N-efficient agronomic practices, the skills and knowledge of farmers are needed to be improved, and the techniques also need to be fine-tuned and made cost-effective, user friendly, and ecofriendly.

2 Causes of Low N-Use Efficiencies

2.1 Loss from the Plant Tissue

Cereal plants release N from plant tissue, chiefly in the form of NH_3 after anthesis (Harper et al. 1987; Francis et al. 1993). Plant N losses have accounted for 52–73 % of the unaccounted N using ^{15}N in corn research (Francis et al. 1993) and between 21 % (Harper et al. 1987) and 41 % (Daigger et al. 1976) in winter wheat. Gaseous plant N loss in excess of 45 kg N/ha/year has also been documented in soybean [*Glycine max* (L.) Merr.] (Stutte et al. 1979).

2.2 N Loss Due to Denitrification

Reported gaseous N losses due to denitrification from applied N fertiliser include 9.5 % in winter wheat (Aulakh et al. 1982), 10 % in lowland rice (De Datta et al. 1991), and 10 % (conventional tillage) to 22 % (no till) in corn (Hilton et al. 1994). Incorporation of straw and/or application of straw on the surface of zero-till plots can double denitrification losses (Aulakh et al. 1984).

2.3 N Losses by Surface Runoff

N fertiliser losses in surface runoff have been reported to range between 1 % (Blevins et al. 1996) and 13 % (Chichester and Richardson

1992) of the total N applied and are generally lower under no tillage. When urea fertilisers are applied to the surface without incorporation, losses of N fertiliser as NH_3 can exceed 40 % (Fowler and Brydon 1989; Hargrove et al. 1977) and are generally greater with increasing temperature, soil pH, and surface residue. From a finger millet plot with 2 % slope, available N loss through sediment loss due to water erosion was 6.7 kg/ha/year in eastern India (Dass and Sudhishri 2010).

2.4 Leaching Losses of N

Leaching of N, mostly in the form of NO_3, causes huge losses of applied N fertiliser. Particularly, when the rate of N fertiliser application is in excess of the crop requirement, NO_3 leaching, is high. In cooler temperate climates, NO_3 through tile drainage has approached 26 kg/ha/year under conventional-tillage corn when only 115 kg N/ha was applied (Drury et al. 1996).

3 Nitrogen-Use Efficiency and Expression

Nitrogen-use efficiency may be defined in various terms and their approximate value over region and crop basis is shown in Table 1.

Terms Used in Describing N-Use Efficiency (Dobermann 2005)

1. *Agronomic efficiency (AE_N):* It may be defined as increase in grain yield kg grain kg^{-1} N applied. Its value ranges from 18 to 24 kg grain kg^{-1} N applied and was the smallest in maize and largest in rice.

2. *Apparent nitrogen recovery (ANR):* It may be defined as per cent increase in the uptake of N in fertilised crop as compared to control where no N was applied. Its value ranges from 10 to 70 % across region and various crops.

3. *Physiological efficiency (PE_N):* It is defined as increase in grain yield Kg grain kg^{-1} N absorbed. Its value ranges from 20 to 52 across various regions and crops.

Table 1 Descriptive statistics of various NUE terms for cereals in various continents

Region/crop	AE_N	RE_{N15}	PE_N	PFP_N
Africa	13.9	0.37	22.9	39.3
Australia	8.0	0.41	–	54.0
Europe	21.3	0.61	27.7	50.4
America	19.6	0.36	28.4	49.6
Asia	21.5	0.44	46.6	53.5
Average/total	19.6	0.44	40.6	51.6
Maize	24.2	0.40	36.7	72.0
Rice	22.0	0.44	52.8	62.4
Wheat	18.1	0.45	28.9	44.5
Average/total	20.6	0.44	40.6	51.6

Agronomic efficiency (AE_N), physiological efficiency index of N (PE_N), partial factor productivity (PFP_N), recovery efficiency (RE_{N15})

Partial Factor Productivity of N (PFP$_N$) It is the gain in grain yield per kg N applied to the crop. Its value ranges between 39 and 72, which means by application of one kg of N, 39–72 kg grain yield can be gained across various continents and crops.

4   Resource Conserving Techniques

Resource conservation means management of the human use of natural resources, such as soil, water, nutrients, etc. to provide the maximum benefit to current generations while maintaining capacity to meet the needs of future generations. Conservation includes both the protection and rational use of natural resources. Resource conserving techniques (RCTs) refer to those practices that conserve resources and ensure their optimal utilisation and enhance resource or input-use efficiency. For example, a cultivar or any plant type which uses plant nutrients, such as nitrogen, phosphorus, potassium, etc. efficiently can be an RCT. There can be a large number of RCTs. The term RCTs should not be confused with conservation agriculture. Conservation agriculture, in fact, refers to the RCTs that involve (1) soil cover, particularly through the retention

of crop residues on the soil surface; (2) sensible, profitable rotations; (3) a minimum level of soil disturbance, e.g. reduced or zero tillage; and (4) the dimension of conservation agriculture in the minimum compaction of soil surface.

The important RCTs usable in agriculture include laser land levelling, zero or minimum tillage (save fuel), direct seeding, permanent or semi-permanent residue cover, new varieties that use plant nutrients more efficiently, furrow irrigated raised bed (FIRB) technology, system of rice intensification (SRI), direct seeded rice (DSR), precision farming techniques, use of leaf colour chart (LCC), chlorophyll meter, GreenSeeker, etc. Some important RCTs useful in enhancing N-use efficiency have been discussed in this chapter.

4.1   Laser Land Levelling

Land levelling is generally done in irrigated agriculture for uniform application or retention of water. Accurate levelling of agricultural land has been a challenge tillage recent past. The advent of laser leveller has made it possible to achieve complete and precise levelling of land. Laser land levelling (LLL) was first introduced in India in 2001 in western Uttar Pradesh. However, the number of laser land levellers rose to 925 and the acreage under LLL grew to 200,000 ha in 2008. Laser land levelling is, basically, a precursor of resource conserving technique and a process of smothering land surface (± 2 cm) from its average elevation using laser-equipped dragged buckets. It levelled the surface having 0–0.2 % slope that leads to uniform distribution of water throughout the field and enhance resource-use efficiency. The potential benefits of laser land levelling are a 3–6 % rise in an area under cultivation due to removal of bunds and channels, saving of 10–30 % water due to uniform distribution, a 3–19 % increase in yield, a 6–7 % increase in nitrogen-use efficiency (Jat et al. 2009), reduction of cost of production, and enhancement of productivity.

4.2 Zero Tillage

Zero tillage is an extreme form of conservation tillage (CT) in which mechanical soil manipulation is reduced to traffic and sowing only. The no-till system is a specialised type of conservation tillage consisting of a one-pass planting and fertiliser operation in which the soil and the surface residues are minimally disturbed (Parr et al. 1990). It helps in paradigm shift in crop production. The current and potential area is 2.0 m ha and 10 m ha under zero tillage in India, respectively. It is very helpful in the area of intensive cultivation where a turn-around period between two crops is really very less and thus it can facilitate timely sowing. Advantages of the zero tillage include saving of fuel and labour cost, reduction of cost of cultivation, and saving of approximately Rs. 3,000/- ha^{-1} towards field preparation; timely planting gave yield advantage approximately 2 q ha^{-1} for a week advanced sowing in wheat crop and reduces soil erosion and improves soil health. Soils in no-till systems tend to be more moist than ploughed soils. Soil moisture ultimately affects nitrogen management. A layer of crop residues on the soil surface reduces evaporation from the soil surface. Alternatively, tillage promotes evaporation and dries out soil to the depth of tillage. A wetter soil also tends to be a cooler soil. Thus, no-till soils tend to be wetter and cooler, so that they reduce microbial activity and the mineralisation of crop residues and organic matter.

The results of a 10-year study showed that N-mineralisation rates were higher in annual cropping systems under no tillage than under conventional tillage (Wienhold and Halvorson 1999). Increased N stored as labile organic forms causes this increased mineralisation. Increased amounts of organic N will supply more nitrogen to crops, which will result in less N required from fertilisers as well as reduced leaching.

In a rice-wheat cropping system, zero tillage gave higher PFP$_N$ as compared to conventional tillage at all levels of fertiliser application. Application of nitrogen beyond 120 kg/ha reduced PFP$_N$, but yield was more at subsequent level under zero-tillage treatment than the conventional tillage (Table 2). This demonstrates that zero

Table 2 Effect of NPK on wheat productivity across the tillage methods

NPK level kg ha^{-1}	Tillage practices			
	Zero tillage		Conventional tillage	
	Yield (q ha^{-1})	PFP$_N$	Yield (q ha^{-1})	PFP$_N$
120:60:40	40.6	33.8	36.0	30.0
150:75:50	48.5	32.3	41.0	27.3
180:90:60	49.6	27.6	43.3	24.1
CD (0.05)	Tillage operations 1.9			
	Fertiliser level 3.8			

Source: Sharma et al. (2005)

Table 3 Effect of residue cover, tillage practices, and N fertilisation methods on residual soil NO$_3^-$

Tillage	Organic matter (g kg^{-1})	Residue cover (%)	Residual soil NO$_3^-$ N	
			0–15 cm kg Nha^{-1}	15–60 cm kg Nha^{-1}
No till (B)	32.0	43.0	29	74
No till (I N)	32.0	41.0	15	58
Chisel plough (B)	32.0	25.0	40	124
Mouldboard plough (I N)	32.0	14.0	52	123

Source: Hilton et al. (1994)

tillage results in better crop responses to higher N doses and also leads to greater PFP$_N$ (Sharma et al. 2005).

In an experiment set to study the effects of methods of tillage on the corn in corn-oat rotation in silty loam soil in South Dakota (USA), N was applied at 112 kg/ha as ammonium nitrate either injected (IN) or broadcasted (B), and in no-till plots N was applied by modified anhydrous ammonia knives prior to planting as urea ammonium nitrate (28 % N). This experiment showed higher residue accumulation on the surface of the no-till plots, which provides greater carbon substrate for microbial activity (Table 3). Residue increases soil-surface wetness and forms the C-source to microbes near the surface where high soil temperature favours denitrification. The zone of denitrifying activity was closer to the soil surface in the no-till treatment than for the other tillage treatments. Even with surface disturbance with ploughing or disking, some residue remains in the surface in chisel plough or mouldboard

Table 4 Effects of lime and N on corn under no tillage in silt loam soil

N treatments	Lime Mg ha^{-1}	Leaf N g kg^{-1}	Yield Mg ha^{-1}
Urea ammonium nitrate – broadcasted	0	10.22c*	27.6b
	1.12	9.93cd	27.3b
Urea ammonium nitrate – incorporated	0	11.55a	30.3a
	1.12	11.21a	30.8a
Urea – broadcasted	0	9.58d	25.4d
	1.12	8.90f	24.4d
Urea – incorporated	0	9.33de	26.2c
	1.12	9.36d	26.7c
Ammonium nitrate – incorporated	0	11.31a	29.4a
	1.12	10.91ab	29.0a

Source: Howard and Essington (1998)
*Within columns, means followed by the same latter are not significantly similar at $\alpha = 0.05$

plough treatment. N loss in wheel track (WT) area was 1.6 times higher than the non-wheel track area (NWT), and this, as shown by regression coefficient which is high in NWT area as compared with WT area due to anaerobiosis, occurs after rainfall (Hilton et al. 1994).

Howard and Essington (1998) conducted an experiment for 12 years in silty loam soil in no-till corn. They applied N at 168 kg ha^{-1} within 5 days after planting of corn as UAN (46 % N). This study showed that NUE as leaf N g kg^{-1} was low in all treatment with application of lime at 1.12 t ha^{-1} as compared to no application (Table 4), as lime application causes N loss by both NH$_3$ volatilisation and immobilisation of N by surface mulch, while in no-lime application there is immobilisation only. Application of fertilisers by broadcasting caused more N loss and less N recovery as compared to incorporation of fertilisers. So, lime application and broadcasting of fertiliser should be avoided to reduce N losses and increase NUE in no-till system.

4.3 Rotary Till Drill

This machine is a combination of rotavator, seed-cum-fertiliser drill and light planker-cum-driving wheel at the back. All the operations like sowing, ploughing, and incorporation of residues and green manure crops can be performed under single operation by using rotary tillage drill. Advantages of the rotary tillage drill are as follows:

- Low energy, time, and labour requirement.
- Useful to incorporate green manure crop, weeds, and residues.
- Simultaneous land preparation and sowing.
- Useful in intercultural operation in horticultural crops.
- Puddling can also be done with single operation only.

4.4 Leaf Colour Chart

Leaf colour chart popularly known as LCC has been used in determination of leaf nitrogen content based on chlorophyll content in the leaves at different growth stages. An LCC value of 4 indicates that there is 1.4–1.5 mg N/g leaf weight. The critical LCC value for rice hybrids and HYVs is 4 and for and basmati rice is 3. These values have to taken from 7–10 DAS or 20–25 DAT to heading. This simple and cost-effective device has been largely used by the south Asian farmers for efficiently managing N fertilisers (Singh et al. 2010). Singh et al. (2007) summarised the data of 350 trials and reported that use of LCC significantly reduced the N requirement without causing any reduction in yield of rice compared to farmers' practice (Fig. 1).

Shukla et al. (2006) computed the agronomic efficiency, recovery efficiency, and physiological efficiency in two varieties of rice and wheat, each under different fertiliser application methods involving LCC at PDFSR, Meerut (Tables 5 and 6), and reported that in sandy loam soil NUE increased with application of N using LCC as compared to the recommended N and farmers' practice (Table 7). However, NUE decreased with increase in critical value of LCC from <3 to 5. This could be due to the fact that use of LCC as an RCT in rice and wheat restricted N leaching losses. Similarly, at several

Fig. 1 Average N fertiliser applied and grain yield of rice in 350 on-farm trials compared (Source: Singh et al. 2007)

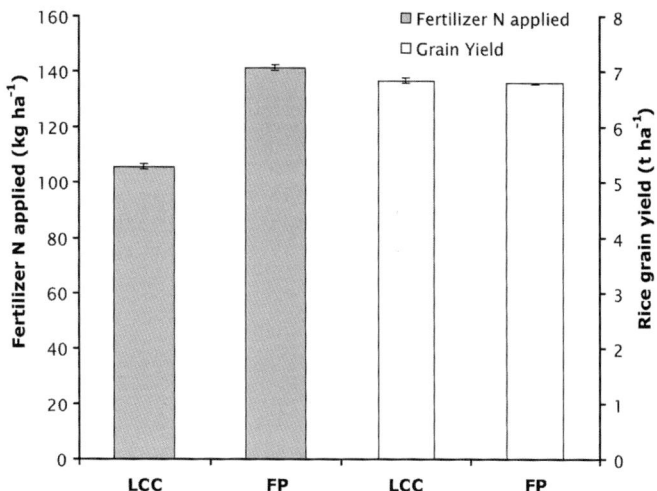

Table 5 Nitrogen-use efficiencies of wheat varieties grown under different N management

N management treatment (kg ha^{-1})	Grain yield (t ha^{-1})	Total N uptake (kg ha^{-1})	AE (kg grain kg^{-1} N applied)	RE (%)	PE (kg grain kg^{-1} N absorbed)
PBW 343 (early sown)					
No N control (zero)	2.15	44.5			
LCC ≤ 3, no basal N (60)	4.20	85.5	34.2	68.3	52.4
LCC ≤ 4, no basal N (120)	5.65	120.0	29.2	62.9	46.4
LCC ≤ 5, no basal N (160)	6.05	132.5	24.4	55.0	44.3
Recommended N (120)	5.05	105.5	24.2	51.0	47.5
Farmers' N practice (150)	5.10	100.0	19.7	41.0	48.0
PBW 226 (late sown)					
No N control (zero)	1.80	35.5			
LCC ≤ 3, no basal N (90)	4.05	84.5	25.0	54.4	45.9
LCC ≤ 4, no basal N (120)	4.45	98.0	22.1	52.1	42.4
LCC ≤ 5, no basal N (160)	4.70	108.0	18.1	45.3	40.0
Recommended N (120)	4.00	88.5	18.3	44.2	41.5
Farmers' N practice (150)	4.05	91.5	15.0	37.3	40.2
CD (0.05) varieties	0.43	5.1	NS	5.2	3.1
N management	0.78	8.3	4.1	8.9	5.6
V × N	1.01	11.4	5.9	12.6	8.4

other places in India, higher NUE was observed when N applications were scheduled using LCC compared to farmers practices (Table 8). Overall, in wheat, applying 30 kg N/ha each time with an LCC score of 4 with a total application of 120 kg N/ha resulted in higher PFP$_N$, N uptake, and NUE than using the same quantity of N in three fixed time splits. The same things were also reported in the case of rice (Ladha et al. 2005).

Table 6 Nitrogen-use efficiencies of rice varieties grown under different N management

N management treatment (kg ha^{-1})	Grain yield (t ha^{-1})	Total N uptake (kg ha^{-1})	AE (kg grain kg^{-1} N applied)	RE (%)	PE (kg grain kg^{-1} N absorbed)
Basmati 370					
No N control (zero)	2.75	60.5			
LCC $\leq$ 2, no basal N (20)	3.30	74.5	27.5	70.0	39.3
LCC $\leq$ 3, no basal N (80)	4.30	108.0	19.3	59.4	32.6
LCC $\leq$ 4, no basal N (100)	4.00	111.5	12.5	50.0	24.5
Recommended N (80)	3.75	98.5	12.5	47.5	26.3
Farmers' N practice (100)	3.65	102.0	9.0	41.5	21.7
PHB 71					
No N control (zero)	3.80	61.5			
LCC $\leq$ 3, no basal N (90)	6.30	114.5	27.8	58.9	47.2
LCC $\leq$ 4, no basal N (130)	7.50	136.5	26.0	53.5	48.0
LCC $\leq$ 4, no basal N (150)	7.70	143.0	24.7	52.0	47.5
Recommended N (150)	6.85	125.5	20.3	42.7	46.9
Farmers' N practice (180)	7.00	129.5	17.8	37.8	47.1
CD (0.05) varieties	0.21	6.1	2.4	4.8	3.8
N management	0.30	11.4	3.9	9.2	7.3
V $\times$ N	0.43	16.7	4.8	11.4	9.8

Table 7 Total N used and grain yield with farmers' practice (FP) and LCC-based N management at different stations, India

Stations	Practice of N management	N used (kg ha^{-1})	Grain yield (t ha^{-1})	PFP$_N$ (kg kg^{-1} N applied)
Haryana, 155 farms, wheat	Farmers' practice	149	6.4	42.9
	LCC 4	124	6.4	51.6
Tamil Nadu, 20 farms, wheat	Farmers' practice	142	5.0	35.2
	LCC 4	108	4.9	45.4
Modipuram, Basmati 370	Farmers' practice	100	3.7	37.0
	LCC 4	100	3.9	39.0
Modipuram, Saket 4	Farmers' practice	150	5.6	37.3
	LCC 4	120	6.1	50.8
Modipuram, Hybrid 6111	Farmers' practice	180	6.9	38.3
	LCC 4	150	7.4	49.3

4.5 Optical Sensors, Like Chlorophyll Meter and GreenSeeker

These sensors are useful tools in guiding precise application of inputs and reducing the GHG emissions from the crop fields. Data in Table 8 shows that SPAD-based N applications resulted in about 50 % N saving in rice and improved agronomic efficiency of N and partial factor productivity by 2.5 times compared to local recommended practice of N application (IRRI-CREMNET 2000). In winter season maize crop, *SPAD-based ($\leq$37)* N application resulted in a saving of 55 kg N ha compared to soil test crop response equation-based N application without any yield reduction. Agronomic efficiency was higher (Dass et al. 2012). In wheat, timely N application at SPAD value $\leq$42 resulted in 9 %

Table 8 Comparison of fixed threshold (SPAD-35) chlorophyll meter value-based N management practice with local or soil test crop response correlation recommendations

Treatment	N used (kg ha^{-1})	Grain yield (t ha^{-1})	AE$_N$[a]	PFP$_N$[b]
India: old Cauvery delta, Padugai soil series (Kuruvai)				
Control	0	4.9 b	–	–
Local recommendation	125	7.3 a	18.6	58.1
SPAD-35	65	7.6 a	41.2	117.2
India: new Cauvery delta, 1996 DS (Kuruvai)				
Control	0	5.3 b	–	–
Local recommendation	125	6.4 a	8.8 b	51.6 b
SPAD-35	60	7.1 a	51.0 a	118.4 a
India: new Cauvery delta, 1998 DS (Kuruvai)				
Control	0	3.6 b	–	–
STCR Recom	142	5.0 a	10.3	35.4
SPAD-35	110	5.0 a	12.9	45.4

Source: IRRI-CREMNET (2000)

[a]AE$_N$: agronomic efficiency of applied N (kg additional grain over control per kg applied N)

[b]PFP$_N$: partial factor productivity for applied N (kg grain yield per kg applied N)

Plate 1 Taking observation with chlorophyll meter in wheat crop

Plate 2 Chlorophyll meter-based N application

higher wheat yield with 20 kg/ha N saving compared to recommended soil-based N supply (Dass et al. 2012) (Plates 1, 2, and 3). Studies on GreenSeeker optical sensor- based N management in rice (cultivar PR 118) revealed that applying 20 kg N as basal, 60 kg at 21 days after transplanting (DAT) and 12 kg at 42 DAT, returned the highest grain yield (6.68 t/ha) and saved 28 kg N/ha compared to recommended schedule of N application (120 kg N /ha in three equal splits) (Gupta 2006).

Plate 3 Soil test-based recommended practice

4.6 Furrow Irrigated Raised Bed (FIRB) Planting

Crops are sown on 40 cm wide raised beds and irrigation providing in 30 cm wide furrows. The number of crop rows to be adjusted on beds is three for wheat; two each for peas, mung bean, soybean, and mustard; and one for maize and pigeon pea. Wheat after pigeon pea, maize, soybean, and peas and mung bean after mustard and wheat can be sown just by reshaping of beds. Advantages of the FIRB planting are as follows:
1. It promotes crop diversification.
2. Saves irrigation water by 25–35 %.
3. Saves N and seed rate up to 25 %.
4. It helps in decreasing weed infestation as well as easy weeding.
5. It provide easy passage for drainage of excess water.
6. It facilitates easy rouging in the field crops.

4.7 Surface Seeding

Surface seeding is the simplest method of zero-tillage systems being promoted in eastern India, Nepal, and Bangladesh. It is mostly suitable in lowland areas where sowing is delayed due to excessive soil moisture. The benefits of surface seeding are given below:
• No equipment needed for surface seeding.
• Very well suitable for heavy textured soils.
• It is boon for the areas where land preparation is very difficult and costly.

4.8 Crop Rotation

Crop rotation is one of the core principles of conservation agriculture. It helps in efficient utilisation of resources and enhances use efficiency of resources like nutrients, water, etc. Chettri and Bandhopadhaya (2005) studied the influence of different fertiliser management practices on yield efficiency, agronomic efficiency, and nutrient recovery efficiency in different rice-based crop rotations in clay loam soil at Kalyani West Bengal, India (Table 9), and reported that the maximum grain yield can be obtained with either green gram or lathyrus incorporated in situ before rice transplanting in addition to 75 % RDN + 10 t FYM/ha. Higher AE is observed in rice crop grown after mung bean or lathyrus as fodder crop, with 75 % recommended dose of NPK + 10 t FYM/ha. This could likely be due to the combined application of all these sources that increased the availability of the nutrients and gave luxurious crop. The inclusion of dual-purpose summer legumes in rice-wheat cropping systems has beneficial effects on the NUE of the system (Jat 2010) (Table 10).

4.9 Efficient N Management

Apart from the rate of application, the time and method of N application play a very important role in utilisation of applied N by crops and consequently govern NUE. Placement of N was more important than time of N application in influencing yields in the semiarid regime. Deep banding N fertiliser (0.1–0.15 m deep) resulted in superior spring wheat yield as compared to broadcast application. The rate of yield increase was determined by partial derivative of yield equation of soil and N fertiliser. Generally it increases with increase in soil water availability. It indicated that each added N is used less and less efficiently and the rate of yield increase with per unit increase in soil N is greater than the N fertiliser. It shows that an NUE increment is directly related to water uptake and generally to soil N but inversely related to N fertiliser. NUE increases with year at <50 kg ha^{-1}, but converse was true at higher dose. This is due to increase in n supplying power of soil with increasing cropping years (Campbell et al. 1993) (Tables 11) The important tools and tactics with their relative benefits and limitations in enhancing the NUE are presented in (Ladha et al. 2005) (Table 12).

Table 9 Evaluation of GreenSeeker based N recommendation in rice at Ludhiana (cultivar PR 118)

Fertilizer N applied (kg N/ha) DAT									Total N applied (kg/ha)	Rice grain yield (t/ha)	Total N uptake (kg/ha)
0	7	15	21	28	35	42	49	56			
									0	3.85	68.1
40			40			40			120	6.19	132.3
20			40			28*			88	6.23	127.8
20			60			12*			92	6.83	131.7
30			30			32*			92	5.63	119.2
30			50			14*			94	6.28	125.7
40			40			24*			104	6.34	131.8
30				50			29*		109	6.29	135.8
	20			40			29*		89	5.97	126.9
	20			60			19*		99	6.59	133.5
	30			30			32*		92	5.66	120.8
	30			50			17*		97	6.25	133.7
40				40			20*		100	6.50	138.0
		40		40			20*		100	5.80	120.3
		40			40			1*	81	4.97	106.9
LSD (P=0.05)										0.774	14.65

Source: Gupta (2006)

Table 10 Agronomic efficiency and recovery fraction of fertiliser use in different cropping sequence with rice at different levels of fertiliser management

Cropping sequence	Yield (kg ha^{-1})	Agronomic efficiency	Recovery fraction
At 100 % of the RD of N, P, and K (urea, DAP, and MOP)			
Kharif rice-oat F – boro rice-green gram GM	8,062	18.75	1.72
Kharif rice-lathyrus GM – boro rice-maize Fallow (F)	8,205	19.08	1.75
Kharif rice-lathyrus F – boro rice-green gram F	7,990	21.03	1.80
Kharif rice-oat F – boro rice-maize F	6,714	17.67	1.69
At 75 % of the RD of N, P, and K (urea, DAP, and MOP)			
Kharif rice-oat F – boro rice-green gram GM	7,590	22.65	2.13
Kharif rice-lathyrus GM – boro rice-maize F	7,620	22.75	2.17
Kharif rice-lathyrus F – boro rice-green gram F	7,345	25.77	2.22
Kharif rice-oat F – boro rice-maize F	6,225	21.84	2.13
At 75 % of the RD of N, P, and K (urea, DAP, and MOP) + FYM at 10 t ha^{-1}			
Kharif rice-oat F – boro rice-green gram GM	9,975	25.91	2.14
Kharif rice-lathyrus GM – boro rice-maize F	10,820	28.10	2.21
Kharif rice-lathyrus F – boro rice-green gram F	9,808	29.28	2.23
Kharif rice-oat F – boro rice-maize F	8,008	23.90	2.10
CD (0.05) for grain yield – at the same level of C	0.219		
CD (0.05) for grain yield – at the same level of F	0.292		

Source: Chettri and Bandhopadhaya (2005)

Table 11 Effect of time, method, and rate of application of N fertiliser on N-use efficiency of zero-tilled spring wheat

Crop year	Time or method of application	Rate of N fertiliser (kg ha^{-1})				
		25	50	75	100	Mean
		NUE Kg grain kg^{-1} N				
1	Band	38	30	31	21	30
	Broadcast	37	26	29	20	28
2	Fall (30–35 DAS)	32	24	20	17	23
	Spring (120–125 DAS)	41	28	21	17	27
3	Mean	15	8	6	5	9
4	Mean	13	8	5	4	8
5	Mean	35	26	22	18	25
6	Band	27	20	17	14	19
	Broadcast	24	17	14	11	17
7	Fall	10	7	5	4	7
	Spring	10	7	5	4	6
8	Fall	31	24	20	17	23
	Spring	34	27	19	16	24
9	Fall	37	27	21	17	25
	Spring	39	27	20	17	26

Table 12 Tools/tactics of enhancing N fertiliser-use efficiency

Tools/tactics	Benefit: cost	Limitations
SSNM	High	Infrastructure for every site
Chlorophyll meter	High	Initial high cost
LCC	Very high	None
Breeding strategy	Very high	Varieties yet to be developed
N fixation in nonlegumes	High	Tech. yet to be developed for field scale
Precision farming technology	High	Tech. need to be fine-tuned
Resource conserving techniques	High	Tech. need to be evaluated for long-term impacts
Integrated crop management	High	Tech. need to be evaluated for long-term impacts

Conclusions

Use of nitrogen in agriculture is indispensable as it is an important constituent of plant material and human food, and its contribution in food production is the largest among all other plant nutrients. More than 50 % of applied N is lost through various processes, and nitrogen-use efficiency remains below 50 % in most crops. Certain resource conserving techniques (RCTs) are now available that can help reduce the N requirement, improve N utilisation, reduce losses, and finally enhance NUE. However, RCTs are more effective in combinations rather than their individual application. Among various RCTs, research work is mostly available on tillage practices, mulching, LCC, and some work on chlorophyll meter and GreenSeeker sensors. Although laser land levelling and rotary tillage seem to be potential technologies, information of research work carried out on these aspects is only meagrely available. LCC and chlorophyll meter could reduce N application rates by 25–50 %. Zero tillage is more responsive to applied N as compared to CT in coarse textured soils and increases NUE. Retention of residues increases N uptake and thereby increases NUE in different cropping systems.

Denitrification rates are high in no tillage in high rainfall area due to creation of anaerobic condition here N is broadcasted. In humid environment and with finer textured soil, potential loss with surface residue appears to be minimised by fertiliser management.

References

Aulakh MS, Rennie DA, Paul EA (1982) Gaseous nitrogen losses from cropped and summer fallowed soils. Can J Soil Sci 62:187–195

Aulakh MS, Rennie DA, Paul EA (1984) The influence of plant residues on denitrification rates in conventional and zero tilled soils. Soil Sci Soc Am J 48:790–794

Beman JM, Arrigo K, Matson PM (2005) Agricultural runoff fuels large phytoplankton blooms in vulnerable areas of the ocean. Nature 434:211–214

Blevins DW, Wilkison DH, Kelly BP, Silva SR (1996) Movement of nitrate fertilizer to glacial till and runoff from a claypan soil. J Environ Qual 25:584–593

Campbell CA, Zentner RP, Selles F, McConkey BG, Dyck FB (1993) Nitrogen management for spring wheat grown annually on zero-tillage: yields and N use-efficiency. Agron J 85:107–114

Cassman KG (1999) Ecological intensification of cereal production systems: yield potential, soil quality, and precision agriculture. Proc Natl Acad Sci U S A 96:5952–5959

Chettri M, Bandhopadhaya P (2005) Effect of integrated nutrient management on fertilizer use efficiency and changes in soil fertility status after rice (*Oryza sativa*) – based cropping system. Indian J Agric Sci 75 (9):596–599

Chichester FW, Richardson CW (1992) Sediment and nutrient loss from clay soils as affected by tillage. J Environ Qual 21:587–590

Daigger LA, Sander DH, Peterson GA (1976) Nitrogen content of winter wheat during growth and maturation. Agron J 68:815–818

Dass A, Sudhishri S (2010) Intercropping in fingermillet (*Eleusine coracana*) with pulses for enhancing productivity, resource-use efficiency and soil fertility in uplands of southern Orissa. Indian J Agron 55(2):89–94

Dass A, Singh DK, Dhar S (2012) Precise supply of nitrogen and irrigation to hybrid maize using plant sensors. In: Proceedings of the international agronomy congress "Agriculture, Diversification, Climate Change management and livelihoods", 26–30 November 2012. ISA, IARI, New Delhi, pp 534–535

De Datta SK, Buresh RJ, Samsom MI, Obcemea WN, Real JG (1991) Direct measurement of ammonia and denitrification fluxes from urea applied to rice. Soil Sci Soc Am J 55:543–548

Dobermann, A.R. 2005. Nitrogen Use Efficiency - State of the Art" (2005). Agronomy & Horticulture – Faculty Publications. Paper 316

Drury CF, Tan CS, Gaynor JD, Oloya TO, Welacky TW (1996) Influence of controlled drainage–subirrigation on surface and tile drainage nitrate loss. J Environ Qual 25:317–324

Fowler DB, Brydon J (1989) No-till winter wheat production fertilizers. Agron J 81:518–524

Francis DD, Schepers JS, Vigil MF (1993) Post-anthesis nitrogen loss from corn. Agron J 85:659–663

Gupta R (2006) Crop canopy sensors for efficient nitrogen management in the Indo-Gangetic plains. Progress report. USDA funded project. The Rice-Wheat Consortium, New Delhi International Maize and Wheat Improvement Center (CIMMYT), Mexico, 28p

Hargrove WL, Kissel DE, Fenn LB (1977) Field measurements of ammonia volatilization from surface applications of ammonium salts to a calcareous soil. Agron J 69:473–476

Harper LA, Sharpe RR, Langdale GW, Giddens JE (1987) Nitrogen cycling in a wheat crop: soil, plant, and aerial nitrogen transport. Agron J 79:965–973

Hilton DR, Fixen PE, Woodward HJ (1994) Effects of tillage, nitrogen placement, and wheel compaction on denitrification rates in the corn cycle of a corn-oats rotation. J Plant Nutr 90:1341–1357

Howard DD, Essington ME (1998) Effect of surface applied limestone on the efficiency of urea containing N sources for no- till corn. Agron J 90:523–528

IRRI-CREMNET (International Rice Research Institute - Crop and Resource Management Network). 2000. Progress Report for 1998 & 1999. IRRI, Los Baños, Philippines.

Jat SL (2010) Effect of dual purpose summer legumes and zinc fertilization on productivity and quality of aromatic hybrid rice and their residual effects on succeeding wheat grown in a summer legume-rice-wheat cropping system. PhD thesis, Indian Agricultural Research Institute (IARI), New Delhi, India

Jat ML, Gupta R, Ramasundaram P, Gathala MK, Sidhu HS, Singh S, Singh RG, Saharawat YS, Kumar V, Chandna P, Ladha JK (2009) Laser assisted precision land leveling: a potential technology for resource conservation in irrigated intensive production systems of Indo-Gangetic plains. In: Ladha JK, Yadvinder Singh O, Erenstein HB, Los B (eds) Integrated crop and resource management in the rice-wheat system of South Asia. International Rice Research Institute, Makati City, pp 223–238

Ladha JK, Pathak H, Krupnik JS, Kessel C (2005) Efficiency of fertilizer nitrogen in cereal production: retrospect and prospects. Adv Agron 87:85–156

London JG (2005) Nitrogen study fertilizes fears of pollution. Nature 433:791

Parr JF, Papendick RI, Hornick SB, Meyer RE (1990) The use of cover crops, mulches and tillage for soil water conservation and weed control. In: Organic-matter management and tillage in humid and sub-humid Africa. IBSRAM proceedings no. 10, IBSRAM, Bangkok, pp 246–261

Ramos C (1996) Effect of agricultural practices on the nitrogen losses in the environment. In: Rodriguez-Barrueco C (ed). Sharma RK, Tripathi SC, Kharub AS, Chhoker RS, Mongia AD, Shoran J, Chauhan DS, Nagrajan S (2005) A decade of research on zero tillage and crop establishment, Research bulletin no. 18. DWR, Karnal, pp 5–8

Sharma, R.K., S.C. Tripathi, A.S. Kharub, R.S. Chhoker, A.D. Mongia, J. Shoran, D.S. Chauhan, and S. Nagrajan. 2005. A decade of research on zero tillage and crop establishment. DWR, Karnal - 132001. Res. Bulletin 18: 5–8.

Shukla AK, Singh VK, Dwivedi BS, Sharma SK, Singh Y (2006) Nitrogen use efficiencies using LCC in rice–wheat cropping system. Indian J Agric Sci 76:651–656

Singh Y, Singh B, Gupta RK, Singh J, Ladha JK, Bains JS, Balasubramanian V (2007) On-farm evaluation of leaf color chart for need-based nitrogen management in irrigated transplanted rice in northwestern India. Nutr Cycl Agroecosyst 78:167–176

Singh V, Singh B, Singh Y, Thind HS, Gupta RK (2010) Need based N management using chlorophyll meter and leaf colour chart in rice and wheat in South Asia: a review. Nutr Cycl Agroecosyst 88:361–380

Stulen I, Perez-Soba M, De Kok LJ, Van Der Eerden L (1998) Impact of gaseous nitrogen deposition on plant functioning. New Phytol 139:61–70

Stutte CA, Weiland RT, Blem AR (1979) Gaseous nitrogen loss from soybean foliage. Agron J 71:95–97

Tilman D (1999) Global environmental impacts of agriculture expansion; the need for sustainable and efficient practices. Proc Natl Acad Sci U S A 96:5995–6000

Wienhold BJ, Halvorson AD (1999) Nitrogen mineralization responses to cropping, tillage, and nitrogen rate in the northern Great Plains. Soil Sci Soc Am J 63:192–196

Strategies for Enhancing Phosphorus Efficiency in Crop Production Systems

Avishek Datta, Sangam Shrestha, Zannatul Ferdous, and Cho Cho Win

Abstract

Phosphorus (P) is the second important key element after nitrogen as a mineral nutrient for crop production. An adequate supply of P during early phases of plant development is essential for the root establishment and growth as well as for laying down the primordia of plant reproductive parts. Although abundant in soils in both organic and inorganic forms, P is the least available to plants due to its high fixation in most soil conditions and slow diffusion. Therefore, P can be a major limiting nutrient for plant growth on many soils across the world. Agricultural productivity will be lower without P, and consequently less food will be produced per unit area of land, especially in the least developed and developing countries where access to P fertilizers are restricted due to the rising costs of P fertilizer. Therefore, P is essential for the intensive agricultural production systems and thus contributes significantly to the present and future global food production and security. P is usually added to soil as chemical P fertilizer to satisfy the nutritional requirements of crop; however, plants can use only a small amount of this P since 75–90 % of added P is precipitated by metal–cation complexes and rapidly becomes fixed in soils. There is an increasing concern about the sustainability of P in agricultural production systems largely due to shortage of inorganic P fertilizer resources and environmental effects of agricultural P use beyond the field in the form of eutrophication. Such environmental concerns have led to the search for sustainable way of P nutrition of crops. Enhancing the efficiency of P in plants can be obtained through improving P acquisition by plants from the

A. Datta (✉) • Z. Ferdous • C.C. Win
Agricultural Systems and Engineering Program, School of Environment, Resources and Development, Asian Institute of Technology, Pathumthani 12120, Thailand
e-mail: datta@ait.ac.th; avishek.ait@gmail.com

S. Shrestha
Water Engineering and Management Program, School of Engineering and Technology, Asian Institute of Technology, Pathumthani 12120, Thailand
e-mail: sangamshrestha@gmail.com

A. Rakshit et al. (eds.), *Nutrient Use Efficiency: from Basics to Advances*,
DOI 10.1007/978-81-322-2169-2_5, © Springer India 2015

soil, internal plant utilization, or both. Hence, this review mainly focuses on three aspects: (1) we provide a brief overview on the holistic understanding of P dynamics in soil, (2) we discuss the role of microorganisms in increasing the availability of P to plants, and (3) we speculate on sustainable management strategies to enhance P use efficiency (PUE) in modern agriculture. Sustainable response strategies are required to improve PUE and recovery in order to cover the losses from the entire food production and consumption chain. An integrated strategy rather than a single strategy should be employed to enhance PUE and to recover unavoidable P losses.

Keywords
Soil P • P efficiency • P acquisition • P utilization • P use efficiency • Plant root development • Soil microorganisms

1   Introduction

The global demand for food is on rise from an ever-decreasing resource base, both in terms of quality and quantity, primarily due to an ever-growing population and increasing consumption of calorie- and meat-intensive diets (Mueller et al. 2012). Crop yield is largely controlled by fertilizer use and irrigation. Phosphorus (P) is by far the most important mineral nutrient for crop production, after nitrogen (N). Compared to other major nutrients, P is the least available to plants due to its high fixation in most soil conditions and slow diffusion (Ramaekers et al. 2010; Shen et al. 2011). Therefore, P can be a major limiting nutrient for plant growth and development on many soils across the world. Agricultural productivity will be lower without P, and consequently less food will be produced per unit area of land, especially in the least developed and developing countries where access to P fertilizers are restricted due to the rising costs of P fertilizer (Lynch 2007; Richardson et al. 2011; Richardson and Simpson 2011). Therefore, P is essential for the intensive agricultural production systems and thus contributes significantly to the present and future global food production and security (Richardson et al. 2011).

While analyzing the flow of P from "mine to fork", Cordell et al. (2009a, 2011) showed that a greater proportion of P is lost along the way when passing from mine to field to fork. The losses are significant from agricultural standpoint. The overall major losses in absolute amounts occur primarily from two main subsystems: arable land and livestock production (Tirado and Allsopp 2012). Inefficiencies in farm management are responsible for the arable land losses, where about 33 % of the P entering the soil is lost by both wind and water erosions. The harvested crops generally uptake between 15 and 30 % of the applied P fertilizer. Improper management of manure causes P loss at the livestock production level. It has been estimated that almost 50 % of the P entering the livestock system is lost into the environment (Tirado and Allsopp 2012).

Overall, P efficiency is divided into P acquisition efficiency (PAE) and P use efficiency (PUE). Enhanced P efficiency in plants can be achieved by improved PAE and/or PUE (Manske et al. 2001; Veneklaas et al. 2012). PAE refers to the ability of crops to uptake P from soils, and PUE is defined as the ability to produce biomass or yield, which is produced per unit of acquired P (Hammond et al. 2009; Wang et al. 2010). However, the contribution from PAE or PUE to crop P efficiency varies with crop species and environmental conditions (Wang et al. 2010). Sustainable response strategies are required to improve PUE and recovery in order to cover the losses from the entire food production and consumption chain (Schroder et al. 2011). An integrated strategy should be employed to enhance PUE and to recover unavoidable P losses. Cordell et al. (2009b) pointed out that the long-term P demand of the world could be possibly met through increased PUE by an order of 70 % and the remaining 30 % could be from high recovery and reuse of P from all resources.

This paper will mainly serve three purposes: (1) we provide a brief overview on the holistic understanding of P dynamics in soil, (2) we discuss the role of microorganisms in increasing the availability of P to plants as microorganisms contribute significantly to plant P nutrition, and (3) we speculate on sustainable management strategies to enhance PUE in modern agriculture. This paper will advance our existing knowledge about P efficiency with respect to crop production and will present better general guidelines to improve PUE in farms (involve both crops and animals) in managing a sustainable environmental system.

2 Holistic Understanding of P Dynamics in Soil

Soil P exists in organic (P_o) and inorganic (P_i) forms, which also differ in their fate and behavior in soils (Hansen et al. 2004). In most soils, the surface horizon contains more P than the subsoil due to the greater activities of microorganism and higher accumulation of organic materials in the surface layer. However, soil P content varies with parent material and management practices followed. These factors also govern the relative amount of P_o and P_i in a soil. Although P_i can vary from 10 to 90 %, it usually accounts for 35–70 % of total P in soil (Harrison 1987). Soils naturally contain 300–1,000 ppm of total P. The P_i forms are dominated by various minerals that include hydrous sesquioxides, amorphous and crystalline aluminum (Al) and iron (Fe) compounds in acidic, noncalcareous soil, and calcium (Ca) compounds in alkaline calcareous substrate. These Al, Fe, and Ca phosphates vary in their dissolution rates, depending on the size of mineral particles and

soil pH (Shen et al. 2011). Hinsinger (2001) mentioned that the solubility of Al and Fe phosphates increases and those of Ca phosphate decreases with increasing pH, except pH values >8. All these P forms are not generally discrete entities but exist in complex equilibrium as transformations between the forms occur continuously with each other, representing from very stable, sparingly available, to plant available pools such as labile P and solution P in soils (Shen et al. 2011).

In general, P_o accounts for 30–65 % of the total P in soils (Harrison 1987). The P from P_o can be released through mineralization facilitated by soil microorganisms and plant roots in association with phosphatase secretion (Shen et al. 2011). P_o mineralization dynamics are largely determined by environmental factors (e.g., soil moisture, temperature) and soil factors (e.g., pH, Eh). Like N, P_i can also be immobilized, which is the reverse reaction of mineralization. During immobilization, microorganisms convert P_i to P_o, which are then assimilated into their living cells. Therefore, the availability of P is greatly complex as it is highly associated with P dynamics and conversion among various P pools (Shen et al. 2011).

3 Microbial Contribution to Enhance P Availability in Soil

3.1 Potential of Plant and Microbial Strategies to Improve PAE

As many soils worldwide have a moderate to high capacity to retain P, most of the fertilizer and manure applied to soil for P are rapidly bound by the soil minerals that are not subject to rapid release. Thus, soil solution P concentration is very low. There is a need to develop plants that are more P efficient at low soil P and P-deficient soil, especially in the least developed and developing countries. P-efficient plants are required to reduce inefficient use of various P inputs and to minimize potential P loss to the environment (Richardson et al. 2011). Soil microorganisms can also influence the availability of P (phosphate) to plant roots as they are involved in a number of processes that affect P transformation in soil (Richardson 2001). Microorganisms can mediate the distribution of P between the available P in soil solution and the total soil P (inorganic and organic pools) through solubilization and mineralization reactions. Microbial processes mediate the availability of P to plants by directly influencing several important processes in the P cycle (Oberson and Joner 2005; Richardson and Simpson 2011). It has been estimated that the microbial biomass of soil contains a P pool ranging from 0.5 to 7.5 % of total P in grassland soils and, in case of arable soils, the range is between 0.4 and 2.4 % of total P (Oberson and Joner 2005). Therefore, soil microorganisms are critical for the transfer of poorly available form of P to plant available form and consequently play an important role in maintaining P in readily available form in soil. These processes are dominant in the rhizosphere of plants (Richardson 2007).

The concept of using soil microorganisms to enhance P availability to plants is actually an old thought (Richardson and Simpson 2011). Indeed, soil and microbial interactions are complex and often difficult to manipulate (Richardson 2001). Therefore, a greater knowledge of the processes and understanding microbial ecology are essential to better exploit soil microorganisms for P mobilization in soils.

3.2 Mechanisms Governing Mobilization of Soil P by Microorganisms

Microorganisms directly influence the ability of plants to acquire P from soil through a number of mechanisms that include (1) an increase in the surface area of roots by either an extension of existing root systems (e.g., mycorrhizal associations) or by hormonal stimulation of root growth, branching, or root hair development, i.e., phytostimulation (Richardson et al. 2009a); (2) by modification of sorption equilibria that may result in increased net transfer of phosphate ions into soil solution or an increase in the mobility of organic P either directly or indirectly

through microbial turnover (Seeling and Zasoski 1993); and (3) through stimulation of metabolic processes that are effective in directly solubilizing and mineralizing P from poorly available forms of inorganic and organic P in soil (Richardson et al. 2009a). These processes include the excretion of protons (hydrogen ions), the release of organic acids, the production of siderophores, and the release of phosphatase and cellulolytic enzymes that are able to hydrolyze organic P (Richardson 2007).

3.3 Interactions Between Plant Roots and Microorganism on P Availability

Microorganisms decompose and mineralize organic material as well as release and transform inorganic nutrients; therefore, they play a dominant role in nutrient cycling (Marschner et al. 2011). Microorganisms can further influence nutrient availability in soil by different reactions such as solubilization, chelation, and oxidation/reduction when they require a specific nutrient. Bacteria and fungi generally release nutrients from their biomass as they undergo turnover or are fed upon by protozoa (Marschner et al. 2011). Most of the soil is low in P availability; thus, microorganisms and plants have developed similar mechanisms for mobilization and uptake of P. Microbes are more competitive than plants for the uptake of P in the rhizosphere (Marschner et al. 2011). In addition, microorganisms can affect plant growth and nutrient uptake by release of growth-stimulating or growth-inhibiting substances that influence root physiology and root system architecture (Govindasamy et al. 2009; Ryu et al. 2005).

Lack of carbon limits the growth and activity of microorganisms in soil (Demoling et al. 2007; De Nobili et al. 2001). This is due to the complex nature of soil carbon resulting in poor decomposability of the soil organic matter. Soil microbial growth can be stimulated by applying a carbon amendment to a soil. However, this process will also result in a concurrent rapid decrease of soil solution P concentration due to the temporary immobilization of P in microbial biomass (Bunemann et al. 2004; Oehl et al. 2001). It is possible for the microbial biomass to grow rapidly and to multiply when readily degradable carbon is available even under low-P conditions (Bunemann et al. 2004). Root exudates, on the other hand, are easy to decompose due to their lower molecular weight. Therefore, microbial density and metabolic activity will be greater in the rhizosphere than bulk soil with the release of exudates by roots (Soderberg and Baath 1998).

Agricultural soil can maintain higher soil organic carbon levels when animal manures are applied to the soil on a regular basis (Leifeld et al. 2009). Soil microbial biomass and activities are also reported to be greater in manure soil compared with soil managed with mineral fertilization (Joergensen et al. 2010). This translates into a higher soil microbial P pool with faster turnover under organic cropping systems (Oehl et al. 2001). Plants and microorganisms require mineralization or hydrolysis of substrates by phosphatase enzymes (either plant or microbial origin) to utilize organic P (Richardson and Simpson 2011). P deficiency results in an increased activity of phosphatases, which is designated as part of P starvation responses (Richardson and Simpson 2011). Extracellular phosphatases are released from roots of plants and are important for capture and recycling of organic P (Richardson et al. 2005). Up to 60 % of the total organic P may be hydrolyzed by phosphatases with the highest amounts being released by phytases (Bunemann 2008).

Higher microbial activity is responsible for the increased mineralization of soil organic matter associated. This occurs in the rhizosphere as a result of a microbial priming effect due to utilization of exudate C with subsequent mineralization of nutrients from soil organic matter (Cheng 2009). Mineralization and immobilization of P depend to a greater extent on C:P or C:organic P ratios as C:N ratios are reasonably constant across different soils having various levels of organic matter (Kirkby et al. 2011). This has implications for competition between plants and microorganisms for nutrient uptake within the rhizosphere, resulting in either a net immobilization or mineralization of P (Richardson and Simpson 2011).

The activity of microorganisms in the soil is different under various management practices. It has been estimated that the recovery of fresh manure P by ryegrass (*Lolium* spp.) (24–35 %) was lower than the recovery of mineral P (37–43 %) (Oberson et al. 2010). The available P content of the soil influenced the recovery of fresh manure P where the recovery was lower in soils with higher plant available P (Simpson et al. 2011). Uptake of residual P was greater from conventionally managed soils than from organically managed soils. It is important to note that the total and available P contents of soil were depleted less in the conventional systems than in organic systems (Oehl et al. 2002).

3.4 Significance of Plant and Microbial Interactions on P Acquisition

The ability of crops to take up P from soil is referred to as P acquisition (Wang et al. 2010). The enhancement of crop P efficiency is dependent on various physiological and biochemical traits. Root morphological characteristics and architecture are the most important traits in P acquisition (Richardson et al. 2009b; Wang et al. 2010). One of the important strategies to improve soil P acquisition is "topsoil foraging" by root systems (Lynch 2011; Ramaekers et al. 2010; Richardson et al. 2011). Surface soil strata or topsoil generally has greater P availability than subsoil strata in most natural soils (Lynch 2011; Ramaekers et al. 2010). Therefore, root traits that enhance topsoil foraging will be able to acquire more P. Plant species greatly differ in their ability for topsoil foraging. Root architectural traits associated with enhanced topsoil foraging include shallower growth angles of axial roots, enhanced adventitious rooting, a greater number of axial roots, and greater dispersion of lateral roots (Lynch 2011). For example, P-efficient common bean (*Phaseolus vulgaris* L.) genotypes produce shallower basal roots under low P availability compared with P-inefficient genotypes, where P acquisition significantly correlates with basal root shallowness (Liao et al. 2001). It has been estimated that variation in root growth angle among closely related genotypes is associated with up to 600 % increase in P acquisition and 300 % increase in yield in common bean (Liao et al. 2001). Studies have also shown that in maize (*Zea mays* L.) and soybean [*Glycine max* (L.) Merr.], shallower growth angles of axial roots (basal roots in legumes, seminal and crown roots in maize) increase topsoil foraging and thereby P acquisition (Lynch 2011; Zhu et al. 2005).

The abundance of root exudates stimulates microbial growth and activity and attracts more soil microorganisms to the root surface in the rhizosphere (Marschner et al. 2011). Root exudates (e.g., sugars, organic acid anions, and amino acids) are released primarily in the root zone immediately behind the root tip and in the distal elongation zone under nutrient deficiency (Marschner et al. 1997a, b). These are also the sites where microorganisms are the least abundant but the most active (Nguyen and Guckert 2001).

Only a small proportion of P is readily available to plants and microorganisms as P is mainly present in forms that are unavailable to them, but the total amount of P in the soil generally is high (Hinsinger 2001; Richardson 2001). The solubility of poorly soluble P_i can be increased by plants and microorganisms through releasing protons, OH^-, or CO_2 and organic acid anions such as citrate, malate, and oxalate (Hinsinger 2001; Roelofs et al. 2001). Plants and microorganisms can mineralize P_o by release of various phosphatase enzymes (Neble et al. 2007).

Crops can also extend their P uptake capacity through a symbiosis with beneficial fungi (Schroder et al. 2011). Association of crop plants with mycorrhizal fungi can increase plant P uptake through extension of plant root system with mycorrhizal hyphae, which increase the soil volume explored (Bucher 2007). Arbuscular mycorrhizal (AM) fungal species colonize the most land plant species (70–90 %) and contribute significantly in the P nutrition, especially in soils with low available P (Parniske 2008). P is taken up by the hyphae and transported to roots, where it is moved across the fungus–plant interface into the host cell. In exchange for P and other nutrients supplied to the plant, the fungus

receives reduced carbon (Smith and Read 2008). AM fungal hyphae extend further from roots than root hairs and are also active in the more mature regions of the root (Richardson et al. 2011; Smith and Read 2008). Therefore, AM plants can exploit larger soil volume, and this is the primary reason for the greater P uptake in low-P soil (Richardson et al. 2011).

4 Management Measures to Improve PUE

4.1 Reducing P Losses from Agricultural Systems

P is lost from arable lands mostly by erosion (both by wind and water), runoff, crop uptake and removal, as well as leaching. Modern agroecosystems not only deplete finite resources (land, "fossil" water, fossil energy, and fossil phosphate) but also contaminate a larger part of our planet with reactive N and P, which damage environmental quality including biodiversity (Correll 1998). Agriculture alone is responsible for the depletion of about 19 Mt year^{-1} of P from phosphate rock for the production of fertilizer (Heffer and Prud'homme 2008).

Loss of P from agricultural soil can be minimized by preventing soil erosion and subsequent losses of P to water bodies. Verheijen et al. (2009) classified erosion as water erosion, wind erosion, tillage erosion, and the erosion resulting from soil particles adhering to lifted crops. It has been estimated that around 12 % of the total European land area suffers from water erosion and 4 % from wind erosion (Louwagie et al. 2009). The amounts of soil lost via erosion for an average European arable soil range from 5–40 Mg ha^{-1} year^{-1} (Verheijen et al. 2009) to 10 Mg ha^{-1} year^{-1} at most for the major part of Europe (Louwagie et al. 2009). Greater proportion of P is contained by the light and more easily erodible soil fraction (clay) compared with the coarse fraction (Quinton 2002). Ruttenberg (2003) estimated that about 20–30 Mg year^{-1} of P is lost via erosion at a global scale, which is equivalent to an annual loss of 15–20 kg P ha^{-1} when erosion from land other than arable land is negligible. Both soluble (dissolved) and particulate (eroded soil particles) forms of soil P can be removed through water moving across the surface or through soils resulting in an increase in the concentration of bioavailable P in surface waters (Schroder et al. 2011). Soils with high concentrations of P will transport high concentrations of P to surface streams if sediments are washed into streams. Surface runoff is the second largest potential cause of dissolved P loss after erosion. Runoff of excess P from agricultural fields can enhance the fertility status of natural waters that can contribute to eutrophication, the increase in the growth of algae and other aquatic plants. Eutrophication can start from a P concentration in water of around 0.10 g total P m^{-3} (Correll 1998). P loss by leaching is usually considered less important than surface runoff.

Conservation practices that can be used to minimize soil erosion and surface runoff include reduced tillage without removal of crop residues, terracing on sloping land, cultivation and planting along the contour, cover crop establishment, agroforestry, or complete reforestation (Schroder et al. 2011). Practices that minimize losses of P by runoff include applying P when soil is best able to capture it. Application of phosphate fertilizers or manures should be avoided to frozen or snow-covered land or to soils that are dry and hard or waterlogged (Schroder et al. 2011). Placement of fertilizer should be made as close as possible to where the roots of the crop plants are growing (Schroder et al. 2011). Moreover, it is also necessary to maximize the use of P in manure for maintaining soil fertility in croplands and pastures and adjusting livestock diets, which will in turn reduce P losses. This will simultaneously work for recovering the P and other nutrients being lost when they are not incorporated into crop plants (Tirado and Allsopp 2012).

can be challenging as the total P content in manure varies widely (depends on manure type) and about 70 % of total P in manure is labile. The P$_i$ contents of manure can range from 50 to 90 % (Dou et al. 2000). Turner and Leytem (2004) mentioned that manure also comprises a substantial amount of P$_o$ (e.g., nucleic acids and phospholipids), which can be released by mineralization to increase soil P$_i$ concentrations. Organic substances like manure can greatly reduce the chance of P adsorption by soil particles as the humic acids contain large number of negative charges (e.g., carboxyl and hydroxyl groups), which strongly compete for the adsorption sites with P$_i$ (Shen et al. 2011). In addition, manure can modify soil P availability by changing soil pH. However, PUE does not depend on the food source of the livestock producing manure; it does not matter whether livestock are fed home-produced feeds or imported feeds (Schroder et al. 2011).

It is important to maintain high soil organic matter levels in order to achieve the best use of P added to the soil (Tirado and Allsopp 2012). Microorganisms play a dominant role in releasing P from the stored P in organic form, which then become available to plants. P application to pastures on low-P soils results in an accumulation of organic matter and consequently organic P in the soil (Simpson et al. 2011). A relationship exists between soil organic carbon accumulation and P fertilizer use (Chan et al. 2010). However, the accumulation of organic P is influenced by the farming system and/or management practices followed (Simpson et al. 2011). Bunemann et al. (2006) estimated that a wheat (*Triticum aestivum* L.)–pasture rotation accumulated ~2 kg P ha^{-1} year^{-1} as organic P in soil, whereas there was zero accumulation of organic P in soil under continuous cropping rotations.

4.2 Application of Manure P in Agricultural Soil

Building P fertility can be accomplished by manure application to soil (Shen et al. 2011). However, matching crop P needs with manure

4.3 Improving P Fertilizer Recommendations and Management

In plant, P is essential and performs a number of physiological functions related to energy transformations. P is a part of many cell constituents and plays a major role in several key physiological processes that include photosynthesis, respiration, energy storage and transfer, cell division, and cell enlargement. An adequate amount of P is required for the early root establishment and growth. Crop P requirement varies depending on the type of crops grown. It is necessary to adopt soil test-based P fertilizer recommendation as the nutrient supplying capacity of the soil varies. Producers may adjust/alter this amount depending on fertilizer costs, yield goals, management intensity, or different expectations in soil P levels (e.g., sufficiency, or maintenance, or buildup). The P fertilization amount is important for improving the PUE. The amount of P fertilization at any one time should be selected in such a way that it matches with P uptake in the harvested crop when the soil tests result in a critical soil P level. Eutrophication is a major and common problem worldwide, caused mostly by the overuse of P and N fertilizers (Tirado and Allsopp 2012). Proper management of P fertilizer applications is essential in order to optimize crop yield and protect water quality. Various management practices that increase P availability include banding (band placement method), increasing organic matter through manure applications, conservation tillage practices, and applying fertilizers as close as possible to meet the peak crop uptake.

McDonald et al. (2011) compared P imbalances across all regions globally and reported that about 29 % of the global cropland area had overall P deficits, whereas 71 % of the cropland area had overall P surpluses. They also mentioned that the present P fertilizer use may be contributing to soil P accumulation in some rapidly developing areas together with relatively low PUE although developing countries, on an average, had P deficits during the mid-twentieth century (Bennett et al. 2001; MacDonald et al. 2011). Cakmak (2002) reported that yields on about 67 % of the global farmland are limited by insufficient soil P levels. Therefore, most producers apply P in their farms in the form of manures, composts, biosolids, or mineral fertilizers, just alone to compensate for the P that is exported to the final produce (Schroder et al. 2011).

4.4 Improving Timing and Placement Methods of P Fertilizer

P fertilizer should be applied at the time of the maximum P uptake by the crop. Banding or deep placement of P near the expanding root system of the crop row is recommended over broadcast applications (Schroder et al. 2011). This is because P banding essentially saturates the soil with P in a small localized area, which provides plant roots an easy access of the P readily available after the initial P–soil reactions take place. In contrast, P will come in contact with much more surface area of soil when P is applied through broadcast method, which will ultimately lead to high levels of P sorption and low levels of available P. Havlin et al. (2011) recommended that the recovery of fertilizer P can be significantly improved by placing fertilizer with the seed or in a band close to it, especially in the year of application.

4.5 Soil Moisture and P Availability

Moisture content of the soil influences the effectiveness and rate of availability of applied P. Soils at field capacity dissolve about 50–80 % of P fertilizer within 24 h compared with 20–50 % for soils at 2–4 % moisture within the same time (Havlin et al. 2011). The response to granular phosphates of high water solubility is greater under wet conditions compared with that from powdered materials. However, powdered materials provide better results under dry conditions (Havlin et al. 2011).

There is an increase in available P after flooding in most soils. This is largely due to the conversion of Fe^{3+}–P minerals to more soluble Fe^{2+}–P minerals. Other mechanisms which are involved for increasing the availability of P after flooding include dissolution of occluded P, release of P from insoluble Fe and Al compounds, some dissolution of Ca phosphates at higher CO_2 levels in the soil solution, and greater P diffusion (Havlin et al. 2011). The release of P by these processes can take several weeks after flooding (Snyder and Slaton 2002). There is a decrease in available P after field draining of the flooded soils. The Fe and Al compounds that became soluble after flooding will react with native and applied fertilizer P to form insoluble P compounds as soils started drying after field draining. They rapidly adsorb soluble phosphates, and the availability of P will decrease (Snyder and Slaton 2002).

4.6 Development of More Efficient Crop Genotypes for P Acquisition and Use

P is one of the most limiting factors for crop production as it is the least mobile and least available mineral nutrient to plant under most soil conditions. Generally, soil and fertilizer P are easily bound by either soil organic matter or chemicals, making it unavailable to plants unless hydrolyzed to release P_i (Akthar et al. 2008). Therefore, successful P management can be obtained by breeding crop varieties that can grow and yield better under low P supply (Shen et al. 2011; Wang et al. 2010). These cultivars or genotypes will be more efficient in P acquisition and use.

The crop P efficiency can be defined as the ability to produce biomass or yield under certain available P supply conditions (Wissuwa et al. 2009). Improving the efficiency of P fertilizer use for crop growth requires enhanced P acquisition by plants from the soil (PAE) and enhanced use of P in processes that lead to faster growth and greater allocation of biomass to the harvestable parts (internal PUE) (Veneklaas et al. 2012). Therefore, it is important to focus on different mechanisms of crop plants that help improve their P acquisition and utilization efficiency. Plants are divided into monocot and dicot members where their root systems are distinct and very diverse. Their root architecture differs quite significantly; however, the main adaptive root traits correlated with enhanced P acquisition are common among all vascular plant species for enhancing P acquisition (Niu et al. 2013; Ramaekers et al. 2010). P is a relatively immobile and the least available element in the soil. Therefore, the foremost strategy used by plants for the greatest P acquisition comprises of maximal and continued soil exploration through proliferation and extension of all root types (Lynch and Ho 2005). The greater preference for P acquisition is for metabolically efficient roots that can obtain P from the soil actively.

Genotypic differences affect the ability of plants to absorb P (Schroder et al. 2011). Genotype differs in its response to limiting P_i availability in terms of its allocation of assimilates to either aboveground parts or roots, extension rates of root growth, branching of root systems, lateral root formation, or distribution of a given root length through the soil profile (Abel et al. 2002; De Willigen and Van Noordwijk 1987; Gaxiola et al. 2011). PUE of crops is governed by the uptake efficiency and the utilization inside the plant once the P is taken up (Schroder et al. 2011). For example, studies with the model plant *Arabidopsis thaliana* have shown that plants grown in low P_i develop root systems with higher numbers of lateral roots and larger root hairs (Lopez-Bucio et al. 2002, 2003). In addition to morphological shifts in the roots, P deficiency also results in the change of the physiological and cellular traits of the roots (Gaxiola et al. 2011). Yan et al. (2002) demonstrated that white lupine (*Lupinus albus* L.), a plant with a very efficient P_i scavenging capacity, upregulates the abundance and activity of the P-ATPase enzyme as a response to limiting P_i.

Breeding crop cultivars or genotypes more efficient for P acquisition and use is one of the important strategies for successful P management (Shen et al. 2011). Researchers from China have identified crop varieties for high PUE by utilizing traditional plant breeding programs. An example of such an efficient genotype was the soybean variety "BX10" that produced superior root traits enabling better adaptation to low-P soils (Yan et al. 2006). Some important root genetic traits have been identified that have the potential to utilize in breeding P-efficient crops, including root exudates, root hair traits, topsoil foraging through basal, or adventitious rooting (Gahoonia and Nielsen 2004).

Conclusion

P is an essential macronutrient for all living organisms. P plays a vital role in different physiological processes of plants including photosynthetic regulation, energy conservation, and carbon metabolism (Abel et al. 2002). Plants absorb orthophosphate (P_i) either as $H_2PO_4^-$ or HPO_4^{2-} depending on pH of the soil. The amounts of $H_2PO_4^-$ and HPO_4^{2-} are similar in soil at pH 7.2; however, the uptake of $H_2PO_4^-$ is much faster in plants than HPO_4^{2-} uptake (Gaxiola et al. 2011). The P_i concentration of average agricultural soil solution is about 0.05 ppm, but this concentration varies widely among soils (Havlin et al. 2011). The deficiency of P_i is critical in highly weathered soils of tropics and subtropics, as well as calcareous/alkaline soils of the Mediterranean basin (Hinsinger 2001). P_i can be adsorbed or precipitated by Ca salts or Fe and Al oxide complexes. These processes greatly reduce the availability of P_i as it becomes trapped in minerals (Holford 1997). Suboptimal levels of P_i result in crop yield losses, although the severity and economical relevance of the losses vary from crop to crop (Gaxiola et al. 2011).

There is an increasing concern about the sustainability of P in agricultural production systems. This is due to shortage of inorganic P fertilizer resources and environmental effects of agricultural P use beyond the field in the form of eutrophication that is caused by runoff from fertilized fields (Gaxiola et al. 2011; Wang et al. 2010). It is important to note that the availability of P in the soils of many developing countries such as India, Spain, Australia, etc., is at a critical low point (Gahoonia and Nielsen 2004). A greater understanding of P dynamics in the soil/rhizosphere/plant continuum is of utmost importance as it can provide a significant basis for optimizing P management to improve PUE in crop production (Shen et al. 2011). It will be

challenging to obtain greater insight on the mechanisms and regulation of P acquisition and internal utilization efficiency. The research thrust should be directed toward the improvement of P efficiency that often features more to PAE under limited P supply and to PUE under high-P conditions (Wang et al. 2010). Therefore, the perfect breeding approach should take into account the improvement of both P acquisition and P utilization efficiency in the given species under different P supply conditions in various soil types (Wang et al. 2010). It is important to establish an integrated P management strategy that involves manipulation of soil and rhizosphere processes, development of P-efficient crops, and improving P recycling efficiency in the future (Shen et al. 2011).

References

Abel S, Ticconi CA, Delatorre CA (2002) Phosphate sensing in higher plants. Physiol Plant 115:1–8. doi:10.1034/j.1399-3054.2002.1150101.x

Akthar MS, Oki Y, Adachi T (2008) Genetic variation in phosphorus acquisition and utilization efficiency from sparingly soluble P-sources by *Brassica* cultivars under P-stress environment. J Agric Crop Sci 194:380–392. doi:10.1111/j.1439-037X.2008.00326.x

Bennett EM, Carpenter SR, Caraco NF (2001) Human impact on erodable phosphorus and eutrophication: a global perspective. BioScience 51:227–234. doi:10.1641/0006-3568(2001)051[0227:HIOEPA]2.0.CO;2

Bucher M (2007) Functional biology of plant phosphate uptake at root and mycorrhizal interfaces. New Phytol 173:11–26. doi:10.1111/j.1469-8137.2006.01935.x

Bunemann EK (2008) Enzyme additions as a tool to assess the potential bioavailability of organically bound nutrients. Soil Biol Biochem 40:2116–2129. doi:10.1016/j.soilbio.2008.03.001

Bunemann E, Bossio DA, Smithson PC, Frossard E, Oberson A (2004) Microbial community composition and substrate use in a highly weathered soil as affected by crop rotation and P fertilization. Soil Biol Biochem 36:889–901. doi:10.1016/j.soilbio.2004.02.002

Bunemann EK, Heenan DP, Marschner P, McNeill AM (2006) Long-term effects of crop rotation, stubble management and tillage on soil phosphorus dynamics. Aust J Soil Res 44:611–618. doi:10.1071/SR05188

Cakmak I (2002) Plant nutrition research: priorities to meet human needs for food in sustainable ways. Plant Soil 247:3–24. doi:10.1023/A:1021194511492

Chan KY, Oates A, Li GD, Conyers MK, Prangnell RJ, Poile JG, Liu DL, Barchia IM (2010) Soil carbon stocks under different pastures and pasture management in the higher rainfall areas of south-eastern Australia. Aust J Soil Res 48:7–15. doi:10.1071/SR09092

Cheng W (2009) Rhizosphere priming effect: its functional relationships with microbial turnover, evapotranspiration, and C-N budgets. Soil Biol Biochem 41:1795–1801. doi:10.1016/j.soilbio.2008.04.018

Cordell D, Drangert JO, White S (2009a) The story of phosphorus: global food security and food for thought. Glob Environ Change 19:292–305. doi:10.1016/j.gloenvcha.2008.10.009

Cordell D, White S, Drangert JO, Neset TSS (2009b) Preferred future phosphorus scenarios: a framework for meeting long-term phosphorus needs for global food demand. In: Ashley K, Mavinic D, Koch F (eds) International conference on nutrient recovery from wastewater streams Vancouver, 2009. IWA Publishing, London. ISBN 9781843392323

Cordell D, Rosemarin A, Schroder JJ, Smit AL (2011) Towards global phosphorus security: a systems framework for phosphorus recovery and reuse options. Chemosphere 84:747–758. doi:10.1016/j.chemosphere.2011.02.032

Correll DL (1998) The role of phosphorus in eutrophication of receiving waters: a review. J Environ Qual 27:261–266. doi:10.2134/jeq1998.00472425002700020004x

De Nobili M, Contin M, Mondini C, Brookes PC (2001) Soil microbial biomass is triggered into activity by trace amounts of substrate. Soil Biol Biochem 33:1163–1170. doi:10.1016/S0038-0717(01)00020-7

De Willigen P, Van Noordwijk M (1987) Roots, plant production and nutrient use efficiency. PhD thesis, Wageningen Agricultural University, Wageningen, The Netherlands, pp 282

Demoling F, Figueroa D, Baath E (2007) Comparison of factors limiting bacterial growth in different soils. Soil Biol Biochem 39:2485–2495. doi:10.1016/j.soilbio.2007.05.002

Dou Z, Toth JD, Galligan DT, Ramberg CF, Ferguson JD (2000) Laboratory procedures for characterizing manure phosphorus. J Environ Qual 29:508–514. doi:10.2134/jeq2000.00472425002900020019x

Gahoonia TS, Nielsen NE (2004) Root traits as tools for creating phosphorus efficient crop varieties. Plant Soil 260:47–57. doi:10.1023/B:PLSO.0000030168.53340.bc

Gaxiola RA, Edwards M, Elser JJ (2011) A transgenic approach to enhance phosphorus use efficiency in crop as part of a comprehensive strategy for sustainable agriculture. Chemosphere 84:840–845. doi:10.1016/j.chemosphere.2011.01.062

Govindasamy V, Senthilkumar M, Mageshwaran V, Annapurna K (2009) Detection and characterization of ACC deaminase in plant growth promoting rhizobacteria. J Plant Biochem Biotechnol 18:71–76. doi:10.1007/BF03263298

Hammond JP, Broadley MR, White PJ (2009) Shoot yield drives phosphorus use efficiency in Brassica oleracea and correlates with root architecture traits. J Exp Bot 60:1953–1968. doi:10.1093/jxb/erp083

Hansen JC, Cade-Menun BJ, Strawn DG (2004) Phosphorus speciation in manure-amended alkaline soils. J Environ Qual 33:1521–1527. doi:10.2134/jeq2004.1521

Harrison AF (1987) Soil organic phosphorus – a review of world literature. CAB International, Wallingford, p 257

Havlin JL, Beaton JD, Tisdale SL, Nelson WL (2011) Soil fertility and fertilizers – an introduction to nutrient management, 7th edn. Pearson Education, Inc., Upper Saddle River, p 515

Heffer P, Prud'homme M (2008) Medium-term outlook for global fertilizer demand, supply and trade 2008–2012. In: 76th IFA annual conference, Vienna, Austria

Hinsinger P (2001) Bioavailability of soil inorganic P in the rhizosphere as affected by root-induced chemical changes: a review. Plant Soil 237:173–195. doi:10.1023/A:1013351617532

Holford ICP (1997) Soil phosphorus, its measurements and its uptake by plants. Aust J Soil Res 35:227–239. doi:10.1071/S96047

Joergensen R, Mader P, Fließbach A (2010) Long-term effects of organic farming on fungal and bacterial residues in relation to microbial energy metabolism. Biol Fertil Soils 46:303–307. doi:10.1007/s00374-009-0433-4

Kirkby CA, Kirkegaard JA, Richardson AE, Wade LJ, Blanchard C, Batten G (2011) Stable soil organic matter: a comparison of C:N:P:S ratios in Australian and other world soils. Geoderma 163:197–208. doi:10.1016/j.geoderma.2011.04.010

Leifeld J, Reiser R, Oberholzer HR (2009) Consequences of conventional versus organic farming on soil carbon: results from a 27-year field experiment. Agron J 101:1204–1218. doi:10.2134/agronj2009.0002

Liao H, Rubio G, Yan X, Cao A, Brown KM, Lynch JP (2001) Effect of phosphorus availability on basal root shallowness in common bean. Plant Soil 232:69–79. doi:10.1023/A:1010381919003

Lopez-Bucio J, Hernandez-Abreu E, Sanchez-Calderon L (2002) Phosphate availability alters architecture and causes changes in hormone sensitivity in the arabidopsis root system. Plant Physiol 129:244–256. doi:10.1104/pp. 010934

Lopez-Bucio J, Cruz-Ramirez A, Herrera-Estrella L (2003) The role of nutrient availability in regulating root architecture. Curr Opin Plant Biol 6:280–287. doi:10.1016/S1369-5266(03)00035-9

Louwagie G, Gay SH, Burrell A (2009) Sustainable agriculture and soil conservation (SoCo); final report. JRC Scientific and Technical Reports EUR 23820EN, Ispra, Italy, pp 171

Lynch JP (2007) Roots of the second green revolution. Aust J Bot 55:493–512. doi:10.1071/BT06118

Lynch JP (2011) Root phenes for enhanced soil exploration and phosphorus acquisition: tools for future crops.

Plant Physiol 156:1041–1049. doi:10.1104/pp. 111. 175414

Lynch JP, Ho MD (2005) Rhizoeconomics: carbon costs of phosphorus acquisition. Plant Soil 269:45–56. doi:10.1007/s11104-004-1096-4

MacDonald GK, Bennett EM, Potter PA, Ramankutty N (2011) Agronomic phosphorus imbalances across the world's croplands. Proc Natl Acad Sci 108:3086–3091. doi:10.1073/pnas.1010808108

Manske GGB, Ortiz-Monasterio JI, van Ginkel M (2001) Importance of P uptake efficiency versus P utilization for wheat yield in acid and calcareous soils in Mexico. Eur J Agron 14:261–274. doi:10.1016/S1161-0301 (00)00099-X

Marschner P, Crowley DE, Higashi RM (1997a) Root exudation and physiological status of a root-colonizing fluorescent pseudomonad in mycorrhizal and non-mycorrhizal pepper (*Capsicum annuum* L.). Plant Soil 189:11–20. doi:10.1023/A:1004266907442

Marschner P, Crowley DE, Sattelmacher B (1997b) Root colonization and iron nutritional status of a *Pseudomonas fluorescens* in different plant species. Plant Soil 196:311–316. doi:10.1023/A:1004236410302

Marschner P, Crowley D, Rengel Z (2011) Rhizosphere interactions between microorganisms and plants govern iron and phosphorus acquisition along the root axis–model and research methods. Soil Biol Biochem 43:883–894. doi:10.1016/j.soilbio.2011.01.005

Mueller ND, Gerber JS, Johnston M, Ray DK, Ramankutty N, Foley JA (2012) Closing yield gaps through nutrient and water management. Nature 490:254–257. doi:10.1038/nature11420

Neble S, Calvert V, Le Petit J, Criquet S (2007) Dynamics of phosphatase activities in a cork oak litter (*Quercus suber* L.) following sewage sludge application. Soil Biol Biochem 39:2735–2742. doi:10.1016/j.soilbio. 2007.05.015

Nguyen C, Guckert A (2001) Short-term utilization of ^{14}C-[U]glucose by soil microorganisms in relation to carbon availability. Soil Biol Biochem 33:53–60. doi:10.1016/S0038-0717(00)00114-0

Niu YF, Chai RS, Jin GL (2013) Responses of root architecture development to low phosphorus availability: a review. Ann Bot 112:391–408. doi:10.1093/aob/mcs285

Oberson A, Joner EJ (2005) Microbial turnover of phosphorus in soil. In: Turner BL, Frossard E, Baldwin DS (eds) Organic phosphorus in the environment. CABI, Wallingford, pp 133–164

Oberson A, Tagmann HU, Langmeier M, Dubois D, Mäder P, Frossard E (2010) Fresh and residual phosphorus uptake by ryegrass from soils with different fertilization histories. Plant Soil 334:391–407. doi:10.1007/s11104-010-0390-6

Oehl F, Oberson A, Probst M, Andreas Fliessbach A, Roth H-R, Frossard E (2001) Kinetics of microbial phosphorus uptake in cultivated soils. Biol Fertil Soils 34:31–41. doi:10.1007/s003740100362

Oehl F, Oberson A, Tagmann HU, Besson JM, Dubois D, Mader P, Roth H-R, Frossard E (2002) Phosphorus budget and phosphorus availability in soils under organic and conventional farming. Nutr Cycl Agroecosyst 62:25–35. doi:10.1023/A:1015195023724

Parniske M (2008) Arbuscular mycorrhiza: the mother of plant root endosymbioses. Nat Rev Microbiol 6:763–775. doi:10.1038/nrmicro1987

Quinton JN (2002) Detachment and transport of particle-bound P: processes and prospects for modelling. In: Chardon WJ, Schoumans OF (eds) Phosphorus losses from agricultural soils: processes at the field scale. Cost Action 832 Quantifying the agricultural contribution to eutrophication. Alterra, Wageningen, pp 61–65

Ramaekers L, Remans R, Rao IM, Blair MW, Vanderleyden J (2010) Strategies for improving phosphorus acquisition efficiency of crop plants. Field Crops Res 117:169–176. doi:10.1016/j.fcr.2010.03.001

Richardson AE (2001) Prospects for using soil microorganisms to improve the acquisition of phosphorus by plants. Aust J Plant Physiol 28:897–906. doi:10.1071/PP01093

Richardson AE (2007) Making microorganisms mobilize soil phosphorus. In: Valazquez E, Rodriguez-Barrueco C (eds) First international meeting on microbial phosphate solubilization, vol 102, Developments in plant and soil science. Springer, Dordrecht, pp 85–90

Richardson AE, Simpson RJ (2011) Soil microorganisms mediating phosphorus availability. Plant Physiol 156:989–996. doi:10.1104/pp. 111.175448

Richardson AE, George TS, Hens M, Simpson RJ (2005) Utilization of soil organic phosphorus by higher plants. In: Turner BL, Frossard E, Baldwin DS (eds) Organic phosphorus in the environment. CABI, Wallingford, pp 165–184

Richardson AE, Barea J, McNeill AM, Prigent-Combaret C (2009a) Acquisition of phosphorus and nitrogen in the rhizosphere and plant growth promotion by microorganisms. Plant Soil 321:305–339. doi:10. 1007/s11104-009-9895-2

Richardson AE, Hocking PJ, Simpson RJ, George TS (2009b) Plant mechanisms to optimize access to soil phosphorus. Crop Pasture Sci 60:124–143. doi:10. 1071/CP07125

Richardson AE, Lynch JP, Ryan PR (2011) Plant and microbial strategies to improve the phosphorus efficiency of agriculture. Plant Soil 349:121–156. doi:10. 1007/s11104-011-0950-4

Roelofs RFR, Rengel Z, Cawthray GR, Dixon KW, Lambers H (2001) Exudation of carboxylates in Australian Proteaceae: chemical composition. Plant Cell Environ 24:891–903. doi:10.1046/j.1365-3040. 2001.00741.x

Ruttenberg KC (2003) The global phosphorus cycle. Treatise on geochemistry, vol 8. Elsevier Ltd, Amsterdam, pp 585–643

Ryu CM, Hu CH, Locy RD, Kloepper JW (2005) Study of mechanisms for plant growth promotion elicited by rhizobacteria in *Arabidopsis thaliana*. Plant Soil 268:285–292. doi:10.1007/s11104-004-0301-9

Schroder JJ, Smit AL, Cordell D, Rosemarin A (2011) Improved phosphorus use efficiency in agriculture: a key requirement for its sustainable use. Chemosphere 84:822–831. doi:10.1016/j.chemosphere.2011.01.065

Seeling B, Zasoski RJ (1993) Microbial effects in maintaining organic and inorganic solution phosphorus concentrations in a grassland topsoil. Plant Soil 148:277–284. doi:10.1007/BF00012865

Shen J, Yuan L, Zhang J, Li H, Bai Z, Chen X, Zhang W, Zhang F (2011) Phosphorus dynamics: from soil to plant. Plant Physiol 156:997–1005. doi:10.1104/pp.111.175232

Simpson RJ, Oberson A, Culvenor RA (2011) Strategies and agronomic interventions to improve the phosphorus-use efficiency of farming systems. Plant Soil 349:89–120. doi:10.1007/s11104-011-0880-1

Smith SE, Read DJ (2008) Mycorrhizal symbiosis, 3rd edn. Academic, London

Soderberg KH, Baath E (1998) Bacterial activity along a young barley root measured by the thymidine and leucine incorporation techniques. Soil Biol Biochem 30:1259–1268. doi:10.1016/S0038-0717(98)00058-3

Synder CS, Slaton N (2002) Effects of soil flooding and drying on phosphorus reactions. News and Views. Potash and Phosphate Institute, Norcross. http://www.ipni.net/ppiweb/ppinews.nsf/0/BFC425FCAE6ACE4E85256B90005BD114/$FILE/Flooded%20Soils.pdf. Accessed 27 Dec 2013

Tirado R, Allsopp M (2012) Phosphorus in agriculture: problems and solutions. Technical report (review) 02-2012, Greenpeace Research Laboratories, Amsterdam, The Netherlands. http://www.greenpeace.to/greenpeace/wp-content/uploads/2012/06/Tirado-and-Allsopp-2012-Phosphorus-in-Agriculture-Technical-Report-02-2012.pdf. Accessed 24 Dec 2013

Turner BL, Leytem AB (2004) Phosphorus compounds in sequential extracts of animal manures: chemical speciation and a novel fractionation procedure. Environ Sci Technol 38:6101–6108. doi:10.1021/es0493042

Veneklaas EK, Lambers H, Bragg J (2012) Opportunities for improving phosphorus-use efficiency in crop plants. New Phytol 195:306–320. doi:10.1111/j.1469-8137.2012.04190.x

Verheijen FGA, Jones RJA, Rickson RJ, Smith CJ (2009) Tolerable versus actual soil erosion rates in Europe. Earth Sci Rev 94:23–38. doi:10.1016/j.earscirev.2009.02.003

Wang X, Shen J, Liao H (2010) Acquisition or utilization, which is more critical for enhancing phosphorus efficiency in modern crops? Plant Sci 179:302–306. doi:10.1016/j.plantsci.2010.06.007

Wissuwa M, Mazzola M, Picard C (2009) Novel approaches in plant breeding for rhizosphere-related traits. Plant Soil 321:409–430. doi:10.1007/s11104-008-9693-2

Yan F, Zhu Y, Muller C, Zorb C, Schubert S (2002) Adaptation of H^+-pumping and plasma membrane H^+ ATPase activity in proteoid roots of white lupin under phosphate deficiency. Plant Physiol 129:50–63. doi:10.1104/pp.010869

Yan X, Wu P, Ling H, Xu XF, Zhang Q (2006) Plant nutriomics in China: an overview. Ann Bot 98:473–48. doi:10.1093/aob/mcl116

Zhu JM, Kaeppler S, Lynch J (2005) Topsoil foraging and phosphorus acquisition efficiency in maize. Funct Plant Biol 32:749–762. doi:10.1071/FP05005

Efficiency of Soil and Fertilizer Phosphorus Use in Time: A Comparison Between Recovered Struvite, FePO$_4$-Sludge, Digestate, Animal Manure, and Synthetic Fertilizer

Céline Vaneeckhaute, Joery Janda, Erik Meers, and F.M.G. Tack

Abstract

The aim of this study was to evaluate the phosphorus use efficiency (PUE) based on the plant reaction and changes in soil P bioavailability status in time by land application of recovered bio-based fertilizers, including struvite, FePO$_4$-sludge, digestate, and animal manure, compared to synthetic triple super phosphate (TSP). First, product characteristics and P fractionations were assessed. Then, a greenhouse experiment was set up to evaluate plant growth and P uptake, as well as changes in P availability on sandy soils with both high and low P status. P soil fractions were determined in extracts with water (Pw), ammonium lactate (PAl), and CaCl$_2$ (P-PAE) and in soil solution sampled with Rhizon samplers (Prhizon). Struvite demonstrated potential as a slow release, mixed nutrient fertilizer, providing a high P availability in the beginning of the growing season, as well as a stock for delayed, slow release. The addition of FePO$_4$-sludge was not interesting in terms of P release, but resulted in the highest PUE regarding biomass yields. The conversion of animal manure by anaerobic (co)digestion and subsequent soil application of digestate improved the PUE. Finally, the additional use of Rhizon samplers is proposed for better understanding and categorization of different inorganic and organic P fertilizers in environmental legislation.

Keywords

Phosphorus use efficiency • Nutrient recycling • Phosphorus bioavailability • Bio-based fertilizers • Chemical extraction • Greenhouse experiment

C. Vaneeckhaute (✉) • J. Janda • E. Meers • F.M.G. Tack
Laboratory of Analytical and Applied Ecochemistry,
Faculty of Bioscience Engineering, Ghent University,
Coupure Links 653, 9000 Ghent, Belgium
e-mail: celinevaneeckhaute@gmail.com;
joery_janda@hotmail.com; Erik.Meers@ugent.be;
Filip.Tack@ugent.be

A. Rakshit et al. (eds.), *Nutrient Use Efficiency: from Basics to Advances*,
DOI 10.1007/978-81-322-2169-2_6, © Springer India 2015

Abbreviations

PAE   Plant available elements, phosphorus
        extracted with $CaCl_2$
PAl   Phosphorus extracted with ammonium
        lactate
PE   Phosphorus efficiency
Pw   Phosphorus extracted with water
TSP   Triple super phosphate

1   Introduction

Phosphorus (P) is an essential nutrient for the sustainability of all life on earth (EFMA 2000). The worldwide demand for P in agriculture is high and still increasing due to the rising world population, increasing meat consumption, and the cultivation of energy crops (Godfray et al. 2010; Syers et al. 2008). Many studies, however, have shown that worldwide P resources are finite on a human timescale (Elser and Bennett 2011; Scholz and Wellmer 2013). It is predicted that the production of high-quality phosphate rock will reach its peak this century, possibly as early as the next few decades, despite growing demand for P fertilizers (Neset and Cordell 2012). This will cause an extreme rise in price for fossil reserve-based mineral P fertilizers, consequently resulting in a decrease of their agricultural use (Oskam et al. 2011). On the other hand, importation of animal feed and the resulting manure excesses, either or not in combination with the (unnecessary use) of mineral fertilizers, have led to surplus fertilization and P accumulation in many soils worldwide. This has caused pollution of terrestrial and aquatic ecosystems, hence stimulating the introduction of continually more stringent fertilization levels for P application to agricultural fields (Kang et al. 2011; Ranatunga et al. 2013). Also the P content in animal feed and consequently in animal manure is decreasing and will continue to decrease in the future (Schröder et al. 2011). Because of this expected decrease in the P supply to agricultural fields, the effective use of soil phosphate, mineral fertilizer, and animal manure, as well as the cradle-to-cradle

recycling by P recovery from municipal, agricultural, and other waste products as green, renewable fertilizer with high PUE, has become highly important (Huang et al. 2012; Ma et al. 2011; Syers et al. 2008). Pathways could be the recuperation of P as struvite (Hong-Duck et al. 2012), iron phosphate sludge (Sano et al. 2012), phosphate ashes (Stark et al. 2006), etc. Moreover, in P-saturated regions the extraction of P from agricultural fields is relevant to export the recovered P towards P-deficient regions and/or for industrial purposes. In this way sustainable alternatives for the use of depleting phosphate rock can be provided.

There are several different P fertilizer types, such as granular vs. liquid fertilizers, water soluble vs. insoluble fertilizers, quick vs. slow release fertilizers, and single vs. mixed nutrient fertilizers (Erro et al. 2011). Their use depends on the P status of the soil, the soil characteristics including the P fractionation in the different soil pools, and the P demand of the agricultural crop. On a daily basis, a rapidly growing crop takes up the P_2O_5 equivalent of about 2.5 kg ha^{-1}. However, it is estimated that only a small proportion (15–20 %) of the total amount of P in the plant is directly provided by the fertilizer applied to that crop. The remainder comes from soil reserves. It is therefore clear that there must be adequate reserves of readily available P in the soil (Syers et al. 2008). The P status of European soils is estimated by routine analysis, and for many countries some 25 % (5–55 %) of soils test as very low and low in readily available P. Such soils require significantly more P to be applied than is removed by the crop to increase soil reserves and thus soil fertility. On the other hand, in many countries some 40 % (15–70 %) of soils test as high and very high in readily available P. On such soils, when crops are grown that have small, inefficient root systems but a large daily intake of P at critical growth stages, it may be necessary to apply more P. On soils with a medium P analysis value, applications need to sustain the P status. This may require a small extra amount of P over and above that removed with the harvested crop (EFMA 2000).

In order to evaluate the agricultural potential and efficiency of new P fertilizers, insights in their P release with time are thus highly important. The performance of a fertilizer can be evaluated via product fractionation, the plant reaction, and/or chemical soil analysis. It can be expressed as bioavailability indices, such as the phosphorus use efficiency (PUE), which can be based on the fresh weight (FW) and dry weight (DW) yield, the growth rate (FW, DW), the P uptake (rate), and the degree and rate in which the P status of the soil changes, as determined by chemical methods (Van Dam and Ehlert 2008). A fractionation of fertilizers based on the solubility of P is, in general, based on using solvents with different strength and selectivity. The most important solvents in the frame of EU legislation are, ranked from strong to weak, mineral acid, neutral ammonium citrate solution, and water (EU 2003). Furthermore, soil measurements can be conducted. The measured P content varies from actual availability to total P reserve in the soil. The analysis can be divided in P capacity and P intensity of the soil, based on the strength of the extraction method. The P intensity gives an indication of the total amount of inorganic P which is directly available for the plant during a short period of time, while the P capacity gives an indication of the amount of P that may be released in the long term (Dekker and Postma 2008).

In some European countries, e.g., the Netherlands, fertilizer recommendations are based on the P status of the soil, measured as PAl and Pw number. These measures are based on an extraction with ammonium lactate and water, respectively (Sissingh 1971). The PAl number is a measure of the P capacity of the soil and is conceived to be P in the soil that can become available for crops over a long period of time. The Pw number is a combination of the capacity and intensity of the soil. However, the PAE method (plant available elements) is internationally receiving more and more attention as a simple alternative for the many extraction procedures that are currently used for single nutrients. It concerns a multielement extraction with 0.01 M $CaCl_2$ (Houba et al. 2000). For phosphorus (P-PAE), this extraction is believed to give an indication of the P intensity.

A current limitation of all these methods is that root formation, soil compaction, and mineralization of organic matter are not taken into account (Ehlert et al. 2006). The use of Rhizon soil moisture samplers (Rhizon SMS) allows to assess the total amount of P in the soil solution, including dissolved organic and inorganic forms (Eijkelkamp 2003). Besides convenience, Rhizon SMS for direct extraction of soil moisture also overcomes disadvantages related to traditional sampling using ceramic cups such as the exchange of (divalent) cations and phosphate (Grossmann and Udluft 1991).

The first aim of this study was to evaluate the phosphorus use efficiency (PUE) based on the plant reaction and changes in soil P bio-availability status (PAl, Pw, P-PAE) in time by application of different bio-based fertilizers as compared to a control and reference fossil reserve-based mineral fertilizer, triple super phosphate (TSP, $Ca(H_2PO_4)_2.H_2O$). To this end a physicochemical characterization and P fractionation of the products was conducted and a greenhouse experiment was set up. The green renewable fertilizers under study were struvite ($MgNH_4PO_4.6H_2O$), iron phosphate sludge ($FePO_4$), digestate from codigestion, and pig manure. Two soils were involved, a nutrient-rich sandy soil with high P status (Pw > 55; Alterra 2012) and a nutrient-poor, P-deficient Rheinsand (Pw < 36; Alterra 2012) to allow to assess the available P effectively provided by the fertilizers. A second aim was to overcome the limitations of the current soil extraction methods by using Rhizon SMS to determine the P delivery to the plant in the short term. The plant under study was maize, which has a high P demand. Based on the findings, the potential use of the renewable P fertilizers in agriculture is evaluated with reference to traditional triple super phosphate fertilizer.

2 Material and Methods

2.1 Experimental Setup

An overview of the experimental setup can be found in Fig. 1. At first, a physicochemical

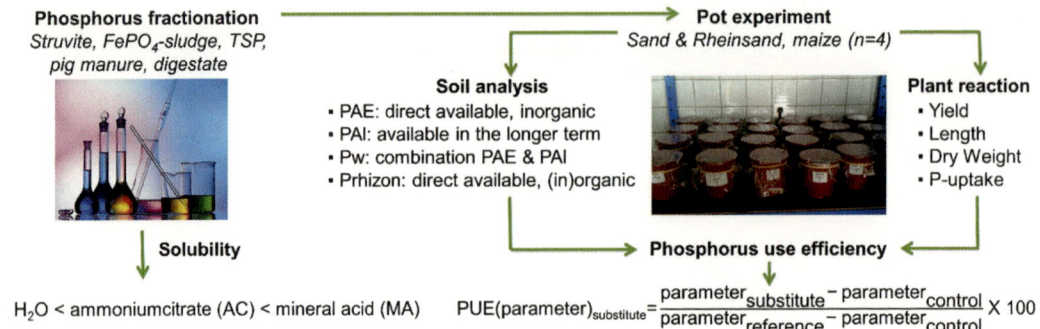

Fig. 1 Overview experimental setup

characterization and P fractionation of the products was conducted (Sect. 2.2). Then, a greenhouse experiment (Sect. 2.3) was set up in order to evaluate the plant reaction and soil P bio-availability status in time (Sect. 2.4). Based on the obtained results, average phosphorus use efficiencies (PUEs) were calculated for use of the different bio-based fertilizers as compared to a control and a reference triple super phosphate (TSP) (Sect. 2.5).

2.2 Product Characterization and P Fractionation

Dry weight (DW) content was determined as residual weight after 72 h drying at 80 °C, while organic carbon (OC) was determined after incineration of the dry samples during 4 h at 550 °C in a muffle furnace (Van Ranst et al. 1999). Conductivity and pH were determined potentiometrically using a WTW-LF537 (GE) electrode and an Orion-520A pH-meter (USA), respectively. The solid samples were first equilibrated for 1 h in deionized water at a 5/1 liquid to dry sample ratio and subsequent filtered (MN 640 m, Macherey–Nagel, GE). Total N content was determined using the Kjeldahl method, and total P was determined using the colorimetric method of Scheel (Van Ranst et al. 1999). Ca, Mg, and K were analyzed using ICP-OES (Varian Vista MPX, USA) (Van Ranst et al. 1999). Ammonium was determined using a Kjeltec-1002 distilling unit (Gerhardt Vapodest, GE) after addition of MgO to the

sample and subsequent titration (Van Ranst et al. 1999). Furthermore, a P fractionation in frame of EU regulation was conducted. The determination of the fraction of P soluble in water, mineral acid (mixture of saltpeter acid and sulfuric acid), and neutral ammonium citrate was determined as described in EU (2003).

2.3 Greenhouse Experiment

Soils used for the greenhouse experiment were a sandy soil from Ranst, Belgium (pH = 5.0, EC = 111 μS cm^{-1}, density = 1.262 kg L^{-1}) and laboratory-grade Rheinsand (pH = 7.9, EC = 67 μS cm^{-1}, density = 1.612 kg L^{-1}). TSP was collected at Triferto (Ghent, Belgium), struvite at the water treatment plant of Clarebout Potatoes (Nieuwkerke-Heuvelland, Belgium), and FePO$_4$-sludge at the piggery of Innova Manure (Ichtegem, Belgium). Animal manure was sampled at the piggery of Ivaco (Gistel, Belgium), and digestate was sampled at the biogas plant SAP Eneco Energy (Merkem, Belgium). The latter concerns an anaerobic codigestion plant with an influent feed consisting of animal manure (30 %), energy maize (30 %), and organic-biological waste from the food industry (40 %). The samples were collected in polyethylene sampling bottles (5 L), stored cool (4 °C), and transported to the laboratory for physicochemical analysis. The data were used to calculate the maximum allowable dosage for the different cultivation scenarios with respect to the Flemish Manure Decree (FMD 2011).

Table 1 Product and macronutrient dosage to soil by bio-based fertilizer application (standardized to 80 kg P_2O_5 ha^{-1}); Differences in N, K, Ca and Mg application were corrected by adding the appropriate amount of a 1 M NH_4NO_3, K_2SO_4, $MgSO_4.7H_2O$ and/or $CaSO_4.2H_2O$ solution

Fertilizer type	Product (t DW[a] ha^{-1})	Total N (kg ha^{-1})	Effective N (kg ha^{-1})	Total P_2O_5 (kg ha^{-1})	Total K_2O (kg ha^{-1})	Total Ca (kg ha^{-1})	Total Mg (kg ha^{-1})	OC[b] (kg ha^{-1})
TSP[c]	0.19	0.09	0.06	80	0.36	26	0.40	3.0
Struvite	0.27	14	9.2	80	3.0	0.16	24	78
FePO$_4$-sludge	3.08	169	92	80	357	29	15	770
Pig manure	1.51	159	112	80	112	44	21	559
Digestate	2.76	185	124	80	160	72	17	938

[a]*DW* dry weight, [b]*OC* organic carbon, [c]*TSP* triple super phosphate

Plastic containers (2 L) were filled with 1 kg of soil, and the soil moisture solution was brought to field capacity (23 % for sand, 19 % for Rheinsand). After 2 days of equilibration (March 16, 2012), an equivalent product dose of 80 kg P_2O_5 ha^{-1} was applied in all containers (Table 1), which is the maximum amount for manure application on a sandy soil in Flanders (FMD 2011). Simultaneously, a control treatment without P fertilization was set up. Differences in N, K, Ca, and Mg application between the different scenarios were corrected by adding the appropriate amount of a 1 M NH_4NO_3, K_2SO_4, $MgSO_4.7H_2O$, and $CaSO_4.2H_2O$ solution, without exceeding the field capacity. Moreover, the Flemish fertilization advice of 135 kg effective N ha^{-1}, 250 kg K_2Oha^{-1}, 50 kg MgO ha^{-1}, and 70 kg Ca ha^{-1} was respected.

Soils were homogenized and soil moisture content was brought to field capacity with deionized water. Each treatment was repeated four times, resulting in a total of 48 containers (5 amendments and 1 control, 2 soil types, 4 replications). After 4 days of equilibration (March 21, 2012), seven energy maize seeds of the species Atletico KWS were sown in each container at a depth of 2 cm. The containers were covered with a perforated plastic in order to reduce evapotranspiration. When the plants reached the height of the plastic, the plastics were removed and the plants were thinned out to five plants per container. In each container a Rhizon SMS was inserted diagonally from the top soil through the soil column. The 48 containers were randomly placed on a greenhouse bench at ± 20 °C. The plants were lightened with Brite-Grow bio-growing lamps (36 W) 50 cm above the plants (LUX 1500) in a day-night cycle (6 AM till 8 PM). Daily the soils were weighed and the soil moisture content was brought to field capacity. After 1 week, leakage of soil solution was visible in two containers: one struvite and one control treatment, both on a sandy soil. These two containers were eliminated from the experiment.

Homogeneous soil samples (10 g) were taken for analysis of PAl, Pw, and P-PAE by means of a soil auger the first 2 weeks and the last 2 weeks of the experiment. Rhizon soil moisture extracts were sampled weekly during the experiment and the P concentration of the soil solution, as well as the pH were each time analyzed. Furthermore, the length of the plants was measured weekly. After 5 weeks of growth, the plants were harvested, their yield was determined, and plant samples were taken for physicochemical analysis. The soils were maintained on the greenhouse bench and were moisturized every 2 weeks up to field capacity. Finally, PAl, P-PAE, and Pw in the soils were measured again after 6 months.

2.4 Plant and Soil Analysis

Dry weight (DW) biomass content was determined as residual weight after 1 week drying at 65 °C. Macronutrients (N, P, K, Ca, Mg) in the biomass were determined in the same way as described for the products (Sect. 2.2). Soil pH and conductivity were determined in the same way as described for the products (Sect. 2.2). Field capacities were determined in accordance with CSA (2012). For the determination of Pw, 4 g of soil and 240 mL of distilled water were mixed in a 250 mL flask, shaken for 1 h, and filtered until colorless (EL&I 2009; Sissingh 1971). For PAl, 2.5 g of soil was mixed with 50 mL of ammonium lactate solution (pH 3.75), shaken for 4 h, and filtered until colorless (CSA 2012). For P-PAE, 1 g of dry soil was mixed with 25 mL 0.01 M $CaCl_2$ in a 40 mL centrifuge tube, shaken for 1 h, centrifuged during 10 min at 4,000 t min$^-$1, and filtered (Van Ranst et al. 1999). Finally, the total P content in the filtered extraction solutions as well as in the Rhizon SMS extracts was determined by the method of Scheel (Van Ranst et al. 1999).

3 Phosphorus Use Efficiency (PUE)

Phosphorus use efficiencies (%) of the bio-based fertilizers were calculated based on the plant reaction and the soil status using the following equation:

$$
\begin{aligned}
&\text{PUE(parameter)}_{\text{bio-fertilizer}} \\
&= \frac{(\text{parameter}_{\text{bio-fertilizer}} - \text{parameter}_{\text{control}})}{(\text{parameter}_{\text{reference}} - \text{parameter}_{\text{control}})} \\
&\quad \times 100
\end{aligned}
$$

where "bio-fertilizer" refers to the bio-based fertilizers under study, "control" to the blank treatment, and "reference" to the TSP treatment and where "parameter" can refer to:
- The plant P uptake and the plant fresh and dry weight yield: PUE(uptake), PUE(FWyield), and PUE(DWyield). Here the PE refers to the percentage of phosphate in the bio-based fertilizers that has the same effectiveness as the reference fossil reserve-based mineral P fertilizer TSP.
- The PAl number, the Pw number, the P-PAE number, and the P concentration in the soil

solution extracted with Rhizon SMS: PUE (PAl), PUE(Pw), PUE(PAE), and PUE (Prhizon). Here the PE refers to the increment in soil P status by application of the bio-based fertilizers as compared to the increment by application of the reference fossil reserve-based mineral P fertilizer TSP.

4 Statistical Analysis

Statistical analysis was conducted using SAS 9.3. A one-way Anova procedure was used to determine the effect of fertilizer type on the different plant and soil parameters per measurement. Furthermore, a two-way Anova was used to determine significant differences of the different plant and soil parameters in time, as well as to determine significant differences between the different treatments over the whole experimental period. Significance of effects was tested by use of an F-test and post hoc pair-wise comparisons were conducted using Tukey's HSD test ($\alpha = 0.05$). The condition of normality was checked using the Kolmogorov Smirnov test and QQ-plots, whereas equality of variances was checked with the Levene test. Significant parameter correlations were determined using the Pearson correlation coefficient (r).

5 Results

5.1 Product Characterization and P Fractionation

At first, it must be noticed that TSP and struvite were dry, granular products, while the other products were liquids. The amount of P soluble in water was low for struvite (1.7 %) and $FePO_4$-sludge (4.0 %) as compared to TSP (96 %), while the amount of P soluble in mineral acid was in the same line as the reference (Table 2). Digestate had approximately the same P solubility's in the different extraction reagents as animal manure (79–85 %). Compared to TSP, the P solubility of these products in water was lower, while it was higher in mineral acid.

Table 2 Product physicochemical characterization and phosphorus fractionation, $n = 2$

Parameter	TSP[a]	Struvite	FePO$_4$-sludge	Pig manure	Digestate
pH	2.6	8.4	4.6	7.7	8.6
Conductivity (mS cm^{-1})	29	547	15	35	37
DW[b] (%)	100	100	2.0 ± 0.0	6.2 ± 0.1	9.8 ± 0.0
OC[c] (% on DW)	1.6 ± 0	29 ± 0	25 ± 0	38 ± 1	34 ± 1
P$_2$O$_5$ (g kg^{-1}FW[d])	430 ± 5	293 ± 3	0.51 ± 0.02	3.3 ± 0.0	2.9 ± 0.0
P$_2$O$_5$ water (g kg^{-1}FW[d])	413 ± 1	5.0 ± 0.0	0.02 ± 0.00	2.8 ± 0.1	2.3 ± 0.0
P$_2$O$_5$ NAC[e] (g kg^{-1} FW[d])	410 ± 1	282 ± 3	0.50 ± 0.01	3.0 ± 0.0	0.50 ± 0.01
P$_2$O$_5$ MA[f] (g kg^{-1}FW[d])	398 ± 1	288 ± 5	0.46 ± 0.00	3.2 ± 0.0	0.46 ± 0.00
N (g kg^{-1}FW[d])	0.49 ± 0.03	52 ± 2	1.1 ± 0.0	6.5 ± 0.0	6.6 ± 0.2
NH$_4$-N (g kg^{-1} FW[d])	0.23 ± 0.06	28 ± 1	0.25 ± 0.00	4.6 ± 0.1	3.8 ± 0.0
Effective N (g kg^{-1}FW[d])	0.31 ± 0.04	34 ± 1	0.59 ± 0.02	4.6 ± 0.1	4.4 ± 0.1
K$_2$O (g kg^{-1} FW[d])	1.9 ± 0.3	11 ± 0	2.3 ± 0.1	4.6 ± 0.4	5.7 ± 0.10
Effective N / P$_2$O$_5$ / K$_2$O	0.00072/1/0.0044	0.12/1/0.038	1.1/1/4.5	1.4/1/1.4	1.5/1/2.0
Ca (g kg^{-1} FW[d])	138 ± 1	0.58 ± 0.00	0.19 ± 0.00	1.8 ± 0.0	2.6 ± 0.0
Mg (g kg^{-1} FW[d])	2.1 ± 0.0	87 ± 1	0.10 ± 0.00	0.90 ± 0.00	0.60 ± 0.00

[a]*TSP* triple super phosphate, [b]*DW* dry weight, [c]*OC* organic carbon, [d]*FW* fresh weight, [e]*NAC* neutral ammonium citrate, [f]*MA* mineral acid

Table 3 Average phosphorus use efficiency (PUE) based on the plant reaction in time (%) for the different bio-based fertilizers; PUE(control) = 0 %; PUE(TSP) = 100 %

PUE (%)	PUE(FWyield) Sand	PUE(FWyield) Rheinsand	PUE(DWyield) Sand	PUE(DWyield) Rheinsand	PUE(uptake) Sand	PUE(uptake) Rheinsand
Struvite	-21[a]	75	10[a]	67	22	42
FePO$_4$-sludge	-68[a]	159	-16[a]	233	16	3.3
Animal manure	-46[a]	-8.9	-8.5[a]	-67[b]	37	80
Digestate	-67[a]	-45[b]	-90[a]	-100[b]	80	63

[a]TSP < control; [b]bio-fertilizer < control

The solubility of P in neutral ammonium citrate was high for all fertilizers (91–100 %). Furthermore, the pH of TSP and FePO$_4$-sludge was low (2.6–4.6), while for struvite and digestate it was alkaline (8.4–8.6). The pH of pig manure was quasi neutral. Finally, struvite, FePO$_4$-sludge, digestate, and pig manure added significantly more organic carbon to the soil as compared to TSP.

5.2 Plant Reaction

On the sandy soil, at the harvest, all treatments had a significantly higher FW biomass yield ($p < 0.0001$), DW biomass yield ($p = 0.0002$), and length ($p = 0.0007$) as compared to the reference TSP, while the DW content and P content

(mg kg^{-1} DW) of the biomass was significantly higher ($p < 0.0001$) for the TSP treatment. The absolute P uptake (mg P) was, however, for the TSP treatment only significantly higher ($p = 0.012$) as compared to the control. The PUE(FWyield) and PUE(DWyield) on the sandy soil were mostly negative as the yield of the reference TSP was lower than the control (Table 3). The best average PUE based on the crop yield was observed for FePO$_4$-sludge and digestate, the latter simultaneously showing the highest PUE(uptake).

On Rheinsand, no significant differences ($p > 0.1$) were observed in the biomass length and DW yield. The DW content was significantly lower ($p < 0.0001$) for TSP and FePO$_4$-sludge as compared to the control and digestate, while FePO$_4$-sludge had a significantly higher FW yield than the control, pig manure, and digestate ($p = 0.003$). The use of TSP, pig manure, and digestate resulted in a significantly higher ($p < 0.0001$) P content (g kg^{-1} DW) and absolute P uptake (mg P) as compared to the control and FePO$_4$-sludge. Also, the plant P uptake for the struvite treatment was significantly lower as compared to the reference TSP. In terms of efficiency, the PUE(FWyield) and PUE(DWyield) were the highest for FePO$_4$-sludge; however, its PUE(uptake) was the lowest (Table 3). The PUE (uptake) for animal manure and digestate were the highest, yet their PUE(FWyield) and PUE (DWyield) were negative as the yields were slightly lower than the control.

5.3 Soil Bio-availability Analysis

5.3.1 Chemical Extractions: P-PAE, PAl, and Pw Number

At first, it must be remarked that P-PAE and Pw could only be detected on the sandy soil, as the values on Rheinsand were lower than the detection limits of the spectrophotometer (0.66 mg P L^{-1}), as well as the continuous flow analyzer (0.05 mg P L^{-1}). Over the whole experimental period, the average P-PAE (mg P kg^{-1} soil) was significantly higher ($p < 0.0001$) for TSP as compared to the other treatments and the control, as well as for struvite compared to the control, digestate, and FePO$_4$-sludge. The effect of FePO$_4$-sludge on the P-PAE number was in average significantly lower ($p < 0.0001$) than that of all other treatments. The two-way Anova for P-PAE indicated a significant ($p < 0.0001$) decrease for all treatments from weeks 2 to 4 and weeks 4 to 5. After 6 months, no more significant differences were observed between the treatments ($p = 0.15$).

Over the whole period of time, the average Pw (mg P$_2$O$_5$ L^{-1} soil) for TSP, digestate, and struvite was significantly higher ($p < 0.0001$) than for the control and FePO$_4$-sludge. There was a significant decrease ($p = 0.0021$) in week 2 for all treatments as compared to the other weeks. After 6 months, the control had a significantly higher Pw number than struvite, pig

manure, and FePO$_4$-sludge ($p = 0.0069$). Overall, the average PAl (mg P$_2$O$_5$ 100 g^{-1} soil) for TSP was significantly higher ($p < 0.0001$) than for the other treatments. On Rheinsand, PAl for TSP was over the whole experimental period significantly higher ($p = 0.030$) as compared to FePO$_4$-sludge, but not to the control. Both on sand and Rheinsand, no significant changes in the weekly average PAl ($p > 0.1$) were found. Also after 6 months the PAl number was not significantly different for the different treatments ($p > 0.05$).

In terms of efficiency, on the sandy soil all fertilizers had a lower PUE(PAE) and PUE(PAl) than the reference TSP during the whole experimental period (Table 4). Struvite had the highest PUE(PAE), while the P-PAE number for FePO$_4$-sludge was even lower than the control. The PUE (Pw) increased in time for struvite and digestate, compared to the reference TSP. For FePO$_4$-sludge, it was negative and decreasing.

5.3.2 Rhizon Soil Solution Extracts: Prhizon

On the sandy soil, the average P$_2$O$_5$ content (mg L^{-1}) over time in the soil solution measured with Rhizon SMS was significantly higher ($p < 0.0001$) for pig manure as compared to struvite, the control, and FePO$_4$-sludge. The latest showed significantly lower values ($p < 0.0001$) than the other treatments and the control. The average values in week 1 were significantly higher ($p < 0.0001$) than in weeks 3, 4, and 5, as well as in week 2 compared to week 4. On Rheinsand, no detectable amount of P$_2$O$_5$ in the soil solution was found for the control, while for FePO$_4$-sludge it was only detectable during the first 2 weeks. For the other treatments, a significant decrease ($p < 0.0001$) in P$_2$O$_5$ concentration over time was observed. The values were significantly higher ($p = 0.0002$) for pig manure as compared to struvite and FePO$_4$-sludge, as well as for TSP compared to struvite. On sand, the average pH over time was significantly lower ($p < 0.0001$) for animal manure as compared to all other treatments, as well as for TSP compared to struvite, FePO$_4$-sludge, the control, and

Table 4 Average phosphorus use efficiency (PUE) based on soil analysis in time (%) for the different bio-based fertilizers; PUE(control) = 0 %; PUE(TSP) = 100 %

PUE (%)	PUE(PAE) Sand	PUE(Pw) Sand	PUE(PAl) Sand	PUE(PAl) Rheinsand	PUE(Prhizon) Sand	PUE(Prhizon) Rheinsand
Struvite	57	374	1.6	-94^b	145	60
FePO$_4$-sludge	-41^a	-46^a	23	-606^b	-131^a	3.2
Animal manure	21	24	34	-215^b	130	114
Digestate	14	212	-3.0^a	453^b	71	81

[a]Bio-fertilizer < control; [b]no significant difference with the control because of high standard error

digestate. The latest showed a significantly higher ($p < 0.0001$) average pH than the other treatments, both on sand and Rheinsand. Moreover, there was a strongly significant ($p < 0.0001$) decrease in pH from weeks 3 to 4 for all treatments.

In terms of efficiency (Table 3), the PUE (Prhizon) was very high for pig manure on both sand and Rheinsand. On sand, the curve for struvite showed a similar pattern as for pig manure up to week 3. However, on Rheinsand the values for struvite were always lower as compared to the reference and pig manure.

6 Discussion

6.1 Evaluation of Biomass Yield and P Uptake in Time

The use of digestate, pig manure, FePO$_4$-sludge, and struvite on an acidic sandy soil with high P status (Pw control > 55) resulted in higher biomass yields and lengths as compared to fossil reserve-based mineral fertilizer TSP. The lower yield and length found for TSP fertilization can be explained by the fact that most of the P was water soluble (96 %) and therefore partly captured by Al and Fe hydroxides in the soil (Van Dam and Ehlert 2008). It should be remarked that if in practice the amount of water soluble P applied is higher than the crop demand on soils low in Fe and Al, the excess supply will cause a high risk of leaching in the field (Kang et al. 2011; Yang et al. 2012). On Rheinsand, which had a low P level (Pw control < 36), the application of FePO$_4$-sludge resulted in similar FW biomass yields as compared to TSP, while

the P uptake was significantly lower. This was likely due to the better implantation of roots in the soil and enhanced growth of mycorrhizal mycelia by poor P availability (Nieminen et al. 2011). The use of pig manure and digestate resulted in a plant P uptake comparable to TSP on a P-deficient soil, indicating that the absolute fertilizer effect in terms of direct available P was similar.

6.2 Evaluation of Soil P Status in Time

The P solubility in water of struvite was much lower as compared to the reference TSP, whereas the solubility in mineral acid was relatively high, indicating that struvite has slow-release properties. This is in line with the slow-release properties of this product for NH$_4$-N found in literature (Hong-Duck et al. 2012) and with the bioavailability curve for Prhizon on P-deficient Rheinsand. In spite of this, struvite demonstrated a relatively high efficiency in terms of direct available P on the P-rich sandy soil, which was also confirmed by the significant correlation between the PAE for struvite and TSP on the sandy soil ($r = 0.625$, $p = 0.030$). The high PUE(PAE) and PUE(Prhizon) may be attributed to the higher amount of NH$_4$-N relative to P$_2$O$_5$ in struvite (Table 3). The uptake of NH$_4^+$ by the roots and the nitrification of NH$_4^+$ into NO$_3^-$ are acidifying processes, which can increase soil P mobilization and uptake in the rhizosphere (Diwani et al. 2007). Indeed, during the first 3 weeks of growth, the pH in the soil solution was the lowest, while the amount of direct available P was the highest. Other potential reasons are the presence of Mg in struvite (Hong-Duck et al. 2012) and/or its high salt content (Hartzell et al. 2010). At the end of the growing season, PUE(PAl) increased, indicating that struvite addition increased the P reserve in the soil for release in the longer term. This phenomenon was also reflected in the high significant correlation between the PAl number on the sandy soil for struvite and $FePO_4$-sludge ($r = 0.86$, $p < 0.0001$).

$FePO_4$-sludge was clearly not interesting for use as starter fertilizer for crop growth, as it had a very low P solubility in water. In agreement to Nieminen et al. (2011), the solubility in neutral ammonium citrate was 100 %. Accordingly, the efficiency of this compound to supply direct available P was low. On the other hand, the P capacity over time was slightly increasing, indicating that the addition of $FePO_4$-sludge increased the amount of P that can be released in the longer term. Although the product's ability to fixate P is not interesting regarding the imminent depletion of P reserves (Scholz and Wellmer 2013), there is a high interest to use $FePO_4$-sludge for forestry on drained peat- and wetlands in order to reduce P leaching and increase P adsorption (Nieminen et al. 2011). Nevertheless, Nieminen et al. (2011) have reported that a long study period will be required because of the slow development of active root/mycorrhiza associations that may be necessary for significant P release and because the duration of the growth response after P fertilization may be over 30 years.

The efficiency of digestate in supplying direct available P was slightly increasing during the greenhouse experiment, indicating that P from digestate was released more slowly than from the reference TSP. The product had, though lower than TSP, a relatively high P solubility in water, while the solubility in mineral acid was 100 %. The PUE(Pw) was therefore high. Pig manure released immediate available P somewhat faster than digestate, as the PUE was higher after 1 week, but equal after 4 weeks. In addition, its P solubility in water was slightly higher as compared to digestate, while the solubility in mineral acid was slightly lower. This is in line with the observed bioavailability indices: P-PAE

and Prhizon were higher for pig manure than for digestate, whereas Pw was slightly lower. All these results correspond to literature data and indicate that anaerobic (co)digestion of animal manure reduces the fraction of immediate inorganic plant available P in the soil solution, whereas it increases the fraction of easily available soil phosphate that can be released in the short term (Möller and Müller 2012). This would be caused by the enhanced formation and precipitation of calcium phosphate, magnesium phosphate, and/or struvite by mineralization of N, P, and Mg in combination with a substantial increase of the manure pH (Möller and Müller 2012). In this perspective, it should also be noticed that the bioavailability curve for direct available P (P-PAE) was highly significant and well correlated for struvite and digestate ($r = 0.90$, $p < 0.0001$), as well as the pH in the soil solution ($r = 0.85$, $p < 0.0001$). The conversion of animal manure by anaerobic (co)digestion and the subsequent use of digestate on agricultural fields may thus offer a solution to control water soluble P in soils, meanwhile supplying sufficient P to support plant growth.

Another interesting remark is that the P intensity of the soil measured as P-PAE was lower for digestate and pig manure than for TSP, while Prhizon was relatively higher, especially for pig manure. It is likely that this extra amount of soluble P for the organic fertilizers, digestate and pig manure is attributed to the release of organic P_2O_5 in the soil solution (Roboredo et al. 2012), which cannot be measured with the PAE method. Indeed, the P-PAE number was significantly correlated for struvite and TSP ($r = 0.63$, $p < 0.0001$), but no significant correlation was found between P-PAE for the other products. On the other hand, the correlations of P in the soil solution on Rheinsand between TSP and pig manure ($r = 0.76$, $p < 0.0001$) and TSP and digestate ($r = 0.73$, $p < 0.0001$) were significant, although only a weak correlation was found between TSP and struvite ($r = 0.59$; $p = 0.01$). Nevertheless, Huang et al. (2012) emphasized that this organic water soluble P in soils also plays a role in plant P utilization. As PUE(Prhizon) was much higher for pig manure than for TSP, both on sand and Rheinsand, and since pig manure is a liquid fertilizer, application of this product might cause a higher risk of leaching in the field. Furthermore, as also the yield and the P uptake on sand were much higher for digestate than for pig manure, treating animal manure by anaerobic (co)digestion before application to the field, and meanwhile producing renewable energy, appears as an interesting option from an environmental point of view.

Finally, an interesting point is that all bio-based fertilizers, especially digestate, added significantly more organic carbon to the soil compared to TSP (Table 1). Application of these products could therefore also contribute to the struggle against organic carbon depletion in many agricultural soils worldwide.

6.3 Practical Implications

In the wastewater and manure treatment industry, Fe salts are often used for P removal. However, results indicate that the production of $FePO_4$-sludge for fertilizer use is not very interesting in terms of P release for crop growth, unless it can be used on drained soils. In the transition from nutrient removal to nutrient recovery, alternative P recovery techniques are therefore recommended, such as anaerobic digestion and/or struvite production. There is evidence that these recovered bio-based products can be used as a sustainable substitute for synthetic P fertilizers in agriculture. However, marketing of these green renewable fertilizers will also depend on the economic viability of the nutrient recovery technique in question and the economic competitiveness of the products as compared to fossil reserve-based mineral fertilizers. Another important bottleneck is that all derivatives produced from animal manure are currently still categorized as animal manure in (European) environmental legislation and can therefore not or only sparingly be returned to agricultural land. The need exists for greater differentiation between soil, crop, and fertilizer types in the advice given on

P fertilizer requirements. For this purpose, a combination of the soil chemical P status and the fertilizer properties, as well as the P demand of the agricultural crop, is recommended. Regarding the aim to reduce P leaching and run-off, the most important parameter will be the measurement of direct available P. As the P-PAE method does not take the release of soluble organic P into account, the measurement with Rhizon SMS appears to be an interesting complementary method. Hence, a combination of these two methods for direct available P is proposed for better categorization of different inorganic and organic P fertilizers in EU legislation.

Conclusions

Results indicate clearly that there are perspectives for reuse of recovered bio-based products as renewable P fertilizers in agriculture. Struvite can be used as a slow-release, mixed-nutrient fertilizer, indirectly providing a high P availability for the plant in the beginning of the growing season, as well as a stock for delayed, slow release. The addition of $FePO_4$-sludge as starter fertilizer was not interesting in terms of P release. Application of this product, however, resulted in the highest efficiency regarding biomass yields. Furthermore, the sustainable use of P from animal manure could be improved by anaerobic (co)digestion in order to create digestate for application to the field. As added benefits, negative environmental impacts of untreated animal manure are avoided, renewable energy is produced, and important amounts of organic carbon are added to the soil. Finally, based on all results, the additional use of Rhizon soil moisture samplers for determination of direct available P is proposed for better understanding and categorization of different inorganic and organic P fertilizers in environmental legislation. This may attribute to an improved differentiation between soil, crop, and fertilizer types in the advice given on P fertilizer requirements, thereby moving towards a more efficient and sustainable use of P in agriculture.

Acknowledgements This work has been funded by the European Commission under the InterregIVb Project Arbor (accelerating renewable energies through valorization of biogenic organic raw material) and by the Environmental and Energy Technology Innovation Platform (MIP) under the project Nutricycle.

References

Alterra (2012) Classification of phosphate categories. Report no. BO-12.12-002-006, Alterra, Wageningen UR

CSA (2012) Compendium for sampling and analysis for the implementation of the waste and soil remediation decree. Flemish Ministerial Order of 18.01.2012 (Jan 18, 2012)

Dekker PHM, Postma R (2008) Verhoging efficiëntie fosfaatbemesting. Report no. PPO-3250061800, Praktijkonderzoek Plant en Omgeving BV, Wageningen UR. (in Dutch)

Diwani GE, Rafie SE, El Ibiari NN, El-Aila HI (2007) Recovery of ammonia nitrogen from industrial wastewater treatment as struvite slow releasing fertilizer. Desalination 214(1–3):200–214

EFMA (2000) Understanding phosphorus and its use in agriculture. European Fertilizers Manufacturers Association, Brussels

Ehlert PAI, Burgers SLGE, Bussink DW, Temminghoff EJM, Van Erp PJ, Van Riemsdijk WH (2006) Dekstudie naar de mogelijkheden van het aanduiden van fosfaatarme gronden op basis van P-PAE. Alterra, Wageningen UR, Report no. 1958. (in Dutch)

Eijkelkamp (2003) Agrisearch equipment. Technical report No. M2.19.21.E, Eijkelkamp, Giesbeek

Elser J, Bennett EMA (2011) Broken biogeochemical cycle. Nature 478:29–31

EL&I (2009) Protocol phosphate differentiation and derogation 2010–2013. Dutch Ministry of Economic Affairs, Agriculture and Innovation, Den Haag (The Netherlands). (in Dutch)

Erro J, Baigorri R, Yvin JC, Garcia-Mina JM (2011) (31) p NMR Characterization and efficiency of new types of water-insoluble phosphate fertilizers to supply plant-available phosphorus in diverse soil types. J Agric Food Chem 59(5):1900–1908

EU (2003) Regulation (EC) No 2003/2003 of the European Parliament and of the Council of 13.10.2003 relating to fertilizers. Official Journal of the European Union, Pub. L. no. 304 (October 13, 2003)

FMD (2011) Flemish Manure Decree of 13.05.2011 concerning the protection of water against nitrate pollution from agricultural sources. Official Belgian Bulletin of Acts, Orders and Decrees, Pub. no. BS13.05.2011-MAP4 (May 13, 2011). (in Dutch)

Godfray HCJ, Beddington JR, Crute IR, Haddad L, Lawrence D, Muir JF (2010) Food security: the challenge of feeding 9 billion people. Science 327:812–818

Grossmann J, Udluft P (1991) The extraction of soil water by the suction-cup method: a review. J Soil Sci 42:83–93

Hartzell JL, Jordan TE, Cornwell JC (2010) Phosphorus burial in sediments along the salinity gradient of the Patuxent river, a subestuary of the Chesapeake Bay (USA). Estuar Coasts 33(1):92–106

Hong-Duck R, Chae-Sung L, Yu Kyung K, Keum-Yong K, Sang-III L (2012) Recovery of struvite obtained from semiconductor wastewater. Environ Eng Sci 29(6):540–548

Houba VJG, Temminghoff EJM, Gaikhorst GA, Van Vark W (2000) Soil analysis procedures using 0.01 M calcium chloride as extraction reagent. Commun Soil Sci Plant 31(9–10):1299–1396

Huang XL, Chen Y, Shenker M (2012) Dynamics of phosphorus phytoavailability in soil amended with stabilized sewage sludge materials. Geoderma 170:144–153

Kang J, Amoozegar A, Hesterberg D, Osmond LD (2011) Phosphorus leaching in a sandy soil affected by organic and inorganic fertilizer sources. Geoderma 161(3–4):194–201

Ma W, Ma L, Wang F, Sisak I, Zhang F (2011) Phosphorus flows and use efficiencies and consumption of wheat, rice, and maize in China. Chemosphere 84:814–821

Möller K, Müller T (2012) Effects of anaerobic digestion on digestate nutrient availability and crop growth: a review. Eng Life Sci 12(3):242–257

Neset TS, Cordell D (2012) Global phosphorus scarcity: identifying synergies for a sustainable future. J Sci Food Agric 92:2–6

Nieminen M, Lauren A, Hokka H, Sarkkola S, Koivusalo H, Pennanen T (2011) Recycled iron phosphate as a fertilizer raw material for tree stands on drained boreal peatlands. For Ecol Manag 261(1):105–110

Oskam A, Meester G, Silvis H (2011) EU policy for agriculture, food and rural areas. Wageningen Academic Publishers, Wageningen

Ranatunga TD, Reddy SS, Taylor RW (2013) Phosphorus distribution in soil aggregate size fractions in a poultry litter applied soil and potential environmental impacts. Geoderma 192:446–452

Roboredo M, Fangueiro D, Lage S, Coutinho J (2012) Phosphorus dynamics in soils amended with acidified pig slurry and derived solid fraction. Geoderma 189–190:328–333

Sano A, Kanomata M, Inoue H, Sugiura N, Xu KQ, Inamori Y (2012) Extraction of raw sewage sludge containing iron phosphate for phosphorus recovery. Chemosphere 89(10):1243–1247

Scholz RW, Wellmer FW (2013) Approaching a dynamic view on the availability of mineral resources: what we may learn from the case of phosphorus? Glob Environ Change 23:11–27

Schröder JJ, Smit AL, Cordell D, Rosemarin A (2011) Improved phosphorus use efficiency in agriculture: a key requirement for its sustainable use. Chemosphere 84:822–831

Sissingh HA (1971) Analytical technique of the Pw method used for the assessment of phosphate status of arable soils of the Netherlands. Plant Soil 34:483–486

Stark K, Plaza E, Hultman B (2006) Phosphorus release from ash, dried sludge and sludge residue from super-critical water oxidation by acid or base. Chemosphere 62(5):827–832

Syers JK, Johnston AE, Curtin D (2008) Efficiency of soil and fertilizer phosphorus use. Report no. 18, FAO Fertilizer and plant nutrition bulletin, Rome

Van Dam AL, Ehlert PAI (2008) Beschikbaarheid van fosfaat in organische meststoffen. Report no. PPO-3236029100, Praktijkonderzoek Plant &omgeving BV, Lisse. (in Dutch)

Van Ranst E, Verloo M, Demeyer A, Pauwels JM (1999) Manual for the soil chemistry and fertility laboratory. University of Ghent, Ghent

Yang JC, Wang ZG, Zhou J, Jiang NM, Zhang JF, Pan R et al (2012) Inorganic phosphorus fractionation and its translocation dynamics in a low-P soil. J Environ Radioact 112:64–69

Strategies for Enhancing Zinc Efficiency in Crop Plants

P.C. Srivastava, Deepa Rawat, S.P. Pachauri, and Manoj Shrivastava

Abstract

Zinc is an essential micronutrient for both plants and animals. Zinc deficiency, widely recorded in many parts of the globe, not only leads to poor yield levels but also causes reduction in the quality of produce and malnutrition in animals and humans. Higher Zn efficiency in crops could be achieved by adopting suitable measures like proper soil and fertilizer management, efficient use of traditional/modified new Zn sources at appropriate time using proper method of application, an appropriate rhizosphere management for harnessing the potential of microbial relationships with host crops, and development of Zn-efficient crop genotypes. In the present chapter, an attempt has been made to encompass each of these options. Wide genotypic variations in Zn efficiency exist in many crops, and a better understanding of the mechanism of Zn tolerance/efficiency and Zn enrichment in edible plant parts of Zn-efficient genotypes could help in identifying key traits/genes which are useful in developing Zn-efficient crop varieties by traditional breeding or genetic engineering methods. More concerted joint efforts of agronomist, soil scientists, plant physiologist, and plant breeders/biotechnologists are required for enhancing Zn efficiency in food crops.

Keywords

Absorption by roots • Biochemical utilization • Chemical fertilization • Nutrient interactions • Rhizosphere • Zinc transporters

P.C. Srivastava (✉) • D. Rawat • S.P. Pachauri
Department of Soil Science, G.B. Pant University of Agriculture & Technology, Pantnagar, Uttarakhand 263 145, India
e-mail: pcsriv@yahoo.com; rawatdeepa291@gmail.com; sp_pachauri@rediffmail.co.in

M. Shrivastava
Centre for Environment Science and Climate Resilient Agriculture, Indian Agricultural Research Institute, New Delhi 110 012, India
e-mail: manojshrivastava@iari.res.in

Zinc is an essential micronutrient for both plants and animals. It is an integral constituent of several important enzymes having role in anabolic and growth processes of the plant. Its deficiency not only leads to poor yield levels but also causes reduction in the quality of produce. Among the micronutrients, Zn deficiency is a widespread micronutrient disorder affecting food production over many countries including Australia, India,

Turkey, and the USA (Sillanpaa 1990; Alloway 2004). Alkaline soil pH, coarse soil texture, low soil organic carbon content, high calcareousness in soil, and application of heavy dose of phosphatic fertilizers to soil are some of the factors which adversely influence the availability of both native and added Zn fertilizers in soil. Zinc deficiency is common under both aerobic and submerged conditions. Soil submergence reduces the availability of Zn to rice crop due the reaction of Zn with free sulfide (S^{2-}) (Mikkelsen and Shiou 1977), increase in soil pH due to gleying process, and also owing to the formation of some insoluble zinc compounds with Mn and Fe hydroxides. Under the submerged soil conditions, Zn (both native soil Zn or Zn applied through fertilizer) is trapped into amorphous sesquioxide precipitates or franklinite ($ZnFe_2O_4$) (Sajwan and Lindsay 1988). The correction of Zn deficiency in soils involves soil or foliar application of Zn fertilizers. Zinc fertilizers applied to soil are subjected to chemical transformation, depending upon the nature of Zn fertilizer and soil characteristics/conditions. These chemical transformations often lead to a reduction in Zn availability with passage of time and consequently result to poor use efficiency of added Zn fertilizer. The necessity of enhancing Zn efficiency in agriculture is, firstly, to achieve twin objectives of sustainable crop production in low-input agriculture and/or in Zn-deficient area and, secondly, to reduce cost of cultivation as Zn fertilizers are one of the costly inputs in agriculture.

Since plant roots occupy only about 1 or 2 % of the soil surface volume, therefore, the amount and proportion of applied nutrients including Zn that reach plant roots determine the efficiency of uptake. Hence, the nutrient absorption efficiency is a function of both the ability of the soil to supply Zn^{+2} and the capacity of plant to absorb Zn^{+2}. The following approaches need to be tried for enhancing Zn efficiency in crop production:

1. Modification of the rhizosphere environment using soil amendments
2. Choice of Zn fertilizer sources and the modifications in method and time of application
3. Utilization of nutrient interactions
4. Increasing the efficiency of crop plants to absorb and utilize Zn

Each of these available options needs to be exploited in a synchronized way to achieve higher use efficiency of this important micronutrient.

1 Modification of the Rhizosphere Environment Using Soil Amendments

Soil factors such as soil texture, nature of soil clays, organic matter content, cation exchange capacity, soil pH, moisture, temperature, aeration, soil compaction, and availability of other plant nutrients in soil influence the availability, transformation, and fixation (sorption) of Zn (Srivastava and Gupta 1996). The bioavailability of Zn in soil is controlled by adsorption-desorption process and/or precipitation-dissolution reactions which in turn are dependent upon pedogenic properties of soils and soil management (Plate 1). Since the mobilization of Zn from soil to plant root is dominantly through diffusion, Barber (1976) pointed out three important soil parameters to be responsible for governing the rate of supply of Zn from the soil to the root: diffusion coefficient, concentration in soil solution, and buffer capacity. The diffusion coefficient is the most important factor, and its magnitude is influenced by volumetric soil water content, the tortuosity of the diffusion path, and the buffer capacity. By increasing the water content of the soil, the tortuosity factor is reduced, while the cross-sectional area available for Zn diffusion is increased to result in higher diffusion coefficient of Zn in soil. Karaman et al. (2013) studied the effect of different matric potentials on the response of Zn doses and Zn uptake of five soybean genotypes (A-3735, A-3127, SA-88, S-4340, and Ilisulu-20) and reported that soil moisture stress significantly decreased physiological responses of soybean genotypes to Zn doses, indicating thereby a close relationship between soil moisture levels and Zn use efficiency as there were significant differences among the soybean genotypes in their ability to accumulate Zn.

Plate 1 Scheme of modifications in rhizospheric environment to enhance Zn availability to growing plants

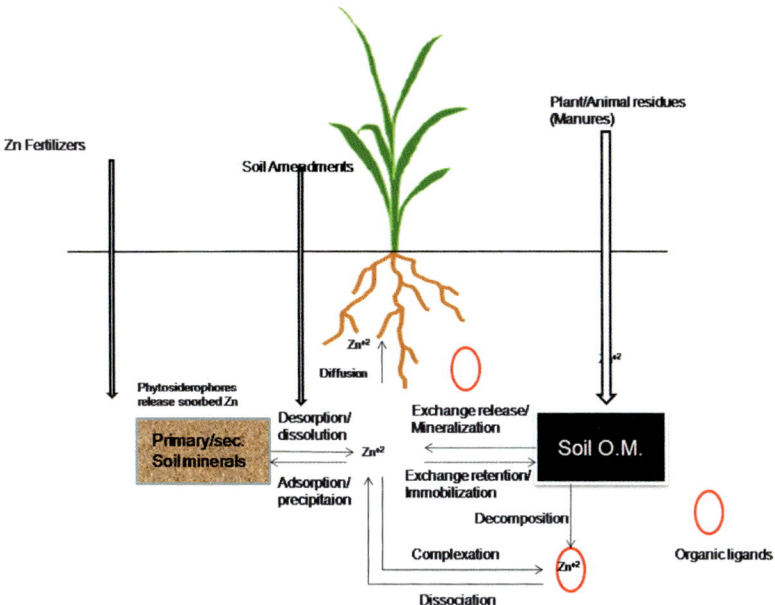

Calcareous and alkaline soils having higher pH values maintain very low solubility of both native and added Zn, and the efficiency of Zn fertilizers on such soils is also poor. In such soils, addition of chemical fertilizers and amendments capable of reducing the soil pH and promoting root growth would certainly help in the enhancement of use efficiency of Zn fertilizers (Mortvedt and Kelsoe 1988). Soil organic matter content has an influence on the exchange capacity of soil and helps to retain ions on the exchange complex at much lower tenacity as compared to soil minerals. In soils dominated by iron oxides and oxyhydroxides and amorphous oxides of iron and aluminum such as Ultisols and Oxisols or in calcareous and alkaline soils, the added Zn fertilizer is irreversibly retained and poor use efficiency of Zn fertilizers could be faced. In such soils, the transformation of added Zn to the chemical fractions of poor availability can be limited by band application of Zn fertilizers to reduce their contact with the soil and by a liberal application of organic manures. It has been demonstrated that the presence of humic substances like humic and fulvic acids pre-sorbed on goethite (α-FeOOH) decreased Zn sorption capacity and increased the desorption of sorbed Zn (Anupama et al. 2005). In a neutral soil, combined application of 2.5 kg Zn + 5 t farmyard manure ha^{-1} to pearl millet-wheat cropping system in alternate years gave significantly higher Zn uptake by crops as compared to application of 10 kg Zn ha^{-1} in alternate years and brought about tenfold increase in the apparent Zn fertilizer use efficiency (Chaube et al. 2007). Similarly, application of 2.5 t press mud compost + 5 kg Zn ha^{-1} to sugarcane increased apparent recovery of applied Zn by the sugarcane ratoon (Siddiqui et al. 2005). Sahai et al. (2006) also evaluated the possibility of further reducing the dose of organics using some readily decomposable matters such as fresh cow dung in place of farmyard manure and observed that the application of a mixture of 2.5 kg Zn with 200 kg of fresh cow dung preincubated for 1 month ha^{-1} to rice crop in rice-wheat rotation gave a total Zn uptake of 517 g Zn ha^{-1} by rice-wheat rotation which was significantly higher than the total Zn uptake obtained with application of 2.5 kg Zn alone ha^{-1} (471 g Zn ha^{-1}); the effect was ascribed to the complexation of Zn by organic acids formed during the decomposition of fresh cow dung. All these researches indicated that the addition of organic matter to soil along with conventional Zn fertilizer like zinc sulfate heptahydrate helps in improving efficiency of Zn applied to soil. However, depending upon the nature of organic manure, some insoluble organic complexes may also form which may strongly bind with Zn and reduce the availability of Zn to plants. Therefore, the effect of manure on the bioavailability of Zn depends on the characteristics of the manure and also on the specific circumstances involved. However, studies on the effect of manure on the bioavailability of Zn and other micronutrients in cereal grain from human nutrition point of view are too rare to support a conclusion.

In acid soils, though the solubility of Zn is not poor yet, these soils are often poor in Zn due to overall poor status of Zn in soils as these soils are developed over highly weathered sandy parent materials. Liming of acidic soils decreases the availability of Zn. The efficiency of applied Zn may be relatively also poor mainly because of poor root growth under toxic levels of Al and Mn; therefore, the use of lime/organic manure in these soils helps in reducing the toxicities and modifying soil conditions for better root growth and could help in achieving higher use efficiency of applied Zn by crops grown in acidic soil.

2   Choice of Zn Fertilizers and the Modifications in Method of Application

Zinc sulfate (20–22 % Zn for heptahydrate form and 35 % Zn for monohydrate form) is the most commonly used water-soluble Zn fertilizer. Some less soluble sources like ZnO, $ZnCO_3$, and $Zn_3(PO_4)_2$ give better performance on acid soils. Insoluble forms like ZnS and Zn frits can also perform well on acidic soils, if these are used in finally divided form. Zinc-EDTA (14 % Zn) is manyfold costlier than $ZnSO_4$; therefore, it is less popular among farmers in the developing countries. Being a chelate compound, Zn-EDTA results relatively higher mobilization efficiency than $ZnSO_4$ in neutral soils (Srivastava et al. 1999). Beside these Zn fertilizers, there are several other organic preparations of Zn in the literature which have been reported to give higher use efficiency of applied Zn fertilizers

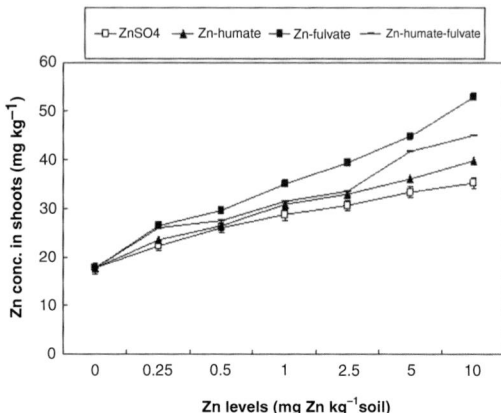

Fig. 1 Effect of Zn oragano-complexes and ZnSO4 levels on Zn concentration in maize plants (35 days after sowing) (Adopted from Kar et al. 2007)

owing to lesser fixation of soluble Zn in soil. Low-yield ammonium-based lignosulfonate Zn complex (5 % Zn) resulted in about more than double recovery of added Zn by beans as compared to $ZnSO_4$ (Singh et al. 1986). Kar et al. (2007) compared some preparations of organo-complexes of Zn like Zn-fulvate, Zn-humate, and Zn-humate-fulvate with $ZnSO_4$ in a glasshouse experiment with maize (*Zea mays* L.) and reported that the near-optimum concentration of Zn in maize tissue ($\approx$30 mg Zn kg^{-1} dry matter) could be attained at 10 mg Zn as $ZnSO_4$, 5.0 mg Zn as Zn-humate, 2.5 mg Zn as Zn-humate-fulvate, and 1.0 mg Zn as Zn-fulvate kg^{-1} soil (Fig. 1). Srivastava et al. (2008) studied the kinetics of desorption, transformation, and availability of Zn applied to soil through ^{65}Zn-tagged zinc-enriched bio-sludge from distillery molasses (ZEMB) or as zinc sulfate heptahydrate (ZSH) to rice crop and subsequently grown wheat. These workers demonstrated that the desorption rate coefficient (K) and desorbed amount of Zn were significantly higher with ZEMB than with ZSH. The ZEMB maintained relatively higher proportion of applied Zn in available forms as compared to ZSH as the former Zn source had major proportion of water-soluble Zn (85.78 % of total water-soluble Zn) in association with the dissolved organic matter which allowed a faster diffusion of Zn to the plant roots. The ZEMB source also maintained

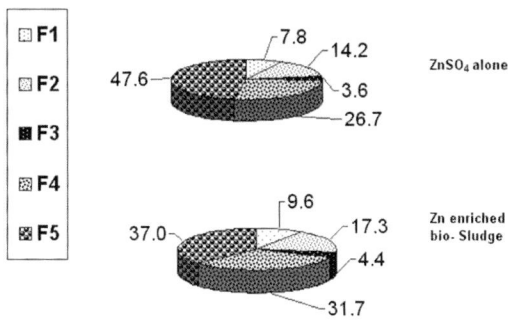

Fig. 2 Percent distribution of added Zn fertilizer among different chemical fractions of Zn (*F1* water-soluble + Exch, *F2* carbonate bound, *F3* organically bound, *F4* reducible, *F5* residual fraction) in Zn-enriched bio-sludge (Adopted from Srivastava et al. 2008)

Table 1 Effect of Zn fertilizer sources on percent utilization of fertilizer Zn by first rice crop and subsequent wheat plants (Srivastava et al. 2008)

Zn fertilizer sources	Percent utilization of fertilizer Zn (%)	
	First rice crop	Subsequent wheat crop
5.00 kg Zn as Zinc sulfate ha^{-1}	0.162	0.184
1.25 kg Zn as ZEMB[a] ha^{-1}	0.610	0.441
2.50 kg Zn as ZEMB ha^{-1}	0.433	0.341
5.0 kg Zn as ZEMB ha^{-1}	0.290	0.293
C.D. ($p \leq 0.05$)	0.047	0.080

[a]Zinc-enriched post-methanation bio-sludge from molasses-based distillery

relatively higher proportion of applied Zn into such chemical fractions of Zn in soil which were likely to release Zn for utilization by plants as compared to ZSH (Fig. 2). The effect could be ascribed to the presence of soluble organic matter in the ZEMB having good complexation or chelation ability to maintain a high content of soluble Zn in the soil solution and suppressing the hydrolysis of Zn^{+2} so as to discourage the strong sorption of Zn^{+2} by soil constituents. In comparison to conventional zinc sulfate fertilizer (ZSH), the use of ZEMB increased the utilization of applied fertilizer Zn by rice and subsequent wheat crops (Table 1). In the follow-up 2 years field experiments with rice-wheat rotation, Srivastava et al. (2009) noted that the apparent Zn utilization efficiency of ZEMB at 5.00 kg Zn ha^{-1} dose was more than twofold higher (7.12 %) than ZSH (3.22 %) (Fig. 3). The values of apparent Zn utilization efficiency of ZEMB at lower doses (1.25 and 2.50 kg Zn ha^{-1}) were still higher than the value observed at 5.00 kg Zn ha^{-1} dose. These findings indicate that Zn applied as organo-complex to soil remains available to crops for a longer period of time than conventional inorganic Zn fertilizer like ZnSO$_4$.

As regards the methods of soil application of Zn fertilizers, only broadcast and band application are common, the latter method ensures the better utilization of applied Zn in soils of high Zn fixation capacity (calcareous and alkaline soils),

and it can be easily practiced in wide row crops. Foliar application of water-soluble Zn fertilizers certainly ensures manyfold higher use efficiency than soil application as it is directly spayed on crop foliage and skips irreversible retention in the soil or chemical transformation of Zn into poorly available chemical fractions of Zn in the soil. Investigations carried out by the authors revealed that foliar application of Zn gave higher apparent Zn utilization efficiency as compared to soil application, and the magnitude of increase varies with the crop and sensitivity of the variety to Zn deficiency and also the level of other critical nutrients supplied to the crop (unpublished data). However, foliar application of Zn should not be treated as an alternative to soil application because it is often resorted practically after the appearance of the deficiency symptoms, a time by which yield damages are already inflicted on the crop.

3 Utilization of Nutrient Interactions for Increasing Zn Use Efficiency

The relationship among some nutrients in plant may be additive or synergistic or antagonistic or nonexistent. In soils of low to medium supply of a critical nutrient, the additive or synergistic relationships can be utilized for higher yields and

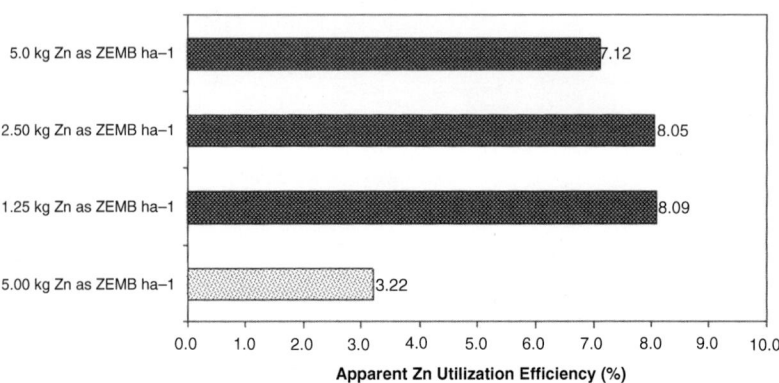

Fig. 3 Apparent Zn utilization efficiency (%) for conventional zinc sulfate heptahydrate (ZSH) and zinc-enriched post-methanation bio-sludge (ZEMB) after two cycles of rice-wheat rotation. The numerical values in front of bars indicate % Zn utilization efficiency (Adopted from Srivastava et al. 2009)

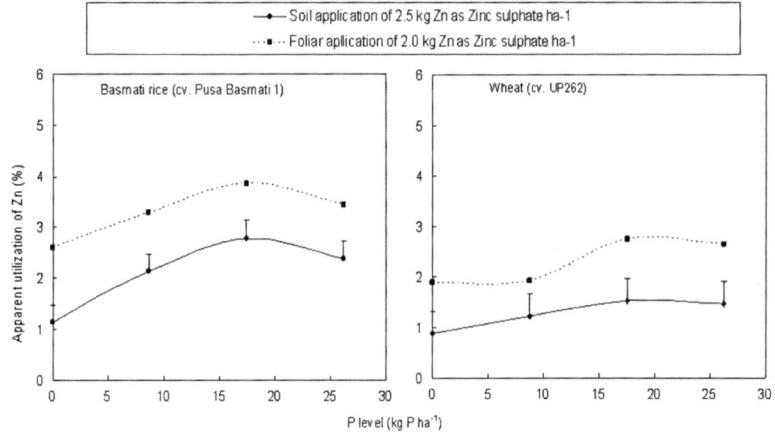

Fig. 4 Apparent utilization of Zn (%) applied through soil or foliar application of Zn by Basmati rice-wheat rotation at varying levels of soil application of phosphatic fertilizer (Pooled data of 2 years) (Adopted from Srivastava et al. 2013b)

acquisition of Zn by crops to enhance the use efficiency of Zn in crop production. Some of such synergistic relationships have been observed in respect of N, P, K, and S. The synergistic relationship of acid producing N fertilizers on Zn utilization is attributed to an improvement in the solubility of Zn in soil and promotion of plant and root growth (Giordano 1979); however, at high doses of ammonium fertilizers, the relationship may turn to be antagonistic due to poor biochemical utilization of Zn (Srivastava and Gupta 1996). In soils deficient in P supply, the application of normal dose of phosphatic fertilizer enhances the absorption of Zn by the crop due to better root growth and results in improved Zn use efficiency. In Basmati rice-wheat rotation, Srivastava et al. (2013b) reported that an increase in P levels

up to 17.5 kg P ha^{-1} increased apparent utilization efficiency of soil or foliar-applied Zn (Fig. 4). However, a very high level of P fertilization may adversely influence Zn efficiency due to reduction in root surface area, absorption, and translocation of Zn (Ali et al. 1990). The interaction of Zn and K also influences utilization efficiency of Zn applied to crops. Srivastava et al. (2013a) reported that soil application of K ensured higher apparent use efficiency of Zn especially, that of foliar-applied Zn in rice-wheat rotation (Fig. 5). A synergistic relationship between Zn and S has been reported in the literature (Kumar and Singh 1980; Bowman and Olsen 1982). In mustard, S fertilization has been reported to increase Zn flux to crop due to increased root surface area and solubilization of Zn in soil (Sharma et al. 1990).

Fig. 5 Apparent utilization of Zn applied through soil or foliar application at varying levels of soil application of potassium fertilizer by rice-wheat rotation (Pooled data of 2 years) (Communicated by Srivastava et al. 2013a)

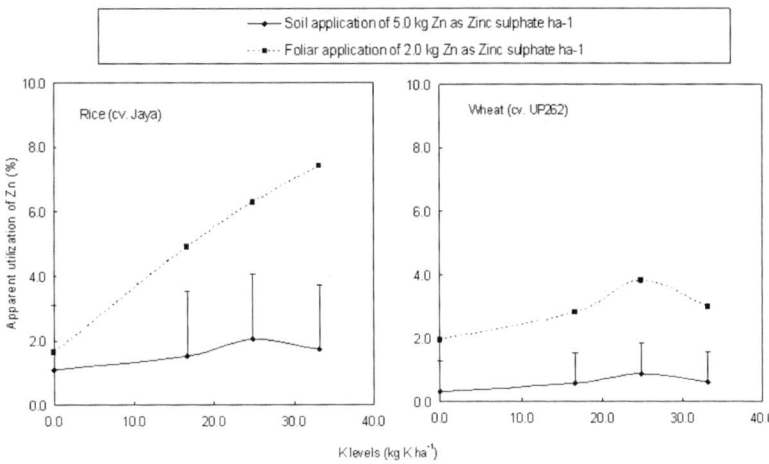

4 Increasing the Efficiency of Crop Plants to Absorb and Utilize Zn

Different crop plants vary in their Zn use efficiency. Fageria et al. (2008) reported that Zn use efficiency for grain production was higher for corn followed by rice and the minimum for soybean. Within a crop, different cultivars of rice (Jiang et al. 2007; Hafeez et al. 2010), wheat (Cakmak et al. 2001), and Chinese cabbage (Wang et al. 2011) have been reported to differ in their Zn efficiency. Differential Zn utilization efficiency of crops and also among different genotypes within a crop can be related to the differences in the "morpho-chemo-socio-physiological" behavior of the plant roots in a Zn-deficient soil environment. Plant roots have different strategies for the enhancement of Zn absorption. These include bestowing special features in root architecture, alterations in the rhizosphere chemistry to effect greater solubilization of Zn in the rhizosphere so as to maintain higher absorption rate even in Zn-deficient soil, maintaining microbial associations in the rhizosphere for higher Zn absorption, and physiological adjustments for remobilization of Zn

and efficient metabolic utilization of Zn (Plate 2). Each of these aspects needs to be understood for breeding Zn-efficient genotypes and also adopting supplementary cultural measures to achieve higher Zn efficiency in crop production.

4.1 Root-Induced Rhizospheric Changes to Increase Labile Pool of Zn

Since the solubility of Zn in soil is governed by pH, any change in pH of the rhizosphere is likely to alter the solubility and ultimate availability of Zn to the growing plants. Lowering of rhizospheric pH induced by plant roots is a result of exudation of H^+ due a cation-anion imbalance in the plant body or formation of HCO_3^- ions upon dissolution of CO_2 released by roots due to respiration of roots or tendency of roots to exude lowmolecular-weight organic acids in the rhizosphere.

Proton exudation or acidification of rhizosphere by plant roots has been reported to mobilize Zn from soil to plant roots. In case of lowland rice plant, acidification of the rhizosphere is possible in two ways: (i) exudation of

Plate 2 Scheme of root-induced and other plant-induced changes to influence Zn availability to growing plants

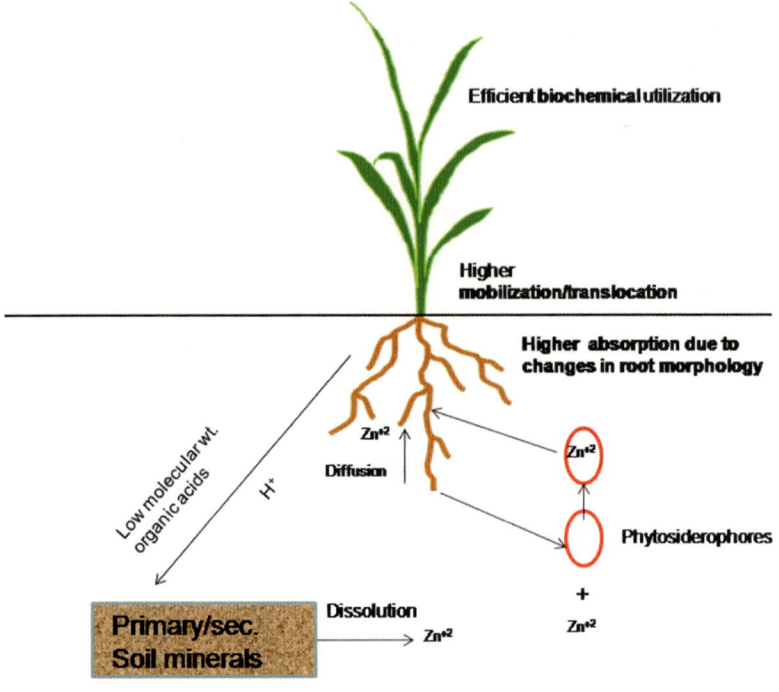

H^+ due to imbalance of cation and anion uptake in rice which preferentially absorbs NH_4^+ and (ii) as a result of radial oxygen loss from roots which causes oxidation of Fe^{2+} to Fe^{+3} with release of two protons (H^+). In rice, we observed that the roots of young seedling (20 days after germination) of NDR359, a cultivar highly susceptible to Zn deficiency, showed proton exudation ability in Zn-deficient growing medium (Plate 3). Proton exudation in the rhizosphere is likely to enhance the solubility of Zn especially in calcareous and alkaline soils; however, whether proton exudation capacity of a genotype can be utilized as a genetic trait for breeding efficient genotypes still remains doubtful.

The roots of certain plants exude low-molecular-weight organic acids (LMWOAs) and phytochelators to solublise. Zn in the rhizosphere. The exudation of several LMWOAs like citrate (Hoffland et al. 2006) or malate (Gao et al. 2009; Rose et al. 2011) or oxalate (Bharti et al. 2014) by plant roots has been reported.

However, the results could not be related to Zn efficiency of the genotypes in many instances. Besides that the phenomenon of exudation of LMWOAs has been also reported to be a result of radical oxygen stress leading to root membrane damage (Rose et al. 2011, 2012) rather than as an adaptive mechanism induced under Zn deficiency. Further, it has also been argued that the concentration of LMWOAs reported in root exudates (0.01–1 mM) may not be sufficient to mobilize the required amount of Zn in the plant rhizosphere (Gao et al. 2009; Rose et al. 2011).

Some cereals release nonprotein amino acids (phytosiderophores), which are capable of chelating micronutrients like Fe and Zn (Marschner 1995). A number of studies have reported the release of phytosiderophores by cereal roots. In solution culture experiment, durum genotypes of wheat which are sensitive to Zn deficiency have been reported to exude relatively smaller amounts of phytosiderophores as compared to

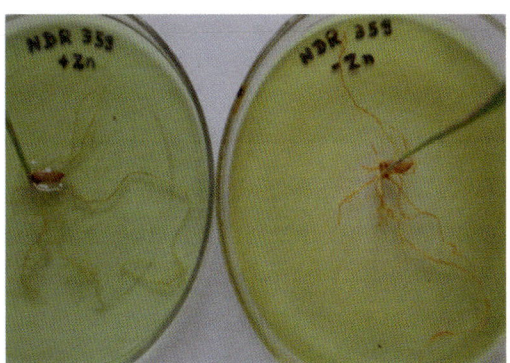

Plate 3 Proton exudation under Zn stress by young (20 days after germination) seedling of rice (cv. NDR359; a variety; susceptible to Zn deficiency). The symbols –Zn and +Zn indicate no Zn (0.00 mg Zn l^{-1}) and the presence of Zn (0.05 mg Zn l^{-1}) in agar medium mixed with bromothymol blue; the initial pH of the medium was adjusted to 7.0 in both –Zn and +Zn treatments before transfer of the seedling. A change from green to yellow color in Zn-deficient medium and relatively high intensity of yellow color near the surface of roots can be seen under –Zn treatment

bread wheat genotypes which are tolerant of Zn deficiency (Walter et al. 1994). Similarly, the secretion of phytosiderophores has been noted under Zn deficiency in wheat (Cakmak et al. 1994) and also in barley (Suzuki et al. 2008). However, other workers failed to notice significant release of phytosiderophores in those cultivars of barley (Gries et al. 1995) and wheat (Pedler et al. 2000) which have already been reported to release phytosiderophores under Zn deficiency. Deoxymugineic acid released by cereals under Fe deficiency (Ishimaru et al. 2011) could also help uptake of Zn in rice genotypes tolerant to Zn deficiency (Ptashnyk et al. 2011). It, therefore, appears that more scientific evidences are still required to prove the utility of deoxymugineic acid as a genetic character in Zn-efficient genotypes of cereals.

Zuo and Zhang (2009) reviewed the potential role of intercropping of dicot plants with cereals for Fe and Zn biofortification and opined that intercropping could be a practical, effective, and sustainable practice in developing countries for enhancing Zn efficiency. In a field experiment conducted on a low Zn soil in Turkey,

Gunes et al. (2007) observed higher concentration of Zn in both wheat and chickpea under intercropping system than in the monocropped system. Similarly, in Chinese peanut/maize intercropping, the excretion of phytosiderophores by maize into the rhizosphere played an important role in improving Fe and Zn nutrition of the peanut crop (Zuo and Zhang 2008).

4.2 Role of Rhizospheric Microorganisms in Enhancing Zn Acquisition

The microorganisms in the rhizosphere play an important role in governing Zn uptake of plants and Zn efficiency (Plate 4). Rengel (1997) observed that Zn deficiency increased the numbers of fluorescent pseudomonads in the rhizosphere of all wheat genotypes tested, and the effect was particularly more pronounced in genotypes tolerant of Zn deficiency. These reports hint at some significant relation between microbial communities in rhizosphere of different genotypes and their tolerance to Zn stress. The effect might be a reaction to the altered root exudation pattern under Zn deficiency. However, whether these changes actively contribute to the acquisition of Zn or passively appear in response to direct tolerance mechanisms of efficient genotypes has to be investigated further. In laboratory and glasshouse conditions, many rhizospheric microorganisms have been reported to stimulate acquisition of Zn by plants (Tariq et al. 2007). The effect could be attributed directly to the production of plant hormones such as indole acetic acid (IAA), gibberellic acid (GA) and cytokinin, phosphate solubilization, Zn solubilization, and nitrogen fixation and indirectly to plant growth promotion through suppression of soilborne or foliar pathogens. Vaid et al. (2013) examined three bacterial strains, namely, *Burkholderia* sp. strain SG1 (BC), *Acinetobacter* sp. strain SG2 (AX), and *Acinetobacter* sp. strain SG3 (AB), isolated from the rhizosphere of rice plant growing in a Zn-deficient Typic Hapludoll for gluconic acid production, Zn solubilization, and IAA

Plate 4 Scheme of
changes induced by
microorganisms to enhance
Zn availability to growing
plants

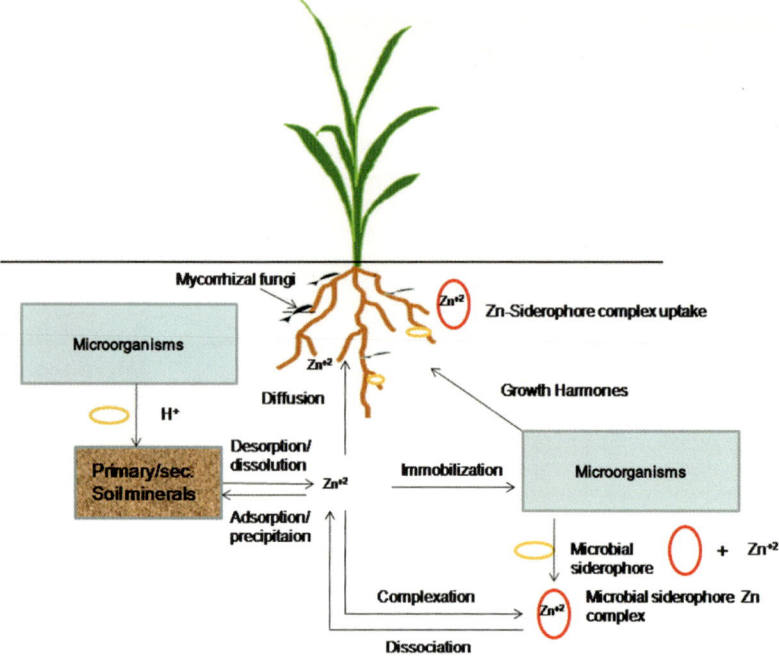

production in vitro cultures. These workers noted that among the three bacterial isolates, the highest gluconic acid production (25.9 mM) after 48 h of inoculation, Zn solubilization, and IAA production (5.79 mg L^{-1}) was recorded with *Acinetobacter* sp. and these effects were relatively lower for the other two strains. In a follow-up greenhouse study, the effect of seed inoculation with these strains alone or in combination was evaluated on yields, and total Zn uptake by two wheat varieties (VL 804, sensitive to Zn deficiency, and WH 1021, tolerant to Zn deficiency) and the highest grain yield was noted with inoculation of AX + AB in WH 1021 and with BC + AX in VL804. These inoculations also increased total Zn uptake significantly over the control (no Zn application).

In nature, mycorrhizal plants are known to take up higher amount of Zn and other nutrients like P and Cu as compared to their non-mycorrhizal counterparts. The beneficial effect of mycorrhizal fungi on Zn uptake of host plants has been reported in pigeon pea (Wellings et al. 1991), wheat (Ryan and Angus 2003), and tomato (Cavagnaro et al. 2010). Mycorrhization brings changes in the root

architecture, and the extramatricular hyphae are likely to extend the effective zone of root exploration further (Kothari et al. 1991). Sharma and Srivastava (1991) demonstrated that vesicular-arbuscular mycorrhizal fungi (AMF) (*Glomus macrocarpum*) inoculation of green gram increased the mobilization of Zn through diffusion process. Cavagnaro (2008) concluded that the improvements in the Zn nutrition of plants colonized by AMF could be attributed to direct uptake of Zn by AMF and/or indirect effects due to alteration in morphological and physiological characteristics of roots. Mycorrhizal inoculation as a tool to improve Zn efficiency of plants holds much promise for vegetable and horticultural crops which can be easily inoculated in the nursery. However, as genetic differences in Zn efficiency are independent of mycorrhizal associations, the role of mycorrhizae in governing Zn efficiency of a genotype remains doubtful.

Despite these claims, the effectiveness of rhizospheric microorganisms under field condition has yet to be proved. However, there lies a possibility that a better understanding of microbial dynamics in the rhizosphere of different genotypes of varying Zn deficiency tolerance might lead to further exploration of opportunities for enhancing the root acquisition of Zn in future.

adaptive trait. This trait is being targeted in breeding for enhanced tolerance to Zn deficiency.

4.3 Role of Root Architecture in Enhancing Zn Acquisition

Since the dominant process of Zn mobilization to the plant roots is a diffusion-controlled process, therefore, root architecture is likely to play an important role in Zn uptake. Dong et al. (1995) opined that root architecture has profound influence on Zn efficiency of plants. Thinner roots with higher surface area explore the soil more thoroughly and may increase the availability of Zn and also of other nutrients (Rengel and Graham 1995). In dryland cereals like wheat and barley, the formation of root hairs and the ability to produce longer and finer roots has also been linked to enhanced Zn uptake (Dong et al. 1995; Genc et al. 2007). Hoffland et al. (2006) observed that the higher number of rice plants per hill showed an improvement in Zn nutrition of crop. The observed effect could be a result of increased localized concentrations of root exudates for affecting Zn solubilization and higher chances of efficient capture of solubilized Zn diffused in the vicinity of intertwined roots. Ptashnyk et al. (2011) noted that root length density in rice was the most important parameter to govern the uptake of Zn solubilized and chelated by the deoxymugineic acid. In nutrient uptake modeling of rice crop, Kirk (2003) envisaged that rice roots having a coarse aerenchymous primary root along with numerous fine, short lateral roots are likely to offer the optimum combination to meet the twin requirements of root aeration and nutrient uptake. It has been also observed that rice cultivars tolerant to Zn deficiency maintained higher number of crown roots as compared to cultivars sensitive to Zn deficiency (Widodo et al. 2010). The difference in crown root emergence among different rice genotypes can be detected as early as 3 days after transplanting, and therefore, this trait could be an independent character responsible for tolerance mechanism rather than being an

4.4 Role of Zn Transporters to Increase Zn Uptake Efficiency

Zinc ions diffuse in the free space of cell wall, and their further passage across the plasma membrane occur through ion transport proteins. Besides that, an alternative mechanism involving uptake of Zn-phytosiderophore complex (Zn-PS) has also been recorded in cereals (Kochian 1993; von Wiren et al. 1996). In cereals, both high-velocity, low-affinity system operational at higher concentrations of Zn and a low velocity, high affinity system functional at low concentrations of Zn are observed (Hacisalihoglu et al. 2001; Hacisalihoglu 2002). The ZIP family transporters are known to facilitate entry of Zn into the root cells. These ZIP transporter genes are reported to be upregulated under Zn deficiency stress (Ishimaru et al. 2011). However, a conclusive evidence proving greater expression of particular Zn transporters in roots of Zn-efficient genotypes is still warranted (Bowen 1987; Hacisalihoglu et al. 2001).

Zinc retranslocation from old parts to the young parts of shoot has been also suggested as a possible mechanism affecting zinc efficiency in common bean (Hacisalihoglu et al. 2004), wheat (Torun et al. 2000), and rice (Hajiboland et al. 2001). Zinc efficient barley genotype has been reported to remobilize greater amounts of Zn from vegetative to reproductive tissues as compared to a Zn-inefficient genotype (Genc and McDonald 2004). Gao et al. (2005) correlated Zn efficiency significantly ($P < 0.05$) with Zn uptake ($R^2 = 0.34$), Zn translocation from root to shoot ($R^2 = 0.19$), and shoot Zn concentration ($R^2 = 0.27$), and these workers could explain only 53 % of variation in zinc efficiency calculated on the basis of Zn uptake and Zn translocation to the shoots. Similarly, a large unexplained variation in Zn efficiency has been reported in wheat (Cakmak et al. 2001). Holloway et al. (2010) also showed that wheat variety "Gatcher" produced 47 % more dry weight of tops and double root length density at maturity as compared to "Excalibur." However, "Excalibur" variety was found to be much more efficient in Zn uptake by roots and sevenfold more efficient than "Gatcher" in partitioning Zn to grain production.

4.5   Efficient Biochemical Utilization of Zn

The unexplained variation in Zn efficiency among different genotypes might also be related to differences in biochemical Zn utilization and Zn retranslocation from older into younger tissues in shoots (Hacisalihoglu and Kochian 2003). Zinc is an essential component of some antioxidant and homeostatic enzymes (Broadley et al. 2007). Zinc efficiency was found to be positively correlated with the activity of the Zn-requiring enzyme like carbonic anhydrase in rice (Rengel 1995) and Cu/Zn superoxide dismutase (SOD) in wheat (Cakmak et al. 1997; Hacisalihoglu et al. 2003) and black gram (Pandey et al. 2002). These findings suggest that Zn-efficient genotypes may be able to maintain a normal functioning of these enzymes under low Zn conditions. These works provide some circumstantial evidence in support of efficient biochemical utilization of Zn in efficient crop genotypes as compared to inefficient genotypes. On the molecular level, it can be interpreted as higher expression of genes responsible for these enzymes. It is also expected that alterations in the Zn-dependent regulation of the expression of these key Zn-requiring enzymes might have a role in Zn efficiency of crop plants.

Besides these enzymes, there are a number of proteins which bind to Zn and are likely to play some role in Zn homeostasis and trafficking. Some of these proteins also appear to be responsible in the regulation of expression of genes involved in Zn metabolism (Berg 1990). A plant homologue of metal response element-binding transcription factor-1 (MTF1) already reported in mammals (Andrews 2001) could possibly regulate the transcription of *MT* genes and

coordinate cellular Zn homeostasis in crop plants. There is need to elucidate further the molecular mechanisms of these possible Zn sensors in Zn homeostasis and efficiency in crop plants.

Conclusion

The objective of enhancing Zn efficiency in agriculture is to achieve sustainability in crop production from Zn-deficient geographical areas and to reduce the cost of cultivation as Zn fertilizers are costly inputs in agriculture. An enhancement in the concentration of Zn in grains and straw of staple food crops is desirable for alleviating Zn malnutrition in human and cattle population. Higher Zn efficiency in crops could be achieved by suitably tailoring in the various available options which include proper soil and fertilizer management, efficient use of traditional and new/modified Zn sources at right time using appropriate method of application, and proper rhizosphere management to harness the potential of microbial relationships with host crops. Use of Zn-efficient genotypes in cultivation is the most simple and economic solution to achieve higher Zn efficiency in agriculture. In view of some genotypic variations which exist in many crops, a better understanding of the mechanism of Zn tolerance and Zn enrichment in edible plant parts in Zn-efficient genotypes could help in identifying key traits/genes which are likely to be useful in developing Zn-efficient crop varieties by employing traditional breeding or genetic engineering methods.

References

Ali T, Srivastava PC, Singh TA (1990) Effect of zinc and phosphorus fertilization on zinc and phosphorus nutrition of maize during early growth. Pol J Soil Sci 23: 79–87

Alloway BJ (2004) Zinc in soil and crop nutrition. International Zinc Association, Brussels

Andrews GK (2001) Cellular zinc sensors: MTF-1 regulation of gene expression. Biometals 14:223–237

Anupama, Srivastava PC, Ghosh D, Kumar S (2005) Zinc sorption-desorption characteristics of goethite

(α-FeOOH) in the presence of pre-sorbed humic and fulvic acids. J Nucl Agric Biol 34:19–26

Barber SA (1976) Efficient fertilizer use. In: Patterson FL (ed) Agronomic research for food. ASA special publication no 26. American Society of Agronomy, Madison

Berg JM (1990) Zinc finger domains: hypotheses and current knowledge. Annu Rev Biophys Biophys Chem 19:405–421

Bharti K, Pandey N, Shankhdhar D, Srivastava PC, Shankhdhar SC (2014) Effect of different zinc levels on activity of superoxide dismutases and acid phosphatases and organic acid exudation on wheat genotypes. Physiol Mol Biol Plant 20:41–48

Bowen JE (1987) Physiology of genotypic differences in Zn and Cu uptake in rice and tomato. In: Gabelman HW, Loughman BC (eds) Genetic aspects of plant mineral nutrition. Martinus Nijhoff Publishers, Dordrecht

Bowman RA, Olsen SR (1982) Effect of calcium sulphate on iron and zinc uptake in Sorghum. Agron J 74: 923–924

Broadley MR, White PJ, Hammond JP, Zelko I, Lux A (2007) Zinc in plants. New Phytol 173:677–702

Cakmak I, Gulut KY, Marschner H, Graham RD (1994) Effect of zinc and iron deficiency on phytosiderophore release in wheat genotypes differing in zinc deficiency. J Plant Nutr 17:1–17

Cakmak I, Ozturk L, Eker S, Torun B, Kalfa H, Yilmaz A (1997) Concentration of Zn and activity of Cu/Zn-SOD in leaves of rye and wheat cultivars differing in sensitivity to Zn deficiency. J Plant Physiol 151:91–95

Cakmak O, Ozturk L, Karanlik S et al (2001) Tolerance of 65 durum wheat genotypes to zinc deficiency in a calcareous soil. J Plant Nutr 24(11):1831–1847

Cavagnaro TR, Dickson S, Smith FA (2010) Arbuscular mycorrhizas modify plant responses to soil zinc addition. Plant Soil 329:307–313

Cavagnaro TR (2008) The role of arbuscular mycorrhizas in improving plant zinc nutrition under low soil zinc concentration: a review. Plant Soil 304:315–325

Chaube AK, Ruhella R, Chakraborty R, Gangwar MS, Srivastava PC, Singh SK (2007) Management of zinc fertilizer under pearl millet-wheat cropping system in a Typic Ustipsamment. J Indian Soc Soil Sci 55: 196–202

Dong B, Rengel Z, Graham RD (1995) Characters of root geometry of wheat genotypes differing in Zn efficiency. J Plant Nutr 18:2761–2773

Fageria NK, Barbosa Filho MP, Santos AB (2008) Growth and zinc uptake and use efficiency in food crops. Commun Soil Sci Plant Anal 39:2258–2269

Gao X, Zou C, Zhang F, van der Zee SEATM, Hoffland E (2005) Tolerance to zinc deficiency in rice correlates with zinc uptake and translocation. Plant Soil 278: 253–261

Gao X, Zhang F, Hoffland E (2009) Malate exudation by six aerobic rice genotypes varying in zinc uptake efficiency. J Environ Qual 38:2315–2321

Genc Y, McDonald GK (2004) The potential of synthetic hexaploid wheats to improve zinc efficiency in modern bread wheat. Plant Soil 262:23–32

Genc Y, Huang CY, Langridge P (2007) A study of the role of root morphological traits in growth of barley in zinc-deficient soil. J Exp Bot 58:2775–2784

Giordano PM (1979) Soil temperature and nitrogen effects on response of flooded and nonflooded rice to zinc. Plant Soil 52:365–372

Gries D, Brunn S, Crowley DE, Parker DR (1995) Phytosiderophore release in relation to micronutrient metal deficiencies in barley. Plant Soil 172:299–308

Gunes A, Inal A, Adak MS, Alpaslan M, Bagci EG, Erol T, Pilbeam DJ (2007) Mineral nutrition of wheat, chickpea and lentil as affected by intercropped cropping and soil moisture. Nutr Cycl Agroecosyst 78: 83–96

Hacisalihoglu G (2002) Physiological and biochemical mechanisms underlying zinc efficiency in monocot and dicot crop plants. PhD thesis, Cornell University, Ithaca, New York, USA

Hacisalihoglu G, Kochian LV (2003) How do some plants tolerate low levels of soil zinc? Mechanisms of zinc efficiency in crop plants. New Phytol 159:341–350

Hacisalihoglu G, Hart JJ, Kochian LV (2001) High- and low-affinity zinc transport systems and their possible role in zinc efficiency in bread wheat. Plant Physiol 125:456–463

Hacisalihoglu G, Hart JJ, Wang Y, Cakmak I, Kochian LV (2003) Zinc efficiency is correlated with enhanced expression and activity of Cu/Zn superoxide dismutase and carbonic anhydrase in wheat. Plant Physiol 131:595–602

Hacisalihoglu G, Ozturk L, Cakmak I, Welch RM, Kochian L (2004) Genotypic variation in common bean in response to zinc deficiency in calcareous soil. Plant Soil 259:71–83

Hafeez B, Khanif YM, Samsuri AW, Radziah O, Zakaria W, Saleem M (2010) Evaluation of rice genotypes for zinc efficiency under acidic flooded condition. In: 19th world congress of soil science, soil solutions for a changing world, Brisbane, Australia, 1–6 August 2010

Hajiboland R, Singh B, Römheld V (2001) Retranslocation of Zn from leaves as important factor for zinc efficiency of rice genotypes. In: Horst WJ (ed) Plant nutrition – food security and sustainability of agro-ecosystems. Kluwer Academic Publishers, Dordrecht

Hoffland E, Wei CZ, Wissuwa M (2006) Organic anion exudation by lowland rice (Oryza sativa L.) at zinc and phosphorus deficiency. Plant Soil 283:155–162

Holloway RE, Graham RD, McBeath TM, Brace DM (2010) The use of a zinc-efficient wheat cultivar as an adaptation to calcareous subsoil: a glasshouse study. Plant Soil 336:15–24

Ishimaru Y, Bashir K, Nishizawa NK (2011) Zn uptake and translocation in rice plants. Rice 4:21–27

Jiang W, Zhao M, Jin L, Fan T (2007) Differences in zinc uptake and use efficiency between different aerobic rice accessions. Acta Metall Sin 13:479–484

Kar D, Ghosh D, Srivastava PC (2007) Efficacy evaluation of different zinc-organo complexes in supplying zinc to maize (*Zea mays* L.) plant. J Indian Soc Soil Sci 55:67–72

Karaman MR, Horuz A, Tuşat E, Adiloğlu A, Fatih E (2013) Effect of varied soil matric potentials on the zinc use efficiency of soybean genotypes (*Glycine Max* L.) under the calcareous soil. Sci Res Essays 8: 304–308

Kirk GJD (2003) Rice root properties for internal aeration and efficient nutrient acquisition in submerged soil. New Phytol 159:185–194

Kochian LV (1993) Zinc absorption from hydroponic solutions by plant roots. In: Robson AD (ed) Zinc in soils and plants. Kluwer Academic Publishers, Dordrecht, pp 45–57

Kothari SK, Marschner H, Romheld V (1991) Contribution of VA mycorrhizal hyphae in acquisition of phosphorus and zinc by maize grown calacareous soil. Plant Soil 131:177–185

Kumar VK, Singh M (1980) Sulphur and zinc interaction in relation to yield, uptake and utilization of sulphur in soybean. Soil Sci 130:19–25

Marschner H (1995) Mineral nutrition of higher plants. Academic, Boston

Mikkelsen DS, Shiou K (1977) Zinc fertilisation and behaviour in flooded soils. Special publication no. 5. Comm. Agric. Bur., Farnham Royal

Mortvedt JJ, Kelsoe JJ (1988) Response of corn to zinc applied with banded acid-type fertilizers and ammonium polyphosphate. J Fertil Issues 5:83–88

Pandey N, Pathac GC, Singh AK, Sharma CP (2002) Enzymic changes in response to zinc nutrition. J Plant Physiol 159:1151–1153

Pedler JF, Parker DR, Crowley DE (2000) Zinc deficiency-induced phytosiderophore release by the Triticaceae is not consistently expressed in solution culture. Planta 211:120–126

Ptashnyk M, Roose T, Jones DL, Kirk GJD (2011) Enhanced zinc uptake by rice through phytosiderophore secretion: a modelling study. Plant Cell Environ 34:2038–2046

Rengel Z (1995) Carbonic anhydrase activity in leaves of wheat genotypes differing in Zn efficiency. J Plant Physiol 147:251–256

Rengel Z (1997) Root exudation and microflora populations in the rhizosphere of crop genotypes differing in tolerance to micronutrient deficiency. Plant Soil 196:255–260

Rengel Z, Graham RD (1995) Wheat genotypes differ in Zn efficiency when grown in chelate-buffered nutrient solution. I. Growth. Plant Soil 173:307–316

Rose MT, Pariasca-Tanaka J, Rose TJ, Wissuwa M (2011) Bicarbonate tolerance of Zn-efficient rice genotypes is not related to organic acid exudation, but to reduced solute leakage from roots. Funct Plant Biol 38: 493–504

Rose TJ, Impa SM, Rose MT, Tanaka PJ, Mori A, Heuer S, Johnson BSE, Wissuwa M (2012) Enhancing phosphorus and zinc acquisition efficiency in rice: a critical review of root traits and their potential utility in rice breeding. Ann Bot 112:331–345

Ryan MH, Angus JF (2003) Arbuscular mycorrhizae in wheat and field pea crops on a low P soil: increased Zn-uptake but no increase in P-uptake or yield. Plant Soil 250:225–239

Sahai P, Srivastava P, Singh SK, Singh AP (2006) Evaluation of organics incubated with zinc sulphate as Zn source for rice-wheat rotation. J Ecofriendly Agric 1: 120–125

Sajwan KS, Lindsay WL (1988) Effect of redox, zinc fertilisation and incubation time on DTPA-extractable zinc, iron and manganese. Commun Soil Sci Plant Anal 19:1–11

Sharma AK, Srivastava PC (1991) Effect of vesicular-arbuscular mycorrhizae and zinc application on dry matter and zinc uptake of greengram (*Vigna radiata* E. Wilczek). Biol Fertil Soils 11:52–56

Sharma UC, Gangwar MS, Srivastava PC (1990) Effect of zinc and sulphur fertilizers on growth, root characteristics, nutrient uptake and yields of mustard (*Brassica juncea* L.). J Indian Soc Soil Sci 38:696–701

Siddiqui A, Srivastava PC, Singh AP, Singh SK (2005) Effect of zinc sulphate and pressmud compost application on yields, zinc concentration and uptake of sugarcane. Indian J Sugarcane Technol 20:35–39

Sillanpaa M (1990) Micronutrient assessment at the country level; an international study. Food and Agriculture Organization of the United Nations, Rome

Singh JP, Karamanos RE, Lewis NG, Stewart JWB (1986) Effectiveness of zinc fertilizer sources on nutrition of beans. Can J Soil Sci 66:183–187

Srivastava PC, Gupta UC (1996) Trace elements in crop production. Science Publishers Inc., New Hampshire

Srivastava PC, Ghosh D, Singh VP (1999) Comparative evaluation of zinc enriched farmyard manure with other common sources for rice. Biol Fertil Soils 30: 168–172

Srivastava PC, Singh AP, Kumar S, Ramachandran V, Shrivastava M, D'souza SF (2008) Desorption and transformation of zinc in a mollisol and its uptake by plants in a rice-wheat rotation fertilized with either zinc-enriched biosludge from molasses or with inorganic zinc. Biol Fertil Soils 44:1035–1041

Srivastava PC, Singh AP, Kumar S, Ramachandran V, Shrivastava M, D'souza SF (2009) Comparative study of a Zn-enriched post-methanation bio-sludge and Zn sulfate as Zn sources for a rice-wheat rotation. Nutr Cycl Agroecosyst 85:195–202

Srivastava PC, Ansari UI, Pachauri SP, Tyagi AK (2013a) Effect of zinc application methods on apparent utilization efficiency of zinc and potassium fertilizers under rice-wheat rotation. J Plant Nutr (in press) (Manuscript no LPLA-2013-0102)

Srivastava PC, Bhatt M, Pachauri SP, Tyagi AK (2013b) Effect of different zinc application methods on apparent utilization efficiency of zinc and phosphorus fertilizers under Basmati rice-wheat rotation. Arch Agron Soil Sci 60:33–48

Suzuki M, Tsukamoto T, Inoue H, Watanabe S, Matsuhashi S, Takahashi M, Nakanishi H, Mori S, Nishizawa NK (2008) Deoxymugineic acid increases Zn translocation in Zn-deficient rice plants. Plant Mol Biol 66:609–617

Tariq M, Hameed S, Malik KA, Hafeez FY (2007) Plant root associated bacteria for zinc mobilization in rice. Pak J Bot 39:245–253

Torun B, Bozbay G, Gültekin I et al (2000) Differences in shoot growth and zinc concentration of 164 bread wheat genotypes in a zinc-deficient calcareous soil. J Plant Nutr 23:1251–1265

Vaid SC, Gangwar BK, Sharma A, Srivastava PC, Singh MV (2013) Effect of zinc solubilizing bioinoculants on zinc nutrition of wheat (*Triticum aestivum* L.). Int J Adv Res 1:805–820

von Wiren N, Marschner H, Romheld V (1996) Roots of iron-efficient maize also absorb phytosiderophore-chelated zinc. Plant Physiol 111:1119–1125

Walter A, Römheld V, Marschner H, Mori S (1994) Is the release of phytosiderophores in zinc-deficient wheat plants a response to impaired iron utilization? Physiol Plant 92:493–500

Wang HX, Guo JY, Xu WH (2011) Response and zinc use efficiency of Chinese cabbage under zinc fertilization [J]. Plant Nutr Fertil Sci 17:154–159

Wellings NP, Wearing AH, Thompson JP (1991) Vesiculararbuscular mycorrhizae (VAM) improve phosphorus and zinc nutrition and growth of pigeonpea in a Vertisol. Aust J Agric Res 42:835–845

Widodo B, Broadley MR, Rose TJ et al (2010) Response to zinc deficiency of two rice lines with contrasting tolerance is determined by root growth maintenance and organic acid exudation rates, and not by zinc-transporter activity. New Phytol 186:400–414

Zuo Y, Zhang F (2008) Effect of peanut mixed cropping with gramineous species on micronutrient concentrations and iron chlorosis of peanut plants grown in a calcareous soil. Plant Soil 306:23–36

Zuo Y, Zhang F (2009) Iron and zinc biofortification strategies in dicot plants by intercropping with gramineous species: a review. Agron Sustain Dev 29:63–71

Nitrification Inhibitors: Classes and Its Use in Nitrification Management

Rajesh Kumar, Balraj S. Parmar, Suresh Walia, and Supradip Saha

Abstract

The explosive expansion of human activity during the last two centuries through industrial and agricultural pursuits has resulted in massive changes in the nitrogen (N) cycle of the planet. Based on the projected population growth and food demand, the N-fertilizer inputs into agricultural systems need to be doubled in the near future which would lead to further increase in the amount of N lost to the environment. If production agriculture continues to move towards high-nitrification agricultural systems with the expansion and intensification of agricultural activities, there is potential for catastrophic consequences to our planet due to the destruction of the ozone layer, global warming, and eutrophication. It is therefore imperative to manage the nitrification in agricultural systems for minimizing N leaks into the environment which are not only a serious economic and energy drain on society but also potentially have long-term ecological and environmental consequences. Currently, more than 60 % of the total N applied to agricultural systems is lost, amounting to an annual economic loss equivalent to US$17 billion worldwide. Wide substrate range of ammonia monooxygenase (AMO), an important enzyme involved in nitrification, has enabled a range of chemicals or chemical formulations that can be effectively deployed as additives to N fertilizers to regulate nitrification. These chemicals by augmenting the efficiency of N-fertilizer use help us to achieve higher food production for catering the ever increasing population and minimize fertilizer-related pollution of the environment. This paper overviews N transformations in agricultural systems and the salient agrochemicals employed for management of nitrification, the most important transformation, in particular.

R. Kumar (✉) • B.S. Parmar • S. Walia • S. Saha
Division of Agricultural Chemicals, Indian Agricultural
Research Institute, New Delhi 110012, India
e-mail: rajesh_agchem@iari.res.in; parmar.
balraj9@gmail.com; suresh_walia@yahoo.com;
s_supradip@yahoo.com

A. Rakshit et al. (eds.), *Nutrient Use Efficiency: from Basics to Advances*,
DOI 10.1007/978-81-322-2169-2_8, © Springer India 2015

103

Keywords

Agricultural chemicals • Nitrification • Nitrification inhibitors • Soils • Agriculture

1   Introduction

Nitrogen (N) is vital for life. It is an essential element for plant growth and reproduction and is one of the most widely distributed elements in nature, with atmosphere as the main reservoir. Of the total naturally available N, 99.96 % is present in the atmosphere. Biosphere contains only 0.005 % out of the remaining 0.04 %. In spite of being present in small proportion in living beings, N is most often the restrictive nutrient for crop production since only a fraction of atmospheric nitrogen is made available to the plants through biological nitrogen fixation. However, the use of chemical nitrogenous fertilizers has resulted in significant increase in crop yields. These fertilizers are one of the key contributors in improving agricultural productivity globally. Ammonium-based fertilizers including urea are the widely used N source for field crops. However, fertilizer N is not an unmixed blessing as the commonly used nitrogenous fertilizers, especially urea, suffer from low nitrogen use efficiency (NUE) and contribute towards environmental pollution and health hazards. Worldwide, the NUE for cereal production (wheat, corn, rice, barley, sorghum, millet, oat, rye etc.) is approximately 33 %. The global annual economic loss due to low NUE is about US$17 billion (Subbarao et al. 2006; Raun and Johnson 1999; Prasad 1998).

The soil accounts for a small fraction of the lithospheric N. It is, however, the main source of plant-available N. Out of the total soil N, only a small portion is available to plants as ammonium and/or nitrate. Although the majority of plants are capable of using both ammonium-N and nitrate-N, the latter is the predominant form utilized by the plants under arable/terrestrial conditions. This is primarily because of compulsion rather than preference due to rapid conversion of most of the ammonium-N to nitrate under favorable conditions. Consequently, ammonium-N is available to the plants only for a limited period of time. Plant roots encounter mainly nitrate-N as source of N in the soil.

Nitrification is a key process in managed agricultural ecosystems because the conversion of ammonium to nitrate can lead to substantial loss of agricultural N by leaching and/or denitrification. The fertilizer N loss is of concern because of economic reasons and associated environmental and health hazards. Some of the hazards of excessive use of nitrogenous fertilizers include: (1) "blue baby" syndrome (methemoglobinemia) in infants and ruminants due to nitrate and nitrite in waters and food; (2) gastric cancer, goiter, birth defects, and heart diseases due to nitrites and nitrosamines; (3) respiratory illness due to nitrate, nitrite, and nitric acid in aerosols; (4) eutrophication due to N in surface waters; (5) accumulation of various oxides of nitrogen in the atmosphere contributing to ozone layer destruction, global warming, and acid rain; (6) plant toxicity due to high levels of nitrite and ammonium in soils; and (7) excessive plant growth due to more available N. Therefore, concerted efforts have been and are being made for improving the use efficiency of N fertilizer and plant N uptake (Prasad 1998; Prasad and Power 1995; Azam and Farooq 2003).

Multidisciplinary approaches followed to increase N use efficiency include: (a) breeding crop varieties with higher fertilizer use efficiency; (b) improved agronomic practices; (c) use of controlled or slow-release fertilizers, urease, and nitrification inhibitors; and (d) supplementation/ integration of fertilizer N with organic manures. These approaches have helped to alleviate the problems arising as a result of fertilizer N use. Fertilizer management through improved formulations, mode and time of application and placement, etc., has also been found helpful to mitigate some of the problems. Likewise, a large variety of chemicals have been tested as potent inhibitors of specific N transformation process including urea hydrolysis and nitrification. These inhibitors improved the fertilizer N use efficiency along with significant reduction in losses due to ammonia volatilization, denitrification, and nitrate leaching. Nitrification inhibition could lead to: (1) increased rhizospheric microbial activities, (2) enhanced mineralization of native soil N, (3) increased fertilizer N use efficiency, and (4) greater photosynthate partitioning to the rhizosphere, thus enriching the soil with organic matter. It is important, therefore, to develop an understanding of the nitrification process, factors affecting nitrification, methods to regulate the process, and its implications to ecosystem functioning (Abalos et al. 2014).

2   The Nitrification Process

Nitrification has been defined as the oxidation of any reduced nitrogen form (organic or inorganic) to nitrate. The microorganisms carry out this oxidation process. Being the only link between reduced and oxidized nitrogen compounds, the nitrification process is of major importance for the nitrogen cycle in aquatic and terrestrial environments. In soils, the nitrification process oxidizes the immobile ammonium to nitrate, a mobile ion.

The biological oxidation of ammonia to nitrate is a two-step process mediated by autotrophic bacteria. It is first oxidized to nitrite and then to nitrate, as follows:

$$NH_3 + O_2 \rightarrow NO_2^- + 3H^+ + 2e^-$$
$$NO_2^- + H_2O \rightarrow NO_3^- + 2H^+ + 2e^-$$

The source of ammonium-N could be soil organic matter (mineralization by soil microorganisms) and/or chemical fertilizers. In the case of soil, organic N is used by the ammonifiers, while chemical fertilizers contain either ammonium as such or its precursors.

The oxidation of reduced organic or inorganic nitrogen to nitrate mediated by heterotrophic organisms has been named heterotrophic nitrification, whereas the oxidation of reduced inorganic nitrogen to nitrate by autotrophic organisms is called autotrophic nitrification.

Several genera and species of ammonium and nitrite-oxidizing heterotrophs including fungi (*Aspergillus flavus*, *Neurospora crassa*, *Penicillium* sp.), actinomycetes (*Streptomycetes* sp., *Nocardia* sp.), and bacteria (*Arthrobacter* sp., *Azotobacter* sp., *Pseudomonas fluorescens*, *Aerobacter aerogenes*, *Bacillus megaterium*, *Proteus* sp.) have been reported (Koops et al. 1991; Purkhold et al. 2000; Regan et al. 2002). However, autotrophic nitrifiers are the main organisms responsible for most of the nitrification. Ammonium-oxidizing autotrophs include *Nitrosomonas*, *Nitrosolobus*, and *Nitrosospira*. These organisms have been isolated from a variety of soil environments with ubiquitous distribution. Nitrite produced by the ammonium-oxidizing autotrophs is rapidly oxidized to nitrate by *Nitrobacter* species (Hovanec and Delong 1996).

All nitrifiers are obligate aerobes and hence a restricted nitrification under waterlogged or aquatic environments can be observed. In addition, these microorganisms, especially *Nitrobacter*, are fairly sensitive to acidic pH. As a result nitrification is inhibited in climax ecosystems like forest soils with thick layer of leaf litter and zones of acidic pH. The process of nitrification itself may lead to lowering of pH of the medium due to release of H^+ as shown in the equation above.

As stated earlier, the autotrophic bacteria mostly carry out the nitrification process. These utilize reduced inorganic nitrogen as energy source and carbon dioxide as carbon source. Hence, this nitrification process may more correctly be called chemolithoautotrophic nitrification. The substrate for the enzyme ammonia monooxygenase (AMO) involved in the first part of the chemolithoautotrophic nitrification process is ammonia (Norton and Stark 2011) rather than ammonium but an acidity-dependent equilibrium always exists between ammonia and ammonium. It is mediated by two distinct groups of bacteria, ammonia-oxidizing bacteria (AOB), and nitrite-oxidizing bacteria. Both groups are dominated by autotrophic metabolism; however, some nitrite oxidizers may also use organic compounds as carbon sources (mixotrophs) (Norton and Stark 2011).

3 Chemolithoautotrophic Ammonia-Oxidizing Bacteria (AOB)

The AOB carry out a specific environmental function of oxidation of ammonia to nitrite. All AOB that have been isolated and characterized are gram-negative, obligate aerobic, and obligate chemolithoautotrophs. Together with the nitrite-oxidizing bacteria, they make up the family *Nitrobacteraceae*. Five different genera for AOB, namely, *Nitrosomonas*, *Nitrosospira*, *Nitrosococcus*, *Nitrosolobus*, and *Nitrosovibrio* were defined on the basis of classical morphological characteristics (Tomiyama et al. 2001). Most AOB described belong to the *ß-Proteobacteria* (ß-AOB), but a few marine isolates of *Nitrosococcus*, not known from soil, belong to the *γ-Proteobacteria* (γ-AOB). Recent research based on sequences of 16S rDNA suggests that the ß-AOB can be divided into only two major phylogenetic lineages, the *Nitrosospira* and the *Nitrosomonas*. New results showed similar, but not identical, evolutionary relationships of ß-AOB when using the 16S rRNA gene or a functional gene (*amo-A*) as marker genes for phylogenetic analysis (Purkhold et al. 2000). Hence, except for a few marine *Nitrosococcus* strains, all known AOB are of monophyletic origin. The high correlation between function and phylogeny is rather unique for AOB when compared to other functional groups of microorganisms in soil, e.g., denitrifiers (Bothe et al. 2000).

Two key enzymes mediate the ammonia oxidation in AOB – ammonia monooxygenase (AMO) and hydroxylamine oxidoreductase (HAO) – and both are codependent because they generate substrate and electrons, respectively, for each other (Bothe et al. 2000).

Oxidation of Ammonia to Hydroxylamine (AMO)

$$NH_3 + O_2 + 2H^+ + 2e^- \rightarrow NH_2OH + H_2O$$

Oxidation of Hydroxylamine to Nitrite (HAO)

$$NH_2OH + H_2O \rightarrow NO_2^- + 5H^+ + 4e^-$$

The AMO enzyme consists of three subunits with different sizes (Bothe et al. 2000) – AMO-A, AMO-B, and AMO-C – and mainly the gene (*amo-A*) encoding the A subunit which carry the active site of AMO has been investigated (Purkhold et al. 2000). The AMO enzyme may catalyze co-oxidation of a broad range of substrates (McCarty 1999). Hence, this enzyme has been in focus when exploiting the role of AOB in bioremediation (Duddleston et al. 2000). Many similarities between AMO of AOB and particulate methane monooxygenase (pMMO) found in methane-oxidizing bacteria have been reported (Bedard and Knowles 1989) and the similarity of nucleotide sequences encoding the enzymes indicate a common evolutionary origin (Holmes et al. 1995).

4 Nitrification Inhibition or Regulation

Nitrification being one of the key N cycle processes under most arable situations on land, a need to inhibit nitrification in order to maintain the economy of agroecosystems has been always felt as discussed below.

Groundwater Pollution Uncontrolled and excessive nitrification may lead to groundwater contamination with nitrate and nitrite as well as increased concentration of the later in eatables, especially vegetables leading to human health hazards. Nitrate itself is not a threat, while nitrite is definitely a potential health hazard and that too when found in places at a wrong time.

Consumption of water and vegetables containing excessive amounts of nitrate may lead to the production of nitrite in the stomach and the later becomes particularly dangerous for the babies. Methemoglobinemia (blue baby syndrome) may occur in 1-year old babies taking diet with too much nitrate. Methemoglobinemia is the condition in the blood which causes infant cyanosis or blue baby syndrome. Methemoglobin is probably formed in the intestinal tract of an infant when bacteria convert the nitrate ion to nitrite ion. One nitrite molecule then reacts with two molecules of hemoglobin to form methemoglobin. In acid mediums, such as the stomach, the reaction occurs quite rapidly. This altered form of blood protein prevents the blood cells from absorbing oxygen which leads to slow suffocation of the infant which may lead to death. Because of the oxygen deprivation, the infant will often take on a blue or purple tinge in the lips and extremities, hence the name, blue baby syndrome. Other signs of infant methemoglobinemia are gastrointestinal disturbances, such as vomiting and diarrhea; relative absence of distress when severely cyanotic but irritable when mildly cyanotic; and chocolate-brown-colored blood. Stomach and gastrointestinal cancer has also been associated with the concentration of nitrate in potable water. Again, it is nitrite that reacts with amines to form N-nitroso compounds, which are reported to cause stomach cancer. Such an illness may result from consumption of vegetables containing high concentrations of nitrate originating from soil or irrigation water.

In water bodies, however, nitrate and other forms of N may encourage the growth of algae and subsequently the bacteria leading to exhaustion of molecular oxygen, thereby affecting animal life. Indeed, whole ecological balance of water bodies may change due to the so-called eutrophication.

Nitrous Oxide Production Nitrous oxide is produced naturally in soils through the microbial processes of nitrification and denitrification. Since 1750, the global atmospheric concentrations of nitrous oxide have risen by approximately 18 % and are continuing to do so at 0.25 % per annum. This increase is attributed mainly to biospheric processes. Flood irrigation leads to rapid nitrification and denitrification resulting in considerable amounts of atmospheric nitrous oxide emission, which may amount to 35–45 % of the applied N. On the global level, >65 % of the atmospheric nitrous oxide comes from the soil, which is twice the amount produced by burning fossil fuels and four times the amount evolved from the oceans. Being a greenhouse gas, nitrous oxide contributes substantially to the destruction of stratospheric ozone (Azam and Farooq 2003). N_2O is approximately 300 times more powerful than CO_2 at trapping heat in the atmosphere.

Nitrification and denitrification are the main contributors (Azam et al. 2002) to atmospheric nitrous oxide. However, since the two processes occur simultaneously, it is difficult to ascertain the real contribution of either to the observed nitrous oxide fluxes. Nevertheless, nitrification is reported to make a substantial contribution to the nitrous oxide emission under aerobic conditions. Higher nitrous oxide emissions are often reported from fertilized than unfertilized soils, rates of emission being greatest following application of ammonium or ammonium-forming fertilizers (Azam et al. 2002). In several studies, using isotope methodology and nitrification inhibitors, this increase is attributed to losses of N_2O occurring during the process of nitrification (Abbasi and Adams 2000). Estimates of the amount of N_2O resulting from nitrification are variable but generally account for <1 % of the fertilizer N applied. In the case of anhydrous NH_3, however, the losses may increase to 6–7 %. In most studies, the onset of N_2O emission is observed very early during the incubation, while nitrification continues for extended periods of time. Williams et al. (1998) reported active nitrification 7–12 days after application of ammonium nitrate, while a flush of N_2O emission from soil was observed around day 1, followed by a decline. These researches (Abbasi and Adams 2000; Azam et al. 2002; Williams et al. 1998) showed very low molar ratios of NO to N_2O and suggested that denitrification was the dominant process involved in N_2O emission.

Contribution of nitrification to nitrous oxide emissions may be high under the semiarid agroclimatic conditions and with the use of urea as major N fertilizer. Urea is rapidly hydrolyzed followed by a quick nitrification of the resultant ammonium especially under relatively warmer conditions. Thus, nitrification not only contributes to nitrous oxide emissions, but the process of denitrification is fairly well supported by sustained availability of nitrate. In most soils, formation and emissions of nitrous oxide to the atmosphere are enhanced by an increase in available mineral N through increased rates of nitrification and denitrification. Therefore, addition of N in organic or inorganic compounds eventually leads to enhanced N_2O emissions.

Nitrification Inhibition and Ecosystem Functioning Nitrification inhibition and consequent accumulation of ammonium would lead to: (1) increased microbial activities including biological nitrogen fixation, (2) greater photosynthate partitioning to the rhizosphere, (3) enhanced mineralization of native soil N, and (4) increased efficiency of fertilizer N use by plants.

Ammonium is preferred over nitrate as a source of N by microorganisms. As not all organisms possess nitrate reductase to enable them to assimilate nitrate, while almost all of them will be able to assimilate ammonium, so this preference is consequential rather than the reason. In addition, assimilation of nitrate is more energy intensive than ammonium. Hence, sufficient easily oxidizable C will be required for efficient assimilation of nitrate. Studies involving the use of glucose as a C source indeed reveal similar assimilation of both ammonium and nitrate by the soil microorganisms. Nevertheless, the presence of ammonium leads to an enhancement in microbial activities in terms of respiratory response. In experiments aimed at studying

the mineralization of native soil N, ammonium-N is reported to have a significantly higher effect as compared to nitrate-N. This so-called "priming" effect or added nitrogen interaction has been found to increase with the amount of applied N. An indirect effect of chemical fertilizers as well as green manures is their positive influence on the mineralization and plant availability of N from the soil organic reserves.

Most of the plants utilize both the ammonium- and nitrate-N with varied preference for one form over another. However, ammonium as an exclusive source of N may cause growth inhibition in many species, particularly in those grown under arable conditions (Marschner 1999). Under these conditions, nitrification is generally quite rapid and hence deleterious effects of ammonium are avoided. Under saline conditions also, ammonium increases the sensitivity of plants whereas nitrate has been reported to moderate the negative effects of salinity (Khan et al. 1994). However, the plants are bound to face higher concentrations of ammonium under saline conditions because of the inhibitory effects of salts on the process of nitrification. Therefore, nitrification inhibition would be a blessing for arable plants grown on normal agricultural soils, whereas it may be an added problem for those grown on salt-affected lands. Several studies indeed show a positive effect of nitrification inhibitors on plant growth and N use efficiency[4] by decreasing the loss of N through denitrification and nitrate leaching and conservation of the applied N through enhanced immobilization.

The form of N plays a significant role in affecting root growth, rhizodeposition, and the concomitant changes in different rhizospheric microbial functions including root-induced N mineralization. In wheat and maize, root growth may be restricted in ammonium compared to nitrate-fed plants and may be attributed to an increased root respiration, greater allocation of photosynthates to nitrogenous than structural component, and increased export of carbon (probably as amino acids) from root to shoot than that occurring under nitrate nutrition

(Azam and Farooq 2003). In addition, ammonium nutrition leads to a higher rhizodeposition, thereby enlarging the below-ground sink for photosynthates, most probably at the expense of plant tops thereby reducing the biomass yield. However, increase in rhizodeposition due to increased/sustained availability of ammonium may also prove beneficial to plants in terms of increased microbial activities, especially the mineralization of native soil N. In laboratory experiments, a significant increase in the mineralization of soil N has been observed[5] following addition of easily oxidizable C.

It has also been suggested above that mineralization of N from soil organic matter is more intense in the presence of ammonium than nitrate. Jenkinson et al. 1985 attributed this to "pool substitution" whereby the native N stands proxy for the applied N giving the impression of enhanced mineralization of the latter. The fact remains, however, that applied N (especially ammonium) leads to an increase in the availability of soil N. Inhibition of nitrification may therefore lead to a higher mineralization of native soil N thereby augmenting N supplies to plants. In addition, microorganisms responsible for the synthesis of aggregation-adhesion macromolecules may be encouraged by higher availability of carbonaceous materials in the rhizosphere. This will result in better soil structure as well as improved moisture-holding capacity of the soil at the root surface. The latter may help the plants withstand drought stress at least temporarily. Thus, in spite of the negative effects of ammonium, inhibition of nitrification may still exert beneficial effects on plant growth. The negative effects can be overcome to a significant extent by developing plant types more efficient in using ammonium, the so-called ammoniphilic plants. Plants like rice and sugarcane growing under high soil moisture conditions can be considered as ammoniphilic plants. Efforts are needed to engineer arable crops (like wheat) for improved tolerance to ammonium while employing nitrification inhibitors.

Another aspect worth consideration is the susceptibility of nitrate to leaching beyond the effective root zone after being converted to calcium nitrate $\{Ca(NO_3)_2\}$ in the presence of ionic calcium (Ca). Hence, in calcareous soils the conditions are quite conducive to this mode of nitrate escape especially following organic amendment that helps in the release of Ca. The leaching is more pronounced in clayey soils at near neutral pH as negative charge on the clays repels nitrate, thereby facilitating the process of leaching. Hence, not only the use efficiency of nitrate will remain low under these conditions but also N economy of the system will be negatively affected.

5   Management Practices to Reduce or Regulate Nitrification

Various approaches have been suggested to improve the use efficiency of nitrogenous fertilizers. These include the improved agronomic practices, use of coatings, chemical additives, and the various chemical and physical modifications.

Improved Agronomic Practices Split application, placement, foliar application, fertigation, etc. are some of the agronomic techniques by which NUE of nitrogenous fertilizers can be increased (Raun and Johnson 1999).

Controlled or Slow-Release Fertilizers By using specific fertilizer formulations to release N in synchrony with plant requirement, it should be possible to provide sufficient N in a single application to satisfy the plant's need, yet maintain low concentrations of mineral N in the soil throughout the growing season. If this could be done, losses would be small because of the limited amount of N in the substrate.

Several slow-release forms of N (Shaviv and Mikkelsen 1993) have been suggested. These include:

- *Coated fertilizers*: Soluble urea is coated with an insoluble, slowly permeable but generally biodegradable material to achieve controlled/delayed release of urea-N. Several organic and inorganic coating materials such as sulfur, gypsum, lac, latexes, polyolefins, resins, plastics, polyurethanes, rock phosphates, etc. have been attempted using three types of coating processes, namely, rolling bed, falling curtain, and fluidized bed.
- *Complex organic N compounds with relatively less solubility in water than urea*: This group of compounds consisting of urea-formaldehyde complexes (38 % N), oxamide (30 % N), isobutylidenediurea (IBDU, 30 % N), urea-Z (35 % N), etc., are only slightly soluble in water. The rate of nitrogen release from these compounds depends upon water solubility, microbiological action, and chemical hydrolysis.
- *Urea supergranules (USG)*: The USG consisting of 1–2 discrete urea particles is not so efficient, but its proper deep placement (1 USG for 4 hills at 7–10 cm soil depth with the hole at the placement site closed) makes it efficient.

Many of these fertilizer formulations have been utilized to grow plants in diverse environments. The influence of slow-release forms on levels of soil mineral N and the recovery of fertilizer N have been assessed for upland crops and lowland rice. The use of these formulations has generally decreased the total loss of fertilizer N.

5.1   Use of Inhibitors

Fertilizer use efficiency could be greatly increased if the hydrolysis of urea to ammonium by soil urease could be retarded by the use of urease inhibitors or if nitrate accumulation during the cropping phase could be regulated by nitrification inhibitors.

Urease Inhibitors These reduce the hydrolysis of urea by inhibiting soil urease activity and thus prevent rapid development of high partial pressure of NH_3 and high pH of floodwater in rice fields and eventually reduce NH_3 volatilization losses. A large number of compounds have been tested for their ability to inhibit soil urease but most are ineffective or do not persist in soil. The phosphoroamides, such as phenylphosphorodiamidate (PPD) and N-(*n*-butyl) thiophosphorictriamide (NBPT), have shown promise for limiting the hydrolysis of urea in laboratory and greenhouse studies when used singly or in combination. Relatively few studies have been done on their ability to reduce NH_3 volatilization and increase grain yield in the field.

Studies using PPD and NBPT as urease inhibitors in flooded rice fields have shown little reduction in NH_3 loss. The reasons for the failure of PPD in flooded soils seem to be its rapid hydrolysis under the alkaline conditions generated in the floodwater by photosynthetic algae and its decomposition due to the high temperatures reached in the floodwater. The reasons for the failure of NBPT in flooded soils have not been completely explained, but the results of laboratory studies with non-flooded soils suggest that it must be converted to the oxygen analogue to inhibit urease activity. Studies with another thiophosphorictriamide, thiophosphoryl triamide, showed it too to be a relatively weak inhibitor of urease activity. Appreciable inhibition was achieved only after it had been converted to the oxon analogue. These studies indicate that the thiophosphorictriamides do not inhibit urease activity, but that the phosphorictriamides are its potent inhibitors.

Field studies in Thailand show that the activity of PPD can be prolonged, and NH_3 loss markedly reduced, by controlling the floodwater pH with the algicide terbutryn. In addition, a mixture of NBPT and PPD in the presence of terbutryn was even more effective than PPD alone. It appears that during the time when the

PPD was effective, NBPT was being converted to its oxygen analogue. This inhibited urease activity when PPD lost its capacity to do so. The combined urease inhibitor-algicide treatment reduced ammonia loss from 10 to 0.4 kg N ha^{-1} (Freney et al. 2011).

In a laboratory study, cyclohexylphosphoric-triamide (CHPT) was found a very effective inhibitor of urease activity and the same was confirmed in a field experiment with flooded rice in Thailand (Freney et al. 2011). The oxon analogue of NBPT, N-(n-butyl) phosphoric-triamide, was compared with CHPT. The two markedly reduced urea hydrolysis, the CHPT being more effective. Its addition maintained the ammoniacal N concentration of the floodwater below 2 g m^{-3} for 11 days, reduced NH$_3$ loss by 90 %, and increased grain yield. Application of NBPT with urea resulted in increase in cotton yield by 14 % and it was also recommended that NBPT cannot be used in combination with DCD (Kawakami et al. 2012). Whereas, in maize crop, NBPT did not significantly increase the grain yield and it was also concluded that effectiveness of NBPT + DCD combination is influenced by management practices (Sanz-Cobena et al. 2012).

Nitrification Inhibitors The nitrification inhibitors (NIs) decrease the availability of nitrate and consequently its vulnerability to escape mechanisms. A lot of work has been reported on the ways to retard/inhibit the rate of nitrification not only to reduce fertilizer N losses (Prasad and Power 1995) but also to prolong the persistence of fertilizer N in ammoniacal form (Prasad and Power 1995; McCarty 1999). Since ammonia or ammonium-producing compounds are the main source of fertilizer N, maintenance of the applied N in the ammonium form should mean that less N is lost by denitrification. One mechanism of maintaining added N as ammonium is to use a nitrification inhibitor with the fertilizer.

Numerous substances have been tested for their ability to inhibit nitrification (Table 1), and several of these have been patented. Only a limited

Table 1 Some commercial and extensively tested synthetic nitrification inhibitors

Nitrapyrin [2-chloro-6-(trichloromethyl)-pyridine]
2-Amino-4-chloro-6-methylpyrimidine (AM)
Dicyandiamide (DCD)
Etridiazole (5-ethoxy-3-trichloromethyl-1,2,4–thiadiazole)
N-(2,5-dichlorophenyl succinamic acid) (DCS)
Potassium azide (KN$_3$)
3-Amino-1,2,4 triazole (ATC)
Thiourea (TU)
MBT (2-mercaptobenzothiazole)
2-Ethynyl pyridine
MPC (3-methylpyrazole-1-carboxamide)
ST (2-sulfanilamidothiazole)
Carbon disulfide
3-Mercapto-1,2,4-triazole
Sodium diethyldithiocarbamate
Acetylene
Gaseous hydrocarbon, e.g., methane, ethane, and ethylene
Ammonium thiosulfate
Thiophosphoryl triamide
4-Mesyl benzotrichloride
4-Nitrobenzotrichloride
Guanyl thiourea
2,4-Diamino-6-trichloromethyl triazine
Potassium trithiocarbonate
Sodium thiocarbonate
2-Amino-4-methyl-6-trichloromethyltriazine (MAST)
2-Benzothiazole sulfone morpholine
3,4-Dimethylpyrazole phosphate

number of chemicals are available commercially for use in agriculture. These include 2-chloro-6-(trichloromethyl) pyridine (nitrapyrin), sulfathiazole, dicyandiamide, 2-amino-4-chloro-6-methyl pyrimidine, 2-mercaptobenzothiazole, thiourea, and 5-ethoxy-3-trichloromethyl-1,2,4-thiadiazole (terrazole). Unfortunately, most of these compounds have limited usefulness. For example, the most commonly used nitrification inhibitor, nitrapyrin, is seldom effective because of sorption on soil colloids, hydrolysis to 6-chloropicolinic acid, and loss by volatilization.

Nitrapyrin Nitrapyrin [2-chloro-6-(trichloromethyl)-pyridine, **I**] was developed by

Dow Chemical Company. It is marketed under the trade name "N-Serve 24 nitrogen stabilizer" (a.i. 240 g L^{-1}) and "N-Serve 24E nitrogen stabilizer" (a.i. 240 g L^{-1}). The rates of application advised by Dow Chemical Company for band and row placement are 1.125–1.25 L ha^{-1} of N-Serve 24E for cotton, maize, sugar beet, sorghum, and wheat and 4.50–6.75 L ha^{-1} for potatoes before or after planting or sowing. For broadcasting, the rate of application has to be increased considerably. When granulated fertilizer is used, it can be applied at 0.2–1.0 % of the amount of fertilizer N (Kawakami et al. 2012). Because of its high vapor pressure, nitrapyrin cannot be granulated with solid-N fertilizer like urea without loss of the inhibitor during processing, storing, and handling. Nitrapyrin sometimes shows poor activity due to sorption on soil colloids, hydrolysis to 6-picolinic acid, and loss by volatilization.

AM 2-Amino-4-chloro-6-methylpyrimidine (**II**) is another well-known nitrification inhibitor developed by Toyo Kaotsu Industries Inc. (now Mitsui Toatsu) of Japan. Pure AM is a white crystalline substance (mp, 182 °C) and is soluble in water but unlike nitrapyrin; it is relatively insoluble in organic solvents. AM is less volatile and less effective than nitrapyrin. AM is effective (Prasad and Power 1995) when applied at 5–6 kg ha^{-1}.

Etridiazole 5-Ethoxy-3-trichloromethyl-1, 2, 4-thiadiazole (Terrazole, Etridiazole, Dwell, **III**) is an effective nitrification inhibitor developed by Olin Corporation, Baltimore, USA. This product is available as a wettable powder or technical grade liquid with 35 % and 95 % a.i., respectively. As a coating on ammonium sulfate and urea, terrazole 95 % a.i. is used up to 1.5 % by weight. The recommended rates of compound for crops like potatoes, sugar beet, lettuce, and onions are 0.6–1 kg ha^{-1} (Slangen and Kerkhoff 1984). Besides this compound, some other thiadiazoles are also known to inhibit ammonia oxidation. Among them 3, 4-dichloro-1, 2, -5-thiadiazole is noteworthy.

Dicyandiamide (DCD, IV) It has been developed both as a slow-release nitrogen source as well as nitrification inhibitor (Slangen and Kerkhoff 1984). In Japan, it is added to mixed fertilizer and a product urea-form plus containing 10 % by weight of DCD is produced. A fertilizer containing urea and DCD in a 4: 1 ratio is commercially available in West Germany. DCD is toxic to plants, but the effect differs with plant species. This compound is effective over a period of 1–3 months. When applied at 10–15 % of applied nitrogen, it remains active for a period of 2 months. Cost of production of DCD is lower than the corresponding cost of nitrapyrin and etridiazole. It has the advantage of completely decomposing in soil to CO_2 and NH_4^+ over several weeks and thereby acts as a high analysis (66.7 % N) slow-release N fertilizer. Compounds containing ammonium sulfate and dicyandiamide are available in granulated and coated form from Suddeutsche Kalksticksoff-Werke AG, Trostberg, Germany, and Sisco Corporation, Japan, and recommended as slow-release fertilizers. Increase in NUE of urea by DCD in field studies on different crops has been reported by various workers (Ma et al. 2013). In a recent study, DCD was found highly effective in reducing N_2O emissions by 58–63 % in a dairy pasture (Ball et al. 2012).

CMP [1-Carbamoyl-3-methylpyrazole, **V**] has been found to be an effective nitrification inhibitor (McCarty 1999). Under flooded conditions, CMP affected almost total inhibition of nitrification and prevented buildup of nitrite and nitrate in floodwater. In pure culture studies, CMP inhibited the growth of *Nitrosomonas* at concentrations as low as 1 ppm.

ST A group of thiazoles such as sulfathiazoles and especially 2-sulfanilamidothiazole (ST, **VI**) were introduced as nitrification inhibitors by Mitsui Toatsu Chemicals Inc., Japan. ST is more stable than AM and can be formulated with both acidic and basic fertilizers. It is apparently more volatile than AM. ST is commercially used in Japan on a limited scale.

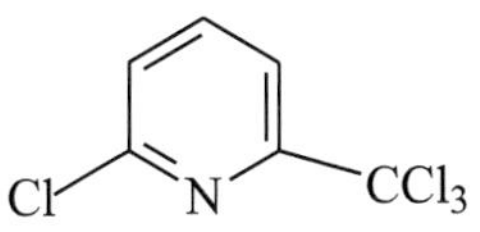

(I)

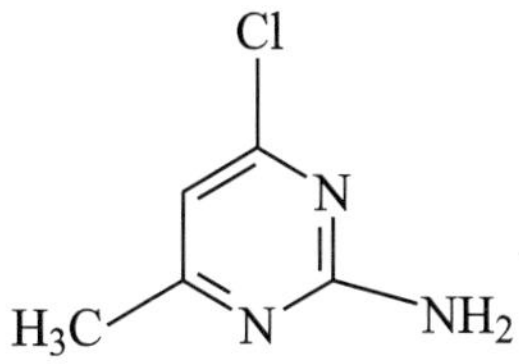

(II)

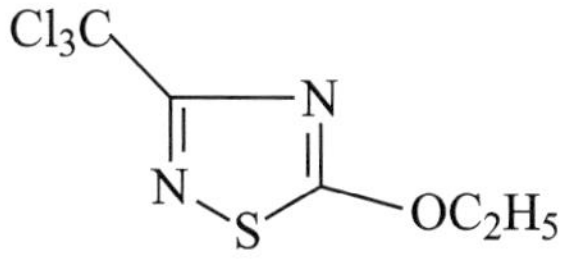

(III)

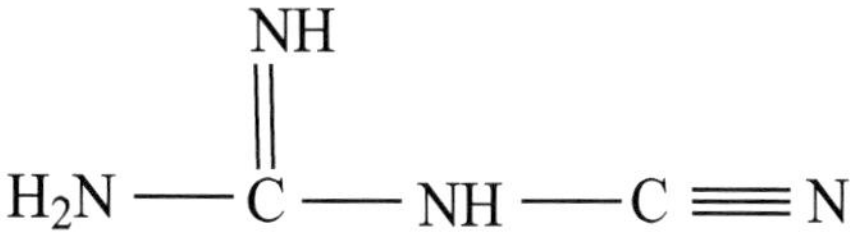

(IV)

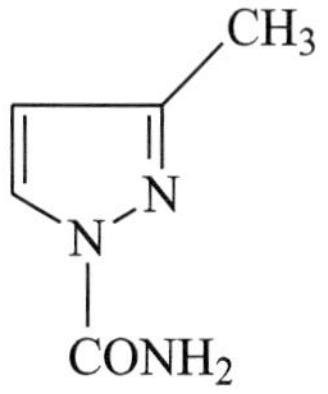

(V)

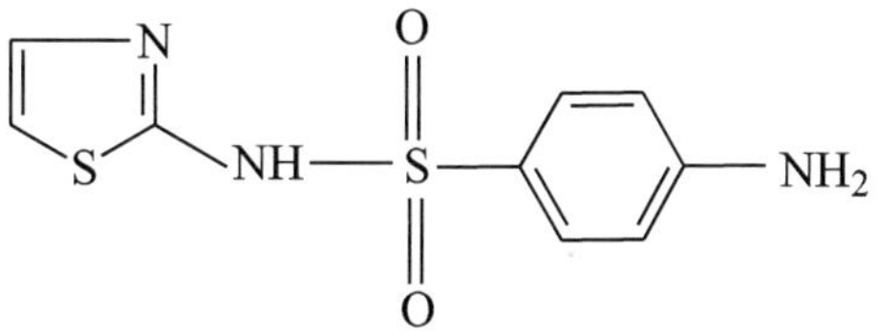

(VI)

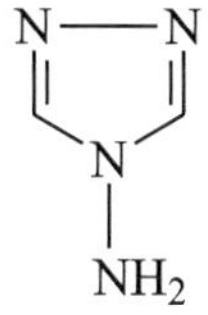

(VII)

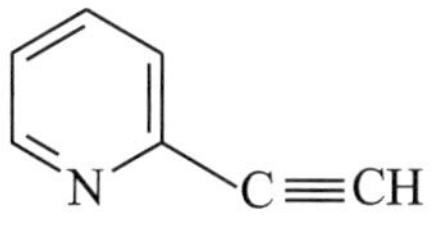

(VIII)

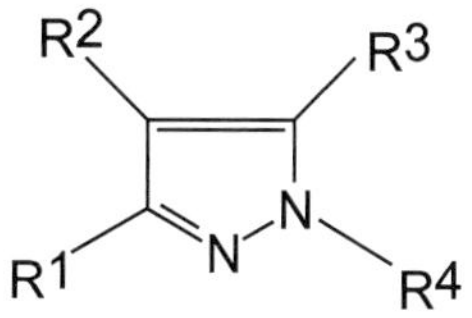

(IX)

Some commercial nitrification inhibitors

ATC Many triazoles particularly 1, 2, 4-triazoles are reported to have nitrification inhibition activity. Among these, 4-amino-1, 2, 4- triazole (ATC, **VII**) is the most potent inhibitor but not as effective as nitrapyrin. ATC was produced from formic acid and hydrazine by Ishihara Industries, Japan. It completely checks nitrification of urea for four weeks at a concentration of 5 % by weight of urea.

Pyridines Among the various pyridines tested, 2-ethynylpyridine (**VIII**) and nitrapyrin (**I**) are the most potent inhibitors of nitrification, but other compounds also possess this activity. Among those, 2-chloropyridine, 2, 6-dichloropyridine, and 6-chloro-2-picoline significantly inhibited ammonia oxidation in soil, whereas compounds containing carboxylic group (6-chloropicolinic acid) had little effect (Ball et al. 2012). In a wheat crop, chlorinated pyridine performed better than DCD in yield enhancement and reduction in N_2O emission (McCarty and Bremner 1989).

3, 4-Dimethylpyrazole Phosphate (DMPP, IX)
It is a new nitrification inhibitor with highly favorable toxicological and ecotoxicological properties and shows several distinct advantages compared to the currently used nitrification inhibitors. Application rates of 0.5–1.5 kg ha^{-1} are sufficient to achieve optimal nitrification inhibition. It can significantly reduce nitrate leaching, without being liable to leaching itself (Zerulla et al. 2001). Significant reduction in N_2O production was observed when DMPP was used as NI (Menendez et al. 2012).

5.2   Indigenous Nitrification Inhibitors

Furan Derivatives Sahrawat and Mukerjee (1977) after observing the effect of furan ring on nitrification inhibition screened some furano compounds like furfural (**X**) and furfural alcohol (**XI**) for possible effect on nitrification. Twenty to 30 % concentrations of these compounds matched 5 % karanjin (the major furanoflavonoid constituent of *Pongamia glabra*) in nitrification inhibition during 45–60 days. The inhibition of nitrification decreased after 45 days, while karanjin remained effective even after this period. Furfural alcohol was a better inhibitor of nitrification of ammonium sulfate than that of urea.

Kuzvinzwa et al. (1984) tested derivatives of furfural along with a natural furanocoumarin – psoralen –for nitrification inhibition in laboratory incubation studies. 5-Nitrofurfural oxime (**XII**), furfural oxime (**XIII**), and furfural semicarbazone (**XIV**) were the most effective followed by 5-nitrofurfural semicarbazone (**XV**), 5′-nitro-3-chloro-2-furanilide (**XVI**), and psoralen (**XVII**). Only 5-nitrofurfural oxime approached nitrapyrin in effectiveness. The nitro derivatives tended to become general bactericides and became effective against *Nitrobacter* species also, thereby causing accumulation of nitrite nitrogen. 3-Chlorofurananilide and furfural oxime caused very little accumulation of nitrite even at the highest concentration (15 %, N-basis).

Datta et al. (2001) examined three series of furfural derivatives, namely, N–O–furfural oxime ethers (**XVIII**), furfural Schiff bases (**XIX**), and furfural chalcones (**XX**), as possible nitrification inhibitors in laboratory incubation study. Furfural oxime ethers and Schiff bases showed potential activity, but furfural chalcones were only mildly active. N-O-Ethyl furfural oxime among the oxime ethers and furfurylidine-4-chloroaniline among Schiff bases performed the best. These two compounds showed more than 50 % nitrification on the 45th day at 5 % dose as compared to 73 % by nitrapyrin.

Activity of the ethers decreased with increase in N-O-alkyl chain length and introduction of chlorine in phenyl ring of furfurylidene anilines increased the activity of Schiff base. Schiff bases derived from 2, 4/2, 6-dichlorobenzaldehyde and 2/3/4-fluoroaniline were also reported as potent nitrification inhibitors (Aggarwal et al. 2009).

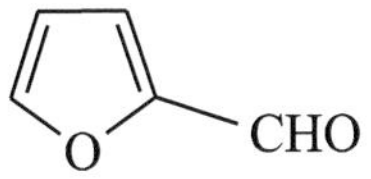

X

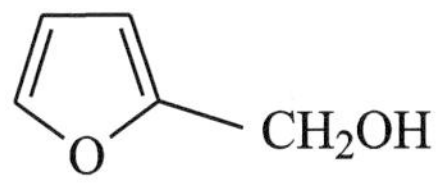

XI

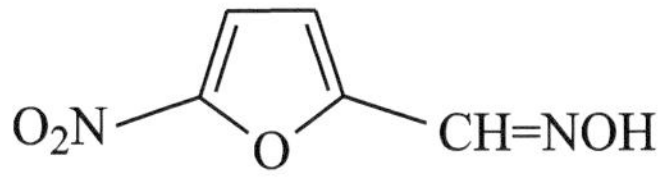

XII

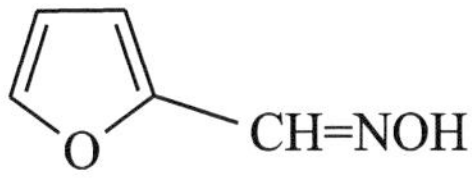

XIII

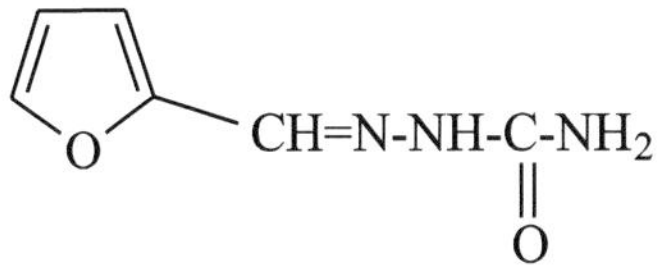

XIV

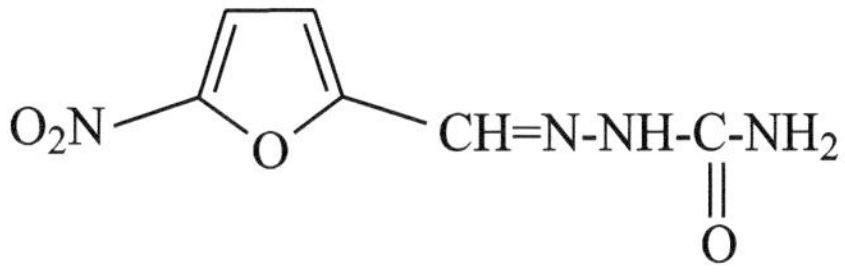

XV

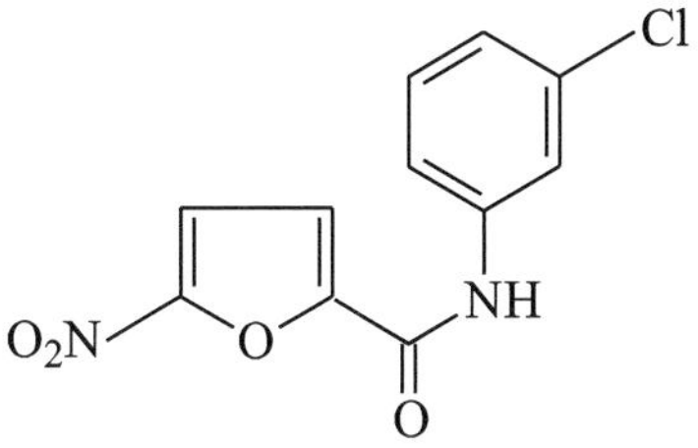

XVI

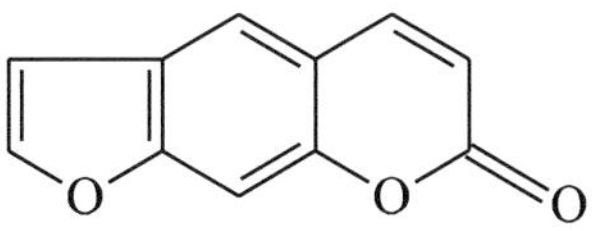

XVII

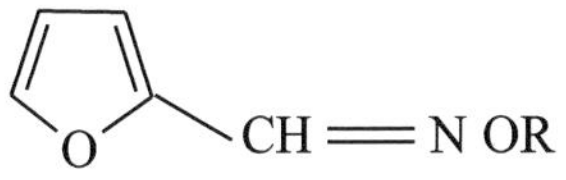

XVIII

R= C_2H_5/$CH(CH_3)_3$/C_4H_9/
C_6H_{11}/CH_2-$CH=CH_2$

XIX

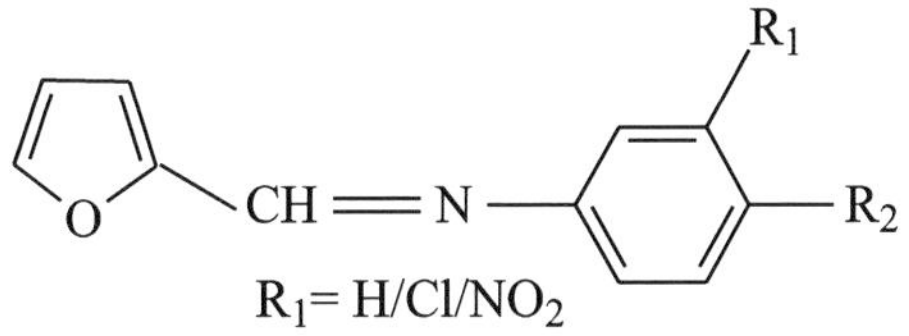

R_1= H/Cl/NO_2
R_2= H/CL

XX

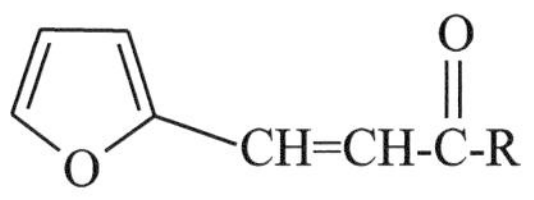

R= CH_3/C_6H_5/4-OH-C_6H_5

Furfural derivatives (X-XX) with nitrification inhibitory property

Acetylenic Compounds Acetylene was first found to inhibit ammonia oxidation in pure cultures of *Nitrosomonas europaea* and then established as a potent inhibitor of nitrification (McCarty 1999). However, because it is a gas, there are problems in introducing it into the soil in field and sustaining its availability during the growing period at a concentration required to limit nitrification. The problem has been overcome by the use of calcium carbide coated with layers of wax and shellac to provide a slow-release source of acetylene. Addition of wax-coated calcium carbide to the fertilized soil has reduced nitrification and increased yield, or recovery of N, in irrigated wheat, maize, cotton, and flooded rice (Banerjee et al. 1990).

Another way of overcoming the problem of applying gaseous acetylene is to use substituted acetylenes such as 2-ethynylpyridine or phenyl acetylene, which are liquids at ambient temperatures. These two compounds have proved as effective inhibitors in laboratory studies. The use of 2-ethynylpyridine in irrigated cotton has resulted in greatly increased recovery of applied N (Freney et al. 2011).

Sulfur Compounds A broad range of S-containing compounds including thiosulfates, thiocarbamates, xanthates, S-containing amino acids, and several pesticides including fungicides inhibit nitrification. Specific compounds include: S-benzyl isothiouronium salts (Kumar et al. 2004; Bhatia et al. 2010), carbon disulfide (CS_2), thiourea, allyl thiourea, guanyl thiourea, 2-mercaptobenzothiazole, 3-mercapto-1, 2, 4-triazole, thioacetamide, sodium diethyl-dithiocarbamate, sodium thiocarbonate, thiosemicarbazide, thiocarbohydrazide, diphenylthiocarbazone, dithiocarbamate, s-ethyl dipropyl thiocarbamate, ethylene-*bis*-dithiocarbamate, and N-methyl dithiocarbamate.

Heterocyclic Compounds Several strong inhibitors of ammonia oxidation in soil can be classified by their heterocyclic ring structures. This class of compounds includes some of the more potent inhibitors of nitrification in soil, namely, nitrapyrin, etridiazole, 2-ethynyl pyridine, 4-amino-1, 2, 4-triazole, 3-methylpyrazole-1-carboxamide and recently reported furan derivatives (Datta et al. 2001), 1, 3, 4-oxa/thiadiazoles (Kumar et al. 2005; Saha et al. 2010), 3, 4-dimethyl pyrazole phosphate (Zerulla et al. 2001), and naphthyridine derivatives (Aggarwal et al. 2010). The heterocycles involved in general are: furan, pyrazole, pyridine, pyridazine, benzotriazole, 1, 2, 4-triazole, thiadiazoles, 1, 3, 5-triazines, and s-tetrazines. Several of the heterocyclic N compounds found to inhibit ammonia oxidation are structurally similar in that they contain chloro (Cl) and/or trichloromethyl (CCl_3) groups substituted on carbon atom(s) adjacent to a ring N [e.g., nitrapyrin {2-chloro-5-(trichloromethyl) pyridine}, etridiazole (5-ethoxy-3-trichloromethyl-1, 2, 4-thiadiazole), 2-chloropyridine, 2,6-dichloropyridine, 6-chloro-2-picoline, and 3,4-dichloro-1,3,4-thiadiazole].

Inhibitors of Natural Origin Synthetic nitrification inhibitors, though expensive, can efficiently inhibit nitrification. Certain allelochemicals released by plants are also reported to have an inhibitory effect. Rice postulated that because inhibition of nitrification results in conservation of both energy and nitrogen, vegetation in late succession or climax ecosystems contains plants that release allelochemicals that inhibit nitrification in soil (Rice 1984). Some natural products from neem (*Azadirachta indica*, A. Juss), karanja (*Pongamia glabra*, Vent.), mint (*Mentha spicata*, *Mentha arvensis* L.), and mahua (*Madhuca longifolia*, L.) are reported to inhibit the activity of nitrifiers (Sahrawat and Parmar 1975; Prasad et al. 1993; Prasad et al. 2002; Saxena et al. 1999; Kumar et al. 2007, 2008, 2010, 2011; Sahrawat 1982; Majumdar 2008; Patra and Chand 2009; Opoku et al. 2014). Among them, neem-based products like Nimin, Neemex, and Neem Gold-A are commercially available in the Indian market.

Nonedible oilseeds like neem *(Azadirachta indica* A. Juss.), karanja *(Pongamia glabra* Vent.), and mahua *(Madhuca indica, M. latifolia)* have been extensively studied for nitrification inhibition properties. Oil cakes in general and nonedible cakes in particular have been known to possess certain minor nonfatty biologically active constituents which make them unsuitable for human consumption. These oilseeds have been traditionally used as slow nutrient release manures or in admixture with manures to regulate the nutrient release.

Neem The utility of neem cake in improving the nitrogen use efficiency of prilled urea in different crops has been exhaustively demonstrated (Sahrawat and Parmar 1975; Prasad et al. 1993). Due to the poor shelf life of neem cake, its industrial production did not merit attention (Prasad et al. 2002). Neem oil-coated urea (NOCU) on the other hand was found to be more suited (Saxena et al. 1999; Kumar et al. 2007, 2008, 2010, 2011). Its efficacy has been demonstrated at two-fertilizer plants in India, *viz.,* KRIBHCO, Hazira, and Shriram Fertilizers and Chemicals, Kota. M/s National Fertilizer Ltd., Panipat, have claimed independently produced neem oil-coated urea. The production of NOCU has been grown to over 2.0 million tonnes per year and neem-coated urea is manufactured by: (1) National Fertilizers Ltd., (2) Shriram Fertilisers and Chemicals Limited, (3) Indo Gulf Fertilisers, (4) Tata Chemicals Ltd, (5) Chambal Fertilisers and Chemicals Ltd, and (6) Mangalore Chemicals and Fertilizers Ltd as per Ministry of Agriculture Notification No: S.O. No. 2073 (E) dated August 10, 2009 (Agricoop 2009).

Karanja A number of physiologically active furanoflavonoids (Sahrawat 1982; Sahrawat et al. 1974; Majumdar 2008, 2002) are found in seeds, bark, and leaves of the tree *Pongamia glabra* Vent. The hot ethanol extract of the defatted seeds (applied at 20 and 30 % of applied N) had maximum nitrification inhibitory activity

followed by bark extract (prepared with 40: 60, v/v, mixture of petroleum ether and acetone), and leaves had negligible effect. The alcohol extract of seeds was effective in retarding nitrification for 60 days when applied at 20 % of the fertilizer nitrogen dose. The percentage inhibition of nitrification of urea in soil was 47–55 % even after 45 days of application of seed extract.

Karanjin, present in karanja seeds, is a potent inhibitor of nitrification. The inhibition of nitrification of urea or ammonium sulfate was around 43 % after 8 weeks of incubation with its dose of 5 % of applied fertilizer nitrogen. Comparative evaluation of karanjin and three commercial inhibitors (nitrapyrin, AM, and DCD) when applied in a sandy loam soil at 5 mg kg^{-1} of soil reduced the nitrification rate as:

Nitrapyrin > Karanjin > AM > DCD

The furan ring of karanjin was responsible for nitrification inhibition property. This information led to the study of several furan derivatives (Sahrawat and Mukherjee 1977; Kuzvinzwa et al. 1984; Datta et al. 2001) as nitrification inhibitors (described above the synthetic compounds).

Mahua Seed cake and extracts of *Bassia latifolia* Roxb., *Madhuca indica* J.F. Gmel, and *Madhuca latifolia* L. are known to possess nitrification inhibition property. The seed cake and extracts contain saponins, responsible for nitrification inhibition. In an incubation study employing clay loam soil, the inhibitory effect of mahua cake extract persisted only for 20 days (Slangen and Kerkhoff 1984).

Miscellaneous Many other plants and plant products from *Citrullus colocynthis* Schrad., Sal *(Shorea robusta)*, *Eucalyptus globosus*, *Ricinus communis*, *Acacia catechu*, *Calotropis gigantea*, *Onosma hispidum*, *Mentha arvensis*, *Mentha spicata*, *Artemisia annua*, *Chrysanthemum cinerariifolium*, *Tagetes erecta*, *Catharanthus roseus*, *Ricinus communis* L., turmeric powder, tea waste, and cashew shell powder have been reported to show varying degree of nitrification inhibition.

6   Biological Nitrification Inhibitor: Concept

Biological nitrification inhibition is a rhizospheric process where different class of compounds released by plant roots act as nitrification inhibitor (Subbarao et al. 2013). It was hypothesized that it can improve N uptake due to its inhibitory effects on nitrification by improving $NUE_{agronomic}$ mostly contributed by the improvement in crop N uptake.

Primary productivity is positively impacted in the tropical savannas dominated by native African grasses such as *Hyparrhenia diplandra* which appear to have a significant ability to suppress nitrification (Boudsocq et al. 2009).

Recent studies by Boudsocq et al. (2011) reported the role of biological nitrification inhibition in controlling nitrification in temperate and tropical grasslands and contrasting preferences for $NH4^+$ or $NO3^-$ between two plant species. The ability of one species to control nitrification (i.e., to stimulate or inhibit) could enhance their ability to compete for mineral N with other species. This is consistent with the results of the studies suggesting that biological nitrification inhibition strongly affects plant invasions (Hawkes et al. 2005; Rossiter-Rachor et al. 2009).

Several compounds belonging to different chemical groups have been successfully isolated and identified from plant tissue or root exudates using bioassay-guided purification approaches and are reported to be biological nitrification inhibitor (Subbarao et al. 2013). The identified compounds from the aerial parts of *Brachiaria humidicola* are the unsaturated free fatty acids, linoleic acid (**XXI**), and α-linolenic acid (**XXII**). They are relatively weak inhibitors of nitrification. In root tissues of *B. humidicola*, two phenyl propanoids, methyl-p-coumarate (**XXIII**) and methyl ferulate (**XXIV**), were identified as major biological nitrification inhibitor (Gopalakrishnan et al. 2007). From root exudates of hydroponically grown sorghum, a phenylpropanoid, methyl 3-(4-hydroxyphenyl) propionate (**XXV**) has been identified as the biological nitrification inhibitor (Zakir et al. 2008). Bachialactone, a cyclic diterpene, was identified from the root exudates of *B. humidicola* (Subbarao et al. 2013). Further, sorgoleone, a p-benzoquinone exuded from sorghum roots, has a strong inhibitory effect on *Nitrosomonas* sp. and contributes significantly to nitrification inhibition capacity in sorghum.

In a recent review (Subbarao et al. 2013), karanjin was also included as the biological nitrification inhibitor. It seems that there is an overlap of compounds known to be biological nitrification inhibitors. Some compounds are secondary metabolites produced and stored in the different parts of plant, and the compounds released as root exudates.

XXI

XXII

XXIII

XXIV

XXV

XXVI

XXVII

Structures of biological nitrification inhibitors

Conclusion

Several nitrification inhibitors have been reported in literature from time to time. Most of these are not fully satisfactory due to one or more of the following disadvantages:

– Complicated synthesis and related preparation steps

– High volatility, leading to low persistence and high losses into the atmosphere unless introduced in soil using technically complicated processes (e.g., by probe)

– High toxicity or ecotoxicity

– Low stability against hydrolysis, reducing the duration of action in soil and the shelf life
– High application rates
– Requirement of further modification of the active molecule
– Addition of costly formulants
– High cost

Therefore, an ideal nitrification inhibitor is still elusive. It needs to be simple, safe, efficient, persistent, specific, and economical in use. It implies that the nitrification inhibitor should have specificity to nitrifying bacteria responsible for conversion of ammonium to nitrite. Inhibition of *Nitrobacter* is not desirable as it leads to accumulation of nitrite. The inhibitor should be nontoxic to other soil organisms, fish, mammals, and crops and be safe to the environment. It should be effective throughout the nitrogen-soil interaction zone and be sufficiently persistent in action so that nitrification is inhibited for an adequate period of time. Furthermore, it should be a low cost additive to the fertilizer.

A major consideration during the selection of nitrification inhibitors (NI) is their high effectiveness at the lowest possible application rate with a minimum of undesirable side effects. The availability of an inhibitor at effective concentration is essential. This can be achieved by coating fertilizer granules with the inhibitor or by incorporating it into granules (Slangen and Kerkhoff 1984). The aim of both the approaches is to ensure an intimate and uniform interaction of the substrate with the inhibitor. The application of an effective concentration of an NI to soil, together with N fertilizer, is a difficult task since it involves different crops, forms, and rates of N application. It leads to different concentrations of NI reaching the nitrifiers, particularly if N is applied as granules.

While application of chemical fertilizers to agricultural crops has resulted in tremendous increase in yield, problems arising due to escape to the environment of different nitrogen species, especially N_2O, nitrite, and nitrate, have raised serious economic and environmental concerns. Of the different processes responsible for these, nitrification and denitrification are of prime importance. Hence, efforts have to be made to regulate the process of nitrification (major source of different N species) as a means to enhancing the use efficiency of N, decreasing environmental/economic concerns, and optimizing the functioning of agroecosystems. Use of nitrification inhibitors has been helpful in mitigating the negative effects of fertilizer application. However, continued efforts need to be made for finding more efficient and environment-friendly products to suit the ever-changing agroclimatic conditions.

References

Abalos D, Jeffery S, Sanz-Cobena A, Guardia G, Vallejo A (2014) Meta-analysis of the effect of urease and nitrification inhibitors on crop productivity and nitrogen use efficiency. Agric Ecosyst Environ 189:136–144

Abbasi MK, Adams WA (2000) Estimation of simultaneous nitrification and denitrification in grassland soil associated with urea-N and ^{15}N and nitrification inhibitor. Biol Fertil Soils 31:38–44

Aggarwal N, Kumar R, Dureja P, Rawat DS (2009) Schiff bases as potential fungicides and nitrification inhibitors. J Agric Food Chem 57:8520–8525

Aggarwal N, Kumar R, Srivastva C, Dureja P, Khurana JM (2010) Synthesis of nalidixic acid based hydrazones as novel pesticides. J Agric Food Chem 58(5):3056–3061

Agricoop (2009) http://agricoop.nic.in/inm/cfqcti9210.pdf. Accessed on 13 Dec 2011

Azam F, Farooq S (2003) Nitrification inhibition in soil and ecosystem functioning – an overview. Pak J Biol Sci 6(6):528–535

Azam F, Mueller C, Weiske A, Benckiser G, Ottow JCG (2002) Nitrification and denitrification as sources of atmospheric N_2O – role of oxidizable C and applied N. Biol Fertil Soils 35:54–61

Ball BC, Cameron KC, Di HJ, Moore S (2012) Effects of trampling of a wet dairy pasture soil on soil porosity and on mitigation of nitrous oxide emissions by a nitrification inhibitor, dicyandiamide. Soil Use Manag 28:194–201

Banerjee NK, Mosier AR, Uppal KS, Goswami NN (1990) Use of encapsulated calcium carbide to reduce denitrification losses from urea-fertilized flooded rice. Mitteilungen der Deutschen Bodenkundlichen Gesellschaft 60:245–248

Bedard C, Knowles R (1989) Physiology, biochemistry, and specific Inhibitors of CH_4, NH_4^+, and CO oxidation by methanotrophs and nitrifiers. Microb Rev 53:68–84

Bhatia A, Sasmal S, Jain N, Pathak H, Kumar R, Singh A (2010) Mitigating nitrous oxide emission from soil under conventional and no-tillage in wheat using nitrification inhibitors. Agric Econ Environ 136:247–253

Bothe H, Jost G, Schloter M, Ward BB, Witzel KP (2000) Molecular analysis of ammonia oxidation and denitrification in natural environments. FEMS Microbiol Rev 24:673–690

Boudsocq S, Lata JC, Mathieu J, Abbadie L, Barot S (2009) Modelling approach to analyse the effects of nitrification inhibition on primary production. Funct Ecol 23:220–230

Boudsocq S, Barot S, Loeuille N (2011) Evolution of nutrient acquisition: When adaptation fills the gap between contrasting ecological theories. Proc R Soc B Biol Sci 278:449–457

Datta A, Walia S, Parmar BS (2001) Some furfural derivatives as nitrification inhibitors. J Agric Food Chem 49:4726–4731

Duddleston KN, Bottomley PJ, Porter AJ, Arp DJ (2000) New insights into methyl bromide cooxidation by *Nitrosomonas europaea* obtained by experimenting with moderately low density cell suspensions. Appl Environ Microbiol 66:2726–2731

Freney JR, Peoples MB, Mosier AR (2011) Efficient use of fertilizer nitrogen by crops. http://www.agnet.org/library/eb/414/. Accessed on 02 Dec 2011

Gopalakrishnan S, Subbarao GV, Nakahara K, Yoshihashi T, Ito O, Maeda I, Ono H, Yoshida M (2007) Nitrification inhibitors from the root tissues of *Brachiaria humidicola*, a tropical grass. J Agric Food Chem 55:1385–1388

Hawkes CV, Wren IF, Herman DJ, Firestone MK (2005) Plant invasion alters nitrogen cycling by modifying the soil nitrifying community. Ecol Lett 8:976–985

Holmes AJ, Costello A, Lindstrom ME, Murrell JC (1995) Evidence that particulate methane monooxygenase and ammonia monooxygenase may be evolutionary related. FEMS Microbiol Lett 132:203–208

Hovanec TA, Delong EF (1996) Comparative analysis of nitrifying bacteria associated with freshwater and marine aquaria. Appl Environ Microbiol 62 (8):2888–2896

Jenkinson DS, Fox RH, Rayner JH (1985) Interactions between fertilizer nitrogen and soil nitrogen – the so-called "priming" effect. J Soil Sci 36:425–444

Kawakami EM, Oosterhuis DM, Snider JL, Mozaffari M (2012) Physiological and yield responses of field grown cotton to application of urea with urease inhibitor NBPT and the nitrification inhibitor DCD. Eur J Agron 43:147–154

Khan MG, Silberbush M, Lips SH (1994) Physiological studies on salinity and nitrogen interaction in alfalfa. I. Biomass production and root development. J Plant Nutr 17:657–668

Koops HP, Bottcher B, Moller UC, Pommerening-Roser-A, Stehr G (1991) Classification of eight new species of ammonia-oxidizing bacteria: *Nitrosomonas communis* sp. nov., *Nitrosomonas ureae* sp. nov., *Nitrosomonas aestuarii* sp. nov., *Nitrosomonas marina* sp. nov., *Nitrosomonas nitrosa* sp. nov., *Nitrosomonas eutropha* sp. nov., *Nitrosomonas oligotropha* sp. nov. and *Nitrosomonas halophila* sp. nov. J Gen Microbiol 137:1689–1699

Kumar R, Anupama, Parmar BS (2004) S-Benzylisothiouronium derivatives as nitrification inhibitors. Pestic Res J 16(1):48–57

Kumar R, Anupama, Parmar BS (2005) Process for the preparation of 5-substituted-1,3,4-oxadiazole-2-thiols as new urease and nitrification inhibitors. Patent application no. 3461/DEL/2005 filed 23 Dec 2005. Publication date: 05 Sept 2008

Kumar R, Devakumar C, Sharma V, Kakkar G, Kumar D, Panneerselvam P (2007) Influence of physico-chemical parameters of neem (*Azadirachta indica* A Juss) oils on nitrification inhibition in soil. J Agric Food Chem 55:1389–1393

Kumar R, Devakumar C, Kumar D, Panneerselvam P, Kakkar G, Arivalagan T (2008) Influence of edaphic factors on the mineralization of neem oil coated urea in four Indian soils. J Agric Food Chem 56 (21):10183–10191

Kumar D, Devakumar C, Kumar R, Das A, Panneerselvam P, Shivay YS (2010) Effect of neem-oil coated prilled urea with varying thickness of neem (*Azadirachta indica* A. Juss)-oil coating and N rates on productivity and nitrogen-use efficiency of lowland irrigated rice under Indo-Gangetic plains. J Plant Nutr 33:1939–1959

Kumar D, Devakumar C, Kumar R, Panneerselvam P, Das A, Shivay YS (2011) Relative efficiency of prilled urea coated with major neem (*Azadirachta indica* A. Juss) oil components in lowland irrigated rice of Indo-Gangetic plains. Arch Agron Soil Sci 57(1):61–74

Kuzvinzwa SM, Devakumar C, Mukherjee SK (1984) Evaluation of furano compounds as nitrification inhibitors. Bull Indian Soc Soil Sci 13:165–172

Ma Y, Sun L, Zhang X, Yang B, Yin B, Yan X, Xiong Z (2013) Mitigation of nitrous oxide emissions from paddy soil under conventional and no-till practices using nitrification inhibitors during the winter wheat growing season. Biol Fertil Soils 49:627–635

Majumdar D (2008) Unexploited botanical nitrification inhibitors prepared from Karanja plant. Nat Prod Radiance 7(1):58–67

Marschner H (1999) Mineral nutrition of higher plants, 2nd edn. Academic Press, London

Mazumdar D (2002) Suppression of nitrification and N_2O emission by Karanjin-a nitrification inhibitor prepared from Karanj (*Pongamia glabra*). Chemosphere 47:845–850

McCarty GW (1999) Modes of nitrification inhibitors. Biol Fertil Soils 29:1–9

McCarty GW, Bremner JM (1989) Inhibition of nitrification in soil by heterocyclic nitrogen compounds. Biol Fertil Soils 8:204–211

Menendez S, Barrena I, Setien I, Gonzalez-Murua C, Estavillo MJ (2012) Efficiency of nitrification inhibitor DMPP to reduce nitrous oxide emissions under different temperature and moisture conditions. Soil Biol Biochem 53:82–89

Norton JM, Stark JM (2011) Regulation and measurement of nitrification in terrestrial systems. In: Klotz MG (ed) Methods in enzymology, vol 486. Academic Press, Burlington, pp 343–368

Opoku A, Chaves B, De Neve S (2014) Neem seed oil: a potent nitrification inhibitor to control nitrate leaching after incorporation of crop residues. Biol Agric Hortic. doi:10.1080/01448765.2014.885394

Patra DD, Chand S (2009) Natural nitrification inhibitors for augmenting nitrogen use efficiency in soil-plant system. In: The proceedings of the international plant nutrition colloquium XVI, Department of Plant Sciences, UC Davis, UC Davis. http://escholarship.org/uc/item/4h30z8tg. Accessed on 03 Dec 2011

Prasad R (1998) Fertilizer urea, food security, health and the environment. Curr Sci 75:677–683

Prasad R, Power JF (1995) Nitrification inhibitors for agriculture, health and the environment. Adv Agron 54:233–281

Prasad R, Devakumar C, Shivay YS (1993) Significance in increasing fertilizer nitrogen efficiency. In: Randhawa NS, Parmar BS (ed) Neem research and development, SPS publication no. 3, Society of Pesticide Science, India, New Delhi, pp 97–108

Prasad R, Sharma SN, Singh S, Devakumar C, Saxena VS (2002) Neem coating of urea for the environment and agriculture. Fertil News 47(5):63–67

Purkhold U, Pommerening-Roser A, Juretschko S, Schmid MC, Koops HP, Wagner M (2000) Phylogeny of all recognized species of ammonia oxidizers based on comparative \6S rRNA and amoA sequence analysis: implications for molecular diversity studies. Appl Environ Microbiol 66:5368–5382

Raun WR, Johnson GV (1999) Improving nitrogen use efficiency for cereal production. Agron J 91:357–363

Regan John M, Harrington GW, Noguera DR (2002) Ammonia- and nitrite-oxidizing bacterial communities in a pilot-scale chloraminated drinking water distribution system. Appl Environ Microbiol 68 (1):73–81

Rice EL (1984) Allelopathy, 2nd edn. Academic Press, New York

Rossiter-Rachor NA, Setterfield SA, Douglas MM, Hutley LB, Cook GD, Schmidt S (2009) Invasive *Andropogon gayanus* (Gamba Grass) is an ecosystem transformer of nitrogen reactions in Australian savanna. Ecol Appl 19:1546–1560

Saha A, Kumar R, Kumar R, Devakumar C (2010) Green synthesis of 5-substituted-1,3,4-thiadiazole-2-thiols as new potent nitrification inhibitors. J Heteroc Chem 47:838–845

Sahrawat KL (1982) Comparative evaluation of Karanjin and extract of Karanj (*Pongamia glabra* Vent) and neem (*Azadirachta indica*) seeds for retardation of nitrification of urea in soil. J Indian Soc Soil Sci 30:156–159

Sahrawat KL, Mukherjee SK (1977) Nitrification inhibitors: I. Studies with furano compounds. Plant Soil 47:687–691

Sahrawat KL, Parmar BS (1975) Alcohol extract of neem (*Azadirachta indica*) as nitrification inhibitor. J Indian Soc Soil Sci 13:131–134

Sahrawat KL, Parmar BS, Mukherjee SK (1974) A note on the nitrification inhibitors in the seeds, bark and leaves of *Pongamia glabra* Vent. Indian J Agric Sci 44:415–418

Sanz-Cobena A, Sanchez-Martin L, Garcia-Torres L, Vallejo A (2012) Gaseous emission of N_2O and NO and NO_3 leaching from urea applied with urease and nitrification inhibitors to a maize crop. Agric Ecosyst Environ 149:64–73

Saxena VS, Devakumar C, Prasad R (1999) Pusaneem-ME coated urea. Indian patent application no. 223/Del/99 dated 5th Feb 1999

Shaviv A, Mikkelsen RL (1993) Controlled release fertilizers to increase efficiency of nutrient use and minimize environmental degradation: a review. Fertil Res 35:1–12

Slangen JHG, Kerkhoff P (1984) Nitrification inhibitors in agriculture and horticulture: a literature review. Fertil Res 5:1–76

Subbarao GV, Ito O, Sahrawat KL, Berry WL, Nakahara K, Ishikawa T, Watanabe T, Suenaga K, Rondon M, Rao IM (2006) Scope and strategies for regulation of nitrification in agricultural systems – challenges and opportunities. Crit Rev Plant Sci 25:305–335

Subbarao GV, Sahrawat KL, Nakahara T, Kishi M, Rao IM, Hash CT, George TS, Srinavasa Rao P, Nardi P, Bonnett D, Berry W, Suenaga K, Lata JC (2013) Biological nitrification inhibition—a novel strategy to regulate nitrification in agricultural systems. In: Sparks D (ed) Advances in agronomy, vol 114. Academic Press, Burlington, pp 249–302

Tomiyama H, Ohshima M, Ishii S, Satoh K, Takahashi R, Isobe K, Iwano H, Tokuyama T (2001) Characteristics of newly isolated nitrifying bacteria from rhizoplane of paddy rice. Microb Environ, Japan 16:101–108

Williams PH, Jarvis SC, Dixon E (1998) Emission of nitric oxide and nitrous oxide from soil under field and laboratory conditions. Soil Biol Biochem 30:1885–1893

Zakir HAKM, Subbarao GV, Pearse SJ, Gopalakrishnan S, Ito O, Ishikawa T, Kawano N, Nakahara K, Yoshihashi T, Ono H, Yoshida M (2008) Detection, isolation and characterization of a root-exuded compound, methyl 3-(4-hydroxyphenyl) propionate, responsible for biological nitrification inhibition by sorghum (*Sorghum bicolor*). New Phytol 180:442–451

Zerulla W, Barth T, Dressel J, Erhardt K, von Loquenghien H, Pasda K, Radle MG, Wissemeier AH (2001) 3,4-Dimethylpyrazole phosphate: a new nitrification inhibitor for agriculture and horticulture. Biol Fertil Soils 34:79–84

Part II

Microbiological Aspects of Nutrient Use Efficiency

Role of Microorganisms in Plant Nutrition and Health

Om Prakash, Rohit Sharma, Praveen Rahi, and Nanjappan Karthikeyan

Abstract

Microbes are the key components of soil nutrient cycling. Status of soil health and richness of soil nutrient pool depend on structure and functions of soil microbial community. Microbes play an important role in nutrient mobilisation and uptake. They promote plant growth and suppress disease by their various activities. Phosphate and sulphate solubilisation, plant growth promotion, siderophore production, nitrogen fixation, denitrification, immune modulation, signal transduction and pathogen control are some of the well-recognised microbial mediated processes which promote the plant growth and protect them from pests. Current chapter starts with a brief introduction of plant nutrients and their classification and mechanisms of nutrient uptake by the plants. After that we discussed the importance of microbes in plant nutrient uptake and mobilisation. In addition, importance of different classes of microorganisms (fungi, bacteria, cyanobacteria) in plant nutrition and health has been discussed in detail under different sections. In the end we conclude the role of microorganisms in sustainable agriculture and environment and suggested for the promotion and use of microbial-based formulations instead of chemical fertilisers. We also emphasised on cultivation and preservation of agriculturally and environmentally important but not yet cultured organisms for sustainable development in agriculture and environment.

O. Prakash (✉) • R. Sharma • P. Rahi
Microbial Culture Collection, National Centre for Cell
Science, Pune 411007, Maharashtra, India
e-mail: prakas1974@gmail.com; rohit@nccs.res.in;
praveen_rahi22@yahoo.co.in

N. Karthikeyan
National Bureau of Agriculturally Important
Microorganisms (NBAIM), Kushmaur, Mau,
Uttar Pradesh, India
e-mail: microkarthi.iari@gmail.com

A. Rakshit et al. (eds.), *Nutrient Use Efficiency: from Basics to Advances*,
DOI 10.1007/978-81-322-2169-2_9, © Springer India 2015

Keywords
Plant nutrients • Microorganisms • Plant growth • Stress tolerance • Disease resistance • Plant immunity

1   Introduction

Plant growth and health depends upon availability of right composition and concentration of nutrients in soil. Inadequate supply, limitations, imbalance and excess presence of these nutrients in soil system affect plant growth, fertility and productivity. A total of 17 nutrients including carbon (C), hydrogen (H), oxygen (O), nitrogen (N), phosphorus (P), potassium (K), sulphur (S), magnesium (Mg), calcium (Ca), iron (Fe), boron (B), manganese (Mn), zinc (Zn), molybdenum (Mo) and copper (Cu) are essential for plant growth and health (soil nutrient). Based on the requirements, these elements are classified as macronutrients and micronutrients or trace elements (Hodges 2010). Plants need macronutrient in large quantity while micronutrients or trace elements are required in minimal amount and work as cofactors for enzymatic activities. Based on the function C, H and O are considered as structural element; N, P and K are primary nutrients; Ca, Mg and S are considered secondary; while others (Mn, B, Cu, Co, Mo, Zn) are micronutrients and trace elements (Hodges 2010; Microbes 2010). Except, carbon, hydrogen and oxygen, plants take these nutrients from soil by their roots in the form of soluble ions (Marschner 1995; Microbes 2010). Availability and uptake of nutrient from soil depends on several factors including chemistry and composition of soil and environmental factors like pH, moisture, soil texture, temperature and microbial community composition. Plant takes nutrient from soil in the form of soluble ions using different mechanisms including direct contact of plant root and nutrient (root interception), movement of dissolved nutrient towards plant root (mass flow) and movement of nutrient from zone of higher concentration to zone of low concentration by the process of diffusion (Morgan and Connolly 2013). It's not the total nutrient pool but availability of bioavailable form of nutrient affects plant growth and productivity. For instance, rich pools of phosphate and sulphate are available in the soil, but the plant only utilised soluble form of these nutrients and the rest are not used for plant utilisation (Solomon et al. 2003). Past data indicated that most of the soil sulphur (>95 %) is present as sulphate ester or sulphonate form while amount of inorganic sulphur (sulphur available for plant nutrient) present in the soil is very little (Solomon et al. 2003). Plant growth follows the Liebig's Law of minimum which explains that plant growth is limited by the short supply of nutrients and cannot be achieved by excess supply or availability of other nutrients (Sinclair 1999) (http://forums2.gardenweb.com/forums/load/contain/msg0921071615772.html). Past study indicated that nutrient-deficient soil produced less nutritious food. Addition of chemical fertilisers, pesticides and herbicides for restoration of soil nutrient for better crop production and yield is in extensive practice. Excess use of chemicals to improve the crop yield disturbed the balance of natural microbial population and consequently geochemical cycling of nature and created environmental pollution. Use of biofertiliser and bipesticides to promote the crop yield and restore the soil nutrients for sustainable agriculture and environment is the better alternative for chemicals and right tool for sustainable environment and agriculture (Bertin et al. 2003).

2   Plant Nutrients and Microbes

Soil microbes are the vital component of soil, health, functionality, sustainable environment and for survival of animals and plants on planet Earth. The notion "Microbes eat, we eat and microbes die, we die" seems absolutely true. According to eminent microbiologist CR Woese, "If we wiped off all the multicellular organisms (plant and animals) from the surface of earth then it would barely affect the microbial community whereas the destruction of microbial community would lead to instant death of all life forms on earth" (http://www.nytimes.com/2013/01/01/science/carl-woese-dies-discovered-lifes-third-domain.html?_r=0). Living soil is a dynamic ecosystem and homes for divers' range of microbes including bacteria, actinobacteria (previously actinomyces), cyanobacteria, fungi, archaea, viruses, protozoa and microalgae. Biodiversity of soil is the measure of soil health. A healthy soil will harbour more diverse population of microorganisms in terms of species richness. It is speculated that 1 gram of soil contains more than 10^8–10^9 cells of bacteria, 10^7–10^8 cells of actinobacteria and 10^5–10^6 cells of fungi with thousands of different species (Microbes 2010).

Microbes are active players of biogeochemical cycling of materials of the nature and affect the composition and concentration of soil nutrients. Carbon, nitrogen, sulphur and iron cycles of the natures are driven and mediated by microorganisms. Microbes create nutrient-like carbon, nitrogen, oxygen, hydrogen, phosphorus, potassium, trace elements, vitamins and amino acids and make them available for plant in right form for their growth and health. Bacteria and fungi are the major decomposer on earth and crucial component for composting and humus formation. Mineralisation of dead and decaying plant and animal residues releases carbon dioxide in the atmosphere and rejuvenates the nutrient pool of the soil. Compost is the best source for trace element and rich source of plant nutrition (Mehta et al. 2013). Compost retains the soil moisture, binds soil particle together and protects plants from disease and maintains soil pH by its buffering action and ideal component for soil restoration (Mehta et al. 2013). In addition to decomposition and mineralisation, microbes provide valuable nutrients to plants by their plant growth-promoting activity, phosphate and sulphate solubilisation, nitrogen fixation, etc. Rhizospheric and rhizoplane bacteria and fungi also produce valuable components like vitamin, amino acids, signalling compounds, growth hormones and many more unknown compounds

for plant growth and disease suppression (Bertin et al. 2003; Bais et al. 2006; Lugtenberg and Kamilova 2009). It has been found that several microorganisms especially mycorrhizal fungi assist in signal transduction and plant to plant communication and act as natural defence system for plant. Microbial community residing in the vicinity of plant roots and on the surface of root is known as rhizospheric and rhizoplane microbial community, respectively. Study of plant microbes' interaction in terms of rhizospheric and rhizoplane microbial community in plant health, nutrient uptake, disease suppression and growth yield is the area of special interest among the plant scientists (Ryu et al. 2003; Bais et al. 2006; Morgan and Connolly 2013). In addition to maintaining the nutrient pool of the soil and promoting and suppressing the plant growth by its various activities, microbes also act as a source of nutrient for plant. A recent study conducted on *Arabidopsis thaliana* and *Lycopersicon esculentum* demonstrated that plant roots take the microbes, digest it and use it as a source of nutrient. Thus it is clear that microbes are essential constituents of soil and crucial for soil health and plant growth (Sparling et al. 1997). Plant microbes' interaction and study of structure of soil microbial community is the area of interest for soil and agricultural microbiologist. In the current chapter we focused on description of different classes of microorganisms and their role in supporting the plant nutrient for sustainable agriculture.

Study of structure and function of soil microbial community using Metagenomic and next-generation sequencing along with plant micro interaction is the area of interest for soil microbiologists across the globe (Prakash et al. 2013b). Due to development of cheap sequencing technologies, good progress towards mapping the soil microbiome has been made (Lou et al. 2013). Now it is essential to cultivate the valuable but not yet cultured microorganism and reintroduce them to the soil for restoration of nutrient and support the sustainable environments and agriculture. Destruction of natural microbial population by excess tilling, overuse of chemicals, addition of manuring, composting, etc. impacted soil health and natural turnover of nutrient in the soil. Now the scientific community across the globe think that microbes are better alternative of chemicals and trying to maintain the natural microflora of soil and also practising for inoculation of beneficial microorganisms isolated from outside for restoration of soil nutrient, promotion of plant growth and suppression of diseases for sustainable agriculture and environment.

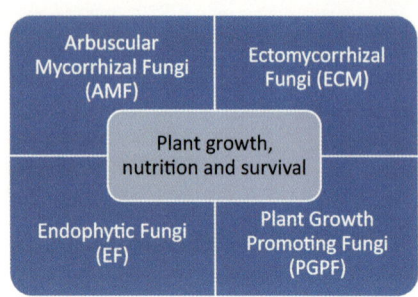

Fig. 1 Different kinds of fungal partners associated with plant growth, survival and nutrition

3 Role of Fungi in Plant Growth and Nutrition

Scientists are trying to increase agricultural production by keeping the concept of sustainable agriculture and environment protection in their mind. Different environmental friendly approaches including the use of natural microorganisms which promote plant growth and provide disease resistance capacity by producing different kinds of new compounds are being regularly used. Various groups of fungi including endophytic, ectomycorrhizal, arbuscular, etc. (Fig. 1) play an important role in increasing plant growth and obtaining nutrition. It also helps in survival by increasing resistance against pathogens and tolerance to different kinds of stresses such as drought and salt. Fossil records indicate that plants have been associated with endophytic (Krings et al. 2007) and mycorrhizal (Redecker et al. 2000) fungi for >400 million years (Myr) more importantly when plants colonised surface of the Earth. These associations helped in survival and adaptation of plants during various ecological changes, benefiting them in various aspects. Thus, fungal symbiotic association is considered to be one of the helpful agents responsible for adaptation of plant from aquatic habitat to land. The plant-fungal interaction has been the topic of study for long. And in the past two decades, the research has been focussed to explore these interactions on physiological, molecular and genetic backgrounds for their better biotechnological exploitation. It can be termed as 'naturopathy for plants', minimising ecological disturbances, reducing dependency on inorganic fertilisers and increasing production in crops and even survival of artificial and natural forest plantations.

We all know that plant in itself is a complex system and multiple factors govern its survival. It is very difficult to study about one factor without considering the others. Hence the role of fungi alone does not govern better plant survival; it is the interaction of many factors put together to bring the final output, i.e. better survival of plant in the environment. In this section we discuss the role of fungi in plant survival by helping them in different ways: induced systemic resistance; plant growth promotion; resistance of hosts to insect feeding; disease resistance; solubilisation of phosphorus; production of plant growth-promoting hormones; increased above ground photosynthesis; plant tolerance to abiotic stresses such as drought, salt, heavy metals; etc. (Fig. 2). Here we talk about various benefits provided by arbuscular, ectomycorrhizal and endophyte fungi to plant.

4 Arbuscular Mycorrhizal Fungi

Obligate symbiosis or association of arbuscular mycorrhizal (AM) fungi particularly members of *Glomeromycota* with angiosperms (excluding *Pinaceae*), bryophytes, pteridophytes and gymnosperms is one of the oldest (>450 Myr) and most common (>80 %) type of terrestrial symbiosis on the Earth (Smith and Read 2008). Most of the crop plants belonging to family

Fig. 2 Beneficial effects of ECM, AM, PGP and endophytic fungi

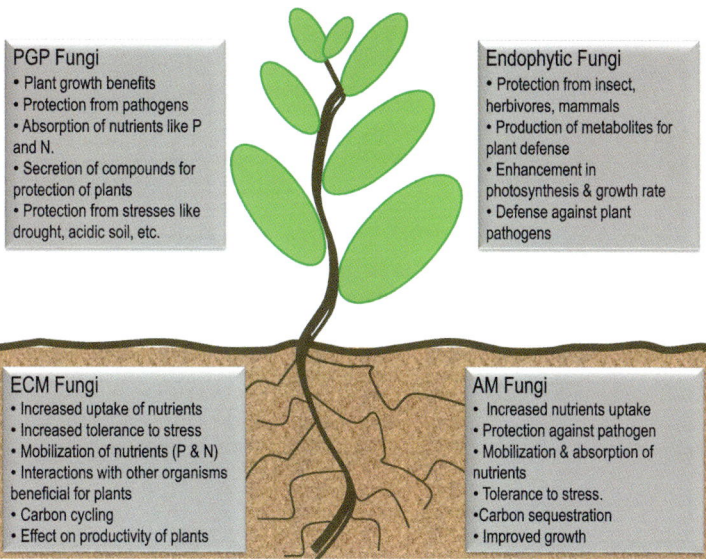

PGP Fungi
• Plant growth benefits
• Protection from pathogens
• Absorption of nutrients like P and N.
• Secretion of compounds for protection of plants
• Protection from stresses like drought, acidic soil, etc.

Endophytic Fungi
• Protection from insect, herbivores, mammals
• Production of metabolites for plant defense
• Enhancement in photosynthesis & growth rate
• Defense against plant pathogens

ECM Fungi
• Increased uptake of nutrients
• Increased tolerance to stress
• Mobilization of nutrients (P & N)
• Interactions with other organisms beneficial for plants
• Carbon cycling
• Effect on productivity of plants

AM Fungi
• Increased nutrients uptake
• Protection against pathogen
• Mobilization & absorption of nutrients
• Tolerance to stress.
• Carbon sequestration
• Improved growth

Gramineae, Palmae, Leguminosae, and *Rosaceae* are frequently associated with this type of symbiosis. Due to widespread occurrence of AM association with crop plant, now we know that it is in fact a rule instead of exception especially for plants growing in fields. It facilitates exchange of nutrients like carbon, phosphorus, nitrogen, etc. (Smith and Smith 2011a, b, 2012), protects the plants from pathogen and insects, provides tolerance to different kinds of environmental stresses and plays a major role in soil biogeochemical cycling of nutrients. Initially researchers did not pay attention to AM fungi due to difficulty in cultivation and maintenance, but in the past decade, techniques have been developed to maintain them in pot and root culture. Consequently, understanding in the area of signalling and cellular interaction between the symbionts, functional studies of AM fungi and gene expression, molecular identification and role of individual members in ecosystem services have improved (Cavagnaro et al. 2003; Karandashov et al. 2004). This has become possible because of physiological experiments, molecular studies and ultrastructural studies using modern equipment, viz., advance microscopes, HPLC, GC-MS, etc. It is understood that there is a perifungal membrane surrounding the arbuscule inside the cytoplasm of plant cell. The extra-radical mycelium gathers nutrition from

vast distances and supplies to the plant cell through cortical cells and perifungal membrane. AM fungi are important because they sequester the nutrients bound with soil matter and particles even in low-nutrient soils and exert beneficial effects (Fig. 3) onto many agricultural crops, viz., maize, potato, sunflower, onion, wheat, soybean, *Jatropha*, etc.

4.1 Role of AM in Plant Nutrient Uptake and Growth

The importance of AM in plant growth and nutrition has been reported elsewhere by several researchers (Bianciotto and Bonfante 2002; Wu et al. 2005; Gosling et al. 2006; Vyas et al. 2007; Rahi et al. 2009), and they considered that AM facilitate mobilisation and uptake of those minerals or compounds to the plants which are less soluble and have low mobility (Baslam et al. 2011a; Alizadeh 2012; Baslam and Goicoechea 2012; Baslam et al. 2013a, b, c). Although most of the work has been done on P (phosphorus) and N (nitrogen), AM also mobilises other minerals such as Zn, Mg, S, Ca, K, etc. (Allen et al. 2003; Hodge 2003). AM are known to supply P (90–100 % of plant P demand) to plants and in return fungi take C

Fig. 3 Various effects of
AM fungi on plant

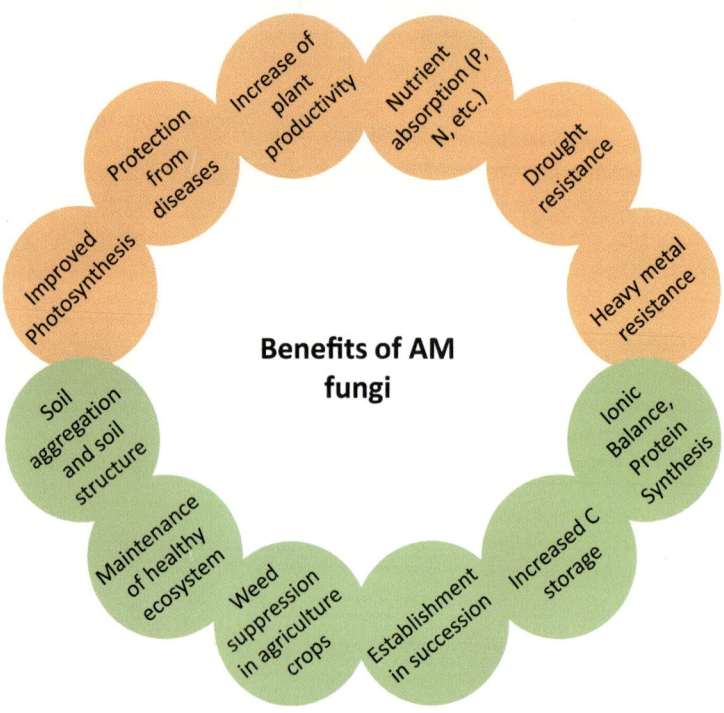

(carbon) from plant in the form of hexose, sucrose and fructose (van der Heijden et al. 2006. The perifungal membrane contains *Pi* and *NH4+* transporter, and it has been shown that the *Pi* intake is much faster with AM-associated plants than NM (non-mycorrhizal) plant. Increased growth of plant is attributed to the absorption of *Pi* by AM fungi. The reason may be that the *Pi* intake by AM fungi is at a distance from the depletion zone formed near the root and root hairs. Studies have shown that the AM fungi help plant to absorb *Pi* and in turn receive C from plant. It is observed that the AM reduces the P absorption if the C flow from plant to AM fungi is reduced. Moreover, the supply of P also depends on the species of AM fungi (Abiala et al. 2013). But the AM is considered to be more effective in absorbing P than an NM plant root (Fig. 5) which can be 3–5 times more than direct root absorption. And even the complete P uptake can be taken by AM only (Smith et al. 2003). Studies have demonstrated the fact that AM-infected plant roots absorb and accumulate more P as compared to NM roots of plant, when grown in

soils low in P (Smith and Read 2008). The effect of AM has been studied in many economically important crops, viz., cassava, potatoes, cocoyam and yam (Nisha and Rajeshkumar 2010; Srinivasan and Govindasamy 2014). AM association is not only important in nutrient-deficient but also essential in nutrient-rich soil as most plant roots are not capable of sequestering the nutrient even from nutrient-rich soil. Generally, extra-radical mycelia of AM fungi absorbed the nutrients and transported them to cortical AM cells and finally transferred to intraradical mycelia. The intracellular mycelia interact with plant cells at apoplast region and transfer nutrients to plant cells through nutrient transporters. It is estimated that AM fungi increases surface of absorption many times by extending the mycelia up to several metres to obtain the same nutrients which plant roots are trying from the vicinity. Due to fine structure of AM hyphae (10 μm diameter) than root (1–3 mm), it penetrates easily within the soil granules and increases the surface area for absorption and helps in growth and development of associated plant community/crop. Cavagnaro and his group have studied AM effect by using a mutant variety of non-mycorrhizal tomato *rmc* (Cavagnaro et al. 2008). Cavagnaro and Martin (2010) showed that the Zn absorption was more in AM plants (243 ± 18 μg/plant) than in NM plants (77 ± 15 μg/plant) highlighting the effect of AM. Research on grapevine has improved shoot and root growth, tissue concentrations of P and water relations.

It has been proved that although P is the most studied nutrient absorption by AM roots, some reviews have also published related to absorption of other nutrients (Mohammadi et al. 2011; Alizadeh 2012). Effect of AM fungi improves in collaboration with other organisms. Bacteria have been reported to enhance spore germination and root colonisation efficiency of AM (Miransari 2011). In their studies Vafadar et al. (2014) have tested several bacterial species, viz., *Azotobacter chrococcum*, *Pseudomonas putida* and *Bacillus polymyxa* along with AM fungus *Glomus intraradices* on *Stevia rebaudiana*. Studies showed that synergistic effect of AM and bacteria improves growth, nutrient absorption, phytohormone production and chlorophyll production and increases biomass (Vasanthakumar 2003; Hemavathi et al. 2006; Aseri et al. 2008). In addition *Glomus fasciculatum*, *G. mosseae* and several other AM fungi have been studied in association with bacteria to enhance plant growth (Das et al. 2007; Singh et al. 2012). In AM plant symbiosis, plants get easy supply of nutrients and invest less energy; consequently it increases the photosynthetic rate and overall biomass of the plant. Improvement in the nutritional quality of greenhouse-grown lettuce by AM fungi is also extensively studied (Baslam et al. 2011b, 2013a, b, c; Baslam and Goicoechea 2012).

In addition to P, AM fungi increase the uptake of nitrogen (Hawkins et al. 2000; Blanke et al. 2005), assist in assimilation of N in plants (Toussaint et al. 2004; Mortimer et al. 2009) and induce better biological N fixation. AM fungi also induces production of chemicals like diterpenoids (Brandle and Telmer 2007; Banchio

et al. 2010; Awasthi et al. 2011) which exert stimulatory effect on plant growth due to enhanced photosynthesis.

4.2 Role of AM in Stress Tolerance, Weed Control and Protection from Insect Pest

AM fungi help plants to overcome and resist various kinds of biotic and abiotic environmental stresses thus protecting plants from metal toxicity. AM fungi acts as a sink for heavy metals and also induces the expression of metal-tolerant genes in plants. Consequently it reduces the metal concentrations near root and root hairs and provides better tolerance, protecting plants from metal toxicity and stress (Andrade et al. 2010). In addition, the positive effect of AM on water stress has been well documented (Wu and Zou 2010; Zhu et al. 2010; Mohammadi et al. 2011). For example, *Glomus intraradices* and *G. claroideum* are known to help plant to overcome drought conditions. It is generally observed that AM fungi symbiosis is specialised. It recognises and establishes symbiosis with all the plants of same population and derecognises others. Thus plants with AM association will be healthy while others (weeds) will be weak, and this concept can be used for weed control programme in the field. It has been found that AM fungi suppresses the competitive ability of weeds in sunflower field (van der Heijden et al. 2008) and can be exploited for biological weed control.

There are several studies demonstrating that AM fungal association protects plants from herbivore insects (Borowicz 2009; Gehring and Bennett 2009; Vanette and Hunter 2011; Roger et al. 2013). It is considered that AM association increases plant nutrition and health providing better protection from insect pests in comparison to NM plant. Hartley and Gange (2009) have shown that the insect herbivores are negatively affected by colonisation of AM fungi to plants while Roger et al. (2013) did not get any significant difference between the NM and mycorrhizal infected plants when attacked by insect herbivore. In our view the AM fungi do not provide any direct protection to the plant from insect herbivory, but stimulate restabilisation of the plant after insect attack due to better nutrition and health.

4.3   Protection from Plant Pathogens

AM fungi protects plant from pathogen invasion using different mechanisms including growth inhibition, enhancing plant nutrition and health, increasing colonisation of plant growth-promoting rhizobacteria (PGPR), improving mineral nutrition and phytohormone production, etc. Among them reduction in mycorrhizospheric proliferation and intraradical replication of pathogenic fungi by AM fungi have been demonstrated and discussed earlier (Lioussanne 2010). Members of *Glomaceae* are more involved in protection from the pathogen (Sikes et al. 2009). During initial colonisation of pathogenic organisms, AM fungi protect plants by preventing the access sites (physical effect) to pathogen. Indirectly it protects by releasing the chemicals like siderophores, salicylic acid (SA), abscisic acid (ABA), jasmonic acid (JA) and ethylene thus enhancing the defence and stress tolerance capacity of host. Additional studies on the effect of AM fungi against fungi, bacteria, virus, nematode and pathogens are required before drawing any substantial conclusion.

Synergistic association of bacteria and AM fungi helps plant to fight against plant pathogenic organisms. *Pseudomonas fluorescence* CHA0 is known to help control plant disease in association with AM fungi (Mukerji and Ciancio 2007). Fiorilli et al. (2011) showed that AM fungi reduce pathogenicity of *Botrytis cinerea* to tomato plants with the help of ABA production and also suppress the symptoms of *Xanthomonas campestris* (Liu et al. 2007). The effect of root-infecting pathogenic bacteria was reduced in apple trees by inoculation of *Glomus fasciculatum* and *G. macrocarpum* (Pal and

Gardener 2006). Many bacterial species live inside AM fungi as endophyte or on the wall of hyphae (mycosphere). For instance, members of *Burkholderia* have been reported as endosymbionts of species of *Gigaspora* and *Scutellospora* (Bonfante 2003). Thus many beneficial effects of AM fungi to the plant can be attributed due to the associated bacteria.

AM fungi reduced the symptoms of *Alternaria alternata*, root rot or wilting caused by fungal pathogens (*Rhizoctonia, Verticillium, Fusarium, Pythium, Aphanomyces, Phytophthora*) and *Phytophthora parasitica* on tomato plant (*Lycopersicon esculentum*) (Vigoa et al. 2000; Fritz et al. 2006). Mechanism known for disease protection by AM is by competitive interactions with pathogenic fungi, anatomical or architectural changes in root structure, change in the microbial community structure in rhizoplane or rhizosphere, activation of plant defence via JA, ET and SA signalling pathways or improved nutrient status of plant (Pozo et al. (2008). Most common among them are enhanced nutrient uptake and compensation of the loss caused by pathogen. It may be due to increased colonisation of AM on plant roots and increased local and systemic resistance. Some researchers group believe that AM fungi alone are not responsible for pathogen protection, but accumulation of defensive plant compounds like phenylpropanoid, chitinases and glucanases and expression of defence-related genes also play an important role in disease suppression (Pozo et al. 2008).

Although the role of AM fungi is well established in plant nutrient uptake, pathogen and insect control and in weed control (Fig. 4), most of the biocontrol agents do not contain AM fungi due to lack of knowledge and non-popularity of mycorrhizal fungi (Whipps 2004). Thus, to exploit them for commercial production and extensive use in agricultural practice, extensive research in field experiments is necessary. However, scientists and farmers should promote exploitation of this symbiotic association for agriculture and its sustainable development (Cardoso and Kuyper 2006; Cavagnaro and Martin 2010).

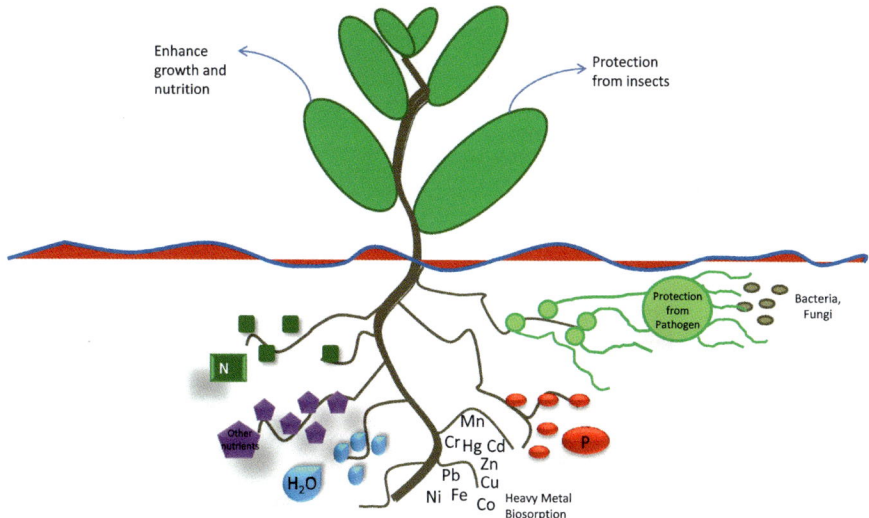

Fig. 4 Diagrammatic depiction of beneficial effects of AM fungi

5 Ectomycorrhizal Fungi

An ectomycorrhiza (ECM) is a mutualistic symbiotic relationship characterised by a root-fungus association in which the fungus grows on root surface and penetrates the cortex intercellularly (in between the cells) to produce a network. Three features are generally recognised during this association: (1) formation of a mantle or sheath of fungal hyphae, (2) development of hyphae between root cells to form Hartig net and (3) hyphae that grow into surrounding soil (extra-radical mycelium). Although the main interface of nutrient exchange in most ECM is Hartig net, the repeated branching of inner mantle hyphae suggests their involvement in bidirectional movement of nutrients. The mantle hyphae may accumulate compounds including lipids, protein and/or phenolics and polyphosphates, while deleterious metals may be bound to polyphosphates and other vacuolar deposits in the mantle, thereby preventing their uptake into roots. This observation is of particular relevance when polluted sites are being planted with tree seedlings inoculated with ECM fungi. The compact nature of mantles of some ECM may contribute to protection of roots from water loss (as soils dry) and ingress of pathogenic organism.

Since the mantle interfaces with the soil, it potentially regulates the transport of water and nutrient ions into the root. The Hartig net is involved in nutrient exchange as the fungal hyphae absorb most of the sugars, minerals and water which are also passed to root cells. Hyphae-forming Hartig net also functions as a depository for soluble and insoluble carbohydrate, lipids, phenolic compounds and polyphosphates. The most obvious function of fine hyphae that comprises much of the extra-radical mycelium is the mobilisation, absorption and translocation of mineral nutrients and water from the soil substrate to plant roots. In species with rhizomorphs, connecting fine hyphae passes water and dissolved nutrients to these structures for more rapid translocation through the wide diameter central hyphae (vessel hyphae) to root (Agerer 2001). Experiments with radioactive isotopes of phosphorus (P) (^{13}P-labelled orthophosphate) have shown that P can be translocated for over 40 cm through rhizomorphs to roots of colonised plants and subsequently to the shoot system (Peterson et al. 2004). Production and final biomass of *Laccaria bicolor* basidiocarps is correlated with the higher rate of photosynthesis of their host *Pinus strobus*. Bacteria are known to associate with extra-radical hyphae to form "biofilms" (layers of bacteria embedded in secreted polysaccharides) and break down petroleum hydrocarbons and other soil pollutants, as bacteria emanating from *Pinus sylvestris-Lactarius rufus* mycorrhizal interface proliferated in patches of soil contaminated with petroleum hydrocarbons (Poole et al. 2001). Studies have shown that several ECM mushrooms are found in India and role of *Cantharellus* has been studied on plant growth and nutrition (Sharma and Rajak 2011). The various aspects for the plant growth enhancement through ECM are discussed in detail below.

5.1 Effect of ECM on Growth and Nutrient Uptake

It has been shown that mycorrhizal fungi contribute to plant diversity, nutrient cycling, acquisition to nutrient sources previously thought not available to plants and finally to ecosystem functioning. As a rule, mycorrhizal infection enhances plant growth by increasing nutrient uptake via increase in the absorbing surface area, by mobilising sparingly available nutrients sources or by excretion of chelating compounds or ectoenzymes. Depending on tree species, a varied proportion of root supply of mineral nutrients from the soil may occur via fungal hyphae (Smith and Read 2008). Some ECM has mycelial cords which transport water and nutrients over long distances and, in some cases, via specialised non-living hyphae. Bidirectional transfer of nutrients between plant and fungus is typical of ECM (including other mycorrhizal types) and is. It is also the basis for prolonged compatible interactions in such symbioses. Many genes are up-regulated exchanging amino acids, oligopeptides and polyamines (Martin and Nehls 2009). At cellular level, interfaces in all types of mycorrhizas are composed of membranes of both partners, separated by an apoplastic region. The interface is simple, intercellular, wall to wall contact in ECM, but fungal partner remains in apoplast outside the plant protoplast. In Hartig net region hyphae branch profusely and septa formation becomes irregular or incomplete to give a characteristic labyrinthine system (Martin et al. 2008). Surface fibrils and acid phosphatase activity present in the mantle disappear as hyphae become tightly pressed against host cell walls. Adjacent fungal and host walls become indistinguishable from each other, forming a homogenous interfacial matrix (Bonfante 2001). Extracellular material is also deposited around hyphae of ECM and accumulates in intercellular spaces of the fungal sheath in ECM of trees. Nature of the compounds transferred between the symbionts has been reviewed previously (Simard and Durall 2004). Sugars are important in carbohydrate transfer. Hydrolysis of sucrose (or trehalose in orchid mycorrhizas) and synthesis of characteristic "nonrecyclable" carbohydrates (e.g. mannitol in ECM) are important steps in polarising transport in favour of one symbiont. ECM has a major influence on N and less effect on P nutrition. There is evidence that inorganic orthophosphate is the major form in which phosphorus is transferred. In ECM, the coexistence of ATPase activity on plant and fungal plasma membrane at Hartig net interface suggests that two systems work cooperatively in bidirectional nutrient exchange. P_i transferred to host tissue in excised beech ECM is through active transport by the sheath hyphae (Bücking and Heyser 2003).

Growth responses following seedling inoculation under controlled conditions with beneficial ECM fungi like *Pisolithus* and *Rhizopogon* species have been repeatedly observed (Menkis et al. 2011). Plant growth responses have been extensively studied using different ECM mushrooms, viz., *Tuber, Lactarius, Laccaria, Scleroderma, Cenoccocum, Thelephora, Cantharellus, Paxillus, Amanita, Hebeloma, Suillus,* etc. (Menkis et al. 2011). Although ECMs are commonly assumed to enhance water uptake by their hosts, few researchers have addressed this experimentally. Some mycorrhizal fungi grown in vitro showed survival at water potentials below the permanent wilting point of their host, and this capacity varies widely among species. As mycorrhizal association is a bidirectional movement of nutrients, the C from the host plant flows to the fungus and nutrients derived

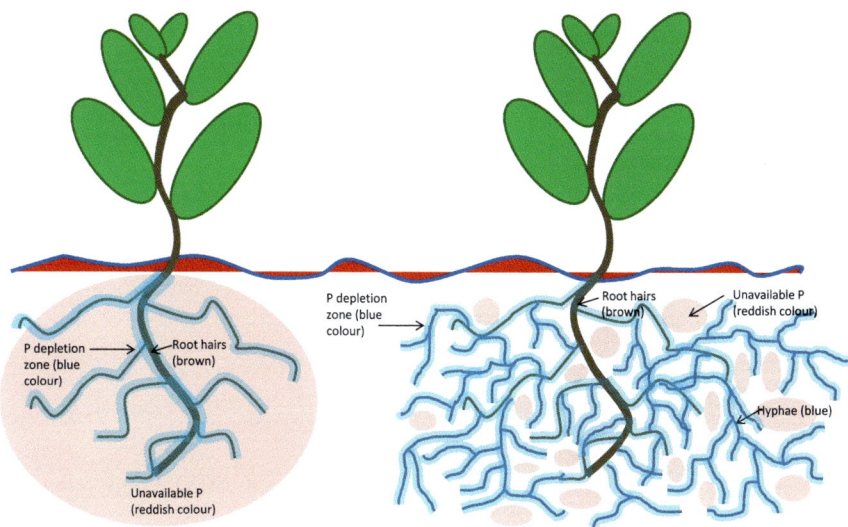

Fig. 5 Different mode of nutrient absorption by non-mycorrhizal (NM) and mycorrhizal (ECM and AM) fungi. Left-non-mycorrhizal and right-mycorrhizal

from fungus flow to the plants. The ECM-infected plants explore the areas beyond the root hair zone by extra-radical hyphae and absorb the nutrients for the host plant.

It has been shown that symbiosis between fungus and tree is essential to P nutrition of the latter in soils of low P availability. Phosphatase activity in ectomycorrhizal fungi (Fig. 5) has been shown to vary with species (Alvarez et al. 2009; Sharma et al. 2010). Although rate of excretion largely depends on composition of growing medium, variation may also occur due to fungal species. Wallander et al. (2013) using particle-induced X-ray emission (PIXE) analysis of element contents suggested that ECM species of *Rhizopogon* has the ability to mobilise P and K to host trees. A large fraction of the P in most temperate forest soils occurs in organic forms, such as inositol phosphates, nucleic acids and phospholipids. Host access to organic P sources may depend on its association with a range of ECM fungi that produce extracellular enzymes capable of acquiring P in organic forms. However, mycorrhizal infection does not always increase plant growth or reproduction. Possible reasons include high efficiency of P acquisition or low P requirement of plants. P acquisition can be reduced when external hyphae are destroyed

by grazing soil animals, soil disturbance or fungicides.

Many studies have shown that ECM fungi can utilise organic N sources through the production of extracellular acid proteinases. Mycorrhizal infection can provide host plants with access to N sources which are normally unavailable to non-mycorrhizal roots (Plassard et al. 2000). Further, proteolytic capacity may vary greatly between fungal isolates. Perhaps all fungi can assimilate ammonia by a combination of the glutamate dehydrogenase and glutamine synthetase pathways. In contrast only a smaller number of species can efficiently reduce nitrate. Ammonia and nitrate is rapidly assimilated in the extra-radical hyphae (Hawkins et al. 2000; Martin and Plassard 2001) and N is transferred to the host primarily as glutamine. Analysis of ECM genomes showed that the free-living mycelium has the potential to import organic and inorganic N sources, including nitrate, ammonium and peptides from the soil (Bonfante and Genre 2010). ECM symbiosis alters metabolic pathways of N assimilation in the fungal symbiont (He et al. 2005). Not much is known on the role of mycorrhiza in uptake of K, Ca, Mg and S. The majority of studies on ECM and micronutrient uptake are focused on the protection from excessive uptake of Cu and Zn on soils high in heavy metals. Production of siderophores is also widespread among ECM. However, boron is essential for the growth of fungi and perhaps ECM may help increase concentrations in the host plants.

Sequencing of genomes of *Lactarius bicolor* and *Tuber melanosporum* has given an opportunity to study the genes expressed after the ECM association in the plant and fungi. In the poplar tree, out of approx. 39,000, nearly 3,000 genes are differentially expressed due to *L. bicolor* association. Genome sequencing has changed the way we look at ECM fungi and has now become the model for plant-microbe interactive study. Now we know the determinants of the factors which are responsible for the way a fungi behaves as saprotroph, pathogen or mycorrhizal. These studies have also highlighted on the mechanism behind fungal nutritional strategies.

5.2 Stress Tolerance

In nature, plants face several stresses like drought, acidification metal tolerance, etc. for which they have evolved several mechanisms to avoid them including ECM formation (Ortega et al. 2004). Various metals are important for plant growth at low concentration but become toxic at high levels (Zn, Cu, Mn, etc.) whereas others (Hg, Cd, Pb, etc.) are not required at all and toxic even in low concentration (Menkis et al. 2011). The hyphae of ECM are known to contain vacuolar polyphosphates which are known to absorb toxic metals and help plants adapt to pollutants. The mechanism by which ECM helps in metal tolerance is extracellular (chelation and cell-wall binding), intracellular (binding to non-protein thiol) and/or detoxification mechanism (Bellion et al. 2005). These include reducing uptake into cytosol by extracellular chelation or binding onto cell-wall components, intracellular chelation of metals in cytosol by a range of ligands (glutathione, metallothioneins) or efflux from cytosol into sequestering compartments. Although it is difficult to study metal tolerance in symbiosis, the

data of full genome of fungi will help to know complete range of genes involved in it. Studies in this regard have been undertaken on various ECM fungi, viz., *Paxillus, Lactarius,* etc. (Jentschkea and Godbold 2000; Gadd 2010). It has been observed by various workers that the water absorbance is more in mycorrhizal plants than NM plants. Moreover, when subjected to sudden drought, mycorrhizal plants behaved better than NM plants.

5.3 Protection from Pathogens

The hyphae of ECM fungi provide increased surface area of their host root system not only for nutrient absorption but also for interactions with other microorganisms and provide an important pathway for translocation of energy-rich plant assimilates (products of photosynthesis) to the soil (Finlay 2004). The interactions may be synergistic, competitive or antagonistic and may have applied significance in areas such as sustainable forestry, biological control or bioremediation. Bacteria with potential to fix nitrogen have been discovered growing in association with tuberculate roots of ECM plants. The extent to which interactions between ECM mycelia and other microorganisms influence different organic or mineral substrates is still unclear. Further experiments are needed to distinguish between the activity of ECM hyphae themselves and facilitated ECM uptake of compounds mobilised by the activities of other organisms.

The microbiota of forest soils is dominated by ECM and saprotrophic decomposer fungi involved in supply of nutrients to trees and decomposition of woody plant litter, respectively. Basidiomycete mycelia (of which many are ECM) are ubiquitous in forest soils where they fulfil a range of key ecological functions (Cairney 2005). Interactions between the two groups of fungi are important in both managed forests and in natural forests. However, there are important differences between the two groups, saprotrophs obtain their C from decaying organic matter while the ECM fungi obtain most of their C directly from their host plants (Leake and

Johnson 2004). ECM mycelia of six different species reduced bacterial activity, estimated as thymidine incorporation, in experiments with sandy soil. Moreover, exudation and reabsorption of fluid droplets at ECM hyphal tips has already been demonstrated by Sun et al. (1999) which showed an important mechanism for conditioning the hyphal environment in the vicinity of tips creating interface for soil environment and other microorganisms. Fungi are known to produce effector molecules which help the plant to form association with the fungi. Hence effector molecules help in the association formation as compared to pathogenic fungi (Martin et al. 2008).

Mycorrhizal fungi also modify the interactions of plants with other soil organisms, both pathogenic (nematodes and fungi) as well as mutualists (nitrogen-fixing bacteria) (van Tichelen et al. 2001). Pathogenic fungi may invade roots but mycorrhizal fungi alter host response to these pathogens. Some bacterial strains isolated from the soils and rhizosphere significantly interacted either positively or negatively with growth of *Rhizopogon luteolus* mycelium along the root surface of *Pinus radiata* seedlings grown in vitro. *Laccaria bicolor* prevented spread of *Fusarium oxysporum* in Douglas-fir roots as a result of flavonoid wall infusions. Role of bacteria in promoting mycorrhizal formation and soil animals grazing external mycelium are among the important areas in functioning of symbiosis. For example, specific bacteria stimulate ECM formation in conifer nurseries and are called mycorrhization helper bacteria as found in *Pinus sylvestris-Lactarius rufus*, *P. sylvestris-Hebeloma* or *Amanita* (Vik et al. 2013; Kozdrój et al. 2007; Kluber et al. 2011). Wild sporocarps of *Laccaria bicolor* and *L. laccata* always contain large populations of bacilli and pseudomonads, while on the other hand species such as *Hebeloma cylindrosporum* do not contain bacteria in nature and form sporocarps in aseptic conditions. Earlier, Danell et al. (1993) have also reported relation of pseudomonads with *Cantharellus cibarius*.

Mycorrhizal fungi colonise feeder roots thereby interacting with root pathogens that parasitise same tissue. In a natural ecosystem where uptake of P is low, a major role of mycorrhizal fungi may be the protection of root system from endemic pathogens such as *Fusarium* spp. Mycorrhizal fungi may reduce the incidence and severity of root diseases. The mechanisms proposed to explain this protective effect include: (1) development of a mechanical barrier – especially the mantle of the ECM; (2) production of antibiotic compounds that suppress the pathogen; (3) competition for nutrients with the pathogen, including production of siderophores; and (4) induction of generalised host defence mechanisms. Over the last 40 years, there has been an increasing interest in the potential role that ECM and AM fungi play in control of plant diseases. Various aspects of this concept have been reviewed by many workers. More than 80 disease biocontrol products are on the market worldwide, but none of these contain mycorrhizal fungi. This is despite ample evidence that both AM and ECM can control a number of plant diseases. A focused approach should now be taken with ECM systems, which seem to have suffered from a lack of funding in this area, which is evident by the paucity of research published. It is possible to exploit these interactions in two ways. First, the effects of other organisms on mycorrhizal fungi may be modified to improve mycorrhizal function. Second, the ability of mycorrhizal fungi to interfere with pathogens might be used as a form of biological control.

6   Endophytic Fungi

Endophytic fungi have been known to benefit plant by secreting several chemicals, producing growth hormones, etc. Nowadays a fresh study has been started to re-evaluate the beneficial effects of plants as it is considered that it may the activity of endophytes. It is a bipartite symbiosis wherein both the members benefit each other (Fig. 6). Here we discuss how endophytes help plants in growth, nutrition and survival in tough conditions.

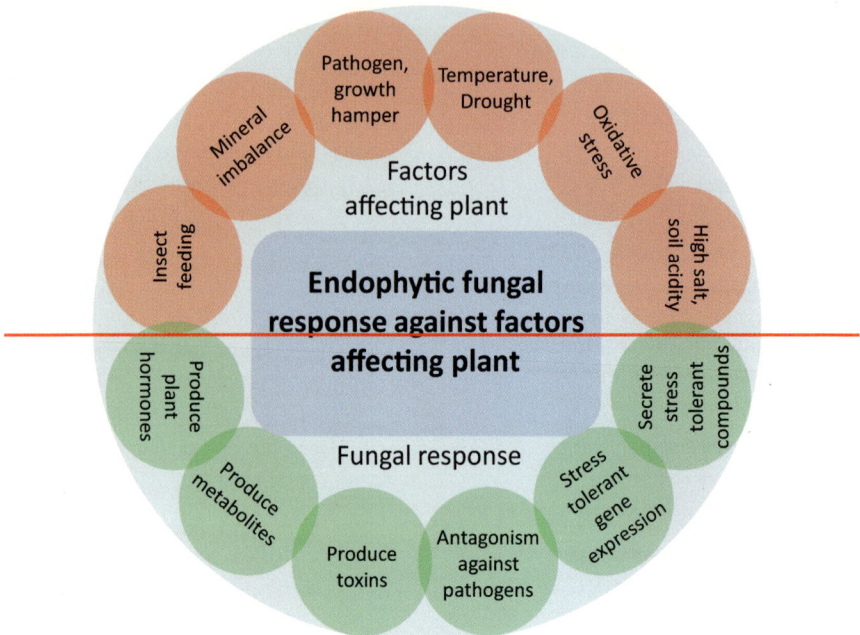

Fig. 6 Endophytic fungi shows various types of response related to factors affecting plant growth

6.1 Plant Growth-Promoting Effect

Endophytic fungi live in symbiosis with plant and play a crucial role in plant growth promotion. Fungi produce plant hormones, viz., gibberellins (GA), abscisic acid and auxin. Gibberellin is known to be produced by *Gibberella fujikuroi, Phaeosphaeria* sp., *Neurospora crassa, Aspergillus* sp., *Penicillium* sp. and *Paecilomyces formosus* (Hamayun et al. 2009c; Khan et al. 2009b, 2012). Screening 35 endophytic fungi for plant growth-promoting hormones found that *Penicillium* sp. produced more gibberellins. *Waito-C* rice mutant variety with deficient gibberellins was used for the study. Khan et al. (2012) observed that the culture filtrate of *Paecilomyces formosus* LHL10 when screened on gibberellin-deficient mutant rice *Waito-C* and normal gibberellin rice *Dongjin-byeo*, it increased the growth as compared to the one on which no filtrate was applied (devoid of other nutrients). The presence of Indole Acetic Acid (IAA) and GA suggested the presence of IAA and GA pathways. Similar effect was observed with cucumber. Waqas et al. (2012) also observed that endophytic fungi *Phoma glomerata* LWL2 and *Penicillium*

sp. LWL# significantly increased plant biomass and other growth-related factors by secreting GA and IAA. *P. indica* is also known to improve the growth rate of various plants. It is also known to activate nitrate reductase responsible for nitrogen acquisition (Waller et al. 2005; Singh et al. 2011). Previous reports have confirmed that endophytes can produce both GA and IAA which improve plant growth and crop production (Yuan et al. 2010). Several other fungi have been reported for gibberellins production, viz., *Sphaceloma manihoticola* (Bomke et al. 2008), *Phaeosphaeria* sp., *Sesamum indicum* (Choi et al. 2005), *Phaeosphaeria* sp. L487, *Penicillium citrinum, Chrysosporium pseudomerdarium* (Hamayun et al. 2009a), *Scolecobasidium tshawytschae* (Hamayun et al. 2009b), *Aspergillus fumigatus* and *Penicillium funiculosum* (Khan et al. 2012). *Beauveria bassiana* and *Metarhizium anisopliae* are two entomopathogenic fungi which are used commercially for biocontrol of insect pests. Researchers have tried to use them on crop plants where it lives endophytically inside host tissue and promote growth apart from harbouring protection against insects. Many other workers have demonstrated similar findings (Ownley et al. 2008). Elena et al. (2011) experimentally showed that *M. anisopliae* had plant growth-promoting effect on tomato plants. Although they were applied from outside, they entered the host plant forming endophytic symbiosis and finally helping in plant growth. This shows that the entomopathogenic fungi do not lose their virulence against target insect when inside plant tissue as endophyte. This capability of fungi can be exploited commercially by introducing potential fungal strains at the time of seed preparation. This kind of study can be of much interest to plant breeders looking for resistance varieties against insect or microbial pests. Plants treated with endophytes are healthier than untreated ones which could be due to the phytohormones secreted by the fungi. Endophytic fungi living in various parts of plant and producing growth-regulating compounds influence development of plant thus increasing survival in tough conditions.

6.2 Resistance to Hosts to Insect Feeding

During the course of evolution, plants have developed mechanism, made structural changes and formed relationship with fungi and other microbes to protect themselves from herbivores. They defend themselves by their defence mechanism or increasing effectivity of insect's natural enemy (Schoonhoven et al. 2005). Plants tend to support a system which promotes maximum growth by microbial interaction and minimum damage due to insects. Studies have shown that fungal endophytes increase hosts' resistance to insect feeding (Van Bael et al. 2009; Saikkonen et al. 2010). It has been reported that endophytes affect herbivores directly, decreasing their survival on host plant or indirectly affecting them by inducing the delays in developmental time, fecundity or foraging behaviours (van Bael et al. 2009; Bittleston et al. 2011). Endophytes also produce metabolites which are toxic to herbivores (Hartley and Gange 2009). Tanaka et al. (2005) have showed that peramine (fungal metabolite) protects plant from herbivores. Endophytes are also known to discourage mammalian herbivores from feeding (Li et al. 2004). Faeth et al. (2006) reported that *Achnatherum robustum* (sleepy grass) produces lysergic acid amide (causing sleepiness in animals) and therefore animals avoid them. *Achnatherum inebrians* is infected by *Neotyphodium gansuense* which is avoided by animals (Li et al. 2004). Schardl et al. (2004) also found that *Neotyphodium* colonises *Festuca arundinacea* (grass) and provides tolerance to the host against environmental factors and also produces alkaloids toxic to cattles. Similar observations have been made by Panaccione et al. (2006) with rabbits showing differential feeding habits for plants with and without alkaloids produced by endophyte. van Bael et al. (2009) have shown that endophytic fungi reduce leaf beetle, *Chelymorpha alternans*, towards plants. Moreover, Bittleston et al. (2011) experimentally demonstrated that Atta ants (*Atta colombica*) feeds on leaves of *Cordia alliodora* plants with less load of endophyte than leaves with more endophytic load. In other words endophytic fungi reduced the likelihood of leaf removal. Recently, Estrada et al. (2013) showed that ants cut about one-third more area of cucumber leaves with lower densities of endophytes. They also concluded that it is the change in the chemical composition of leaves (caused by endophytic *Colletotrichum tropicale*) that are responsible for the above different choice by ants. It removed 20 % more paper disks impregnated with the extracts of those leaves compared with leaves and disks from plants hosting the fungus. However, *C. tropicale* colonisation did not cause detectable changes in the composition of volatile compounds, cuticular waxes, nutrients or leaf toughness. van Bael et al. (2011) had earlier demonstrated that different compounds of plant leaves (metabolites, volatile compounds, toxins) determine the host selection by insects. Moreover, these compounds can be produced in different quantities as single fungus produces different metabolites in different media. Multiple species and strains present inside tissues complicate the issue. Analyses of compounds correlated with increased fungal biomass will help to identify the fungi acting as deterrent factor against insects. Moreover, chitin of fungal cell wall may also play an important role in this, making it more tough for insects to break. It is possible that for ants, leaves hosting fungal endophytes take longer to process than those free of the symbiont (van Bael et al. 2012). That means digesting the leaves with endophytic fungi may be difficult for insects and ants. We think the difference is same as organic and inorganic food and young and old plant. Since most of the metabolites are volatile, compounds with smell may also play a major role in the same. Mycotoxin compounds released from endophytic fungi may also play a role. Toxicity of metabolites released by endophyte to specific insect or herbivore or mammal may tell the exact reason for the deterrence of animals from plants, just as naphthalene, an insect repellent, is produced by *Muscodor vitigenus*, a novel endophytic fungus (Daisy et al. 2002). *Muscodor albus* produces volatile organic compound which is effective against insects (Riga et al. 2008; Lacey et al. 2009). Detailed studies in these aspect are lacking, but many other factors affect (positive or negative) the role of beneficial fungi. Plant developmental stage may also influence plant defences against herbivores. Gange et al. (1994) had shown that different combinations of *Glomus* species had different effects on host acceptance by a leaf-mining insect (*Chromatomyia syngenesiae*) and seed-feeding insect (*Tephritis neesii* and *Ozirhincus leucanthemi*). It has been observed that mycorrhizal fungi have positive or neutral effect on phloem feeders & specialist chewers and negative effect on mesophyll feeders & general chewing insects (Gehring and Bennett 2009; Hartley and Gange 2009; Koricheva et al. 2009; Pineda et al. 2010). Thus species composition and abundance also plays a key role in it. Environmental stress also changes the effects of fungal endophytes on mortality of plant-feeding insects, for example, endophytic fungus *Acremonium strictum* enhanced the mortality of whitefly (*Trialeurodes vaporariorum*) feeding on tomato plants during drought (Vidal 1996). Physiology and genetic mechanisms related to the effect of various fungi on plant

growth promotions and defence against plant pests need to be investigated in detail. Studies on effects of endophytic fungi on sap composition of plants and in tissue regeneration are understudied (Pineda et al. 2010). The outcome of the plant-fungi-insect interaction depends on several points, viz., type of plant, type of insect (generalist or specialist in terms of host preference), feeding habit of insect (piercing-cutting of host or chewing insect), abiotic and biotic environmental factors, amount of symbiosis/mutualism and mechanism of action (toxin production/ enhanced growth/volatile metabolite production, etc.). It still needs exploitation for biotechnological purpose in agriculture and forestry. These endophytic fungi can provide an effective solution to pest control and breeding-resistant varieties of crops or forest tree varieties apart from stress tolerance.

6.3 Disease Resistance

Another reason of endophytes becoming popular especially in the field of plant pathology, is their ability to resist pathogens (bacteria, fungi or nematodes) thus keeping plant healthy. It has been reported that plants harbouring endophytes show more resistance to pathogens in comparison to plant without or with low density of endophytes. Many endophytic fungi have shown antagonistic actions against plant pathogens in dual plate assays. Although it is difficult to relate the in vitro inhibition of pathogens by endophytes with actual in vivo conditions, studies have demonstrated that endophytes give resistance to plant against plant pathogens. Several studies have shown that endophytic fungi can help plants in protection against pathogens (Park et al. 2005a, b; Kim et al 2007). Clark et al. (2006) and Bonos et al. (2005) have shown that endophyte-infected turf grass showed more resistance to *Sclerotinia homoeocarpa* (dollar spot disease) and *Laetisaria fuciformis* (red thread disease). *Acremonium strictum* isolated from *Dactylis glomerata* (grass) is found to be a mycoparasite of *Helminthosporium solani* causing disease in potato (Rivera Varas et al. 2007). *Epichloe festucae* provides resistance against *Sclerotinia homoeocarpa* (causing dollar spot disease) in *Festuca rubra* (Clark et al. 2006). Barley inoculated *Piriformospora indica* shows resistance to *Fusarium culmorum* (vascular pathogen), *Blumeria graminis* (leaf pathogen) (Waller et al. 2005) and *Phytophthora* sp. (Arnold et al. 2003). Many entomopathogenic fungi, viz., *Beauveria bassiana* and *Lecanicillium* spp., found as endophytes (Vega et al. 2008) also suppress plant disease (Goettel et al. 2008; Ownley et al. 2008). However, it has been questioned by several researchers about reason behind resistance to disease, whether it should be attributed to antifungal compounds of endophyte, compounds produced by the plant in response to the endophyte or combination of other factors. Researchers from several different laboratories have already demonstrated the effect of fungal endophytes to protection against nematode to host plant. Studies in the past demonstrated that endophytes utilise various mechanisms to protect the plant from invasion of pathogen. It includes secretion of antibiotic or antifungal metabolites, increased host immune response, competition, antibiosis, mycoparasitism, providing of systemic resistance, etc. *Muscodor albus* have shown to produce volatile organic compounds against microbes (Strobel 2006). Many endophytes produce antibiotics which are effective against bacterial pathogens (Wang et al. 2007). Studies have also shown that if extract of an already present endophytic fungi is sprayed to plants, it colonises plant tissues and provides protection against pathogens. An important beneficial aspect of these endophytes is that they are mostly not host specific; however, some are reported to be one. As these endophytes are transferred vertically, seeds infected with them can be produced which can reduce the seed treatment practice by inorganic fertilisers. However, the mechanism is still not known completely as there are only few studies in this regard (Bordallo et al. 2002; Vu et al. 2006).

It seems that inside environment of plant is unique and harbours various kinds of fungi together. Moreover, many fungal species can live together protecting their host tissue

"livelihood" from plant pathogens. Antibacterial and antivirus effects of endophytes have also been observed by researchers (Lehtonen et al. 2006; Wang et al. 2007). Systematic studies looking for diversity of endophytes in crop species and varieties at various geographic and climatic regions are necessary. Evaluation of their potential against pathogens (fungi, bacterial, nematodes, virus) will help the plant breeders to come up with resistant varieties and reduce load on inorganic pesticides. With cost of pesticides increasing and many pesticides failing to control pathogens, it can be a cost-effective solution for farmers too and a step towards sustainable agriculture.

6.4 Stress Tolerance of Endophytes

Endophytes are known to help plants by various means. One of the beneficial effects of endophytes is the stress tolerance caused by biotic and abiotic factors. Plant faces various kinds of stress in the environment, viz., drought, heat, cold, oxidative stress, salinity, heavy metal toxicity, etc. With time and evolution, plant has developed various means and mode to overcome the same. For long time, plants were considered a single organism, but research in past few decades has highlighted the fact that it contains populations of multiple organisms which have helped them to adapt to various changes in environmental conditions in evolution. Abiotic stresses are a serious threat to crop plants and one of the major causes of loss of crop productivity around the world (Singh et al. 2011). These stresses cause considerable damage to plants in the form of disruption of cellular function, metabolic pathways, damaging structure of proteins, etc. (Wang et al. 2000). It is a common observation in crop fields that water stress at the time of branching, flowering and/or seed formation affects crop production (e.g. chickpea). In previous studies, *Piriformospora indica* has been experimentally shown to protect host plant from various stresses (Waller et al. 2005; Schäfer et al. 2007). We think that any kind of associations which had occurred during the evolution may be due to some or other reasons. Stress environments, viz., drought, high salt, etc. have forced plants and fungi to form associations to come out of the situation. Kuldau and Bacon (2008) have studied such beneficial effect of clavicipitaceous endophytes on host plants. In cacao plant, *Trichoderma* spp. occurs as endophyte and act against abiotic stress by inducing stress-tolerant gene expression (Bailey et al. 2006; Bae et al. 2009). Wilberforce et al. (2002) observed that some non-mycorrhizal fungi are found in roots in more quantities than mycorrhizal during stress environment indicating their role in tolerance. *Fusarium culmorum* and *Curvularia protuberata* (Redman et al. 2002). Rodriguez et al. (2008) observed that the plant and endophyte does not survive temperature above 40 °C and when in symbiotic association can tolerate up to 65 °C. Similar effect is observed between *F. culmorum* and host plant in context to high salt tolerance. The role of individual component of endophytic diversity in overcoming stress by a plant will help to elucidate this fact. It is possible that different component of endophytic diversity may be delivering different benefits. *P. indica* is an unusual fungus which has been shown to provide multiple benefits to host plants including stress tolerance (Waller et al. 2005). Redman et al. (2002) have shown that some plants do not tolerate the stress condition in the absence of endophytic fungi. *Dichanthelium lanuginosum* shows increased tolerance to heat due to *Curvularia* sp. Fungi are also known to provide tolerance to multiple hosts (Waller et al. 2005). Endophytes of grasses also provide protection to abiotic stresses. Fungi have also been involved in tripartite symbiotic association, where presence of bacteria/virus inside the endophytic fungi triggers its beneficial effect to host. In the absence of the third partner (bacteria/virus), it fails to provide protection to abiotic stress (Singh et al. 2011). Similar observation is observed with *Piriformospora* (Nautiyal et al. 2010). The molecular mechanism for the same is not much and needs further study. Yuan et al. (2010) has very nicely reviewed the stress tolerance benefits of endophytic fungi on plants. Studies have shown that the plants secrete various stress-tolerant chemicals/compounds like antioxidant enzymes, stress-tolerant hormones, metabolites, etc. Endophytic fungi help in stress-tolerating gene expression, viz., DREB2A, CBL1, ANAC072, RD29A, CAS protein, etc. The kind of mechanism adapted by the plant will depend on the plant and fungi involved and kind of stress. Trehalose an antioxidant, is one of the main stress-tolerant molecule produced by most plants which forms adaptive mode of stress tolerance mechanism. The other one is avoidance by growing roots to deeper region. The in vivo production of trehalose is considered to be an effect of endophytic fungal association. Trehalose is part of stress-tolerant mechanism for most plants. The kind of signal crosstalking which goes inside plant tissue against any stress condition is amazing to understand the complexity that goes during the process.

These can be crucial for drought-tolerant crop varieties for water scarce areas, an important aspect which plant breeders should look into and exploit. The ability of certain endophytic fungi originally isolated from grasses providing stress tolerance to tomato plants (genetically distant group of plants) (Chaw et al. 2004) is an important aspect in crop protection. With vertical transmission being a yes, these endophytic fungi can be exploited while developing new varieties. Moreover, different varieties for different regions (having different stress) with different compositions of endophytic diversity may help to solve regional environmental problems of crops.

7 Role of Bacteria in Soil Nutrient Mobility and Uptake

The effects of microorganisms on plants are well established for several microorganism-plant pairs, and interference with plant health and growth has been reported (Babalola 2010; Berlec 2012). The mechanisms behind beneficial plant-microbe interactions are complex phenomena involving a combination of direct and indirect mechanisms (Kloepper et al. 1989; Rodríguez et al. 2008; Son et al. 2009; Lugtenberg and Kamilova 2009; Compant et al. 2005; Berlec 2012). The direct beneficial effects of PGPB strains include enhancing phosphorus availability fixing atmospheric nitrogen (Bashan et al. 2004); mobilisation of potassium (Singh et al. 2010); sequestering iron for plants by production of siderophores (Bakker et al. 2007); producing plant hormones (Gutierrez-Manero et al. 2001; Spaepen et al. 2007) such as gibberellins, cytokinins, and auxins; and synthesising the enzyme ACC-deaminase, which lowers plant levels of ethylene, thereby reducing environmental stress on plants (Glick et al. 2007). The indirect mechanisms of plant growth promotion by PGPB include antibiotic production, depletion of iron from the rhizosphere, synthesis of antifungal metabolites, production of fungal cell-wall lysing enzymes, competition for sites on roots and induced systemic resistance (Sayyed and Chincholkar 2009). Many PGPB possess multiple plant growth-promoting attributes which influence plant growth at different developmental stages (Naik et al. 2008; Poonguzhali et al. 2008; Gulati et al. 2009).

7.1 Nitrogen

The nitrogen cycle is an essential and complex biogeochemical cycle that has a great impact on soil fertility (Jetten 2008). The cycle is dominated by four major microbial processes: N fixation, nitrification, denitrification, and N mineralisation (Ogunseitan 2005). Microbial inoculants have demonstrated significant roles in N cycling and plant utilisation of fertiliser N in the plant-soil system (Briones et al. 2003; Adesemoye et al. 2009). The biological nitrogen fixation (BNF) has a great practical importance because the use of nitrogenous fertilisers has resulted in unacceptable levels of water pollution (increasing concentrations of toxic nitrates in drinking water supplies) and the eutrophication of lakes and rivers (Sprent and Sprent 1990). It has been reported that fertiliser is usually applied in large doses, up to 50 % of which may be

leached, while BNF can be tailored to the needs of the organism (Sprent and Sprent 1990). A wide range of organisms have the ability to fix nitrogen including 87 species in 2 genera of archaea, 38 genera of bacteria and 20 genera of cyanobacteria (Sprent and Sprent 1990; Zahran et al. 1995).

In legumes the bacteria (rhizobia) reside in small growths on the roots called nodules and fix nitrogen which is absorbed by the plant. Rhizobia-legume symbioses have been reported to provide well over half of the biological source of fixed nitrogen and are the primary source of fixed nitrogen in land-based systems (Tate 1995). A renewable source of N for agriculture has been represented by the atmospheric nitrogen fixed symbiotically by the association between *Rhizobium* species and legumes (Peoples et al. 1995). Impressive values ranging from 200 to 300 kg N ha^{-1} $year^{-1}$ have been estimated for various legume crops and pasture species (Peoples et al. 1995). Inputs of fixed N for alfalfa, red clover, pea, soybean, cowpea and vetch have been estimated about 65–335 kg of N ha^{-1} $year^{-1}$ (Tate 1995). Preference for cheap and sensitive acetylene reduction assay for measuring nitrogen fixation has been suggested over the accurate and expensive ^{15}N isotopic method (Hardy et al. 1973; Sprent and Sprent 1990). Nodulation and nitrogenase activity have been employed as major traits for the evaluation of rhizobia-legume symbiosis and to select potential strains of rhizobia (Younis 2007).

In addition to symbiotic nitrogen fixation, bacteria also have nitrogen-fixing ability under free-living condition. The free-living nitrogen-fixing bacteria are widely distributed among phylogenetically diverse bacteria such as *Acetobacter, Arthrobacter, Azoarcus, Azospirillum, Azotobacter, Bacillus, Burkholderia, Enterobacter, Herbaspirillum, Klebsiella* and *Pseudomonas* associated with some agronomically important crops. Although the root-nodulating bacteria are known for their capacity to fix atmospheric nitrogen, they are not considered as plant growth-promoting rhizobacteria due to their highly specific symbiotic interactions. Free-living nitrogen-fixing bacteria have been considered as an alternative for inorganic nitrogen fertiliser for promoting plant growth (Park et al. 2005a, b). An increasing supply of N through dinitrogen fixation has been reported to increase crop production in saline habitats (Yao et al. 2010).

7.2   Phosphorus

Phosphorus is another plant growth-limiting nutrient as it affects plant structure at cellular level and stimulates growth and hastens maturity. Plants with P deficiency exhibit stunted growth, wilting of leaves, delayed maturity and reduced yield (Mallarino et al. 2002; Loria and Sawyer 2005). Although most agricultural soils have large amounts of inorganic and organic P, these are immobilised and mostly unavailable. Hence, only a very low concentration of P is available to plants, and many soils are actually P deficient (Fernández et al. 2007). One major reason that P is not readily available to plants is because of the high reactivity of P with some metal complexes such as iron (Fe), aluminium (Al), and calcium (Ca) leading to the precipitation or adsorption of between 75 and 90 % of P in the soil (Igual et al. 2001; Gyaneshwar et al. 2002). Even upon the application of P fertilisers to soils, 75 % of the soluble phosphate may be bound in soil or become sparingly soluble in form by reaction with the free Ca^{2+} ions in high pH soils or with Fe^{3+} or Al^{3+} in low pH soils, resulting in less than sufficient amount of P available for crop growth and yield (Gyaneshwar et al. 2002).

PGPB play significant roles in the solubilisation of inorganic phosphate and mineralisation of organic phosphates. Several strains have been reported from different environments with the capacity to solubilise mineral phosphate (Nautiyal et al. 2000; Gyaneshwar et al. 2002; Gulati et al. 2008; Patel et al. 2008; Khan et al. 2009a, b; Park et al. 2009; Son et al. 2009; Hariprasad and Niranjana 2009; Singh et al. 2010a). Bacteria belonging to *Achromobacter*, *Aerobacter*, *Agrobacterium*, *Bacillus*, *Burkholderia*, *Erwinia*, *Flavobacterium*, *Gluconacetobacter*, *Micrococcus*, *Pseudomonas*, *Ralstonia*, *Rahnella*, *Rhizobium*, *Serratia* and others have been reported for the

conversion of insoluble inorganic phosphates into soluble forms (Pandey et al. 2006; Pérez et al. 2007; Gulati et al. 2008; Poonguzhali et al. 2008; Linu et al. 2009; Vyas et al. 2010; Zabihi et al. 2011).

The principal mechanism for the phosphate solubilisation capacity has been reported as production of organic acids (Patel et al. 2008; Park et al. 2009). Gluconic acid has been the major organic acid produced by most of phosphate-solubilising bacteria. Direct periplasmic oxidation of glucose to gluconic acid is considered as the metabolic basis of inorganic phosphate solubilisation by many gram-negative bacteria as a competitive strategy to transform the readily available carbon sources into less readily utilisable products by other microorganisms (Goldstein and Krishnaraj 2007). Other organic acids including 2-ketogluconic, acetic, citric, glycolic, isovaleric, isobutyric, lactic, malonic, oxalic, propionic and succinic acids have also been detected during phosphate solubilisation (Vyas and Gulati 2009). The organic acids produced by the microorganisms reduce pH and act as chelating agents, forming complexes with Ca, Fe or Al and thereby releasing the phosphates to solution. Other mechanisms of solubilisation comprise the release of other chelating substances and inorganic acids such as sulphuric, nitric and carbonic acids. Secretion of phosphatase enzymes (acid and alkaline phosphatase, phytase, phosphohydrolase) by phosphobacteria has also been recorded as a common mode of conversion of insoluble forms of P to available forms and thus enhances plant P uptake and growth (Kohler et al. 2007).

The solubilisation of phosphates has been found in several species of root-nodulating bacteria of different legumes (Alikhani et al. 2006; Rivas et al. 2007). The process of formation of the nitrogen-fixing nodule has been reported to be limited by the availability of P (MacDermott 1999). High positive response has been recorded to P supplementation in legumes like alfalfa, clover, common bean, cow pea and pigeon pea (Al-Niemi et al. 1997; Deng et al. 1998). The nitrogen-fixing potential of aquatic legumes *Sesbania rostrata*, an important constituent of the green-manure technology for rice, has also been limited by P (Ladha et al. 1992; Ventura and Ladha 1997). The root-nodulating bacteria with phosphate-solubilising ability have been proved to be good plant growth-promoting bacteria for non-legumes (Yanni et al. 2001). In addition to the beneficial effects of rhizobia on legume and non-legume plants, inoculation and inoculant production technologies are already available, and they have been used with legumes for many years without causing harm to the environment or to farmers.

The rhizosphere bacteria pose the beneficial effect on plant growth by their potential for phosphate solubilisation. The phosphate-solubilising bacteria isolated from the rhizosphere of various plants have been known to be metabolically more active than those isolated from sources other than rhizosphere (Baya et al. 1981; Gyaneshwar et al. 2002). The studies on the diversity analysis of phosphate-solubilising microorganisms in rhizosphere and bulk soils collected from rock phosphate in Tachira, Venezuela, have concluded that numbers of phosphate-solubilising microorganisms were higher in the rhizosphere than in the bulk soil (Reyes et al. 1999). Increased plant growth, biomass and yield of different crops and plants have been recorded upon the inoculation of phosphate-solubilising rhizobacteria (Hariprasad and Niranjana 2009; Vyas and Gulati 2009). Although phosphorus solubilisation is one of the important mechanisms through which PGPB promote plant growth, many workers have suggested the consideration of other ways of plant growth promotion for the selection of PGPB simultaneously to phosphate solubilisation (Vassilev et al. 2006; Naik et al. 2008; Hariprasad and Niranjana 2009; Gulati et al. 2009).

7.3 Potassium

Potassium (K) is essential macronutrient for plant growth and plays significant roles in activation of several metabolic processes including protein synthesis, photosynthesis and enzymes, as well as in resistance to diseases and insects (Rehm and Schmitt 2002). Though it is present in soil as an abundant element or is also applied to fields as natural or synthetic fertilisers, only 1–2 % of it is available to plants, the rest being bound with other minerals and therefore unavailable to plants. The most common soil components of potassium, 90–98 %, are feldspar and mica (McAfee 2008). Composting of mica along with rice straw and rock phosphate inoculated with *Aspergillus awamori* has been used as a viable technology, where significant amount of insoluble K present in waste mica are mobilised into plant available form of K and used as a source of potassium in crop production, which could help to reduce the reliance on costly chemical fertilisers. Several soil microorganisms are able to solubilise unavailable form of K-bearing minerals, like mica, illite and orthoclases, by excreting organic acids which either directly dissolve mineral K or chelate silicon ions to bring the K into the solution (Friedrich et al. 1991; Ullman et al. 1996; Bennett et al. 1998). Three PGPR *Bacillus mucilaginosus*, *Azotobacter chroococcum* and *Rhizobium* sp. (specific to sunn hemp) exhibited their potential in mobilisation of K from waste mica under hydroponic cultivation of maize and wheat (Singh et al. 2010b). Significant increase in plant assimilation of K has been reported by the use of potassium-solubilising microorganisms (Goldstein and Liu 1987; Sheng 2005).

7.4 Iron

Iron (Fe) is a structural component of many of the proteins involved in important processes such as photosynthesis and nitrogen fixation. A large portion of the soil Fe is in highly insoluble form of ferric hydroxide which acts as a limiting factor for plant growth even in iron-rich soils. Although the total Fe content in soils usually far exceeds plant requirement for Fe, its bioavailability in the soil, especially in calcareous soils, is often severely limited.

Under conditions of Fe limitation, microorganisms and plants commonly rely on chelating agents to solubilise and transport inorganic Fe. The siderophores produced by microorganisms and phytosiderophores produced by few Fe-efficient grasses are among the most important naturally occurring biosynthetic chelates (Crowley et al. 1991). An important role of soil microbial activity in favouring Fe uptake has been suggested by plant growth experiments on sterile and non-sterile soils (Masalha et al. 2000; Jin et al. 2006). Variation in the rhizosphere microbial community has been observed with plant's Fe nutritional status (Yang and Crowley 2000). It has been found that, when the soil solution of a calcareous soil was incubated on an agar plate containing phenolic root exudates from Fe-deficient red clover, only a few microbial species thrived while growth of the rest is inhibited, and the majority of the microbes which thrived can secrete siderophores under Fe-deficient conditions (Jin et al. 2006, 2010).

Siderophores produced by soil microbes are seen as one of the microbial functions most supportive of Fe acquisition by plants, because siderophores have a high affinity for chelating Fe (III), and the resulting chelates have been proven to be an efficient bioavailable Fe source for plants. Higher Fe accumulation by plants supplemented with siderophore-Fe than by those supplemented with EDTA-Fe has been recorded in both shoots and roots suggesting the incorporation of siderophore-Fe in the roots in a more efficient way (Jin et al. 2010). Previously, higher Fe concentration has been observed in the Arabidopsis plants fed with Fe-pyoverdine than in those fed with EDTA-Fe (Vansuyt et al. 2007). From the studies on the plant Fe status on microbial communities, it has been concluded that the phenolic compounds exuded from plant roots under Fe-deficient conditions may selectively modify the microbial community structure in favouring more siderophore-secreting microbes, which helps to improve the solubility of insoluble iron and plant iron nutrition via microbial siderophores (Jin et al. 2010).

A total of approximately 500 different siderophore structures have been described so far which mainly consist of one or more ligand structures (Fig. 7) (Haselwandter and Winkelmann 2007). The three common functional groups that coordinate to Fe^{3+} in siderophores are hydroxamic acids, catechols and α-hydroxy-carboxylic acids (Zawadzka et al. 2006). Based on the functional groups, siderophores are classified into hydroxamate, catecholate or carboxylate types (Baakza et al. 2004). Catecholate- and hydroxamate-type siderophores have been widely reported in bacteria including *Aeromonas hydrophila*, *Azotobacter*, *Methylobacterium* spp., *Pseudomonas fluorescens*, *P. aeruginosa*, *P. pseudomallei*, *P. putida*, *P. stutzeri* and *Pseudomonas* sp. (Thirumurugan et al. 2006; Storey et al. 2006; Lacava et al. 2008). Carboxylate-type siderophores have been reported in a few bacteria including *Pseudomonas mediterranea*, *Pseudomonas* sp. and *Rhizobium meliloti* (Baakza et al. 2004; Tian et al. 2008, 2009). Catecholate-type siderophores have been reported for stronger binding to iron than hydroxamate-type siderophores (Matzanke 1991).

The ability to acquire iron by siderophore mediation is greatly advantageous to rootnodulating bacteria as iron is the structural component to many proteins involved in nitrogen fixation including nitrogenase, leghemoglobin and hydrogenase. Strains belonging to the group rhizobia have been reported to produce a variety of siderophores; some of these siderophores are unique in their functional group (Smith et al. 1985; Patel et al. 1988; Persmark et al. 1993). Nine rhizobia have been found positive for siderophore production out of 84 strains of rhizobia isolated from different legumes (Derylo and Skorupska 1992). Stimulated Fe uptake and shoot transport has been observed in clover plant on the application of purified rhizobial siderophore isolated for *Rhizobium leguminosarum* bv. *trifolii* (Derylo and Skorupska 1992). Strong antagonism has been observed against *Macrophomina phaseolina* by the siderophore producing strains RMP3 and RMP5 of *Rhizobium meliloti* (Arora et al. 2001).

Several microorganisms including *Alcaligenes*, *Azotobacter*, *Azospirillum*, *Bacillus*, *Enterobacter*, *Pseudomonas* and *Rhizobium* produce siderophores under low-iron conditions (Yang et al. 2009; da Silva and de Almeida 2006; Storey et al. 2006; Sayyed and Chincholkar 2009). Siderophore-producing rhizobacteria suppress fungal pathogens by making iron unavailable for fungal growth (Mahmoud and Abd-Alla 2001; Sharma and Johri 2003; Sayyed et al. 2007; Sayyed and

Fig. 7 Types of
siderophores

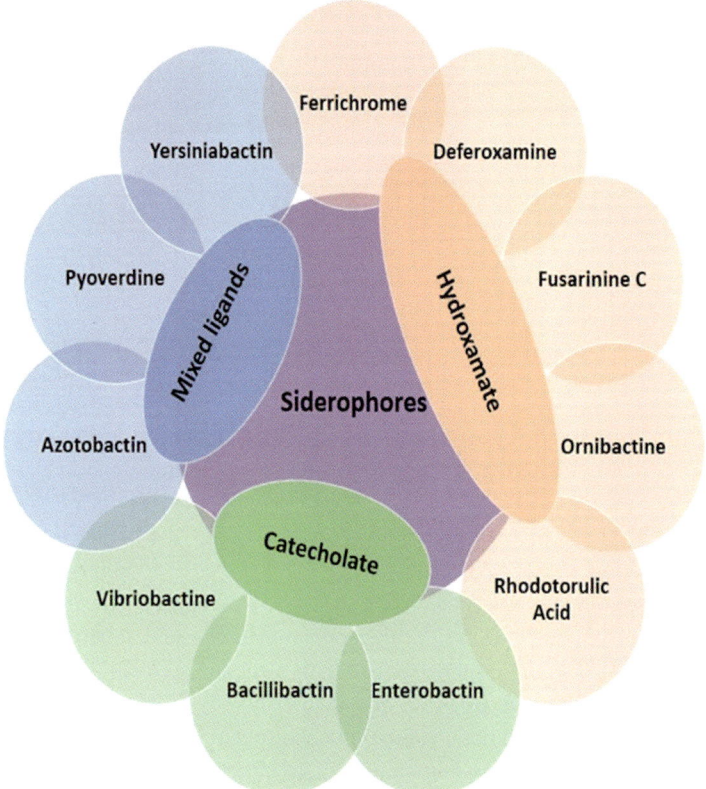

Chincholkar 2009). These bacteria also enhance availability of P to the plants through the solubilisation of iron-bound phosphorus in the soil (Duponois et al. 2006; Jing et al. 2007). Siderophore-mediated competition for iron is also a major factor in determining the interaction among bacterial strains during rhizosphere competence by PGPR (Loper and Henkels 1999; Jing et al. 2007).

8 Role of Cyanobacteria in Crop Nutrition

Cyanobacteria are ancient, large and diverse group of prokaryotic autotrophs, which exhibit oxygenic photosynthesis. Some are also capable of a photoheterotrophic mode of nutrition (Rippka 1972). In addition to their oxygenic photosynthesis, many cyanobacteria also perform biological nitrogen fixation, thereby using atmospheric N_2 as their sole nitrogen source, although they readily switch over to other

sources of nitrogen (combined nitrogen), if available. In the rice fields, cyanobacterial genera such as *Anabaena*, *Nostoc*, *Scytonema*, *Calothrix*, *Gloeotrichia* and *Fischerella* are abundant, apart from unicellular forms (Roger and Kulasooriya 1980). Cyanobacteria such as *Nostoc*, *Anabaena*, *Calothrix*, *Aulosira* and *Plectonema* are found to be ubiquitous while *Hapalosiphon*, *Scytonema* and *Cylindrospermum* are found to be localised in distribution (Venkataraman 1972; Kannaiyan 1985).

Contribution of cyanobacteria in agriculture is not restricted to its ability to carry out biological nitrogen fixation; it also helps in solubilisation of phosphorous and improves soil physical properties by exopolysaccharide production, addition of organic matter in the soil and production of plant growth-promoting substances. Ability of cyanobacteria in mobilising the inorganic phosphates in laboratory condition as well as soil is well documented (Roychoudhury and Kaushik 1989; Natesan and Shanmugasundaram 1989;

Yandigeri and Pabbi 2005). Kleiner and Harper (1977) reported that soils with cyanobacterial cover showed more extractable phosphates than in nearby soils without cover. The production of phthalic acid as a possible mode of P-solubilisation by cyanobacteria was proposed by Yandigeri et al. (2011). Application of cyanobacteria to rice crop was investigated extensively even though their application to other crops has also been shown interest in recent years.

8.1 Role of Cyanobacteria in Paddy Cultivation

As early as in 1939, attributing cyanobacteria for the maintenance of soil fertility in rice fields by De was the first report recognising the agronomic potential of cyanobacteria. Venkataraman (1972) termed 'algalization' as the inoculation of algae in rice fields at various localities and demonstrated the positive effect on grain yield of rice. Nitrogen-fixing cyanobacteria are known to play an important role in maintaining and improving the productivity of rice fields (Kaushik 1994). The application of cyanobacteria with reduced nitrogenous fertiliser gave yield equivalent to that of full dose nitrogenous fertiliser (Swarnalakshmi et al. 2006). Venkataraman (1981) reported the estimate of N_2-fixed as a result of cyanobacterial inoculation in the range of 20–30 kg ha^{-1}. Wet land flooded condition of soil supports sustenance of nitrogen fertility than dry land conditions (Watanabe and Roger 1984). Since the rice field ecosystem supplies all the vital requirements for cyanobacterial growth such as light, moisture, and mineral nutrients; the nitrogen-fixing cyanobacteria dominate in the Indian paddy fields. The favourable conditions of rice fields for the fixation of atmospheric nitrogen are one of the reasons for maintaining higher yields of paddy crop under low land conditions. Algalization increases the fertiliser use efficiency of crop plants and reduces the loss of nitrogen and above all provides these benefits for a longer duration (Goyal 1993).

Out of the various symbiotic associations between plants and the cyanobacteria, *Azolla-Anabaena azollae* symbiosis is the most important with respect to application in agriculture. The free floating fern *Azolla* is endowed with microsymbiont *Anabaena azollae*, a heterocystous nitrogen-fixing cyanobacteria. *Azolla* is widely applied as an organic fertiliser in the tropical rice fields of India, China, Vietnam etc. It not only supplies the fixed nitrogen but also valuable organic matter to the soil which in turn boost the growth of other plant growth-promoting bacteria in the soil (Gaur 2006). When applied in suitable conditions, *Azolla* is able to fix nitrogen up to 1.1 kg N h^{-1} day^{-1}. Becking (1979) showed that the maximum biomass produced by Azolla ranged from 0.8 to 5.2 t dry matter per day. Under reduced nitrogen levels of nitrogen, Azolla application maintains the higher rice yield (Yanni 1992).

Various researchers reported the cyanobacterial exudates showed increase in seed germination, root and shoot growth, weight of rice grains and their protein content (Venkataraman and Neelakantan 1967; Mishra and Kaushik 1989). There are different views on the nature of bioactive agents in the cyanobacterial exudates. The presence of both IAA and amino acids in the culture filtrates of cyanobacteria has already been reported (Karthikeyan 2006; Karthikeyan et al. 2009).

8.2 Wheat Production and Cyanobacteria

In the recent years, interest on utilising nitrogen-fixing cyanobacteria for wheat crop has gained importance as wheat requires large quantities of fertiliser nitrogen to achieve good yields. Spiller and Gunasekaran (1990, 1991) reported that cyanobacteria could enhance the growth of wheat crop under nitrogen-free hydroponic system. Under Saudi Arabian soil conditions, Kabli et al. (1997) conducted experiment to test the efficiency of cyanobacteria in wheat crop and concluded that *Nostoc commune* and *Anabaena cylindrica* could supply atmospheric nitrogen to wheat to maintain its vegetative growth. At Indian Agricultural Research Institute, New Delhi, the beneficial effects of cyanobacterial application to other crops notably wheat have been addressed by many workers. It has been found that yield of wheat crop with treatment including 1/3 N + full dose of P and K plus cyanobacterial culture was equivalent to wheat crop treated with full dose of NPK (Karthikeyan et al. 2007). They also reported the soil fertility improvement by the application of cyano-bacteria. Investigation on performance of cyanobacteria and bacterial combination for wheat crop (variety HD 2687) under pot culture condition was evaluated by Nain et al. (2010). They reported that combination of cyanobacteria and bacteria isolated from wheat rhizosphere exhibited plant growth promotion and increase in yield. The soil microbiological properties like dehydrogenase activity, FDA hydrolase, alkaline phosphatase and microbial biomass were also found to be enhanced by the application of cyanobacteria and bacteria combinations. Simi-lar results in wheat crop were also reported (Manjunath et al 2011; Rana et al. 2012). Swarnalakshmi et al. (2013) evaluated biofilmed preparation of cyanobacterium *Anabaena torulosa* with agriculturally beneficial bacteria such as *Azotobacter*, *Mesorhizobium*, *Serratia* and *Pseudomonas* in wheat crop in which single super phosphate (SSP) was supplied as P source. Results indicated that nitrogen fixation in terms of acetylene reduction (ARA) was enhanced even 14 weeks after inoculation. Application of SSP showed beneficial role in terms of increasing nitrogen fixation as well as increased P uptake by plant. Cyanobacterial con-sortium for the organic cultivation of wheat-rice cropping sequence was also investigated for their agronomic potential (Prakash et al. 2013a).

8.3   Other Crops

The attempt to use cyanobacteria for vegetable crops was done by Dadhich et al. (1969). They studied the efficiency of cyanobacterium *Calothrix anomala* on *Capsicum annum* and *Lactuca sativa* and found that yield and percent-age nitrogen was increased significantly in the treatment that included the combination of cyanobacteria and urea. Application of culture filtrate of *Aulosira fertilissima* to tomato crops enhanced vegetative growth of tomato

significantly than the control (Rizvi and Sharma 1994). Similar enhancement in growth para-meters of tomato seedlings was reported by Al-khiat (2006). Increase in vitamin C content of tomato by the application of cyanobacteria to tomato crop was reported by Kaushik and Venkataraman (1979).

Kumar et al. (2013) evaluated the potential of eight thermotolerant bacteria (seven *Bacillus* spp. and one actinobacterium *Kocuria* sp.) and two cyanobacteria (*Anabaena laxa* and *Calothrix elenkinii*) as plant growth-promoting (PGP) agents for seed spices, such as coriander, cumin and fennel, under controlled conditions in potting mix fortified with microbial cultures and found that germination percentage of cumin was enhanced by 25 % over control in treatment containing *A. laxa* while *Calothrix* enhanced root/shoot length significantly in all the three crops, especially fennel. Plant dry weight and peroxidase activity of shoots and roots were enhanced by five- to tenfold in all the microbe-inoculated treatments. Enhancement of soil ferti-lity is imperative to achieve maximum as well as sustainable crop yields to meet the growing demand for food. Cyanobacteria are proven biofertilisers for wetland crop like rice. But its utility is expanding to other crops such as wheat, vegetables and spice crops. They form an in-expensive farm grown input which helps in better crop nutrient management while working in per-fect harmony with nature. Cyanobacteria provide a holistic approach to improve soil physical, chemical and biological properties when com-bined with other plant growth-promoting microorganisms.

Conclusion

Supply of adequate amount of food and provi-sion of healthy environment for expanding global human population without changing the quality of natural ecosystem are the major challenges for agricultural as well as environmental scientists. Good status of soil health and adequate supply of bioavailable nutrients in the soil are crucial for plant dis-ease suppression, production of nutritious food and to improve the crop yield. In order to protect the crops from diseases and to improve their yield, chemical fertilisers and pesticides have been extensively used in the recent past. Although use of chemical fertilisers and pesticides has increased the crop yield, it has simultaneously destroyed the quality of soil, disturbed the natural community structure of soil microbial population and the natural cycling of materials and hampered the concept of sustainability. In addition, excess use of chemicals not only affected the diversity of soil microflora but also indirectly distracted the quality of underground water table as well as surface water bodies by the processes of leaching and surface run-off, respectively. Data from past research indicated that microbes play valuable role in plant nutrition and disease suppression by their diverse activities like phosphate and sulphate solubilisation, production of signalling compounds and growth hormones, nitrogen fixation, etc. It is an established fact that microbial-based biofertilisers and biopesticides are the better alternative of chemicals and beneficial for human health and environment. Unfortunately microbial-based formulation lacks universality in their effects and activity and gives different results in different geographical and climatic conditions. More research and extra efforts are required to discover the better microbes for effective formulations. Recent advances in sequencing technology and soil metagenomics have revealed that vast majority of microbial population of the soil are unidentified and functionally not characterised. Despite all the effort microbiologists are only able to cultivate the 1–10 % of microbes, and the rest 90–99 % are not yet cultured (Rappe and Giovannoni 2003). Study of soil microbiome using next-generation sequencing further widened the windows of uncultured soil microbial diversity. Here it is advisable to first of all trace out the black box of microbial diversity either using the soil metagenomics or by soil microbiome study. After that, formulate the strategy for isolation of ecologically and agriculturally important organisms under laboratory conditions using modern cultivation approaches (Prakash et al. 2013a) and preserve them for commercial formulations (Prakash et al. 2013b). After cultivation and functional characterisation, select the ecologically beneficial microorganisms like PGPR, diazotroph phosphate solubiliser and pesticide degraders for the formulation of biopesticide, biofertilisers, biocomposting and/or biodegradation of environmental pollutants for sustainable agriculture and environment (Lugtenberg and Kamilova 2009; Chandler et al. 2008).

Initially it was considered that due to small size and universal distribution, ecological disturbances do not affect the microbial diversity and its conservation. Therefore, the concept of microbial diversity conservation got less attention than plant and animal diversity conservation. Now it is established that ecological disturbances like addition of chemicals, climatic variations and tillage affect the structure and functions of soil microbial community and even cause the permanent loss of valuable microorganisms from the habitats. Progress in microbial ecology has proved that microbial diversity is more valuable and needs conservation. Preservation of cultivable organisms under culture collection or bioresource centres has been optimised, but conservation of not yet cultured but ecologically valuable microorganisms is a challenging task and needs more attention. Conservation of intact habitats is a good option for conservation of not yet cultured microorganism. Therefore, not only cultivation, screening and preservation of ecologically and agriculturally important microorganisms are important but conservation of valuable but not yet cultured organisms are equally important for better future and sustainable development.

Acknowledgments This work was supported by the Department of Biotechnology (DBT; Grant no. BT/PR/ 0054/NDB/52/94/2007), Govt. of India, under the project "Establishment of microbialculture collection."

References

Abiala MA, Popoola OO, Olawuyi OJ, Oyelude JO, Akanmu AO, Killani AS, Osonubi O, Odebode AC (2013) Harnessing the potentials of Vesicular Arbuscular Mycorrhizal (VAM) fungi to plant growth – a review. Int J Pure Appl Sci Technol 14(2):61–79

Adesemoye AO, Torbert HA, Kloepper JW (2009) Plant growth-promoting rhizobacteria allow reduced application rates of chemical fertilizers. Microb Ecol 54:921–929

Agerer R (2001) Exploration types of ectomycorrhizae: a proposal to classify ectomycorrhizal mycelial systems according to their patterns of differentiation and putative ecological importance. Mycorrhiza 11:107–114

Al-khiat SHA (2006) Effect of *Cyanobacteria* as a soil conditioner and biofertilizer on growth and some biochemical characteristics of tomato (*Lycopersicon esculentum* L.) Seedlings. Dissertation, King Saud University

Al-Niemi TS, Summers ML, Elkins JG, Kahn ML, Mcdermott TR (1997) Regulation of the phosphate stress response in *Rhizobium meliloti* by PhoB. Appl Environ Microbiol 63:4978–4981

Alikhani HA, Saleh-Rastin N, Antoun H (2006) Phosphate solubilization activity of rhizobia native to Iranian soils. Plant Soil 287:35–41

Alizadeh O (2012) A critical review on the nutrition role of arbuscular mycorrhizal fungi. ELBA Bioflux 4:1–7

Allen MF, Swenson W, Ouerejeta JI, Egerton-Warburton LM, Treseder KK (2003) Ecology of mycorrhizae: a conceptual framework for complex interactions among plants and fungi. Ann Rev Phytopathol 41:271–303

Alvarez M, Huygens D, Olivares E, Saavedra I, Alberdi M, Valenzuela E (2009) Ectomycorrhizal fungi enhance nitrogen and phosphorus nutrition of *Nothofagus dombeyi* under drought conditions by regulating assimilative enzyme activities. Physiol Plant 136(4):426–36. doi:10.1111/j.1399-3054.2009.01237.x

Andrade SA, Silveira AP, Mazzafera P (2010) Arbuscular mycorrhiza alters metal uptake and the physiological response of *Coffea arabica* seedlings to increasing Zn and Cu concentrations in soil. Sci Total Environ 408 (22):5381–5391. doi:10.1016/j.scitotenv.2010.07.064

Arnold AE, Mejía LC, Kyllo D, Rojas EI, Maynard Z, Robbins N, Herre EA (2003) Fungal endophytes limit pathogen damage in a tropical tree. PNAS 100: 15649–15654. doi:10.1073/pnas.2533483100

Arora NK, Kang SC, Maheshwari DK (2001) Isolation of siderophore producing strains of *Rhizobium meliloti* and their biocontrol potential against *Macrophomina phaseolina* that causes charcoal rot of groundnut. Curr Sci 81:673–677

Aseri GK, Jain N, Panwar J, Rao AV, Meghwal PR (2008) Biofertilizers improve plant growth, fruit yield, nutrition, metabolism and rhizosphere enzyme activities of Pomegranate (*Punica granatum* L.) in Indian Thar desert. Sci Hortic 117:130–135

Awasthi A, Bharti N, Nair P, Singh R, Shukla AK, Gupta MM, Darokar MP, Kalra A (2011) Synergistic effect of *Glomus mosseae* and nitrogen fixing *Bacillus subtilis* strain Daz26 on artemisinin content in *Artemisia annua* L. Appl Soil Ecol 49:125–130

Baakza A, Vala AK, Dave BP, Dube HC (2004) A comparative study of siderophore production by fungi from marine and terrestrial habitats. J Exp Mar Biol Ecol 311:1–9

Babalola OO (2010) Beneficial bacteria of agriculture importance. Biotechnol Lett 32:1559–1570

Bae H, Sicher RC, Kim MS, Kim S-H, Strem MD, Melnick RL, Bailey BA (2009) The beneficial endophyte *Trichoderma hamatum* isolate DIS 219b promotes growth and delays the onset of the drought response in *Theobroma cacao*. J Exp Bot 60: 3279–3295. doi:10.1093/jxb/erp165

Bailey BA, Bae H, Strem MD, Roberts DP, Thomas SE, Samuels GJ, Choi I-Y, Holmes KA (2006) Fungal and plant gene expression during the colonization of cacao seedlings by endophytic isolates of four *Trichoderma* species. Planta 224:1449–1464

Bais HP, Weir TL, Perry LG, Gilroy S, Vivanco JM (2006) The role of root exudates in rhizosphere interactions with plants and other organisms. Annu Rev Plant Biol 57:233–266

Bakker PAHM, Pieterse CMJ, van Loon LC (2007) Induced systemic resistance by fluorescent *Pseudomonas* spp. Phytopathology 97:239–243

Banchio E, Bogino PC, Santoro M, Torres L, Zygadlo J, Giordano W (2010) Systemic induction of monoterpene biosynthesis in *Origanumxmajoricum* by soil bacteria. J Agric Food Chem 58:650–654

Baslam M, Goicoechea N (2012) Water deficit improved the capacity of arbuscular mycorrhizal fungi (AMF) for inducing the accumulation of antioxidant compounds in lettuce leaves. Mycorrhiza 22:347–359. doi:10.1007/s00572-011-0408-9

Baslam M, Garmendia I, Goicoechea N (2011a) Arbuscular mycorrhizal fungi (AMF) improved growth and nutritional quality of greenhouse-grown lettuce. J Agric Food Chem 59:5504–5515. doi:10.1021/jf200501c

Baslam M, Pascual I, Sanchez-Díaz M, Erro J, García-Mina JM, Goicoechea N (2011b) Improvement of nutritional quality of greenhouse-grown lettuce by arbuscular mycorrhizal fungi is conditioned by the source of Phosphorus nutrition. J Agric Food Chem 59:11129–11140. doi:10.1021/jf202445y

Baslam M, Esteban R, García-Plazaola JI, Goicoechea N (2013a) Effectiveness of arbuscular mycorrhizal fungi (AMF) for inducing the accumulation of major carotenoids, chlorophylls and tocopherol in green and red leaf lettuces. Appl Microbiol Biotechnol 97:3119–3128. doi:10.1007/s00253-012-4526-x

Baslam M, Garmendia I, Goicoechea N (2013b) Enhanced accumulation of vitamins, nutraceuticals and minerals in lettuces associated with arbuscular mycorrhizal fungi (AMF): a question of interest for both vegetables and humans. Agriculture 3:188–209. doi:10.3390/agriculture3010188

Baslam M, Garmendia I, Goicoechea N (2013c) The arbuscular mycorrhizal symbiosis can overcome reductions in yield and nutritional quality in greenhouse-lettuces cultivated at inappropriate growing seasons. Sci Hortic 164:145–154

Baya MA, Boehhing RS, Ramos-Cormenzana A (1981) Vitamin production in relation to phosphate solubilization by bacteria. Soil Biol Biochem 13:527–531

Becking JH (1979) Environmental requirements of Azolla for use in tropical rice production. Nitrogen and Rice, IRRI, Manila, pp 354–374

Bellion M, Courbot M, Jacob C, Blaudez D, Chalot M (2005) Extracellular and cellular mechanisms sustaining metal tolerance in ectomycorrhizal fungi. FEMS Microbiol Lett 254(2):173–181. doi:10.1111/j.1574-6968.2005.00044.x

Bennett PC, Choi WJ, Rogers JR (1998) Microbial destruction of feldspars. Miner Manag 8:149–150

Berlec A (2012) Novel techniques and findings in the study of plant microbiota: search for plant probiotics. Plant Sci 2(193):96–102

Bertin C, Yang XH, Weston LA (2003) The role of root exudates and allelochemicals in the rhizosphere. Plant Soil 256:67–83

Bianciotto V, Bonfante P (2002) Arbuscular mycorrhizal fungi: a specialized niche for rhizospheric and endocellular bacteria. Anton van Leeuwenhoek 81: 365–371

Bittleston LS, Brockmann F, Wcislo W, van Bael SA (2011) Endophytic fungi reduce leaf-cutting ant damage to seedlings. Biol Lett 7(1):30–32. doi:10.1098/rsbl.2010.0456

Blanke V, Renke C, Wagner M, Fuller K, Held M, Kuhn AJ, Bruscot F (2005) Nitrogen supply affects arbuscular mycorrhizal colonization of *Artemisia vulgaris* in a phosphate polluted field sites. New Phytol 166:981–992

Bomke C, Rojas MC, Gong F, Hedden P, Tudzynski B (2008) Isolation and characterization of the gibberellin biosynthetic gene cluster in *Sphaceloma manihoticola*. Appl Environ Microbiol 74:5325–5339

Bonfante P (2001) At the interface between mycorrhizal fungi and plants: the structural organization of cell wall, plasma membrane and cytoskeleton. In: Hock B (ed) Mycota, IX: fungal associations. Springer, Berlin

Bonfante P (2003) Plants, mycorrhizal fungi and endobacteria: a dialog among cells and genomes. Biol Bull 204:215–220

Bonfante P, Genre A (2010) Mechanisms underlying beneficial plant-fungus interactions in mycorrhizal symbiosis. Nat Commun 1:48. doi:10.1038/ncomms1046

Bonos SA, Wilson MM, Meyer WA, Funk CR (2005) Suppression of red thread in fine fescues through endophyte-mediated resistance. Appl Turfgrass Sci 10:1094

Bordallo JJ, López Llorca LV, Jannson HB, Salinas J, Persmark L, Asensio L (2002) Colonization of plant roots by egg-parasitic and nematode-trapping fungi. New Phytol 154:491–499

Borowicz VA (2009) Organic farm soil improves strawberry growth but does not diminish spittlebug damage. J Sustain Agric 33:177–188

Brandle JE, Telmer PG (2007) Steviol glycoside biosynthesis. Phytochemistry 68:1855–1863

Briones AM, Okabe S, Umemiya Y, Ramsing N, Reichardt W, Okuyama H (2003) Ammonia-oxidizing bacteria on root biofilms and their possible contribution to N use efficiency of different rice cultivars. Plant Soil 250:335–348

Bücking H, Heyser W (2003) Uptake and transfer of nutrients in ectomycorrhizal associations: interactions between photosynthesis and phosphate nutrition. Mycorrhiza 13:59–68. doi:10.1007/s00572-002-0196-3

Cairney JWG (2005) Basidiomycete mycelia in forest soils: dimensions, dynamics and roles in nutrient distribution. Mycol Res 109(1):7–20

Cardoso IM, Kuyper TW (2006) Mycorrhizas and tropical soil fertility. Agric Ecosyst Environ 116:72–84

Cavagnaro TR, Martin AW (2010) The role of mycorrhizas in plant nutrition: field and mutant based approaches. In: 19th world congress of soil science, soil solutions for a changing world, Brisbane, Australia, 1–6 August 2010

Cavagnaro TR, Smith FA, Ayling SM, Smith SE (2003) Growth and phosphorus nutrition of a *Paris*-type arbuscular mycorrhizal symbiosis. New Phytol 157: 127–134

Cavagnaro TR, Langley AJ, Jackson LE, Smukler SM, Koch GW (2008) Growth, nutrition, and soil respiration of a mycorrhiza-defective tomato mutant and its mycorrhizal wild-type progenitor. Funct Plant Biol 35:228–235

Chandler D, Davidson G, Grant W, Greaves J, Tatchell MG (2008) Microbial biopesticides for integrated crop management: an assessment of environmental and regulatory sustainability. Trends Food Sci Technol 19:275–283. ISSN 0924-2244

Chaw S, Chang C, Chen H, Li W (2004) Dating the monocot–dicot divergence and the origin of core eudicots using whole chloroplast genomes. J Mol Evol 58:424–441

Choi WY, Rim SO, Lee JH, Lee JM, Lee IJ, Cho KJ, Rhee IK, Kwon JB, Kim JG (2005) Isolation of gibberellins producing fungi from the root of several *Sesamum indicum* plants. J Microbiol Biotechnol 15:22–28

Clark MM, Gwinn KD, Ownley BH (2006) Biological control of *Pythium myriotylum*. Phytopathology 96: S25

Compant S, Duffy B, Nowak J, Climent C, Barka EA (2005) Use of plant growth promoting bacteria for biocontrol of plant diseases: principles, mechanism of action, and future prospects. Appl Environ Microbiol 71:4951–4959

Crowley DE, Wang YC, Reid CPP, Szanislo PJ (1991) Mechanisms of iron acquisition from siderophores by microorganisms and plants. In: Chen Y, Hadar Y (eds) Iron nutrition and interactions in plants. Kluwer Academic Publishers, Dordrecht

da Silva GA, de Almeida EA (2006) Production of yellow-green fluorescent pigment by *Pseudomonas fluorescens*. Braz Arch Biol Technol 49:411–419

Dadhich KS, Varma AK, Venkataraman GS (1969) The effect of *Calothrix* inoculation on vegetable crops. Plant Soil 31:377–379

Daisy BH, Strobel GA, Castillo U, Ezra D, Sears J, Weaver DK, Runyon JB (2002) Naphthalene, an insect repellent, is produced by *Muscodor vitigenus*, a novel endophytic fungus. Microbiology 148: 3737–3741

Danell E, Alstrom A, Ternstrom A (1993) *Pseudomonas fluorescens* in association with fruit bodies of the ectomycorrhizal mushroom *Cantharellus cibarius*. Mycol Res 97:1148–1152

Das K, Dang R, Shivananda T, Sekeroglu N (2007) Influence of bio-fertilizers on the biomass yield and nutrient content in *Stevia rebaudiana* Bert. grown in Indian subtropics. J Med Plant Res 1:005–008

Deng SP, Summers M, Kahn ML, McDermott TR (1998) Cloning and characterization of a *Rhizobium meliloti* non-specific acid phosphatase. Arch Microbiol 170: 18–26

Derylo M, Skorupska A (1992) Rhizobial siderophore as an iron source for clover. Physiol Plant 85:549–553

Duponois R, Kisa M, Plenchette C (2006) Phosphate-solubilizing potential of the nematophagous fungus *Arthrobotrys oligospora*. J Plant Nutr Soil Sci 169: 280–282

Elena GJ, Beatriz PJ, Alejandro P, Roberto L (2011) *Metarhizium anisopliae* (Metschnikoff) sorokin promotes growth and has endophytic activity in tomato plants. Adv Biol Res 5:22–27

Estrada C, Wcislo WT, van Bael SA (2013) Symbiotic fungi alter plant chemistry that discourages leaf-cutting ants. New Phytol 198(1):241–251. doi:10.1111/nph.12140

Faeth SH, Gardner DR, Hayes CJ, Jani A, Wittlinger SK, Jones TA (2006) Temporal and spatial variation in alkaloid levels in *Achnatherum robustum*, a native grass infected with the endophyte *Neotyphodium*. J Chem Ecol 32:307–324

Fernández LA, Zalba P, Gomez MA, Sagardoy MA (2007) Phosphate-solubilization activity of bacterial strains in soil and their effect on soybean growth under greenhouse conditions. Biol Fertil Soils 43: 805–809

Finlay RD (2004) Mycorrhizal fungi and their multi-functional role. Mycologist 18(2):91–96

Fiorilli V, Catoni M, Francia D, Cardinale F, Lanfranco L (2011) The arbuscular mycorrhizal symbiosis reduces disease severity in tomato plants infected by *Botrytis cinerea*. J Plant Pathol 93(1):237–242

Friedrich S, Platonova NP, Karavaiko GI, Stichel E, Glombitza F (1991) Chemical and microbiological solubilization of silicates. Acta Biotechnol 11(3): 187–196

Fritz M, Jakobsen I, Langkjaer MF, Thordal-Christensen-H, Pons-Kühnemann J (2006) Arbuscular mycorrhiza reduces susceptibility of tomato to *Alternaria solani*. Mycorrhiza 16:413–419

Gadd GM (2010) Metals, minerals and microbes: geomicrobiology and bioremediation. Microbiology 156:609–643

Gange AC, Brown VK, Sinclair GS (1994) Reduction of black vine weevil larval growth by vesicular–arbuscular mycorrhizal infection. Entomol Exp Appl 70:115–119

Gaur AC (2006) *Azolla* act as green manure and production. Biofertilizers in sustainable agriculture. ICAR, New Delhi

Gehring C, Bennett A (2009) Mycorrhizal fungal-plant-insect interactions: the importance of a community approach. Environ Entomol 38:93–102

Glick BR, Cheng Z, Czarny J, Duan J (2007) Promotion of plant growth by ACC deaminase-producing soil bacteria. Eur J Plant Pathol 119:329–339

Goettel MS, Koike M, Kim JJ, Aiuchi D, Shinya R, Brodeur J (2008) Potential of Lecanicillium spp. for management of insects, nematodes and plant diseases. J Invertebr Pathol 98:256–261

Goldstein AH, Krishnaraj PU (2007) Phosphate solubilizing microorganisms vs. phosphate mobilizing microorganisms: what separates a phenotype from a trait? In: Velázquez E, Rodríguez-Barrueco C (eds) Proceedings of 1st international meeting on microbial phosphate solubilization, vol 102. Springer, Dordrecht

Goldstein AH, Liu ST (1987) Molecular cloning and regulation of a mineral phosphate solubilizing gene from *Erwinia herbicola*. Biotechnology 5:72–74

Gosling P, Hodge A, Goodlass G, Bending GD (2006) Arbuscular mycorrhizal fungi and organic farming. Agric Ecosyst Environ 113:17–35

Goyal SK (1993) Algal biofertilizer for vital soil and free nitrogen. Proc Indian Natl Sci Acad 59:295–301

Gulati A, Rahi P, Vyas P (2008) Characterization of phosphate-solubilizing fluorescent pseudomonads from the rhizosphere of seabuckthorn growing in the cold deserts of Himalayas. Curr Microbiol 56:73–79

Gulati A, Vyas P, Rahi P, Kasana RC (2009) Plant growth promoting and rhizosphere competent *Acinetobacter rhizosphaerae* strain BIHB 723 from the cold deserts of Himalayas. Curr Microbiol 58:371–377

Gutierrez-Manero FJ, Ramos-Solano B, Robanza A, Mehouachi J, Tadeo FR, Talon M (2001) The plant-growth-promoting rhizobacteria *Bacillus pumilus* and *Bacillus licheniformis* produce high amounts of physiologically active gibberellins. Physiol Plant 111: 206–211

Gyaneshwar P, Kumar GN, Parekh LJ, Poole PS (2002) Role of soil microorganisms in improving P nutrition of plants. Plant Soil 245:83–93

Hamayun M, Khan SA, Iqbal I, Hwang YH, Shin DH, Sohn EY, Lee BH, Na CI, Lee IJ (2009a) *Chrysosporium pseudomerdarium* produces gibberellins and promotes plant growth. J Microbiol 47: 425–430

Hamayun M, Khan SA, Kim HY, Chaudhary MF, Hwang YH, Shin DH, Kim IK, Lee BH, Lee IJ (2009b) Gibberellins production and plant growth enhancement by newly isolated strain of *Scolecobasidium tshawytschae*. J Microb Biotechnol 19:560–565

Hamayun M, Khan SA, Khan MA, Khan AL, Kang S-M, Kim S-K, Joo G-J, Lee I-J (2009c) Gibberellin production by pure cultures of a new strain of *Aspergillus fumigatus*. World J Microbiol Biotechnol 25:1785–1792

Hardy RWF, Burns RC, Holsten RD (1973) Applications of the acetylene-ethylene assay for measurement of nitrogen fixation. Soil Biol Biochem 5:47–81

Hariprasad P, Niranjana SR (2009) Isolation and characterization of phosphate solubilizing rhizobacteria to improve plant health of tomato. Plant Soil 316:13–24

Hartley SE, Gange AC (2009) Impacts of plant symbiotic fungi on insect herbivores: mutualism in a multitrophic context. Annu Rev Entomol 54:323–342

Haselwandter K, Winkelmann G (2007) Siderophores of symbiotic fungi. In: Varma A, Chincholkar SB (eds) Microbial siderophores, vol 12, Soil biology. Springer, Berlin/Heidelberg

Hawkins HJ, Johansen A, George E (2000) Uptake and transport of organic and inorganic nitrogen by arbuscular mycorrhizal fungi. Plant Soil 226:275–285

He X, Critchley C, Ng H, Bledsoe C (2005) Nodulated N_2-fixing *Casuarina cunninghamiana* is the sink for net N transfer from non-N_2-fixing Eucalyptus maculate via an ectomycorrhizal fungus *Pisolithus* sp. using $^{15}NH_4^+$ or $^{15}NO_3^-$ supplied as ammonium nitrate. New Phytol 167:897–912. doi:10.1111/j.1469-8137.2005.01437.x

Hemavathi VN, Sivakumar BS, Suresh CK, Earanna N (2006) Effect of *Glomus fasciculatum* and plant growth promoting rhizobacteria on growth and yield of *Ocimum basilicum*. Karnataka J Agric Sci 19:17–20

Hodge A (2003) Plant nitrogen capture from organic matter as affected by spatial dispersion, interspecific competition and mycorrhizal colonization. New Phytol 157(2):303–314

Hodges SC (2010) Soil fertility basics. Soil science extension. North Carolina State University, Raleigh, pp 4927–4932

Igual JM, Valverde A, Cervantes E, Velázquez E (2001) Phosphate-solubilizing bacteria as inoculants for agriculture: use of updated molecular techniques in their study. Agronomie 21:561–568

Jentschkea G, Godbold DL (2000) Metal toxicity and ectomycorrhizas. Physiol Plant 109:107–116

Jetten MSM (2008) The microbial nitrogen cycle. Environ Microbiol 10(11):2903–2909

Jin CW, He YF, Tang CX, Wu P, Zheng SJ (2006) Mechanisms of microbial enhanced iron uptake in red clover. Plant Cell Environ 29:888–897

Jin CW, Li GX, Yu XH, Zheng SJ (2010) Plant Fe status affects the composition of siderophore-secreting microbes in the rhizosphere. Ann Bot 105:835–841

Jing YD, He ZL, Yang XE (2007) Role of soil rhizobacteria in phytoremediation of heavy metal contaminated soils. J Zhejiang Univ Sci B 8:192–207

Kabli SA, Al-Garni SM, Al-Fassi FA (1997) Efficiency of cyanobacteria from soil of western region in the Kingdom of Saudi Arabia as biofertilizer for wheat. Arab Gulf J Sci Res 15:481–503

Kannaiyan S (1985) Studies on the algal application for lowland rice crop. Tamil Nadu Agric Univ Bull, Coimbatore

Karandashov V, Nagy R, Wegmüller S, Amrhein N, Bucher M (2004) Evolutionary conservation of a phosphate transporter in the arbuscular mycorrhizal symbiosis. Proc Natl Acad Sci U S A 101:6285–6290

Karthikeyan N (2006) Characterization of cyanobacteria from the rhizosphere of wheat. Dissertation, Indian Agricultural Research Institute

Karthikeyan N, Prasanna R, Lata KBD (2007) Evaluating the potential of plant growth promoting cyanobacteria as inoculants for wheat. Eur J Soil Biol 43:23–30

Karthikeyan N, Prasanna R, Sood A, Jaiswal P, Nayak S, Kaushik BD (2009) Physiological characterization and electron microscopic investigations of cyanobacteria associated with wheat rhizosphere. Folia Microbiol 54:43–51

Kaushik BD (1994) Algalization of rice fields in salt affected soils. Ann Agric Res 15:105–106

Kaushik BD, Venkataraman GS (1979) Effect of algal inoculation on the yield and vitamin C content of two varieties of tomato. Plant Soil 52:135–137

Khan AA, Jilani G, Akhtar MS, Naqvi SMS, Rasheed M (2009a) Phosphorus solubilizing bacteria: occurrence, mechanisms and their role in crop production. J Agric Biol Sci 1:48–58

Khan SA, Hamayun M, Kim HY, Yoon HJ, Seo JC, Choo YS, Lee IJ, Kim SD, Rhee IK, Kim JG (2009b) A new strain of *Arthrinium phaeospermum* isolated from *Carex kobomugi* Ohwi is capable of gibberellin production. Biotechnol Lett 31:283–287

Khan AL, Hamayun M, Kang SM, Kim YH, Jung HY, Lee JH, Lee IJ (2012) Endophytic fungal association via gibberellins and indole acetic acid can improve plant growth under abiotic stress: an example of *Paecilomyces formosus* LHL10. BMC Microbiol 12:3

Kim HY, Choi GJ, Lee HB, Lee SW, Kim HK, Jang KS, Son SW, Lee SO, Cho KY, Sung ND, Kim JC (2007) Some fungal endophytes from vegetable crops and their anti-oomycete activities against tomato late blight. Lett Appl Microbiol 44:332–337

Kleiner KT, Harper KT (1977) Soil properties in relation to cryptogamic ground cover in Canyon-lands National park. J Range Manag 30:202–205

Kloepper JW, Lifshitz R, Zablotowicz RM (1989) Free living bacterial inocula for enhancing crop productivity. Trends Biotechnol 7:39–44

Kluber LA, Smith JE, Myrolda DD (2011) Distinctive fungal and bacterial communities are associated with mats formed by ectomycorrhizal fungi. Soil Biol Biochem 43:1042–1050

Kohler J, Caravaca F, Carrasco L, Roldan A (2007) Interactions between a plant growth-promoting rhizobacterium, an AM fungus and a phosphate-solubilising fungus in the rhizosphere of *Lactuca sativa*. Appl Soil Ecol 35:480–487

Koricheva J, Gange AC, Jones T (2009) Effects of mycor-rhizal fungi on insect herbivores: a meta-analysis. Ecology 90:2088–2097

Kozdrój J, Piotrowska-Seget Z, Krupa P (2007) Mycor-rhizal fungi and ectomycorrhiza associated bacteria isolated from an industrial desert soil protect pine seedlings against Cd (II) impact. Ecotoxicology 16 (6):449–456

Krings M, Taylor TN, Hass H, Kerp H, Dotzler N, Hermsen EJ (2007) Fungal endophytes in a 400-mil-lion-yr-old land plant: infection pathways, spatial dis-tribution, and host responses. New Phytol 174(3): 648–657

Kuldau G, Bacon C (2008) Clavicipitaceous endophytes: their ability to enhance resistance of grasses to multi-ple stresses. Biol Control 46:57–71

Kumar M, Prasanna R, Bidyarani N, Babu S, Mishra BK, Kumar A, Adaka A, Jauharia S, Yadav K, Singh R, Saxena AK (2013) Evaluating the plant growth pro-moting ability of thermotolerant bacteria and cyano-bacteria and their interactions with seed spice crops. Sci Hortic 164:94–101

Lacava PT, Silva-Stenico ME, Araújo WL, Simionato AVC, Carrilho E, Tsai SM, Azevedo JL (2008) Detec-tion of siderophores in endophytic bacteria *Methylo-bacterium* spp. associated with *Xylella fastidiosa* subsp. p*auca*. Pesq Agrop Brasileira 43:521–528

Lacey LA, Horton DR, Jones DC, Headrick HL, Neven LG (2009) Efficacy of biofumigant fungus *Muscodor albus* (Ascomycota: Xylariales) for control of codling moth (Lepidoptera: Tortricidae) in stimulated storage conditions. J Econ Entomol 102:43–49

Ladha JK, Pareek RP, Beeker N (1992) Stem-nodulating legume-*Rhizobium* symbiosis and its agronomic use in lowland rice. In: Stewart BA (ed) Advances in soil science, vol 20. Springer, New York

Leake JR, Johnson D (2004) Networks of power and influence: the role of mycorrhizal mycelium in controlling plant communities and agroecosystem functioning. Can J Bot 82(8):1016–1045

Lehtonen PT, Helander M, Siddiqui SA, Lehto K, Saikkonen K (2006) Endophytic fungus decreases plant virus infections in meadow ryegrass (*Lolium pratense*). Biol Lett 2:620–623

Li CJ, Nan ZB, Paul VH, Dapprich PD, Liu Y (2004) A new *Neotyphodium* species symbiotic with drunken horse grass (*Achnatherum inebrians*) in China. Myco-taxon 90:141–147

Linu MS, Stephen J, Jisha MS (2009) Phosphate solubi-lizing *Gluconacetobacter* sp., *Burkholderia* sp. and their potential interaction with cowpea (*Vigna unguiculata* (L.) Walp.). Int J Agric Res 4:79–87

Lioussanne L (2010) The role of the arbuscular mycorrhiza-associated rhizobacteria in the biocontrol of soilborne phytopathogens. Span J Agric Res 8(S1): S51–S61

Liu J, Maldonado-Mendoza I, Lopez-Meyer M, Cheung F, Town CD, Harrison MJ (2007) Arbuscular

mycorrhizal symbiosis is accompanied by local and systemic alterations in gene expression and an increase in disease resistance in the shoots. Plant J 50:529–544

Loper JE, Henkels MD (1999) Utilization of heterologous siderophores enhances levels of iron available to *Pseudomonas putida* in the rhizosphere. Appl Environ Microbiol 65:5357–5363

Loria ER, Sawyer JE (2005) Extractable soil phosphorus and inorganic nitrogen following application of raw and anaerobically digested swine manure. Agron J 97: 879–885

Lou DI, Hussmann JA, McBee RM, Acevedo A, Andino R, Press WH, Sawyer SL (2013) High-throughput DNA sequencing errors are reduced by orders of magnitude using circle sequencing. Proc Nat Acad Sci U S A 110:19872–19877

Lugtenberg B, Kamilova F (2009) Plant growth-promoting rhizobacteria. Ann Rev Microbiol 63:541–556

MacDermott TR (1999) Phosphorus assimilation and regulation in rhizobia. In: Triplett EW (ed) Nitrogen fixation in prokaryotes: molecular and cellular bio-logy. Horizon Scientific Press, Norfolk

Mahmoud ALE, Abd-Alla MH (2001) Siderophores pro-duction by some microorganisms and their effect on *Bradyrhizobium*-mung bean symbiosis. Int J Agric Biol 3:157–162

Mallarino AP, Stewart BM, Baker JL, Downing JD, Sawyer JE (2002) Phosphorus indexing for cropland: overview and basic concepts of the Iowa phosphorus index. J Soil Water Conserv 57:440–447

Manjunath M, Prasanna R, Sharma P, Nain L, Singh R (2011) Developing PGPR consortia using novel genera *Providencia* and *Alcaligenes* along with cyano-bacteria for wheat. Arch Agron Soil Sci 57:873–887

Marschner H (1995) Mineral nutrition of higher plants, 2nd edn. Academic, London, p 889

Martin F, Nehls U (2009) Harnessing ectomycorrhizal genomics for ecological insights. Curr Opin Plant Biol 12:508–515

Martin F, Plassard C (2001) Nitrogen assimilation by ectomycorrhizal symbiosis. In: Morot-Gaudry JF (ed) Nitrogen assimilation by plants: physiological, biochemical and molecular aspects. Science Publishers, Enfield

Martin F, Aerts A, Ahren D, Brun A, Danchin EG, Duchaussoy F, Gibon J, Kohler A, Lindquist E, Pereda V et al (2008) The genome of *Laccaria bicolor* provides insights into mycorrhizal symbiosis. Nature 452:88–92

Masalha J, Kosegarten H, Elmaci O, Mengel K (2000) The central role of microbial activity for iron acquisi-tion in maize and sunflower. Biol Fertil Soils 30: 433–439

Matzanke BF (1991) Structures, coordination chemistry and functions of microbial iron chelates. In: Winkelmann G (ed) CRC handbook of microbial iron chelates. CRC Press, Boca Raton

McAfee J (2008) Potassium, a key nutrient for plant growth. Department of Soil and Crop Sciences, http://jimmcafee.tamu.edu/files/potassium

Mehta CM, Palni U, Franke-Whittle IH, Sharma AK (2013) Compost: its role, mechanism and impact on reducing soil-borne plant diseases. Waste Manag 34: 607–622

Menkis A, Bakys R, Lygis V, Vasaitis R (2011) Mycorrhization, establishment and growth of out-planted *Picea abies* seedlings produced under different cultivation systems. Silv Fenn 45(2):283–289

Microbes S (2010) Understanding soil microbes and nutrient recycling. Actinomycetes 107:40–50

Miransari M (2011) Interactions between arbuscular mycorrhizal fungi and soil bacteria. Appl Microbiol Biotechnol 89:917–930

Mishra S, Kaushik BD (1989) Growth promoting substances of cyanobacteria. I. Vitamins and their influence on rice plant. Proc Indian Nat Sci Acad B55:295–300

Mohammadi K, Khalesro S, Sohrabi Y, Heidari G (2011) Beneficial effects of the mycorrhizal fungi for plant growth. J Appl Environ Biol Sci 1(9):310–319

Morgan JB, Connolly EL (2013) Plant-soil interactions: nutrient uptake. Nat Educ Knowl 4(8):2

Mortimer PE, Pérez-Fernández MA, Valentine AJ (2009) Arbuscular mycorrhiza affects the N and C economy of nodulated *Phaseolus vulgaris* (L.) during NH4 nutrition. Soil Biol Biochem 41:2115–2121

Mukerji KG, Ciancio A (2007) Mycorrhizae in the integrated pest and disease, section-2. In: Ciancio A, Mukerji KG (eds) Management general concepts in integrated pest and disease management. Springer, Dordrecht

Naik PR, Sahoo N, Goswami D, Ayyadurai N, Sakthivel N (2008) Genetic and functional diversity among fluorescent pseudomonads isolated from the rhizosphere of banana. Microb Ecol 56:492–504

Nain L, Rana A, Joshi M, Shrikrishna JD, Kumar D, Shivay YS, Paul S, Prasanna R (2010) Evaluation of synergistic effects of bacterial and cyanobacterial strains as biofertilizers for wheat. Plant Soil 331: 217–230

Natesan R, Shanmugasundaram S (1989) Extracellular phosphate solubilization by the cyanobacterium *Anabaena* ARM 310. J Biosci 14:203–208

Nautiyal CS, Bhadauria S, Kumar P, Lal H, Mondal R, Verma D (2000) Stress induced phosphate solubilization in bacteria isolated from alkaline soils. FEMS Microbiol Lett 182:291–296

Nautiyal CS, Chauhan PS, DasGupta SM, Seem K, Varma A, Staddon WJ (2010) Tripartite interactions among *Paenibacillus lentimorbus* NRRLB-30488, *Piriformospora indica* DSM 11827 and *Cicer arietinum* L. World J Microbiol Biotechnol 26: 1393–1399. doi:10.1007/s11274-010-0312-z

Nisha MC, Rajeshkumar S (2010) Effect of arbuscular mycorrhizal fungi on growth and nutrition of *Wedelia chinensis* (Osbeck) Merril. Indian J Sci Technol 3(6): 676–678

Ogunseitan O (2005) Microbial diversity: form and function in prokaryotes. Blackwell, Malden

Ortega U, Dunabeitia M, Menendez S, Gonzalez-Murua-C, Majada J (2004) Effectiveness of mycorrhizal inoculation in the nursery on growth and water relations of *Pinus radiata* in different water regimes. Tree Physiol 24:65–73

Ownley BH, Griffin MR, Klingeman WE, Gwinn KD, Moulton JK, Pereira RM (2008) *Beauveria bassiana*: endophytic colonization and plant disease control. J Invertebr Pathol 98:267–270

Pal KK, Gardener BM (2006) Biological control of plant pathogens. Plant Health Instr. doi:10.1094/PHI-A-2006-1117-02

Panaccione DG, Cipoletti JR, Sedlock AB, Blemings KP, Schardl CL, Machado C, Seidel G (2006) Effects of ergot alkaloids on food preference and satiety in rabbits as assessed with gene-knockout endophytes in perennial ryegrass (*Lolium perenne*). J Agri Food Chem 54:4582–4587

Pandey A, Trivedi P, Kumar B, Palni LMS (2006) Characterization of a phosphate solubilizing and antagonistic strain of *Pseudomonas putida* (B0) isolated from a sub-alpine location in the Indian central Himalaya. Curr Microbiol 53:102–107

Park JH, Choi GJ, Lee HB, Kim KM, Jung HS, Lee SW, Jang KS, Cho KY (2005a) Griseofulvin from *Xylaria* sp. strain F0010, and endophytic fungus of *Abies holophylla* and its antifungal activity against plant pathogenic fungi. J Microbiol Biotechnol 15(1): 112–117

Park M, Kim C, Yang J, Lee H, Shin W, Kim S, Sa T (2005b) Isolation and characterization of diazotrophic growth promoting bacteria from rhizosphere of agricultural crops of Korea. Microbiol Res 160:127–133

Park KH, Lee CY, Son HJ (2009) Mechanism of insoluble phosphate solubilization by *Pseudomonas fluorescens* RAF15 isolated from ginseng rhizosphere and its plant growth-promoting activities. Lett Appl Microbiol 49: 222–228

Patel HN, Chakraborty RN, Desai SB (1988) Isolation and partial characterization of phenolate siderophore from *Rhizobium leguminosarum* IARI 102. FEMS Microbiol Lett 56:131–134

Patel DK, Archana G, Kumar GN (2008) Variation in the nature of organic acid secretion and mineral phosphate solubilization by *Citrobacter* sp. DHRSS in the presence of different sugars. Curr Microbiol 56:168–174

Peoples MB, Herridge DF, Ladha JK (1995) Biological nitrogen fixation: an efficient source of nitrogen for sustainable agricultural production? Plant Soil 174: 3–28

Pérez E, Sulbarán M, Ball M, Yarzabal LA (2007) Isolation and characterization of mineral phosphate-solubilizing bacteria naturally colonizing a limonitic crust in the south-eastern Venezuelan region. Soil Biol Biochem 39:2905–2914

Persmark M, Pittman P, Buyer JS, Schwyn B, Gill PR, Neilands JB (1993) Isolation and structure of rhizobactin 1021, a siderophore from alfalfa symbiont *Rhizobium meliloti* 1021. J Am Chem Soc 115: 3950–3956

Peterson RL, Massicotte HB, Melville LH (2004) Mycorrhizas: anatomy and cell biology. CABI Publishing/CAB International, Wallingford/Oxon

Pineda A, Zheng S-J, van Loon JJA, Pieterse CMJ, Dicke M (2010) Helping plants to deal with insects: the role of beneficial soil-borne microbes. Trends Plant Sci 15:507–514. doi:10.1016/j.tplants.2010.05.007

Plassard C, Bonafos B, Touraine B (2000) Differential effects of mineral and organic N sources, and of ectomycorrhizal infection by *Hebeloma cylindrosporum*, on growth and N utilization in *Pinus pinaster*. Plant Cell Environ 23:1195–1205

Poole EJ, Bending GD, Whipps JM, Read DJ (2001) Bacteria associated with *Pinus sylvestris–Lactarius rufus* ectomycorrhizas and their effects on mycorrhiza formation *in vitro*. New Phytol 151(3):743–751

Poonguzhali S, Madhaiyan M, Sa T (2008) Isolation and identification of phosphate solubilizing bacteria from Chinese cabbage and their effect on growth and phosphorus utilization of plants. J Microbiol Biotechnol 18:773–777

Pozo MJ, Verhage A, García-Andrade J, García JM, Azcón-Aguilar C (2008) Priming plant defence against pathogens by arbuscular mycorrhizal fungi. In: Azcon-Aguilar C et al (eds) Mycorrhizas – functional processes and ecological impact. Springer, Berlin/Heidelberg

Prakash O, Nimonkar Y, Shouche YS (2013a) Practice and prospects of microbial preservation. FEMS Microbiol Lett 339:1–9

Prakash O, Shouche Y, Jangid K, Kostka JE (2013b) Microbial cultivation and the role of microbial resource centers in the omics era. Appl Microbiol Biotechnol 97:51–62

Rahi P, Vyas P, Sharma S, Gulati A (2009) Plant growth promoting potential of the fungus Discosia sp. FIHB 571 from tea rhizosphere tested on chickpea, maize and pea. Indian J Microbiol 49:128–133

Rana A, Joshi M, Prasanna R, Shivay YS, Nain L (2012) Biofortification of wheat through inoculation of plant growth promoting rhizobacteria and cyanobacteria. Eur J Soil Biol 50:118–126

Redecker D, Morton JB, Bruns TD (2000) Ancestral lineages of arbuscular mycorrhizal fungi (Glomales). Mol Phylogenet Evol 14:276–284

Redman RS, Sheehan KB, Stout RG, Rodriguez RJ, Henson JM (2002) Thermotolerance conferred to

plant host and fungal endophyte during mutualistic symbiosis. Science 298:1581

Rehm G, Schmitt M (2002) Potassium for crop production. Retrieved February 2, 2011, from Regents of the University of Minnesota website: http://www.extension.umn.edu/distribution/cropsystems/dc6794.html

Reyes I, Bernier L, Simard R, Antoun H (1999) Effect of nitrogen source on solubilization of different inorganic phosphates by an isolate of *Penicillium rugulosum* and two UV-induced mutants. FEMS Microbiol Ecol 28:281–290

Riga K, Lacey LA, Guerra N (2008) The potential of the endophytic fungus, *Muscodor albus*, as a biocontrol agent against economically important plant parasitic nematodes of vegetable crops in Washington State. Biol Control 45:380–385

Rippka R (1972) Photoheterotrophy and chemoheterotrophy among unicellular blue green algae. Arch Microbiol 87:94–98

Rivas R, Peix A, Mateos PF, Trujillo ME, Martínez-Molina E, Velázquez E (2007) Biodiversity of populations of phosphate solubilizing rhizobia that nodulates chickpea in different Spanish soils. In: Velazquez E, Rodriguez-Barrueco C (eds) First international meeting on microbial phosphate solubilization. Springer, Dordrecht

Rivera Varas VV, Freeman TA, Gusmestad NC, Secor GA (2007) Mycoparasitism of *Helminthosporium solani* by *Acremonium strictum*. Phytopathology 9(97):1331–1337

Rizvi Z, Sharma VK (1994) Algae as biofertilizer for tomato plants. Rec Adv Phycol, conference paper Vol. NA:221–223

Rodríguez H, Vessely S, Shah S, Glick BR (2008) Effect of a nickel-tolerant ACC deaminase-producing *Pseudomonas* strain on growth of nontransformed and transgenic canola plants. Curr Microbiol 57: 170–174

Rodriguez RJ, Henson J, Van Volkenburgh E, Hoy M, Wright L, Beckwith F, Kim Y, Redman RS (2008) Stress tolerance in plants via habitat-adapted symbiosis. ISME J 2:404–416. doi:10.1038/ismej.2007.106

Roger PA, Kulasooriya SA (1980) Blue green algae and rice. International Rice Research Institute, Manila

Roger A, Gétaz M, Rasmann S, Sanders IR (2013) Identity and combinations of arbuscular mycorrhizal fungal isolates influence plant resistance and insect preference. Ecol Entomol 38(4):330–338. doi:10.1111/een.12022

Roychoudhury P, Kaushik BD (1989) Solubilization of Mussorie rock phosphate by cyanobacteria. Curr Sci 58:569–570

Ryu CM, Farag MA, Hu CH, Reddy MS, Wei HX et al (2003) Bacterial volatiles promote growth in

Arabidopsis. Proc Natl Acad Sci U S A 100: 4927–4932

Saikkonen K, Saari S, Helander M (2010) Defensive mutualism between plants and endophytic fungi? Fungal Divers 41:101–113

Sayyed RZ, Chincholkar SB (2009) Siderophore-producing *Alcaligenes faecalis* exhibited more bio-control potential vis-à-vis chemical fungicide. Curr Microbiol 58:47–51

Sayyed RZ, Naphade BS, Chincholkar SB (2007) Siderophore producing *A. feacalis* promoted the growth of Safed musali and Ashwagandha. J Med Arom Plant 29:1–5

Schäfer P, Khatabi B, Kogel KH (2007) Root cell death and systemic effects of *Piriformospora indica*: a study on mutualism. FEMS Microbiol Lett 275:1–7

Schardl CL, Leuchtmann A, Spiering MJ (2004) Symbioses of grasses with seed borne fungal endophytes. Annu Rev Plant Biol 55:315–340

Schoonhoven LM, van Loon JJA, Dicke M (2005) Insect-plant biology. Oxford University Press, Oxford/New York

Sharma A, Johri BN (2003) Growth promoting influence of siderophore-producing *Pseudomonas* strains GRP3A and PRS$_9$ in maize (*Zea mays* L.) under iron limiting conditions. Microbiol Res 158:243–248

Sharma R, Rajak RC (2011) Ectomycorrhizal interaction between *Cantharellus* and *Dendrocalamus*. In: Rai M, Varma A (eds) Diversity and biotechnology of ectomycorrhizae, vol 25, Soil biology. Springer, Berlin/Heidelberg

Sharma R, Baghel RK, Pandey AK (2010) Dynamics of acid phosphatase production of the ectomycorrhizal mushroom *Cantharellus tropicalis*. Afr J Microbiol Res 4(20):2072–2078

Sheng XF (2005) Growth promotion and increased potassium uptake of cotton and rape by a potassium releasing strain of *Bacillus edaphicus*. Soil Biol Biochem 37:1918–1922

Sikes BA, Cottenie K, Klironomos JN (2009) Plant and fungal identity determines pathogen protection of plant roots by arbuscular mycorrhizas. J Ecol 97:1274–1280

Simard SW, Durall DM (2004) Mycorrhizal networks: a review of their extent, function, and importance1. Can J Bot 82:1140–1165

Sinclair TR (1999) Limits to crop yield. In: Fedroeff NV, JE cohen (eds) Plants and population: is there time? Colloquium. National Academy of Sciences, Washington, DC

Singh AV, Shah S, Prasad B (2010a) Effect of phosphate solubilizing bacteria on plant growth promotion and nodulation in soybean (*Glycine max* (L.) Merr.). J Hill Agric 1(1):35–39

Singh G, Biswas DR, Marwaha TS (2010b) Mobilization of potassium from waste mica by plant growth promoting and its assimilation by maize (*Zea mays*) and wheat (*Triticum aestivum* L.): a hydroponics study

under phytotron growth chamber. J Plant Nutr 33(8): 1236–1251

Singh LP, Gill SS, Tuteja N (2011) Unraveling the role of fungal symbionts in plant abiotic stress tolerance. Plant Signal Behav 6(2):175–191

Singh R, Soni SK, Kalra A (2012) Synergy between *Glomus fasciculatum* and a beneficial Pseudomonas in reducing root diseases and improving yield and forskolin content in *Coleus forskohlii* Briq. under organic field conditions. Mycorrhiza 23(1):35–44. doi:10.1007/ s00572-012-0447-x

Smith SE, Read DJ (2008) Mycorrhizal symbiosis, 3rd edn. Elsevier, New York. ISBN 978-0-12-370526-6

Smith FA, Smith SE (2011a) What is the significance of the arbuscular mycorrhizal colonisation of many economically important crop plants? Plant Soil 348: 63–79. doi:10.1007/s11104-011-0865-0

Smith SE, Smith FA (2011b) Roles of arbuscular mycorrhizas in plant nutrition and growth: new paradigms from cellular to ecosystem scales. Annu Rev Plant Biol 62:227–250

Smith SE, Smith FA (2012) Fresh perspectives on the roles of arbuscular mycorrhizal fungi in plant nutrition and growth. Mycologia 104(1):1–13. doi:10.3852/11-229

Smith MJ, Schoolery JN, Schwyn B, Neilands JB (1985) Rhizobactin, a structurally novel siderophore from *Rhizobium meliloti*. J Am Chem Soc 107:1739–1743

Smith S, Smith A, Jakobsen I (2003) Mycorrhizal fungi can dominate phosphorus supply to plant irrespective of growth response. Plant Physiol 133:16–20

Solomon D, Lehmann J, Martinez CE (2003) Sulphur K-edge XANES spectroscopy as a tool for understanding sulphur dynamics in soil organic matter. Soil Sci Soc Am J 67:1721–1731

Son SH, Khan Z, Kim SG, Kim YH (2009) Plant growth-promoting rhizobacteria, *Paenibacillus polymyxa* and *Paenibacillus lentimorbus* suppress disease complex caused by root-knot nematode and *Fusarium* wilt fungus. J Appl Microbiol 107:524–532

Spaepen S, Vanderleyden J, Remans R (2007) Indole-3-acetic acid in microbial and microorganism-plant signaling. FEMS Microbiol Rev 31(4):425–448

Sparling GP, Pankhurst C, Doube BM, Gupta VVSR (1997) Soil microbial biomass, activity and nutrient cycling as indicators of soil health. Biol Indic Soil Health :97–119

Spiller H, Gunasekaran M (1990) Ammonia excreting mutant strain of the cyanobacterium *Anabaena variabilis* supported growth of wheat. Appl Microbiol Biotechnol 33:477–480

Spiller H, Gunasekaran M (1991) Simultaneous oxygen production and nitrogenase activity of an ammonia excreting mutant strain of the cyanobacterium *Anabaena variabilis* in a coculture with wheat. Appl Microbiol Biotechnol 35:798–804

Sprent JI, Sprent P (1990) Nitrogen fixing organisms: pure and applied aspects. Chapman and Hall, London

Srinivasan R, Govindasamy C (2014) Influence of native arbuscular mycorrhizal fungi on growth, nutrition and phytochemical constituents of *Catharanthus roseus* (L.) G. Don. J Coast Life Med 2(1):31–37

Storey EP, Boghozian R, Little JL, Lowman DW, Chakraborty R (2006) Characterization of 'Schizokinen'; a dihydroxamate-type siderophore produced by *Rhizobium leguminosarum* IARI 917. Biometals 19:637–649

Strobel G (2006) Microbial gifts from rain forests. Curr Opin Microbiol 9:240–244. doi:10.1016/j.mib.2006.04.001

Sun YP, Unestam T, Lucas SD, Johanson KJ, Kenne L, Finlay R (1999) Exudation-reabsorption in a mycorrhizal fungus, the dynamic interface for interaction with soil and soil microorganisms. Mycorrhiza 9:137–144

Swarnalakshmi K, Dhar DW, Singh PK (2006) Blue green algae: a potential biofertilizer for sustainable rice cultivation. Proc Indian Natl Sci Acad 72:135–143

Swarnalakshmi K, Prasanna R, Kumar A, Pattnaik S, Chakravarty K, Shivay YS, Singh R, Saxena AK (2013) Evaluating the influence of novel cyanobacterial biofilmed biofertilizers on soil fertility and plant. Eur J Soil Biol 55:107–116

Tanaka A, Tapper BA, Popay A, Parker EJ, Scott B (2005) A symbiosis expressed nonribosomal peptide synthetase from a mutualistic fungal endophyte of perennial ryegrass confers protection to the symbiotum from insect herbivory. Mol Microbiol 57:1036–1050

Tate RL (1995) Soil microbiology (symbiotic nitrogen fixation). Wiley, New York

Thirumurugan R, Murugappan RM, Rekha S (2006) Characterization and quantification of siderophores produced by *Aeromonas hydrophila* isolated from *Cyprinus carpio*. Pak J Biol Sci 9:437–440

Tian F, Ding Y, Zhu H, Yao L, Jin F, Du B (2008) Screening, identification and antagonistic activity of a siderophore-producing bacteria G-229-21T from rhizosphere of tobacco. Wei Sheng Wu Xue Bao 48:631–637

Tian F, Ding Y, Zhu H, Yao L, Du B (2009) Genetic diversity of siderophore-producing bacteria of tobacco rhizosphere. Braz J Microbiol 40:276–284

Toussaint JP, ST-Arnaud M, Charest C (2004) Nitrogen transfer and assimilation between the arbuscular mycorrhizal fungus *Glomus intraradices* Schenck and Smith and Ri T-DNA roots of *Daucus carota* L. in an in vitro compartment system. Can J Microbiol 50:251–260

Ullman WJ, Kirchman DL, Welch SA, Vandevivere P (1996) Laboratory evidence for microbially mediated silicate mineral dissolution in nature. Chem Geol 132(1):11–17

Vafadar F, Amooaghaie R, Otroshy M (2014) Effects of plant-growth-promoting rhizobacteria and arbuscular mycorrhizal fungus on plant growth, stevioside, NPK, and chlorophyll content of *Stevia rebaudiana*. J Plant Interact 9(1):128–136. doi:10.1080/17429145.2013.779035

van Bael SA, Fernández-Marín H, Valencia M, Rojas E, Wcislo W, Herre EA (2009) Two fungal symbioses collide: endophytic fungi are not welcome in leaf-cutting ant gardens. Proc R Soc B 276:2419–2426. doi:10.1098/rspb.2009.0196

van Bael SA, Estrada C, Wcislo WT (2011) Fungal–fungal interactions in leaf cutting ant agriculture. Psyche. doi:10.1155/2011/ 617478

van Bael SA, Seid MA, Wcislo WT (2012) Endophytic fungi increase the processing rate of leaves by leaf-cutting ants (Atta). Ecol Entomol 37:318–321

van der Heijden MGA, Streitwolf-Engel R, Riedl R, Siegrist S, Neudecker A, Ineichen K, Boller T, Wiemken A, Sanders IR (2006) The mycorrhizal contribution to plant productivity, plant nutrition and soil structure in experimental grassland. New Phytol 172:739–752

van der Heijden MGA, Rinaudo V, Verbruggen E, Scherrer C, Bàrberi P, Giovannetti M (2008) The significance of mycorrhizal fungi for crop productivity and ecosystem sustainability in organic farming systems. In: 16th IFOAM organic world congress, Modena, Italy, 16–20 June 2008

van Tichelen KK, Colpaert JV, Vangronsveld J (2001) Ectomycorrhizal protection of *Pinus sylvestris* against copper toxicity. New Phytol 150(1):203–213

Vanette RL, Hunter MD (2011) Plant defence theory re-examined: nonlinear expectations based on the cost and benefits of resource mutualisms. J Ecol 99: 36–45

Vansuyt G, Robin A, Briat JF, Curie C, Lemanceau P (2007) Iron acquisition from Fe-pyoverdine by *Arabidopsis thaliana*. Mol Plant Microbe Interact 20:441–447

Vasanthakumar SK (2003) Studies on beneficial endorhizosphere bacteria in solanaceous crop plants. M. Sc. (Agri) thesis, Dharwad, Karnataka, India: University of Agricultural Sciences

Vassilev N, Vassileva M, Nikolaeva I (2006) Simultaneous P-solubilizing and biocontrol activity of microorganisms: potentials and future trends. Appl Microbiol Biotechnol 71:137–144

Vega FE, Posada F, Aime MC, Pava-Ripoll M, Infante F, Rehner SA (2008) Entomopathogenic fungal endophytes. Biol Control 46:72–82

Venkataraman GS (1972) Algal biofertilizer and rice cultivation. Today and Tomorrow's Printer and Publishers, New Delhi

Venkataraman GS (1981) Blue green algae for rice production- a manual for its promotion. FAO Soil Bull No 46, Rome

Venkataraman GS, Neelakantan S (1967) Effect of the cellular constituents of the nitrogen fixing bluegreen algae *Cylindrospermum muscicola* on the root growth of rice seedlings. J Gen Appl Microbiol 13:53–61

Ventura W, Ladha JK (1997) Sesbania phosphorus requirements when used as biofertilizers for long term rice cultivation. Soil Sci Soc Am J61:1240–1244

Vidal S (1996) Changes in suitability of tomato for whiteflies mediated by a non-pathogenic endophytic fungus. Entomol Exp Appl 80:272–274

Vigoa C, Normana JR, Hookerb JE (2000) Biocontrol of the pathogen *Phytophthora parasitica* by arbuscular mycorrhizal fungi is a consequence of effects on infection loci. Plant Pathol 49:509–514

Vik U, Logares R, Blaalid R, Halvorsen R, Carlsen T, Bakke I, Kolstø A-B, Økstad OA, Kauserud H (2013) Different bacterial communities in ectomycorrhizae and surrounding soil. Sci Rep 3:3471. doi:10.1038/srep03471

Vu T, Hauschild R, Sikora RA (2006) *Fusarium oxysporum* endophytes induced systemic resistance against *Radopholus similis* on banana. Nematology 8:847–852

Vyas P, Gulati A (2009) Organic acid production *in vitro* and plant growth promotion in maize under controlled environment by phosphate-solubilizing fluorescent *Pseudomonas*. BMC Microbiol 9:174

Vyas P, Rahi P, Chauhan A, Gulati A (2007) Phosphate solubilization potential and stress tolerance of Eupenicillium parvum from tea soil. Mycol Res (Now Fungal Biol) 111:931–938

Vyas P, Joshi R, Sharma KC, Rahi P, Gulati A, Gulati A (2010) Cold-adapted and rhizosphere-competent strain of *Rahnella* sp. with broad-spectrum plant growth-promotion potential. J Microbiol Biotechnol 20(12):1724–1734

Wallander H, Ekblad A, Godbold DL, Johnson D, Bahr A, Baldrian P, Björk RG, Kieliszewska-Rokicka B, Kjøller R, Kraigher H, Plassard C, Rudawska M (2013) Evaluation of methods to estimate production, biomass and turnover of ectomycorrhizal mycelium in forest soils – a review. Soil Biol Biochem 57:1034–1042

Waller F, Achatz B, Baltruschat H, Fodor J, Becker K, Fischer M, Heier T, Hückelhoven R, Neumann C, von Wettstein D, Franken P, Kogel KH (2005) The endophytic fungus *Piriformospora indica* reprograms 6 barley to salt-stress tolerance, disease resistance, and higher yield. Proc Natl Acad Sci U S A 102(7):13386–13391

Wang WX, Barak T, Vinocur B, Shoseyov O, Altman A (2000) Abiotic resistance and chaperones: possible physiological role of SP1, a stable and stabilizing protein from Populus. In: Vasil IK (ed) Plant biotechnology 2000 and beyond. Kluwer, Dordrecht

Wang FW, Jiao RH, Cheng AB, Tan SH, Song YC (2007) Antimicrobial potentials of endophytic fungi residing in *Quercus variabilis* and brefeldin A obtained from *Cladosporium* sp. World J Microbiol Biotechnol 23:79–83

Waqas M, Khan AL, Kamran M, Hamayun M, Kang SM, Kim YH, Lee IJ (2012) Endophytic fungi produce gibberellins and indole acetic acid and promote host-plant growth during stress. Molecules 17:10754–10773

Watanabe I, Roger PA (1984) Nitrogen fixation in wetland rice fields. In: Subba Rao NS (ed) Current developments in biological nitrogen fixation. Oxford IBH, New Delhi, pp 237–276

Whipps JM (2004) Prospects and limitations for mycorrhizas in biocontrol of root pathogens. Can J Bot 82:1198–1227

Wilberforce EM, Griffith GW, Boddy L, Griffiths R (2002) The widespread occurrence of dark septate endophyte fungi in grassland communities. In: The 7th international mycological congress, Oslo, 11–17 August 2002

Wu Q, Zou Y (2010) Beneficial roles of arbuscular mycorrhizas in citrus seedlings at temperature stress. Sci Hortic 125:289–29

Wu SC, Cao ZH, Li ZG, Cheung KC, Wong MH (2005) Effects of biofertilizer containing N-fixer, P and K solubilizers and AM fungi on maize growth: a greenhouse trial. Geoderma 125:155–166

Yandigeri MS, Pabbi S (2005) Response of diazotrophic cyanobacteria to alternative sources of phosphorus. Indian J Microbiol 45:131–134

Yandigeri MS, Kashyap S, Yadav AK, Srinavasan R, Pabbi S (2011) Studies on mineral phosphate solubilization by cyanobacteria *Westiellopsis* and *Anabaena*. Microbiology 80:552–559

Yang CH, Crowley DE (2000) Rhizosphere microbial community structure in relation to root location and plant iron nutritional status. Appl Environ Microbiol 66:345–351

Yang J, Kloepper JW, Ryu CM (2009) Rhizosphere bacteria help plants tolerate abiotic stress. Trends Plant Sci 14:1–4

Yanni YG (1992) The effect of cyanobacteria and *Azolla* on the performance of rice under different levels of fertilizer nitrogen. World J Microbiol Biotechnol 8:132–136

Yanni YG, Rizk RY, El-Fattah FKA, Squartini A, Corich V, Giacomini A, de Bruijn F, Rademaker J, Maya-Fores J, Ostrom P, VegaHernandez M, Hollingsworth RI, Martinez-Molina E, Mateos P, Velaquez E, Woperis J, Triplett E, Umali-Garcia M, Anarna JA, Rolfe BG, Ladha JK, Hill J, Mujoo R, Nag PK, Dazzo FB (2001) The beneficial plant-growth promoting association of *Rhizobium leguminosarum* bv. *trifolii* with rice roots. Aust J Plant Physiol 28:845–870

Yao L, Wu Z, Zheng Y, Kaleem I, Li C (2010) Growth promotion and protection against salt stress by *Pseudomonas putida* Rs-198 on cotton. Eur J Soil Biol 46:49–54

Younis M (2007) Responses of *Lablab purpureus-Rhizobium* symbiosis to heavy metals in pot and field experiments. World J Agri Sci 3(1):111–122

Yuan ZL, Zhang CL, Lin FC (2010) Role of diverse non-systemic fungal endophytes in plant performance and response to stress: progress and approaches. J Plant Growth Reg 29:116–126

Zabihi HR, Savaghebi GR, Khavazi K, Ganjali A, Miransari M (2011) *Pseudomonas* bacteria and phosphorous fertilization, affecting wheat (*Triticum aestivum* L.) yield and P uptake under greenhouse and field conditions. Acta Physiol Plant 33:145–152

Zahran HH, Ahmad MS, Afkar EA (1995) Isolation and characterization of nitrogen-fixing moderate halophilic bacteria from saline soils of Egypt. J Basic Microbiol 35:269–275

Zawadzka AM, Vandecasteele FPJ, Crawford RL, Paszczynski AJ (2006) Identification of siderophores of *Pseudomonas stutzeri*. Can J Microbiol 52:1164–1176

Zhu XC, Song FB, Xu HW (2010) Arbuscular mycorrhizae improves low temperature stress in maize via alterations in host water status and photosynthesis. Plant Soil 331:129–137

Role of Cyanobacteria in Nutrient Cycle and Use Efficiency in the Soil

Manish Kumar, D.P. Singh, Ratna Prabha, and Arun K. Sharma

Abstract

Cyanobacteria are ancient key photosynthetic prokaryotic organisms playing critical role in the biological nutrient cycling in different habitats. They have tremendous capabilities for the management of agroecosystem. The organism possesses various attributes that directly or indirectly not only improve nitrogen (N), phosphorus (P), potassium (K), iron (Fe), and other mineral content in the soils but facilitate plants to make better use of such minerals in plant growth promotion for enhanced crop production. Although much work has been carried out on the nitrogen fixation mechanisms of cyanobacteria, its direct implication in the field is still awaited. Similarly continuous research work is required on the identification of efficient strains of cyanobacteria to make better utilization of them in improving macro- and micronutrients use efficiency by the plants.

Keywords

Cyanobacteria • Nitrogen cycle • Nutrient use efficiency • PGPRs • Microbes • Agroecology

1 Introduction

During the last several decades, agriculture is heavily dependent of the increasing doses of chemical fertilizers and synthetic nutrient supplements for enhancing soil fertility and thereby, increasing crop production although this has many times negatively affected the complex biogeochemical cycle of the agroecosystem (Perrott et al. 1992; Steinshamn et al. 2004). Due to the excessive usage of chemical fertilizers, problems like leaching, segregation, or runoff of nutrients, particularly nitrogen and phosphorus, are critical leading to the degradation of the agroecology and the underground water quality (Gyaneshwar et al. 2002). The most important reasons behind this problem are the low use efficiency of the chemical fertilizers by the plants and low retention of these fertilizers by the soils to make it available to the crops for a longer duration. Since the world population is growing at a faster pace, there will remain a high demand of chemical fertilizers in the future to cope up

M. Kumar • D.P. Singh (✉) • R. Prabha • A.K. Sharma
National Bureau of Agriculturally Important
Microorganisms, Kushmaur, Maunath Bhanjan 275101,
Uttar Pradesh, India
e-mail: dpsfarm@rediffmail.com; nbaimicar@gmail.com

A. Rakshit et al. (eds.), *Nutrient Use Efficiency: from Basics to Advances*,
DOI 10.1007/978-81-322-2169-2_10, © Springer India 2015

with the increased food production for the increasing population. This will again stimulate high quantities of chemical fertilizers for the forthcoming agriculture and keep on excessive pressure on the environment and sustainability of the agroecosystem (Adesemoye and Kloepper 2009). In the last several years, the rate of the application of N, P, and K has increased tremendously, and although this has helped in increasing agriculture production, it has also created severe problems to the fertility of soils, native flora, and fauna including microbial communities and *belowground* water. The challenge, therefore, is to keep on increasing agricultural productivity without creating harm to the agroecology and disturbing environment due of the excessive use of chemical fertilizers (Vitousek et al. 2002).

Nutrients remain in the environment in three basic forms including gaseous form (e.g., N_2 and CO_2), minerals (such as apatite, P-containing mineral $Ca_5(PO_4)_3$, iron, etc.), and organic form (nutrients bound with carbon (C)-based compounds in living or dead organisms and their products like composts). Enhanced crop productivity either in low fertility soils or in the soils with plenty of nutrients but not being available to the plants due to various reasons is a challenge to cover increasing food demand. Nutrient use efficiency (NUE), i.e., the capacity of a plant to acquire or utilize nutrients, can be considered to emphasize productivity or internal nutrient requirement of the cells (Gourley et al. 1994). Plant nutrient uptake usually occurs in ionic forms and animals through the consumption of living or dead organic forms, while microbes can use both minerals and organic materials. The proper interconversion between various nutrient forms generally reflects a sign of balanced ecosystem function which predominantly requires the role of microorganisms. Nutrient cycling is the rotation of various minerals and nutrients among various abiotic and biotic biological entities from the water and soil to organisms (plants, animals, or microbes) to the environment. During the course of the plant growth and development in a particular soil habitat, elements are regularly extracted by the plants from the soils or fixed from the environment within plants and soils and ultimately return to the same environment. Huge diversity of organisms greatly influences nutrient cycling by creating various physical structures and mechanisms to regulate fluxes of nutrients among different compartments of the environment. These structures and environmental processes limit losses and excessive transfer of nutrients by acting as buffers (Lavelle et al. 2005). During the past several decades, the cycling of the key elements like phosphorus, nitrogen, carbon, potash, zinc, iron, silicon, and other micronutrients also has seen substantial deterioration with less positive and more negative impact over a range of ecosystem and their functioning because of the ongoing industrialization, chemicalization, and human interferences in the agricultural ecosystems. Imbalances in the nutrient cycling have largely affected the quality and fertility of soils, diminished agroecosystems, affected crop productivity, and made poor availability of nutrients to the crops.

In general, NUE covers diverse physiological processes including relation between nutrient content of the plant and its growth rate, availability of nutrients and minerals in the soil, soil processes and biological factors to make existing nutrients available to the plants, partitioning of nutrients between the litter fall from plants and resorption pathways, and role of microbes in regulating nutrient cycling in the soils (Aerts 1990).

The principles of nutrient use efficiency are rooted in the judicious availability of nutrients in agroecosystem, proper nutrient cycling, and utilization of the nutrients by the crop plants. Nutrients in any form required by the crops are not in plenty as is largely believed by the public. Because of the rising prices of the chemical fertilizers and imbalances in the prices of the produced crops in return, farming fraternity is also becoming conscious about the economics of the fertilizer used and the crop yield. This has actually put pressure on the fertilizer industries on one hand to develop better fertilizers with high nutrient release and loss efficiency in the soils and scientists on the other hand to develop strategies to search for the viable methods for better NUE. Since the issue of the plant nutrition is related to the biological component of the soils where microbes have to play a critical role in improving nutrient availability to the plants and, thus, enhancing the nutrient use efficiency of the plants especially in the nutrient-poor soil, it is aimed to discuss the role of microbes and especially cyanobacteria in the nutrient cycling and use efficiency. Cyanobacteria are the most natural inhabitants of the agroecosystem and are dominant phototrophic organisms in water (fresh and marine water, hot springs) and soil. Their long evolutionary history in the environment and abilities for adaptation in almost all kinds of natural habitats make these organisms ubiquitous in nature. Their presence has been reported from a wide range of soils with a variety of characteristics (normal, saline, alkaline, acidic soils, etc.). These organisms have largely been considered for possessing various characteristics to contribute to the productivity of the agricultural crops, fertility of soils, and balance of the ecosystems.

2   Microbial Role in Nutrient Use Efficiency

Like soil and water, a great diversity of microbial resources available in the soils also play a critical role in the uptake and acquisition of the nutrients by the plants. It is usually believed that the huge diversity of the microbial communities associated with the rhizoplane in the rhizosphere and phylloplane helps plants to acquire minerals, organic substances, and many other small molecule metabolites including amino acids, phytohormones, etc. to improve plant productivity. The role of the microbes in the soils has been emphasized as nitrogen fixers (Bashan et al. 2004), P-solubilizers (Rodriguez and Fraga 1999), potassium solubilizer and mobilizers (Basak and Biswas 2009), siderophore production for iron sequestration in plants (Bakker et al. 2006), plant hormone producers (Prasanna et al. 2011), and 1-aminocyclopropane-1-carboxylate

(ACC) deaminase producers that alleviate biotic stresses in the plants (Adesemoye and Kloepper 2009). Their interactions with the plants have been shown to improve plant growth and impart biological control against biotic and abiotic stresses, and much work is available on these aspects (Bhardwaj et al. 2014). The basic concept of organic farming has its roots in the role of microbial indigenous communities that silently work to improve biogeochemical cycle in the natural ecosystem. Plant growth-promoting rhizobacteria (PGPRs) and the arbuscular mycorrhizal fungi (AMF) are known to improve nutrient status of the soils that will ultimately be available to the crop plants (Adesemoye and Kloepper 2009), and their formulations applied as seed treatment or soil inoculants are supposed to multiply manyfolds in the soils and participate in the nutrient cycling to benefit crop improvement (Singh et al. 2011). Inoculations with PGPR and AMF have been shown to enhance nutrient use efficiency of fertilizers (Bhardwaj et al. 2014), and various combinations of chemical fertilizers with PGPRs and AMF have been reported to have impact on fertilizer usage and management in the soils (Han and Lee 2005; Adesemoye et al. 2008). Application of soybean residues and *Azospirillum brasilense* inoculation with or without inorganic N-fertilizer in poor fertility sandy soils showed a significant increase in N-accumulation in wheat shoots or grains. Improved phosphorus use efficiency is reported in French bean when rock phosphate was applied with farmyard manure (FYM) or vermicompost (1:2 ratio) along with phosphate-solubilizing bacteria (Manjunath et al. 2006). It is shown that the application of phosphate-solubilizing bacteria (PSB) significantly increased available phosphorus in an acidic soil (oxisol) (Setiawati and Handayanto 2010). Mycorrhizal association may help to improve early season P-acquisition in crops. The AM endophytes are not host-specific in general, although preferential associations are observed with some host plants (Jansa et al. 2013).

It has largely been considered that tripartite interactions of plant-PGPR-AMF may facilitate nutrient uptake by the plants (Barea et al. 1998) and such interaction seems promising although the association between PGPRs and AMF may become synergistic or antagonistic, and therefore, investigations in this direction are required (Adesemoye and Kloepper 2009). The mechanisms behind plant-microbe (PGPR or AMF) interaction is a complex phenomenon involving various direct and indirect molecular and biochemical cellular processes (Berg 2009), but a clearer understanding of such processes will definitely help in identifying how such interactions can benefit nutrient uptake by the plants.

3   Cyanobacteria and Nitrogen Fixation

Cyanobacteria are considered to have a direct impact on improving soil fertility and consequently enhance growth and production of rice crop (Song et al. 2005). They play important roles in increasing soil physical structure, porosity and retaining soil moisture due to their filamentous structure and production of mucilaginous substances, production and release into the soil phytohormones, vitamins, amino acids, secondary metabolites like phenolics, flavonoids, etc. (Roger and Reynaud 1982; Rodriguez et al. 2006), improving mineral sequestration (N and P) in the soils and accumulating organic biomass in the soil which improves soil organic level after their decomposition (Wilson 2006; Saadatnia and Riahi 2009). Besides all, they have an indigenous potential to fix atmospheric nitrogen into the habitat they belong and thereby increase direct fertility of the soils. In the early ecosystems, fixed nitrogen was limited due to bacteria nitrification and anaerobic ammonia oxidation, while oxygen levels were still comparatively low in the atmosphere. Cyanobacteria, being the primary producers on the Earth, were able to provide their own fixed nitrogen continued to outcompete other microorganisms and to oxygenate the Earth (Antonia 2008). Since nitrogen limitation is among the major factors limiting crop production, a large proportion of the energy budget of the developed and developing world is being spent on the production of N-based fertilizers. At the same time, it is also a fact that the overuse of N-based fertilizers is a serious matter as it not only affect the cost of production but create problems of polluting drinking and recreational water bodies and low fertility of the soils. Plenty of nitrogen in the gaseous form (78 % by volume, 3.9×10^{15} tonnes) is available in Earth's atmosphere, but only a few number of prokaryotic organisms are known to fix atmospheric nitrogen. These organisms are called nitrogen fixers or diazotrophs, and many of them have been applied in enhancing agricultural productivity since long back.

Many of the cyanobacterial species have the intrinsic ability to fix atmospheric nitrogen with the help of a very specialized cell called *heterocyst* (Fay et al. 1968). These specialized cells are relatively very enlarged than the usual vegetative cells and are present terminally or intercalary in the filamentous cyanobacterial organisms (Sah 2008). Cyanobacterial nitrogen fixation is a process coupled by the generation of hydrogen gas (H_2) at the same time, and 40 % of the evolved H_2 is recycled with the help of hydrogen-uptake gene (*hup* gene) (Margheri et al. 1990), while the rest 60 % can be used as a source of green fuel (Dutta et al. 2005). Heterocysts are tri-layered specialized cellular structures in which the outer fibrous and middle homogenous layers are comprised of noncellulosic polysaccharide, while the inner laminated layer is composed of glycolipids (Lang 1968). The specialized cell wall of the heterocyst permits atmospheric nitrogen gas to diffuse inside but restricts the atmospheric oxygen to come inside. The enzyme cascade nitrogenase catalyzes biological N_2 fixation inside the heterocysts, and since this enzyme is highly sensitive to oxidation in the presence of oxygen, the cell wall of the heterocyst checks the diffusion of the oxygen inside the cells. Heterocysts are so specialized in their physiology that they lack photosystem II (PS II) in it because in PS II photolysis occurs with the generation of free oxygen. Heterocyst possesses photosystem I (PS I) which generates ATP that facilitates in nitrogen fixation. Even in the conditions of the external oxygen entering inside the hetero-cystous cells through polar plugs, these cells make a reaction of entering oxygen with that of hydrogen molecules present inside the cells to make molecules of water and, thereby, maintain the internal environment of the cells as reducing and not oxygenating to facilitate nitrogen fixation (Sah 2008). Some filamentous cyanobacteria (e.g., of the genera *Anabaena* and *Nostoc*) confine the nitrogen fixation activity to the specialized heterocysts which is containing nitrogenase enzyme; other unicellular and filamentous forms express nitrogenase activity in the dark periods of light-dark growth cycle (Herrero et al. 2001). Another example is *Trichodesmium* sp. which fixes nitrogen aerobically and expresses nitrogenase activity in the light period of light-dark growth cycle (Capone et al. 1997). Symbiotic biological nitrogen fixation also occurs with the water fern *Azolla* living symbiotically with a cyanobacterium *Anabaena azollae* which colonizes the base of *Azolla* fronds. These symbiotic cyanobacteria fix significant amount of nitrogen in their heterocystous cells and remained prominent as biofertilizers in the wetland paddy fields over so many years in the Southeast Asia (Wagner 2011). In many areas, paddy fields are typically covered with the blooms of *Azolla* that fix as high as 600 kg N per ha per year during the rice-growing season (Postgate 1982; Fattah 2005).

4 Cyanobacteria to Improve Nutrient Use Efficiency

Nitrogen (N_2) is essential for all life forms in which it helps the biosynthesis of amino acids, proteins, and other nitrogenous organic biomolecules and formation of chlorophyll to support photosynthesis and, thus, plant growth. Although chemical fertilizers, especially urea, are extensively being used as a source of nitrogen for the crops, farmyard manure (FYM), compost and vermicompost, legume cover crops, and off-farm fertilizers like fish emulsion, poultry wastes, etc. are largely being used by organic practitioners to meet out the demand of nitrogen

in organic crops. Limitations of the FYM or other manures or composts lie in their nutrient composition which varies widely depending on the source and method of production, plenty of availability at the farms, and, simultaneously, the cost of transportation. In this regard, biological nitrogen fixation (BNF) usually performed by a number of free-living or symbiotic microorganisms, especially by bacteria in the soil ecosystem, is again an alternative option to cover up the demand of nitrogen fertilizer, although to a limited extent (Shridhar 2012). Cyanobacteria offer an ecologically sound and economically impressive alternative to the ever-increasing usage of chemical fertilizers for increasing crop productivity especially in the rice cultivation (Mishra and Pabbi 2004). They play an important role in the maintenance of soil fertility in the rice fields to yield enhanced crop production (Song et al. 2005). The application of N_2-fixing cyanobacteria (as individual organism or in consortia) as a potential N_2-biofertilizer source in the field, especially paddy as an alternative to the commonly used organic fertilizers, can be of great value and is becoming popular day by day (Thajuddin and Subramanian 2005; Choudhary 2011; Prasanna et al. 2011; Hasan 2012; Sahu et al. 2012). Most of the rice fields inhabit a natural population of cyanobacterial strains that act as a potential source of nitrogen fixation in the field. Ammonia produced within the cells as a product of nitrogen fixation is taken up by passive diffusion or as ammonium ion by specific uptake mechanism. Alternatively amino acid alanine, asparagine, and glutamine can also act as sources of nitrogen for the plants. There was a study on the performance of selected bacterial strains, namely, *Providencia* sp., *Brevundimonas* sp., and *Ochrobactrum* sp. in combination with two species of *Anabaena* and one *Calothrix* sp. in pot experiment with rice variety Pusa 1460 comprising 51 treatments and recommended fertilizer as control (Prasanna et al. 2011). This study illustrated positive impact of co-inoculation of bacterial and cyanobacterial strains for integrated nutrient management of rice crop.

Soil incubation and greenhouse studies were conducted to evaluate the impact of nitrogen availability from mineralizable nitrogen on different types of soil textures on the growth of lettuce crop. N-mineralization potential of cyanobacterial biofertilizers was compared with the traditionally used fertilizers in two soils with contrasting textures in an experiment at constant temperature (25 °C) and 60 % moisture content for 140 days (Sukor 2013). Results indicated that in comparison to the commonly used organic fertilizers (compost and fish emulsion), N availability facilitated by cyanobacterial biofertilizers was fairly efficient in sandy soils, and it also enhanced lettuce growth and lettuce root response on N use efficiency. The use of foliar cyanobacterial biofertilizer in combination with soil applied, solid and liquid cyanobacterial biofertilizers showed higher percentage of fertilizer recovery and NUE at 56 kg N ha^{-1} compared to the composted manure that correspond to lettuce yield component which was higher in fish emulsion compared to the composted manure (Sukor 2013).

It was demonstrated at Central Rice Research Institute, Cuttack, India, that application of cyanobacterium *Aulosira* sp. with soil at a rate of 60 kg ha^{-1} (fresh weight) caused significant change in the content of soil nitrogen content. Nitrogen content in the soil was increased by 13–14 % under field conditions by the incorporation of cyanobacteria, and the soil amended with cyanobacteria releases almost 50 % of its ammonium nitrogen at 50 days of flooding (Syiem 2005). The rate of nitrogen released by cyanobacteria was recorded to be 12 and 35 % after 7 and 35 days of flooding in field, respectively. Apart from *Aulosira*, *Nostoc muscorum*, *Nostoc commune*, and *Anabaena* sp. also enhanced the release of inorganic nitrogen into the soil. N content in the soil incorporated with cyanobacteria was found to be higher when it was exposed to light (due to N gain from cyanobacteria) than that of the unexposed soil. A significant increase in total inorganic nitrogen content was also reported in pots inoculated with *Tolypothrix tenuis* (Mishra and Pabbi 2004). Cyanobacteria potentially secrete substantial

amount of extracellular nitrogenous compounds into the medium. Different cyanobacterial strains (*Calothrix ghosei*, *Westiellopsis*, *Hapalosiphon intricatus*, and *Nostoc* sp.) isolated from the rhizosphere of wheat were analyzed for their physiological attributes (Karthikeyan et al. 2007). Soil enriched with cyanobacterial isolates helps in enhancing soil fertility and makes it suitable for cultivation.

Cyanobacteria were also used as biofertilizer in wheat crops (a major staple crop) (Abd-Alla et al. 1994), lettuce (Sukor 2013), and maize to enhance the nutrient use efficiency by increasing the nitrogenase activity. There are some reports in which cyanobacterial biofertilizers along with some chemical fertilizers ($N_{40}P_{30}K_{30}$) had interesting and promising results treated with wheat crop (var HD 2687). Cyanobacterial inoculants along with vermicompost lead to the increase in nitrogenase activity, and *Azotobacter* with cyanobacteria increases the extent of chlorophyll. *Calothrix*, *Anabaena*, *Oscillatoria*, and *Phormidium* are abundant cyanobacterial community reported in wheat crop soil ecosystem (Prasanna et al. 2009). In some study, root sections of wheat seedlings cocultured with *Calothrix ghosei* showed short filaments inside the root hairs and cortical region (Karthikeyan et al. 2007). The higher crop yield in clayey soil types, which are expected to absorb more available nutrients than sandy soil, can be obtained by the application of cyanobacteria as biofertilizers. Usually, crops seldom utilize more than 40 % of the applied N, and the increasing use of N-fertilizer decreases nutrient use efficiency (Dobermann 2005). Nitrogen use efficiency depends on the time of application of fertilizer, keeping in view the demand of plant growth. It has also been observed that nutrient use efficiency initially increases when the supplied rate of N reaches up to a maximum and then declines when nutrient transport in plants is regulated to ensure sufficient absorbing rate of nutrient to cover plant growth demand (Bloom et al. 2006). Cyanobacteria, being the producers of mucilage and enhancers of organic matter content in the soils, also help in the slow release process of the nutrients and minerals present in the soils or being absorbed by themselves and made available to the plants after the decay of their cells.

5 Other Attributes of Cyanobacteria

The use of cyanobacterial biofertilizer led to increase in microbial diversity in soil and subsequently improve the fertility of soils because it was found to increase organic matter content and enzymatic activity carried out by microbes (dehydrogenase and nitrogenase activities) (El-Gaml 2006). Cyanobacteria also excrete indole-3-acetic acid (IAA) and amino acids which can stimulate the growth of microbial populations in soil (Song et al. 2005; Karthikeyan et al. 2007). An increased microbial activity in the rhizosphere is reported to increase microbial colonization of plant roots and help plants to acquire N from soils. Phytohormones (IAA) are important factors that contribute to enhance root system in the plants (Spaepen et al. 2007) and, thus, increase the possibility of enhanced nutrient acquisition by the roots that can support plant growth. This could be the indirect impact of cyanobacterial biofertilization in the soils to improve plant growth and crop yield.

There are reports of possible increase in the NUE due to co-inoculation of cyanobacteria with bacteria on crop yield. Prasanna et al. (2011) explained C–N sequestration in soil under rice crop after co-inoculation. The synergistic effect of rhizobacteria and cyanobacteria in promoting yield of rice was demonstrated to result in 30–40 kg nitrogen savings in terms of chemical fertilizers (Prasanna et al. 2011). Moreover positive influence on carbon sequestration in soil was also reported. Enhanced agronomic efficiency and utility in integrated nutrient management of rice–wheat cropping sequence was documented using cyanobacteria and rhizobacteria together as inoculants. Significant enhancement was reported in organic carbon in plants inoculated with microbes and correlated it with MBC (microbial biomass carbon) values (Prasanna et al. 2011). Several

enzyme activities were also significantly correlated with soil microbial carbon. These enzymes play important roles in nutrient cycling. Treatment of cyanobacteria and bacterial strains to crop plants not only enhanced C/N ratio but provided an indication that in combination, the use of plant growth-promoting bacteria and cyanobacteria can aid in carbon sequestration in soils and adjust nutrient imbalance in soils with mineral deficiencies (Prasanna et al. 2009). Cyanobacterial carbon-sequestering ability leads to a significant role in the carbon enrichment of soils (Jaiswal et al. 2010). Biomineralization of carbon dioxide by $CaCO_3$ precipitation using cyanobacteria is demonstrated to offer strategies for point-source carbon capture and sequestration (Tabita and Colletti 1979; Jansson and Northen 2010). Carbon sequestration is an important process in carbon cycling and involves capturing and storing of carbon and its subsequent removal from the global carbon cycle. *Nostoc muscorum* is supposed to act as an ecosystem manager by creating an environment within the soil system that is suitable for plant and microbial growth. *N. muscorum* leads the formation of stable soil aggregates and, thus, help in increasing nutrient use efficiency.

Conclusion

Cyanobacteria, being a key photosynthetic microorganism in the biological nutrient cycling, have tremendous capability for the agroecosystem management. The organism possesses various attributes that directly or indirectly not only improve nitrogen, phosphorus, iron, and other mineral content in the soils but facilitate plants to make better use of such minerals in plant growth promotion for enhanced crop production. Although much work has been carried out on the nitrogen fixation mechanisms of cyanobacteria, its direct implication in the field is still awaited. Similarly continuous research work is required on the identification of efficient strains of cyanobacteria to make better utilization of them in improving P, K, and Fe use efficiency by the plants in the soils.

References

Abd-Alla MH, Mahmoud ALE, Issa AA (1994) Cyanobacterial biofertilizer improve growth of wheat. Phyton 34:11–18

Adesemoye AO, Kloepper JW (2009) Plant–microbes interactions in enhanced fertilizer-use efficiency. Appl Microbiol Biotechnol 85:1–12. doi:10.1007/s00253-009-2196-0

Adesemoye AO, Torbert HA, Kloepper JW (2008) Enhanced plant nutrient use efficiency with PGPR and AMF in an integrated nutrient management system. Can J Microbiol 54:876–886

Aerts R (1990) Nutrient use efficiency in evergreen and deciduous species from heathlands. Oecologie 84:391–397

Antonia H (2008) The cyanobacteria. Caister Academic Press, Northforlk, pp 13–20

Bakker PAHM, Pieterse CMJ, van Loon LC (2006) Induced systemic resistance by fluorescent *Pseudomonas* spp. Phytopathology 97:239–243

Barea JM, Andrade G, Bianciotto V, Dowling D, Lohrke S, Bonfante P, O'Gara F, Azcon-Anguilar C (1998) Impact on arbuscular mycorrhiza formulation of *Pseudomonas* strains used as inoculants for biocontrol of soil-borne fungal plant pathogens. Appl Environ Microbiol 64:2304–2307

Basak BB, Biswas DR (2009) Influence of potassium solubilizing microorganism (*Bacillus mucilaginosus*) and waste mica on potassium uptake dynamics by sudan grass (*Sorghum vulgare* Pers.) grown under two Alfisols. Plant Soil 317:235–255

Bashan Y, Holguin G, de-Bashan LE (2004) Azospirillum-plant relations: physiological, molecular, agricultural and environmental advances (1997–2003). Can J Microbiol 50:521–577

Berg G (2009) Plant-microbe interactions promoting plant growth and health: perspectives for controlled use of microorganisms in agriculture. Appl Microbiol Biotechnol 84:11–18

Bhardwaj D, Ansari MW, Sahoo RK, Tuteja N (2014) Biofertilizers function as key player in sustainable agriculture by improving soil fertility, plant tolerance and crop productivity. Microb Cell Factories 13:66. doi:10.1186/1475-2859-13-66

Bloom AJ, Fresnsch J, Taylor AR (2006) Influence of inorganic nitrogen and pH on the elongation of maize seminal roots. Ann Bot 97:867–873

Capone DG, Zehr JP, Paerl HW, Bergman B, Carpenter EJ (1997) *Trichodesmium*, a globally significant marine cyanobacterium. Science 276:1221–1229

Choudhary KK (2011) Occurrence of nitrogen-fixing cyanobacteria during different stages of paddy cultivation. Bangladesh J Plant Taxon 18:73–76

Dobermann AR (2005) Nitrogen use efficiency – state of the art. Agronomy & Horticulture-Faculty Publications. Paper 316. http://digitalcommons.unl.edu/agronomy facpub/316

Dutta D, De D, Chaudhari S, Bhattacharya SK (2005) Hydrogen production by cyanobacteria. Microb Cell Factories 4:36

El-Gaml M (2006) Studies on cyanobacteria and their effect on some soil properties. Thesis, Benha University. Enhanced-Efficiency Fertilizers, Frankfurt, Germany. Digital Commons, University of Nebraska

Fattah QA (2005) Plant resources for human development. In: Third international botanical conference 2005. Bangladesh Botanical Society, Dhaka, Bangladesh

Fay P, Stewart WDP, Walsby AE, Fogg GE (1968) Is the heterocyst the site of nitrogen fixation in blue-green algae? Nature (London) 220:810–812

Gourley CJP, Allan DL, Russelle MP (1994) Plant nutrient efficiency: a comparison of definitions and suggested improvements. Plant Soil 158:29–37

Gyaneshwar P, Kumar GN, Parekh LJ, Poole PS (2002) Role of soil microorganisms in improving P nutrition of plants. Plant Soil 245:83–93

Han HS, Lee KD (2005) Phosphate and potassium solubilizing bacteria effect on mineral uptake, soil availability and growth of egg plants. Res J Agric Biol Sci 1:176–180

Hasan MA (2012) Investigation of the nitrogen fixing cyanobacteria (BGA) in rice fields of North-West region of Bangladesh. I: Nonfilamentous. J Environ Sci Nat Resour 5:185–192

Herrero A, Muro-Pastor AM, Flores E (2001) Nitrogen control in cyanobacteria. J Bacteriol 183:411–425

Jaiswal P, Kashyap AK, Prasanna R, Singh PK (2010) Evaluating the potential of *N. calcicola* and its bicarbonate resistant mutant as bioameliorating agents for 'Usar' soil. Indian J Microbiol 50:12–18

Jansa J, Bukovská P, Gryndler M (2013) Mycorrhizal hyphae as ecological niche for highly specialized hypersymbionts–or just soil free-riders? Front Plant Sci 4:134. doi:10.3389/fpls.2013.00134

Jansson C, Northen T (2010) Calcifying cyanobacteria–the potential of biomineralization for carbon capture and storage. Curr Opin Biotechnol 21:365–371. doi:10.1016/j.copbio.2010.03.017

Karthikeyan N, Prasanna R, Nain L, Kaushik BD (2007) Evaluating the potential of plant growth promoting cyanobacteria as inoculants for wheat. Eur J Soil Biol 43:23–30

Lang N (1968) The fine structure of blue-green algae. Ann Rev Microbiol 22:15–42

Lavelle P, Dugdale R, Scholes R, Berhe AA, Carpenter E, Codispoti L, Izac A-M, Lemoalle J, Luizao F, Scholes M, Tréguer P, Ward B (2005) Nutrient cycling. In: Hassan R, Scholes R, Ash N (eds) Ecosystems and human well-being: current state and trends, vol I. Island Press, London, pp 321–351

Manjunath MN, Patil PL, Gali SK (2006) Effect of organics amended rock phosphate and P solubilizer on P use efficiency of French bean in a Vertisol of Malaprabha Right Bank command of Karnataka. Karnataka J Agric Sci 19:36–39

Margheri MC, Tredici MR, Allotta G, Vagnoli L (1990) Heterotrophy metabolism and regulation of uptake hydrogenase activity in symbiotic cyanobacteria. In: Polsinelli M, Materassi R, Vincenzin M (eds) Developments in plant and soil sciences - biological nitrogen fixation. Kluwer Academic Publishers, Dordrecht, pp 481–486

Mishra U, Pabbi S (2004) Cyanobacteria – a potential biofertilizer for rice. Resonance 9(6):6–10

Perrott KW, Sarathchandra SU, Dow BW (1992) Seasonal and fertilizer effects on the organic cycle and microbial biomass in a hill country soil under pasture. Aust J Soil Res 30:383–394

Postgate JR (1982) The fundamentals of nitrogen fixation. Cambridge University Press, New York

Prasanna R, Nain L, Ancha R, Srikrishna J, Joshi M, Kaushik BD (2009) Rhizosphere dynamics of inoculated cyanobacteria and their growth-promoting role in rice crop. Egypt J Biol 1:26–36

Prasanna R, Joshi M, Rana A, Shivay YS, Nain L (2011) Influence of co-inoculation of bacteria-cyanobacteria on crop yield and C–N sequestration in soil under rice crop. World J Microbiol Biotechnol. doi:10.1007/s11274-011-0926-9

Rodriguez H, Fraga R (1999) Phosphate solubilizing bacteria and their role in plant growth promotion. Biotechnol Adv 17:319–339

Rodriguez AA, Stella AA, Storni MM, Zulpa G, Zaccaro MC (2006) Effects of cyanobacterial extracellular products and gibberellic acid on salinity tolerance in *Oryza sativa* L. Saline Syst 2:7

Roger PA, Reynaud PA (1982) Free-living blue-green algae in tropical soils. Martinus Nijhoff Publisher, La Hague

Saadatnia H, Riahi H (2009) Cyanobacteria from paddy fields in Iran as a biofertilizer in rice plants. Plant Soil Environ 55:207–212

Sah P (2008) Understanding the physiology of heterocysts and nitrogen fixation in cyanobacteria or blue-green algae. Nat Sci 6:28–33

Sahu D, Priyadarshini I, Rath B (2012) Cyanobacteria – as potential biofertilizer. CIB Tech J Microbiol 1:20–26

Setiawati TC, Handayanto E (2010) Role of phosphate solubilising bacteria on availability phosphorus in Oxisols and tracing of phosphate in corn by using ^{32}P. In: 19th world congress of soil science, soil solutions for a changing world, Brisbane, Australia, 1–6 August 2010

Shridhar BS (2012) Review: nitrogen fixing microorganisms. Int J Microbiol Res 3:46–52

Singh JS, Pandey VC, Singh DP (2011) Efficient soil microorganisms: a new dimension for sustainable agriculture and environmental development. Agric Ecosyst Environ 140:339–353

Song T, Martensson L, Eriksson T, Zheng W, Rasmussen U (2005) Biodiversity and seasonal variation of the cyanobacterial assemblage in a rice paddy field in Fujian, China. Fed Eur Mat Soc Microbiol Ecol 54: 131–140

Spaepen S, Vanderkyden J, Remans R (2007) Indole-3-acetic acid in microbial and micro organism – plant signaling. FEMS Microbiol Rev 31:425–448

Steinshamn H, Thuen E, Bleken MA, Brenoe UT, Ekerholt G, Yri C (2004) Utilization of nitrogen (N) and phosphorus (P) in and organic dairy farming system in Norway. Agric Ecosyst Environ 104:509–522

Sukor A (2013) Effect of cyanobacterial fertilizers compared to commonly used organic fertilizers on nitrogen availability, lettuce growth and nitrogen use efficiency on different soil textures. MS thesis, Colorado State University

Syiem MB (2005) Entrapped cyanobacteria: Implications for biotechnology. Indian J Biotechnol 4:209–215

Tabita FR, Colletti C (1979) Carbon dioxide assimilation in cyanobacteria: regulation of ribulose, 1,5-bisphosphate carboxylase. J Bacteriol 140:452–458

Thajuddin N, Subramanian G (2005) Cyanobacterial diversity and potential applications in biotechnology. Curr Sci 89:47–57

Vitousek PM, Cassman K, Cleveland C, Crews T, Field CB, Grimm NB, Howarth RW, Marino R, Martinelli L, Rastetter EB, Sprent JI (2002) Towards an ecological understanding of biological nitrogen fixation. Biogeochemistry 57–58:1–45

Wagner SC (2011) Biological nitrogen fixation. Nat Educ Knowl 3:15

Wilson LT (2006) Cyanobacteria: a potential nitrogen source in rice fields. Texas Rice 6:9–10

Trichoderma Improves Nutrient Use Efficiency in Crop Plants

Sayaji T. Mehetre and Prasun K. Mukherjee

Abstract

Trichoderma spp. are better known as suppressor of plant diseases. *Trichoderma*-based formulations thus dominate the biofungicide market. Intense researches however identified many traits in *Trichoderma* that extends the applications beyond plant protection. Various species of *Trichoderma* are capable of enhancing root growth and development, imparting tolerance to abiotic stresses, and improving uptake and use efficiency of micro- and macronutrients, culminating in enhancing crop productivity. Microbe-mediated improvement of nutrient use efficiency is gaining a lot of importance in the context of gradual loss of soil fertility/ productivity that has resulted from intensive agriculture. *Trichoderma* spp. can enhance crop productivity by virtue of both enhanced decomposition of biomass and improving uptake of inorganic fertilizers, the topics of discussion in the current review.

Keywords

Trichoderma • Growth promotion • Nutrient use efficiency • Composting • Biodegradation

1 Introduction

Nutrients play important role in crop productivity. These nutrients are categorized as primary, secondary, and micronutrients depending on the requirement to the crops. Nutrient use efficiency in crop is yield per unit of nutrient supplied (from the soil and/or fertilizer). This is usually divided into two components: uptake, or the ability of the plant to extract the nutrient from the soil, and utilization/use efficiency, the ability of plant to convert the absorbed nutrient into yield. It also covers transport, storage, mobilization, and usage of nutrients within the plant and depends on the environment. The nutrients that most commonly limit plant growth are N, P, K, S, and micronutrients.

Nutrient availability to the plant is influenced by microorganism present in the soil and the rhizosphere. The omnipresent *Trichoderma* spp. play significant role in nutrients recycling and

S.T. Mehetre (✉) • P.K. Mukherjee
Nuclear Agriculture and Biotechnology Division, Bhabha Atomic Research Centre, Trombay, Mumbai 400085, Maharashtra, India
e-mail: smehetre@gmail.com

A. Rakshit et al. (eds.), *Nutrient Use Efficiency: from Basics to Advances*,
DOI 10.1007/978-81-322-2169-2_11, © Springer India 2015

nutrient availability to plants. Most *Trichoderma* spp. are rhizosphere competent and hence have competitive advantages as far as colonization of roots/rhizosphere is concerned (Chet et al. 1981; Elad 2000; Harman 2006). Apart from its ability to accelerate the nutrient uptake, by virtue of being prolific producers of plant cell wall degrading enzymes, *Trichoderma* has also shown to enhance the process of composting in soil. This in turn enhances the nutrient availability to the crops. Root colonization by *Trichoderma* spp. frequently enhances root growth and development, thus directly contributing to improved nutrient uptake.

2 Enhancement of Plant Growth by *Trichoderma*

Use of *Trichoderma* spp. has very often resulted in higher growth of plant and improved the yield of several crops. This has been well studied and documented over the period of time, but the exact mechanism is not clearly understood. A possible mechanism for increased plant growth is an increase in nutrient transfer from soil to root, which is supported by the fact that *Trichoderma* can colonize the interior of roots (Kleifield and Chet 1992). Several mechanisms, by which *Trichoderma* spp. influences plant development, were suggested, such as production of growth hormones (Windham et al. 1986), solubilization of insoluble minor nutrients in soil (Altomare et al. 1999), and increased uptake and translocation of less-available minerals, in addition to suppression of minor pathogens (Baker 1989). Growth stimulation is evidenced by increases in biomass, productivity, stress resistance, and increased nutrient absorption (Hoyos-Carvajal et al. 2009). Certain *Trichoderma* strains have been shown to stimulate plant growth through the production of plant-growth-promoting (PGP) compounds (Chang et al. 1986; Ousley et al. 1994; Contreras-Cornejo et al. 2009; Vinale et al. 2009).

There are many reports of enhanced plant growth and yield due to application of *Trichoderma* spp. Application of *Trichoderma*

spp. under glasshouse conditions increased seedling height by 26–61 %, root exploration by 85–209 %, leaf area 27–38 %, and root dry weight by 38–62 % within 15 days after the treatments in bitter gourd (Lo and Lin 2002). Chang et al. (1986) demonstrated that *T. harzianum* strains improved tomato seedling growth. There were differences between the untreated control and the treatments for all of the growth parameters at 4 weeks after inoculation with the exception of root fresh and dry weight. In another study, *T. harzianum* was applied to cucumber and pepper seedlings as a peat-bran preparation incorporated into the propagative mixture in a commercial production nursery. Significant increase of 23.8 and 17.2 % in seedling height, 96.1 and 50 % in leaf area, and 24.7 and 28.6 % in plant dry weight was observed in cucumber and pepper seedlings, respectively, as compared to their non-treated counterparts. *Trichoderma*-treated seedlings were much more developed and vigorous and had higher chlorophyll contents (Inbar et al. 1994). Windham et al. (1986) showed the production of growth regulating factors for enhanced plant productivity due to application of *Trichoderma* spp.

Effect of *T. harzianum* on seed germination was studied in maize and beans. Maize seeds coated with *Trichoderma* inoculum and planted in soils without fertilizer addition recorded the highest germination rate of 82.7 % followed by seeds coated with the inoculum and planted in soils treated with manure (82.2 %). Combination of the inoculum and fertilizer performed better at improving maize seed germination compared with fertilizers applied singly (Okoth et al. 2011). The yield was also enhanced significantly at the end of experiment. Adams et al. (2007) investigated if *T. harzianum* Rifai (1295-22) could be used to enhance the establishment and growth of crack willow (*Salix fragilis*) in a soil containing no organic or metal pollutants and in a metal-contaminated soil by comparing this fungus with non-inoculated controls and an ectomycorrhizal formulation. Results showed saplings grown with *T. harzianum* (T22) produced shoots and roots that were 40 and 20 % longer than those grown with ectomycorrhiza. *T. harzianum* (T22) treated saplings produced more than double the dry biomass of controls.

Several species of *Trichoderma* promoted growth and development of seedlings of vegetable and non-vegetable crops as well as strawberry (Bal and Altintas 2006, 2008; Elad et al. 2006). Zhang et al. (2012) showed that the putative *T. harzianum* mutant (T-E5) enhanced the production of IAA and plant colonization ability in cucumber and thus enhanced the yield over control.

Recently, some researchers have reported the effect of *Trichoderma* isolates directly on the plant growth parameters in some commercial crops (Shanmugaiah et al. 2009; Bal and Altintas, 2008; Babeendean et al. 2000; Zheng and Shetty 2000; Phuwiwat and Soytong 1999, Lynch et al. 1991). Chacon et al. (2007) showed that *T. harzianum* is able to promote tomato plant growth by colonizing the roots, increasing the foliar area and secondary roots, as well as changing the root architecture under sterile condition (Bjorkman et al. 1998). In contrast, Bal and Altintas (2006) demonstrated that application of *T. harzianum* did not increase yield in tomato, thus revealing a strain specificity. De facto, the effect of *Trichoderma* on plant growth improvement is not the result of *Trichoderma* isolate and plant species but also the complex interaction of many factors that may have an influence on the *Trichoderma*-plant interaction such as environmental parameters, soil microorganisms, and soil-plant interactions (Harman et al. 2004).

Mehetre et al. (2008) studied the value addition of the biogas manure with microbial enrichments with *Azotobacter*, *Rhizobium*, and *Trichoderma* and evaluated its performance under field condition on Mungbean (*Vigna radiata* L.). The results revealed that there was significant increase in microbial population of bacteria, fungi, rhizobium, and azotobacter during entire period of crop growth in biogas manure treated plot as compared to urea treated and control plots. All the biological and agronomical parameters were significantly stimulated in biogas manure enriched with *Trichoderma* as compared to only manure treatment. Barakat and

Al-Masri (2009) studied the effect of *T. harzianum* (Jn14) in combination with an amendment of sheep manure on the soil suppressiveness of *Fusarium* wilt of tomato. In addition to disease control, the treatment also increased tomato plant fresh weights by 52 % after 28 months, and the 10 % amendment increased fresh weights by 56, 40, and 63 %, after 18, 24, and 28 months, respectively, over controls.

In addition of being root colonizers, several *Trichoderma* isolates (many are new species) have been reported to be "true" endophytes (Druzhinina et al. 2011). Some of these isolates not only promoted growth of plants but also imparted tolerance to biotic and abiotic stresses (Bae et al. 2009, 2011; Bailey et al. 2009). This is an emerging area and needs to be investigated thoroughly to explore the discovery of more novel strains/ species of endophytic *Trichoderma* and making use of them in improving crop health.

3 Nutrient Use Efficiency

3.1 Nitrogen

Nitrogen is a primary nutrient and is the most ephemeral of the key plant nutrients. Nitrogen is critical for plant growth as it is a fundamental part of chlorophyll, and when leaves contain sufficient nitrogen, photosynthesis occurs at high rates. Application of *Trichoderma* spp. increases nitrogen use efficiency (NUE) in plants. Simple seed treatment with *Trichoderma* can reduce the requirement of nitrogen application to the extent of 30–50 % (Harman 2011; Shoresh et al. 2010; Harman and Mastouri 2010; Shoresh and Harman 2008). Singh et al. (2010) studied the effect of a formulation based on *T. harzianum* (Th 37) (applied @ 20 kg/ha) on the stubbles at the ratoon initiation stage of sugarcane and found that the treatment increased the availability of nitrogen (N), phosphorus (P), and potassium (K) by 27, 65, and 44 %, respectively. Improvements in uptake of nutrients and growth due to application of *Trichoderma* were also noticed in sugarcane (Srivastava et al. 2006;

Yadav et al. 2008; Shukla et al. 2008). A combined inoculation of *T. harzianum* and *Pseudomonas fluorescens* recorded the maximum nitrogen uptake (61.28 mg plant^{-1}) in vanilla (Sandheep et al. 2013).

3.2 Phosphorus

Phosphorus is one of the essential nutrients for plant growth. Its functions cannot be performed by any other nutrient, and an adequate supply of P is required for optimum growth and reproduction. Altomare et al. (1999) showed for the first time the ability of a *T. harzianum* Rifai 1295-22 (T-22) to solubilize insoluble or sparingly soluble minerals via three possible mechanisms including acidification of the medium, production of chelating metabolites, and redox activity. This strain was able to solubilize MnO_2, metallic zinc, and rock phosphate (mostly calcium phosphate) in a liquid sucrose yeast extract medium, as determined by inductively coupled plasma emission spectroscopy. Solubilization of tricalcium phosphate and other forms of phosphorus by *Trichoderma* spp. has been well studied in different crops, and the results of such trials clearly demonstrated enhanced phosphorus availability to plants (Rudresh et al. 2005, Anil and Lakshmi 2010, Azarmi et al. 2011, Saravanakumar et al. 2013). *Trichoderma* also has been shown to increase the root length, total biomass, and seed production of wheat, probably as a consequence of an increased efficiency in the absorption of phosphorus and nitrogen from the soil (Behl et al. 2003; Gupta and Baig 2001). Cuevas (2006) showed that the treatment with *Trichoderma* at 0 N fertilizer yielded grains significantly higher than that treated with 90 kg N/ha but without the fungus. Soil chemical analyses after rice harvest showed that yield was negatively correlated with available soil P and Zn, significant at 1 % level. Treatment plots with high yields had lower available P and Zn. These two studies showed that presence of the fungus in the soil in sufficient population resulted in more mineral nutrient availability especially P and Zn for plant use that increased crop growth and yield.

3.3 Potassium

Potassium is an essential plant nutrient and is required in large amounts for proper growth and reproduction. Potassium regulates the opening and closing of the stomata by a potassium ion pump. Several recent reports indicate that the fungi enhances tolerance to abiotic stresses during plant growth (Yildirim et al. 2006), in part due to improved root growth, improvement in water-holding capacity of plants, or enhancement in nutrient uptake (i.e., potassium), whereas, in the absence of stress, plant growth may or may not be enhanced. Mohammadi et al. (2010) applied *T. harzianum* in combination with other biofertilizers and showed significant increase in N, P_2O_5, K_2O, Fe, and Mg content in leaves and grains of chick pea.

3.4 Micronutrients

Micronutrients though required in small quantity have major role in crop productivity. Slight decrease in the availability of such nutrients hampers the yield of the crop significantly. Santiago et al. (2011) studied the effect of *T. asperellum* inoculation on the uptake of Fe, Cu, Mn, and Zn by wheat (*Triticum aestivum* L.) grown in a calcareous medium. Results showed the inoculation with T34 increased Fe concentration in Fe-deficient media, thus revealing a positive effect of this microorganism on Fe nutrition in wheat. Harman (2000, 2001) reported that *T. harzianum* (1295-22) could also solubilize a number of poorly soluble nutrients, such as Mn^{4+}, Fe^{3+}, Cu^{2+} etc. As mentioned before, in addition to N, P, and K, *T. harzianum* improved the availability of some of the micronutrients, viz., Cu, Fe, Mn, and Zn, that were enhanced, respectively by 6, 100, 79, and 66 % in ratoon sugarcane (Singh et al. 2010).

4 Enhanced Composting and Carbon Recycling

Compost contains humus or humified organic matter, which serves as a "bank" or "reserve" for important plant nutrients. Composting is a very complex process and requires lot of time and involves variety of microorganisms. The benefit from increased soil organic matter and composting includes:

i. Nutrients being available to crops during times when there are minimal or zero external inputs

ii. Reduced need for commercial fertilizers

iii. Improved plant health, which is an important line of defense against pests, diseases, and environmental stress

iv. Better retention of fertilizers and reduced runoff

v. Soil buffering, because the organic matter in compost neutralizes both acidic and alkaline soils and brings pH levels to the optimum range for nutrient availability to plants

Thus, composting is an important process of soil organic matter degradation and has role in nutrient availability to the different crops. *Trichoderma* spp. have the ability to accelerate the composting process and play positive role in the process of humification of compost. There are several reports on the use of *Trichoderma* alone or in combination with different organic manure and compost for better availability of nutrients. Haque et al. (2010) studied the effects of *Trichoderma* enriched biofertilizer on mustard (*Brassica campestris*) under field condition. Results showed increased seed yield per plant (by 5.34 %) with *Trichoderma* enriched biofertilizer. Mokhtar et al. (2013) studied the effect of *T. harzianum* and some essential oils alone or in combination with compost on the peanut crown rot disease under field conditions. All the treatments significantly reduced the peanut crown rot disease. The highest reduction was obtained with combined treatments (compost + *T. harzianum* + thyme and compost + *T. harzianum* + lemongrass) which reduced the disease incidence at both pre- and post-emergence growth stages, respectively. Espiritu (2011) studied the effect of compost prepared from coconut coir dust/chicken manure along with *T. harzianum* and applied to mung bean. The combination treatment gave highest plant fresh biomass and number of nodules over control. *Trichoderma hamatum* (GD12) isolated from soil in Devon, UK, promoted plant growth in low pH and nutrient-poor peat and displayed biological protection against pre- and postemergence diseases of lettuce seedlings, caused by *Sclerotinia sclerotiorum* and *Rhizoctonia solani* under the same conditions (Thornton 2005, 2008; Ryder et al. 2012).

Conclusions

Trichoderma species are free-living fungi that are common in soil and root ecosystems and have been known for decades as biocontrol fungi (Mukherjee et al. 2013). Apart from its biocontrol activity, recent studies have demonstrated that *Trichoderma* has many other useful attributes including enhanced nutrient uptake and availability and plant growth promotion. There are clear indications and experimental evidences for increased nutrient up take by application of *Trichoderma* spp. The nutrients include nitrogen, phosphorus, potassium, and micronutrients. Nutrient use efficiency has gained lot of significance recently due to loss of soil productivity. There is also an increase in demand for organic inputs in agriculture, and nutrient availability plays a major role in organic farming. Compared to the attention that *Trichoderma* spp. received as biofungicides, little in-depth research has been done on the other attributes of these plant-beneficial fungi, like the mechanisms of nutrients uptake as well as plant growth promotion. Recent advances in genetics and genomics would unravel some of these mechanisms and which, in turn, will enhance our capability to fully realize the potential of *Trichoderma* beyond biocontrol (Mukherjee et al. 2013).

Acknowledgements The authors thank the Head of Nuclear Agriculture and Biotechnology Division, Bhabha Atomic Research Centre, Mumbai, for encouragement and support.

References

Adams P, De-Leij FA, Lynch JM (2007) *Trichoderma harzianum* Rifai 1295-22 mediates growth promotion of Crack willow (*Salix fragilis*) saplings in both clean and metal contaminated soil. Microb Ecol 54:306–313

Altomare C, Norvell WA, Bjorkman T, Harman GE (1999) Solubilization of phosphate and micronutrients by the plant growth promoting and biocontrol fungus *Trichoderma harzianum* Rifai (1295 22). Appl Environ Microbiol 65:2926–2933

Anil K, Lakshmi T (2010) Phosphate solubilization potential and phosphatase activity of rhizospheric *Trichoderma* spp. Braz J Microbiol 41(3):787–795

Azarmi R, Hajieghrari B, Giglou A (2011) Effect of *Trichoderma* isolates on tomato seedling growth response and nutrient uptake. Afr J Biotechnol 10 (31):5850–5855

Babeendean N, Moot DJ, Jones EE, Stewart A (2000) Inconsistent growth promotion of Cabbage and Lettuce from *Trichoderma* isolates. New Z Plant Prot 53:143–146

Bae H, Sicher RC, Kim MS, Kim SH, Strem MD, Melnick RL, Bailey BA (2009) The beneficial endophyte *Trichoderma hamatum* isolate DIS 219b promotes growth and delays the onset of the drought response in Theobroma cacao. J Exp Bot 60:3279–3295

Bae H, Roberts DP, Lim HS, Strem MD, Park SC, Ryu CM, Melnick RL, Bailey BA (2011) Endophytic Trichoderma isolates from tropical environments delay disease onset and induce resistance against *Phytophthora capsici* in hot pepper using multiple mechanisms. Mol Plant-Microbe Interact 24:336–351

Bailey BA, Strem MD, Wood D (2009) Trichoderma species form endophytic associations within *Theobroma cacao* trichomes. Mycol Res 113:1365–1376

Baker R (1989) Improved *Trichoderma* spp. for promoting crop productivity. Trends Biotechnol 7(34):38

Bal U, Altintas S (2006) A positive side effect from *Trichoderma harzianum*, the biological control agent: increased yield in vegetable crops. J Environ Prot Ecol 7:383–387

Bal U, Altintas S (2008) Effects of *Trichoderma harzianum* on lettuce in protected cultivation. J Cent Eur Agric 1:63–70

Barakat RM, Al-Masri MI (2009) *Trichoderma harzianum* in combination with sheep manure amendment enhances soil suppressiveness of *Fusarium* wilt of tomato. Phytopathol Mediterr 48:385–395

Behl RK, Sharma H, Kumar V, Narula N (2003) Interactions amongst mycorrhiza, *Azotobacter chroococcum* and root characteristics of wheat varieties. J Agron Crop Sci 189:15–19

Bjorkman T, Blanchard LM, Harman GE (1998) Growth enhancement of shrunken-2 (sh2) sweet corn by *Trichoderma harzianum* 1295–22: effect of environmental stress. J Am Soc Hortic Sci 123:35–40

Chacon MR, Rodriguez-Galan O, Beritez T, Sousa S, Rey M, Llobell A, Delgado-Jarana J (2007) Microscopic and transcriptome analyses of early colonization of tomato roots by *Trichoderma harzianum*. Int Microbiol 10:19–27

Chang YC, Baker R, Kleifeld O, Chet I (1986) Increased growth of plants in the presence of the biological control agent *Trichoderma harzianum*. Plant Dis 70:145–148

Chet I, Harman G, Baker R (1981) *Trichoderma hamatum*: its hyphal interactions with *Rhizoctonia solani* and *Pythium* spp. Microb Ecol 7:29–38. doi:10.1007/BF02010476

Contreras-Cornejo H, Macias-Rodriguez L, Cortes-Penagos C, Lopez-Bucio J (2009) *Trichoderma virens*, a plant beneficial fungus, enhances biomass production and promotes lateral root growth through an auxin-dependent mechanism in Arabidopsis. Plant Physiol 149:1579–1592

Cuevas VC (2006) Soil inoculation with *Trichoderma pseudokoningii* Rifai enhances yield of rice. Philip J Sci 135(1):31–37

de Santiago A, Quintero JM, Avilés M, Delgado A (2011) Effect of *Trichoderma asperellum* strain T34 on iron, copper, manganese, and zinc uptake by wheat grown on a calcareous medium. Plant Soil 342(1–2):97–104

Druzhinina IS, Seidl-Seiboth V, Herrera-Estrella A, Horwitz BA, Kenerley CM, Enrique M, Mukherjee PK, Zeilinger S, Grigoriev IV, Kubicek CP (2011) Trichoderma: the genomics of opportunistic success. Nat Rev Microbiol 9:749–759

Elad Y (2000) Biological control of foliar pathogens by means of *Trichoderma harzianum* and potential modes of action. Crop Prot 19:709–714. doi:10.1016/S0261-2194(00)00094-6

Elad Y, Chet I, Henis Y (2006) Biological control of *Rhizoctonia solani* in strawberry fields by *Trichoderma harzianum*. Plant Soil 60:245–254

Espiritu BM (2011) Use of compost with microbial inoculation in container media for mungbean (*Vigna radiata* L. Wilckzek) and pechay (*Brassica napus* L.). J ISSAAS 17(1):160–168

Gupta N, Baig S (2001) Evaluation of synergistic effect of phosphate solubilizing *Penicillium* spp, AM fungi and rock phosphate on growth and yield of wheat. Philip J Sci 130:139–143

Haque MM, Haque AM, Ilias GNM, Molla HA (2010) *Trichoderma*-enriched biofertilizer: a prospective substitute of inorganic fertilizer for mustard (*Brassica campestris*) production. Agriculturists 8(2):66–73

Harman EG (2000) The dogmas and myths of biocontrol. Changes in perceptions based on research with *Trichoderma harzianum* T22. Plant Dis 84:377–393

Harman EG (2001) Microbial tools to improve crop performance and profitability and to control plant diseases. In: Tzeng DS, Huang JW (eds) Proceeding of international symposium on biological control of plant diseases for the new century-mode of action and application technology. NCHU, Taichung, pp 71–81

Harman GE (2006) Overview of mechanisms and uses of *Trichoderma* spp. Phytopathology 96:190–194. doi:10.1094/PHYTO- 96-0190

Harman GE (2011) Multifunctional fungal plant symbionts: new tools to enhance plant growth and productivity. New Phytol 189:647–649

Harman GE, Mastouri F (2010) Enhancing nitrogen use efficiency in wheat using *Trichoderma* seed inoculants, vol 7. International Society for Plant Microbe Interaction, St. Paul, pp 1–4

Harman GE, Howell CR, Viterbo A, Chet I, Lorito M (2004) *Trichoderma* species-opportunistic, avirulent plant symbionts. Nat Rev Microbiol 2:43–56

Hoyos-Carvajal L, Orduz S, Bissett J (2009) Growth stimulation in bean (Phaseolus vulgaris L.) by *Trichoderma*. Biol Control 51(3):409–416

Inbar J, Abramsky M, Cohen D, Chet I (1994) Plant growth enhancement and disease control by *Trichoderma harzianum* in vegetable seedlings grown under commercial conditions. Eur J Plant Pathol 100:337–346

Kleifield O, Chet I (1992) *Trichoderma* plant interaction and its effect on increased growth response. Plant Soil 144(267):272

Lo C-T, Lin C-Y (2002) Screening strains of *Trichoderma* spp for plant growth enhancement in Taiwan. Plant Pathol Bull 11:215–220

Lynch JM, Wilson KL, Ousley MA, Wipps JM (1991) Response of lettuce to *Trichoderma* treatment. Lett Appl Microbiol 12:59–61

Mehetre ST, Shrivastava M, Kale SP (2008) Influence of different bio-fertilizers amended biogas manure on soil microbial population and growth of Mungbean. J Ecofriendly Agric 3(2):112–115

Mohammadi K, Ghalavand A, Aghaalikhani M (2010) Study the efficacies of green manure application as Chickpea pre plant. World Acad Sci Eng Technol 4:10–20

Mokhtar MA-K, Abdel-Kareem F, El-Mougy NS, El-Mohamady RS (2013) Integration between compost, *Trichoderma harzianum* and essential oils for controlling peanut crown rot under field conditions. J Mycol 2013:1–7

Mukherjee PK, Horwitz BA, Herrera-Estrella A, Schmoll M, Kenerley CM (2013) Trichoderma research in the genome era. Annu Rev Phytopathol 51:105–129

Okoth S, Jane O, James O (2011) Improved seedling emergence and growth of maize and beans by *Trichoderma harziunum*. Trop Subtrop Agroecosyst 13:65–71

Ousley MA, Lynch JM, Whipps JM (1994) Potential of *Trichoderma* spp. as consistent plant growth stimulators. Biol Fertil Soils 17:85–90. doi:10.1007/BF00337738

Phuwiwat W, Soytong K (1999) Growth and yield response of Chinese radish to application of *Trichoderma harzianum*. Thammasat Int J Sci Technol 4(1):68–71

Rudresh DL, Shivaprakash MK, Prasad RD (2005) Tricalcium phosphate solubilizing abilities of *Trichoderma* spp. in relation to P uptake and growth and yield parameters of chickpea (*Cicer arietinum* L.). Can J Microbiol 51(3):217–222

Ryder LS, Harris BD, Soanes DM, Kershaw MJ, Talbot NJ, Thornton CR (2012) Saprotrophic competitiveness and biocontrol fitness of a genetically modified strain of the plant-growth-promoting fungus *Trichoderma hamatum* GD12. Microbiology 158:84–97. doi:10.1099/mic.0.051854-0

Sandheep AR, Asok AK, Jisha MS (2013) Combined inoculation of Pseudomonas fluorescens and Trichoderma harzianum for enhancing plant growth of vanilla (Vanilla planifolia). Pak J Biol Sci 16(12):580–584

Saravanakumar K, Shanmuga Arasu V, Kathiresan K (2013) Effect of *Trichoderma* on soil phosphate solubilization and growth improvement of *Avicennia marina*. Aquat Bot 104:101–105

Shanmugaiah V, Balasubramanian N, Gomathinayagam S, Monoharan PT, Rajendran A (2009) Effect of single application of *Trichoderma viride* and *Pseudomonas fluorescens* on growth promotion in cotton plants. Afr J Agric Res 4(11):1220–1225

Shoresh M, Harman GE (2008) The molecular basis of maize responses to *Trichoderma harzianum* T22 inoculation: a proteomic approach. Plant Physiol 147:2147–2163

Shoresh M, Mastouri F, Harman GE (2010) Induced systemic resistance and plant responses to fungal biocontrol agents. Annu Rev Phytopathol 48:21–43

Shukla SK, Yadav RL, Suman A, Singh PN (2008) Improving rhizospheric environment and sugarcane ratoon yield through bioagents amended farm yard manure in *udic ustochrept* soil. Soil Tillage Res 99:158–168

Singh V, Singh PN, Yadav RL, Awasthi SK, Joshi BB, Singh RK, Lal RJ, Duttamajumder SK (2010) Increasing the efficacy of *Trichoderma harzianum* for nutrient uptake and control of red rot in sugarcane. J Hortic For 2(4):66–71

Srivastava SN, Singh V, Awasthi SK (2006) *Trichoderma* induced improvement in growth, yield and quality of sugarcane. Sugar Technol 8:166–169

Thornton CR (2005) Use of monoclonal antibodies to quantify the dynamics of alpha-galactosidase and endo-1, 4-beta-glucanase production by *Trichoderma hamatum* during saprotrophic growth and sporulation in peat. Environ Microbiol 7:737–749. doi:10.1111/j.1462-2920.2005.00747.x

Thornton CR (2008) Tracking fungi in soil with monoclonal antibodies. Eur J Plant Pathol 121:347–353. doi:10.1007/s10658-007-9228-3

Vinale F, Flemattoi G, Sivasith AK, Lorito M, Marra R, Skelton BW (2009) Harzianic acid, an antifungal and plant growth promoting metabolite from *Trichoderma harzianum*. J Nat Prod 72:2032–2035. doi:10.1021/np900548p

Windham MT, Elad Y, Baker R (1986) A mechanism for increased plant-growth induced by *Trichoderma* spp. Phytopathology 76:518–521

Yadav RL, Singh V, Srivastav SN, Lal RJ, Sangeeta S, Awasthi SK, Joshi BB (2008) Use of *Trichoderma harzianum* for the control of red rot disease of sugarcane. Sugarc Int (UK) 26:28–33

Yildirim E, Taylor AG, Spittler TD (2006) Ameliorative effects of biological treatments on growth of squash plants under salt stress. Sci Hortic (Amst) 111:1–6

Zhang F, Yuan J, Yang X, Cui Y, Chen L, Ran W, Shen Q (2012) Putative *Trichoderma harzianum* mutant promotes cucumber growth by enhanced production of indole acetic acid and plant colonization. Plant Soil 012-1519-6

Zheng Z, Shetty K (2000) Enhancement of pea (*Pisum sativum*) seedling vigor and associated phenolic content by extraction of apple pomace fermented with *Trichoderma* spp. Proc Biochem 36:79–84

Bio-priming Mediated Nutrient Use Efficiency of Crop Species

Amitava Rakshit, Kumai Sunita, Sumita Pal, Akanksha Singh, and Harikesh Bahadur Singh

Abstract

Soil contamination and environmental hazard from the indiscriminate and excessive application of agrochemicals on crops have been key issues for the present-day agriculture. Additionally, the risk to human health has also led to stringent regulatory framework around the use of synthetic chemicals in agriculture. Bio-inoculants have emerged as the most feasible eco-friendly solution to these issues and have been gaining considerable consumer acceptance since the time they were first introduced. Bioagents are substances containing living microorganisms which promote plant growth and maintain the soil and crop health by increasing the supply or availability of primary nutrients to the host plant. Bio-priming which involves seed priming in combination with low dosage of beneficial microorganisms is becoming a potentially prominent technique to induce profound changes in versatility of plant performance, encourage desired attributes in crop growth, and stabilize the efficacy of biological agents in the present fragile setup of agriculture by reducing dependency on chemical inputs and offers an attractive option for resource-poor farmers being an easy and cost effective method. The most prominent contributors in fungi and bacteria which are used extensively in bio-priming include *Trichoderma*, *Pseudomonas*, *Glomus*, *Bacillus*, *Agrobacterium*, and *Gliocladium*. Here in this review, we discuss the potential of

A. Rakshit (✉) • K. Sunita
Department of Soil Science and Agricultural Chemistry,
Institute of Agricultural Science, BHU, Varanasi 221005,
UP, India
e-mail: amitavar@bhu.ac.in; sumeena66@gmail.com

S. Pal • A. Singh • H.B. Singh
Department of Mycology and Plant Pathology, Institute of
Agricultural Science, BHU, Varanasi 221005, UP, India
e-mail: sumitapal99@gmail.com; bhuaks29@gmail.com;
hbs1@rediffmail.com

A. Rakshit et al. (eds.), *Nutrient Use Efficiency: from Basics to Advances*,
DOI 10.1007/978-81-322-2169-2_12, © Springer India 2015

bio-priming for improving crop growth and nutrient use efficiency and provide an assessment of bioagents currently used with crop species and key limitations involved.

Keywords

Biological agents • Bio-priming • Crop growth improvement

1   Introduction

The fertilizer industry presents one of the most energy-intensive sectors within the Indian economy and is therefore of particular interest in the context of both local and global environmental discussions (GOI 2011). During fertilizer production energy is consumed in the form of natural gas, associated gas, naphtha, fuel oil, low sulfur heavy stock, and coal (Phylipsen et al. 1998). In the present-day scenario, there are more than 57 large-sized and 64 medium- and small-sized chemical fertilizer production units in India which have performed a significant role in enabling the increased supply of essential nutrients to plants to achieve the objective of being self-sufficient in the production of food grains.

There is no denying the fact that over the years increased usage of fertilizer has played a significant role in the increase of agriculture productivity. Current trends in agricultural output, however, depict that the marginal productivity of soil in relation to the application of fertilizers is declining. Decline in soil organic matter and deficiency of secondary nutrients and micronutrients were major issues which led to yield stagnation. The comparatively high usage of straight fertilizers as against the complex fertilizers (NPK) which are considered to be agronomically better including low or non-usage of secondary nutrients and micronutrients has also probably contributed towards slowdown in fertility and growth of productivity. The declining fertilizer use efficiency is also one of the factors for low productivity. The chemical and synthetic fertilizers, particularly nitrogen, phosphorous, and potassium (NPK), are highly subsidized (Table 1), and the amount have quadrupled over the past 10 years.

Further the fertilizer sector needs Rs 40,000 crore to address the current payment crisis due to subsidy over dues. The pricing of subsidized fertilizers is also probably responsible for higher usage of straight fertilizers and skewed usage of nutrients. Huge amount of subsidy allocation provided directly to the industry has led to indiscriminate production and availability while neglecting the locally available knowledge on soil nutrient management. Widespread usage of such fertilizers has resulted in the degradation of natural resource base, especially soil.

In recent years environmental pollution is a serious global problem, and plants are more and more subjected to a variety of stresses. Among these stresses, biotic and abiotic factors like low soil fertility, drought, and temperature extremes are very common. And with this, present context low-input agricultural systems have gained attention due to rising interest for the conservation of natural resources, reduction of environmental degradation, and escalating price of inorganic fertilizers. Conventional farming systems with lower application of fertilizers and pesticides have been developed and perfected for many crops under varied agroecological conditions. Moreover, advances in science and technology enabled us to apply the potential of biological diversity for pollution abatement which is termed as bioresource management. A more appropriate management of microorganism in agriculture is expected to allow a substantial reduction in the amount of minerals used without losses in productivity while permitting a more sustainable production system.

The negative consequences of environmental damage, land constraints, population pressure, and institutional deficiencies have been reinforced by a limited understanding of the

Table 1 Input statistics in agriculture

Input	Usage (10^6 tonnes)		Subsidy (Rs. billion)		Size of the industry (Rs. billion)		Energy involvement (MJ kg^{-1})
	India	Global	India	Global	India	Global	
Fertilizer	24.5	170	750	–	30	5,000	78.2(N);17.5(P);13.8(K)
Pesticide	0.85	2.6	–	–	180	2,500	215 (herbicide), 238 (insecticide), and 92 (fungicide)
Biopesticide	0.25	25	–	–	2	200	–
Biofertilizer	0.28	200	–	–	4	180	0.01(liquid); 0.3(solid)

Source: Mihov and Ttringovska (2010), www.bccresearch.com, http://www.nic.in/agri

biological processes necessary to optimize nutrient cycling, minimize use of external inputs, and maximize input use efficiency and plant protection particularly in tropical and subtropical agriculture. The overall strategy for increasing crop yields and sustaining them at a high level must include an integrated approach to the management of soil nutrients, along with other complementary measures of plant protection. Integration of chemicals, plant extracts, biotic agents along with priming agents are some of the novel approach for a holistic management reducing the cost and pollution hazards while causing minimum interference with biological equilibrium (Papavizas 1973).

Research on the use of bioagents for use in agriculture has a history of more than 70 years, and considerable attention has been directed to biological seed and transplant treatments (Lewis and Lumsden 2001). Bioagents, especially beneficial bacteria and fungi, have shown promise in many seed enhancement studies, and its use is consistent with the development of integrated crop management systems.

2 Seed: A Basic and Critical Input for Agricultural Production

Seed is an important facet of agriculture, contributing greatly to the successful production of food and feed crops worldwide, and yet it faces challenges like production, storage, and quality control of seeds. Such challenges along with inadequate soil moisture and soil salinity lead to poor and unsynchronized seedling emergence, poor establishment of crop stand, and a reduction in crop yield and/or total crop failure.

To overcome these challenges, a number of seed technologies (priming, pelleting, coating, etc.) that enhance germination and synchronization of seedling emergence under adverse environmental conditions such as drought have been developed (Rakshit et al. 2013; Roy and Srivastava 2000; Ashraf et al. 2011; Basra et al. 2005; Tzortzakis 2009). One seed treatment method has proven successful after Heydecker (1973) attempted initial experimentation to improve germination and emergence under stressful conditions. The technique is known as priming. Priming, by definition, is the process by which the physiological processes of seed germination are partially activated to point that is inhibitive for radical emergence (Ashraf and Foolad 2005). The technique involves pre-soaking seeds in aerated solutions or solid matrices for a specific period of time and then re-drying before field planting. Currently, it is an important tool for accelerating seed germination rate, ensuring uniform seedling emergence, and improving stand establishment and seedling vigor (McDonald 2000; Nascimento and Pereira 2007). Priming is becoming an extremely widely used method in agriculture. In general, most kinds of seeds experimented with so far have shown an overall advantage over seeds that are not primed. Seed priming can be accomplished through different methods such as hydropriming (water), solid matrix priming or matripriming (hydrated sand, peat, and vermiculite), osmopriming (soaking in osmotic solutions such as PEG or inorganic salts), thermopriming (treatment with low or high temperatures), and plant growth inducers.

Several investigations across varied agro ecological situations confirmed that seed priming has many benefits including break of seed dormancy, uniform emergence, deeper roots, higher resource use efficiency, better competition with weeds, early flowering and maturity and grow faster under stress conditions (Bajehbaj 2010). This method can be useful to farmers because it saves them the money and time spent for fertilizers, reseeding, and weak plants. It is a form of seed planting preparation, in which seeds are soaked before planting. However, because of environmental concerns with chemical control strategies and a potential for toxic residues to accumulate in soil, biological control strategies with antagonistic fungi and rhizobacteria have been developed. However, among the viable options available with priming, bio-priming is the most evolved process of resource conservation by mutual interaction of plant/plant propagules and suitable microbial flora.

3 Bio-priming

Bio-priming is a process of biological seed treatment that refers to combination of seed hydration (physiological aspect of disease control) and inoculation (biological aspect of disease control) of seed with beneficial organism to protect seed. Seed treatments with biocontrol agents along with priming agents may serve as important means of managing many of the soil and seed-borne pathogens and diseases and improving nutrient use efficiency, the process often known as bio-priming. Bio-priming is a relatively new and emerging seed and/or seedling treatment tool that can be used to induce systemic resistance in treated crops against abiotic and biotic stresses (Rakshit et al. 2014). In most cases, microbial inoculants such as plant growth-promoting rhizosphere or endophytic microbes (bacteria or fungi) are used. Like other seed priming techniques, this technology has proven to be of paramount importance in improving seed quality and performance as well as in promoting plant growth (Aliye et al. 2008; Rajkumar et al. 2010, 2012). Seeds may be treated with microorganisms in a specific concentration for a specific duration or by coating with microbes. The use of bioagents or botanicals with priming agents has become an inevitable method of disease control, particularly in crops and in the absence of resistant cultivars.

Some biocontrol agents applied as seed dressers are capable of colonizing the rhizosphere, potentially providing benefits to the plant beyond the seedling emergence stage (Nancy et al. 1997). As a result, bio-priming holds a tremendous scope most likely to be exploited by seed companies and organic farmers in the format of sustainable agriculture where seed-microbe association will be in a position to survive and thrive in stressful man-made environments.

4 Bio-priming-Mediated Nutrient Use Efficiency

Improving NUE is an important goal to harvest better crop yield on sustained basis (Rakshit et al. 2002). According to statistics, the worldwide transaction amount of fertilizer is roughly US$40 billion. Of this, 135 million metric tons of chemical fertilizer is applied each year, with sales volume of about US$30 billion (http://www.fertilizer.org/Statistics). Although there are no clear application statistics for biofertilizer, however, its sales volume is estimated to be as much as US$3 billion. Microbial inoculants play a critical role in this by taking part in regulating enzyme activities followed by nutrient dynamics in the rhizosphere. Further, the protective effect of microbes against a broad range of stress has been well documented and is the reason for their multifaceted use in sustainable agriculture. Different mechanisms have been reported to explain alleviation of stress by microbes including biochemical changes in plant tissues, microbial changes in rhizosphere, nutrient status, anatomical changes to cells, and changes to root system morphology.

Studies have shown that the growth-promoting ability of microbes may be highly specific to certain plant species, cultivar, and genotype (Table 2) (Bashan 1998; Gupta et al. 2000; Lucy et al. 2004). PGPR can affect plant growth by different direct and indirect mechanisms (Glick 1995; Gupta et al. 2000). Some examples of these mechanisms (Table 3), which can probably be active simultaneously or sequentially at different stages of plant growth, are (1) increased mineral nutrient solubilization and nitrogen fixation,

Table 2 Experiments on bio-priming of seed carried out in series of crop species

Primer	Effective against	Mechanism/responsible metabolites	References
Pseudomonas fluorescens	Alternaria blight (*Alternaria helianthi*) of sunflower	Production of bacterial allelochemicals, antibiosis (antibiotics)	Rao et al. (2009)
Pseudomonas aureofaciens AB254	Damping off (*Pythium ultimum*) of sweet corn	Production of bacterial allelochemicals, antibiosis (antibiotics), lytic enzymes	Callan et al. (1990)
Enterobacter cloacea (Jordan) + *Trichoderma* spp.	Damping off (*Pythium ultimum*) of tomato and cucumber	Antibiosis (antibiotics), biocidal volatiles, lytic enzymes, detoxification enzymes	Harman and Taylor (1988)
Pseudomonas fluorescens strains: UTPf76 and UTPf86	Sunflower seed germination and promotion of seedling growth	Production of bacterial allelochemicals, indirect promotion through induced systemic resistance	Moeinzadeh et al. (2010)
Pseudomonas fluorescens AB254	Sweet corn	Production of bacterial allelochemicals, antibiosis (antibiotics), indirect promotion through induced systemic resistance	Callan et al. (1991)
Rhizopseudomonads strain 7NSK2	Maize and barley	Biocidal volatiles, antibiosis (antibiotics), lytic enzymes, detoxification enzymes	Iswandi et al. (1987)
Pseudomonas putida	Canola rapeseed	Production of bacterial allelochemicals, lytic enzymes, detoxification enzymes	Lifshitz et al. (1987)
Pseudomonas fluorescens	Growth of pearl millet	Production of bacterial allelochemicals, antibiosis (antibiotics), indirect promotion through induced systemic resistance	Raj et al. (2004)
Fluorescent pseudomonads	Take-all of wheat (*Gaeumannomyces graminis*)	Production of bacterial allelochemicals, antibiosis (antibiotics), indirect promotion through induced systemic resistance	Weller and Cook (1983)
Serratia plymuthica (strain HRO-C48) + *Pseudomonas chlororaphis* (strain MA 342)	Blackleg disease (Leptosphaeria maculans) in oilseed rape	Production of bacterial allelochemicals, including iron-chelating siderophores, antibiosis (antibiotics)	Ruba et al. (2011b)
Serratia plymuthica + *Pseudomonas chlororaphis*	*Verticillium longisporum* infection in oilseed rape	Production of bacterial allelochemicals, including iron-chelating siderophores, antibiosis (antibiotics)	Ruba et al. (2011a)
Pseudomonas	Corn, sorghum, and wheat	Production of bacterial allelochemicals, antibiosis (antibiotics), indirect promotion through induced systemic resistance	El-Meleigi (1989)
Enterobacter cloacae	Rots caused by *Pythium* in pea and cucumber	Biocidal volatiles, lytic enzymes, and detoxification enzymes	Hadar et al. (1983)
T. harzianum, T. viride, T. hamatum, B. subtilis, B. cereus, and *P. fluorescens*	Faba bean root rot by *Rhizoctonia solani, Fusarium solani*, and *Sclerotium rolfsii*	Production of bacterial allelochemicals, including iron-chelating siderophores, antibiosis (antibiotics), biocidal volatiles	El-Mougy and Abdel-Kader (2008)

(continued)

Table 2 (continued)

Primer	Effective against	Mechanism/responsible metabolites	References
Clonostachys rosea	Carrot seed infected with seed-borne *Alternaria* spp.	Production of bacterial allelochemicals, including iron-chelating siderophores, antibiosis (antibiotics), biocidal volatiles, lytic enzymes, detoxification enzymes, indirect promotion through induced systemic resistance	Jensen et al. (2004)
Pseudomonas cepacia or *P. fluorescens*	*Pythium* damping off and *Aphanomyces* root rot of peas	Production of bacterial allelochemicals, antibiosis (antibiotics), indirect promotion through induced systemic resistance	Parke (1991)
Agrobacterium rubi strain A 16 + *Burkholderia gladii* strain BA 7 + *Pseudomonas putida* strain BA 8 + *Bacillus subtilis* strain BA 142, and *Bacillus megaterium* strain M 3	Radish under salinity	Production of bacterial allelochemicals, antibiosis (antibiotics), biocidal volatiles, lytic enzymes	Kaymak et al. (2009)
Beneficial microorganisms	Carrot and onion	Production of bacterial allelochemicals, including iron-chelating siderophores, antibiosis (antibiotics), biocidal volatiles, lytic enzymes, detoxification enzymes, indirect promotion through induced systemic resistance	Bennett et al. (2009)
PGPR like *Azotobacter chroococcum* + *Azospirillum lipoferum* + *Azotobacter chroococcum* + *Azospirillum lipoferum*	Hybrid maize	Production of bacterial allelochemicals, including iron-chelating siderophores, antibiosis (antibiotics), indirect promotion through induced systemic resistance	Sharifi Raouf Seyed (2011)
Trichoderma harzianum + *T. virens* and + *Pseudomonas aeruginosa*	Damping off (*Colletotrichum truncatum*) in soybean	Production of bacterial allelochemicals including iron-chelating siderophores, antibiosis (antibiotics), biocidal volatiles	Begum et al. (2010)

making nutrients available for the plant; (2) repression of soilborne pathogens (by the production of hydrogen cyanide, siderophores, antibiotics, and/or competition for nutrients); (3) improved plant stress tolerance to drought, salinity, and metal toxicity; and (4) production of phytohormones such as indole-3-acetic acid (IAA) (Gupta et al. 2000). Moreover, some PGPR have the enzyme 1-aminocyclopropane-1-carboxylate (ACC) deaminase, which hydrolyses ACC, the immediate precursor of ethylene in plants (Glick et al. 1995). By lowering ethylene concentration in seedlings and thus its inhibitory effect, these PGPR stimulate seedlings root length (Glick et al. 1999).

Nonetheless, these research findings suggest that bio-priming with different beneficial microbes may not only enhance seed quality but also boost seedling vigor and ability to withstand abiotic and biotic stressors and thus offer an innovative crop protection tool for the sustainable improvement of crop yield.

Table 3 Bio-priming-mediated nutrient use efficiency

S. No.	Crop	Bioagent	Nutrient use efficiency Primary (N, P, K)	Secondary (Ca, Mg)	Micro (Cu, Fe, Zn, Mn)	References
1.	Rice (*Oryza sativa*)	*A. amazonense*	N (3.5–18.5 %)			Rodrigues et al. (2008)
2.	*Wheat-rice* and *wheat-black gram* rotations	Natural mycorrhiza consortium + *Pseudomonas fluorescens* (strains R62 + R81)	0.695 PUE [kg P grain kg^{-1} P fertilizer]			Mäder et al. (2011)
3.	Maize (*Zea mays*)	*T. harzianum*	8.8–9.76 % N in root; 3.5 % N in shoot			Akladious and Abbas (2012)
4.	Sugarcane (*Saccharum officinarum*)	Fluorescent *Pseudomonas* strains R62 + R81	0.719 PUE [kg P grain kg^{-1} P fertilizer]			Yadav et al. (2013)
5.	Soybean (*Glycine max*)	Trichoderma harzianum AS19-2	N (15.8 %)		Zn (8.24 %); Fe (57.82 %)	Entesari et al. (2013)
		Trichoderma virens As10-5	N (5.2 %)		Zn (21.6 %); Fe (14.81)	
		Trichoderma atroviride As18-5	N (11 %)		Zn (37.25 %); Fe (14.6 %)	
6.	Cucumber (*Cucumis sativus*)	*Trichoderma asperellum* strain T 34			Cu (25 %); Zn (11.4 %); Zn (29.5 %); Mn (58.6 %); Cu (10.5 %); Fe (85.7 %)	Santiago et al. (2012)
7.	Cucumber (*Cucumis sativus*)	*T. harzianum*	N (13 %); P (12 %); K (11.7 %)	Ca (13.5 %); Mg (3.7 %)	Fe (9 %); Mn (8.2 %); Cu (35 %); Zn (5.7 %)	Moharam and Negim (2012)
		Trichoderma viride Tv2	N (5.9 %); P (1.2 %); Ca (5.3 %)		Fe (7.5 %); Mn (1.1 %); Cu (13.8 %); Zn (1.4 %)	
8.	Cucumber (*Cucumis sativus*)	*Trichoderma harzianum*	P (30 %)		Zn (25 %); Mn (70 %)	Yedidia et al. (2001)
9.	Tomato (*Lycopersicon esculentum*)	*T. harzianum* T22	K (9.7 %); P (38 %); N (2.5 %)	Ca (22 %); Mg (20 %)	Fe (46 %); Zn (27 %)	Molla et al. (2012)
		BioF/liquid (broth of spores suspension of *T. harzianum* T22)	K (15.3 %); P (24.7 %)	Ca (18.2 %); Mg (24.4 %)	Fe (64.6 %); Zn (45 %)	
10.	Tomato (*Lycopersicon esculentum*)	*T. harzianum* T969	P (65.85 %); K (324.35 %)			Azarmi et al. (2011)
		T. harzianum T447	P (359.53 %); K (782.97 %)	Ca (528.63 %); Mg (220.86 %)		
		T. harzianum T969	P (42.98 %); K (162.82 %)	Ca (31.46 %); Mg (38.98 %)		

(*continued*)

Table 3 (continued)

S. No.	Crop	Bioagent	Nutrient use efficiency Primary (N, P, K)	Secondary (Ca, Mg)	Micro (Cu, Fe, Zn, Mn)	References
11.	Broccoli (*Brassica oleracea*)	AM fungi	N (102.08 %); P (53.33 %)			Tanwar et al. (2013)
		P. fluorescens	N (235.42 %); P (163.33 %)			
		T. viride	N (735 %); P (210 %)			
12.	Melon (*Cucumis melo*)	*T. harzianum*	N (27.03 %), P (137.8 %); K (27.96 %)			Martínez-Medina et al. (2009)
		Glomus constrictum under reduced fertilization dosage	N (11.05 %); K (32.2 %)			
		Glomus mosseae under reduced fertilization dosage	N (31.05 %), P (67.56 %), and K (46.6 %)			
		Glomus claroideum under reduced fertilization dosage	N (9.47 %); P (27.02 %); K (27.96 %)			
		Glomus intraradices under reduced fertilization dosage	N (7.89 %); P (21.62 %); K (13.55 %)			
		Trichoderma harzianum under conventional fertilization dosage	N (20.6 %); K (30 %)			
		Glomus constrictum under conventional fertilization	K (30 %)			
		Glomus mosseae under conventional fertilization	N (1.26 %); K (29.16 %)			
		Glomus claroideum under conventional fertilization	N (9.47 %); P (27 %); K (27.96 %)			
		Glomus intraradices under conventional fertilization	N (7.89 %); P (21.62 %); K (13.56 %)			
13.	Tea (*Camellia sinensis*)	*Trichoderma harzianum*	N (44 %); P (50 %); K (16 %)			Thomas et al. (2010)
		Azospirillum brasilense	N (65 %); P (25 %); K (14 %)			
		Pseudomonas fluorescens	N (52 %); P (67 %); K (18 %)			

5   Present Status

Apart from the general issues relating to the need to use integrated approaches to promotion, local access to the materials and up-front cost of some of the materials used can be a bottleneck to adoption in rural areas. On the contrary measuring the correct quantities, repackaging and selling the small quantities of bio inoculums needed by farmers offers opportunities as well for disadvantaged people to generate income. However, a holistic approach to rural development is necessary, which is something that sectoral line agencies may find difficult. A second-approach more readily available materials are being evaluated as an alternative source for priming. The need for commercialization of the supply of biological agents is a problem that will need to be addressed. There is also an opportunity for seed producers to develop a value-added process if the technology can be refined to allow longer-term storage of bio-primed seeds.

Conclusions

Bio-priming is a smart practice that is consistent with sustainable agriculture goals. Although chemical seed treatment constitutes a low-volume pesticide use, today's growers are considering ways to continue to combat stress while reducing synthetic chemical inputs. Biological seed treatment using naturally occurring soil microorganisms has the potential to provide safe, nonpolluting, and environmentally sound disease control. The development of a delivery system for biological seed protectants that reliably furnishes the grower with a healthy seedling stand aids in making bio-priming competitive with chemical seed treatment. Findings suggest that bio-priming with different beneficial microbes may not only enhance seed quality but also boost seedling vigor and ability to withstand abiotic and biotic stresses and thus offer an innovative crop protection tool for the sustainable improvement of crop yield. Private and public sector research scientists are making valuable improvements in the efficacy, reliability, and utility of bio-priming for various crop species with reference to long shelf life, high density of viable propagules, stability under stress, ease of application to seed, and economy of production. Many areas are yet to be explored further looking at the work and progress with bioagents applied to seed as a subset of microbial ecology.

References

Akladious SA, Abbas SM (2012) Application of *Trichoderma harzianum* T22 as a biofertilizer supporting maize growth. Afr J Biotechnol 11 (35):8672–8683

Aliye N, Fininsa C, Hiskias Y (2008) Evaluation of rhizosphere bacterial antagonists for their potential to bio-protect potato (*Solanum tuberosum*) against bacterial wilt (*Ralstonia solanacearum*). Biol Control 47:282–288

Ashraf M, Foolad MR (2005) Pre-sowing seed treatment – a shotgun approach to improve germination, plant growth, and crop yield under saline and non-saline conditions. Adv Agron 88:223–271

Ashraf MA, Rasool M, Mirza MS (2011) Nitrogen fixation and indole acetic acid production potential of bacteria isolated from rhizosphere of sugarcane (*Saccharum officinarum* L.). Adv Biol Res 5 (6):348–355

Azarmi R, Hajieghrari B, Giglou A (2011) Effect of *Trichoderma* isolates on tomato seedling growth response and nutrient uptake. Afr J Biotechnol 10 (31):5850–5855

Bajehbaj AA (2010) The effects of NaCl priming on salt tolerance in sunflower germination and seedling grown under salinity conditions. Afr J Biotechnol 9:1764–1770

Bashan Y (1998) Inoculants for plant growth-promoting bacteria in agriculture. Biotechnol Adv 16:729–770

Basra SMA, Farooq M, Tabassum R, Ahmed N (2005) Evaluation of seed vigour enhancement techniques on physiological and biochemical basis in coarse rice. Seed Sci Technol 34:741–750

Begum MM, Sariah M, Puteh AB, Zainal Abidin MA, Rahman MA, Siddiqui Y (2010) Field performance of bio-primed seeds to suppress *Colletotrichum truncatum* causing damping-off and seedling stand of soybean. Biol Control 53(1):18–23

Bennett AJ, Mead A, Whipps JM (2009) Performance of carrot and onion seed primed with beneficial microorganisms in glasshouse and field trials. Biol Control 51(3):417–426

Callan NW, Mathre DE, Miller JB (1990) Bio-priming seed treatment for biological control of *Pythium ultimum* pre-emergence damping off in sh2 sweet corn. Plant Dis 74:368–372

Callan NW, Mathre DE, Miller JB (1991) Yield performance of sweet corn seed bioprimed and coated with *Pseudomonas fluorescens* AB254. Hortic Sci 26:1163–1165

El-Meleigi MA (1989) Effect of soil *Pseudomonas* isolates applied to corn, sorghum and wheat seeds on seedling growth and corn yield. Can J Plant Sci 69:101–108

El-Mougy NS, Abdel-Kader MM (2008) Long-term activity of bio-priming seed treatment for biological control of faba bean root rot pathogens. Australas Plant Pathol 37:464–471

Entesari M, Sharifzadeh F, Ahmadzadeh M, Farhangfar M (2013) Seed biopriming with *Trichoderma* species and *Pseudomonas fluorescent* on growth parameters, enzymes activity and nutritional status of soybean. Int J Agron Plant Prod 4(4):610–619

Glick BR (1995) The enhancement of plant growth by free-living bacteria. Can J Microbiol 41:109–117

Glick BR, Karaturovic DM, Newell PC (1995) A novel procedure for rapid isolation of plant growth promoting *Pseudomonas*. Can J Microbiol 41:533–536

Glick BR, Patten CL, Holguin G, Penrose DM (1999) Biochemical and genetic mechanisms used by plant growth promoting bacteria. Imperial College Press, London

Government of India (2011) Fertilizer situation in INDIA, Department of Agriculture and Co-operation, http://www.nic.in/agri. National Informatics Centre (NIC)

Gupta A, Gopal M, Tilak KV (2000) Mechanism of plant growth promotion by rhizobacteria. Indian J Exp Biol 38:856–862

Hadar Y, Harman GE, Taylor AG, Norton JM (1983) Effects of pre-germination of pea and cucumber seeds and of seed treatment with *Enterobacter cloacae* on rots caused by *Pythium*. Phytopathology 73:1322–1325

Harman GE, Taylor AG (1988) Improved seedling performance by integration of biological control agents at favorable pH levels with solid matrix priming. Phytopathology 78:520–525

Heydecker W (1973) The priming of seeds. University of Nottingham School of Agriculture Report, pp 50–67

Iswandi A, Bossier P, Vandenabeele J, Verstraete W (1987) Effect of seed inoculation with the rhizopseudomonads strain 7NSK2 on root microbiota of maize (*Zea mays*) and barley (*Hordeum vulgare*). Biol Fertil Soils 3:153–158

Jensen B, Knudsen IMB, Madsen M, Jensen DF (2004) Biopriming of infected carrot seed with an antagonist, *Clonostachys rosea*, selected for control of seed borne *Alternaria* spp. Phytopathology 94:551–560

Kaymak HC, İsmail G, Faika Y, Mesude FD (2009) The effects of bio-priming with PGPR on germination of radish (*Raphanus sativus* L.) seeds under saline conditions. Turk J Agric For 33:173–179

Lewis JA, Lumsden RD (2001) Biocontrol of damping-off of greenhouse-grown crops caused by Rhizoctonia solani with a formulation of *Trichoderma* spp. Crop Prot 20:49–56

Lifshitz R, Kloepper JW, Kozlowksi M, Simon C, Carlson J, Tipping B, Zaleska I (1987) Growth promotion of canola rapeseed seedling by a strain of *Pseudomonas putida* under gnotobiotic condition. Can J Microbiol 33:390–395

Lucy M, Reed E, Glick BR (2004) Application of free living plant growth-promoting rhizobacteria. Antonie van Leeuwenhoek Int J Gen Mol Microbiol 86:1–25

Mäder P, Kaiser F, Adholeya A, Singh R, Uppal HS, Sharma AK, Srivastava R, Sahai V, Aragno M, Wiemken A, Johri BN, Fried PM (2011) Inoculation of root microorganisms for sustainable wheat-rice and wheat-black gram rotations in India. Soil Biol Biochem 43:609–619

Martínez-Medina A, Roldán A, Pascual JA (2009) Performance of a Trichoderma harzianum bentonite-vermiculite formulation against Fusarium wilt in seedling nursery melon plants. Hort Sci 44:2025–2027

McDonald MB (2000) Seed priming. In: Black M, Bewley MJD (eds) Seed technology and its biological basis. Sheffield Academic Press, Sheffield, pp 287–325

Mihov M, Ttringovska I (2010) Energy efficiency improvement of greenhouse tomato production by applying new biofertilizers. Bulg J Agric Sci 16(4):454–458

Moeinzadeh A, Sharif-Zadeh F, Ahmadzadeh M, Heidari Tajabadi F (2010) Biopriming of sunflower (Helianthus annuus L.) seed with Pseudomonas fluorescens for improvement of seed invigoration and seedling growth. Aust J Crop Sci 4(7):564–570

Moharam MH, Negim OO (2012) Biocontrol of *Fusarium* wilt disease in cucumber with improvement of growth and mineral uptake using some antagonistic formulations. Commun Agric Appl Biol Sci 77(3):53–63

Molla AH, Haque M, Haque A, Ilias GNM (2012) *Trichoderma*-enriched biofertilizer enhances production and nutritional quality of tomato (*Lycopersicon esculentum* Mill.) and minimizes NPK fertilizer use. Agric Res 1(3):265–272

Nancy W, Don Mathre E, James B, Charles S (1997) Biological seed treatments: factors involved in efficacy. Hortic Sci 32:179–183

Nascimento WM, Pereira RS (2007) Preventing thermo inhibition in carrot by seed priming. Seed Sci Technol 35:503–506

Papavizas GC (1973) Status of applied biological control of soil borne plant pathogens. Soil Biol Biochem 5:709–720

Parke JL (1991) Root colonization by indigenous and introduced microorganisms. In: Keister DL, Gregan PB (eds) The rhizosphere and plant growth. Kluwer Academic Publishers, Dordrecht, pp 33–42

Phylipsen GJM, Blok K, Worrell E (1998) Handbook on international comparisons of energy efficiency in the manufacturing industry in commission of the Ministry of Housing, Physical Planning and the Environment. Department of Science, Technology and Society, Utrecht University, Utrecht

Phylipsen GJM (2002) International comparisons of energy efficiency. In: Meyers RA (ed) The encyclopedia of physical science and technology. Academic, San Diego

Raj N, Shetty N, Shetty H (2004) Seed biopriming with *Pseudomonas fluorescens* strains enhances growth of pearl millet plants and induces resistance against downy mildew. Int J Pest Manag 50(1):41–48

Rajkumar M, Prasad MNV, Freitas H (2010) Potential of siderophore-producing bacteria for improving heavy metal phytoextraction. Trends Biotechnol 28:142–149

Rajkumar M, Sandhya S, Prasad MNV, Freitas H (2012) Perspectives of plant-associated microbes in heavy metal phytoremediation. Biotechnol Adv 30(6):1562–1574

Rakshit A, Bhadoria PBS, Mittra BN (2002) Nutrient use efficiency for bumper harvest. Yojona 56:28–30

Rakshit A, Pal S, Rai S, Rai A, Bhowmick MK, Singh HB (2013) Micronutrient seed priming: a potential, tool in integrated nutrient management. Satsa Mukkhapatra 17:77–89

Rakshit A, Pal S, Meena S, Manjhee B, Preeti P, Rai S, Rai A, Bhowmik MK, Singh HB (2014) Bio-priming: a potential tool in the integrated resource management. Satsa Mukkhapatra 18:94–103

Rao MSL, Kulkarni S, Lingaraju S, Nadaf HL (2009) Bio priming of oilseeds: a potential tool in the integrated management of *Alternaria* blight of sunflower. Helia 32(50):107–114

Rodrigues EP, Rodrigues LS, de Oliveira ALM, Baldani VL, Teixeira KRS, Urquiaga S, Reis VM (2008) *Azospirillum amazonense* inoculation: effects on growth, yield and N$_2$ fixation of rice (*Oryza sativa* L.). Plant Soil 302:249–261

Roy NK, Srivastava AK (2000) Adverse effect of salt stress conditions on chlorophyll content in wheat (*Triticum aestivum* L.) leaves and its amelioration through pre-soaking treatments. Indian J Agric Sci 70:777–778

Ruba A, Salman M, Ehlers R-U (2011a) Differential resistance of oilseed rape cultivars (*Brassica napus*) to *Verticillium longisporum* infection is affected by rhizosphere colonisation with antagonistic bacteria, *Serratia plymuthica* and *Pseudomonas chlororaphis*. BioControl 56(1):101–112

Ruba A, Salman M, Ehlers R-U (2011b) Effect of seed priming with *Serratia plymuthica* and *Pseudomonas chlororaphis* to control *Leptosphaeria maculans* in different oilseed rape cultivars. Eur J Plant Pathol 130(3):287–295

Santiago AD, García-López AM, Quintero JM, Avilés M, Delgado A (2012) Effect of *Trichoderma asperellum* strain T34 and glucose addition on iron nutrition in cucumber grown on calcareous soils. Soil Biol Biochem 57:598–605

Sharifi RS (2011) Grain yield and physiological growth indices in maize (*Zea mays* L.) hybrids under seed biopriming with plant growth promoting rhizobacteria (PGPR). J Food Agric Environ 9(3&4):393–397

Tanwar A, Aggarwal A, Kaushish S, Chauhan S (2013) Interactive effect of AM fungi with *Trichoderma viride* and *Pseudomonas fluorescens* on growth and yield of broccoli. Plant Prot Sci 49:137–145

Thomas J, Ajay D, Kumar RR, Mandal AK (2010) Influence of beneficial microorganisms during in vivo acclimatization of in vitro-derived tea (*Camellia sinensis*) plants. Plant Cell Tissue Organ Cult 101:365–370

Tzortzakis NG (2009) Effect of pre-sowing treatment on seed germination and seedling vigor in endive and chicory. Hortic Sci (Prague) 36(3):117–125

Weller DM, Cook RJ (1983) Suppression of take-all of wheat by seed treatment with *fluorescent Pseudomonads*. Phytopathol 73:463–469

Yadav SK, Dave A, Sarkar A, Singh HB, Sarma BK (2013) Co-inoculated biopriming with Trichoderma, Pseudomonas and Rhizobium improves crop growth in Cicer arietinum and Phaseolus vulgaris. Inter J Agric Environ Biotechnol 6(2):255–259

Yedidia I, Srivastva AK, Kapulnik Y, Chet I (2001) Effect of *Trichoderma harzianum* on microelement concentrations and increased growth of cucumber plants. Plant Soil 235:235–242

Unrealized Potential of Seed Biopriming for Versatile Agriculture

Kartikay Bisen, Chetan Keswani, Sandhya Mishra, Amrita Saxena, Amitava Rakshit, and H.B. Singh

Abstract

Seeds are the crucial input in agriculture as most of the world food crops are grown from seeds and they are circulated at large scale in international trade. However, many plant pathogens can be seed transmitted, and seed distribution is an extremely capable way of introducing plant pathogens into fresh areas as well as a means of endurance of the pathogen between growing seasons. In past decades, chemicals are widely used for seed treatment as a potent approach towards disease control; however, rising concern about their negative impact on the environment and human health minimizes their use and promotes biological control for plant pathogens. Biopriming is a currently popular approach of seed treatment which includes inoculation of seed with beneficial microorganisms (biological aspect) and seed hydration (physiological aspect) to protect the seed from various seed- and soilborne diseases. Biopriming treatment is able to incite changes in plant characteristics and facilitate uniform seed germination and growth associated with microorganism inoculation. Seed priming and osmo-priming are commonly being used in many horticultural crops to amplify the growth and uniformity of germination. However, it may be used alone or in combination with biocontrol agents to advance the rate of seed emergence and minimize soilborne diseases. On the other hand, some biocontrol agents are used as seed dressers and are able to colonize the rhizosphere, helping seeds to resist various abiotic stresses

K. Bisen • C. Keswani • S. Mishra •
A. Saxena • H.B. Singh (✉)
Department of Mycology and Plant Pathology, Institute of
Agricultural Sciences, Banaras Hindu University,
Varanasi 221005, Uttar Pradesh, India
e-mail: kartikaybisen@rediffmail.com;
chetankeswani@rediffmail.com; smnbri@gmail.com;
amrita.bhu08@gmail.com; hbs1@rediffmail.com

A. Rakshit
Department of Soil Science and Agricultural Chemistry,
Institute of Agricultural Science, BHU, Varanasi 221005,
India
e-mail: amitavar@bhu.ac.in

A. Rakshit et al. (eds.), *Nutrient Use Efficiency: from Basics to Advances,*
DOI 10.1007/978-81-322-2169-2_13, © Springer India 2015

such as salinity, drought, low fertility and heavy metal stress, etc. Therefore, biopriming is becoming a viable alternative for inorganic chemicals.

Keywords

Seed • Biopriming • Bioinoculants • Plant growth • Biotic stress • Abiotic stress • PGPR

1 Introduction

Seed priming technique is used for improving the vigor, establishment, and efficiency of seedlings in the fields. The early stage of seed germination requires suitable conditions; however, various biotic and abiotic factors hinder the process of germination. Therefore, the new concept of seed biopriming emerged that amalgamates biological and physiological aspects together. Seed biopriming using beneficial and eco-friendly biological agents could lead to improved physiology of seeds resulting into enhanced vigor of the seedlings (Ghassemi-Golezani 2008).

Plethora of plant pathogens ranging from viroids to parasitic higher plants is responsible for various diseases in crops. The detrimental effect of diseases may range from placid symptoms to disaster in which large planted areas of food crops are damaged. Control of plant pathogens is difficult because their populations are unpredictable in time, space, and genotype. Additionally, pathogen-infected seeds also contribute to the establishment of diseases and make the control strategy tedious. As the agricultural production is largely based on the seed (essential input), until and unless the purity, superiority, and seed standards are maintained, any production program cannot be deemed successful. Most crops are raised from seed each year, producing more than 2.3 billion tons of grains (Reddy et al. 2011). Seeds must germinate and seedlings emerge, speedily and constantly throughout the field so that water, light, and soil nutrients can be used with utmost efficiency. If seeds emerge and grow slowly after germination, they often become weak and stunted and more vulnerable to pathogens which may result in low production. Apart from dealing with plant disease control, seed

biopriming also helps in alleviating various abiotic stress conditions, viz., salinity, drought, low fertility, heavy metals, etc.

As a consequence of both industrialization and increasing population, the earth's atmosphere and ecosystems are no longer enough for absorption and breakdown of waste that we produce. As a result, the atmosphere is gradually more fouled with various toxic metals and compounds. The global increases in both human population demands and environmental damages have a significant outcome that worldwide production of food may soon become unsatisfactory to feed all of the global population (de Rosa et al. 2006). It is therefore necessary that agricultural productivity be considerably boosted within the next few decades. In this context, agricultural practices are shifting towards a more sustainable approach of using transgenic plants, plant growth-promoting bacteria, nanoformulations, biofertilizer, and biocontrol agents for enhancing crop productivity (Berg et al. 2010; Adesemoye and Egamberdieva 2013; Mishra et al. 2014).

2 Consequences of Seed-Borne and Soilborne Diseases

Approximately 90 % of all the worlds' food crops are grown from seeds (Schwinn 1994), and seeds are widely disseminated in national and international trade. Germplasm is also dispersed and traded in the form of seeds for breeding purpose. However, many plant pathogens can be seed transmitted, and seed distribution is an extremely capable way of introducing plant pathogens into fresh areas as well as a means of endurance of the pathogen between growing seasons. Plant diseases causing micro-organisms (usually fungi, bacteria, viruses, and nematodes) may be carried with, on or in seeds and, in appropriate environmental conditions and may be transmitted to cause disease in developing seedlings or plants. Currently, two viable options for increasing crop productivity are by firstly introducing high-yielding varieties and secondly avoiding crop disease incidence. However, the benefits of using high-yielding variety may be nullified by seed-borne diseases. These diseases may be guilty for about 10 % losses in chief crops in India (Neergaard 1979).

Interactions between pathogenic soil organisms and plants may result in death of young seedlings and even adult trees. Many organisms target younger plants, but others appear as problems at mature plant. Other pathogens are able to cause disease in many different plant species. The soil organisms that have the potential to be plant pathogens include fungi, bacteria, viruses, nematodes, and protozoa. Some pathogens of the aboveground parts of plants survive in the soil at various stages in their life cycles. Regardless of having a severe infection caused by soilborne pathogens, mostly host plants do not exhibit symptoms of the disease. Usually, plant diseases occur in unfavorable condition or when a new soilborne pathogen is unexpectedly introduced in the vicinity of susceptible crop.

3   History of Seed Treatment

Most primitive documented use of seed treatment dates back to 60 A.D. when wine and crushed cypress leaves were used to maintain seed free from storage insects. Recently, scientists have given credibility to this practice because hydrogen cyanide evolves under these conditions which kill insects. Coincidently in the seventeenth century, a ship transporting large amount of food grains met with a catastrophic accident causing the grains be soaked with seawater. Seed recovered from the ship produced a crop that showed significantly lesser infection of stinking smut than in neighboring fields planted with unsoaked seed (https://www.apsnet.org/edcenter/advanced/topics/Pages/CerealSeedTreatment.aspx). Although this observation was described and acknowledged at that time, it was not until 1750 that Tillet demonstrated scientifically the advantage of using salt and lime in controlling common bunt in wheat. Seed treatment has since evolved into a more complex science. The arrival of the organic mercurial in 1920s started a new era in seed treatment that has resulted in the many commercial contact and systemic fungicides (Goggi 2011). However, recent awareness about the environmental hazard of mercurial compounds has led to their global injunction.

4   Worldwide Status of Seed Treatment

Seed treatment market is categorized under fungicides, insecticides, biocontrol seed treatments, and other seed treatment chemicals. It is further divided as chemical and nonchemical nature of products. Bio-based seed treatment, a part of nonchemical share that includes natural active ingredients, is likely to be one of the greatest emerging seed treatment segments in the near future (Nautiyal 1999). Chemical-based seed treatments are further divided as fungicides and insecticides. Globally, about 51.7 % of the total required seed treatment in 2012 was covered by only insecticides. Growth rate of the international fungicide seed treatment market is of 9.2 % CAGR (compound annual growth rate) and is projected to attain $1,367.8 million by 2018. The global insecticide seed treatment market, rising at a CAGR of 10.8 %, is expected to reach $2,182.8 million by 2018 (http://www.marketsandmarkets.com). One of the significant success issues in the market is the ability of a market contestant to innovate an absolute protection solution against various abiotic and biotic stresses in a single product.

5   Status of Seed Treatment in India

Biological seed treatment has grown from 3 to 5 % for agricultural and horticultural crops in 2012–2013 (http://nsai.co.in/). However, chemical treatment still accounts for 90–95 % of the total seed treatment component. The seeds of self-pollinated crops are treated to the extent of 10–15 % on an average. Seed treatment estimated to enhance productivity by 8–10 % with minimal cost.

6   Seed Priming

Priming could be defined as "controlling the hydration level within seeds so that the metabolic activity necessary for germination can occur but radicle emergence is prevented." Different physiological activities within the seed occur at different moisture levels (Leopold and Vertucci 1989; Taylor 1997). The last physiological activity in the germination process is radicle emergence. The commencement of radicle emergence necessitates high seed water content. By restrictive seed water content, the entire metabolic steps essential for germination can occur without the unalterable act of radicle emergence. Prior to radicle emergence, seeds are considered dehydration tolerant, thus the primed seed moisture content can be reduced by drying. Primed seeds can be stored until time of sowing after drying. For practical purposes, seeds are primed for the following reasons:

- To triumph over or improve phytochrome-induced dormancy in plants
- To reduce the time required for germination and for successive emergence to take place
- To promote and stand uniformity in order to simplify production management

6.1   Seed Biopriming

The prime objective behind seed treatment is to increase seed performance in many of the following ways: (1) eliminate any threat posed by seed- and soilborne phytopathogens; (2) meet the needs of thinning seedlings, especially in case of mechanical planters; and (3) enhance germination rates. In general crop production, chemical fungicides are often used which reduce seed and seedling losses due to seed-borne and soil-borne diseases. For organic growers, most of the

seed protectants are not a viable alternative; however, there are some seed treatments, such as priming, pelletizing, and the use of hot water treatment, that can be used by organic farmers to improve seed performance.

Biopriming is a new fangled technique of seed treatment (Fig. 1) that incorporates biological inoculation of seed with beneficial microorganisms to guard seeds and regulate seed hydration for biotic and abiotic stress management. It is the modern method for controlling major seed- and soilborne pathogens. Seed priming was used commercially in many horticultural crops as a means to boost speed and uniformity of germination and advance final stand. Seed priming alone or in combination with low dosage of fungicides and/or biocontrol agents has been used to improve the rate and uniformity of seed emergence and to diminish diseases.

Biological control agents as seed treatments are being vigorously developed by a number of companies across the globe. These products may also offer a limited solution for protection against specific pathogens. It should be strained that effectiveness of biological seed treatment is currently distant in replacing chemical seed treatment; nevertheless, biocontrol agents offer dual advantage of crop protection and plant growth promotion simultaneously. If the global demand for organic crops elevates, then these products may have an ecological advantage over their chemical counterparts for crop protection.

Biological management of plant diseases relies on both potential biocontrol organisms and methods of introducing the organism in high-incidence regions. Regardless of the activity of the biocontrol agents, the methods used to produce, formulate, and deliver these organisms may profoundly affect their efficacy under field conditions. One of the popular methods of introducing biological control agents is seed treatment (Fig. 1). Applying microorganisms to seed is an attractive proposition because of the combination of specific effect and limited environmental impact. In the familiar saying, seed treatment has the potential to deliver agents "in the right amount, at the right place, and at the right time" (McQuilken et al. 1998).

Fig. 1 General procedure of seed biopriming

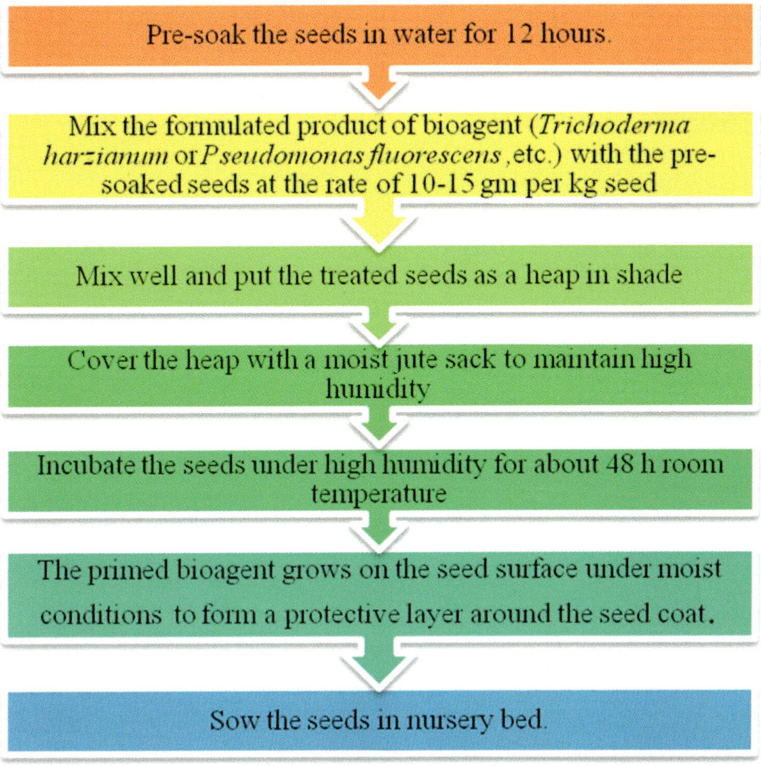

Pre-soak the seeds in water for 12 hours.

Mix the formulated product of bioagent (*Trichoderma harzianum* or *Pseudomonas fluorescens*, etc.) with the pre-soaked seeds at the rate of 10-15 gm per kg seed

Mix well and put the treated seeds as a heap in shade

Cover the heap with a moist jute sack to maintain high humidity

Incubate the seeds under high humidity for about 48 h room temperature

The primed bioagent grows on the seed surface under moist conditions to form a protective layer around the seed coat.

Sow the seeds in nursery bed.

6.2 Benefits of Seed Priming with Bioinoculants

6.2.1 Disease Suppression

Bioinoculants help in disease suppression by utilizing different mechanisms such as siderophore production, antimicrobial secondary metabolite, and secretion of lytic enzymes (Keswani et al. 2014). Even though contest between bacterial and fungal plant pathogens for space or nutrients has been known to exist as a mechanism of biocontrol for many years, the furthermost attention recently has involved competition for iron. Siderophores (Greek: "iron carrier") are small, high-affinity iron-chelating compounds secreted by microorganisms such as bacteria, fungi, and grasses. Siderophores are among the strongest soluble Fe^{3+}-binding agents known (Leong 1986). Certain strains of fluorescent *Pseudomonas* increase yield or control biologically one or more soilborne pathogens when applied as seed or seed inoculants to agricultural crops (Burr and Caesar 1984). It is well documented that *Pseudomonas*

siderophores improve biocontrol and simultaneously enhance plant growth by the production of fluorescent siderophores that chelate molecular iron in rhizosphere, making it less assessable for other competing microorganisms (Singh et al. 2011, 2014; Jain et al. 2012). Biopriming treatment is potentially prominent to induce profound changes in plant characteristics and to encourage more uniform seed germination and plant growth associated with fungi and bacteria coatings (Entesari et al. 2013).

6.2.1.1 Parasitism and Production of Extracellular Enzymes

The use of specific mycolytic soil microorganisms to control plant pathogens is an ecological approach to overcome the problems caused by standard chemical methods of plant protection. The ability to produce lytic enzymes is a widely distributed property of rhizosphere-competent fungi and bacteria (Viterbo et al. 2002). *Trichoderma* spp. are active mycoparasites against a range of economically important soilborne plant pathogens and are successfully used as a biocide in greenhouse and field applications (Chet 1987; Papavizas 1985; Nayak et al. 2009). This is well complimented by the secretion of extracellular hydrolytic enzymes such as chitinases (Carsolio et al. 1994; de la Cruz et al. 1992; Harman et al. 1993), β-glucanases (Haran et al. 1995; Lora et al. 1995; Lorito et al. 1994), and proteases (Geremia et al. 1993). The effect of these compounds in phytopathogenic fungi includes degradation of the cell wall (Harman et al. 2004). Biopriming propagules with some plant growth-promoting rhizobacteria (PGPR) can also offer systemic resistance against broad-spectrum plant pathogens.

6.2.1.2 Systemic Acquired Resistance (SAR)

Certain bacterial species induce a systemic response in host plant known as induced systemic resistance (ISR) that is phenotypically alike systemic acquired resistance (SAR). SAR occurs when plants activate their defense mechanism against primary infection by a pathogen, particularly when a pathogen induces a hypersensitive reaction and becomes limited in a local necrotic lesion (Van Loon et al. 1998). PGPR-elicited ISR was first studied on carnation with declined susceptibility to *Fusarium* wilt (Van Peer et al. 1991) and on cucumber to *Colletotrichum orbiculare* (Wei et al. 1991).

6.2.2 Plant Growth Promotion

The amount of total phosphorus in the soil is generally high (400 and 1,200 mg kg^{-1} of soil), and most of this phosphorus is insoluble and not accessible to the plant. The insoluble phosphorus is found as an inorganic mineral and in organic forms including inositol phosphate, phospho-monoesters, and phosphotriesters (Khan et al. 2007). The limited accessibility of phosphorus combines with the fact that this element is crucial for plant growth (Feng et al. 2004). Therefore, phosphorus mineralization and solubilization are an important trait in almost all plant growth-promoting microbes (Richardson 2001; Rodríguez and Fraga 1999).

The positive effects of siderophores on the plants' growth have been documented in different types of experiments. For example, (i) the use of labeled iron siderophores as a sole source of iron demonstrated that plants are able to obtain the labeled iron (Lope and Buyer 1991; Duijff et al. 1994; Jin et al. 2006; Siebner-Freibach et al. 2003), (ii) plants inoculated with the siderophore-producing *Pseudomonas* and raised under iron-limiting conditions expressed lesser chlorotic symptoms and an increased chlorophyll level (Sharma et al. 2003), and (iii) *Arabidopsis* plants showed an enhanced iron concentration in plant tissues and better growth when treated with *Pseudomonas fluorescens* which synthesized Fe-pyoverdine complex (Vansuyt et al. 2007). Overall, the seed biopriming results in enhanced plant growth through the abovementioned mechanisms adopted by microbes (Fig. 2).

6.2.3 Effect on Abiotic Stresses

Abiotic stresses are responsible for reduced agricultural production and slow microbial activity in the soil. Microorganisms may play a significant role in adaptation strategies and increase of forbearance to abiotic anxiety in agricultural crop plants. Different microorganisms are united with plant roots and alleviate the force of abiotic stresses on plants by formation of biofilm and exopolysaccharide production (Selvakumar et al. 2012). Induction of osmoprotectors and heat shock proteins is also an important mechanism by which microorganisms help plants to stand firmly when exposed to stress (Hasanuzzaman et al. 2013). Studies have shown that certain microbial species induced tolerance to plants against different abiotic stresses such as drought, salinity, nutrient deficiency, and high concentration of heavy metals (Rajapaksha et al. 2004; Grover et al. 2010; Milošević et al. 2002; Dimkpa et al. 2009).

6.2.3.1 Drought/Excessive Moisture Stress

Drought stress bounds crop growth and productivity, particularly in arid and semiarid areas. Some microbes that inhabit the plant rhizosphere use different methods to restrain the side effects of drought on plants (Table 1). According to

Fig. 2 Process of seed biopriming using *Trichoderma harzianum* and its effect on plant health

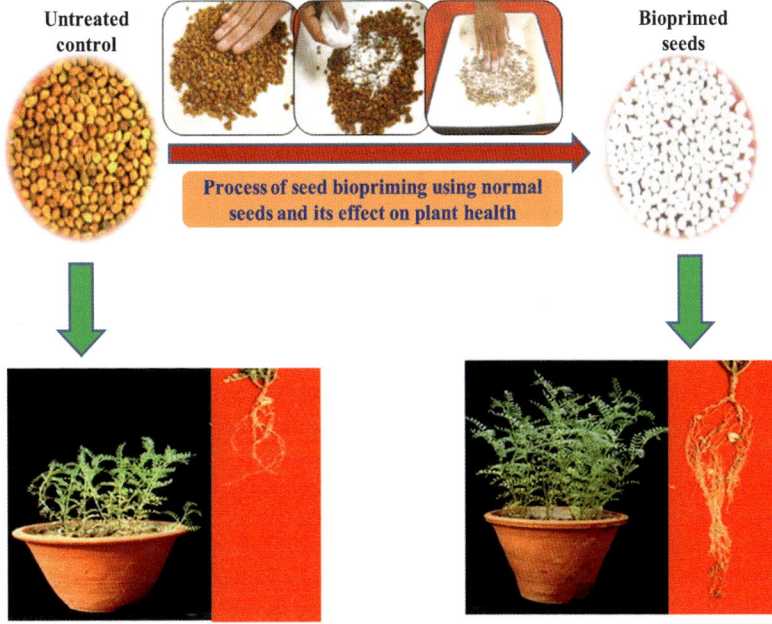

Table 1 Effect of microorganisms on drought mitigation in crops

Microorganisms	Crop	Mechanism	References
Pantoea agglomerans	Wheat	Production of EPS which affects the structure of rhizospheric soil	Amellal et al. (1998)
Rhizobium sp.	Sunflower	Production of EPS which affects the structure of rhizospheric soil	Alami et al. (2000)
Pseudomonas sp.	Pea	ACC deaminase activity	Arshad et al. (2008)
Phyllobacterium brassicacearum STM196	*Arabidopsis*	Induction of physiological changes and reproductive delay	Bresson (2013)
Pseudomonas mendocina	Lettuce	Enhanced phosphatase activity and proline accumulation	Kohler et al. (2008)
Paenibacillus polymyxa and *Rhizobium tropici*	Bean	Hormonal balance and stomatal conductance	Figueiredo et al. (2008)

Grover et al. (2010), certain microbial species may mitigate the effect of drought by several mechanisms such as inducing resistance genes, improving water movement within the plant, production of exopolysaccharides, and synthesizing ACC deaminase, indoleacetic acid (IAA), and proline.

Seed inoculation with *Bacillus amyloliquefaciens* results in the production of exopolysaccharides (EPS) which have the ability to improve soil structure by aiding the formation of macroaggregates. This leads to amplified plant

resistance to stress due to scarcity of water (Milošević et al. 2002). Macroaggregates are custodians of soil fertility, because they preserve equilibrium between aerobic and anaerobic states and make sure a steady uptake of nutrients from soil reserves. Bacteria like *Pseudomonas* keep on surviving under stress by the production of EPS, which protects microbes from hydric stress, and by enhancing water withholding and diffusing carbon in environment (Sandhya et al. 2009). EPS own elite water holding and cementing properties and play an imperative role in the

Table 2 Effect of microorganisms on imparting salinity stress tolerance to crops

Crop	Microorganisms	Effect	Reference
Sorghum	AM fungi	Increased water circulation	Feng et al. (2002) and Grover et al. (2010)
Corn		Improved osmoregulation and proline accumulation	
Clover			
Wheat	Rhizobium sp., Pseudomonas sp.	Restricted Na+ uptake	Grover et al. (2010)
Sorghum	AM fungi	Improved water relation	Cho et al. (2006)
Tomato	Achromobacter piechaudii	Reduced levels of ethylene and improved plant growth	Mayak et al. (2004)
Groundnut	Pseudomonas fluorescens	Enhanced ACC deaminase activity	Saravanakumar and Samiyappan (2007)
Cotton	Pseudomonas putida	Reduced uptake of Na^{2+} from the soil	Yao et al. (2010)

formation and stabilization of soil aggregates and mobilization of nutrients and water flow across plant roots through biofilm formation (Roberson and Firestone 1992; Tisdall and Oadea 1982).

6.2.3.2 High-Temperature Stress

Temperature is a crucial factor for agricultural production as it affects most of the physiological function in plants. Extreme high and low temperatures are limiting factor to the production and geographical distribution of agricultural plants. Some bacterial species and strains affect plant tolerance to high temperature. *Pseudomonas* sp. induces thermotolerance in sorghum seedlings by synthesizing high molecular weight proteins and enhancing plant biomass (Grover et al. 2010).

6.2.3.3 Salinity Stress

Soil salinity has an insightful effect on seed germination, which is the most imperative phase of thriving crop production. Under such circumstances, it is necessary to encourage seed germination and growth. The most relevant solution in such condition is to use salt-tolerant bacterial inoculants that generate auxins and gibberellins and encourage plant growth in salinity conditions (Mayak et al. 2004) (Table 2). Giri and Mukerji (2004) demonstrated that, in saline soil, higher absorption of P in inoculated plants may advance their growth rate and salt tolerance and restrain the unfavorable effect of salinity stress.

6.2.3.4 Low Fertility Stress

Soil microbes play an essential role in determining soil fertility by recycling soil nutrients (Glick 2012). Therefore, this is an important aspect because soil fertility directly affects the agronomic efficiency by serving as nutrient inputs. Nowadays major emphasis is being given on plant nutrient management system by integrating biological components in the form of agriculturally important microbes. Previous findings have clearly reported the significant contribution of plant growth-promoting microbes in enhancing the supply of major nutrients N, P, and K and other elements leading to enhanced plant growth (Joo et al. 2004; Sheng and He 2006; Glick et al. 2007).

These studies enlighten the use of beneficial microorganisms in combating low fertility stress in soil system. Soil is the major source of nutrients, and water required for plant growth and soil microbes have great influence in nutrient cycling (Nannipieri et al. 2003). Hence, seed biopriming using such nutrient-supplying microbes could yield better crop growth. For example, seed inoculation with nitrogen-fixing bacteria facilitate supply of nitrogen to the plants (Deaker et al. 2012; Kloepper et al. 1992; Bashan et al. 2004). Likewise, phosphate-solubilizing bacteria such as *Azotobacter*, *Bacillus*, *Pseudomonas*, and *Serratia* supply P to the plants to stimulate the efficiency of nitrogen-fixing microbes and enhance the availability of other trace elements (Zaidi et al. 2009; Bhattacharyya and Jha 2012). In another report by Entesari

Table 3 Effect of bioinoculants in heavy metal remediation

Bioinoculants	Heavy metals	References
Pseudomonas spp.	Cr, Cu, Cd, Ni, Zn	Hussein et al. (2004)
Trichoderma atroviride	Zn, Ba, Fe	Kacprzak and Malina (2005)
Trichoderma harzianum	Zn, Cd, Hg	Krantz-Rülcker et al. (1993)
Arbuscular mycorrhiza	Cu, Cd	Liao et al. (2003)

et al. (2013), seed biopriming with *Trichoderma* sp. and *Pseudomonas* sp. increased uptake of Fe, Zn, and N. In the present scenario, reduced soil fertility has become major problem and thus recently concept of nutrient use efficiency is gaining much attention. In this context, seed biopriming using beneficial microbes could fulfill the increasing demand of nutrient availability.

6.2.3.5 Heavy Metal Stress

Heavy metal toxicity in cultivated land has become a serious environmental concern due to their possible adverse effects on ecology. Large part of agricultural soils throughout the world is contaminated by heavy metal toxicity. These metals include Cd, Cu, Zn, Ni, Co, Cr, Pb, and As. Several factors such as industrial waste, sewage sludge, domestic waste materials, and long-term use of phosphatic fertilizers in agricultural lands are responsible for heavy metal stress (Bell et al. 2001; Schwartz et al. 2001). Heavy metal toxicity in plants results in poor plant growth, leaf chlorosis, restricted nutrient uptake, disturbed plant metabolism, reduced ability to fix atmospheric nitrogen in legume plants, and ultimately yield reduction (Bazzaz et al. 1974; Chaudri et al. 2000; Broos et al. 2005).

A wide range of soil fungi including *Trichoderma* are important tools for bioremediation of heavy metal-stressed soil (Errasquín and Vazquez 2003). *Trichoderma* can remove the heavy metal ions, and sorption was acknowledged as the main mechanism of uptake (Kacprzak and Malina 2005; Yazdani et al. 2009; Srivastava et al. 2011). Several microbes are reported (Table 3) in bioremediation of heavy metal-contaminated soil and induction of resistance in plants against heavy metals.

7 Key Challenges in Production and Application of Inoculants to Seed

There are several barriers and problems that must be overcome before the registration and commercialization of microbes and microbial products (Mathre et al. 1999; Gardener and Fravel 2002). These include maintaining stability, efficacy, and quality of the product during upgrading and production of the organism under commercial condition. The key factors that must be for formulation development are compatibility of microbial product with application methods, shelf life, expenditure, and easiness of application.

7.1 Seed

Seed is a capable release vehicle for the test application of microbes to soil; it is a very composite substrate. Water relationships between the seed and coatings are not well understood. Seeds are stored under clean and dry conditions as rise in moisture content may lead to increased seed respiration rate and consequently heat, thus reducing their shelf life. Many microbial inoculants have a need of some hydration to stay viable, and this may be unfavorable to the seed during storage. Other physical restrictions include the nature of the seed coat and size of seed where smaller seeds carry lesser priming agent in comparison to larger seeds. This is further challenged by natural seed microflora that may fight for space on the surface of the seed. Some legume seeds release inhibitory exudates that can cause restriction in endurance of bioinoculants on seed (Deaker et al. 2004).

7.2 Shelf Life of Bioinoculants

Limited understanding of the survival of biocontrol agents on seed surface and also during seed storage is a major challenge limiting commercial success of bioprimed seed. The causes responsible for cell deaths during storage are poorly understood. Microorganisms' survival is enhanced on seed surfaces at low temperatures (Callaghan et al. 2006), but this is not often commercially practicable.

Conclusion

Rising concern among scientists and general public regarding serious health hazards on human health associated with the use of chemical in food supplies has propelled research for eco-friendly alternative approaches for plant disease management and overall growth promotion and performance (Wilson and Wisniewski 1994; Gerhardson 2002). Seed biopriming using biological control agents and growth promoter may be an appropriate alternate choice of fungicides to control soil- and seed-borne fungi (Harman and Taylor 1988). Biopriming creates a complimentary environment for seed bioinoculants by increasing nutrient uptake from seed exudates and initial moisture of the seeds which can contribute to the proliferation of microbes on the seed surface (Wright et al. 2003). Seed biopriming is a phenomenal technique used globally for the management of seed- and soil-borne phytopathogens of many economically important crops. This practice is being used since past decades effectively in the field and offers better or equal results over conventional fungicides (Callen et al. 1990; Callen and Mathre 2000; Niranjan Raj et al. 2004). Seed biopriming through beneficial microorganisms can be regarded as direct substitute of seed treatment with agrochemicals. Moreover, this technique could be an efficient part of an integrated system, mingling both

microorganisms and pesticides for successful plant disease management.

Advances have been achieved through a greater understanding of the control mechanisms used by biocontrol agents against seed and soil borne pathogens (Glare et al. 2011). Interestingly, seed biopriming can act as a model system for the delivery of dense population of beneficial microorganisms to soil, where they can colonize emerging roots of crop plants. However, the major constrain associated with seed biopriming is maintaining the high numbers of microorganisms on seed surface during seed treatment and storage. Therefore, better understanding of the interactions between microorganisms, seed, and formulation components is required for developing functional microbial inoculants.

Acknowledgments SM is highly grateful to the University Grants Commission, New Delhi, India, for providing Dr. D. S. Kothari postdoctoral fellowship. CK and KB thank Banaras Hindu University, Varanasi, Uttar Pradesh, India, for providing financial support.

References

Adesemoye AO, Egamberdieva D (2013) Beneficial effects of plant growth promoting rhizobacteria on improved crop production: the prospects for developing economies. In: Maheshwari DK (ed) Bacteria in agrobiology: crop productivity. Springer, Berlin/Heidelberg

Alami Y, Achouak W, Marol C, Heulin T (2000) Rhizosphere soil aggregation and plant growth promotion of sunflowers by exopolysaccharide producing *Rhizobium* sp. strain isolated from sunflower roots. Appl Environ Microbiol 66:3393–3398

Amellal N, Burtin G, Bartoli F, Heulin T (1998) Colonization of wheat rhizosphere by EPS producing *Pantoea agglomerans* and its effect on soil aggregation. Appl Environ Microbiol 64:3740–3747

Arshad M, Sharoona B, Mahmood T (2008) Inoculation with *Pseudomonas* spp. containing ACC deaminase partially eliminate the effects of drought stress on growth, yield and ripening of pea (*Pisum sativum* L.). Pedosphere 18:611–620

Bashan Y, Holguin G, de-Bashan LE (2004) Azospirillum-plant relationships: physiological, molecular, agricultural, and environmental advances (1997–2003). Can J Microbiol 50:521–577

Bazzaz FA, Carlson RW, Rolfe GL (1974) The effect of heavy metals on plants: Part I. Inhibition of gas exchange in sunflower by Pb, Cd, Ni and Tl. Environ Pollut 7:241–246

Bell FG, Bullock SET, Halbich TFJ, Lindsay P (2001) Environmental impacts associated with an abandoned mine in the Witbank Coalfield, South Africa. Int J Coal Geol 45:195–216

Berg G, Egamberdieva D, Lugtenberg B, Hagemann M (2010) Symbiotic plant-microbe interactions: stress protection, plant growth promotion and biocontrol by stenotrophomonas. In: Seckbach JMG, Grube M (eds) Symbiosis and stress. Springer, Dordrecht/Heidelberg/London/New York, pp 445–460

Bhattacharyya PN, Jha DK (2012) Plant growth-promoting rhizobacteria (PGPR): emergence in agriculture. World J Microbiol Biotechnol 28:1327–1350

Bresson J (2013) The PGPR strain *Phyllobacterium brassicacearum* STM196 induces a reproductive delay and physiological changes that result in improved drought tolerance in *Arabidopsis*. New Phytol 200:558–569

Broos K, Beyens H, Smolders E (2005) Survival of rhizobia in soil is sensitive to elevated zinc in the absence of the host plant. Soil Biol Biochem 37:573–579

Burr TJ, Caesar AJ (1984) Beneficial plant bacteria. CRC Crit Rev Plant Sci 2:1–20

Callaghan M, Swaminathan J, Lottmann J, Wright D (2006) Seed coating with biocontrol strain *Pseudomonas fluorescens* F113. N Z Plant Prot 59:80–85

Callen NW, Mathre DE (2000) Biopriming seed treatment. Encyclopedia of plant pathology. John Wiley and Sons, New York

Callen NW, Mathre DE, Miller JB (1990) Biopriming seed treatment for biological control of *Pythium ultimum* pre emergence damping-off in sh2 sweet corn. Plant Dis 74:368–372

Carsolio C, Gutierrez A, Jimenez B, Van Montgu M, Herrera Estrells A (1994) Characterization of ech 42 and *Trichoderma harzianum* endochitinase gene expressed during mycoparasitism. Proc Natl Acad Sci USA 91(23):10903–10907

Chaudri AM, Allain CMC, Barbosa-Jafferson VL, Nicholson FA, Chambers BJ, McGrath SP (2000) A study of the impacts of Zn and Cu on two rhizobial species in soils of a long term field experiment. Plant Soil 22:167–179

Chet I (1987) *Trichoderma*: application, mode of action and potential as a biocontrol agent of soil borne plant pathogenic fungi. In: Chet I (ed) Innovative approaches to plant disease control. Wiley, New York, pp 137–160

Cho K, Toler H, Lee J, Ownley B, Stutz JC, Moore JL, Auge RM (2006) Mycorrhizal symbiosis and response of sorghum plants to combined drought and salinity stresses. J Plant Physiol 163:517–528

De la Cruz J, Hidalgo-Gallego A, Lora JM, Benitez T, Pintor-Toro JA, Llobell A (1992) Isolation and characterization of three chitinases from *Trichoderma harzianum*. Eur J Biochem 206:859–867

de Rosa CT, Johnson BL, Fay M, Hansen H, Mumtaz MM (2006) Public health implications of hazardous waste sites: findings assessment and research. Food Chem Toxicol 34:1131–1138

Deaker R, Roughly RJ, Kennedy IR (2004) Legume seed inoculation technology – a review. Soil Biol Biochem 36:1275–1288

Deaker R, Hartley E, Gemell G (2012) Conditions affecting shelf-life of inoculated seed. Agriculture 2:38–51

Dimkpa C, Weinand T, Asch F (2009) Plant-rhizobacteria interactions alleviate abiotic conditions. Plant Cell Environ 32:1682–1694

Duijff BJ, Bakker PAHM, Schippers B (1994) Ferric pseudobactin 358 as an iron source for carnation. J Plant Nutr 17:2069–2078

Entesari M, Sharifzadeh F, Ahmadzadeh M, Farhangfar M (2013) Seed biopriming with *Trichoderma* species and *Pseudomonas fluorescent* on growth parameters, enzymes activity and nutritional status of soybean. Int J Agron Plant Prod 4:610–619

Errasquín EL, Vazquez C (2003) Tolerance and uptake of heavy metals by *Trichoderma atroviride* isolated from sludge. Chemosphere 50(1):137–143

Feng G, Zhang FS, Li XL, Tian CY, Tang C, Renegal Z (2002) Improved tolerance of maize plants to salt stress by arbuscular mycorrhiza is related to higher accumulation of leaf P-concentration of soluble sugars in roots. Mycorrhiza 12:185–190

Feng K, Lu HM, Sheng HJ, Wang XL, Mao J (2004) Effect of organic ligands on biological availability of inorganic phosphorus in soils. Pedosphere 14:85–92

Figueiredo MVB, Burity HA, Martinez CR, Chanway CP (2008) Alleviation of drought stress in common bean (*Phaseolus vulgaris* L.) by co-inoculation of *Paenibacillus polymyxa* and *Rhizobium tropici*. Appl Soil Ecol 40:182–188

Gardener BBMS, Fravel DR (2002) Biological control of plant pathogens: research, commercialization, and application in the USA. Plant Health Prog. doi:10.1094/PHP-2002-0510-01-RV

Geremia RA, Goldman GH, Jacobs D, Ardiles W, Vila SB, Van Montagu M, Herrera-Estrella A (1993) Molecular characterization of the proteinase encoding gene, prb1, related to mycoparasitism by *Trichoderma harzianum*. Mol Microbiol 8:603–613

Gerhardson B (2002) Biological substitute for pesticides. Trends Biotechnol 20:338–343

Ghassemi-Golezani K, Sheikhzadeh-Mosaddeg P, Valizadeh M (2008) Effect of hydropriming duration and limited irrigation on field performance of chickpea. Res J Seed Sci 1(1):34–40

Giri B, Mukerji KG (2004) Mycorrhizal inoculant alleviates salt stress in *Sesbania aegyptiaca* and *Sesbania grandiflora* under field conditions: evidence for reduced sodium and improved magnesium uptake. Mycorrhiza 14:307–312

Glare T, Caradus J, Gelernter W, Jackson T, Keyhani N, Kohl J, Marrone P, Morin L, Stewart A (2012) Have biopesticides come of age? Trends Biotechnol 30:250–258

Glick BR (2012) Plant growth-promoting bacteria: mechanisms and applications. Hindawi Publishing Corporation, Scientifica, Article ID 963401, p 15. http://dx.doi.org/10.6064/2012/963401

Glick BR, Cheng Z, Czamy J, Duan J (2007) Promotion of plant growth by ACC deaminase-containing soil bacteria. Crit Rev Plant Sci 26:227–242

Goggi AS (2011) Evolution, purpose and advantages of seed treatments. III seed congress of the Americas, Santiago, Chile, pp 27–29

Grover M, Ali SZ, Sandhya V, Rasul A, Venkateswarlu B (2010) Role of microorganisms in adaptation of agriculture crops to abiotic stress. World J Microbiol Biotechnol 30:312–321

Haran S, Schickler H, Oppenheim A, Chet I (1995) New components of the chitinolytic system of *Trichoderma harzianum*. Mycol Res 99(4):441–446

Harman GE, Taylor AG (1988) Improved seedling performance by integration of biological control agents at favorable pH levels with solid matrix priming. Phytopathology 78:520–525

Harman GE, Hayes CK, Lorito M, Broadway RM, DiPietro A, Peterbauer CK, Tronsmo A (1993) Chitinolytic enzymes of *Trichoderma harzianum*: purification of chitobiosidase and endochitinase. Phytopathology 83:313–318

Harman GE, Howell CR, Viterbo A, Chet I, Lorito M (2004) *Trichoderma* species-opportunistic, avirulent plant symbionts. A reviews. Nat Rev Microbiol 2:43–56

Hasanuzzaman M, Nahar K, Alam M, Roychowdhury R, Fujita M (2013) Physiological, biochemical, and molecular mechanisms of heat stress tolerance in plants. Int J Mol Sci 14:9643–9684

Hussein H, Moawad H, Farag S (2004) Isolation and characterization of *Pseudomonas* resistant to heavy metals contaminants. Arab J Biotechnol 7(1):13–22

Jain A, Singh S, Sarma BK, Singh HB (2012) Microbial consortium–mediated reprogramming of defence network in pea to enhance tolerance against *Sclerotinia sclerotiorum*. J Appl Microbiol 112(3):537–550

Jin CW, He YF, Tang P, Zheng SJ (2006) Mechanisms of microbially enhanced Fe acquisition in red clover (*Trifolium pratense* L.). Plant Cell Environ 29:888–897

Joo GJ, Kim YM, Lee IJ, Song KS, Rhee IK (2004) Growth promotion of red pepper plug seedlings and the production of gibberellins by Bacillus cereus, Bacillus macroides and Bacillus pumilus. Biotechnol Lett 26:487–491

Kacprzak M, Malina G (2005) The tolerance and Zn^{2+}, Ba^{2+} and Fe^{3+} accumulation by *Trichoderma atroviride* and *Mortierella exigua* isolated from contaminated soil. Can J Soil Sci 85(2):283–290

Keswani C, Mishra S, Sarma BK, Singh SP, Singh HB (2014) Unraveling the efficient applications of secondary metabolites of various *Trichoderma* spp. Appl Microbiol Biotechnol 98(2):533–544

Khan MS, Zaidi A, Wani PA (2007) Role of phosphate-solubilizing microorganisms in sustainable agriculture—a review. Agron Sustain Dev 27:29–43

Kloepper JW, Schippers B, Bakker PAHM (1992) Proposed elimination of the term endorhizosphere. Phytopathology 82:726–727

Kohler J, Herna´ndez JA, Caravaca F, Rolda´n A (2008) Plant-growth promoting rhizobacteria and arbuscular mycorrhizal fungi modify alleviation biochemical mechanisms in water-stressed plants. Funct Plant Biol 35:141–151

Krantz-Rülcker C, Allard B, Schnürer J (1993) Interactions between a soil fungus, *Trichoderma harzianum*, and IIb metals—adsorption to mycelium and production of complexing metabolites. Biometals 6:223–230

Leong J (1986) Siderophores: their biochemistry and possible role in the biocontrol of plant pathogens. Annu Rev Phytopathol 24:187–209

Leopold AC, Vertucci CW (1989) Moisture as a regulator of physiological reactions in seeds. In: Stanwood PC, McDonald MB (eds) Seed moisture, CSSA special publication number 14. Crop Science Society of America, Madison, pp 51–69

Liao JP, Lin XG, Cao ZH, Shi YQ, Wong MH (2003) Interactions between arbuscular mycorrhizae and heavy metals under sand culture experiment. Chemosphere 50:847–853

Lope JE, Buyer JS (1991) Siderophore in microbial interaction on plant surfaces. Mol Plant Microbe Interact 4:5–13

Lora JM, De La Cruz J, Benitez T, Pintor-Toro JA (1995) A putative catabolite-repressed cell wall protein from the mycoparasitic fungus *Trichoderma harzianum*. Mol Gen Genet 247:639–645

Lorito M, Harman CK, DiPietro A, Woo SL, Harman GE (1994) Purification, characterization and synergistic activity of a glucan-1, 3-β glucosidase and an N-acetylglucosaminidase from *Trichoderma harzianum*. Phytopathology 84:398–405

Mathre DE, Cook RJ, Callan NW (1999) From discovery to use: traversing the world of commercializing biocontrol agents for plant disease control. Plant Dis 83:972–983

Mayak S, Tirosh T, Glick BR (2004) Plant growth-promoting bacteria confer resistance in tomato plants to salt stress. Plant Physiol Biochem 42:565–572

McQuilken MP, Halmer P, Rhodes DJ (1998) Application of microorganisms to seeds. Microbiol Rev Can Microbiol 44:162–167

Milošević N, Govedarica M, Kastori R, Petrović N (2002) Effect of nickel on wheat plants, soil microorganisms and enzymes. Biologia XLVII:177–181

Mishra S, Singh BR, Singh A, Keswani C, Naqvi AH et al (2014) Biofabricated silver nanoparticles act as a strong fungicide against bipolaris sorokiniana causing spot blotch disease in wheat. PLoS One 9(5):e97881

Nannipieri P, Ascher J, Ceccherini MT, Landi L, Pietramellara G, Renella G (2003) Microbial diversity and soil functions. Eur J Soil Sci 54:655–670

Nautiyal CS (1999) Bioinoculants for sustainable agriculture: recent status and constraints. In: Rajak RC (ed) Microbial biotechnology for sustainable development and productivity, Scientific Publisher, Jodhpur, pp 1–11

Nayaka SC, Niranjana SR, Uday Shankar AC, Niranjan Raj S, Reddy MS, Prakash HS, Mortensen CN (2009) Seed biopriming with novel strain of *Trichoderma harzianum* for the control of toxigenic *Fusarium verticillioides* and fumonisins in maize. Arch Phytopathol Plant Protect 43(3):264–282

Neergaard P (1979) Seed pathology. The MacMillan Press, London

Niranjan Raj S, Shetty NP, Shetty HS (2004) Seed biopriming with *Pseudomonas fluorescens* isolates enhance growth of pearl millet plants and induces resistance against downy mildew. Int J Pest Manag 50:41–48

Papavizas GC (1985) *Trichoderma* and *Gliocladium* biology, ecology, and potential for biocontrol. Annu Rev Phytopathol 23:23–54

Rajapaksha RM, Tobor – Kapłon MA, Baath E (2004) Metal toxicity affects fungal and bacterial activities in soil differently. Appl Environ Microbiol 70:2966–2973

Reddy ASR, Madhavi GB, Reddy KG, Yellareddygari SK, Reddy MS (2011) Effect of seed biopriming with *Trichoderma viride* and *Pseudomonas fluorescens* in chickpea (*Cicer arietinum*) in Andhra Pradesh, India. In: Reddy MS, Wang Q, Li Y, Zhang L, Du B, Yellareddygari SKR (eds) Plant growth-promoting rhizobacteria (PGPR) for sustainable agriculture, Proceedings of the 2nd Asian PGPR conference, Beijing, China, pp 324–429

Richardson AE (2001) Prospects for using soil microorganisms to improve the acquisition of phosphorus by plants. Funct Plant Biol 28:897–906

Roberson E, Firestone M (1992) Relationship between desiccation and exopolysaccharide production in soil *Pseudomonas* sp. Appl Environ Microbiol 58:1284–1291

Rodríguez H, Fraga R (1999) Phosphate solubilizing bacteria and their role in plant growth promotion. Biotechnol Adv 17:319–339

Sandhya V, Ali SKZ, Grover M, Reddy G, Venkateswarlu B (2009) Alleviation of drought stress effects in sunflower seedlings by exopolysaccharides producing *Pseudomonas putida* strain P45. Biol Fertil Soil 46:17–26

Saravanakumar D, Samiyappan R (2007) Effects of 1-aminocyclopropane-1-carboxylic acid (ACC) deaminase from *Pseudomonas fluorescence* against saline stress under in vitro and field conditions in groundnut (*Arachis hypogea*) plants. J Appl Microbiol 102:1283–1292

Schwartz C, Gerard E, Perronnet K, Morel JL (2001) Measurement of in situ phytoextraction of zinc by spontaneous metallophytes growing on a former smelter site. Sci Total Environ 279:215–221

Schwinn F (1994) Seed treatment – a panacea for plant protection? In: Martin TJ (ed) Seed treatment: progress and prospects, BCPC monograph no. 57. British Crop Protection Council, Farnham, pp 3–14

Selvakumar G, Panneerselvam P, Ganeshamurthy AN (2012) Bacterial mediated alleviation of abiotic stress in crops. In: Maheshwari DK (ed) Bacteria in agrobiology: stress management. Springer, Berlin/Heidelberg, pp 205–224

Sharma A, Johri BN, Sharma AK, Glick BR (2003) Plant growth-promoting bacterium *Pseudomonas* sp. strain GRP3 influences iron acquisition in mung bean (*Vigna radiata* L. Wilczek). Soil Biol Biochem 35:887–894

Sheng XF, He LY (2006) Solubilization of potassium-bearing minerals by a wild-type strain of *Bacillus edaphicus* and its mutants and increased potassium uptake by wheat. Can J Microbiol 52(1):66–72

Siebner-Freibach H, Hadar Y, Chen Y (2003) Siderophores sorbed on Ca-montmorillonite as an iron source for plants. Plant Soil 251:115–124

Singh BN, Singh A, Singh SP, Singh HB (2011) *Trichoderma harzianum*-mediated reprogramming of oxidative stress response in root apoplast of sunflower enhances defense against Rhizoctonia solani. Eur J Plant Pathol 131:121–134

Singh A, Jain A, Sarma BK, Upadhyay RS, Singh HB (2014) Rhizosphere competent microbial consortium mediates rapid changes in phenolic profiles in chickpea during *Sclerotium rolfsii* infection. Microbiol Res 169:353–360

Srivastava PK, Vaish A, Dwivedi S, Chakrabarty D, Singh N, Tripathi RD (2011) Biological removal of arsenic pollution by soil fungi. Sci Total Environ 409:2430–2442

Taylor AG (1997) Seed storage germination and quality. In: Wien HC (ed) The physiology of vegetable crops. CAB International, Wallingford, pp 1–36

Tisdall JM, Oadea JM (1982) Organic matter and water stable aggregates in soils. J Soil Sci 33:141–163

Van Loon LC, Bakker PAHM, Pieterse CMJ (1998) Systemic resistance induced by rhizosphere bacteria. Annu Rev Phytopathol 36:453–483

Van Peer R, Niemann GJ, Schippers B (1991) Induced resistance and phytoalexin accumulation in biological control of Fusarium wilt of carnation by *Pseudomonas* sp. strain WCS417r. Phytopathology 81:728–734

Vansuyt G, Robin A, Briat JF, Curie C, Lemanceau P (2007) Iron acquisition from Fe-pyoverdine by *Arabidopsis thaliana*. Mol Plant-Microbe Interact 20:441–447

Viterbo A, Ramot O, Leonid C, Chet I (2002) Significance of lytic enzymes from *Trichoderma* spp. in the biocontrol of fungal plant pathogen. Antonie Van Leeuwenhoek 81(4):549–556

Wei L, Kloepper JW, Tuzun S (1991) Induction of systemic resistance of cucumber to *Colletotrichum orbiculare* by select strains of plant growth-promoting rhizobacteria. Phytopathology 81:508–1512

Wilson CL, Wisniewski ME (1994) Biological control of postharvest diseases of fruits and vegetables– theory and practices. CRC Press, Boca Raton

Wright B, Rowse HR, Whipps JM (2003) Microbial populations on seeds during drum and steeping priming. Plant Soil 255:631–640

Yao L, Wu Z, Zheng Y, Kaleem I, Li C (2010) Growth promotion and protection against salt stress by *Pseudomonas putida* Rs-198 on cotton. Eur J Soil Biol 46:49–54

Yazdani M, Yap CK, Abdullah F, Tan SG (2009) *Trichoderma atroviride* as a bioremediator of Cu pollution: an in vitro study. Toxicol Environ Chem 91:1305–1314

Zaidi A, Khan MS, Ahemad M, Oves M (2009) Plant growth promotion by phosphate solubilizing bacteria. Acta Microbiol Immunol Hung 56:263–284

Part III

Molecular and Physiological Aspects of Nutrient Use Efficiency

Improving Nutrient Use Efficiency by Exploiting Genetic Diversity of Crops

S.P. Trehan and Manoj Kumar

Abstract

Low nutrient use efficiency by crops needs to be increased to have lower unit crop production costs, higher economic returns and minimal negative environmental impact. In the twenty-first century, nutrient-efficient plants will play a major role in increasing crop yields compared to the twentieth century, mainly due to limited land and water resources available for crop production, higher cost of inorganic fertiliser inputs, declining trends in crop yields globally and increasing environmental concerns. Many studies have reported large genetic variation for nutrient efficiency in different crop species and cultivars. The efficient cultivars gave higher yield under nutrient stress and had higher agronomic use efficiency (AUE) than less efficient cultivars. The variation in nutrient efficiency of different crop cultivars was due to both their capability to use absorbed nutrients to produce yield and to their capacity to take up more nutrient per unit soil. The indication of higher expression of ammonium transporter, cytochrome oxidase and asparagine synthetase in potato leaves had been reported to be used as parameters to screen potato genotypes for high metabolism, utilisation, transport and storage of N. There is a need to establish linkage of nutrient efficiency with root and shoot parameters/DNA markers/genes to have/breed multi-nutrient-efficient crop varieties.

Keywords

Agronomic use efficiency (AUE) • Nitrogen • Nutrient uptake efficiency (NUE) • Phosphorus • Potassium • Physiological use efficiency (PUE) • Potato • Yield • Varieties

S.P. Trehan (✉)
Central Potato Research Station, P.B. No. 1, P.O. Model Town, Jalandhar 144 003, Punjab, India
e-mail: sptrehan2@yahoo.co.in

M. Kumar
Central Potato Research Station, Sahaynagar, Patna 801 506, Bihar, India
e-mail: manojkumar0105@gmail.com

1 Introduction

The crop yield gap is observed at several stages: between the genetic potential and the yield maximisation on station research, between yields on station experiments and in on-farm trials, between on-farm trials and farmers' fields in a

A. Rakshit et al. (eds.), *Nutrient Use Efficiency: from Basics to Advances*,
DOI 10.1007/978-81-322-2169-2_14, © Springer India 2015

given environment and between farmers' fields in a given environment or in comparable environments. Large-yield gaps are observed not only between on-station and on-farm trials, but also at the farm level. The differences in yield are due to many factors. These include differences in the crops/cultivars, the soil, the crop production inputs including seed and nutrients, poor water management and the level of crop husbandry. The farmer's yield is often low due to neglect of many factors. In addition, many farmers may not have access to adequate and timely resources to make full use of the production potentials made available by research. The aim of technology transfer and crop management including efficient nutrient supply/use should be to enable farmers to come as close as possible to on-station yields, keeping in mind the economics of the operation. Increasing nutrient use efficiency can reduce these yield gaps to some extent.

Adequate crop nutrition is a key component of improved production technology. The gap between nutrients required by the crops and amounts expected to be made available from soil nutrient supplies has to be bridged through external nutrient application. This can be done through a number of organic, microbial and mineral sources, often in an integrated manner. Fertilisers are by far the most important source of plant nutrients and have made a place for themselves in intensive cropping by helping obtain and sustain high crop yields. Balanced and efficient use of fertilisers, organics and bio-fertilisers is the key step which results in lower unit crop production costs, higher economic returns and minimal negative environmental impact. The four watch words around which the plant nutrition needs and strategies should be built are optimum, balance, efficiency and effectiveness. This is possible by putting proven scientific findings into practices on individual farm holdings and revitalising support services such as soil and plant testing, fertility monitoring, matching product supply with soil nutrient deficiencies and taking care of all production inputs.

The definition of fertiliser use efficiency (FUE) is using all nutrients applied without leaving any residue in the soil. So it reduces leaching and run off losses and also avoid luxury consumption of particularly K thereby reducing nutrient imbalance in crop like K-Mg and K-Ca interactions and saving wastage of fertilisers. So to reduce pollution of ground water with NO_3 and save quality of produce due to excessive accumulation of NO_3 in produce, nutrient use efficiency needs to be increased.

2 Techniques for Increasing Fertiliser/Nutrient Use Efficiency

Many techniques and methods can be used to enhance nutrient/fertilisers efficiency for improved nutrient management in crops. Nutrient use efficiency can be increased by applying the right source (which supply nutrients in plant-available forms, suit soil physical and chemical properties, recognise synergisms among nutrient elements and sources, recognise blend compatibility, recognise benefits and sensitivities to associated elements and control effects of non-nutritive elements), at the right rate (by using adequate methods to assess soil nutrient supply, assessing all indigenous nutrient sources available to the crop, assessing crop demand for nutrients, predicting fertiliser use efficiency, considering soil resource impacts and considering rate-specific economics), at the right time (by assessing timing for crop uptake, assessing dynamics of soil nutrient supply, recognising timing of weather factors influencing nutrient loss and evaluating logistics of field operations) and at the right place (by recognising root-soil dynamics, managing spatial variability within fields and among farms, fitting needs of tillage system and limiting potential off-field transport of nutrients). In addition, nutrient use efficiency can also be improved by exploiting genetic diversity of crops.

3 Need for Exploiting Genetic Diversity of Crops for Better Nutrient Use Efficiency

In the twenty-first century, nutrient-efficient plants will play a major role in increasing crop yields compared to the twentieth century, mainly due to limited land and water resources available for crop production, higher cost of inorganic fertiliser inputs, declining trends in crop yields globally and increasing environmental concerns. These factors will enhance the importance of nutrient-efficient cultivars that are also higher producers (Fageria et al. 2008).

A nutrient-efficient cultivar, in an agronomic sense, is a cultivar that is able to grow and yield well in soils too deficient for a standard cultivar (Graham 1984). Different crop cultivars may have different capability to use nutrients from soil/fertilisers/organic manure/waste and show differential response to them. Differential responses among cultivars can provide some insight into the mechanism of tolerance to nutrient deficiency. More importantly, characteristics of efficient cultivars may ultimately be useful as criteria for selecting tolerance to nutrient deficiency in crops.

4 Genomics Approaches to Improve Nutrient Use Efficiency

Recent progress in plant molecular biology has imposed both opportunity and challenge to plant biology research, resulting in the emergence of many new areas of studies. Among these root biology is a new frontier of plant biology for systematic studies of the many processes involved in plant nutrient mobilisation, uptake, transport, translocation as well as their effects on plant growth, development and adaptability to adverse soil conditions. The ultimate objective of root biology is to facilitate the development of nutrient-efficient, stress-tolerant and high-quality crop varieties that will contribute to the agricultural sustainability, food security and environmental safety in the world.

Nitrogen Nitrogen is a crucial plant macronutrient and is needed in the greatest amount of all mineral elements required by plants. Plants consume much less than half of the N fertilisers applied, while majority of N fertilisers are lost to the environment, which causes increasingly severe pollutions. Excess of nitrogen in the plant also has negative effects on the eating and cooking quality of produce. Moreover, fertiliser application has now become the major cost in the production, which greatly affects the income of the farmers. Thus, developing crops that are less dependent on the heavy application of N fertilisers is essential for the sustainability of agriculture.

Many studies have reported large genetic variation for N efficiency in different crop species and cultivars (Trehan 2006, 2007, 2009a, b; Kumar et al. 2009; Trehan and Singh 2013). The potato cv. Kufri Pukhraj was the most N-, P- and K-efficient cultivar among ten cultivars tested in the absence as well as presence of green manure (Trehan 2009a). The efficient cultivars gave higher tuber yield under N, P and K stress (i.e. with less dose of N, P and K fertiliser) and had higher AUE than less efficient cultivars. Mean AUE of N of different cultivars varied between 62 and 97 kg tubers/kg N without green manure and between 68 and 100 kg tubers/kg N with green manure. Mean AUE of Kufri Pukhraj was 97 and 100 kg tubers/kg N without and with green manure, respectively, which was significantly higher than all other cultivars. The main cause of higher nitrogen efficiency in the presence of green manure was the capacity of a genotype to use/absorb more N per unit green manured soil, i.e. the ability of the root system of a genotype to acquire more N from green manured soil (NUE). Most P-efficient cv. K. Pukhraj produced yield of 300 q/ha without P, whereas K. Badshah and K. Ashoka needed 100 kg P_2O_5/ha to produce yield of 270 and 304 q/ha, respectively, in the same field. Similarly most K-efficient cv. K. Pukhraj produced yield of 364 q/ha without K, whereas K. Badshah and K. Sutlej needed 80 kg K_2O/ha to produce yield of 361 and 370 q/ha, respectively, in the same field. The variation in phosphorus and potassium efficiency of different potato cultivars was due to both their capability to use absorbed P and K to produce potato tubers and to their capacity to take up more P and K per unit soil. In the presence of green manure, nitrogen efficiency of different potato cultivars was mainly related to their capability to use nitrogen from green manured soil. More N-efficient cultivars took up higher nitrogen per unit soil than less efficient cultivars. The mean nitrogen uptake by more N-efficient cvs., K. Pukhraj, K. Sutlej, K. Bahar, K. Ashoka and K. Sindhuri was significantly higher (171–214 kg/ha) than other less N-efficient cultivars (139–149 kg/ha)

In a recent study, the potato hybrid JX 576, now released as cv. Kufri Gaurav, has been shown to be more N efficient than other cultivars (Trehan and Singh 2013). The hybrid JX 576 produced significantly higher tuber yield than other five cultivars tested particularly under nitrogen stress (Fig. 1). For example, in 2008–2009, it gave 6.5–12.5 t/ha higher tuber yield under N stress (No N applied), than other five cultivars. It required lower doses of N than other cultivars to produce particular fixed tuber yield in the same field. For example, It gave a yield of 45.9 t/ha with 160 kg N/ha in 2008–2009, whereas Kufri Pushkar and Kufri Pukhraj needed 240 kg N/ha to produce a yield of 44.1 and 46.2 t/ha, respectively. Mean AUE of different cultivars/hybrid varied between 51 and 90 kg tubers/kg N for nitrogen (Fig. 2). The hybrid JX 576 had the highest AUE of N among the six cultivars tested at all rates of nutrient application, whereas cv. Kufri Jyoti had the least AUE. The hybrid JX 576 showed about 10–24 % greater AUE of N than best control Kufri Pukhraj and Kufri Pushkar during different years. The higher efficiency of hybrid JX 576 was mainly because of its better utilisation of absorbed nitrogen for potato production (i.e. higher PUE) than other cultivars (Fig. 2).

A large spectrum of organisms are able to directly take up urea and subsequently utilise it as a nitrogen source for growth. In agricultural crop production, urea is quantitatively the most important N fertiliser worldwide. However, due to processes that cause N losses, such as ammonia volatilisation and denitrification, which might occur after urea hydrolysis into ammonium by urease that is released from soil microorganisms, the N use efficiency (NUE) of urea-based fertilisers in crop production was reported to be less than 50 % of the total N applied. Although ammonium and nitrate can be effectively taken up via a variety of transport system indicating AMTs and NRTs (Gazzarrini et al. 1999; Forde 2000), higher plants are likely to be able to directly absorb or transport urea through protein-mediated pathways such as the active urea transporter AtDUR3 and some MIP proteins (Liu et al. 2003a, b). Therefore, the exploration of urea transport mechanisms at the molecular level in crops will not only provide a fundamental knowledge to understand urea movement into and/or within the crop plants, but may also promise a significant strategy to manipulate crop urea transport pathways, enhancing effective uptake of urea prior to its external degradation.

Root and Shoot Parameters of Crops/ Cultivars for High N Efficiency Sattelmacher et al. (1990) stressed the importance of root length and surface area for nitrogen acquisition by potato cultivars from soils. Information on nitrate uptake kinetic parameters, and their variation among cultivars, is lacking for potato. Mehdi and Bernie (2006) suggested that I_{max} in combination with root morphology may be important in controlling differences in N uptake efficiency among potato cultivars.

In another study, results showed that potato cv. Kufri Pukhraj was more N efficient than other cultivars, Kufri Jyoti and Kufri Jawahar, because it produced higher tuber and total dry matter at lower N rates (Trehan 2009b). The cause/mechanism of its higher N efficiency was its ability to maintain higher N influx (N uptake rate per unit root) than other cultivars (Table 1). At 3.58 m mol N kg^{-1}, the N influx in to the roots of more N-efficient cv. Kufri Pukhraj was 119 × 10^{-14} mol cm^{-1} S^{-1} in comparison to 39 and 81 × 10^{-14} mol cm^{-1} S^{-1} in less N-efficient cvs., Kufri Jyoti and Kufri Jawahar, respectively. Total

Fig. 1 Tuber yield of different potato cultivars/hybrid as affected by N application (The CD (0.05) was 1.60, 1.22 and 3.20 in 2007–2008 and 1.85, 1.50 and 3.71 in 2008–2009 and 1.76, 14.3 and 3.53 in 2009–2010 for cultivar mean, N rate mean and cultivar × N rate, respectively)

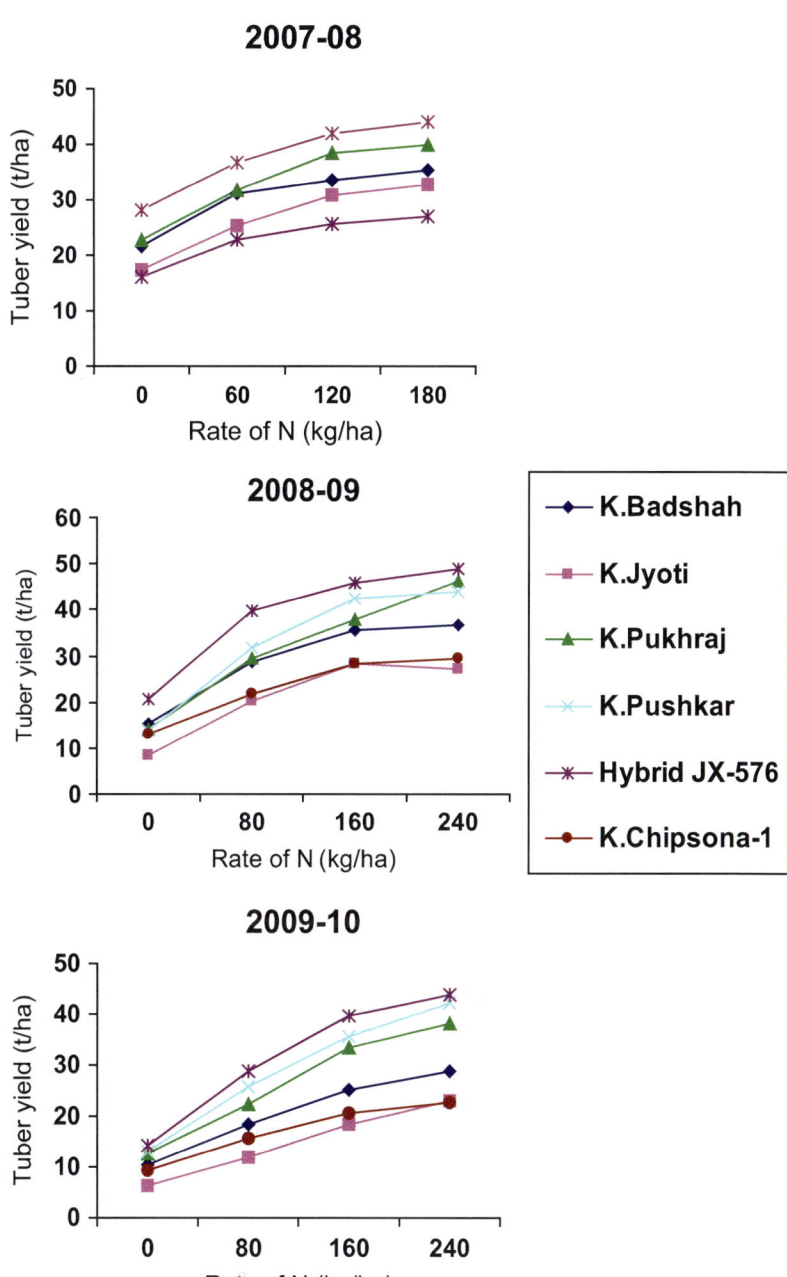

uptake of N by all the cultivars was higher than the amount of N applied at all N rates (Table 2). Kufri Pukhraj was able to maintain higher N influx than other cultivars because of its capacity to use higher amount of N from the soil. At 3.58 m mol N kg^{-1}, Kufri Pukhraj used 13.3 % of total soil organic N in comparison to 6.4 and 10.0 % used by Kufri Jyoti and Kufri Jawahar, respectively.

Phosphorus The development of P-efficient crop varieties that grow and yield better with low P availability is key to the crop production. Many studies have demonstrated substantial

Fig. 2 Mean (3 years) agronomic use efficiency (kg tubers produced/kg nutrient supply), uptake efficiency (kg nutrient uptake/kg nutrient supply) and physiological use efficiency (kg tuber produced/kg nutrient absorbed) of N by six potato cultivars/hybrid CD (0.05): 4.4, 0.028 and 20.9 for AUE, UE and PUE, respectively

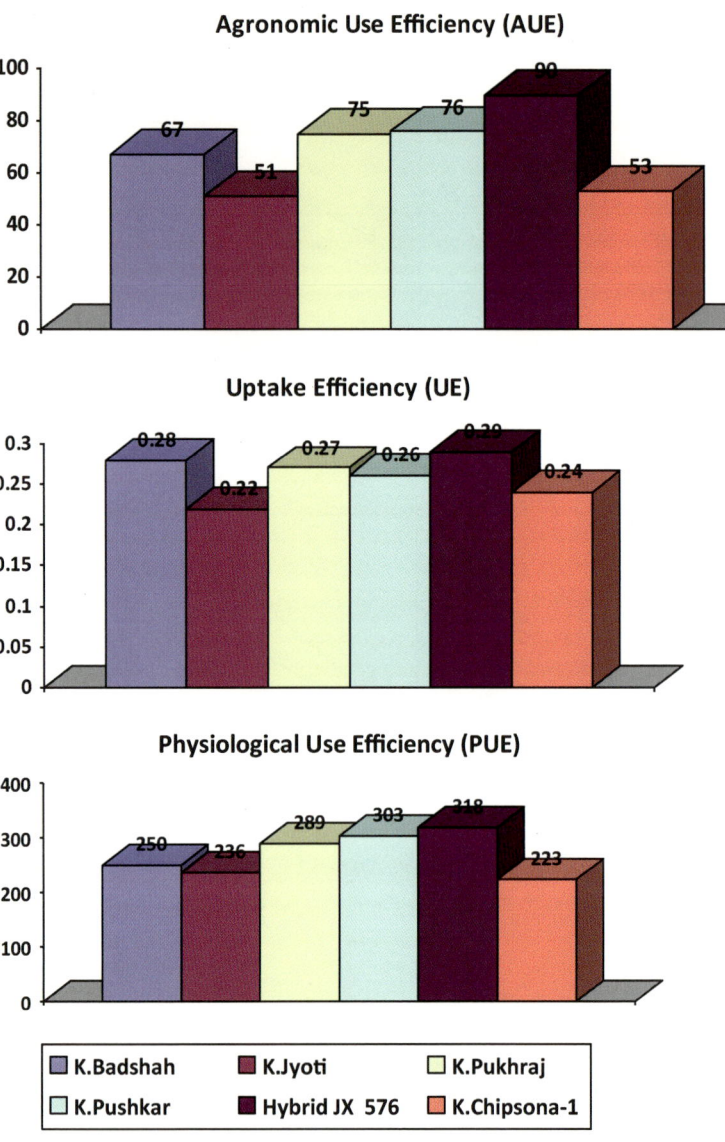

Table 1 Influx of nitrogen (uptake rate per unit root) in to roots of three potato cultivars

Rate of N applied (μ mol kg^{-1})	N influx (10^{-14} mol cm^{-1} s^{-1})		
	Potato cultivars		
	K. Pukhraj	K. Jyoti	K. Jawahar
0	12	16	12
1,790	54	30	42
3,580	119	39	81
7,160	319	116	141

genetic variation for P efficiency in potato and soybean, and these crops have been bred for adaptation to low P soils.

Liu et al. (2005) discovered that sugars and photosynthates are critical effectors for enhanced expression of P-deficiency-induced genes in cluster roots. P-deficiency-induced genes and genes involved in sugar metabolism and/or sensing seem to be co-regulated by P status as well as by light/dark conditions.

Plant growth and P uptake by a diverse range of cereal genotypes differed considerably, indicating that breeding for P uptake efficiency in wheat is possible (Liao Mingtan et al. 2005). The close correlations between shoot biomass and plant P uptake suggest that shoot biomass

Table 2 Utilisation of soil nitrogen by different potato cultivars at 63 days after emergence

Rate of N applied (μ mol kg^{-1})	Total uptake of N (μ mol kg^{-1}) (B)			Extra uptake of N from soil (μ mol kg^{-1}) (B−A)			Percent of soil organic N used $\frac{[(B-A)-210] \times 100^a}{22,500}$		
(A)	K. Pukhraj	K. Jyoti	K. Jawahar	K. Pukhraj	K. Jyoti	K. Jawahar	K. Pukhraj	K. Jyoti	K. Jawahar
0	1,565	1,621	1,832	1,565	1,621	1,832	6.0	6.3	7.2
1,790	3,990	3,763	3,726	2,200	1,973	1,936	8.8	7.8	7.7
3,580	6,790	5,218	6,034	3,210	1,638	2,454	13.3	6.4	10.0
7,160	11,925	9,038	8,437	4,765	1,878	1,277	20.2	7.4	4.7

[a]Where 210 μ mol N kg^{-1} is NO$_3$-N in the soil at the beginning and 22,500 μ mol N kg^{-1} is organic N assuming C:N ratio of 20: 1 in 0.63 % organic carbon present in the soil at the beginning

is a reliable parameter for screening wheat genotypes in soils for P-uptake efficiency.

It would be far more reliable and preferable to identify and select specific traits that are directly related to P efficiency. In recent years systematic experiments in the field, greenhouse and laboratory were set out to investigate the physiological, genetic and molecular aspects of important root traits conferring better P efficiency in soybean and potato.

P efficiency of potato genotype CGN 17903 was related to higher P utilisation efficiency and that of CIP 384321.3 to both higher P uptake efficiency in terms of root-shoot ratio and intermediate utilisation efficiency (Schenk and Balemi 2009).

The number of leaves of P utilisation-inefficient genotype CGN 18233 was reduced by 40 %, whereas that of the P utilisation-efficient genotype CGN 17903 was not affected by P deficiency. This might be due to the ability of the P utilisation-efficient genotype to maintain cell division at shoot meristems to optimum level under lower P concentration leading to maintenance of optimum leaf number per plant (Lynch et al. 1991; Chiera et al. 2002). On the other hand, leaf area of the P-inefficient genotype, was reduced by 70 %, whereas for the two P-efficient genotypes, it was reduced by only 15 % due to low P supply (Balemi 2009). Thus, in P stress-sensitive genotype, the leaf area was the most severely affected plant morphological parameters by P deficiency. Significant differences in root number, length and surface area were found. There was significant correlation between glasshouse and field measures; stolon root number from the glasshouse screen could be used to indicate total root length in the field (Wishart et al. 2009). There is therefore increased interest in using root traits as a selection criterion for improving yields (White et al. 2005; Lynch 2007).

P Efficiency and Root Exudates In sand culture, uptake of P from sparingly soluble rock phosphate was higher in P-efficient plants than in P-inefficient maize. In sand, but not in solution culture, higher citrate concentrations were detected in the rhizosphere of P-efficient than of P-inefficient maize. Quartz sand amended with rock phosphate was a better substrate than nutrient solution for revealing the varietal differences in P acquisition efficiency in short-term experiments (Corrales et al. 2007). Organic acids in root exudates can enhance phosphorus availability. Two methods to identify and quantify twelve aliphatic and eleven aromatic organic acids in the rhizosphere, using chromatographic methods with UV/Vis detection, were developed. Dechassa and Schenk (2004) concluded that the high P efficiency of cabbage could be due to its ability to exude large amounts of citrate, which mobilises soil P for plant uptake.

Iron To meet iron demand for growth and development, two effective iron acquisition systems known as strategy I and strategy II

(Roemheld and Marchner 1986) have been evolved in higher plants. Tomato uses the strategy I mechanism to acquire iron from the soil. The cores of this strategy are acidification of rhizosphere by enhanced extrusion of proton to increase solubility of ferric iron, activation of ferric-chelate reductase reducing Fe^{3+} to Fe^{2+} on root surface in the subapical region, induction of the high-affinity Fe^{2+} transporter system to absorb ferrous iron from the soil into the roots as well as morphological changes of roots, such as thickening of the subapical root zone, increased formation of root hairs and so on. These iron-deficiency responses must be carefully regulated because excess iron can be toxic.

Prom-uthai et al. (2005) investigated the variation in total extractable phenol and tannin content among five rice varieties and their relationship with Fe bioavailability. Total extractable phenols and tannin contents are inhibitory to Fe bioavailability. The bioavailability of Fe in rice grain may be enhanced by eliminating organic compounds through storage and processing.

Zinc Complexation of Zn by phytic acid, the storage form of P in seeds, has a strong negative effect on Zn absorption by the human body (Lonnerdal 2000). It can reduce the bioavailability of Zn to less than 3 % of total Zn in the grain (Bosscher et al. 2001). Therefore, the phytate/Zn molar ratio of seeds has often been used as a predictor for Zn bioavailability in foods. Usually, ratios above 20 are supposed to induce Zn deficiency (Cakmak et al. 1999).

The seed phytate content has been shown highly dependent on P uptake and transport and, therefore, on P availability in the soil. Consequently, phytate-Zn molar ratios are highly sensitive to P fertilisation. Application of P can increase phytate-zinc molar ratios in millet grain from the critical level of about 20 to almost 30 (Buerkert et al. 1998).

Mechanism of High Zinc Efficiency Tolerance to Zn deficiency is still poorly understood and many potential mechanisms have been proposed. Tolerant rice cultivars may be better capable of taking up Zn from Zn-deficient soils. Citrate exudation could increase the availability and uptake of Zn in two ways; (i) it is exuded by plant roots together with protons, for reasons of electro-neutrality. These protons reduce the rhizosphere pH, which increase Zn bioavailability, especially in alkaline soils where Zn deficiency problems are most severe (Kirk and Bajita 1995), and (ii) citrate can form a weak complex with Zn and complexation of Zn may also increase Zn bioavailability. Rice plants are known to exude low-molecular-weight organic anions, including citrate (Aulakh et al. 2001). Synthesis and exudation of citrate are increased under P deficiency in rice (Kirk et al. 1999).

In a study with potato to find mechanism of high Zn efficiency, three cultivars, Kufri Badshah, Kufri Jyoti and Kufri Chandramukhi, were grown in low Zn soil in the absence and presence of Zn (Trehan and Sharma 2003). Two harvests were taken to obtain final dry matter accumulation (DMA), rates of shoot and root growth and Zn uptake rate per unit root length per unit time (Zn influx). Taking the yield of the unfertilised relative to the fertilised plant as a measure of Zn efficiency, i.e. capacity to grow under low soil Zn supply, cv. Kufri Chandramukhi was less Zn efficient producing 71 % total dry matter (shoot + tubers) as compared to cv. Kufri Jyoti and Kufri Badshah producing 81 and 102 %, respectively. Low Zn uptake efficiency of Kufri Chandramukhi was due to its low root length-shoot weight (DMA) ratio (4.7 m/g) than Kufri Badshah (13.4 m/g) and Kufri Jyoti (7.7 m/g). Zinc influx did not differ significantly among cultivars in the absence of Zn. The results showed that response to Zn differed greatly among potato cultivars. The cv. Kufri Chandramukhi had a lower Zn uptake efficiency than other cultivars because of its lower capacity to absorb Zn at limiting Zn supply due to its lower root-shoot (DMA) ratio. This shows that root-shoot ratio was an important factor of Zn efficiency of the potato cultivars since their Zn influx did not differ significantly in the absence of Zn. The root-shoot (DMA) ratio among the cultivars varied by a factor of more than 2. Fohse et al. (1988) have also reported low P uptake efficiency of some crop plants due to their low root-shoot ratio.

Potassium The variation in potassium efficiency of different potato cultivars has been reported due to both their capability to use absorbed K to produce potato tubers and to their capacity to take up more K per unit soil (Trehan 2009a). Most K-efficient cv. K. Pukhraj took up highest K (109 kg K/ha), and less K-efficient cvs. Jyoti, Bahar and Badshah took up less K (81–88 kg K/ha) from the soil under K-deficiency conditions, i.e. in the absence of K fertiliser. The cv. K. Sutlej also took up the same amount of K (109 kg K/ha) as K. Pukhraj in the absence of K fertiliser, but it was less K efficient than K. Pukhraj because it produced less tuber yield per kg of K absorbed (2.9 q tubers/kg K) than K. Pukhraj (3.4 q tubers/kg K).

The potassium-efficient cultivars like Kufri Pukhraj give high potassium efficiency but deplete the soil and create imbalance particularly of K. On the other hand, the addition of organic material to the soil do improve OC status to maintain soil health, productivity and nutrient balance, but the problem is of slow release of nutrients from OM/waste which do not match with the required rate of uptake of nutrients by plants. So use of both efficient cultivars along with OM/waste and fertilisers can maintain nutrient balance and soil health as well as increase nutrient use efficiency. Plants do not recognise boundaries of inorganic and organic manures because plants take nutrients in ionic form. So we need to tailor our varieties which can convert nutrients from organic to ionic form and increase rate of release/uptake of nutrients from organic material or from the soil which can eventually be compensated by release of nutrients from organic materials over a period of time.

5 Genes Controlling Nutrient Efficiency

Identification of characters/parameters/genes and their variation in available biodiversity (plant species and cultivars – Fauna and Flora) is essential for developing nutrient-efficient cultivars, since variety improvement is not possible until we have identified characters/genes. In the past decade, use of molecular markers has emerged as a powerful tool in plant science. Molecular understanding of the inheritance of agriculturally important traits has created new opportunities to streamline the process of altering plant genotypes. Molecular markers permit the study of any morphological, physiological or developmental process in which genetic variants exist with a minimum of prior information. The first towards marker-assisted selection of a trait of interest, be it monogenic or polygenic, is gene mapping through DNA markers. Now when a particular genotype is saturated with DNA markers, one can look forward to locating genes influencing a trait along the chromosomes. The utility of DNA markers in plant breeding is based on finding tight linkages between the markers and genes of interest. Such linkage permits one to infer the presence of a desirable gene by assaying for the marker. Disease resistance traits have been frequently transferred from one genetic background to another. Traditionally, progenies are being screened for the presence of disease resistance genes by inoculation with the pathogen. In contrast, detecting disease resistance genes by their linkage to DNA marker makes it practical to screen for many different disease resistance genes simultaneously without the need to inoculate the pathogen. Of particular potential value to the plant breeder is DNA marker-assisted diagnostic of traits which are difficult to measure like nutrient efficiency, nematode resistance, etc. The introduction of DNA marker-aided selection promises to overcome major limitations of backcross breeding. If the genes to be transferred are marked by tightly linked DNA markers, segregating populations of plants can be screened at the seedling stage, before the trait is expressed, for the presence of the gene(s) of interest. Since the DNA markers can be used to mark quantitative trait loci (QTL) as well as major genes, there are no limitations to the types of characters that can be manipulated by marker-based selection. Without linked

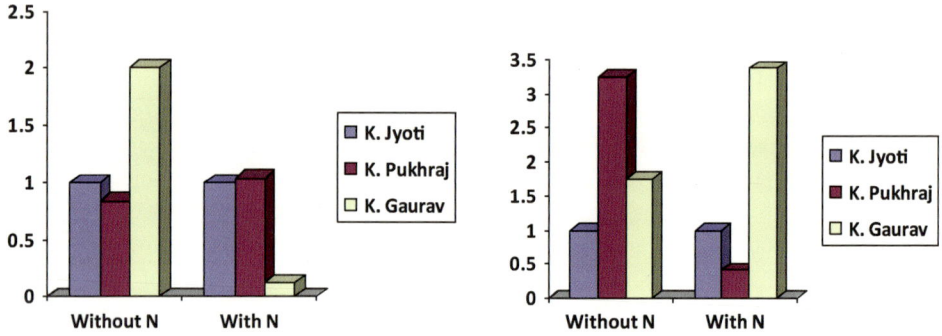

Fig. 3 Ammonium transporter (AMT) (*left side*) and Asparagine synthetase (AS) (*right side*) activities in leaves of three cultivars without and with N application

markers it would be extremely difficult and time consuming to monitor the flow of such genes in a breeding programme. DNA marker-aided backcrossing has been very widely used by commercial firms – saving of even 1–2 generations means that an improved cultivar reaches the market place sooner. Smith (1997) showed that uptake rate of a particular nutrient was controlled by a specific gene or group of genes and also identified several plant genes responsible for sulphate uptake and located the plant genes responsible for phosphate transport. Recent study done at the institute on gene expression analysis gave encouraging results.

Gene Expression Analysis The real-time gene expression of ammonium transporter (AMT), cytochrome oxidase (COX1), asparagines synthetase (AS), nitrate reductase (NR) and nitrite reductase (NIR) was analyzed in leaves of potato cultivars under N stress and other field conditions at the appropriate level (Trehan and Singh 2013). Results showed that gene expression of AMT, AS and COXI in leaf tissues particularly under N stress successfully explained the variation of N efficiency in three cultivars, Kufri Gaurav, Kufri Pukhraj and Kufri Jyoti, which differ widely in N efficiency (Figs. 3 and 4). Ammonium is taken up by plant cells via ammonium transporters in the plasma membrane and distributed to intracellular compartments such as chloroplasts, mitochondria and vacuoles probably via different transporters in each case. Ammonium is also produced by plant cells

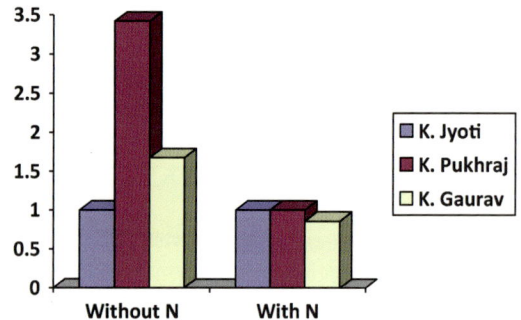

Fig. 4 Cytochrome oxidase (COXI) activity in leaves of three cultivars without and with N application

during normal metabolism, and ammonium transporters enable it to be moved from intracellular sites of production to sites of consumption. Results showed that most N-efficient cv. Kufri Gaurav had the highest activity of this enzyme than Kufri Pukhraj and Kufri Jyoti (Fig. 3). It indicated that Kufri Gaurav is better equipped for ammonium utilisation in the plant cells than Kufri Pukhraj and Kufri Jyoti. The higher expressions of AMT, COX1 and AS in leaves can be used as parameters to screen potato genotypes for high metabolism, utilisation, transport and storage of N.

References

Aulakh MS, Wassmann R, Bueno C, Kreuzwieser J, Rennenberg H (2001) Characterization of root exudates at different growth stages of ten rice (Oryza sativa L.) cultivars. Plant Biol 3:139–148

Balemi T (2009) Effect of phosphorus nutrition on growth of potato genotypes with contrasting phosphorus efficiency. Afr Crop Sci J 17:199–212

Bosscher D, Lu ZL, Janssens G, Van Caillie-Bertrand M, Robberecht H, De Rycke H, De Wilde R, Deelstra H (2001) In vitro availability of zinc from infant foods with increasing phytic acid contents. Br J Nutr 86:241–247

Buerkert A, Haake C, Ruckwied M, Marchner H (1998) Phosphorus application affects the nutritional quality of millet grain in the Sahel. Field Crops Res 57:223–235

Cakmak I, Kalayei M, Ekiz H, Braun HJ, Kiline Y, Yilmaz A (1999) Zinc deficiency as a practical problem in plant and human nutrition in Turkey: a NATO science for stability project. Field Crops Res 60:175–188

Chiera J, Thomas J, Rufty T (2002) Leaf initiation and development in soybean under phosphorus stress. J Exp Bot 53:473–481

Corrales I, Montserrat A, Charlotte P, Juan B (2007) Phosphorus efficiency and root exudates in two contrasting tropical maize varieties. J Plant Nutr 30:887–900

Dechassa N, Schenk MK (2004) Exudation of organic anions by roots of cabbage, carrot, and potato as influenced by environmental factors and plant age. J Plant Nutr Soil Sci 167(5):623–629

Fageria NK, Baligar VC, Li YC (2008) The role of nutrient efficient plants in crop yields in the twenty first century. J Plant Nutr 31:1121–1157

Fohse D, Claassen N, Jungk A (1988) Phosphorus efficiency of plants. I. External and internal P requirement and P uptake efficiency of different plant species. Plant Soil 110:101–109

Forde BG (2000) Nitrate transporters in plants: structure, function and regulation. Biochim Biophys Acta 1465:219–235

Gazzarrini S, Lejay L, Gojon A, Ninnemann O, Frommer WB, von Wiren N (1999) Three functional transporters for constitutive, diurnally regulated, and starvation-induced uptake of ammonium into Arabidopsis roots. Plant Cell 11:937–948

Graham RD (1984) Breeding for nutritional characteristics in cereals. Adv Plant Nutr 1:57–102

Kirk GJD, Bajita JB (1995) Root-induced iron oxidation, pH changes and zinc solubilisation in the rhizosphere of lowland rice. New Phytol 131:129–137

Kirk GJD, Santos EE, Findenegg GR (1999) Phosphate solubilisation by organic anion excretion from rice (Oryza sativa L.) growing in aerobic soil. Plant Soil 211:11–18

Kumar M, Trehan SP, Jatav MK, Lal SS (2009) Efficacy of potato (Solanum tuberosum) cultivars under varying levels of nitrogen and growth duration in eastern Indo-Gangetic plains. Indian J Agron 54(1):63–68

Liao Mingtan T, Hocking Peter J, Dong Bei Delhaize, Richardson E, Alan E, Ryan Peter R (2005) Screening for genotypic variation in phosphorus-uptake efficiency in cereals on Australian soils. In: Li CJ et al (eds) Plant nutrition for food security, human

health and environmental protection. Tsinghua University Press, Beijing, pp 114–115

Liu LH, Ludewig U, Frommer WB, von Wirén N (2003a) AtDUR3 encodes a new type of high-affinity urea/H+ symporter in Arabidopsis. Plant Cell 15:790–800

Liu LH, Ludewig U, Gassert B, Frommer WB, von Wirén N (2003b) Urea transport by nitrogen-regulated tonoplast in- trinsic proteins in Arabidopsis. Plant Physiol 133:1220–1228

Liu J, Yamagishi M, Uhde-Stone C, Bucciarelli B, Samac D, Allan D, Vance C (2005) Shoot signals and molecular regulation of P-deficiency induced genes in cluster roots of white lipin. In: Li CJ et al (eds) Plant nutrition for food security, human health and environmental protection. Tsinghua University Press, Beijing, pp 94–95

Lonnerdal B (2000) Dietary factors influencing zinc absorption. J Nutr 130:1378S–1383S

Lynch JP (2007) Roots of the second green revolution. Aust J Bot 55:493–512

Lynch J, Läuchli A, Epstein E (1991) Vegetative growth of common bean in response to phosphorus nutrition. Crop Sci 31:380–387

Mehdi S, Bernie Z (2006) Nitrate influx kinetic parameters of five potato cultivars during vegetative growth. Plant Soil 288:91–99

Prom-uthai C, Huang L, Fukai S, Rerkasem B (2005) Effect of organic compounds in rice grain on iron bioavailability. In: Li CJ et al (eds) Plant nutrition for food security, human health and environmental protection. Tsinghua University Press, Beijing, pp 434–435

Römheld V, Marschner H (1986) Evidence for a specific uptake system for iron phytosiderophores in roots of grasses. Plant Physiol 80:175–180

Sattelmacher B, Klotz F, Marschner H (1990) Influence of the nitrogen level on root growth and morphology of two potato varieties differing in nitrogen acquisition. Plant and Soil 121:131–137

Schenk MK, Balemi T (2009) Mechanisms of phosphorus efficiency in potato genotypes. In: The proceedings of the international plant nutrition colloquium XVI, Department of Plant Sciences, University of California, Davis

Smith F (1997) Scientists discover how plants eat. Biotechnol Lab Int 2:1–4

Trehan SP (2006) Genetic control of different potato cultivars in the manipulation of nitrogen uptake from green manured soil. Adv Hortic Sci 20:199–207

Trehan SP (2007) Efficiency of potassium utilization from soil as influenced by different potato cultivars in the absence and presence of green manure (Sesbania aculeata). Adv Hortic Sci 21(3):156–164

Trehan SP (2009a) Improving nutrient use efficiency by exploiting genetic diversity of potato. Potato J 36:121–135

Trehan SP (2009b) Mechanism of high nitrogen efficiency in potato cultivars. Adv Hortic Sci 23 (3):179–184

Trehan SP, Sharma RC (2003) Root-shoot ratio as an indicator of zinc uptake efficiency of different potato cultivars. Commun Soil Sci Plant Anal 34:919–932

Trehan SP, Singh BP (2013) Nutrient efficiency of different crop species and potato varieties – in retrospect and prospect. Potato J 40(1):1–21

White PJ, Broadley MR, Hammond JP, Thompson AJ (2005) Optimising the potato root system for phosphorus and water acquisition in low-input growing systems. Asp Appl Biol 73:111–118

Wishart J, George TS, Brown LK, Thompson JA, Ramsay G, Bradshaw JE, White PJ, Gregory PJ (2009) Variation in rooting habit of potatoes: potential for improving resource capture. In: International symposium "Root Research and Applications", 2–4 September 2009, Boku, Vienna, Austria

MicroRNA-Based Approach to Improve Nitrogen Use Efficiency in Crop Plants

Subodh K. Sinha, R. Srinivasan, and P.K. Mandal

Abstract

Nitrogen is one of the most important plant nutrients, which is made available for agricultural crops other than legumes. Most of the N added to the soil is lost to the environment, with an average of only 30–50 % being taken up by the plants depending on the species and cultivar. One of the current strategies at improving crop nitrogen use efficiency (NUE) using modern genetic manipulation techniques and transgenic approaches involves manipulating the expression of genes involved in N and N/C metabolism. In this chapter, we have presented an overview of the recent advances in understanding the regulatory roles of microRNAs (miRNA) in nitrogen metabolism with a view toward engineering crops with increased NUE. We also discuss how miRNAs that influence N metabolism could be engineered. Finally the future prospect of involving miRNA approach for NUE in crop plants has been discussed.

Keywords

Nitrogen • Use efficiency • Metabolism • NUE • Uptake • Transporter • MicroRNA • Genetic manipulation

1 Introduction

The ever-growing world population, which is supposed to reach around 11 billion by 2050 as per United Nation estimates (Anonymous 2011), poses a great challenge to feed them in a most economical and environment-friendly ways. The recent trend of rise in price index of food grains

S.K. Sinha • R. Srinivasan • P.K. Mandal (✉)
National Research Centre on Plant Biotechnology,
New Delhi 110012, India
e-mail: subsinha@gmail.com; srinivasan53@gmail.com;
pranabkumarmandal@gmail.com

and fertilizer at global level reflects significant increase in the cost of cultivation. The remarkable increase in crop productivity observed over the last 50 years is linked with a 20-fold increase in the global use of N fertilizer applications (Glass 2003). The global nitrogen fertilizer use is likely to increase further, at least threefold by 2050 (Good et al. 2004). As nitrogen (N) is a key nutrient for the growth of any crops, the primary goal of using nitrogenous fertilizers, these are applied to the economic optimum level to increase the yield output of a crop per unit of land area (Firbank 2005). Because of the very

A. Rakshit et al. (eds.), *Nutrient Use Efficiency: from Basics to Advances*,
DOI 10.1007/978-81-322-2169-2_15, © Springer India 2015

mobile nature of the available form of N compounds, i.e., nitrate and ammonium in the soil, the percentage of N fertilizer actually used by crops has been very low, and crop plants are able to utilize only 30–40 % of the applied N (Raun and Johnson 1999). Thus, more than 60 % of the soil N is lost through a combination of processes like leaching, surface runoff, denitrification, volatilization, and microbial consumption, which further causes considerable impact on environments throughout the world. The environmental concerns together with the increasing cost of N fertilizer has prompted the plant biologists throughout the world to look for crops that are better able to uptake, utilize, and remobilize the nitrogen available to them. There have been several approaches adopted to develop nitrogen use efficient crop plants. One among them is transgenic approach using candidate gene of nitrogen metabolism. As nitrogen use efficiency is widely believed to be a complex trait including various subcomponents in it, regulatory factors involved in this metabolic pathway have always been considered as one of the favorable target genes for manipulation. One of the key regulatory factors is small RNAs, specifically the microRNAs, which play significant role in gene regulation in almost all aspects of plant life.

2   Why Do We Need Nitrogen Use Efficient Plants: The Rationale?

Nitrogen is taken up from the soil and utilized by plants for various metabolic processes such as biosynthesis of proteins, nucleic acids, cofactors, as well as signaling and storage molecules. Nitrogen is therefore quantitatively one of the most important nutrients and limiting factors for the growth and development of plants (Kraiser et al. 2011). It is mainly absorbed from the soil in the form of nitrate (NO_3^-), ammonium (NH_4^+), or urea. Insufficient nitrogen severely affects yields of crops, while surplus has no significant effect on yield and mainly contributes in N pollution (Liao et al. 2012; Amiour

et al. 2012). In field condition, however, it has been widely observed that on an average only 30–50 % of applied N is actually taken up by plants depending on species and cultivars and the rest of the applied N to the soil is lost to the environment (Garnett et al. 2009). Leaching and surface runoff of inorganic nitrogen in freshwater can cause algal blooms that in turn result in eutrophication of aquatic ecosystems. On the other hand, denitrification of previously fixed N carried out by soil microbes results into the emission of nitrous oxide (N_2O, ~300 times more potent greenhouse gas than CO_2) which has led to a steady increase in the level of this gas at the rate of 5–7 % per decade in the atmosphere since 1979 (Montzka et al. 2011). In addition, the energy required to produce much of the N in commercial fertilizers, through the Haber–Bosch process, is estimated to utilize approximately 1 % of the worlds' annual energy supply, adding to the cost of food production (Smith 2002). Apart from energy cost and environmental issues, the intake of nitrate-contaminated water leads to the incidence of stomach cancer in humans, particularly infants (Abrol et al. 1999; Hill et al. 1973; Weisenburger 1991). The unaccounted percentage of N represents loss of N fertilizer worth Rs. 720 billion annually (NAAS 2005). For the country like India, manufacturing of nitrogenous fertilizers involves huge amounts of foreign exchange and considerable consumption of large quantities of nonrenewable energy resources such as naphtha, natural gas, coal, etc. (NAAS 2005). Whatever may be the chemical nature of N fertilizer, in most of the cases the applied nitrogen finally gets converted to nitrate form. This form of N is not held tightly by soil particles and hence leaches out from the soil during excessive rains, especially in light-textured soil. Furthermore, the strategies to boost yields using more and more nitrogen fertilizers have plateaued as there is a limit for every crop to metabolize nitrogen and that limit we probably have crossed. Naturally, there is a growing interest in reducing fertilizer N inputs by improving plant N use efficiency.

3 A Brief Idea of Nitrogen Metabolism

NUE in plants is a complex trait that depends on nitrogen availability in the soil and the way plants use nitrogen throughout their life span. The use of nitrogen by plants involves several steps, including uptake, assimilation, transloca- tion, and, at the time of plant aging, recycling, and remobilization. Soil nitrogen availability generally fluctuates greatly with both space and time due to factors such as precipitation, temper- ature, wind, soil type, and pH. Plants that are adapted to low pH and reducing soils as found in mature forests or Arctic tundra tend to take up ammonium or amino acids, whereas those adapted to higher pH and more aerobic soils prefer nitrate (Maathuis 2009). Apart from this, some higher plants also take up amino acids under particular conditions of soil composition (Bick et al. 1998; Neelam et al. 1999; Schwacke et al. 1999).

3.1 Nitrogen Uptake

For nitrate uptake which occurs at the root level, two nitrate transport systems have been shown to coexist in plants which work coordinately to take up nitrate from the soil and distribute it within the whole plant. It is generally assumed that the nitrate transporter 1 (NRT1) gene family mediates the root low-affinity transport system (LATS), with the exception of the AtNRT1.1, which is both a dual-affinity transporter (Wang et al. 1998; Liu et al. 1999) and a nitrate sensor (Ho et al. 2009). In *Arabidopsis*, 53 genes belong to the NRT1family. *AtNRT1.1* (formerly known as*Chl1*) is the most extensively studied gene as far as nitrate transporters are concerned. The protein is located on the plasma membrane, and the gene is expressed in the epidermis of the root tips and in the cortex and endodermis in the more mature part of the root (Huang et al. 1999). *AtNRT1.2* is constitutively expressed only in the root epidermis and participates in the constitutive low-affinity system (Huang et al. 1999). Once

taken up by root cells, nitrate must be transported across several cell membranes and distributed in various tissues. The *AtNRT1.4* gene is only expressed in the leaf petiole, and in the mutant the level of nitrate content in the petiole is half that in the wild type (Chiu et al. 2004). *AtNRT1.5*, located on the plasma membrane of root pericycle cells close to the xylem, is involved in long-distance transport of nitrate from the root to the shoot (Lin et al. 2008). The *AtNRT1.6* gene, expressed in the vascular tissue of the silique and funiculus, is thought to deliver nitrate from maternal tissue to the developing embryo (Almagro et al. 2008).

The high-affinity transport system (HATS), acting when the external nitrate concentration is low, relies on the activity of the so-called NRT2 family genes (Williams and Miller 2001). *AtNRT2.1*, in interaction with a nitrate assimilation-related 2 (NAR2) protein (Orsel et al. 2006), has been found as a major compo- nent of the HATS in *Arabidopsis*, as evidenced by the fact that a mutant disrupted for the *AtNRT2.1* gene has lost up to 75 % of the high- affinity NO_3^- uptake activity and showed a lower leaf nitrate content (Filleur et al. 2001). Another *AtNRT2* gene, i.e., *AtNRT2.7*, is expressed in aerial organs and also highly induced in dried seeds. In two allelic *atnrt2.7* mutants, less amount of nitrate is accumulated in the seeds, while seeds from plants overexpressing *AtNRT2.7* accumulate more nitrate (Chopin et al. 2007).

With functional complementation of a yeast mutant defective in methylammonium uptake and recent efforts in sequencing the genome of model species, 6 genes belonging to the same family of ammonium transporters were found in *Arabidopsis* (Gazzarrini et al. 1999); 10 in rice (Sonoda et al. 2003), a species adapted to take up nitrogen in the form of ammonium (Yoshida 1981); and 14 in poplar (Couturier et al. 2007). The physiological and ammonium influx studied carried out on single, double, triple, and quadru- ple mutants showed the involvement of several ammonium transporters (AMT) with varying capacities in *Arabidopsis*, e.g., AMT1.1, AMT1.2, AMT1.3, and AMT1.5 (Yuan et al. 2007). Forward and reverse genetic approaches resulted in the identification of transporters involved in amino acid uptake in the roots (Hirner et al. 2006; Svennerstam et al. 2007). For instance, LHT1 (lysine/histidine transporter), belonging to the ATF (amino acid transport) family, is crucial for the uptake of acidic and neutral amino acids in the roots. The AAP1 (amino acid permease 1) protein was also shown to transport uncharged amino acids, but only when they are supplied at high concentrations in the external medium (Lee et al. 2007). In *Arabidopsis*, the uptake of cationic amino acids such as L-lysine or L-arginine is mediated by AAP5 within the concentration range relevant for field conditions (Svennerstam et al. 2008).

3.2   Nitrogen Assimilation

Nitrogen assimilation requires the reduction of nitrate to ammonium, followed by ammonium assimilation into amino acids. Nitrate reduction takes place in both roots and shoots but spatially separated between the cytoplasm and plastids/chloroplasts where nitrate and nitrite reduction, respectively, takes place. Nitrate reduction into nitrite is catalyzed in the cytosol by the enzyme nitrate reductase (NR) (Meyer and Stitt 2001). After nitrate reduction, nitrite is translocated to the chloroplast where it is reduced to ammonium by the second enzyme of the pathway, the nitrite reductase (NiR). Ammonium, originating from nitrate reduction, is mainly assimilated in the plastid/chloroplast by the so-called GS/GOGAT cycle (Lea and Miflin 1974; Lea and Forde 1994). The glutamine synthetase (GS) fixes ammonium on a glutamate molecule to form glutamine. This glutamine reacts subsequently with 2-oxoglutarate to form two molecules of glutamate, and this step is catalyzed by the glutamine   2-oxoglutarate   aminotransferase (or glutamate synthase, GOGAT). Two classes of nuclear genes code for GS: the *GLN2* and *GLN1* genes. *GLN2*, presented as a single nuclear gene in all the species studied so far, codes for the chloroplastic*GS2*, thought to be involved in the primary assimilation of ammonium coming from nitrate reduction in both C3 and C4 plants.

Conversely, the *GLN1* gene family codes for cytosolic *GS1* isoforms, present in different organs such as roots or stems and thought to be involved in ammonium recycling during particular developmental steps such as leaf senescence and in glutamine synthesis for transport into the phloem sap (Bernard and Habash 2009). Two different forms of glutamate synthase are present in plants: Fd-GOGAT and NADH-GOGAT use ferredoxin and NADH as the electron donors, respectively (Vanoni et al. 2005). Fd-GOGAT is predominantly localized in leaf chloroplasts, whereas NADH-GOGAT is primarily located in plastids of non-photosynthetic tissues, such as roots, etiolated leaf tissues, and companion cells. In addition to the GS/GOGAT cycle, three enzymes probably participate in ammonium assimilation. Cytosolic asparagine synthetase (AS) catalyzes the ATP-dependent transfer of the amido group of glutamine to a molecule of aspartate to generate glutamate and asparagine (Lam et al. 2003). Arabidopsis has three genes which encode AS (*ASN1*, *ASN2*, and *ASN3*). Asparagine has a higher N/C ratio than glutamine and can be used as a nitrogen compound for long-range transport and storage, especially in legumes (Rochat and Boutin 1991). AS in certain situations could compensate for the reduced GS-dependent ammonium assimilatory activity. Apart from asparagine and glutamine, citrulline and   arginine   are   also   synthesized   from carbamoyl phosphate by the action of carbamoyl phosphate synthase (CPSase) within plastids using bicarbonate, ATP, and ammonium or the amide group of glutamine. Another enzyme the mitochondrial NADH-glutamate dehydrogenase can alternatively incorporate ammonium into glutamate in response to high levels of ammonium under stress (Skopelitis et al. 2006). However, the major catalytic activity for GDH in plant cells has been reported to be deamination of glutamate (Masclaux-Daubresse et al. 2006; Purnell   and   Botella   2007).   For   ammonium assimilation, carbon skeletons and especially keto-acids are essential for the synthesis of organic nitrogen as amino acids. Therefore metabolic processes such as photosynthesis, respiration, and photorespiration are essential as these pathways provide the carbon skeletons, ATP, Fdx (ferredoxin), and NADH for ammonium condensation.

4 The Role of MicroRNA in N Starvation

MicroRNAs (miRNAs) are a class of short, single-stranded, noncoding RNAs which negatively regulate gene expression at the posttranscriptional level in plants (Bartel 2007; Filipowicz et al. 2008; He and Hannon 2004). It is now well known that miRNAs are involved in regulating almost all biological and metabolic processes, such as stem cell maintenance and differentiation, organ development, signaling pathways, disease resistance, and response to environmental stress (Bushati and Cohen 2007; Jones-Rhoades et al. 2006; Leung and Sharp 2010). Apart from its role in the regulation of developmental processes, recent functional analyses have demonstrated that several plant miRNAs play vital roles in plant resistance to biotic as well as abiotic stresses including nutrient stresses.

Majority of miRNAs are 21–23 nucleotides long; however, 16–29-nucleotide-long microRNAs have also been observed (Bartel 2007; Zhang et al. 2007). In contrast to animal miRNA genes, which are located anywhere in genomes including coding regions, majority of plant miRNA genes are predominantly located at intergenic regions (Millar and Waterhouse 2005; Zhang et al. 2007). miRNA genes are transcribed by RNA polymerase II (RNA Pol II) (Jones-Rhoades et al. 2006; Lee et al. 2004), and in some cases, they can be transcribed by RNA polymerase III also (Faller and Guo 2008). The initial miRNA transcripts thus produced are called primary miRNAs (pri-miRNAs). RNA Pol II generates capped and polyadenylated pri-miRNAs in both plants and animals (Lee et al. 2004). Based on the current findings of several homologues to animal genes coding for enzymes involved in miRNA biogenesis in plants, it is believed that the mechanism of production of mature miRNA from pri-miRNA is somewhat similar. Once pri-miRNAs are transcribed, it is stabilized by an

RNA-binding protein known as DAWDLE (DDL), to prevent its degradation which also enhances its conversion from pri-miRNAs to miRNA precursors (pre-miRNAs) and further to mature miRNAs. The dicer-like 1 (DCL1) enzyme cuts off the imperfectly folded ends of pri-miRNAs to generate pre-miRNAs with stem-loop hairpin secondary structures (Kurihara et al. 2006; Kurihara and Watanabe 2004). This process is ongoing in a nuclear processing center called the D-body or SmD3/SmB body, which requires the concerted action and physical interactions of several enzymes and/or proteins, including the double-stranded RNA-binding protein hyponastic leaves1 (HYL1), the C2H2-zinc finger protein serrate (SE), the nuclear cap-binding complex (CBC), and DCL1 (Voinnet 2009; Xie 2010; Zhu 2008). Pre-miRNAs are once again subjected for cleavage by DCL1 and HYL1 into an miRNA:miRNA* duplex in the nucleus (Kurihara et al. 2006; Song et al. 2007). A small RNA methyltransferase, called Hua enhancer 1 (HEN1), immediately methylates the $3'$ terminal nucleotides of each strand of the duplexes to prevent their uridylation and subsequent degradation by a class of exonucleases called small RNA degrading nuclease (SDN) (Ramachandran and Chen 2008), and thereby the miRNA:miRNA* duplexes are stabilized (Yu et al. 2005). After it is released from the pre-miRNA, the miRNA:miRNA* duplex is transported from the nucleus into the cytoplasm by HASTY, the nuclear pore, in an ATP-dependent manner (Park et al. 2005). HASTY is the plant homologue of exportin-5 that transports pre-miRNAs from the nucleus into the cytoplasm in animals (Murchison and Hannon 2004). In the cytoplasm, the miRNA:miRNA* duplex is separated; the miRNA strand is incorporated into the RNA-induced silencing complex (RISC) and forms an miR-RISC complex (Chen 2005), in which argonaute 1 (AGO1) cleaves the target mRNA in the middle of the mRNA–miRNA duplex. On the contrary, the miRNAs* are degraded by an unknown mechanism. However, some miRNA* may also function as regular miRNA sequences to target the expression of specific genes (Guo and Lu 2010). Unlike miRNAs in animals, the majority of plant miRNAs cleave their target mRNAs instead of inhibiting protein translation (Zhang et al. 2007).

Recent studies have shown the changes in the expression level of some of the miRNA level in nitrogen limitation/starvation condition in crop plants such as maize, rice, *Arabidopsis*, common bean, etc. (Trevisan et al. 2012; Xu et al. 2011; Zhao et al. 2011, 2012; Jeong et al. 2011; Nischal et al. 2012). The identity and expression profile of different miRNAs involved in such situation reveal several common features among these species. In majority of the cases, it has been observed that some miRNAs are upregulated, while others are downregulated under N starvation condition, commonly to all these species. For instance, miR156 in maize, rice, and *Arabidopsis* and miR160 and miR447 in rice and *Arabidopsis* get upregulated, while miR169, miR397, miR399, and miR408 in maize, common bean, and *Arabidopsis*; miR398 in common bean, rice, and Arabidopsis; and miR827 in maize and *Arabidopsis* are downregulated. However, miR164 shows peculiar expression pattern. It gets downregulated in bean, whereas upregulated in maize (Fischer et al. 2013).

4.1 MicroRNA-Based Engineering: Potential and Prospects

Current trends of improving crop NUE through transgenic approaches focus mainly on the alternation of N uptake, translocation, assimilation, and remobilization processes. These include transgenic expression of genes that encode the autophagy-related factor 8c (Xia et al. 2012), isopentenyl transferase (Rubio-Wilhelmi et al. 2011), the early nodulin factor ENOD93-1 (Bi et al. 2009), sucrose non-fermenting-1-related protein kinase 1 (Wang et al. 2012), alanine aminotransferase (Shrawat et al. 2008), transcription factor Dof1 (Yanagisawa et al. 2004; Kurai et al. 2011), glutamine synthetase (Brauer et al. 2011), NADH-glutamate synthase (Chickova et al. 2001), NADP(H)-glutamate dehydrogenase (Abiko et al. 2010), aspartate aminotransferase (Zhou et al. 2009), asparagine synthetase (Lam et al. 2003), NAD kinase

2 (Takahashi et al. 2009), type I H^+-PPase (Paez-Valencia et al. 2013), and nitrogen transporter homologue PTR9 (Fang et al. 2012). Apart from utilizing these genes, the expression of additional factors involved in carbon–nitrogen balance and signaling may also serve the purpose of improving NUE (McAllister et al. 2012). However, transgene-targeted metabolic engineering approaches pose great limitation by complexity and interconnectedness of many related metabolic processes. Transgene expression probably has less effect than it is expected on plant metabolism due to various feedback mechanisms, such as gene silencing that maintains homeostasis within a plant cell (Brosnan and Voinnet 2011; Kasai et al. 2012). Despite continuous efforts, the regulatory aspects of genes involved in N metabolism and the N starvation response are not fully characterized so far. Here in this section, we have attempted to discuss recent advances in understanding the regulatory role of miRNAs in plant N metabolism with a view toward finding more effective means to engineer nitrogen use efficiency in crops. The different miRNAs which get differentially expressed under nitrogen starvation condition in major crops have been given in Table 1.

4.1.1 miRNAs Downregulated Under N Starvation

4.1.1.1 miR397, miR398, miR408, and miR857

miR397 targets members of the laccase family (*LAC2, LAC4,* and *LAC17*), which are copper-containing enzymes that catalyze the oxidation of phenolic compounds and the concomitant reduction of oxygen to water (Jeon et al. 2012). Laccases are involved in a diverse range of functions related to defense and cell wall lignification (Jeon et al. 2012; Mayer and Staples 2002). miR397 and miR398 are involved in both N and Cu homeostasis; however, in the former case, they get downregulated by N starvation, while in the later conditions, they get upregulated in response to Cu deficiency (Abdel-Ghany and Pilon 2008). miR408 and miR857 also target members of the laccase family in which condition they get downregulated

Table 1 Expression pattern of different miRNAs under nitrogen starvation condition

Expression pattern	Maize (Ref. Trevisan et al. 2012; Xu et al. 2011; Zhao et al. 2012)	Rice (Ref. Cai et al. 2012; Jeong et al. 2011)	Common bean (Ref. Valdés-López et al. 2010)	Arabidopsis (Ref. Zhao et al. 2011; Pant et al. 2009; Gifford et al. 2008; Liang et al. 2012)	Root/shoot
Upregulated	miR164a–h miR167e, f,j miR171h,k miR172a–f miR394a, b miR827 miRC1:1–3 miRC7:1–6 miRC19 miRC11:1–6 miRc37		miR396	–	Shoot
	miR162 miRC5	miR156e,g miR157d miR399i miR447	–	miR156e,g miR157d miR160 miR169d–g miR447c miR780 miR826 miR842 miR846	Both (Liang et al. 2012)
	miR156g, miR159c,d miR160a–i,f, miR164e–h,l, miR167a–d, miR168a–c miR169f–h, miR171g, miR319a–e, miR393b, miR395a,b,d–j,n, p miR399d,j miRC12, miRC71:p,q		miR1508 miR1509 miR1511		Root
Downregulated	miR164f miR169d,e,f–h, m,n miR397a,b miR398a,b,c miR827 miRC23		miR164 miR168 miR390 miR393 miR398 miR399 miR1509 miR1510 miR1511 miR1513 miR1515 miR1516 miR1526 miR2119		Shoot
	miR164h miR169 miR397b miR399d,j miR408 miR528a,b	miR398a miR530	miR319 miR397 miR408 miR156 miR157 miR159 miR169	miR167d miR169a–c,h–n miR171 miR395 miR397 miR398 miR399 miR408 miR827 miR857	Both (Liang et al. 2012)
	miR166j,k,n miR167a–d, n,o–t miR169b,j–l, p miR169i,j,k miR395a, b,d–l,n–p miR396e,f miR399a,c,h,g miR408b miR827 miRC16 miRC50	miR3979	miR1514a		Root

under N starvation in *Arabidopsis* (Liang et al. 2012). These miRNAs target plantacyanin, *LAC3*, *LAC12*, *LAC13* (miR408), and *LAC7* (miR857) mRNAs (Abdel-Ghany and Pilon 2008). Plantacyanin is a Cu-containing regulatory factor involved in plant reproduction, and overexpression of this protein in *Arabidopsis* caused defects in the reproductive organ development (Dong et al. 2005). The expression of plantacyanin under N stress would therefore inhibit reproduction. As far as the mechanism of regulation of laccase through miR397, miR408, and miR857 is concerned, it has been hypothesized that they are involved in maintaining C:N homeostasis. Under N starvation condition, it is presumed that excess fixed C gets incorporated into lignin through increased laccase activity. For instance, it has been observed that tobacco plants grown in N-limiting conditions are highly lignified (Fritz et al. 2006). Further another report suggests that sugarcane laccase *SofLAC* which can complement *Arabidopsis LAC17* mutants is coordinately expressed with several phenylpropanoid biosynthesis genes (Cesarino et al. 2013), suggesting a potential role for laccases in phenylpropanoid and perhaps anthocyanin production too, whereas miR398 targets Cu/Zn superoxide dismutases (*CSD1* and *CSD2*) mRNA which imparts oxidative stress tolerance (Jagadeeswaran et al. 2009; Sunkar et al. 2006), as well as a cytochrome C oxidase subunit (*COX5b-1*) and copper chaperone (*CCS1*) (Zhu et al. 2011; Bouché 2010). Overexpression of *CSD2* in *Arabidopsis* confers tolerance to high light and oxidative stress (Sunkar et al. 2006), which suggests that the downregulation of miR398 under N starvation may provide tolerance to ROS generated by N stress. miR398 downregulation may therefore protect photosynthetic machinery for long-term N remobilization.

4.1.1.2 miR399

miR399 has been reported to target the *PHO2* gene that encodes a ubiquitin-conjugating E2 enzyme *UBC24* (Fujii et al. 2005) which gets upregulated in phosphate-limiting condition. However, N starvation conditions cause downregulated expression of miR399. The ubiquitin-conjugating activity of *UBC24* (*PHO2*) is responsible for the degradation of PHO1, a membrane-associated putative Pi transporter (Liu et al. 2012). The downregulated expression of miR399 presumably permits higher-level expression of *UBC24* which may enhance proteasome-mediated N remobilization of other unidentified targets such as Rubisco. Alternatively, decreased phosphate transport could represent an additional mechanism that conserves plant resources in the form of high-energy phosphate compounds. The cross talk between N and P metabolism is not well characterized, but recent data indicates that the

N:P link is regulated by *PHO2*, *NLA*, miR399, and miR827 (Kant et al. 2011).

4.1.1.3 miR827

Apart from its role in phosphate homeostasis, miR827 also targets the nitrogen limitation adaptation gene *NLA* in *Arabidopsis*, which encodes an E3 ubiquitin ligase, though this is not a conserved target in rice (Kant et al. 2011; Lin et al. 2010). The disruption of *NLA* in *Arabidopsis* caused plants to undergo early senescence under limiting N conditions, impaired anthocyanin production, and reduced photosynthetic capacity (Peng et al. 2007, 2008). Both N remobilization and anthocyanin biosynthesis are key features of N starvation tolerance. Thus, the downregulation of miR827 under N stress may be crucial in both increasing anthocyanin levels and N remobilization in conjunction with miR399 through the ubiquitin-mediated 26S proteasome degradation pathway (Kant et al. 2011).

4.1.2 miRNAs Upregulated Under N Starvation

4.1.2.1 miR156/157

The squamosa promoter binding-protein (SBP) or SBP-like (SPL) gene family (Gandikota et al. 2007) functions to regulate flowering, vegetative phase change, fertility, and leaf formation (Wu and Poethig 2006; Wang et al. 2008, 2009; Wu et al. 2009; Xing et al. 2010) which are found to be targeted by miR156/157. The effect of enhanced expression of miR156 resulted into prolonged juvenile phase, stunted growth, and delayed flowering in transgenic plants of switchgrass, rice, *Arabidopsis*, tomato, torentia, and maize (Wu and Poethig 2006; Xie et al. 2006; Chuck et al. 2007; Zhang et al. 2011; Fu et al. 2012; Shikata et al. 2012). In case of transgenic tomato, the enhanced expression of miR156 resulted into reduction in yield (Xie et al. 2006; Zhang et al. 2011). It has further been observed that the expression of miR156 is linked to increased anthocyanin biosynthesis (Gou et al. 2011), which is a key feature of the plant N starvation response. Relatively modest upregulation of miR156 expression has been speculated to have a positive effect on NUE in grain crops by promoting anthocyanin production. It would be of interest to examine whether mild ectopic expression of miR156 would have a positive effect on plant growth under N starvation. The expression patterns of miR156 and miR172 are interrelated in which the former downregulates the expression of the latter via the expression of transcription factors SPL9 and SPL10. Consequently members of the *APETALA2-like* (*AP2-like*) family of transcription factors, such as *TOE1* and *TOE2*, get regulated (Wu et al. 2009). *TOE1* reportedly functions as a negative regulator of flowering in *Arabidopsis* (Glazińska et al. 2009). Therefore, as miR172 levels decrease under N starvation, the resulting changes in AP2-like factor expression would further delay plant development.

4.1.2.2 miR160

Transgenic overexpression of miR160 in *Arabidopsis* plants resulted in reduced sensitivity to abscisic acid (ABA) as well as abnormal root morphology, including increased production of adventitious roots and a lack of gravitropic response in root tips (Liu et al. 2007; Gutierrez et al. 2009; Wang et al. 2005). However, the transgenic expression of miR160-resistant form of ARF10 gives rise to abnormal leaves, siliques, and flowers and demonstrated impaired seedling growth (Liu et al. 2007). Furthermore, such plants also had increased levels of mRNAs that encode seed storage protein, seed maturation, and ABA-responsive genes (Liu et al. 2007). Likewise, *Arabidopsis* plants that demonstrated increased levels of ARF17 mRNA due to impaired miR160 regulation had severe developmental defects, including the production of fewer lateral and adventitious roots (Mallory et al. 2005; Du et al. 2012). These findings suggest that under N starvation conditions, miR160 upregulation may have a role in promoting lateral root growth in order to access additional N. As with the expression of miR156, strong overexpression of miR160 in *Arabidopsis* plants under the control of the *CaMV 35S* promoter produces undesirable traits, while a modest

upregulation of miR160 levels may promote root exploration for other nutrients in addition to nitrogen. Interestingly, the expression of miR160 gets suppressed by high sucrose concentrations in *Arabidopsis* (Ren and Tang 2012), suggesting a key regulatory role for miR160 in regulating the plant C:N balance. miR160 that gets upregulated in N starvation has also been postulated to target R2R3-MYB family of transcription factors in maize (Du et al. 2012).

4.1.2.3 miR164/miR167

miR164 controls lateral root growth in *Arabidopsis* by regulating the expression of NAM/ATAF/CUC (NAC) transcription factor 1 (Guo et al. 2005). Plants with downregulated expression of miR164 (and thus upregulated NAC1 expression) produced more lateral roots compared to wild-type plants (Guo et al. 2005). Whereas miR167, which targets ARF6 and ARF8, downregulates these positive regulators of adventitious root growth (Mallory et al. 2005), miR167 also targets the geneIAR3, which encodes a factor that hydrolyzes the inactive auxin derivative indole-3-acetic acid-alanine into bioactive auxin (Kinoshita et al. 2012). Further it has been observed that there exists a complex spatial regulation of miR160 and miR164 which may be responsible for specific root architecture. Interestingly, miR167 precursors get either up- or downregulated under N stress in a tissue-specific manner. For example, miR167a is downregulated in *Arabidopsis* under N starvation (Liang et al. 2012).

4.1.2.4 miR447

miR447 has been found to target mRNAs that encode a 2-phosphoglycerate kinase (2PGK) (Allen et al. 2005). This enzyme catalyzes the nucleotide-dependent phosphorylation of 2-phosphoglycerate to form 2,3-bisphosphoglycerate. N starvation condition would therefore downregulate this gene (*2PGK*) and consequently alters the production of pyruvate which will affect downstream energy metabolism, including the production of carbon skeletons for amino acid biosynthesis.

4.1.2.5 miR826

miR826 reportedly targets *AOP2* which encodes 2-oxoglutarate-dependent dioxygenase involved in glucosinolate biosynthesis (Liang et al. 2012). Under N starvation condition, the expression of miR826 is dramatically induced in roots and shoots (Liang et al. 2012). Glucosinolates are a group of plant secondary metabolites produced mainly in *Brassica*, and these compounds are rich in nitrogen and sulfur. They are involved in the production of the nitrogenous glucosinolate defense compounds and are highly expressed in photosynthetic tissue (Neal et al. 2010). Therefore, the suppression of *AOP2* by miR826 could decrease the production of glucosinolates, which results in the decrease of N demand. The downregulation of secondary biosynthetic pathways such as glucosinolate production may represent yet another means of tolerating N starvation.

4.1.3 miRNAs Up- and Downregulated Under N Starvation

4.1.3.1 miR169

miR169 targets nuclear transcription factor Y subunit alpha (NFYA) gene family like *NFYA1, NFYA2, NFYA3, NFYA8, NFYA9*, and *NFYA10*, some of which encode nitrate transporter in *Arabidopsis NRT1.1* and *NRT2.1*. In N starvation condition, the expression of miR169 gets downregulated which in turn enhances the expression of its target gene (Zhao et al. 2011). Transgenic overexpression of miR169 caused *Arabidopsis* plants to accumulate less N, and plants showed more severe symptoms when grown under N-limiting conditions (Zhao et al. 2011) by reducing the production of nitrate transporters. Similar expression pattern of miR169 and its corresponding target gene have also been observed in maize under N starvation (Xu et al. 2011; Zhao et al. 2012). However, some members of the miR169 family get upregulated during N starvation, while others show little differential expression in *Arabidopsis* (Liang et al. 2012). For example, miR169a, b, and c are downregulated under N, P, and S starvation, whereas miR169d–g are significantly upregulated only by N starvation and miR169h–n show little change either way (Liang et al. 2012), whereas overexpression of miR169c in tomato conferred drought resistance in plants. These observations suggest that endogenous promoter-specific regulation might be playing a significant role in regulating these miRNAs (Fischer et al. 2013).

5 Future Prospects

As far as biochemical function is concerned, miRNAs seem to act as transcription factors as it also controls the expression of its target genes. Additionally, transcription and processing of miRNAs are presumably less energy intensive compared to transcriptional and translational activation (Zhang et al. 2012). Therefore, microRNA-based engineering for modification of miRNA expression is an attractive alternative to overexpression of mRNAs for engineering NUE in plants, particularly in light of recent studies on miRNA expression patterns in response to N starvation (Fischer et al. 2013). However, the expression pattern of different microRNAs associated with a particular phenotype under certain circumstances differs from organism to organism. For instance, in *Arabidopsis*, N starvation causes increase in the expression of miR160 and miR171, while miR167 is repressed (Liang et al. 2012). In maize roots, expressions of miR167a–d, g–i, and miR164e–f are upregulated (Zhao et al. 2012). miR160a–e, g–i, and m expressions are upregulated under transient N starvation in maize roots, while miR167 is downregulated under chronic N stress (Xu et al. 2011). Considering this fact, miRNA-based engineering under such circumstances may limit its application in NUE improvement as these miRNAs exhibit complex temporal regulation of root development in species-specific manner. Furthermore, high-level expression of miRNAs under the control of the *CaMV 35S* and *Ubi* promoters can result in undesirable traits such as stunted growth, reduced yield, delayed flowering, and infertility. NUE may be enhanced but at the cost of adverse plant development.

It has been demonstrated that NUE improvement could be achieved by involving the expression of miRNAs that are upregulated by N starvation along with overexpression of factors that are involved in the N deprivation response (Fischer et al. 2013). For instance, the simultaneous overexpression of miR156 and NLA (the target of miR827) could provide increased NUE (Fu et al. 2012; Gou et al. 2011). Another example would be to upregulate miR156 in rice that already expresses the maize transcription factor Dof1 (Yanagisawa et al. 2004; Kurai et al. 2011), which may increase biomass while improving NUE.

As we discussed above that moderate overexpression of miR156 in switchgrass improved N uptake and increased biomass, whereas high expression levels resulted in severely stunted plant growth (Fu et al. 2012). Experiments in which miRNAs are significantly overexpressed under the control of the *CaMV 35S* or *Ubi* promoters produce plants with severe undesirable developmental phenotypes. Distinct differences in miRNA expression levels under N starvation have been observed based on deep sequencing results, with miR156 clearly being the most abundantly expressed species (Liang et al. 2012). Constitutive and inducible tissue-specific expression, as well as promoter strength, must therefore be carefully evaluated, taking into account the variability of endogenous plant miRNA responses. Thus, in order to manipulate nitrogen use efficiency, expression levels of various miRNAs have to be carefully selected and very finely tuned with the use of specific promoters.

References

Abdel-Ghany S, Pilon M (2008) MicroRNA-mediated systemic down-regulation of copper protein expression in response to low copper availability in Arabidopsis. J Biol Chem 283:15932–15945

Abiko T, Wakayama M, Kawakami A, Obara M, Kisaka H, Miwa T, Aoki N, Ohsugi R (2010) Changes in nitrogen assimilation, metabolism, and growth in transgenic rice plants expressing a fungal NADP(H)-dependent glutamate dehydrogenase (gdhA). Planta 232:299–311

Abrol YP, Chatterjee SR, Kumar PA, Jain V (1999) Improvement in nitrogenous fertilizer utilization – physiological and molecular approaches. Curr Sci 76:1357–1364

Allen E, Xie Z, Gustafson A, Carrington J (2005) microRNA-directed phasing during trans-acting siRNA biogenesis in plants. Cell 121:207–221

Almagro A, Lin S, Tsay Y (2008) Characterization of the Arabidopsis nitrate transporter NRT1.6 reveals a role of nitrate in early embryo development. Plant Cell 20:3289–3299

Amiour N, Imbaud S, Clément G, Agier N, Zivy M, Valot B, Balliau T, Armengaud P, Quilleré I, Cañas R, Tercet-Laforgue T, Hirel B (2012) The use of metabolomics integrated with transcriptomic and proteomic studies identifying key steps involved in the control of nitrogen metabolism in crops such as maize. J Exp Bot 63:5017–5033

Anonymous (2011) U.N. Department of Social and Economic Affairs Population Division, World population prospects: the 2010 revision, highlights and advance tables, 2011, Working paper no. ESA/P/WP.220

Bartel DP (2007) MicroRNAs: genomics, biogenesis, mechanism, and function (Reprinted from Cell, vol 116, pp 281–297, 2004). Cell 131:11–29

Bernard SM, Habash DZ (2009) The importance of cytosolic glutamine synthetase in nitrogen assimilation and recycling. New Phytol 182:608–620

Bi YM, Kant S, Clarke J, Gidda S, Ming F, Xu J, Rochon A, Shelp BJ, Hao L, Zhao R, Mullen RT, Zhu T, Rothstein SJ (2009) Increased nitrogen-use efficiency in transgenic rice plants over-expressing a nitrogen-responsive early nodulin gene identified from rice expression profiling. Plant Cell Environ 32:1749–1760

Bick JA, Neelam A, Hall JL, Williams LE (1998) Amino acid carriers of *Ricinus communis* expressed during seedling development: Molecular cloning and expression analysis of two putative amino acid transporters, RcAAP1 and RcAAP2. Plant Mol Biol 36:377–385

Bouché N (2010) New insights into miR398 functions in Arabidopsis. Plant Signal Behav 5:684–686

Brauer E, Rochon A, Bi YM, Bozzo G, Rothstein S (2011) Reappraisal of nitrogen use efficiency in rice overexpressing glutamine synthetase 1. Physiol Plant 141:361–372

Brosnan C, Voinnet O (2011) Cell-to-cell and long-distance siRNA movement in plants: mechanisms and biological implications. Curr Opin Plant Biol 14:580–587

Bushati N, Cohen SM (2007) MicroRNA functions. Annu Rev Cell Dev Biol 23:175–205

Cai H, Lu Y, Xie W, Zhu T, Lian X (2012) Transcriptome response to nitrogen starvation in rice. J Biosci 37:731–747

Cesarino I, Araújo P, Sampaio Mayer JL, Vicentini R, Berthet S, Demedts B, Vanholme B, Boerjan W,

Mazzafera P (2013) Expression of SofLAC, a new laccase in sugarcane, restores lignin content but not S:G ratio of Arabidopsis lac17 mutant. J Exp Bot 64:1769–1781

Chen XM (2005) microRNA biogenesis and function in plants. FEBS Lett 579:5923–5931

Chickova S, Arellano J, Vance C, Hernández G (2001) Transgenic tobacco plants that over express alfalfa NADH-glutamate synthase have higher carbon and nitrogen content. J Exp Bot 52:2079–2087

Chiu CC, Lin CS, Hsia AP, Su RC, Lin HL, Tsay YF (2004) Mutation of a nitrate transporter, AtNRT1:4, results in a reduced petiole nitrate content and altered leaf development. Plant Cell Physiol 45:1139–1148

Chopin F, Orsel M, Dorbe MF et al (2007) The Arabidopsis ATNRT2.7 nitrate transporter controls nitrate content in seeds. Plant Cell 19:1590–1602

Chuck G, Cigan A, Saeteurn K, Hake S (2007) The heterochronic maize mutant *Corngrass1* results from the over expression of a tandem microRNA. Nat Genet 39:544–549

Couturier J, Montanini B, Martin F, Brun A, Blaudez D, Chalot M (2007) The expanded family of ammonium transporters in the perennial poplar plant. New Phytol 174:137–150

Dong J, Kim S, Lord E (2005) Plantacyanin plays a role in reproduction in Arabidopsis. Plant Physiol 138:778–789

Du H, Feng BR, Yang S, Huang YB, Tang YX (2012) The R2R3-MYB transcription factor gene family in maize. PLoS ONE 7:e37463

Faller M, Guo F (2008) MicroRNA biogenesis: there's more than one way to skin a cat. Biochim Biophys Acta Gene Regul Mech 1779:663–667

Fang Z, Xia K, Yang X, Grotemeyer MS, Meier S, Rentsch D, Xu X, Zhang M (2012) Altered expression of the PTR/NRT1 homologue OsPTR9 affects nitrogen utilization efficiency, growth and grain yield in rice. Plant Biotechnol J 11:446–458

Filipowicz W, Bhattacharyya SN, Sonenberg N (2008) Mechanisms of post-transcriptional regulation by microRNAs: are the answers in sight? Nat Rev Genet 9:102–114

Filleur S, Dorbe M, Cerezo M et al (2001) An arabidopsis T-DNA mutant affected in Nrt2 genes is impaired in nitrate uptake. FEBS Lett 489:220–224

Firbank LG (2005) Striking a new balance between agricultural production and biodiversity. Ann Appl Biol 146:163–175

Fischer JJ, Beatty PH, Good AG, Muench DG (2013) Manipulation of microRNA expression to improve nitrogen use efficiency. Plant Sci 210:70–81

Fritz C, Palacios-Rojas N, Feil R, Stitt M (2006) Regulation of secondary metabolism by the carbon–nitrogen status in tobacco: nitrate inhibits large sectors of phenylpropanoid metabolism. Plant J 46:533–548

Fu C, Sunkar R, Zhou C, Shen H, Zhang JY, Matts J, Wolf J, Mann DG, Stewart CN Jr, Tang Y, Wang ZY (2012) Overexpression of miR156 in switchgrass (*Panicum virgatum* L.) results in various morphological alterations and leads to improved biomass production. Plant Biotechnol J 10:443–452

Fujii H, Chiou TJ, Lin SI, Aung K, Zhu JK (2005) A miRNA involved in phosphate starvation response in Arabidopsis. Curr Biol 15:2038–2043

Gandikota M, Birkenbihl RP, Höhmann S, Cardon GH, Saedler H, Huijser P (2007) The miRNA156/157 recognition element in the 3 UTF of the Arabidopsis box gene SPL3 prevents early flowering by translational inhibition in seedlings. Plant J 49:683–693

Garnett T, Conn V, Kaiser BN (2009) Root based approaches to improving nitrogen use efficiency in plants. Plant Cell Environ 32:1272–1283

Gazzarrini S, Lejay L, Gojon A, Ninnemann O, Frommer WB, von Wiren N (1999) Three functional transporters for constitutive, diurnally regulated, and starvation induced uptake of ammonium into Arabidopsis roots. Plant Cell 11:937–947

Gifford M, Dean A, Gutierrez R, Coruzzi G, Birnbaum K (2008) Cell-specific nitrogen responses mediate developmental plasticity. Proc Natl Acad Sci U S A 105:803–808

Glass ADM (2003) Nitrogen use efficiency of crop plants: physiological constraints upon nitrogen absorption. Crit Rev Plant Sci 22:453–470

Glazińska P, Zienkiewicz A, Wojciechowski W, Kopcewicz J (2009) The putative miR172 target gene In APETALA2-like is involved in the photoperiodic flower induction of Ipomoea nil. J Plant Physiol 166:1801–1813

Good AG, Shrawat AK, Muench DG (2004) Can less yield more? Is reducing nutrient input into the environment compatible with maintaining crop production? Trends Plant Sci 9:597–605

Gou J, Felippes F, Liu C, Weigel D, Wang J (2011) Negative regulation of anthocyanin biosynthesis in Arabidopsis by a miR156-targeted SPL transcription factor. Plant Cell 23:1512–1522

Guo L, Lu ZH (2010) The fate of miRNA* strand through evolutionary analysis: implication for degradation as merely carrier strand or potential regulatory molecule? PLoS One 5:e11387

Guo H, Xie Q, Fei J, Chua N (2005) MicroRNA directs mRNA cleavage of the transcription factor NAC1 to down regulate auxin signals for Arabidopsis lateral root development. Plant Cell 17:1376–1386

Gutierrez L, Bussell JD, Pacurar DI, Schwambach J, Pacurar M, Bellini C (2009) Phenotypic plasticity of adventitious rooting in Arabidopsis is controlled by complex regulation of AUXIN RESPONSE FACTOR transcripts and microRNA abundance. Plant Cell 21:3119–3132

He L, Hannon GJ (2004) MicroRNAs: small RNAs with a big role in gene regulation. Nat Rev Genet 5:522–531

Hill MJ, Hawksworth G, Tatterstall G (1973) Bacteria, nitrosamines and cancer of the stomach. Br J Cancer 28:562–567

Hirner A, Ladwig F, Stransky H et al (2006) Arabidopsis LHT1 is a high affinity transporter for cellular amino acid uptake in both root epidermis and leaf mesophyll. Plant Cell 18:1931–1946

Ho C, Lin S, Hu H, Tsay Y (2009) CHL1 functions as a nitrate sensor in plants. Cell 18:1184–1194

Huang NC, Liu KH, Lo HJ, Tsay YF (1999) Cloning and functional characterization of an Arabidopsis nitrate transporter gene that encodes a constitutive component of low-affinity uptake. Plant Cell 11:1381–1392

Jagadeeswaran G, Saini A, Sunkar R (2009) Biotic and abiotic stress down-regulate miR398 expression in Arabidopsis. Planta 229:1009–1014

Jeon J, Baldrian P, Murugesan K, Chang Y (2012) Laccase-catalysed oxidations of naturally occurring phenols: from in vivo biosynthetic pathways to green synthetic applications. Microb Biotechnol 5:318–332

Jeong DH, Park S, Zhai J, Gurazada SG, De Paoli E, Meyers BC, Green PJ (2011) Massive analysis of rice small RNAs: mechanistic implications of regulated microRNAs and variants for differential target RNA cleavage. Plant Cell 23:4185–4207

Jones-Rhoades MW, Bartel DP, Bartel B (2006) MicroRNAs and their regulatory roles in plants. Annu Rev Plant Biol 57:19–53

Kant S, Peng M, Rothstein S (2011) Genetic regulation by NLA and microRNA827 for maintaining nitrate-dependent phosphate homeostasis in Arabidopsis. PLoS Genet 7:e1002021

Kasai M, Koseki M, Goto K, Masuta C, Ishii S, Hellens RP, Taneda A, Kanazawa A (2012) Coincident sequence-specific RNA degradation of linked transgenes in the plant genome. Plant Mol Biol 78:259–273

Kinoshita N, Wang H, Kasahara H, Liu J, MacPherson C, Machida Y, Kamiya Y, Hannah M, Chua NH (2012) IAA-Ala Resistant3, an evolutionarily conserved target of miR167, mediates Arabidopsis root architecture changes during high osmotic stress. Plant Cell 24:3590–3602

Kraiser T, Gras DE, Gutiérrez AG, González B, Gutiérrez RA (2011) A holistic view of nitrogen acquisition in plants. J Exp Bot 62:1455–1466

Kurai T, Wakayama M, Abiko T, Yanagisawa S, Aoki N, Ohsugi R (2011) Introduction of the ZmDof1 gene into rice enhances carbon and nitrogen assimilation under low-nitrogen conditions. Plant Biotechnol J 9:826–837

Kurihara Y, Watanabe Y (2004) Arabidopsis micro-RNA biogenesis through Dicer-like 1 protein functions. Proc Natl Acad Sci U S A 101:12753–12758

Kurihara Y, Takashi Y, Watanabe Y (2006) The interaction between DCL1 and HYL1 is important for efficient and precise processing of pri-miRNA in plant microRNA biogenesis. RNA 12:206–212

Lam M, Wong P, Chan HK, Yam KM, Chen L, Chow CM, Coruzzi GM (2003) Overexpression of the ASN1 gene enhances nitrogen status in seeds of Arabidopsis. Plant Physiol 132:926–935

Lea PJ, Forde BG (1994) The use of mutants and transgenic plants to study amino acid metabolism. Plant Cell Environ 17:541–556

Lea P, Miflin B (1974) Alternative route for nitrogen assimilation in higher plants. Nature 251:614–616

Lee Y, Kim M, Han J, Yeom KH, Lee S, Baek SH, Kim VN (2004) MicroRNA genes are transcribed by RNA polymerase II. EMBO J 23:4051–4060

Lee YH, Foster J, Chen J, Voll LM, Weber AP, Tegeder M (2007) AAP1 transports uncharged amino acids into roots of Arabidopsis. Plant J 50:305–319

Leung AKL, Sharp PA (2010) MicroRNA functions in stress responses. Mol Cell 40:205–215

Liang G, He H, Yu D (2012) Identification of nitrogen starvation-responsive microRNAs in Arabidopsis thaliana. PLoS ONE 7:e48951

Liao C, Peng Y, Ma W, Liu R, Li C, Li X (2012) Proteomic analysis revealed nitrogen-mediated metabolic, developmental, and hormonal regulation of maize (Zea mays L.) ear growth. J Exp Bot 63:5275–5288

Lin SH, Kuo HF, Canivenc G, Lin CS, Lepetit M, Hsu PK, Tillard P, Lin HL, Wang YY, Tsai CB (2008) Mutation of the ArabidopsisNRT1.5 nitrate transporter causes defective root-to-shoot nitrate transport. Plant Cell 20:2514–2528

Lin SI, Santi C, Jobet E, Lacut E, El Kholti N, Karlowski WM, Verdeil JL, Breitler JC, Périn C, Ko SS, Guiderdoni E, Chiou TJ, Echeverria M (2010) Complex regulation of two target genes encoding SPX-MFS proteins by rice miR827 in response to phosphate starvation. Plant Cell Physiol 51:2119–2131

Liu KH, Huang CY, Tsay YF (1999) CHL1 is a dual-affinity nitrate transporter of Arabidopsis involved in multiple phases of nitrate uptake. Plant Cell 11:865–874

Liu PP, Montgomery TA, Fahlgren N, Kasschau KD, Nonogaki H, Carrington JC (2007) Repression of AUXIN RESPONSE FACTOR10 by microRNA160 is critical for seed germination and post-germination stages. Plant J 52:133–146

Liu TY, Huang TK, Tseng CY, Lai YS, Lin SI, Lin WY, Chen JW, Chiou TJ (2012) PHO2-dependent degradation of PHO1 modulates phosphate homeostasis in Arabidopsis. Plant Cell 24:2168–2183

Maathuis F (2009) Physiological functions of mineral nutrients. Curr Opin Plant Biol 12:250–258

Mallory AC, Bartel DP, Bartel B (2005) MicroRNA-directed regulation of Arabidopsis AUXIN RESPONSE FACTOR17 is essential for proper development and modulates expression of early auxin response genes. Plant Cell 17:1360–1375

Masclaux-Daubresse C, Reisdorf-Cren M, Pageau K et al (2006) Glutamine synthetase-glutamate synthase pathway and glutamate dehydrogenase play distinct roles in the sink-source nitrogen cycle in tobacco. Plant Physiol 140:444–456

Mayer A, Staples R (2002) Laccase: new functions for an old enzyme. Phytochemistry 60:551–565

McAllister C, Beatty P, Good A (2012) Engineering nitrogen use efficient crop plants: the current status. Plant Biotechnol J 10:1011–1025

Meyer C, Stitt M (2001) Nitrate reductase and signalling. In: Lea PJ, Morot-Gaudry J-F (eds) Plant nitrogen. Springer, New York, pp 37–59

Millar AA, Waterhouse PM (2005) Plant and animal microRNAs: similarities and differences. Funct Integr Genomics 5:129–135

Montzka S, Dlugokencky E, Butler J (2011) Non-CO_2 greenhouse gases and climate change. Nature 476:43–50

Murchison EP, Hannon GJ (2004) miRNAs on the move: miRNA biogenesis and the RNAi machinery. Curr Opin Cell Biol 16:223–229

NAAS (2005) Policy option for efficient nitrogen use. Policy Paper No. 33, National Academy of Agricultural Sciences, New Delhi

Neal C, Fredericks D, Griffiths C, Neale A (2010) The characterisation of AOP2: a gene associated with the biosynthesis of aliphatic glucosinolates in Arabidopsis thaliana. BMC Plant Biol 10:170

Neelam A, Marvier AC, Hall JL, Williams LE (1999) Functional characterization and expression analysis of the amino acid permease RcAAP3 from castor bean. Plant Physiol 120:1049–1056

Nischal L, Mohsin M, Khan I, Kardam H, Wadhwa A, Abrol YP, Iqbal M, Ahmad A (2012) Identification and comparative analysis of microRNAs associated with low-N tolerance in rice genotypes. PLoS ONE 7:e50261

Orsel M, Chopin F, Leleu O et al (2006) Characterization of a two-component high-affinity nitrate uptake system in Arabidopsis. Physiology and protein–protein interaction. Plant Physiol 142:1304–1317

Paez-Valencia J, Sanchez-Lares J, Marsh E, Dorneles LT, Santos MP, Sanchez D, Winter A, Murphy S, Cox J, Trzaska M, Metler J, Kozic A, Facanha AR, Schachtman D, Sanchez CA, Gaxiola RA (2013) Enhanced proton translocating pyrophosphatase activity improves nitrogen use efficiency in romaine lettuce. Plant Physiol 161:1557–1569

Pant B et al (2009) Identification of nutrient-responsive Arabidopsis and rapeseed microRNAs by comprehensive real-time polymerase chain reaction profiling and small RNA sequencing. Plant Physiol 150:1541–1555

Park MY, Wu G, Gonzalez-Sulser A, Vaucheret H, Poethig RS (2005) Nuclear processing and export of microRNAs in arabidopsis. Proc Natl Acad Sci U S A 102:3691–3696

Peng M, Hannam C, Gu H, Bi YM, Rothstein S (2007) A mutation in NLA, which encodes a RING-type ubiquitin ligase, disrupts the adaptability of Arabidopsis to nitrogen limitation. Plant J 50:320–337

Peng M, Hudson D, Schofield A, Tsao R, Yang R, Gu H, Bi YM, Rothstein SJ (2008) Adaptation of Arabidopsis to nitrogen limitation involves induction of anthocyanin synthesis which is controlled by the NLA gene. J Exp Bot 59:2933–2944

Purnell M, Botella J (2007) Tobacco isozyme 1 of NAD (H)-dependent glutamate dehydrogenase catabolizes glutamate in vivo. Plant Physiol 143:530–539

Ramachandran V, Chen X (2008) Degradation of microRNAs by a family of exoribonucleases in Arabidopsis. Science 321:1490–1492

Raun WR, Johnson GV (1999) Improving nitrogen use efficiency for cereal production. Agron J 91:357–363

Ren L, Tang G (2012) Identification of sucrose-responsive microRNAs reveals sucrose-regulated copper accumulations in and SPL7-dependent and independent manner in Arabidopsis thaliana. Plant Sci 187:59–68

Rochat C, Boutin JP (1991) Metabolism of phloem-borne amino acids in maternal tissues of fruit of nodulated or nitrate-fed pea plants (Pisum sativum L.). J Exp Bot 42:207–214

Rubio-Wilhelmi MM, Sanchez-Rodriguez E, Rosales MA, Blasco B, Rios JJ, Romero L, Blumwald E, Ruiz JM (2011) Cytokinin-dependent improvement in transgenic PSARK::IPT tobacco under nitrogen deficiency. J Agric Food Chem 59:10491–10495

Schwacke R, Grallath S, Breitkreuz KE, Stransky E, Stransky H, Frommer WB, Rentsch D (1999) LeProT1, a transporter for proline, glycine betaine, and γ-amino butyric acid in tomato pollen. Plant Cell 11:377–391

Shikata M, Yamaguchi H, Sasaki K, Ohtsubo N (2012) Overexpression of Arabidopsis miR157b induces bushy architecture and delayed phase transition in Torenia fournieri. Planta 236:1027–1035

Shrawat A, Carroll R, DePauw M, Taylor G, Good A (2008) Genetic engineering of improved nitrogen use efficiency in rice by the tissue-specific expression of alanine aminotransferase. Plant Biotechnol J 6:722–732

Skopelitis D, Paranychianakis N, Paschalidis K, Pliakonis E, Delis I, Yakoumakis D, Kouvarakis A, Papadakis A, Stephanou E, Roubelakis-Anfelakis K (2006) Abiotic stress generates ROS that signal expression of anionic glutamate dehydrogenases to form glutamate for proline synthesis in tobacco and grapevine. Plant Cell 18:2767–2781

Smith BE (2002) Nitrogenase reveals its inner secrets. Science 297:1654–1655

Song L, Han MH, Lesicka J, Fedoroff N (2007) Arabidopsis primary microRNA processing proteins HYL1 and DCL1 define a nuclear body distinct from the Cajal body. Proc Natl Acad Sci U S A 104:5437–5442

Sonoda Y, Ikeda A, Saiki S, Yamaya T, Yamaguchi J (2003) Feedback regulation of the ammonium transporter gene family AMT1 by glutamine in rice. Plant Cell Physiol 44:1396–1402

Sunkar R, Kapoor A, Zhu JK (2006) Posttranscriptional induction of two Cu/Zn superoxide dismutase gene in Arabidopsis is mediated by down regulation of miR398 and important for oxidative stress tolerance. Plant Cell 18:2051–2065

Svennerstam H, Ganeteg U, Bellini C, Nasholm T (2007) Comprehensive screening of Arabidopsis mutants suggests the lysine histidine transporter 1 to be involved in plant uptake of amino acids. Plant Physiol 143:1853–1860

Svennerstam H, Ganeteg U, Nasholm T (2008) Root uptake of cationic amino acids by Arabidopsis depends on functional expression of amino acid permease 5. New Phytol 180:620–630

Takahashi H, Takahara K, Hashida SN, Hirabayashi T, Fujimori T, Kawai-Yamada M, Yamaya T,

Yanagisawa S, Uchimiya H (2009) Pleiotropic modulation of carbon and nitrogen metabolism in Arabidopsis plants overexpressing the NAD kinase2 gene. Plant Physiol 151:100–113

Trevisan S, Nonis A, Begheldo M, Manoli A, Palme K, Caporale G, Ruperti B, Quaggiotti S (2012) Expression and tissue-specific localization of nitrate-responsive miRNAs in roots of maize seedlings. Plant Cell Environ 35:1137–1155

Valdés-López O et al (2010) MicroRNA expression profile in common bean (Phaseolus vulgaris) under nutrient deficiency stresses and manganese toxicity. New Phytol 187:805–818

Vanoni M, Dossena L, van den Heuvel R, Curti B (2005) Structure–function studies on the complex iron-sulfur flavoprotein glutamate synthase: the key enzyme of ammonia assimilation. Photosynth Res 83:219–238

Voinnet O (2009) Origin, biogenesis, and activity of plant microRNAs. Cell 136:669–687

Wang R, Liu D, Crawford NM (1998) The Arabidopsis CHL1 protein plays a major role in high-affinity nitrate uptake. Proc Natl Acad Sci U S A 95:15134–15139

Wang J, Wang LJ, Mao YB, Cai WJ, Xue HW, Chen XY (2005) Control of root cap formation by microRNA-targeted auxin response factors in Arabidopsis. Plant Cell 17:2204–2216

Wang J, Schwab R, Czech B, Mica E, Weigel D (2008) Dual effects of miR156-targeted SPL genes and CYP78A5/KLUH on plastochron length and organ size in Arabidopsis thaliana. Plant Cell 20:1231–1243

Wang J, Czech B, Weigel D (2009) miR156-regulated SPL transcription factors define an endogenous flowering pathway in Arabidopsis thaliana. Cell 138:738–749

Wang X, Peng F, Li M, Yang L, Li G (2012) Expression of a heterologous SnRK1 in tomato increases carbon assimilation, nitrogen uptake and modifies fruit development. J Plant Physiol 169:1173–1182

Weisenburger DD (1991) Potential health consequences of ground-water contamination by nitrates in Nebraska. In: Bogorad I, Kuzerka RD (eds) Nitrate contamination: exposure consequences and control, NATO ASI series G: Ecological sciences 30. Springer Verlag, Berlin, pp 309–331

Williams L, Miller A (2001) Transporters responsible for the uptake and partitioning of nitrogenous solutes. Annu Rev Plant Physiol Plant Mol Biol 52:659–688

Wu G, Poethig R (2006) Temporal regulation of shoot development in Arabidopsis thaliana by miR156 and its target SPL3. Development 133:3539–3547

Wu G, Park MY, Conway SR, Wang JW, Weigel D, Poethig RS (2009) The sequential action of miR156 and miR172 regulates developmental timing in Arabidopsis. Cell 138:750–759

Xia T, Xiao D, Liu D, Chai W, Gong Q, Wang NN (2012) Heterologous expression of ATG8c from soybean confers tolerance to nitrogen deficiency and increases yield in Arabidopsis. PLoS ONE 7(5):e37217

Xie Z (2010) Piecing the puzzle together: genetic requirements for miRNA biogenesis in Arabidopsis thaliana. Methods Mol Biol 592:1–17

Xie K, Wu CC, Xiong L (2006) Genomic organization, differential expression, and interaction of SQUAMOSA promoter-binding-like transcription factors and microRNA156 in rice. Plant Physiol 142:280–293

Xing S, Salinas M, Höhmann S, Berndtgen R, Huijser P (2010) miR156-targeted and non targeted SBP-box transcription factors act in concert to secure male fertility in Arabidopsis. Plant Cell 22:3935–3950

Xu Z, Zhong S, Li X, Li W, Rothstein SJ, Zhang S, Bi Y, Xie C (2011) Genome-wide identification of microRNAs in response to low nitrate availability in maize leaves and roots. PLoS ONE 6:e28009

Yanagisawa S, Akiyama A, Kisaka H, Uchimiya H, Miwa T (2004) Metabolic engineering with Dof1 transcription factor in plants: improved nitrogen assimilation and growth under low-nitrogen conditions. Proc Natl Acad Sci U S A 101:7833–7838

Yoshida S (1981) Fundamentals of rice crop science. IRRI, Los Banos, 110pp

Yu B, Yang ZY, Li JJ, Minakhina S, Yang MC, Padgett RW, Steward R, Chen XM (2005) Methylation as a crucial step in plant microRNA biogenesis. Science 307:932–935

Yuan L, Loque D, Ye F, Frommer WB, von Wiren N (2007) Nitrogen-dependent posttranscriptional regulation of the ammonium transporter AtAMT1;1. Plant Physiol 143:732–744

Zhang BH, Wang QL, Pan XP (2007) MicroRNAs and their regulatory roles in animals and plants. J Cell Physiol 210:279–289

Zhang X, Zou Z, Zhang J, Zhang Y, Han Q, Hu T, Xu X, Liu H, Li H, Ye Z (2011) Over-expression of sly-miR156a in tomato results in multiple vegetative and reproductive trait alterations and partial phenocopy of the sft mutant. FEBS Lett 585:435–439

Zhang Z, Qin Y, Brewer G, Jing Q (2012) MicroRNA degradation and turnover: regulating the regulators. Wiley Interdiscip Rev RNA 3:593–600

Zhao M, Ding H, Zhu JK, Zhang F, Li WX (2011) Involvement of miR169 in the nitrogen-starvation responses in Arabidopsis. New Phytol 190:906–915

Zhao M, Tai H, Sun S, Zhang F, Xu Y, Li WX (2012) Cloning and characterization of maize miRNAs involved in responses to nitrogen deficiency. PLoS ONE 7:e29669

Zhou Y, Cai H, Xiao J, Li X, Zhang Q, Lian X (2009) Over-expression of aspartate aminotransferase genes in rice resulted in altered nitrogen metabolism and increased amino acid content in seeds. Theor Appl Genet 118:1381–1390

Zhu JK (2008) Reconstituting plant miRNA biogenesis. Proc Natl Acad Sci U S A 105:9851–9852

Zhu C, Ding Y, Liu H (2011) MiR398 and plant stress responses. Physiol Plant 143:1–9

Biofortification for Selecting and Developing Crop Cultivars Denser in Iron and Zinc

Sushil Kumar, Nepolean Thirunavukkarasu, Govind Singh, Ramavtar Sharma, and Kalyani S. Kulkarni

Abstract

Deficiency in minerals especially iron and zinc is a global burden chiefly in developing countries due to poverty and lack of awareness. Among current interventions such as food dietary diversification, supplementation, and fortification available for incorporating micronutrients into diet, breeding-based biofortification is the most feasible and best alternative. It involves the exploitation of genetic diversity present in the mineral-dense germplasm, land races, and wild species to create micronutrient denser lines/variety. In present genomics era, molecular breeding approaches employing molecular markers are being extensively utilized for marker-assisted selection (MAS) to develop mineral-denser lines mainly for iron and zinc. Currently, the focus of plant science is on quantitative trait locus (QTL) detection followed by MAS for the development of mineral-dense crops predominantly wheat, rice, maize, and pearl millet through biofortification. Several QTLs have been mapped for micronutrient concentration in grain/leaf using various mapping population and different marker systems. However, the success of this strategy requires long time and trials as increasing the mineral can cause yield penalty. Thus, a

S. Kumar (✉) · K.S. Kulkarni
Department of Agricultural Biotechnology, Anand
Agricultural University, Anand 388 110, Gujarat, India
e-mail: sushil254386@yahoo.com

N. Thirunavukkarasu
Division of Genetics, Indian Agricultural Research
Institute, New Delhi 110 012, India
e-mail: tnepolean@iari.res.in

G. Singh
Plant Biotechnology Centre, S.K. Rajasthan Agricultural
University, Bikaner 334 006, Rajasthan, India
e-mail: govindsingh10@rediffmail.com

R. Sharma
Division of Plant Improvement, Propagation and Pest
Management, Central Arid Zone Research Institute,
Jodhpur 342 003, India
e-mail: ras_rau@rediffmail.com

A. Rakshit et al. (eds.), *Nutrient Use Efficiency: from Basics to Advances*,
DOI 10.1007/978-81-322-2169-2_16, © Springer India 2015

combinatorial approach encompassing the identification and introgression of micronutrient-rich line into locally adapted variety, detection of allergenicity/toxicity, withstanding of nutrient during postharvest processing, and acceptance of new variety by farmers and consumers for a cost-effective intervention is required for the successful development of micronutrient-rich cultivars/lines. Biofortification strategies should be further enhanced with the support from governments for the popularization of varieties through extension workers to reach to the farmers and ultimately acceptance in market.

Keywords

Biofortification • Breeding • Hunger • Malnutrition • Mineral deficiency • Poverty • QTL

1   Introduction

Micronutrients deficiency is a global burning health problem pervasive in both urban and rural areas. About three billion people in world are deficient in key vitamins and minerals, particularly vitamin A, iodine (I), iron (Fe), and zinc (Zn) (Dahiya et al. 2008). Poverty, lack of affordability to diverse and balanced foods, lack of awareness about optimal dietary practices, and high incidence of infectious diseases are some of the factors leading to micronutrient deficiency. In the developing countries, micronutrient deficiency is the major underlying causes of numerous human health problems, and therefore, the situation of nutrient deficiencies is more drastic industrialized countries (Welch and Graham 2004). Moreover, modern agricultural practices including the improved cultivars of crops following the green revolution have further contributed and aggravated the malnutrition in the resource-poor populations by greater removal and exhaustion of major- and micro-plant nutrients in soil. Hence, in the present genomic era, plant nutrition research needs a new paradigm for agriculture and nutrition to meet the global demand for sufficient food production with enhanced nutritional value (Cakmak 2002).

The consumption of less diverse and monotonous food leads to deficiencies in micronutrients, especially iron (Fe), zinc (Zn), iodine (I), selenium (Se), and vitamin A. Among trace elements, Fe and Zn are essential for a variety of metabolic processes (Underwood 1977; Prasad 1978). The main sources of Zn in poor population are staple cereals, starchy roots, tubers, and legumes which are low either in quantity or bioavailability of Zn (Gibson 1994). Cereals contribute up to 50 % of the Fe intake in the poorest households. This means that doubling the Fe or Zn density of food staples could increase total intakes by $\geq$50 % (Ruel and Bouis 1998).

Fe deficiency is estimated to affect about 30 % of the world population, making Fe by far the most deficient nutrient worldwide (Lucca et al. 2001). Zn deficiency has subsequently been reported from all over the world and could be ascribed to the removal of high amounts of Zn from the soil due to the intensive cultivation of high yielding varieties (Takkar and Walker 1993). The intake of Fe and Zn appears to be below the recommended dietary allowance for an average Indian adult; this was observed in particularly low-income rural households in the pearl millet-consuming regions (ICMR 2002; Parthasarathy Rao et al. 2006). Fe deficiency is often accompanied by Zn deficiency as both of these nutrients are derived from similar sources in the diet (Welch 2001). In addition, minerals are essential for plant growth and reproduction, and nutrient deficiencies can limit yield potential and plant products that represent an important source of minerals in the human diet.

2 Importance of Fe and Zn in Human Nutrition

Fe has several vital functions in the human metabolism, *viz.*, synthesis of the oxygen transport proteins (hemoglobin and myoglobin) and formation of heme enzymes and other Fe-containing enzymes, which are particularly important for energy production, immune defense, and thyroid function (Roeser 1986). The other key functions for the Fe-containing enzymes include the synthesis of steroid hormones and bile acids, the detoxification of foreign substances in the liver, and signal controlling in some neurotransmitters such as the dopamine and serotonin systems in the brain. Zinc is involved in the functioning of more than 300 enzymes and is an essential component of many Zn-dependent enzymes. Zn plays a major role in gene expression and acts as a stabilizer of membrane structures and cellular components (Palmgren et al. 2008). Although body Zn homeostasis can be maintained over a wide range of Zn intakes by increasing or decreasing both intestinal Zn absorption and endogenous intestinal Zn excretion; ultimately low Zn intake and/or bioavailability results in Zn deficiency. Meat and seafood are good sources of Zn (Sanstead 1995). However, in many parts of the developing world, most Zn is provided by cereals and legume seeds. These plant foods are high in phytic acid, which is a potent inhibitor of Zn absorption (Navert et al. 1985).

3 Micronutrient Deficiency in Soil and Its Implication on Human Nutrition

In enhancing agricultural productivity and quality, micronutrient supply is of critical importance as both agricultural production and quality are constrained by the deficiencies of plant nutrients and nutrient imbalances. Therefore, the information on the micronutrient status of soil and crop edible tissues is crucial (Mahnaz et al. 2010;

Sahrawat et al. 2010). Research has been conducted to address the relationship between soil micronutrient status and crop yield and quality (Welch and Graham 2004; Gupta 2005).

In fact, intensified land use, without the addition of fertilizers, has apparently resulted in substantial removal of minerals (Sahrawat et al. 2007). Instead of judicious application, the imbalanced use of fertilizers has the problems, especially in the developing countries. Furthermore, it is reported that soils are becoming Zn and Fe deficient worldwide (Ghorbani et al. 2009; Sahrawat et al. 2007). The low availability of Fe, Zn, and copper (Cu) in calcareous or alkaline soils is also considered as the cause for the low mineral concentrations in edible plant parts (White and Brown 2010; Sahrawat et al. 2008). Thus, a successful breeding program for the biofortification of crops with grains denser in minerals will very much depend on the size of plant available mineral pool in soil (Cakmak 2008).

Zinc deficiency is an increasingly important risk factor to the global agriculture and human health, especially in the arid and semiarid regions of world (Nayyar et al. 1990; Sahrawat et al. 2007). Among the cereals, wheat and rice in particular suffer from Zn deficiency. Duffy (2007) reported that 30 % yield loss was common in wheat, rice, maize, and other staple crops grown on Zn-deficient soils. Hence, the widespread deficiency of Zn has serious implications for human health in countries where dominant diet is cereal based and also equally important for all forms of life including plants and animals. Moreover, low solubility of Zn in soils rather than the total amount of Zn is the major reason for the widespread occurrence of Zn-deficiency problem in crops (Cakmak 2008; Sahrawat et al. 2007). Total mineral concentrations in many infertile soils are often sufficient to support mineral-dense crops, if only the minerals are in the plant available form (Graham et al. 1999).

It is indicated that soils on which cereals are regularly grown for human rations are actually low in native nutrient reserves, and thus, it may lead to a situation in which nutrient deficient crops will be food of poor people. Moreover, the availability of Fe is lower in soils of the arid and semiarid regions, and as a result, grains produced on these soils have lower Fe content (Singh 2009). Frequent application of herbicide glyphosate could lead to the shortage of energy needed to maintain root growth and initiate ferric-reductase activity, and this may lead to Fe deficiency. A likely reason for this is that glyphosate interferes with root uptake of Fe by inhibiting ferric-reductase activity in plant roots, required for Fe acquisition by dicot and non-grass species (Ozturk et al. 2008).

Eventually, we have come to understanding of micronutrient deficiency in human is derived from the deficiencies of trace elements in soils and foods. Therefore, it is a multifaceted vicious cycle among the soil-plant-human system. Soil is the base medium for all living things; thus, sick soil means sick plants, sick animal, and sick people (http://www.ecoorganics.com/sick-soil/). It is simpler to cure the sick soils than the sick people. Nevertheless, not all soils are nutritionally sick; in such cases, an improvement in plant uptake and efficiency by genetic modulation is an imperative strategy to combat the mineral deficiency in plant as well as humans.

4 Bioavailability of Fe and Zn and Factors Affecting It

The nutritional quality of a diet can be determined based on the concentration of individual nutrients as well as by the interactions of other elements, promoters, and antinutrients, which affect the bioavailability of micronutrients (Khoshgoftarmanesh et al. 2010). Bioavailability is a term used to describe the digestion, absorption, and subsequent utilization of dietary compounds (Linder 1991). Not all ingested minerals are completely absorbed and utilized in humans and livestock, leading to certain segments of vegetarian population at risk for Fe and Zn and other trace element deficiency (Grusak and Cakmak 2004). Thus, just producing the mineral-dense food does not mean an improved nutrient status of people as the bioavailability of micronutrients needs also to be

improved. That is why, for effective biofortification of food, the understanding of bioavailability of minerals to humans is a prerequisite (Dahiya et al. 2008).

The levels of bioavailable mineral in staple food crop seeds and grains are as low as 5 % and 25 %, respectively; thus, breeder should consider the bioavailability of micronutrients while considering breeding program (Bouis and Welch 2010). The micronutrients interact with various types of biochemical substances which promote or inhibit the bioavailability of minerals (Khoshgoftarmanesh et al. 2010). Inhibitory substances, called antinutrients, reduce, whereas promotive substances, called promoters, enhance/stimulate micronutrient bioavailability to humans (Graham et al. 2001). Amounts of both antinutrients and promoters in grains depend on genetic and environmental factors (Welch and Graham 2004; White and Broadley 2005).

The bioavailability of dietary Fe and Zn is generally impaired by the phytic acid, fiber, and possibly other constituents of some plant foods (Hunt 2002; Mendoza 2002), while oxalate (Sotelo et al. 2010), polyphenolics (Ma et al. 2010), and to certain extent calcium (Zamzam et al. 2005) inhibit Fe absorption. Dietary phytate can influence the bioavailability of several minerals, because of its capacity to form insoluble precipitates which cannot cross the membrane transporters on the surface of enterocytes, making nutrients unavailable (Wise 1995). Negatively charged phytate is the primary storage form of phosphorus in most mature seeds and grains and complexes with positively charged Fe and Zn ions, inhibiting their uptake (Zhou and Erdman 1995). Being a monogastric creature, humans do not synthesize the phytate-degrading enzyme, phytase, as a result digestive tract cannot absorb, but can excrete (Lott et al. 2000).

Zn bioavailability can be predicted by considering phytate-to-Zn molar ratios in foods and has been widely used for zinc bioavailability (International Zinc Nutrition Consultative Group IZiNCG (IZiNCG) 2004; Gargari et al. 2007). Zinc absorption in the intestine is reduced at ratios above a value of around 20 (Frossard et al. 2000). A bioavailability model should be used to screen a large number of promising lines of micronutrient-enriched genotypes identified in breeding programs before advancing them as it is impractical to test the bioavailability of micronutrients in genotypes of staple plant foods generated in plant-breeding programs (Welch and Graham 2004).

Earlier reports in humans revealed that cysteine had a positive effect on mineral absorption, particularly Zn (Snedeker and Greger 1981, 1983; Martinez-Torres and Layrisse 1970), while Fe and copper were less affected by the sulfur-amino acids. Further research is required to focus on the effects of protein and sulfur-containing amino acids on Zn and non-heme Fe bioavailability in diets.

5   Biofortification: A Vital Device for Alleviating Micronutrient Malnutrition

Nutrition-related research has reached to micronutrient level, yet translation of these achievements into macro level community action is limited only to the supplementation of Fe and folic acid (IFA) tablets or fortification of some of the food items. For addressing micronutrient malnutrition, a combination of strategies involving food fortification and pharmaceutical supplementation and food diversification has been emphasized. The successful implementation of these exogenous fortification strategies requires safe delivery systems, stable policies, appropriate social infrastructures, and continued financial support (White and Broadley 2005). However, dietary modifications are promising but require behavioral changes that depend on education, communication, social marketing, and investments. Furthermore, fortification is difficult for each micronutrient, especially for Fe as the fortification of Fe leads to its rapid oxidation as well as increases the loss of iodine (I).

Unfortunately, none of the strategies have been successful against hidden hunger (Bouis 2003; Lyons et al. 2003). Agriculture is a vital tool for ameliorating micronutrient malnutrition

(Singh 2009). Therefore, alternatively, problem can be tackled through agricultural methods of crop cultivation by adding fertilizers – agronomic fortification – in farming system (White and Brown 2010) known as fertifortification (Prasad 2010). Fertifortification depends upon sufficient amount of available minerals in the soil (Cakmak 2008). Despite its success in Finland and Turkey, fertifortification is not practicable in the developing countries because of financial and ecological considerations (Ju et al. 2009) as well as it requires specific agricultural practices with regular application of nutrients. Additionally, they are not effective for Zn and Fe due to their limited mobility in phloem (Marschner 1995) and do not always increase mineral concentrations in edible or economic parts to the desired level and increase the cost of cultivation (Dai et al. 2004; White and Broadley 2005). Complementarily, agronomic fortification can be used as an approach to increase the mineral content in edible plant parts. A substitute approach, endogenous fortification is used by the accumulation of trace minerals directly in cereal grains using breeding. This complimentary solution termed "biofortification" by Bouis (2003).

Crop improvement through breeding has been the key in the past successes of agricultural production (Beddington 2010). Although, breeding-based strategy for biofortification is unproven as yet, it has the potential to become sustainable and cost-effective and to reach remote rural populations (Mannar and Sankar 2004; Genc et al. 2005). It is argued that once mineral-dense lines have been developed, there will be little additional cost in incorporating them into ongoing breeding programs (Welch and Graham 2004). It has been reported that the seed of mineral-dense crops produce more vigorous seedlings on infertile soils (Rengel and Graham 1995). High trace mineral density in seed produces more viable and vigorous seedlings in the next generation, and the efficiency in the uptake of trace minerals improves disease resistance (Welch 1999; Yilmaz et al. 1997). Variety, land races, and wild species are being explored for their mineral levels, and this knowledge is further used to create new varieties with higher micronutrient content (Ghandilyan et al. 2009). Hence, plant-breeding approaches utilize existing genetic variation coupled with marker-/genomics-assisted selection.

6 Relationships Among Grain Minerals and Yield

Biofortification to address nutrient deficiencies is an enticing concept, but there is much to understand about the potential impact on other important traits. For instance, it is not clear whether selection for increased mineral micronutrient content negatively affects yield or other important agronomic and end-use characters. This could occur if genes that increase mineral content are linked with genes that have a deleterious effect on other desired traits, or it could occur as a consequence of trait associations. Correlation between grain Fe and Zn has been studied in several crops, with results showing similar trends. For instance, positive and highly significant correlation between Fe and Zn concentrations had been observed in many crops (Gregorio et al. 2000; Ozkan et al. 2007; Velu 2013). Such correlations among micronutrients indicate that improvement in one element may simultaneously improve the concentration of other element (Ozkan et al. 2007). However, in few studies, negative correlations between the concentrations of Zn in grain and grain yield were reported in wheat (Oury et al. 2006; Zhao et al. 2009) and indicate the difficulty to breed wheat with high Zn concentration and high grain yield. Positive correlations among micronutrients suggest that similar transport and chelation process affect the accumulation of elements in seeds (Ding et al. 2010). The correlations among different minerals implement pleiotrophy for genes controlling the accumulation of these minerals or have close linkage of genes (Wu et al. 2008). Moreover, the positive correlation between Fe and Zn concentrations in grain is less affected by environment and can be combined with other agronomic traits (Banziger and Long 2000; Welch 2005).

7 Exploiting Existing Genetic Variation: Prerequisite for Biofortification

Genetic variation in wild, landraces, and cultivated species is the most important basic resource to generate new plant types with desirable traits for effective crop improvement programs (Vreugdenhil et al. 2004). Observed variation among crop plants can either be qualitative, caused by one or two major loci, or quantitative, caused by the combined effects of multiple loci (Salt et al. 2008). Germplasm of crops differs in the grain mineral content, and the selection followed by utilization of mineral-rich germplasm for breeding is an important component of research for increasing the grain mineral content. Thus, genetic resources enable plant breeders to create novel plant gene combinations and select crop varieties more suited to the needs of diverse agricultural systems (Glaszmann et al. 2010). With the aim to improve nutritional value of food for human beings, researchers in the past decade have shown much interest in developing cultivars of staple food with higher mineral content (Graham et al. 1999; Grusak and DellaPenna 1999; White and Broadley 2005; Cakmak 2008; Tiwari et al. 2009; Norton et al. 2010), but very little attention has been paid in breeding for grain mineral content (Vreugdenhil et al. 2004). The identification of "left behind" valuable alleles in the wild ancestors of crop plants and their reintroduction into cultivated crops is the target of modern plant breeding (Tanksley and McCouch 1997; Chatzav et al. 2010).

Dissecting the variation is prerequisite to utilize the natural diversity through molecular breeding for crop improvement. Therefore, research on the screening of natural genetic variability for seed mineral concentrations in various crop species in order to use selected lines for breeding has also been conducted (White and Broadley 2009). Identification of genotypes with differing nutrient efficiencies generally includes investigation of the potential morphological, physiological, and biochemical mechanisms involved therein (Khoshgof-tarmanesh et al. 2010). Growing evidences indicate that the wild and primitive genotypes show large and useful genetic variation for grain concentrations of Zn and Fe (Ghandilyan et al. 2006). The genetic variations for Fe and Zn in major food crops are explained as followed.

Wheat (*Triticum* spp.), a major staple food crop having significant impact on human health, contributes 28 % of the world's edible dry matter and up to 60 % of the daily calorie intake in several developing countries (Grusak and Cakmak 2005; FAOSTAT 2008). To examine genetic variation for Fe and Zn with other trace minerals, 132 wheat germplasm accessions at the CIMMYT were screened (Monasterio and Graham 2000). The variability in grain Fe ranged from 28.8 to 56.5 mg kg^{-1} and from 25.2 to 53.3 mg kg^{-1} for Zn. In a set of 30 *T. tauschii*, Monasterio and Graham (2000) reported mean Fe concentration of 76 mg kg^{-1} and a maximum value of 99 mg kg^{-1}. Similarly, Oury et al. (2006) identified wheat cultivars with Zn concentration ranging from 15 to 35 mg kg^{-1}, but the grain Zn increased to 43 mg kg^{-1} in selected germplasm. Fe concentration ranged from 20 to 60 mg kg^{-1} and was 88 mg kg^{-1} in non-adapted material. A total of 154 genotypes, including wild emmer accessions were evaluated for Fe and Zn by Chatzav et al. (2010) and reported that Fe ranged from 36 to 69 mg kg^{-1} with a mean of 52 mg kg^{-1}. Similarly, grain Zn concentrations ranged from 35 to 90 mg kg^{-1} with a mean of 58 mg kg^{-1}. The results of other studies (Balint et al. 2001; Morgounov et al. 2007) clearly showed the existence of alleles for mineral diversity within wheat germplasm to improve the food value.

Rice is a dominant cereal crop accounting for 50 % of the worldwide consumption in many developing countries (Lucca et al. 2001). However, currently polished rice is a poor source of essential micronutrients such as Fe and Zn (Bouis and Welch 2010) and contains average of only 2 parts per million (ppm) iron (Fe) and 12 ppm of zinc (Zn). Experts estimate that a rice-based diet should contain 14.5 μg g^{-1} Fe in endosperm,

the main constituent of polished grain, but breeding programs have failed to achieve even half of that value. Low mineral concentration in rice may be attributed to low level of minerals in endosperm and the loss during grain polishing as well. Since 1992, genetic difference for grain Fe has been explored by researchers at the IRRI (Gregorio et al. 2000; Graham et al. 1999). Gregorio et al. (2000) evaluated 1,138 brown rice genotypes for Fe and Zn content and reported that grain Fe and Zn contents ranged between 6.3 and 24.4 mg kg^{-1} and 13.5–58.4 mg kg^{-1}, respectively. On the other hand, aromatic rice exhibited consistently more grain Fe (range 18–22 mg kg^{-1}) and Zn (24–35 mg kg^{-1}) content than the nonaromatic rice genotypes. Research at the International Rice Research Institute (IRRI) showed that local varieties had iron content up to 2.5 times higher than that of the common high yielding varieties (Kennedy and Burlingame 2003). Glahn et al. (2002) evaluated 15 selected Fe-dense and normal genotypes of unpolished rice from the IRRI and reported that the Fe concentration ranged from 14 to 39 mg kg^{-1}. These results indicated that "aromatic and brown rice germplasm" as a potential reservoir of micronutrients which can be harnessed to improve existing micronutrient levels in rice.

Maize is the world's leading staple food along with rice and wheat due to its diverse functionality as a food source for both humans and animals (Grusak and Cakmak 2005; Nuss and Tanumihardjo 2010). Unfortunately, even though maize kernels supply many macro- and micronutrients necessary for human metabolic needs, the amounts of some essential nutrients with phytic acid are ill balanced or inadequate for consumers who rely on maize as a major food source (Grusak and Cakmak 2005; Nuss and Tanumihardjo 2010). The range of Fe and Zn in maize kernels is not as high as in other cereals, but considerable variation in the grain micronutrient content has been reported (Welch and Graham 2004). Menkir (2008) evaluated 149 lowland and 129 mid-altitude maize inbred lines at the IITA, Nigeria, and showed that the lines varied between 11 and 34 mg kg^{-1} in Fe and 14 and 45 mg kg^{-1} in Zn. The best-inbred line in each trial had a kernel Fe concentration that exceeded the average of all the inbred lines by 37% in trial-1, 32% in trial-2, 52% in trial-3, 39% in trial-5, 42% in trial-6 and 78% in trial-7. Similarly, the best-inbred line in each trial had 14–180 % greater concentrations of Zn and other mineral elements than the average of all inbred lines. This represents a broad range of variability in adapted maize germplasm available in the maize breeding program at the IITA. During an F4-mapping population, Simic et al. (2009) reported good range of Fe (17–34 mg kg^{-1}) and Zn (17–28 mg kg^{-1}).

Jambunathan (1980) reported an average Fe concentration of 59 mg kg^{-1} with a range of 26–96 mg kg^{-1}, while grain Zn varied between 19 and 57 mg kg^{-1} with an average of 33 mg kg^{-1} in the samples of 100 varieties of sorghum. At the ICRISAT, Reddy et al. (2005) screened 84 accessions of sorghum for grain Fe and Zn content. The grain Fe and Zn varied from 20.1 to 37 mg kg^{-1}, and grain Zn content varied from 13.4 to 31 mg kg^{-1}. Kayode et al. (2006) evaluated 76 farmers' varieties of sorghum for Fe and Zn concentrations. The Fe and Zn concentration of the grains ranged from 30 to 113 mg kg^{-1} and 11 to 44 mg kg^{-1}, respectively. These varieties exhibited fourfold range in grain Fe and Zn concentrations. In most genotypes, grain Fe was higher than Zn, the difference being one- to fivefold. The level of Fe found in the Kayode et al. (2006) study is in agreement with values reported in the literature. Waters and Pedersen (2009) also reported a wide range in grain Fe (24–73 mg kg^{-1}) and Zn (15–59 mg kg^{-1}) in sorghum.

Pearl millet is an important staple food in arid and semiarid regions of Asia and Africa and serving as a major source of dietary energy in these regions (Velu et al. 2006). Like other cereals, no much work has been done on the genetic variation of Fe and Zn content and the potential to improve it through plant breeding in pearl millet. Preliminary studies were conducted by Jambunathan and Subramanian (1988) in 27 pearl millet genotypes and Hulse et al. (1980) which reported as high as 38 mg kg^{-1} of Fe and 16 mg kg^{-1} of Zn. Similar studies for genetic

variation for grain Fe and Zn content have been reported by Khetarpaul and Chauhan (1990), Kumar and Chauhan (1993), and Abdalla et al. (1998). Higher micronutrient densities in African pearl millet landraces were comparable to those reported in improved varieties and hybrid lines. This demonstrated the potential of landraces for breeding pearl millet with grains denser in Fe and Zn (Buerkert et al. 2001). The genetic variation is presently being exploited in breeding program at different CGIAR centers under HarvestPlus program coordinated by IFPRI and CIAT (Bouis 2003).

8   Genetics of Fe and Zn Content in Grain

Understanding the nature of gene action and inheritance of seed mineral content is crucial to develop effective breeding strategies for micronutrients (Cichy et al. 2005). Very limited information has been generated on the inheritance of grain Fe and Zn content in crops. The genetic bases responsible for the uptake of some micronutrients, especially Fe uptake, in crop plants is now much better understood. Research on the genetics of kernel micronutrient density of maize described additive gene action in the 1960s and 1970s (Gorsline et al. 1964; Arnold and Bauman 1976). The recurring feature of micronutrient efficiency characters are single, major-gene inheritance (Epstein 1972). Weiss (1943) demonstrated this by detecting single major dominant gene while working with Fe efficiency in soybeans. Since Weiss's pioneering study, another study in soybean indicated that several minor additive genes contributed to Fe efficiency (Fehr 1982). Cichy et al. (2005) reported a single dominant gene controlling the high seed Zn in navy bean. Velu et al. (2006) found the prevalence of additive gene action in pearl millet, controlling grain Fe and Zn content. Based on the inheritance study, selection during breeding should be undertaken in a later generation (such as F5), where the dominance effect (unfixable genes) is not present.

9   Molecular Breeding: Maximizing the Exploitation of Genetic Variation

Genetic diversity offers opportunity to utilize various genomic sources and technologies in an effort to manipulate mineral levels in crop edible parts (Grusak and Cakmak 2005). But characterization of genetic variation within natural populations and among breeding lines is crucial for effective conservation and exploitation of genetic resources for crop improvement programs (Varshney and Tuberosa 2007). The development of molecular marker techniques has lead to a great increase in our knowledge of cereal genomics and our understanding of structure and behavior of the cereal genomes (Gupta and Varshney 2000). Renewed interest in the use of markers was generated when studies with maize and tomato demonstrated that some markers explained much of the phenotypic variance of complex traits (Anderson et al. 1993). DNA-based molecular markers having no known effects on phenotype, unaffected by environmental conditions and gene interactions, proved to be powerful and ideal tools for examining quantitative traits and genetic research (Beckmann and Soller 1986). A variety of genetic models and designs including the analysis of mating designs in segregating population are being used to study the quantitative traits to estimate the effective factors applying biometrical or molecular marker methods (Lynch and Walsh 1998; Zeng et al. 1990).

In this genomic era, molecular markers have been proven to be useful in characterization of the available germplasm and estimation of genetic diversity with the aim of using this information for the selection of parents for hybridization programs (Roy et al. 2002; Kalia et al. 2011). Furthermore, the recent development in quantitative genetics by employing of molecular markers allow the development of linkage map to determine the map position and effect of different loci/genes of metric characters known as quantitative trait loci (QTL). This development is to expedite the use of markers for tagging genes/QTLs for qualitative and quantitative traits and for marker-assisted selection (MAS) (Sharma 2001; Yadav et al. 2002). Thus, molecular breeding can enhance the pace of genetic variation exploitation.

10   Mapping QTLs Associated with Grain Mineral (Fe and Zn) Concentrations

Quantitative trait locus (QTL) analysis provides a powerful approach to understand the genetic factors and to unravel the genes underlying the natural variation for Fe and Zn concentrations (Ghandilyan et al. 2006). The identification and tagging of major QTLs for grain micronutrients with large effects would be helpful in the selection of the QTLs in early generations with MAS technique and will greatly accelerate wheat cultivar development for improving mineral concentration in grain (Ortiz-Monasterio et al. 2007). Using various populations, many QTLs for micronutrient concentration in grain/leaf have been mapped in recent years (Table 1). Brief results of various QTL studies in major staple crops are described in brief in the later sections.

10.1 Rice

In a rice double haploid (DH) population, two QTLs for phytate concentration (explaining 24 % and 15 % of total phenotypic variation), three QTLs for Fe concentration (explaining 17 %, 18 %, and 14 % of total phenotypic variation), and two QTLs for Zn concentration (explaining 15 % and 13 % of total phenotypic variation) were identified by Stangoulis et al. (2007) and reported that Zn concentration QTL co-localized with the Fe QTL. Garcia-Oliveira et al. (2009) reported 31 putative QTLs for eight mineral elements (Fe, Zn, Mn, Cu, Ca, Mg, P, and K) in seeds of introgression lines (IL) by single-point analysis, out of which, 17 QTLs were observed during both years. QTLs associated with Zn and Si content in rice was identified by Biradar

Table 1 QTL/s associated with concentrations of essential mineral elements in various crop species

Crop species	Tissue	Elements	Mapping population	Number of lines	Number of markers	Number of QTLs	References
Rice (*Oryza sativa*)	Grain	Fe, Zn	DH	129	582	3 Fe, 2 Zn	Stangoulis et al. (2007)
		Fe, Zn,	RIL	241	221	3 Zn, 2 Fe	Lu et al. (2008)
		Fe, Zn	BIL	85	179	2 Fe, 3 Zn	Garcia-Oliveira et al. (2009)
		Fe, Zn	RIL	79	164	4 Fe, 4 Zn	Norton et al. (2010)
		Zn	DH	127	243	2	Zhang et al. (2011)
		Zn	DH	93	254	6	Biradar et al. (2007)
		Zn, Fe	BC1F1	115	93	2 Zn, 3 Fe	Susanto (2009)
		Fe, Zn	RIL	168	110	7 Fe, 6 Zn	Anuradha et al. (2012)
Wheat (*Triticum* spp.)	Grain	Zn	DH	119	39	11	Shi et al. (2008)
		Fe, Zn	DH	90	470	4 Zn, 1 Fe	Genc et al. (2009)
		Fe, Zn	RIL	152	690	6 Zn, 11 Fe	Peleg et al. (2009)
		Fe, Zn	RIL	93	169	3 Fe, 2 Zn	Tiwari et al. (2009)
		Fe, Zn, Mn	RIL	168	477	1 Fe, 2 Zn, 2 Mn	Ozkan et al. (2007)
Barley (*Hordeum vulgare*)	Grain	Zn	DH	150	417	5	Lonergan et al. (2009)
		Zn	DH	150	302	2	Sadeghzadeh et al. (2010)
Maize (*Zea mays*)	Grain	Fe	RIL	232	1,338	3	Lungaho et al. (2011)
		Fe, Zn	F4	294	121	3 Fe, 1 Zn	Simic et al. (2012)
		Fe, Zn	RIL	113	47	7 Fe, 11 Zn	Beebe et al. (2000)
		Fe Zn	F2:3	218	240	4 Zn, 1 Fe	Jin et al. (2013)
Bean (*Phaseolus vulgaris*)	Seed	Zn	RIL	73	5	Two markers associated with Zn	Gelin et al. (2007)
		Fe, Zn	RIL	87	236	13 Fe, 13 Zn	Blair et al. (2009)
		Fe, Zn	RIL	77		6 Fe, 4 Zn	Cichy et al. (2009)
		Fe, Zn	RIL	110	114	8 Fe, 9 Zn	Blair et al. (2010)
		Fe, Zn	RIL	100	122	6 Fe, 3 Zn	Blair et al. (2011)
		Fe, Zn	F2:3	120	57	1 Zn, 2 Fe	Guzman-Maldonado et al. (2003)
Soybean (*Glycine max*)	Seed	Ca	F2:3	178	148	4	Zhang et al. (2009)
Oilseed Rape (*Brassica napus*)	Seed	Fe, Zn	RIL	124	553	10 Zn, 9 Fe	Ding et al. (2010)
B. oleracea	Leaf	Ca, Mg	DH	90	547	11 Mg, 17 Ca	Broadley et al. (2008)
B. rapa	Leaf	Fe, Zn	DH	183	287	2 Zn, 1Fe	Wu et al. (2008)
Pearl millet (*Pennisetum glaucum*)	Grain	Fe, Zn	RIL	106	305	1 Fe, 1 Zn	Kumar (2011)
		Fe, Zn	RIL	317	234	11 Fe, 8 Zn	Kumar (2011)

DH Double haploid, *RIL* Recombinant inbred line, *BIL* Backcross inbred lines, *F2:3* F2 derived F3, *BC1* First Backcross generation et al. (2007) in DH population. Based on the interval mapping results, one QTL was detected for Si. Similarly, a total of 6 QTLS were detected for Zn content using SMA explaining 1–10 % of total phenotypic variation.

Lu et al. (2008) reported ten QTLs for Cu, Ca, Mn, Zn, and Fe in a RIL population in grains and reported three QTLs for Zn content. Among these QTLs, the major QTL accounted for 19 % of phenotypic variation, whereas two QTLs for Fe accounted for 37 % phenotypic variation. Gregorio et al. (2000) and Avendano (2000) also detected QTL for Fe and Zn, respectively, on same chromosomes.

Using a mapping population consisting of 85 backcross inbred lines (BIL), two QTLs for increasing grain cadmium (Cd) concentration were detected by Ishikawa et al. (2010). A major effect QTL accounted for 35 % of all phenotypic variation. A putative QTL for grain Fe concentration explained 15 % of the phenotypic variation, whereas no QTL for grain Zn concentration was found. Three QTLs for straw Fe concentration and two QTLs for straw Zn concentration were found. Grain concentration QTL was not genetically related to any QTL for other mineral concentration or those for agronomic trait, suggesting that QTL was specific for Cd.

10.2 Wheat

Peleg et al. (2009) identified 82 significant QTLs for nine grain mineral nutrient concentrations including four secondary mineral nutrients and proteins. GEI was exhibited by 38 QTLs. A total of six significant QTLs were associated with Zn explaining 1–13 % of variance with three QTLs showing significant GEI, while a total 11 significant QTLs were associated with Fe, explaining 2–18 % variance with GEI for five QTLs, out of which three QTLs for Zn were in agreement with the results reported in previous studies (Ozkan et al. 2007; Shi et al. 2008; Distelfeld et al. 2007; Genc et al. 2009). Similarly, two out of 11 QTLs have been mapped (Ozkan et al. 2007; Distelfeld et al. 2007). In another study, Tiwari et al. (2009)

detected Zn concentration QTL on same region as reported by Shi et al. (2008). The QTL for grain Fe and Zn mapped in the study conducted by Tiwari et al. (2009) explained 25–30 % of the total phenotypic variation with significant correlation between both elements.

10.3 Maize

Three modest QTLs for grain Fe concentration (FeGC) were detected by Lungaho et al. (2011), indicating that FeGC was controlled by many small QTLs. Ten QTLs for FeGB were identified 54 % of the variance observed in samples from a single year/location. Three of the largest FeGB QTLs were isolated in sister derived lines, and their effect was observed in three subsequent seasons in New York. The results indicated that iron biofortification of maize grain is achievable using specialized phenotyping tools and conventional plant breeding techniques. The analysis of variance indicated that environment played a strong role in influencing grain Fe concentration.

By using 294 F4 lines of a biparental population taken from field trials of over 3 years, Simic et al. (2009) revealed 32 significant QTLs (three for Fe and one for Zn). Significant additive effects with no significant dominant effects suggested that biofortification traits in maize were predicted by a simple additive model and mostly controlled by numerous small-effect QTLs.

10.4 Pearl Millet

Using 106 RILs (ICMB 841-P3 × 863B-P2), two co-localized QTLs for Fe and Zn concentrations on LG 3 were identified in pearl millet by Kumar (2011). Fe and Zn QTLs explained 19 % and 36 % of observed phenotypic variation, respectively. Likewise, Kumar (2011) also detected 19 putative QTL for grain Fe and Zn concentration in ICMS 8511B × AIMP 92901-derived-08 RIL population (317 RILs) on the base of single environment data, of which 11 were for Fe (66 % of phenotypic variation) and eight were for Zn (60 % of phenotypic variation). LG 1 harbored two co-localized main effect putative QTLs for Fe and Zn concentrations.

11   Issues and Conclusions

Micronutrient deficiency is prevalent in populations depending on nondiversified plant-based diets eventually leading to hidden hunger. Breeding-based crop biofortification is a feasible and most economical approach for overcoming "hidden hunger" catalyzed through hastening the breeding efficiency aided by identification of the genes/loci responsible for mineral uptake and translocation to the economic part of plant. However, success of this long, multistage strategy will depend on local dietary patterns as well as technology efficacy. For successful and cost-effective biofortification strategy, few facts must be satisfied like:

1. Identification of micronutrient-rich germplasm/line(s) to introgress gene/QTL(s) into locally adapted varieties through breeding methods for the establishment of nutritional efficacy (biological impact under controlled conditions) and effectiveness (biological impact in real life) of a biofortified crop
2. Detection of allergenicity and toxicity
3. Withstanding of nutrient during postharvest processing
4. Acceptance of new variety by farmers and consumers for a cost-effective intervention

Simultaneously, to enhance the effectiveness of biofortification strategies, governments should recognize the benefits and consider providing structure through nutrition and agricultural policies. Furthermore, like health consequences of malnutrition, the effect of biofortified staple crops should be quick to make a difference. Even after the development of biofortified varieties, it will be essential to address various socioeconomical and sociopolitical challenges to popularize their cultivation by farmers (market price premium) and ultimately their public acceptance to combat malnutrition by biofortification.

Euphorically, emerging evidences of this fast-evolving practice supporting the endurable and cost-effective breeding-based strategy will answer the many important unresolved problems to meet the nutritional needs of malnourished communities throughout the world. Eventually, a multi-tiered network and interdisciplinary research will play a pivotal role for successful breeding-based biofortification strategy to address mineral malnutrition in humans and other animals.

References

Abdalla AA, El Tinay AH, Mohamed BE, Abdalla AH (1998) Proximate composition, starch, phytate and mineral contents of 10 pearl millet genotypes. Food Chem 63:243–246

Anderson JA, Churchill GA, Autrique JE, Tanksley SC, Sorrells ME (1993) Optimizing parental selection for genetic linkage maps. Genome 36:181–186

Anuradha K, Agarwal S, Rao YV, Rao KV, Viraktamath BC, Sarla N (2012) Mapping QTLs and candidate genes for iron and zinc concentrations in unpolished rice of Madhukar × Swarna RILs. Gene 508 (2):233–240

Arnold JM, Bauman LF (1976) Inheritance and interrelationships among maize kernel traits and elemental contents. Crop Sci 16:439–440

Avendano BS (2000) Tagging high zinc content in the grain, and zinc deficiency tolerance genes in rice (Oryza sativa L.) using simple sequence repeats (SSR). Dissertation, Laguna College, Los Banos

Balint AF, Kovacs G, Erdei L, Sutka J (2001) Comparison of the Cu, Zn, Fe, Ca and Mg contents of the grains of wild, ancient and cultivated species. Cereal Res Commun 29:375–382

Banziger M, Long J (2000) The potential for increasing the iron and zinc density of maize through plant-breeding. Food Nutr Bull 21:397–400

Beckmann JS, Soller M (1986) Restriction fragment length polymorphisms and genetic improvement of agricultural species. Euphytica 35:111–124

Beddington J (2010) Food security: contributions from science to a new and greener revolution. Philos Trans R Soc B 365(1537):61–71

Beebe S, Gonzalez AV, Rengifo J (2000) Research on trace minerals in the common bean. Food Nutr Bull 21:387–391

Biradar H, Bhargavi MV, Sasalwad R, Parama R, Hittalmani S (2007) Identification of QTL associated with silicon and Zn content in rice (Oryza sativa L.) and their role in blast disease resistance. Indian J Genet 67:105–109

Blair MW, Astudillo C, Grusak MA, Graham R, Beebe SE (2009) Inheritance of seed iron and zinc concentrations in common bean (Phaseolus vulgaris L.). Mol Breed 23:197–207

Blair MW, Gonzalez LF, Kimani M, Butare L (2010) Genetic diversity, inter-gene pool introgression and nutritional quality of common beans (*Phaseolus vulgaris* L.) from Central Africa. Theor Appl Genet 121(2):237–248

Blair MW et al (2011) Gene-based SSR markers for common bean (*Phaseolus vulgaris* L.) derived from root and leaf tissue ESTs: an integration of the BMc series. BMC Plant Biol 11:50

Bouis HE (2003) Micronutrient fortification of plants through plant breeding: can it improve nutrition in man at low cost? Proc Nutr Soc 62(2):403–411

Bouis HE, Welch RM (2010) Biofortification-a sustainable agricultural strategy for reducing micronutrient malnutrition in the global south. Crop Sci 50:S20–S32

Broadley MR, Hammond JP, King GJ (2008) Shoot calcium and magnesium concentrations differ between subtaxa, are highly heritable, and associate with potentially pleiotropic loci in *Brassica oleracea*. Plant Physiol 146:1707–1720

Buerkert A, Bationo A, Piepho HP (2001) Efficient phosphorus application strategies for increase crop production in Sub-Saharan West Africa. Field Crop Res 72:1–15

Cakmak I (2002) Plant nutrition research: priorities to meet human needs for food in sustainable way. Plant Soil 247:3–24

Cakmak I (2008) Enrichment of cereal grains with zinc: agronomic or genetic biofortification? Plant Soil 302:1–17

Chatzav M, Peleg Z, Ozturk L, Yazici A, Fahima T, Cakmak I, Saranga Y (2010) Genetic diversity for grain nutrients in wild emmer wheat: potential for wheat improvement. Ann Bot 105:1211–1220

Cichy KA, Shana F, Kenneth LG, George LH (2005) Inheritance of seed zinc accumulation in navy bean. Crop Sci 45:864–870

Cichy KA, Caldas GV, Snapp SS, Blair MW (2009) QTL analysis of seed iron, zinc, and phosphorus levels in an Andean bean population. Crop Sci 49:1742–1750

Dahiya S, Chaudhary D, Jaiwal R, Dhankher O, Singh R et al (2008) Elemental biofortification of crop plants. In: Jaiswal P, Singh R, Dhankar OP (eds) Plant membrane and vacuolar transporters. CABI International, Wallingford/Cambridge, pp 345–371

Dai JL, Zhu YG, Zhang M, Huang YZ (2004) Selecting iodine-enriched vegetables and the residual effect of iodate application to soil. Biol Trace Elem Res 101:265–276

Ding G, Yang M, Hu Y, Liao Y, Shi L, Xu L, Meng J (2010) Quantitative trait loci affecting seed mineral concentrations in *Brassica napus* grown with contrasting phosphorus supplies. Ann Bot 105:1221–1234

Distelfeld A, Cakmak I, Peleg Z (2007) Multiple QTL-effects of wheat Gpc-B1 locus on grain protein and micronutrient concentrations. Physiol Plant 129:635–643

Duffy B (2007) Zinc and plant disease. Miner Nutr Plant Dis 35:155–175

Epstein E (1972) Mineral nutrition of plants: principles and perspectives. Wiley, New York

Fehr WR (1982) Control of iron deficiency chlorosis in soybeans by plant breeding. J Plant Nutr 5:611–621

Food and Agriculture Organization of the United Nations, FAOSTAT database (FAOSTAT, 2008). Available at http://faostat.fao.org/site/362/DesktopDefault.aspx? PageID=362

Frossard E, Bucher M, Mächler F, Mozafar A, Hurrell R (2000) Potential for increasing the content and bioavailability of Fe, Zn and Ca in plants for human nutrition. J Sci Food Agric 80:861–879

Garcia-Oliveira AL, Tan L, Fu Y, Sun C (2009) Genetic identification of quantitative trait loci for contents of mineral nutrients in rice grain. J Integr Plant Biol 51:84–92

Gargari BP, Mahboob S, Razavieh SV (2007) Content of phytic acid and its mole ratio to zinc in flours and breads consumed in Tabriz, Iran. Food Chem 100:1115–1119

Gelin JR, Forster S, Grafton KF, McClean P, Rojas-Cifuentes GA (2007) Analysis of seed-zinc and other nutrients in a recombinant inbred population of navy bean (*Phaseolus vulgaris* L.). Crop Sci 47:1361–1366

Genc Y, Humphries JM, Lyons GH, Graham RD (2005) Exploiting genotypic variation in plant nutrient accumulation to alleviate micronutrient deficiency in populations. J Trace Elem Med Biol 18:319–324

Genc Y, Verbyla AP, Torun AA, Cakmak I, Willsmore K, Wallwork H, McDonald GK (2009) Quantitative trait loci analysis of zinc efficiency and grain zinc concentration in wheat using whole genome average interval mapping. Plant Soil 314:49–66

Ghandilyan A, Vreugdenhil D, Aats MGM (2006) Progress in the genetic understanding of plant iron and zinc nutrition. Physiol Plant 126:407–417

Ghandilyan A, Barboza L, Tisné S, Granier C, Reymond M, Koornneef M, Schat H, Aarts MG (2009) Genetic analysis identifies quantitative trait loci controlling rosette mineral concentrations in *Arabidopsis thaliana* under drought. New Phytol 184:180–192

Ghorbani R, Wilcockson S, Koocheki A, Leifert C (2009) Soil management for sustainable crop disease control: a review. Org Farm Pest Control Remediat Soil Pollut Sustain Agric Rev 1:177–201

Gibson R (1994) Zinc nutrition in developing countries. Nutr Res Rev 7:151–173

Glahn RP, Chen SQ, Welch RM, Gregorio GB (2002) Comparison of iron bioavailability from 15 rice genotypes. J Agric Food Chem 50(12):3586–3591

Glaszmann JC, Kilian B, Upadhyaya HD, Varshney RK (2010) Accessing genetic diversity for crop improvement. Curr Opin Plant Biol 13:1–7

Gorsline GW, Thomas WI, Baker DE (1964) Inheritance of P, K, Mg, Cu, B, Zn, Mn, Al and Fe concentrations by corn (*Zea mays* L.) leaves and grain. Crop Sci 4:207–210

Graham RD, Senadhira C, Beebe SE, Iglesias C, Monasterio I (1999) Breeding for micronutrient density in edible portions of staple food crops: conventional approaches. Field Crops Res 60:57–80

Graham RD, Welch RM, Bouis HE (2001) Addressing micronutrient malnutrition through enhancing the nutritional quality of staple foods: principles, perspectives and knowledge gaps. Adv Agron 70:77–142

Gregorio GB, Senadhira D, Htut H, Graham RD (2000) Breeding for trace mineral density in rice. Food Nutr Bull 21:382–386

Grusak MA, Cakmak I (2004) Methods to improve the crop delivery of minerals to humans and livestock. Plant Nutr Genomics 22:13

Grusak MA, Cakmak I (2005) Methods to improve the crop-delivery of minerals to humans and livestock. In: Broadley MR, White PJ (eds) Plant nutritional genomics. Blackwell, Oxford, pp 265–286

Grusak MA, DellaPenna D (1999) Improving the nutrient composition of plants to enhance human nutrition and health. Annu Rev Plant Physiol Plant Mol Biol 50:133–161

Gupta AP (2005) Micronutrient status and fertilizer use scenario in India. J Trace Elem Med Biol 18 (4):325–331

Gupta PK, Varshney RK (2000) The development and use of microsatellite markers for genetic analysis and plant breeding with emphasis on bread wheat. Euphytica 113:163–185

Guzman-Maldonado SH, Martinez O, Acosta-Gallegos JA et al (2003) Putative quantitative trait loci for physical and chemical components of common bean. Crop Sci 43:1029–1035

Hulse JH, Laing EM, Pearson OE (1980) Sorghum and millets: their composition and nutritive value, International Development Research Centre (IDRC). Academic, London, p 997

Hunt JR (2002) Moving toward a plant-based diet: are iron and zinc at risk? Nutr Rev 60(5 Pt 1):127–134

Indian Council of Medical Research (ICMR) (2002) Nutrient requirements and recommended dietary allowances for Indians. Indian Council of Medical Research, New Delhi, p 83

International Zinc Nutrition Consultative Group (IZiNCG) (2004) Assessment of the risk of zinc deficiency in populations and options for its control. Hotz C, Brown KH (eds) Food Nutr Bull 25(Suppl 2):S91–S204

Ishikawa S, Abe T, Kuramata M, Yamaguchi M, Ando T, Yamamoto T, Yano M (2010) A major quantitative trait locus for increasing cadmium-specific concentration in rice grain is located on the short arm of chromosome 7. J Exp Bot 5:923–934

Jambunathan R (1980) Improvement of nutritional quality of sorghum and pearl millet. U N Univ Press Food Nutr Bull 2(1):39–53

Jambunathan R, Subramanian V (1988) Grain quality and utilization in sorghum and pearl millet. In: de Wet JMJ, Preston TA (eds) Proceedings of the workshop on biotechnology for tropical crop improvement. ICRISAT, Patancheru, pp 133–139

Jin T, Zhou J, Chen J, Zhu L, Zhao Y, Huang Y (2013) The genetic architecture of zinc and iron content in maize grains as revealed by QTL mapping and meta-analysis. Breed Sci 63(3):317–324

Ju XT, Xing GX, Chen XP, Zhang SL, Zhang LJ, Liu XJ, Cui ZL, Yin B, Christie P, Zhu ZL, Zhang FS (2009) Reducing environmental risk by improving N management in intensive Chinese agricultural systems. PNAS 106:3041–3046

Kalia RK, Rai MK, Kalia S, Singh R, Dhawan AK (2011) Microsatellite markers: an overview of the recent progress in plants. Euphytica 177:309–334

Kayode APP, Linnemann AR, Hounhouigan JD (2006) Genetic and environmental impact on iron, zinc, and phytate in food sorghum grown in Benin. J Agric Food Chem 54:256–262

Kennedy G, Burlingame B (2003) Analysis of food composition data on rice from a plant genetic resources perspective. Food Chem 80:589–596

Khetarpaul N, Chauhan BM (1990) Improvement in HCl-extractability of minerals from pearl millet by natural fermentation. J Food Chem 37:69–75

Khoshgoftarmanesh AH, Schulin S, Chaney RL, Daneshbakhsh B, Afyuni M (2010) Micronutrient-efficient genotypes for crop yield and nutritional quality in sustainable agriculture a review. Agron Sustain Dev 30:83–107

Kumar S (2011) Development of new mapping population and marker-assisted improvement of iron and zinc grain density in pearl millet [Pennisetumglaucum (L.) R. Br]. PhD thesis, SK Rajasthan Agricultural University, Bikaner, Rajasthan, India

Kumar A, Chauhan BM (1993) Effects of phytic acid on protein digestibility (in vitro) and HCl-extractability of minerals in pearl millet sprouts. Cereal Chem 70 (5):504–506

Linder MC (1991) Nutritional biochemistry and metabolism: with clinical applications, 2nd edn. Elsevier, New York

Lonergan PF, Pallotta MA, Lorimer M, Paull JG, Barker SJ, Graham RD (2009) Multiple genetic loci for zinc uptake and distribution in barley (Hordeum vulgare). New Phytol 184:168–179

Lott JNA, Ockenden I, Raboy V, Batten GD (2000) Phytic acid and phosphorus in crop seeds and fruits: a global estimate. Seed Sci Res 10:11–33

Lu K, Li L, Zheng X (2008) Quantitative trait loci controlling Cu, Ca, Zn, Mn and Fe content in rice grains. J Genet 87(3):305–310

Lucca P, Hurrell R, Potrykus I (2001) Genetic engineering approaches to improve the bioavailability and the level of iron in rice grains. Theor Appl Genet 102:392–397

Lungaho MG, Mwaniki AM, Szalma SJ (2011) Genetic and physiological analysis of iron biofortification in maize kernels. PLoS One 6(6):e20429

Lynch M, Walsh B (1998) Genetics and analysis of quantitative traits. Sinauer Associates, Sunderland

Lyons G, Stangoulis J, Graham R (2003) High-selenium wheat: biofortification for better health. Nutr Res Rev 16(1):45–60

Ma Q, Kim E-Y, Han O (2010) Bioactive dietary polyphenols decrease heme iron absorption by decreasing basolateral iron release in human intestinal Caco-2 cells. J Nutr 140(6):1117–1121

Mahnaz P, Afyuni M, Khoshgoftarmanesh A, Schulin R (2010) Micronutrient status of calcareous paddy soils and rice products: implication for human health. Biol Fertil Soils 46(4):317–322

Mannar MGV, Sankar R (2004) Micronutrient fortification of foods rationale, application and impact. Indian J Pediatr 71(11):997–1002

Marschner H (1995) Mineral nutrition of higher plant, 2nd edn. Academic, New York

Martinez-Torres C, Layrisse M (1970) Effect of amino acids on iron absorption from a staple vegetable food. Blood 35:669–682

Mendoza C (2002) Effect of genetically modified low phytic acid plants on mineral absorption. Int J Food Sci Technol 37:759–767

Menkir A (2008) Genetic variation for grain mineral content in tropical-adapted maize inbred lines. Food Chem 110:454–464

Monasterio I, Graham RD (2000) Breeding for trace minerals in wheat. Food Nutr Bull 21(4):392–396

Morgounov A, Gómez-Becerra HF, Abugalieva A (2007) Iron and zinc grain density in common wheat grown in Central Asia. Euphytica 155:193–203

Navert B, Sandsteraöm B, Cederblad Å (1985) Reduction of the phytate content of bran by leavening in bread and its effect on zinc absorption in man. Br J Nutr 53:47–53

Nayyar VK, Takkar PN, Bansal RL (1990) Amelioration of micro and secondary nutrient deficiencies. In: Micronutrients in soils and crops of Punjab. Research Bulletin, Department of Soils, Punjab Agricultural University, Ludhiana, p 148

Norton GJ, Deacon CM, Xiong L, Huang S, Meharg AA, Price AH (2010) Genetic mapping of the rice ionome in leaves and grain: identification of QTLs for 17 elements including arsenic, cadmium, iron and selenium. Plant Soil 329:139–153

Nuss ET, Tanumihardjo SA (2010) Maize: a paramount staple crop in the context of global nutrition. Compr Rev Food Sci Food Saf 9:417–436

Ortiz-Monasterio JI, Palacios-Rojas N, Meng E (2007) Enhancing the mineral and vitamin content of wheat and maize through plant breeding. J Cereal Sci 46:293–307

Oury F-X, Leenhardt F, Remesy C (2006) Genetic variability and stability of grain magnesium, zinc and iron concentrations in bread wheat. Eur J Agron 25:177–185

Ozkan H, Brandolini A, Torun A, AltIntas S, Eker S, Kilian B, Braun HJ, Salamini F, Cakmak I (2007) Natural variation and identification of microelements content in seeds of Einkorn wheat (Triticum monococcum). Dev Plant Breed 12:455–462

Ozturk L, Yazici A, Eker S, Gokmen O, Romheld V, Cakmak I (2008) Glyphosate inhibition of ferric reductase activity in iron deficient sunflower roots. New Phytol 177:899–906

Palmgren MG, Clemens S, Williams LE, Krämer U, Borg S, Schjørring JK, Sanders D (2008) Zinc biofortification of cereals: problems and solutions. Trends Plant Sci 13:464–473

Parthasarathy Rao P, Birthal PS, Reddy BVS, Rai KN, Ramesh S (2006) Diagnostics of sorghum and pearl millet grains-based nutrition in India. Int Sorghum Millets Newsl 46:93–96

Peleg Z, Cakmak I, Ozturk L, Yazici A, Jun Y, Budak H, Korol AB, Fahima T, Saranga Y (2009) Quantitative trait loci conferring grain mineral nutrient concentrations in durum wheat × wild emmer wheat RIL population. Theor Appl Genet 119:353–369

Prasad AS (1978) Trace elements and iron in human metabolism. Wiley, New York/Chichester

Prasad R (2010) Zinc biofortification of food grains in relation to food security and alleviation of zinc malnutrition. Curr Sci 98(10):1300–1304

Reddy BVS, Ramesh S, Longvah T (2005) Prospects of breeding for micronutrients and carotenedense sorghums. Int Sorghum Millets Newsl 46:10–14

Rengel Z, Graham RD (1995) Importance of seed Zn content for wheat growth on Zn deficient soil. 2. Grain yield. Plant Soil 173:267–274

Roeser HP (1986) Iron. J Food Nutr 42:82–92

Roy JK, Balyan HS, Prasad M, Gupta PK (2002) Use of SAMPL for a study of polymorphism, genetic diversity and possible gene tagging. Theor Appl Genet 104:465–472

Ruel MT, Bouis HE (1998) Plant breeding: a long-term strategy for the control of zinc deficiency in vulnerable populations. Am J Clin Nutr 68(Suppl 2):488S–494S

Sadeghzadeh B, Rengel Z, Li C, Yang H (2010) Molecular marker linked to a chromosome region regulating seed Zn accumulation in barley. Mol Breed 25 (1):167–177

Sahrawat KL, Wani SP, Rego TG, Pardhasaradhi G, Murthy KVS (2007) Widespread deficiencies of sulphur, boron and zinc in dryland soils of the Indian semi-arid tropics. Curr Sci 93:1428–1432

Sahrawat KL, Rego TG, Wani SP, Pardhasaradhi G (2008) Sulfur, boron and zinc fertilization effects on grain and straw quality of maize and sorghum grown on farmers' fields in the semi-arid tropical region of India. J Plant Nutr 31:1578–1584

Sahrawat KL, Wani SP, Pardhasaradhi G, Murthy KVS (2010) Diagnosis of secondary and micronutrient deficiencies and their management in rainfed agro ecosystems: case study from Indian semi-arid tropics. Commun Soil Sci Plant Anal 41:346–360

Salt D, Baxter I, Lahner B (2008) Ionomics and the study of the plant ionome. Annu Rev Plant Biol 59:709–733

Sanstead HH (1995) Is zinc deficiency a public health problem? Nutrition 11:87–92

Sharma A (2001) Marker-assisted improvement of pearl millet (*Pennisetum glaucum*) downy mildew resistance in elite hybrid parental line H 77/833-2. PhD thesis, Chaudhary Charan Singh Haryana Agricultural University, Hisar, Haryana, India

Shi R, Li H, Tong Y, Jing R, Zhang F, Zou C (2008) Identification of quantitative trait locus of zinc and phosphorus density in wheat (*Triticum aestivum* L.) grain. Plant Soil 306:95–104

Simic D, Sudar R, Ledencan T, Jambrovic A, Zdunic Z, Brkic I, Kovacevic V (2009) Genetic variation of bioavailable iron and zinc in grain of a maize population. J Cereal Sci 50(3):392–397

Simic D, Drinic SM, Zdunic Z, Jambrovic A, Ledencan T, Brkic J, Brkic A, Brkic I (2012) Quantitative trait loci for biofortification traits in maize grain. J Hered 103:47–54

Singh MV (2009) Effect of trace element deficiencies in soil on human and animal health. Bull Indian Soc Soil Sci 27:75–101

Snedeker SM, Greger JL (1981) Effect of dietary protein, sulfur amino acids, and phosphorus on human trace element metabolism. Nutr Rep Int 23:853–863

Snedeker SM, Greger JL (1983) Metabolism of zinc, copper, and iron as affected by dietary protein, cysteine, and histidine. J Nutr 113:644–652

Sotelo A, González-Osnaya L, Sánchez-Chinchillas A, Trejo A (2010) Role of oxate, phytate, tannins and cooking on iron bioavailability from foods commonly consumed in Mexico. Int J Food Sci Nutr 61(1):29–39

Stangoulis JCR, Huynh BL, Welch RM (2007) Quantitative trait loci for phytate in rice grain and their relationship with grain micronutrient content. Euphytica 154:289–294

Susanto U (2009) Mapping QTLs controlling iron and zinc contents in polished rice grains using SSR. Poster presentation on international rice genetic symposium, Manila, Philippines

Takkar PN, Walker CD (1993) The distribution and correction of zinc deficiency. In: Robson AD (ed) Zinc in soils and plants. Kluwer Academic Press, Dordrecht, pp 151–165

Tanksley SD, McCouch SR (1997) Seed banks and molecular maps: unlocking genetic potential from the wild. Science 277:1063–1066

Tiwari VK, Rawat N, Chhuneja P (2009) Mapping of quantitative trait loci for grain iron and zinc concentration in diploid a genome wheat. J Hered 100:771–776

Underwood EJ (1977) Trace elements in human nutrition, 4th edn. Academic, New York, p 545

Varshney RK, Tuberosa R (2007) Genomics-assisted crop improvement, vol. 1: genomics approaches and platforms, p. 386 and vol. 2: genomics applications in crops, p. 516. Springer, Houten

Velu G (2013) Biofortification strategies to increase grain zinc and iron concentrations in wheat. J Cereal Sci 59(3):365–372

Velu G, Rai KN, Sahrawat KL, Sumalini K (2006) Variability for grain iron and zinc contents in pearl millet hybrids. SAT eJournal 6:1–4

Vreugdenhil D, Aarts MGM, Koornneef M, Nelissen H, Ernst WHO (2004) Natural variation and QTL analysis for cationic mineral content in seeds of *Arabidopsis thaliana*. Plant Cell Environ 27:828–839

Waters BM, Pedersen JF (2009) Sorghum germplasm profiling to assist breeding and gene identification for biofortification of grain mineral and protein concentration. In: The proceedings of the international plant nutrition colloquium XVI, UC Davis, 26–30 Aug 2009

Weiss MG (1943) Inheritance and physiology of efficiency in iron utilization in soybeans. Genetics 28:253–268

Welch RM (1999) Effects of nutrient deficiencies on seed production and quality. Adv Nutr Res 2:205–247

Welch RM (2001) Micronutrients, agriculture, and nutrition: linkages for improved health and well-being. In: Singh K et al (eds) Perspectives on the micronutrient nutrition of crops. Scientific Publishers, Jodhpur, pp 247–289

Welch RM (2005) Biotechnology, biofortification, and global health. Food Nutr Bull 26:419–421

Welch RM, Graham RD (2004) Breeding for micronutrients in staple food crops from a human nutrition perspective. J Exp Bot 55:353–364

White PJ, Broadley MR (2005) Biofortifying crops with essential mineral elements. Trends Plant Sci 10:586–593

White PJ, Broadley MR (2009) Biofortification of crops with seven mineral elements often lacking in human diets – iron, zinc, copper, calcium, magnesium, selenium and iodine. New Phytol 182:49–84

White PJ, Brown PH (2010) Plant nutrition for sustainable development and global health. Ann Bot 105:1073–1080

Wise A (1995) Phytate and zinc bioavailability. Int J Food Sci Nutr 46:53–63

Wu J, Yuan YX, Zhang XW (2008) Mapping QTLs for mineral accumulation and shoot dry biomass under different Zn nutritional conditions in Chinese cabbage (Brassica rapa L. ssp. pekinensis). Plant Soil 310:25–40

Yadav RS, Hash CT, Bidinger FR, Cavan GP, Howarth CJ (2002) Quantitative trait loci associated with traits determining grain and stover yield in pearl millet under terminal drought stress conditions. Theor Appl Genet 104:67–83

Yilmaz A, Ekiz H, Torun B, Gultekin I, Karanlik S, Bagci SA, Cakmak I (1997) Effect of different zinc application methods on grain yield and zinc concentration in wheat grown on zinc-deficient calcareous soils in Central Anatolia. J Plant Nutr 20:461–471

Zamzam K (Fariba) Roughead, Zito CA, Hunt JR (2005) Inhibitory effects of dietary calcium on the initial uptake and subsequent retention of heme and nonheme iron in humans: comparisons using an intestinal lavage method. Am J Clin Nutr 82 (3):589–597

Zeng ZB, Houle D, Cockerham CC (1990) How informative is Wright's estimator of the number of genes affecting a quantitative character? Genetics 126:235–247

Zhang B, Chen P, Shi A, Hou A, Ishibashi T, Wang D (2009) Putative quantitative trait loci associated with calcium content in soybean seed. J Hered 100:263–269

Zhang X, Guoping Z, Longbiao G, Huizhong W, Dali Z, Guojun D, Qian Q, Dawei X (2011) Identification of quantitative trait loci for Cd and Zn concentrations of brown rice grown in Cd-polluted soils. Euphytica 180:173–179

Zhao FJ, Su YH, Dunham SJ et al (2009) Variation in mineral micronutrient concentrations in grain of wheat lines of diverse origin. J Cereal Sci 49:290–295

Zhou JR, Erdman JW Jr (1995) Phytic acid in health and disease. Crit Rev Food Sci Nutr 35:495–508

Understanding Genetic and Molecular Bases of Fe and Zn Accumulation Towards Development of Micronutrient-Enriched Maize

H.S. Gupta, F. Hossain, T. Nepolean, M. Vignesh, and M.G. Mallikarjuna

Abstract

Micronutrient malnutrition is a global problem afflicting billions of people worldwide. The effects are more prevalent in developing countries where people rely upon cereal-based diets that are inherently deficient in micronutrients. Micronutrients are required in less quantity but play critical role in the growth and development of humans. Since human body cannot synthesize micronutrients, they must be made available through diet. Among micronutrients, deficiency of iron (Fe) and zinc (Zn) has profound effects and require urgent attention. Development of micronutrient-rich staple plant foods through plant breeding, a process referred to as "biofortification," holds promise for sustainable food-based solutions to combat micronutrient deficiency. Maize is the third most important crop of the world, serving as staple food to billions of people in sub-Saharan Africa, Latin America and Asia. The development of Fe- and Zn-rich maize cultivar(s) would therefore have positive effects on health and well-being of humans. Wide variability has been reported for Fe and Zn in maize, which can be explored for genetic improvement of the trait. Genetics of Fe and Zn has been well elucidated, and genes/QTLs governing high Fe and Zn accumulation in maize have been identified. Moreover, by targeting the genes involved in Fe and Zn uptake, transportation and translocation, concentration of the same can be increased in the maize endosperm. Further, manipulating genes for promoter and antinutritional factors, bioavailability of Fe and Zn can be enhanced. Quality protein maize (QPM) genotype reported to have higher

H.S. Gupta (✉) • F. Hossain • T. Nepolean • M. Vignesh •
M.G. Mallikarjuna
Indian Agricultural Research Institute, New Delhi 110
012, India
e-mail: hsgupta.53@gmail.com; fh_gpb@yahoo.com;
tnepolean@gmail.com; pmvignesh@yahoo.co.in;
mgrpatil@gmail.com

A. Rakshit et al. (eds.), *Nutrient Use Efficiency: from Basics to Advances*,
DOI 10.1007/978-81-322-2169-2_17, © Springer India 2015

concentration of Fe and Zn provides opportunity to develop multinutrient-rich maize through a systematic breeding approach. We discussed here available genetic variation for Fe and Zn and their interactions with environments, relationship among micronutrients and grain yield, summary of research efforts with specific emphasis on mechanism of uptake and translocation, genetic and molecular basis of Fe and Zn accumulation, and the strategies that can be explored to breed for high Fe and Zn maize.

Keywords

Micronutrient deficiency • Maize • Fe and Zn • Variability • Accumulation • Bioavailability • Genetic improvement

1 Global Status of Micronutrient Malnutrition

Significant progress in agricultural research worldwide has resulted in manifold increase of food grain productions, yet over half of the global population is afflicted by micronutrient deficiency popularly phrased as "hidden hunger" (Khush et al. 2012; Stein 2010). Micronutrient malnutrition is a global problem but is particularly prevalent in developing countries where people rely upon cereal-based diets that are inherently deficient in micronutrients (Bouis and Welch 2010; Pfeiffer and McClafferty 2007). Micronutrients play a critical role in cellular and humoral immune responses, cellular signaling and function, work capacity, reproductive health, learning and cognitive functions (Guerrant et al. 2000; Kapil and Bhavna 2008). Since human body cannot synthesize micronutrients, they must be made available through diet. Inadequate consumption of these nutrients leads to adverse metabolic conditions resulting in poor health, impaired growth and socioeconomic losses, besides having profound effects on cognitive development, reproductive performance and work productivity (Bouis 2002; Welch and Graham 2004). Among the various micronutrients, the effects of deficiency caused by iron (Fe) and zinc (Zn) have been quite prominent (Dalmiya and Schultink 2003). It is estimated that over 60 % of the world's six billion people are Fe deficient, while it is 30 % for Zn

(White and Broadley 2009). These deficiencies affect people of all ages, but their effects appear more devastating in pregnant women and children especially infants. Considering the importance of the widespread micronutrient deficiency, Millennium Development Goals (MDGs) were adopted by the General Assembly of the UN, and 'child mortality' and 'maternal health' were included among the problems affecting the world population the most; and currently micronutrient malnutrition is considered to be one of the major public health challenges to humankind (Black et al. 2008; UNSCN 2004).

2 Role of Fe and Zn in Humans and Deficiency Symptoms

2.1 Iron

Humans require Fe for basic cellular functions and proper functioning of the muscle, brain and red blood cells (Roeser 1986). It serves as oxygen carrier to the tissues from the lungs by red blood cell haemoglobin and transport medium for electrons within cells and is an integral part of important enzyme systems in various tissues (Brock et al. 1994; Hallberg 1982). Most of the Fe in the body is present in the erythrocytes as haemoglobin, and it comprises of four units, each containing one haem group and one protein chain. Myoglobin, the Fe-containing oxygen storage protein and several other Fe-containing enzymes, like cytochromes, has one haem group and one globin protein chain. These enzymes act as electron carriers within the cell and transfer energy within the cell during the oxidative metabolism. Fe-containing enzymes such as cytochrome P450 plays a vital role in the synthesis of steroid hormones and bile acids and are responsible for detoxification of foreign substances in the liver (Mascotti et al. 1995).

Anaemia is considered as one of the hallmarks of Fe deficiency in human (DeMaeyer and Adiels-Tegman 1985). Fe-deficiency anaemia is present in all age groups and, thus, is a public health problem in most regions of the world. Worldwide, the highest prevalence of Fe deficiency is found in infants, children, adolescents and especially pregnant women (Lozoff et al. 1991). About one-fifth of perinatal mortality and one-tenth of maternal mortality in developing countries, particularly in Africa and Asia, are mainly due to Fe deficiency. Fe-deficiency anaemia in early childhood reduces intelligence in mid-childhood, and under most severity, it will also lead to mild mental retardation (Scrimshaw 1984). Besides, Fe deficiency also causes goitre and eye problems along with profound effects on cognitive development, growth, reproductive performance and work productivity (Bouis 2002).

2.2　Zinc

Zn is an essential mineral for humans, animals and plants for many biological functions. It plays a crucial role in more than 300 enzymes in the human body, for the synthesis and degradation of carbohydrates, lipids, proteins and nucleic acids (Sandstorm 1997). Zn is found in all parts of the body, *viz.*, organs, tissues, bones, fluids and cells. Muscles and bones contain most of the body's Zn; particularly high concentrations of Zn are in the prostate gland and semen (Frossard et al. 2000). Zn stabilizes the structure of cellular components and membranes and plays a critical role in the maintenance of cell and organ integrity. Furthermore, Zn plays an essential role in transcription and thus in the process of gene expression (Sandstorm 1997). It also plays a central role in the immune system, affecting a number of aspects of cellular and humoral immunity (Hambidge 1987; Shankar and Prasad 1998).

The clinical symptoms of severe Zn deficiency in humans are growth retardation, delayed sexual and bone maturation, skin lesion, diarrhoea, impaired appetite, increased susceptibility to infections mediated via defects in the immune system and the appearance of behavioural changes (Prasad 1996). A reduced growth rate and impairments of immune defence are so far the only clearly demonstrated signs of mild form of Zn deficiency in humans. Severe form of Zn deficiency is characterized by short stature, hypogonadism, impaired immune function, skin disorders, cognitive dysfunction and anorexia (Brown et al. 1998; Goldenberg et al. 1995). Besides, depression, psychosis, altered reproductive biology and gastrointestinal problems are some of the important symptoms that are caused due to Zn deficiency (Solomons 2003).

Thus, micronutrient deficiency has become one of the challenging problems affecting people across age groups causing significant socioeconomic losses. Though strategies such as fortification, supplementation and dietary diversification have been in place worldwide to ameliorate the micronutrient deficiency in humans, development of micronutrient-enriched or biofortified crops holds immense promise due to its sustainability, cost-effectiveness and the ability of micronutrient to reach the target group in pure form (Banziger and Long 2000; Bouis and Welch 2010). In this context, understanding the extent of genetic variability, nature of inheritance, interactions with the environment, genetic factors affecting accumulation and bioavailability of micronutrients holds importance to develop an effective breeding strategy towards development of micronutrient-rich maize cultivars.

3　Mechanism of Uptake, Transportation and Translocation

Increased uptake of minerals by roots or leaves, effective redistribution of minerals within the plants and accumulation of minerals in edible portion of the plants are the three important regulations that are mainly responsible for enhancement of micronutrients (Welch and Graham 2005). The molecular mechanisms involved in uptake, transportation and translocation are tightly regulated and governed by complex regulation of multiple genes (Ghandilyan et al. 2006). A brief overview of mechanisms involved in the uptake and distribution of mineral elements in plants are described below.

3.1 Uptake of Minerals from Rhizosphere

The minerals present in the rhizosphere are required to be mobilized by plants so that they become available for uptake by plants. Among the various molecular and physiological mechanisms, membrane-bound transport systems such as ATP-powered pump, channel proteins and co-transporter are responsible for uptake of minerals from the soil. While ATP-powered pump in the plant root cell generates proton motive force across the plasma membrane leading to the ion uptake, channel proteins facilitate diffusion of water and ions. In case of co-transporter, the different type of membrane transport proteins can move solutes either up or down gradients (Chrispeels et al. 1999). The basic mechanisms of uptake of Fe and Zn by plants are described below.

3.1.1 Iron

Mechanism of Fe uptake differs quite significantly in dicots and monocots in the plant system (Frossard et al. 2000; Graham and Stangoulis 2003). In case of dicots and non-graminaceous monocots, a membrane-bound ferric reductase that is linked to a divalent ion transporter or channel and ATP-driven proton extrusion pump plays a pivotal role. Reduction of ferric (Fe^{3+}) to ferrous (Fe^{2+}) form on the root surface is the most important process for Fe absorption from soil in this group of plants (Yi and Guerinot 1996). The ferric reductase reduces the Fe^{3+} to readily absorbable Fe^{2+} form, and with the help of ATP-driven proton extrusion pump, the Fe^{2+} is

absorbed within the root cells. The Fe^{2+} is taken up by the Fe transporter (IRT), a member of ZIP-like transporter (Hell and Stephan 2003). Genes, responsible for ferric chelate reductase (*FRO2*) and an ion transport protein (IRT1), have been cloned and once expressed in *Arabidopsis*; increased uptake of Fe was reported in an Fe-deficient soil (Eide et al. 1996; Guerinot and Yi 1994: Robinson et al. 1999). The expression of *FRO3* like *FRO2* is strongly induced upon Fe deficiency, which suggests that the latter gene has a similar function (Wu et al. 2005). This strategy acts as the constitutive system of supply of Fe in a well-aerated healthy and Fe-sufficient soil.

In case of graminaceous monocots, plant synthesizes and releases non-proteinaceous amino acids known as phytosiderophores (structural derivatives of mugineic acid). Phytosiderophores once released in the soil chelate with the Fe^{3+} and form stable Fe in soil (Roberts et al. 2004). A highly specific Fe transport system then transports Fe^{3+}- phytosiderophores across the plasma membrane of the root cell (Graham and Stangoulis 2003). This highly specific Fe transport system is unique, and genes encoding this transporter belong to *yellow stripe1* (YS1)-like protein family (Curie et al. 2001) or natural resistance-associated macrophage protein (NRAMP) family (Curie et al. 2000; Thomine et al. 2003) or the interferon-y-responsive transcript (IRT-1) family (Eide et al. 1996). The first gene *YS1* encoding Fe^{3+}-phytosidophores has been identified in maize (Curie et al. 2001). *YS1* belongs to the oligopeptide transporter (OPT) family, and 8 and 19 homologs (*yellow stripe* like: YSL) have been found in *Arabidopsis* and rice, respectively (Curie et al. 2001; Jean et al. 2005). Although phytosiderophores predominantly bind to Fe, it can also transport Zn, Cu and Mn across the cell membrane (Ueno et al. 2009).

3.1.2 Zinc

Zn prefers to enter the graminaceous system as divalent cation (Zn^{2+}) rather than phytosiderophore-mediated chelated complex (Bell et al. 1991; Halvorson and Lindsay 1977; Norvell and Welch 1993). However, Von-Wiren et al. (1996) demonstrated that roots of

Fe-efficient maize do absorb Zn in the form of Zn-phytosiderophores. Higher uptake rates of free Zn compared to its chelated species imply that free Zn^{2+} remains the preferential form for Zn uptake even in the presence of Zn^{2+}-phytosiderophores (Frossard et al. 2000). This could be attributed to the higher transport capacity of the Zn^{2+} transporter over phytosiderophore transporter for Zn^{2+}-phytosiderophores. Two Zn transporters (ZIP1 and ZIP3) from A. thaliana have recently been cloned, and they express in roots in response to Zn deficiency (Grotz et al. 1998). As a response to low Zn availability, graminaceous plants can induce the release of phytosiderophores (Cakmak 2008; Walter et al. 1994), which may accumulate in the rhizosphere to concentrations of up to 1 μm (Shi et al. 2008). Von-Wiren et al. (1996) argued with regard to the coincidence of Fe and Zn deficiencies in calcareous soils, the direct uptake of Zn^{2+}-phytosiderophores via the phytosiderophore transporter might be an ecological advantage for graminaceous plants to cover their demand for Zn.

3.2 Transportation of Minerals from Root to Shoots

Once Fe and Zn are absorbed within the root cells, they are transferred to xylem vessels for transportation to shoot portions. There are several transporters identified that mediates translocation of Fe and Zn in plants (Clemens et al. 2002; Hall and Williams 2003; Maser et al. 2001). The heavy metal ATPase (HMA) genes like AtHMA2, AtHMA3 and AtHMA4 expressed in the vascular tissue play a role in transportation of Zn (Axelsen and Palmgren 2001; Eren and Arguello 2004; Hussain et al. 2004). The FRD3 gene plays significant role in transportation of Zn, and in the absence of functional FRD3 protein, the root to shoot transportation of Fe is severely inhibited (Green and Rogers 2004). However, not all the amount of Fe and Zn absorbed by the roots are transported to the shoot portion. Some amount of the minerals are stored in the root itself, probably

in the vacuoles, and once the demand from the shoot increases, the stored Fe and Zn are then again transported to the shoot portion (Ghandilyan et al. 2006). The natural resistance-associated macrophage protein (NRAMP) transporters are most likely controlling the mobilization of metals to and from the vacuolar pool (Thomine et al. 2003). Interestingly, while being transported through the xylem vessels, the positively charged cations can be withdrawn from the xylem vessels by binding the negatively charged cell wall components along the stem. This is very evident in case of polyvalent Ca^{2+} which is mostly withdrawn midway of the travel to the leaf blades. On the contrary, Fe and Zn ions are generally complexed with low-molecular-weight organic compound which protects them from being withdrawn midway and their concentration remains high in the leaf blade.

3.3 Partitioning of Minerals to Different Plant Parts

Though minerals are distributed over all the tissues, certain plant parts accumulate more minerals than the others. Partitioning of minerals also depends upon the stage of the crop growth (Akman and Kara 2003) and the position of the storage organ in the plant (Calderini and Ortiz-Monasterio 2003). Carbohydrates produced in the green leaves due to the activity of photosynthesis are loaded into the phloem for transportation to all parts of the plants either for direct usage or for storage for future purposes. During loading into the phloem vessels, the minerals such as Fe and Zn are also loaded along with the carbohydrates for their transportation to different parts. Fe is found to make complex with nicotianamine and move as a chelate (Von-Wiren et al. 1999), and loading of Fe in the phloem is conditioned by the availability of the chelator rather than the availability of Fe ions (Grusak 2000). However, the dependency of the Zn loading on the Zn-chelator is yet to be established, although both the Fe and Zn are easily transported through phloem. Interestingly, during the deficiency, remobilization of Fe and Zn from the rich older leaves takes time till the senescence is initiated in the older tissue, and thus, the deficiency symptoms become more prominent in the younger leaves rather than the older leaves.

In general mineral concentration in the leaf blades on a dry matter basis is higher than in other tissues (Frossard et al. 2000). Minerals after uptake from soil move from root to shoot via xylem tissue that ends at leaf blades. Accumulation of minerals such as Fe is mainly determined by transpiration rate of the leaves (Marschner 1995). The leaf blades are the sites of high metabolic activity pertaining to photosynthesis and transpiration and therefore require higher amount of Fe and Zn. Leaf chloroplasts contains nearly 80 % of the Fe which is mainly utilized for photosynthetic redox reactions (Terry and Low 1982). Zn acts as cofactor in many of the important enzymes that are involved in photosynthetic processes and detoxification of oxygen-free radicals produced when light energy is absorbed and photosynthetic redox activity is high (Frossard et al. 2000). The concentration of Fe and Zn in the seeds is less as compared to the leaf; however, consumption of larger amount of cereal grains makes it important as a source of Fe and Zn. Accumulation of Fe in grain involves several genes that are involved in the processes such as chelation, membrane transport and deposition of phytate (Graham and Stangoulis 2003). The deposition of Fe and Zn in the roots is probably least due to the fact that minerals are supplied by phloem rather than xylem and the metabolic activity is also much less as compared to the green leaves.

4   Distribution of Fe and Zn in Kernel

Maize kernel is constituted of embryo (12 %), endosperm (82 %) and pericarp (6 %) (Watson and Ramstad 1987). Concentration of micronutrients like Fe and Zn in maize is reported to be highest in the seed coat and scutellum as compared to endosperm (Bityutskii et al. 2002). However, since endosperm occupies greater proportion of the maize kernel, the total content of these micronutrients are much higher in the endosperm (60–80 %), followed by scutellum (15–35 %) and seed coat (8–12 %). In contrast, seed coat contains more than 50 % of the micronutrient in the wheat kernels (Moussavi-Nik et al. 1998). Hence, endosperm and scutellum are the major micronutrient reserves in the maize kernel as compared to other cereal grains (Bityutskii et al. 2001, 2002).

Major portion of Fe and Zn is located in the aleurone, the outer most layer of the endosperm just beneath the pericarp and an integral part of maize bran (Banziger and Long 2000; Ortiz-Monasterio et al. 2007; Puga and Kutka 2012; Wang et al. 2013). Aleurone layer in maize kernel is made up of single-layered cells; however, occasional doubling of individual aleurone layer has been reported in some of the yellow dent maize (Wolf et al. 1972). Multiple aleurone layers (MAL) with an average of 2.0–3.7 layers per kernel and maximum of six layers in some of the kernels have been reported in Coroico maize landrace found in Bolivia, Peru, Ecuador and Brazil (Wolf et al. 1972). Cells of the outer most aleurone layer were quite large, while cell size in the inner layers was progressively smaller in size. Shen et al. (2003) later reported the presence of seven layers of aleurone cells in *supernumerary aleurone-1* (*sal-1*) maize mutant. While *sal-1-1* showed up to seven layers of aleurone cells, *sal-1-2* had 2–3 layers of aleurone cells. MAL appeared to be inherited by a single gene with partial dominance (Wolf et al. 1972; Nelson and Chang 1974). The cloning of *sal1* gene suggested that it is a homolog of human chromatin-modifying protein-1 (*Chmp1*) that is responsible for vacuolar protein sorting responsible in membrane vesicle trafficking.

Since majority of Fe and Zn is located in the aleurone layers, increase in proportion of aleurone layers in relation to starchy endosperm is advantageous in increasing the micronutrient concentration (Wolf et al. 1972). Welch et al. (1993) studied maize genotypes with single and multiple aleurone layers for various micronutrients in kernels. Genotypes with single aleurone layer had 22.3 mg/kg of Fe, while the same was found to be 26.5 mg/kg in MAL with an increase of 19 % Fe. In case of Zn, 31.3 mg/kg was observed in MAL as compared to 22.5 mg/kg in single aleurone genotypes, with a 39 % increase of Zn. Similar observations of significant increase in concentration in MAL were also observed in case of Mn, Cu and Ca (Welch et al. 1993).

5 Variability of Fe and Zn

Based on the estimated average requirement (EAR) of 1,460 μg/day of Fe in nonpregnant and non-lactating women, the target level of Fe in maize on dry weight basis (having 90 % retention after processing and 5 % bioavailability) has been fixed at 60 μg/g (Bouis and Welch 2010). In case of Zn, with EAR of 1,860 μg/day and having 90 % retention and 25 % bioavailability, the same has been fixed at 38 μg/g. It is therefore important to look for existing natural variations for kernel Fe and Zn in maize that can be potentially utilized in the development of cultivar enriched with kernel micronutrients. Various research groups worldwide have studied the extent of genetic variation for kernel Fe and Zn among diverse maize genotypes that included core collection, landraces, populations, varieties, inbreds and hybrids (Table 1).

5.1 Kernel Fe

Banziger and Long (2000) at CIMMYT, Mexico, evaluated 1,814 maize germplasm under 13 different trials in Zimbabwe and Mexico. These germplasm included white landraces from CIMMYT's core collection, white and yellow pools and populations, active breeding germplasm of CIMMYT-Zimbabwe breeding program and released white cultivars of South Africa. Kernel Fe among these diverse germplasm varied from 9.6 to 63.2 mg/kg during 1994–1999. Similar range of wide variability (0–71 mg/kg) among 2019 diverse inbred lines from Brazil was also observed by Guimaraes et al. (2004). In India, Prasanna et al. (2011) and Agrawal et al. (2012) reported significant variation for kernel Fe with a

range of 11.28–60.11 mg/kg and 20.38–54.29 mg/kg, respectively, while evaluating inbred lines and local landrace collections. Chen et al. (2007) generated 36 maize hybrids from high Fe inbreds of China and reported a range of 45.9–69.1 mg/kg. However, much broader range of 13.60–159.43 mg/kg was found by Maziya-Dixon et al. (2000) among 109 mid-altitude and lowland inbred lines evaluated at Ibadan, Nigeria. In contrast, much narrower range (~10–40 mg/kg) of kernel Fe was observed among inbred lines, hybrids and OPVs evaluated in Africa (Long et al. 2004; Menkir 2008; Oikeh et al. 2003a, b, 2004a), Ethiopia and Mexico (Pixley et al. 2011), Brazil (Queiroz et al. 2011), Croatia (Brkic et al. 2003) and India (Chakraborti et al. 2009, 2011a, b).

5.2 Kernel Zn

Kernel Zn among elite inbred lines developed at IITA for the mid-altitude and lowland agro-ecologies of West and Central Africa varied from 11.65 to 95.62 mg/kg (Maziya-Dixon et al. 2000). Banziger and Long (2000) reported that kernel Zn concentration ranged from 12.9 to 57.6 mg/kg among diverse set of heterogeneous and homogeneous maize germplasm in Zimbabwe and Mexico. Similar range of 15.14–52.95 mg/kg, 4–63 mg/kg, 15–47 mg/kg, 13.44–46.39 mg/kg, 17.57–49.14 mg/kg, 21.85–40.91 mg/kg, 14–45 mg/kg and 17.5–42.0 mg/kg were observed by Prasanna et al. (2011), Guimaraes et al. (2004), Ortiz-Monasterio et al. (2007), Chakraborti et al. (2009, 2011a, b), Menkir (2008) and Queiroz et al. (2011), respectively, among maize genotypes. However, much smaller range of 19.3–30.9 mg/kg (Pixley et al. 2011), 16.0–23.6 mg/kg (Brkic et al. 2003), 16.5–20.5 mg/kg (Oikeh et al. 2003a); 16.5–24.6 mg/kg (Oikeh et al. 2003b), 19.4–24.6 mg/kg (Oikeh et al. 2004a), 18.1–29.8 mg/kg (Long et al. 2004), 3.81–35.83 mg/kg (Guleria et al. 2013) and 7.01–29.88 mg/kg (Agrawal et al. 2012) were also observed.

The presence of considerable variability for kernel Fe and Zn among diverse maize genotypes suggests that using natural variations, it is

Table 1 Details of variability of kernel Fe and Zn in selected studies since 2000

| | Range | | | No. of | Place of | |
S. No.	Fe (mg/kg)	Zn (mg/kg)	Type of germplasm	germplasm	evaluation	References
1.	9.6–63.2	12.9.1–57.6	Landrace, pools, population, active breeding germplasm, released cultivar	1,814	Zimbabwe and Mexico	Banziger and Long (2000)
2.	13.60–159.43	11.65–95.62	Inbred lines	109	Nigeria	Maziya-Dixon et al. (2000)
3.	13.6–30.3	16.0–23.6	Hybrids	28	Croatia	Brkic et al. (2003)
4.	15.5–19.1	16.5–20.5	Varieties	20	Nigeria	Oikeh et al. (2003a)
5.	16.8–24.4	16.5–24.6	Varieties	49	Nigeria	Oikeh et al. (2003b)
6.	19.2–24.4	19.4–24.6	Varieties	20	Nigeria	Oikeh et al. (2004a)
7.	15.9–28.1	18.1–29.8	Inbreds	14	Zimbabwe	Long et al. (2004)
8.	0–71	4–63	Inbreds	2009	Brazil	Guimaraes et al. (2004)
9.	45.9–69.1	–	Inbreds, hybrids	9, 36	China	Chen et al. (2007)
10.	11–39	15–47	Improved genotypes, core accessions	1,400	-	Ortiz-Monasterio et al. (2007)
11.	11–34	14–45	Inbreds	310	Nigeria	Menkir (2008)
12.	13.23–40.09	13.44–46.39	Inbreds	25	India	Chakraborti et al. (2009)
13.	12.02–38.46	17.57–49.14	Inbreds, hybrids	7, 42	India	Chakraborti et al. (2011a)
14.	13.95–39.31	21.85–40.91	Inbreds	31	India	Chakraborti et al. (2011b)
15.	12.9–26.5	19.3–30.9	Hybrids	42	Mexico and Ethiopia	Pixley et al. (2011)
16.	11.28–60.11	15.14–52.95	Inbreds, landraces	30	India	Prasanna et al. (2011)
17.	12.2–36.7	17.5–42.0	Inbreds	22	Brazil	Queiroz et al. (2011)
18.	20.38–54.29	7.01–29.88	Inbreds, landraces	67	India	Agrawal et al. (2012)
19.	–	3.81–35.83	Inbreds, landraces	81	India	Guleria et al. (2013)

possible to achieve the target levels set by HarvestPlus in maize (Bouis and Welch 2010).

5.3 Ear-Leaf Fe and Zn

Mineral concentration in the leaves has been found to be much higher as compared to grain (Frossard et al. 2000). Fewer studies on variability for leaf Fe and Zn in maize have been reported worldwide. Chen et al. (2007) found that ear-leaf Fe concentration varied from 219 to 636 mg/kg among the inbred lines, while the same for hybrids were 287–653 mg/kg. Kovacevic et al. (2004) evaluated 20 maize

hybrids in Croatia during 2000–2001. Ear-leaf Fe among the hybrids varied from 137 to 222 mg/kg during 2 years, with a mean of 184 mg/kg and 147 mg/kg during 2000 and 2001, respectively. In case of Zn, much smaller range of 16.6–30.0 mg/kg in ear leaf was observed, with a mean of 22.5 mg/kg across trials.

The genetic control of mechanism pertaining to accumulation of Fe and Zn in leaf blade and grain may be quite different (Chen et al. 2007). This is also evident from the fact that Gorsline et al. (1964) found no association between grain and ear-leaf Fe and Zn concentration in maize and hypothesized that genetic factors controlling the accumulation of micronutrients in two different tissues were independent. It is also interesting to note that although QTLs for Fe and Zn in maize kernel and cob are co-localized, there was no significant correlation observed between Fe and Zn concentration of kernel and cob. This emphasizes the existence of complex molecular mechanism in mineral mobilization and accumulation in various tissues (Qin et al. 2012).

6   Genetics of Accumulation of Fe and Zn in Kernel

The polygenic nature of gene action on accumulation of kernel Fe and Zn in maize was observed by Gorsline et al. (1964) and Arnold and Bauman (1976). Recently, QTL mapping experiments by Baxter et al. (2013), Lungaho et al. (2011), Qin et al. (2012) and Simic et al. (2011) provided further insight to the genetics of accumulation of kernel Fe and Zn in maize, where it was found that it is under the control of numerous genetic loci. Partitioning of genetic variance has direct bearing upon the choice of breeding procedures to be followed. Several studies have been aimed at exploring the nature and magnitude of gene action worldwide. Gorsline et al. (1964) and Arnold and Bauman (1976) in their diallel studies reported that additive gene action was more important than nonadditive gene action for both kernel Fe and Zn concentration. Long et al. (2004) generated 91 hybrids and evaluated them at six locations in Zimbabwe during 1999–2000. Significant GCA effects were observed in flour Fe and Zn concentration, while SCA effects were nonsignificant, indicating the importance of additive gene action. Brkic et al. (2003) evaluated 28 diallel-based hybrids and also reported preponderance of additive gene action for kernel Fe and Zn. The importance of additive gene action was similarly observed by Pixley et al. (2011) and Simic et al. (2011). Preponderance of additive gene action has also been the feature of ear-leaf Fe and Zn (Chen et al. 2007; Gorsline et al. 1964). Significant contribution of GCA effects in majority of the studies suggested that in general per se performance of inbreds should be a good indicator of hybrid performance.

In contrast, Qin et al. (2012) in QTL mapping experiment reported that genetic effects for kernel Fe and Zn was predominantly of partially dominant and overdominant nature. On the other hand, Chakraborti et al. (2010) observed that additive gene action was of higher magnitude as compared to dominance for kernel Fe, while dominance was relatively higher in case of kernel Zn. Besides, epistatic interaction such as additive × dominance component was significant for kernel Fe, whereas additive × additive component was predominant for kernel Zn. Thus, the best inbred lines selected from different trials can be used as parents to exploit both additive and nonadditive gene actions to increase the concentration of minerals in kernels (Menkir 2008).

Heritability is one of the important parameters that determine the success of selection for the target trait in the breeding program. Qin et al. (2012) found that heritability for kernel Fe was <0.60, while the same for kernel Zn was >0.70. Simic et al. (2011) could find heritability for kernel Fe and Zn as 0.64 and 0.60, respectively. Similar values of heritability for kernel Zn (0.59) and much lesser value for kernel Fe (0.46) were observed by Simic et al. (2009). Pixley et al. (2011) and Lungaho et al. (2011) found heritability of Fe as 0.73 and 0.74, respectively. Baxter et al. (2013) found that heritability for Fe and Zn were 0.55 and 0.69, respectively, from the QTL mapping experiments with IBM populations derived from B73 × Mo17. The extent of heritability found in different studies suggests considerable influence of genes/QTLs in determining kernel Fe and Zn in maize. Transgressive segregants were also observed in $F_{2:3}$ families primarily due to segregation and recombination of minor QTLs (Qin et al. 2012). Lungaho et al. (2011) also detected the same for kernel Fe from a cross between B73 × Mo17. Transgressive segregants for kernel Fe and Zn caused primarily due to additive effects have also been recovered in different crosses attempted in India under the Maize Biofortification Program (Personal communication by Dr. P.K. Agrawal, VPKAS, Almora, India). Segregants from high

× high crosses hold considerable promise in developing lines with enhanced Fe and Zn concentration.

7 Relationship of *opaque2* Allele with Kernel Fe and Zn

Maize storage proteins, zeins are deficient in two essential amino acids, lysine and tryptophan. The recessive *opaque2* allele was found to alter the amino acid composition of the endosperm protein, resulting in enhanced concentration of lysine and tryptophan (Mertz et al. 1964). This finding led to the development of high lysine maize, popularly called as quality protein maize (QPM) (Bjarnason and Vasal 1992; Gupta et al. 2009, 2013a, b; Prasanna et al. 2001). Besides, *opaque2* mutant was found to have pleotropic effect on the accumulation of micronutrients in maize. Arnold et al. (1977) compared normal and *opaque2* kernels of a heterozygous population of maize for various micronutrients. Kernel Fe and Zn in normal kernels were 15.64 mg/kg and 19.40 mg/kg, while in *opaque2* kernels, the same were 17.71 mg/kg and 23.90 mg/kg, respectively. Besides, the correlation between lysine and kernel Zn was significantly positive in both homozygous *opaque2* and heterozygous population segregating for *opaque2* allele. Welch et al. (1993) compared five maize lines, B8, NY821, OH51A, W23 and W64A with their *opaque2* (*o2*) version for several micronutrients including Fe and Zn. B8-*o2* showed 19 % and 35 % increase in Fe and Zn concentration, respectively, over B8, while the increase for W64A-o2 was 3 % and 16 %, respectively. Other inbreds OH51A-o2 had 16 % increase in Zn, but it did not show any change for kernel Fe. Chakraborti et al. (2009) reported significant difference between normal and QPM inbreds for kernel Zn concentration, although the same could not be found for kernel Fe. The enhanced concentration of Zn in maize kernel could be due to the direct influence of *opaque2* and other closely linked genes (Arnold et al. 1977). Besides, modifier genes that are influencing the *opaque2* gene

in lysine concentration may also be influencing the Zn concentration. The greater concentration of kernel Fe was attributed to lower kernel density of *opaque2* kernels (Arnold et al. 1977). Interestingly, Welch et al. (1993) found that some of the *opaque2* version of the inbreds showed less kernel Fe and Zn concentration as compared to normal version, thereby suggesting that although *opaque2* may play an important role, there might be other favourable loci that are required to be present along with the *opaque2* gene for the enhancement of micronutrients.

8 Association of Kernel Fe, Zn and Grain Yield

Several studies have reported significant positive correlation between kernel Fe and Zn (Arnold et al. 1977; Baxter et al. 2013; Brkic et al. 2003; Chakraborti et al. 2009; Guimaraes et al. 2004; Lungaho et al. 2011; Maziya-Dixon et al. 2000; Menkir 2008; Oikeh et al. 2003a, b). This could be possibly due to linkage between the genes affecting the accumulations or pleotropic effects of the genes governing the accumulation of micronutrients. A large number of genes encode metal transporter proteins and some of which transport multiple metals (Qin et al. 2012). Positive correlation was observed between kernel Fe and Zn by Qin et al. (2012), and interestingly, QTLs for both the traits were found to co-localize on bin 2.08 and 9.07, thereby suggesting the feasibility of simultaneous improvement of the both. In contrast, Simic et al. (2009) reported weak association between kernel Fe and Zn, while no association was observed in studies by Agrawal et al. (2012), Arnold and Bauman (1976) and Prasanna et al. (2011). This suggests that different sets of genes governing the accumulation of micronutrients are there, and genetic improvement could be undertaken independent of each of the traits. This is evident from the fact that majority of the QTLs identified for kernel Fe and Zn are unique (Simic et al. 2011). This contrast could be attributed to the inherent nature of the specific type of germplasm used in these studies. Pixley et al. (2011) could find significant positive correlation only in testcross hybrid trial, while the same was not observed in other two hybrid trials.

Grain yield is one of the most important characters that determine the acceptability of a cultivar among the farming community, and it is thus imperative to understand the genetic relationship of grain yield with kernel micronutrients. Banziger and Long (2000) observed negative correlation between grain yield and kernel micronutrients (Fe and Zn) in various experiments conducted between 1994 and 1998–1999. The negative association resulted into many low yielding maize germplasm having high kernel Fe and Zn. Increased carbohydrate content in the grain of high yielding genotypes possibly dilutes the concentration of the micronutrients (Banziger and Long 2000). Three differential doses of N fertilizer were applied in soil, and negative correlation was observed between doses of N fertilizer with accumulation of Zn in kernel (Feil et al. 2005). Dilution effect caused due to higher grain yield as a result of fertilizer application contributed to the inverse relationship. The extent of relationship may also be affected by specific environments (Simic et al. 2009). For example, correlation coefficients of grain yield with kernel Fe and Zn varied from −0.60 to 0.16 and −0.44 to −0.03, respectively, across 12 maize trials (Banziger and Long 2000). Besides, nature of germplasm is also an important factor in determining the degree and extent of association. Pixley et al. (2011) reported negative correlation between kernel Fe and grain yield in only one of the three trials, while kernel Zn did not show any correlation with grain yield across all three trials. Long et al. (2004) found negative correlation between grain yield and kernel Fe; however, no association was found between grain yield and kernel Zn. Baxter et al. (2013) found negative but weak association between kernel weight and kernel Fe and Zn. Despite negative relationship of grain yield with one or both micronutrients, it is possible to potentially select for high yielding maize germplasm that also possess high Fe and Zn concentration (Banziger and Long 2000).

On the other hand, Brkic et al. (2003), Lungaho et al. (2011), Menkir (2008) and Simic et al. (2009) detected no association between grain yield and that of kernel Fe and Zn. Chakraborti et al. (2009) and Mi et al. (2004) also observed no association between grain yield and kernel Fe. This suggested that improvement of kernel Fe and Zn can be undertaken without reducing the yield potential. In contrast, Chakraborti et al. (2009) observed positive correlation between grain yield and kernel Zn. This could be due to the fact that micronutrient dense seeds are associated with greater seedling vigour and in turn would produce more yield (Ruel and Bouis 1998). Besides, micronutrient efficient variety normally has better and deeper root system in micronutrient-deficient soil and is better able to tap subsoil water and mineral. This would ensure better absorption of minerals and water from the deep into the soil, thereby making the variety more disease resistant and drought tolerant. Zn-efficient variety would increase the disease resistance due to higher uptake of Zn and would in turn reduce the cost of fungicides (Ruel and Bouis 1998).

9 Stability of Kernel Fe and Zn Expression

The extent of interaction of genotypes with environments is one of the key factors that determine the course of breeding program. A promising genotype with stable trait expression can effectively be utilized as common donor or directly used as parent in the crossing program, across environments. Menkir (2008) reported no significant genotype × location (G × L) interaction for kernel Fe (except one out of eight trials) and Zn concentration in maize. Similar stable nature of kernel Fe and Zn across locations was also observed by Pixley et al. (2011). However, Oikeh et al. (2003a, b), while evaluating 20 early and 49 late-maturing maize varieties, respectively, at three locations detected significant G × L interactions for both kernel Fe and

Zn. Long et al. (2004) also detected significant G × L interactions in a set of 91 maize hybrids evaluated at six sites of Zimbabwe during 1999–2000. Significant G × L interactions for both kernel Fe and Zn concentration in maize have also been reported by Chakraborti et al. (2009) while experimenting with a set of inbred lines at two locations. Oikeh et al. (2004b) further evaluated 20 maize hybrids in three locations representing main growing agro-ecologies of West and Central Africa during 2000 and 2001. Significant effects of genotypes × year (G × Y) and G × L interactions for kernel Fe, and genotype × year × location (G × Y × L) interactions for kernel Zn, were detected. Qin et al. (2012) observed that there were some of the QTLs for both kernel Fe and Zn that could be detected in either of the two locations, suggesting strong QTL × location interaction. Interestingly, significant effects of year on kernel Fe and Zn was observed by Prasanna et al. (2011) and Agrawal et al. (2012) in a set of maize genotypes evaluated during 2006, 2007 and 2008. While Simic et al. (2009) reported significant G × Y kernel Fe, Guleria et al. (2013) detected the same for kernel Zn. Significant QTL × year interaction for kernel Zn in maize was also reported by Simic et al. (2011).

Oikeh et al. (2003b) found that proportion of sum of square of G × E over total sum of square (TSS) was 8 % for kernel Fe and 28 % for kernel Zn, respectively. Oikeh et al. (2004b) observed that same for kernel Fe was 28 %, while it was 35 % for kernel Zn. Similar trend with much higher proportion of G × E variance (kernel Fe: 31.77 % of TSS; kernel Zn: 58.37 % of TSS) was observed by Prasanna et al. (2011). Greater proportion of G × E variance in case of kernel Zn than the kernel Fe is perhaps indicative of the sensitivity of kernel Zn to the soil and microclimatic conditions. In contrast, Chakraborti et al. (2011b) reported that the sum of square for G × E interaction for Fe consisted of 47.71 % of the TSS, while the same was 35.35 % for the Zn concentration. Agrawal et al. (2012) also found that proportion of G × E variance for kernel Fe concentration was 38.62 % of TSS, while for the kernel Zn concentration, it was 27.67 %. Oikeh et al. (2003a), on

the other hand, reported almost equal proportion of G × E variance for both kernel Fe (17 % of TSS) and Zn (19 % of TSS). Among various environments, location was found to have more effects on kernel Fe and Zn than the effects of years/seasons. Oikeh et al. (2004b) observed that in case of kernel Fe, G × L constituted 55 % of G × E variance, while the same for G × Y and G × Y × L were 23 % and 22 %, respectively. For kernel Zn, variance for G × L, G × Y and G × Y × L were 38 %, 28 % and 34 % of total G × E variance, respectively. This suggested that although location remains the major contributing factor for G × E, variations in different year also contributes considerably towards total variations.

Micronutrient concentration is mainly affected by various factors such as soil type and fertility, soil moisture, environmental factors and interactions among nutrients (Arnold and Bauman 1976; Arnold et al. 1977; Feil et al. 2005; Gorsline et al. 1964; Qin et al. 2012). Feil et al. (2005) conducted experiments with four maize varieties of Thailand and evaluated them in two water regimes (irrigated and drought) and three N fertilizer regimes (0, 80 and 160 N ha^{-1}). The study revealed significant effect of doses of N fertilizer application on kernel Zn micronutrients, while water regimes did not show any effects. Higher doses of N fertilizer resulted in indirect reduction of kernel Zn, primarily due to dilution effects caused due to higher grain yield achieved in high N soil. Chiripa et al. (1999) evaluated four Romanian maize hybrids under four water regimes (optimum, 60 % of optimum, 30 % of optimum and nonirrigated) and recorded kernel Fe and Zn concentration of plant samples in 3–4 leaf stage, flowering stage and maturity stage. Interestingly kernel Fe registered higher concentration with the increase of water supply, and in contrast kernel Zn was higher with increase of water stress. Agrawal et al. (2012) evaluated maize genotypes under similar soil profile during 2006–2008 in Northern Himalayan hilly region of India. Year 2006 experienced 548.5 mm of total rainfall during the crop growth period, while it increased substantially during 2007 (613.5 mm) and 2008 (748.5 mm). Mean kernel

Fe among genotypes was found to be the highest during 2008, while 2006 experienced highest mean for kernel Zn. Ferreira et al. (2012) reported similar observation among 10 corn cultivars grown at Rolandia County, Parana State, Brazil, during 2006 and 2007. Kernel Fe concentration was significantly higher during 2007, a year that experienced higher rainfall than the preceding year, while Zn concentration in maize kernel was higher during 2006 which experienced comparatively dry. Temporary water logging could have favoured reduction reactions thereby increasing available Fe to the plants during 2007 (Ferreira et al. 2012). Prasanna et al. (2011) and Guleria et al. (2013) also reported the effect of dry season on higher accumulation of kernel Zn in maize. The study emphasized significant effects of soil water availability on grain nutrient concentration.

The performance of a genotype is a function of genotypic constitution, its interaction with environment, and the possible involvement of a complex network of diverse factors related to soil dynamics and microclimates (Feil et al. 2005; Gorsline et al. 1964; House et al. 1996). Even minor changes in one factor in combination with other factors may also lead to significant variation in micronutrient traits. It is important to note that even microenvironmental variations, such as spatial and temporal variation, and system variations caused due to differential management practices can significantly affect the accumulation of micronutrients (Pfeiffer and McClafferty 2007). Planting seasons and planting dates within a specific season in crops such as rice, pearl millet and wheat have been found to have the significant effects on the kernel mineral concentrations. Despite considerable G × E interactions for kernel Fe and Zn in maize, major proportion of variation is caused due to genetic factors, and it is possible to identify the genotypes with stable mineral concentrations across environments (Agrawal et al. 2012; Guleria et al. 2013; Prasanna et al. 2011). Besides, it is also feasible to combine high micronutrient traits with high yield (Gregorio 2002).

10 Bioavailability of Fe and Zn in Human

Bioavailability is defined as the amount of a micronutrient that is available for absorption in the gastrointestinal tract and in turn utilized for normal metabolic function (Welch and Graham 2004). Different bioavailability assays are available that reliably estimate the amount of micronutrients available for absorption in the gut. Among in vitro systems, Caco-2 assay is one of the most abundantly used systems to measure bioavailable Fe. This assay typically mimics absorption of non-haem Fe from plant-based foods in human body and has been found to be rapid and cost-effective (Glahn et al. 1998; Jovani et al. 2001). Caco-2 cell reduces Fe^{+3} to Fe^{+2} during the uptake and causes synthesis of intracellular Fe storage protein, ferritin, that is regarded as an indicator of bioavailable Fe. Besides, it also behaves in similar manner in response to antinutritional and promoting factors. Various researchers (Lungaho et al. 2011; Oikeh et al. 2003a, b, 2004a; Pixley et al. 2011; Tako et al. 2013) have studied Fe bioavailability in maize using Caco-2 cell line. Besides, molar ratios of P/Fe and Zn/P have been used as indicator of bioavailability (Simic et al. 2011). Phytic acid has been found to be the major antinutritional factors that reduce the bioavailability of Fe and Zn. Since more than 80 % of total cellular P in cereals such as maize is present as phytate, so P can be used as reliable indicator of bioavailability of Fe and Zn. Bioavailability of Fe and Zn decreases when dietary phytate/Fe and phytate/ Zn molar quotients achieve value >1 and >6, respectively (Lonnerdal 2002). Researchers such as Simic et al. (2009) and Simic et al. (2011) estimated Fe and Zn bioavailability in maize using P as the indicator. In vivo poultry (*Gallus gallus*) model of Fe bioavailability has also been found to be useful due to its faster growth rate, simple anatomy, smaller size and low cost to maintain the birds (Tako et al. 2010). Tako et al. (2013) compared Fe bioavailability in maize using poultry model and Caco-2 cell line. These results indicate that this model exhibits the appropriate responses to Fe deficiency and has potential to serve as a model for Fe bioavailability.

Various factors that determines the level of bioavailability includes food type, inherent status of micronutrients in the body, pH level of the gut, state of health, transit time of food in the gut, competition between different ions for the common transporters and levels of different antinutritional and promoting factors (Bohn et al. 2004; Frossard et al. 2000; Lopez et al. 2002; Welch and Graham 2004). The major factors that determines the bioavailability in human is described below.

10.1 Type of Food

The mechanism of Fe uptake in the gut of human differs with (i) haem and (ii) non-haem Fe (Lynch et al. 1985). While, haemoglobin and myoglobin in animal products provide haem Fe, non-haem Fe is taken predominantly from plant sources. The absorption of haem Fe is high, primarily due to the lack of inhibition by the chelating agent such as phytate (Frossard et al. 2000; Lynch et al. 1985). Its contribution varies from 10 to 20 % of the dietary Fe but may contribute up to 50 % (Carpenter and Mahoney 1992). On the other hand, antinutritional factors bind with the non-haem Fe, thereby reducing the bioavailability to a great extent. Interestingly, on an average, staple food grains such as maize contain Fe having only 5 % bioavailability in human gut (Bouis and Welch 2010). However, absorption of non-haem Fe may vary from <1 % to >50 %, depending on the composition of dietary components, Fe status of the individual, gastric acidity, transit time, mucus secretion and overall health status (Bohn et al. 2008; King et al. 2000).

In the developed countries, seafood, meat and dairy products are the primary sources of Zn in the diet (Sandstead 1995). However, in the developing countries, where intake of animal-based food is limited, plant-based foods like cereals and pulses provide the most of dietary Zn (King et al. 2000). In cereals including maize, 25 % of Zn has been found to be bioavailable to humans (Bouis and Welch 2010). Among various factors, presence of phytic acid is regarded as one of the important factors that determine its bioavailability. King et al. (2000) reported that at low Zn level with no inhibitory agents, Zn absorption can be as high as 50 %. Since the diet pattern of the developed countries is mainly based on animal food, it is likely to be sufficient in micronutrients, while countries where considerable sections of the people are dependent on vegetarian food are likely to have deficiency.

10.2 Physiological Status of Body

The gut is the site of the absorption of minerals in humans, and any abnormalities or diseases that influence the normal functioning of intestine and stomach could affect iron status (Bohn et al. 2008). Fe levels in the body besides, blood oxygen concentration and inflammation in the body regulates the synthesis hormone hepcidin from the liver and in turn affects Fe uptake in the gut (Atanasiu et al. 2007; Nicolas et al. 2002). Level of acidity in the human gut is also one of the important factors for absorption of minerals. Lower pH is required for solubilization of mineral ions for their absorption, while higher pH causes synthesis of insoluble salts and lower rate of absorption (Bohn et al. 2008; Lopez et al. 2002). At intestinal pH, Fe^{2+} is more soluble than Fe^{3+} (Salovaara et al. 2003). Besides, proteins get denatured due to high acidity, and Fe is released from protein complexes due to the activity of pepsin. During absorption in the gut, carrier proteins undergo competitions among ions for binding. Copper (Cu) competes with Fe to bind with protein transfer molecule, transferrin. Higher concentration of Cu may displace already bound Fe and thereby reduces the absorption (Lopez et al. 2002). Similarly, cobalt (Co) competes with Zn for binding with carboxypeptide enzyme in the body and causes decrease in activity, that in turn affects overall mineral absorption in the gut.

10.3 Antinutritional Factors

Among various antinutritional components, phytic acid/phytate plays the major role in reducing the bioavailability of minerals such as Fe and Zn. Other substances such as oxalic acids, haemagglutinins, goitrogen, heavy metals (Cd, Hg, Pb, etc.), tannins and polyphenols, and fibres act as antinutrients and considerably reduce the bioavailability of Fe and Zn in the human body (Bouis and Welch 2010; Ortiz-Monasterio et al. 2007; Welch and Graham 2004). The basic mechanism of reduction of bioavailability of micronutrients mainly lies with chelating or reducing the solubility of the nutrients. Phytic acid is one of the important metabolites that are found to accumulate in cereal grains in significant amount, and in maize more than 80 % of the total phosphorus is present as phytic acids (Frossard et al. 2000; Raboy 1997). Although the primary function of the phytate in seeds is to store phosphorus as energy source and antioxidants essentially required for the germinating seeds, the negative charge of phytic acids significantly chelates positively charged minerals and forms insoluble complexes in the gut (Raboy 2001). Among various phenolic compounds, tannins are one of the abundantly found phenolics in plant (Klopfenstein and Hoseney 1995). Tannins form insoluble complexes with Fe^{3+} and makes the minerals unavailable for absorption in the gastrointestinal tract (Gillooly et al. 1984). Dietary fibres, such as insoluble polysaccharides and lignin, bind to mineral ions and in turn reduce the bioavailability of minerals (Charalampopulos et al. 2002). It binds directly with minerals and may form complex interaction with substance, such as phytate, tannin and oxalate, and reduces the availability of minerals (Harland 2006). Besides, it speeds up the passage time through gastrointestinal tract and thereby reduces the time available for nutrient absorption. Oxalic acids form insoluble complexes with cations and reduce their absorption (Savage 2002).

10.4 Promoting Factors

Carotenoids such as β-carotene, lutein and zeaxanthin; organic acids like ascorbic acid, fumarate, malate and citrate; certain amino acids like methionine, cysteine, histidine and lysine; and inulin and other non-digestible carbohydrates enhance the bioavailability of both Fe and Zn (Bouis and Welch 2010; Graham et al. 2001; Ortiz-Monasterio et al. 2007; Welch and Graham 2004). Besides this, while haemoglobin increases the absorption of Fe, long-chain fatty acids like palmitic acids enhance the bioavailability of Zn (Bouis and Welch 2010).

Carotenoids especially β-carotene, lutein and zeaxanthin have been found to enhance the absorption of Fe in human (Ortiz-Monastario et al. 2007). Enhancement of bioavailability could be due to formation of soluble complex or indirect effects of carotenoids on Fe absorption (Walezyk et al. 2003). Besides, β-carotene also have the capacity to somewhat negate the effects of certain inhibitors like phytates and polyphenols (Garcia-Casal et al. 2000; Layrisse et al. 2000). Vitamin C or ascorbic acid has been identified as the important factor that enhances the absorption of minerals (Welch and Graham 2004). It increases bioavailability of Fe mainly by reducing Fe^{3+} to Fe^{2+} and thereby preventing precipitation of Fe as ferric hydroxides. It also increases the acidity in the gut and creates favourable conditions for absorption. Besides Fe^{3+}-ascorbate, chelate remains stable and soluble in higher pH (Teucher et al. 2004). Since the seed of major cereals including maize is devoid of ascorbic acids, additional intake of fruits and vegetables rich in vitamin C is required to increase the bioavailability of Fe of cereal-based diets (Frossard et al. 2000). Meat is one of the important factors for increasing the bioavailability of non-haem Fe. The mechanism of enhancement is possibly due to myosin peptides generated upon degradation by pepsin in the gut, which binds Fe in solution and/or reduction of

Fe^{3+} to Fe^{2+} ions by sulphydryl group of cysteine amino acids that induce the gastric juice production thereby increasing the acidic condition (Carpenter and Mahoney 1992; Mulvihill and Morrissey 1998; Storcksdieck et al. 2007). Around 15 % of the plants including cereals such as rice, wheat and maize store starch as fructans, while the majority of the plant stores starch in the form of carbohydrates (Brinch-Pedersen et al. 2007). Oligofructans and inulins are the non-digestible oligosaccharides and have been found to increase the absorption of minerals (Gibson et al. 1995). Short-chain fatty acids produced due to fermentation lowers the pH of the intestine, thereby increases the solubility of the minerals. Besides, it also forms complexes with the minerals and in turn enhances the absorption (Scholz-Ahrens and Schrezenmeir 2002).

Since the humans lack the phytase enzyme responsible of hydrolysis of phytate, the addition of phytase enzyme increases the bioavailability of minerals. Phytase hydrolyzes the phytate and thereby separates minerals which can be readily absorbed in the gastrointestinal tract of the humans. Phytase is found in diverse organism such as bacteria, fungi and plant. Supplemental microbial phytase has been found to increase Zn bioavailability in the growing pigs and reduced the demand of Zn by one-third (Revy et al. 2006). Among plants phytase have been purified from cereals including maize (Bohn et al. 2008). Thus, increasing of phytase activity either by natural selection or transgenic approach is becoming a tool for increasing mineral bioavailability.

11 Genetics of Kernel Fe and Zn Bioavailability in Maize

Significant variation for bioavailable Fe and Zn was observed among diverse maize genotypes. Oikeh et al. (2003a, 2004a) while working with Caco-2 cell assay found bioavailable Fe ranged from 4 % and 14 % below and 49 % and 43 % above the reference variety, respectively. On the other hand, Oikeh et al. (2003b) found that mean bioavailable Fe varied between 30 % below and 88 % above the reference control variety. Similar

observation of significant variation has also been reported by Lungaho et al. (2011) and Pixley et al. (2011). In case of Zn, bioavailability as measured by relative proportion of P and Zn showed significant variations among maize genotypes (Simic et al. 2009, 2011). Thus, it is possible to identify maize genotypes that can provide more bioavailable Fe and Zn than other genotypes having low bioavailability. Studies on the evaluation of maize genotypes at multiple locations revealed that bioavailable Fe may vary with the environments. While Oikeh et al. (2003a, 2004a) and Simic et al. (2011) did not find any significant G × E interaction, Pixley et al. (2011) observed lesser importance of G × E interaction, thereby suggesting that Fe bioavailability is more or less consistent across environments. On the contrary, Simic et al. (2009) reported significant G × E interactions for P/Fe and P/Zn molar quotients, while Simic et al. (2011) did not detect any G × E interactions for Fe/P and Zn/P.

Bioavailable Fe in the maize kernel did not show any association with kernel Fe concentration (Lungaho et al. 2011; Oikeh et al. 2003b, 2004a; Pixley et al. 2011). The lack of correlation suggested that the two traits are possibly under different genetic control as there was not much agreement between the traits in relation to having common QTLs (Lungaho et al. 2011; Simic et al. 2011). Besides, P/Fe and P/Zn displayed very weak association, suggesting that genetic control of bioavailability of Fe and Zn is quite different (Simic et al. 2009). However, Simic et al. (2011) found that QTLs for Fe/P and Zn/P were co-localized on chromosome 3. SSR marker, *bnlg1456* linked to these QTLs on chromosome 3, is present close enough to the location of *phys1* and *phys2* genes that encode phytase which in turn release available P, Fe and Zn. Significant negative relationship was detected between kernel P concentration and bioavailable Fe, indicating that Fe became less available due to increased phytate content (Oikeh et al. 2003a). On the contrary, Lungaho et al. (2011) found that phytate was not a significant determinant for differences in Fe bioavailability in IBM RI population. Kernel Zn did not show any association with Fe bioavailability (Lungaho et al. 2011; Oikeh et al. 2003b). However, negative correlation between kernel Zn with bioavailable Fe was observed in some studies, thereby suggesting that Zn may act as inhibitor of Fe bioavailability (Glahn et al. 2002; Hunt 2005; Pixley et al. 2011). Besides, Fe and Zn bioavailability did not show any correlation with grain yield, a feature that suggests that bioavailability can be increased without compromising the grain yield potential (Pixley et al. 2011; Simic et al. 2009).

Heritability of Fe bioavailability has been found to be moderate in degree. Simic et al. (2011) found that heritability for Fe/P and Zn/P was 66.3 % and 52.9 %, respectively, while it was 0.53 each for the P/Fe and P/Zn molar ratio (Simic et al. 2009). Pixley et al. (2011) could find heritability ranging from 0.18 to 0.65 for various ferritin-related parameters that were used to assess bioavailable Fe among maize hybrids. The IBM (B73 × Mo17) population used by Lungaho et al. (2011) identified transgressive segregants for bioavailable Fe, a feature that is indicative of preponderance of additive gene action. Pixley et al. (2011) found significant GCA effects for bioavailable Fe, while SCA effects were nonsignificant.

Three QTLs were detected for kernel Fe, while 10 QTLs were found for Fe bioavailability in maize (Lungaho et al. 2011). Simic et al. (2011) also found that seven QTLs governed Fe bioavailability, while three QTLs could be related to the accumulation of Fe in kernel, thereby suggesting that bioavailable Fe is controlled by more loci than the Fe in maize kernel. However, Lungaho et al. (2011) concluded that Fe bioavailability is more simply inherited than the kernel Fe, as greater number of larger QTLs was detected for bioavailable Fe than kernel Fe, where QTLs were having minor effects. Backcross progenies derived from crosses between RI lines of IBM populations (B73 × Mo17) were genotyped with SSR markers that were linked to three QTLs on chromosome 3, 6 and 9. Progenies with high Fe bioavailability selected based on Caco-2 assay were found to possess superior alleles, thereby

suggesting the modest effects of QTLs pertaining to Fe bioavailability. However, for Zn, the number of loci detected for Zn/P and Zn was one each, and each of them was located on different chromosomes (Simic et al. 2011).

12   Strategies to Develop Micronutrient-Rich Cultivar

The success of biofortified crop in order to ameliorate hidden hunger mainly depends on the adaptation of nutritionally improved cultivars by farmers and acceptance among target population (Ruel and Bouis 1998). Micronutrient-rich crops should possess acceptable grain yield that is at par with the popular variety; it should show stable performance across the environments for micronutrient concentration and finally the variety should be free from any sort of undesirable characters. An effective breeding strategy for the development of biofortified crop includes (i) search for genetic variation, (ii) identification of genes/QTLs, (iii) their effective utilization in breeding program, (iv) generation of elite inbreds and (v) development of heterotic maize hybrids with high kernel Fe and Zn. Strategies depicting diverse options to breed maize cultivars with high kernel Fe and Zn have been shown in Fig. 1.

Some of the strategies where progress/lead is available in relation to improvement of kernel Fe and Zn, are discussed here.

12.1 QTLs for Kernel Fe and Zn

Quantitative trait loci (QTL) analysis provides a greater insight to unravel the genes and genetic basis of the accumulation of Fe and Zn in kernels (Simic et al. 2011; Xu et al. 2011). Modest QTLs (for kernel Fe and Zn) that are consistently expressed under multiple environments can be used in marker-assisted selection (MAS) and transferred into desirable genetic backgrounds. In maize, several researchers (Baxter et al. 2013; Lungaho et al. 2011; Qin et al. 2012; Simic et al. 2011) have reported

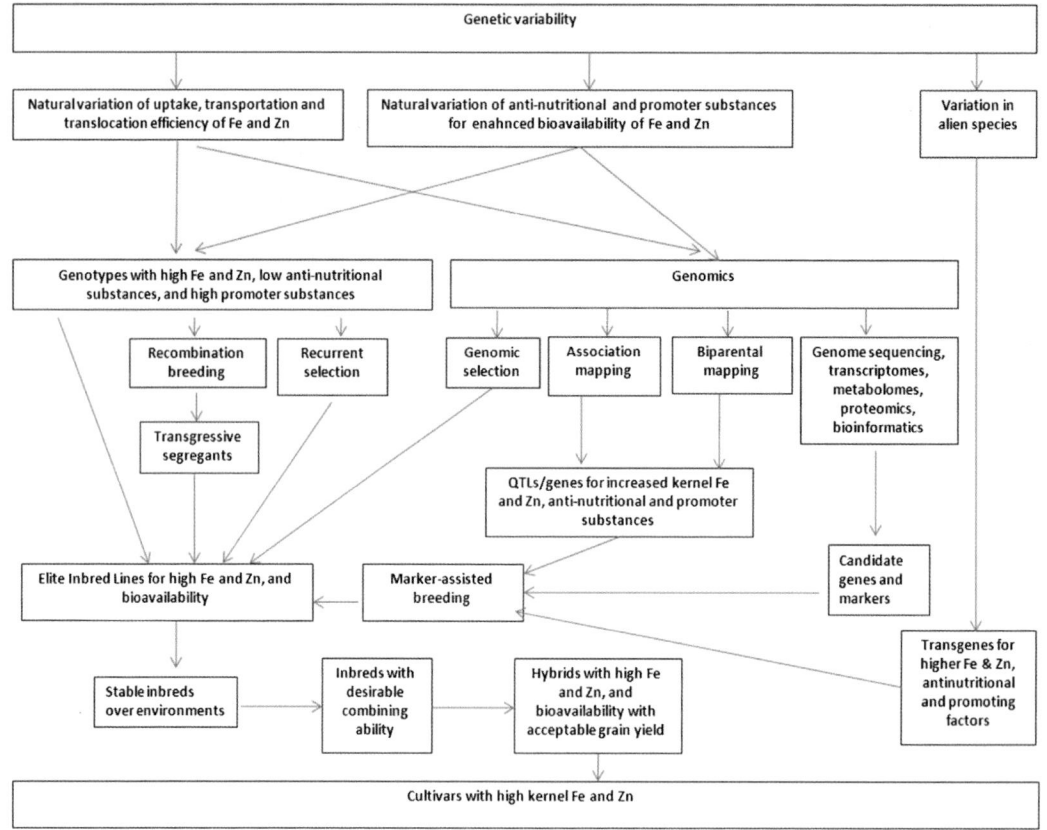

Fig. 1 Strategies to develop Fe- and Zn-enriched maize cultivar

several QTLs for kernel Fe and Zn concentration (Table 2). QTLs for leaf micronutrient traits have also been mapped in maize (Soric et al. 2011). The QTL studies also unravel the possible reasons behind the correlation between kernel Fe and Zn concentration either by their co-localization or pleotropic effect of QTLs which control the network of metal uptake, transportation and accumulation mechanisms (Clemens 2001; Qin et al. 2012). The report of different QTLs for kernel Fe and Zn in terms of their position and numbers further indicates complex nature of trait that is under the influence of genetic background and environment. The significantly large effect QTLs are required to be fine mapped and cloned to understand the mechanism of accumulation of Fe and Zn in kernel.

12.2 Candidate Genes for Kernel Fe and Zn

Isolation of gene(s) for a target trait is one of the prominent objectives in molecular genetics in order to put them in use via various breeding methodologies. Classical methods of gene isolation including positional cloning and insertional mutagenesis have been used successfully for isolation of major genes (Prioul et al. 1999; Tanksley et al. 1995). At this juncture, the candidate gene approach serves as one of the potential alternative strategies. Plant genome codes for large number of metal transporters that vary in their substrate specificities, expression patterns, mineral uptake, mobilization and redistribution in plant and accumulation in seeds; each of these steps is under the control of several genes,

Table 2 Details of mapping populations used and QTLs detected for kernel Fe and Zn

S. no.	Parents	Type of mapping population used	Environments used for evaluation	Fe		Zn		References
				No. of significant QTLs detected	R^2	No. of significant QTLs detected	R^2	
1	B84 × Os6-2	F_4	4 years, one location	3 (K)	21.1 %	1 (K)	4.2 %	Simic et al. (2011)
				7 (BA)	33.2 %	1 (BA)	3.6 %	
2	B73 × Mo17	IBM-RI	5 years, two locations	3 (K)	26.1 %	–	–	Lungaho et al. (2011)
				10 (BA)	54 %			
3	Mu6 × SDM and Mo17 × SDM	$F_{2:3}$	1 year, two locations	4 (K)	10.02–21.12 %	7 (K)	6.34–21.33 %	Qin et al. (2012)
4	B73 × Mo17	IBM-RI	3 years, three locations	2 (K)	9–11 %	3 (K)	5–10 %	Baxter et al. (2013)

K kernel, *BA* bioavailability, *IBM-RI* Intermated B73 × Mo17 recombinant inbreds

making the accumulation of minerals including Fe and Zn in seeds as a complex polygenic trait (Colangelo and Guerinot 2006; Qin et al. 2012; Sharma and Chauhan 2008). Recent developments in genomic approaches coupled with bioinformatic tools have enabled the sequencing of various grass genomes; these sequence data provide greater insight to identify genes for various traits including mineral accumulation, *viz.*, cesium (Payne et al. 2004) and selenium (Zhang et al. 2006). Using rice and maize genome sequence data, candidate genes were predicted for Fe and Zn transporters in maize (Chauhan 2006; Sharma and Chauhan 2008).

12.3 *Omics* of Fe and Zn Homeostasis

The breakthrough developments in molecular genetics and biological research lead to foundation of various *omics* approaches, *viz.*, genomics, transcriptomics, proteomics, metabolomics and ionomics. These approaches can be employed to determine the underlying genetic, physiological and molecular mechanisms and gene(s) responsible for expression of any trait of interest. Genomics research is critical to any major crop improvement program. Genes function for Fe and Zn homeostasis can be tackled by various genomics-based approaches. With the availability of voluminous sequence information, genes can be predicted by using comparative genomics, by assigning homology to genes with functions that are better known in related species employing novel bioinformatics tools and genomic databases of various species (Chauhan 2006; Sharma and Chauhan 2008). Transcriptomics is another important approach, which emphasizes on the experiments that are framed to monitor and manipulate the dynamics of gene expression events that occur for a particular treatment as against the control (Coram et al. 2008). The most powerful applications of transcriptomics are the study of gene expression patterns, for various treatments under question that survey a wide array of cellular responses and phenotypes in addition to the development of gene expression markers that can be used to predict tolerance or adaptation for a particular treatment (Coram et al. 2008; Pollock 2002). Genome-wide transcriptome analysis enabled understanding of interaction among minerals and detection of candidate genes engaged in homeostasis. The existence of negative correlation between Fe and P interaction was confirmed at genome level in terms of regulation of genes engaged in homeostasis of these elements (Zheng et al. 2009). The results from root proteomics and transcriptomics studies in maize for Fe starvation not only yielded the differential protein profiles and genes but it also accounted for allelic diversity leading to qualitative peptide differences that might have an impact on functional variation (Urbany 2012). In maize, transcriptomic analyses identified candidate genes for the *ys3* mutant, where unspliced introns of *ZmTOM1* is found in *ys3* mutant but not in its wild counterpart, suggesting that *ZmTOM1* may be involved in the *ys3* phenotype (Nozoye et al. 2013). Metabolomics and ionomics approaches are performed as potential postgenomic tools in quantitative trait loci (mQTL) analysis, metabolite or ionome profiling of plants, identification and assignment of functions to genes and evaluation of genetically modified crops (Keurentjes et al. 2006; Riedelsheimer et al. 2012; Schauer et al. 2006; Weckwerth et al. 2004). Ionome profiling was successfully employed in maize RILs (B73 × Mo17) to detect QTLs for kernel minerals including Fe and Zn (Baxter et al. 2013).

12.4 Transgenics for High Kernel Fe and Zn

Transgenic breeding approach possesses unique advantage over conventional breeding methods, in terms of no limitation of gene pool and its direct application to improve cultivar using targeted expression of genes in desired organs (Zhu et al. 2007). Maize *yellow stripe1* (*YS1*) gene in yeast Fe uptake mutant restores growth specifically on Fe^{3+}-phytosiderophore-sufficient media and translocate Fe that is bound either by phytosiderophore or by the related compound like nicotianamine (Curie et al. 2001). Thus, ZmYS1 is involved in both primary Fe acquisition and intracellular transport of Fe and other metals in maize (Roberts et al. 2004; Ueno et al. 2009). Thus, overexpression of genes for these several transporters and chelating compounds leads to higher accumulation of Fe and Zn in sink. For instance, over expression of nicotianamine synthase (NAS) and nicotianamine aminotransferase (NAAT) leads to increased phytosiderophore synthesis (Zheng et al. 2010). *Aspergillus niger* phytase in maize resulted in decrease in phytate by up to 95 % while increase in Fe content by 20–70 % (Drakakaki et al. 2005). Expression of soybean ferritin transgene in maize endosperm altered the expression of native Fe homeostasis genes and accumulated significantly higher concentrations of calcium and magnesium in addition to Fe (Kanobe et al. 2013). The above approaches can be coupled with expression of phytase which breakdown the antinutritional factor and facilitate the Fe absorption during digestion.

12.5 Breeding for Reduced Level of Antinutritional Factors

Phytic acid is one of the potential antinutritional factors present in cereal grains including maize, which hinders the absorption of Fe and Zn by digestive system. Low-phytic acid (*lpa*) mutations were isolated in several cereal grains: maize (Raboy et al. 2000), barley (*Hordeum vulgare*) (Larson et al. 1998) and rice (*Oryza sativa*) (Larson et al. 2000). In maize more than 20 *lpa*, mutants have been isolated, which resulted in reductions in seed phytic acid-P ranging from 55 to 66 % (*lpa 1–1 and lpa 2–1*), but in some lines up to 90 % (*lpa 241*) reduction was observed (Pilu et al. 2003; Raboy 2001). However, the usage of mutations such as *lpa 241* has been limited by severe negative effects on seed viability and germination and plant growth, resulting in various levels of yield penalty (Pilu et al. 2003, 2005). To develop a commercial product, it is necessary to obtain plants that have low-phytic acid along with robust agronomically desirable characteristics. Recent advances in genomics and genetic engineering helped in the development of transgenic line in several crops which fulfils these requirements by expressing exogenous phytase genes (Brinch-Pedersen et al. 2006; Chen et al. 2008; Kuwano et al. 2009; Nunes et al. 2006). Maize transgenics with *phyA2* gene from *Aspergillus niger* resulted in 50-fold increase in phytase activity as compared to nontransgenic maize seeds without affecting agronomic performance of a line (Chen et al. 2008). Drakakaki et al. (2005) also reported expression of *Aspergillus niger* phytase gene in maize resulted in decreases in phytate by up to 95 %. Hence, transgenic technology in breeding will serve as one of the potential tools in breeding low kernel phytic acid maize cultivars.

However, the effects of lowering antinutritional factors in plants are not devoid of problems and required to be addressed with cautions (Bouis and Welch 2010; Welch and Graham 2004). In plants, many of the factors such as phytate and polyphenols are part of plant metabolism and play a major role in plants' response to different abiotic stresses, besides possessing positive effects on imparting resistance against pathogens and insect pests (Graham et al. 2001; Welch and Graham 2004). Phytic acid is also required for higher seedling vigour and reduced aflatoxin development in grain (Morris 1995). Besides, phytate has been found to protect seeds against oxidative stress during the seed's life span (Doria et al. 2009).

12.6 Breeding for Increased Level of Bioavailability Promoting Factors

Dietary substances that promote/enhance the absorption of plant minerals by the human digestive system can be increased using genotypes having suitable genes (Bouis and Welch 2010; Gibson 2007; Graham et al. 2001). Reports by several researchers suggest the synergistic effect of *opaque2* allele in enhancing the concentration of micronutrients especially Zn (Arnold et al.

1977; Chakraborti et al. 2009; Welch et al. 1993). Besides, Zn-deficient rats showed increase in absorption of Zn from 64 to 69 % with lysine supplementation (House et al. 1996). Thus, breeding for quality protein maize (QPM) which possesses higher lysine has the potential to enhance the kernel Zn concentration and its absorption in digestive system. Yellow/orange maize contains high amount of carotenoids as compared to other cereals and thus possesses built-in system of enhancing bioavailability of minerals (Harjes et al. 2008). The utilization of rare natural mutants of *crtRB1* gene in maize has led to the significant increase of β-carotene by nearly 10–24 folds (Babu et al. 2013; Gupta et al. 2013b; Vignesh et al. 2012; Yan et al. 2010; Zhang et al. 2012). In maize, the presence of β-carotene increased the Fe absorption level up to 1.8-fold (Garcia-Casal et al. 2000). Besides, the addition of lutein in maize-based diet increased the bioavailability of Fe by twofolds (Garcia-Casal 2006). Maize transgenics developed by overexpressing bacterial genes *crtB* (for phytoene synthase) and *crtI* (for the four desaturation step) under the control of a γ-zein endosperm-specific promoter, resulted in 34-fold increase in total carotenoids with a preferential accumulation of β-carotene in the maize endosperm (Aluru et al. 2008). Zhu et al. (2008) used genetic engineering approach to develop carotenoid-rich maize line in a genetic background of WTM37W which resulted in the accumulation of 57.35 μg/g β-carotene. Maize variety M37W (white endosperm) lines transformed with (i) maize phytoene synthase (*psy1*) cDNA driven by wheat LMW glutenin promoter, (ii) *Pantoea ananatis* (formerly *Erwinia uredovora*) *crtI* gene (encoding carotene desaturase) driven by barley D-hordein promoter (to increase carotene level), (iii) rice dehydroascorbate reductase (*dhar*) cDNA driven by barley D-hordein promoter (to increase ascorbate level), (iv) *E. coli folE* gene encoding GTP cyclohydrolase (GCH1) driven by barley D-hordein promoter (to increase folate level) and (v) selectable marker *bar* led to the synthesis of β-carotene (~60 μg/g) with nearly 169-fold increase as compared to non-transformed lines (0.35 μg/g). It also enhanced the levels of ascorbate (106.94 µg/g) and folate (1.95 µg/g) in the transformed lines (Naqvi et al. 2009).

Addressing antinutrients and promoter substances in crops could be an effective approach for biofortification of Fe and Zn, as fewer genes with profound effects have been found to operate in their biosynthesis and metabolism as compared with the complex mechanism of uptake, transport and deposition of Fe and Zn in kernels, where it involves large number genes with minor effects. An increase of bioavailable Fe from 5 to 20 % would in turn relate to fourfold increase in total Fe (Bouis and Welch 2010). A minor change in the levels of antinutrients and promoter may lead to significant effects on the bioavailability of micronutrients.

Thus, with the availability of (i) wide genetic variations for kernel Fe and Zn, (ii) understanding of the genetic behaviour of QTLs/genes underlying the traits, (iii) scope of manipulating genes for antinutritional and promoting factors to increase their bioavailability and (iv) availability of suitable breeding and biotechnological tools, it is possible to develop Fe- and Zn-enriched maize cultivars.

References

Agrawal PK, Jaiswal SK, Prasanna BM, Hossain F, Saha S, Guleria SK, Gupta HS (2012) Genetic variability and stability for kernel iron and zinc concentration in maize (*Zea mays* L.) genotypes. Indian J GenetPlant Breed 72:421–428

Akman Z, Kara B (2003) Genotypic variations for mineral content at different growth stages in wheat (*Triticum aestivum* L.). Cereal Res Commun 31:459–466

Aluru M, Xu Y, Guo R, Wang Z, Li S, White W, Wang K, Rodermel S (2008) Generation of transgenic maize with enhanced provitamin A content. J Exp Bot 59:3551–3562

Arnold JM, Bauman LF (1976) Inheritance of and interrelationships among maize kernel traits and elemental contents. Crop Sci 16:439–440

Arnold JM, Bauman LF, Aycock HS (1977) Interrelations among protein, lysine, oil, certain mineral element concentrations, and physical kernel characteristics in two maize populations. Crop Sci 17:421–425

Atanasiu V, Manolescu B, Stoian I (2007) Hepcidin: central regulator of iron metabolism. Eur J Haematol 78:1–10

Axelsen KB, Palmgren MG (2001) Inventory of the superfamily of P-type ion pumps in Arabidopsis. Plant Physiol 126:696–706

Babu R, Rojas NP, Gao S, Yan J, Pixley K (2013) Validation of the effects of molecular marker polymorphisms in *lcyE* and *crtRB1* on provitamin A concentrations for 26 tropical maize populations. Theor Appl Genet 126:389–399

Banziger M, Long J (2000) The potential for increasing the iron and zinc density of maize through plant-breeding. Food Nutr Bull 21:397–400

Baxter IR, Gustin JL, Settles AM, Hoekenga OA (2013) Ionomic characterization of maize kernels in the intermated B73 × Mo17 population. Crop Sci 53:208–220

Bell PF, Chaney RL, Angle JS (1991) Determination of the copper Zt activity required by maize using chelator-buffered nutrient solutions. Soil Sci Soc Am J 55:1366–1374

Bityutskii N, Magnitski S, Lapshina I, Lukina E, Soloviova A, Patsevitch V (2001) Distribution of micronutrients in maize grains and their mobilization during germination. In: Horst WJ et al (eds) Plant nutrition-food security and sustainability of agro-ecosystems. Kluwer Academic Publishers, Netherlands, pp 218–219

Bityutskii NP, Magnitskiy SV, Korobeynikova LP, Lukina EI, Soloviova AN, Patsevitch VG, Lapshina IN, Matveeva GV (2002) Distribution of iron, manganese and zinc in mature grain and their mobilization during germination and early seedling development in maize. J Plant Nutr 25:635–653

Bjarnason M, Vasal SK (1992) Breeding of quality protein maize (QPM). Plant Breed Rev 9:181–216

Black RE, Allen LH, Bhutta ZA, Caulfield LE, de Onis M, Ezzati M, Mathers C, Rivera J, Maternal Child Under nutrition Study Group (2008) Maternal and child under nutrition: global and regional exposures and health consequences. Lancet 371:243–260

Bohn T, Davidsson L, Walczyk T, Hurrell RF (2004) Phytic acid added to white-wheat bread inhibits fractional apparent magnesium absorption in humans. Am J Clin Nutr 79:418–423

Bohn L, Meyer AS, Rasmussen SK (2008) Phytate: impact on environment and human nutrition. A challenge for molecular breeding. J Zhejiang Univ (Sci) 9:165–191

Bouis H (2002) Plant breeding: a new tool for fighting micronutrient malnutrition. J Nutr 132:S491–S494

Bouis HE, Welch RM (2010) Biofortification-a sustainable agricultural strategy for reducing micronutrient malnutrition in the global south. Crop Sci 50:S20–S32

Brinch-Pedersen H, Hatzack F, Stoger E, Arcalis E, Pontopidan K, Holm PB (2006) Heat-stable phytases in transgenic wheat (*Triticum aestivum* L.): deposition pattern, thermostability, and phytate hydroslysis. J Agric Food Chem 54:4624–4632

Brinch-Pedersen H, Borg S, Tauris B, Holm PB (2007) Molecular genetic approaches to increasing mineral availability and vitamin content of cereals. J Cereal Sci 46:308–326

Brkic I, Simic D, Zdunic C, Jambrovic A, Ledencan T, Kovacevic V, Kadar I (2003) Combining abilities of corn-belt inbred lines of maize for mineral content in grain. Maydica 48:293–297

Brock JH, Halliday JW, Pippard MJ, Powell LW (1994) Iron metabolism in health and disease. W.B. Saunders Company Ltd., London

Brown K, Peerson JM, Allen LH (1998) Effects of zinc supplementation on children's growth. In: Sandstrom B, Walter P (eds) Role of trace elements for health promotion and disease prevention, Bibliotheca Nutritioet Dieta, 54. Karger, Basel, pp 76–83

Cakmak I (2008) Enrichment of cereal grains with zinc: agronomic or genetic biofortification? Plant Soil 302: 1–17

Calderini DF, Ortiz-Monasterio I (2003) Grain position affects grain macronutrient and micronutrient concentrations in wheat. Crop Sci 43:141–151

Carpenter CE, Mahoney AW (1992) Contributions of heme and nonheme iron to human nutrition. Crit Rev Food Sci Nutr 31:333–367

Chakraborti M, Prasanna BM, Hossain F, Singh AM, Guleria SK (2009) Genetic evaluation of kernel Fe and Zn concentrations and yield performance of selected Maize (Zea mays L.) genotypes. Range Manag Agrofor 30:109–114

Chakraborti M, Prasanna BM, Singh A, Hossain F (2010) Generation mean analysis of kernel iron and zinc concentrations in maize (Zea mays L.). Indian J Agric Sci 80:956–959

Chakraborti M, Prasanna BM, Hossain F, Mazumdar S, Singh AM, Guleria SK, Gupta HS (2011a) Identification of kernel iron- and zinc-rich maize inbreds and analysis of genetic diversity using microsatellite markers. J Plant Biochem Biotechnol 20:224–233

Chakraborti M, Prasanna BM, Hossain F, Singh AM (2011b) Evaluation of single cross Quality Protein Maize (QPM) hybrids for kernel iron and zinc concentrations. Indian J Genet Plant Breed 71:312–319

Charalampopulos D, Wang R, Pandiella SS, Webb C (2002) Application of cereals and cereal components in functional foods: a review. Int J Food Microbiol 79:131–141

Chauhan RS (2006) Bioinformatics approach towards identification of candidate genes for zinc and iron transporters in maize. Curr Sci 91:510–515

Chen F, Chun L, Song J, Mi G (2007) Heterosis and genetic analysis of iron concentration in grains and leaves of maize. Plant Breed 126:107–109

Chen R, Xue G, Chen P, Yao B, Yang W, Ma Q, Fan Y, Zhao Z, Tarczynski MC, Shi J (2008) Transgenic maize plants expressing a fungal phytase gene. Transgenic Res 17:633–643

Chiripa V, Cracium M, Cracium I, Condei G, Fluiera V (1999) Ionic interactions in maize grown under water stress conditions. Rom Ind Res 11:65–74

Chrispeels MJ, Crawford NM, Schroeder JI (1999) Proteins for transport of water and mineral nutrients across the membranes of plant cells. Plant Cell 11: 661–675

Clemens S (2001) Molecular mechanisms of plant metal tolerance and homeostasis. Planta 212:475–486

Clemens S, Palmgren MG, Kramer U (2002) A long way ahead: understanding and engineering plant metal accumulation. Trends Plant Sci 7:309–315

Colangelo EP, Guerinot ML (2006) Put the metal to the petal: metal uptake and transport throughout plants. Curr Opin Plant Biol 9:322–330 Choose Destination

Coram TE, Brown-Guedira G, Chen XM (2008) Using transcriptomics to understand the wheat genome. CAB Rev Perspect Agric Vet Sci Nutr Nat Resour 3:1e9

Curie C, Alonso JM, Le Jean M, Ecker JR, Briat JF (2000) Involvement of NRAMP1 from Arabidopsis thaliana in iron transport. Biochem J 347:749–755

Curie C, Panaviene Z, Loulergue C, Dellaporta SL, Briat JF, Walker EL (2001) Maize yellow stripe1 encodes a membrane protein directly involved in Fe (III) uptake. Nature 409:346–349

Dalmiya N, Schultink W (2003) Combating hidden hunger: the role of international agencies. Food Nutr Bull 24: S69–S77

DeMaeyer E, Adiels-Tegman M (1985) The prevalence of anaemia in the world. World Health Stat Q 38: 302–316

Doria L, Galleschi L, Calucci L, Pinzino C, Pilu R, Cassani E, Nielsen E (2009) Phytic acid prevents oxidative stress in seeds: evidence from a maize (Zea mays L.) low phytic acid mutant. J Exp Bot 60: 967–978

Drakakaki G, Marcel S, Glahn RP, Lund EK, Pariagh S, Fisher R, Christou P, Stoger E (2005) Endosperm specific coexpression of recombinant soybean ferritin and Aspergillus phytase in maize results in significant increases in the levels of bioavailable iron. Plant Mol Biol 59:869–880

Eide D, Broderius M, Fett J, Guerinot ML (1996) A novel iron regulated metal transporter from plants identified by functional expression in yeast. Proc Natl Acad Sci U S A 93:5624–5628

Eren E, Arguello JM (2004) Arabidopsis HMA2, a divalent heavy metal-transporting P IB-type ATPase, is involved in cytoplasmic Zn 2þ homeostasis. Plant Physiol 136:3712–3723

Feil B, Moser SB, Jampatong S, Stamp P (2005) Mineral composition of the grains of tropical maize varieties as affected by pre-anthesis drought and rate of nitrogen fertilization. Crop Sci 45:516–523

Ferreira CF, VargasMotta AC, Prior SA, Reissman CB, Santos dos NZ, Gabardo J (2012) Influence of corn (Zea mays L.) cultivar development on grain nutrient concentration. Int J Agron doi:10.1155/2012/842582

Frossard E, Bucher M, Machler F, Mozafar A, Hurrell R (2000) Potential for increasing the content and bioavailability of Fe, Zn and Ca in plants for human nutrition. J Sci Food Agric 80:861–879

Garcia-Casal M (2006) Carotenoids increase iron absorption from cereal based food in the human. Nutr Res 26:340–344

Garcia-Casal MN, Leets I, Layrisse M (2000) β-carotene and inhibitors of iron absorption modify iron uptake by Caco-2 cells. J Nutr 130:5–9

Ghandilyan A, Vreugdenhil D, Aarts MGM (2006) Progress in the genetic understanding of plant iron and zinc nutrition. Physiol Plant 126:407–417

Gibson RS (2007) The role of diet- and host-related factors in nutrient bioavailability and thus in nutrient-based dietary requirement estimates. Food Nutr Bull 28:77–100

Gibson GR, Beatty ER, Wang X, Cummings JH (1995) Selective simulation of Bifidobacteria in the human colon by oligofructase and inulin. Gastroenterology 108:975–982

Gillooly M, Bothwell TH, Charlton RW, Torrance JD, Bezwoda WR, MacPhail AP, Derman DP, Novelli L, Morrall P, Mayet F (1984) Factors affecting the absorption of iron from cereals. Br J Nutr 51:37–46

Glahn RP, Lai C, Hsu J, Thompson JF, Guo M, Van-Campen DR (1998) Decreased citrate improves iron availability from infant formula: application of an in vitro digestion/Caco-2 cell culture model. J Nutr 128:257–264

Glahn RP, Cheng Z, Ross MW, Gregorio GB (2002) Comparison of iron bioavailability from 15 rice genotypes: studies using an in vitro digestion/Caco-2 cell culture model. J Agric Food Chem 50: 3586–3591

Goldenberg RL, Tamura T, Neggers Y, Copper RL, Johnston KE, DuBard MB, Hauth JC (1995) The effect of zinc supplementation on pregnancy outcome. JAMA 274:463–468

Gorsline GW, Thomas WI, Baker DE (1964) Inheritance of P, K, Mg, Cu, B, Zn, Mn, Al, and Fe concentrations by corn (Zea mays L.) leaves and grain. Crop Sci 4: 207–210

Graham RD, Stangoulis JCR (2003) Trace element uptake and distribution in plants. J Nutr 133:S1502–S1505

Graham RD, Welch RM, Bouis HE (2001) Addressing micronutrient malnutrition through enhancing the nutritional quality of staple foods: principles, perspectives and knowledge gaps. Adv Agron 70:77–142

Green LS, Rogers EE (2004) FRD3 controls iron localization in Arabidopsis. Plant Physiol 136:2523–2531

Gregorio GB (2002) Progress in breeding for trace minerals in staple crops. J Nutr 132:S500–S502

Grotz N, Fox T, Connolly E, Park W, Guerinot ML, Eide D (1998) Identification of a family of zinc transporter genes from Arabidopsis that respond to zinc deficiency. Proc Natl Acad Sci U S A 95:7220–7224

Grusak MA (2000) Strategies for improving the iron nutritional quality of seed crops: lessons learned from the study of unique iron hyper accumulating pea mutants. Pisum Genet 32:1–5

Guerinot ML, Yi Y (1994) Iron: nutritious, noxious and not readily available. Plant Physiol 104:815–820

Guerrant RL, Lima AAM, Davidson F (2000) Micronutrients and infection: interactions and implications with enteric and other infections and future priorities. J Infect Dis 182:S134–S138

Guimaraes PEO, Schaffert RE, Ribeiro PEA, Sena MR, Costa LP, Paes MCD, Alves VMC, Coelho AM, Nutti MR, Carvalho JLM, Pixley K, Nogueira ARA, Souza GB (2004) Mineral grain content and association among Zn and other mineral in yellow QPM and normal endosperm maize lines. HarvestPlus MILHO biofortificado (brochure), Washington

Guleria SK, Chahota RK, Kumar P, Kumar A, Prasanna BM, Hossain F, Agrawal PK, Gupta HS (2013) Analysis of genetic variability and genotype × year interactions on kernel zinc concentration in selected Indian and exotic maize (Zea mays L.) genotypes. Indian J Agric Sci 83:836–841

Gupta HS, Agrawal PK, Mahajan V, Bisht GS, Kumar A, Verma P, Srivastava A, Saha S, Babu R, Pant MC, Mani VP (2009) Quality protein maize for nutritional security: rapid development of short duration hybrids through molecular marker-assisted breeding. Curr Sci 96:230–237

Gupta HS, Raman B, Agrawal PK, Mahajan V, Hossain F, Nepolean T (2013a) Accelerated development of quality protein maize hybrid through marker-assisted introgression of opaque-2 allele. Plant Breed 132: 77–82

Gupta HS, Vignesh M, Hossain F, Nepolean T (2013b) Enrichment of nutritional qualities in maize through marker-assisted selection. In: National Seminar on Genomics for Crop Improvement. IBAB, Bangalore, pp 69–70

Hall JL, Williams LE (2003) Transitional metal transporters in plants. J Exp Bot 54:2601–2613

Hallberg L (1982) Iron absorption and iron deficiency. Hum Nutr Clin Nutr 36:259–278

Halvorson AD, Lindsay WL (1977) The critical $Zn2^+$ concentration for corn and the nonabsorption of chelated zinc. Soil Sci Soc Am J 41:531–534

Hambidge KM (1987) Zinc. In: Mertz W (ed) Trace elements in human and animal nutrition, 5th edn. Academic, Orlando, pp 1–137

Harjes CE, Rocheford TR, Bai L, Brutnell TP, Kandianis CB, Sowinski SG, Stapleton AE, Vallabhaneni R, Williams M, Wurtzel ET, Yan J, Buckler ES (2008) Natural genetic variation in lycopene epsilon cyclase tapped for maize biofortification. Science 319:330–333

Harland BF (2006) Dietary fibre and mineral bioavailability. Nutr Res Rev 2:133–147

Hell R, Stephan UW (2003) Iron uptake, trafficking and homeostasis in plants. Planta 216:541–551

House WA, Van-Campen DR, Welch RM (1996) Influence of dietary sulfur-containing amino acids on the bioavailability to rats of zinc in corn kernels. Nutr Res 16:225–235

Hunt JR (2005) Dietary and physiological factors that affect the absorption and bioavailability of iron. Int J Vitam Nutr Res 75:375–384

Hussain D, Haydon MJ, Wang Y, Wong E, Sherson SM, Young J, Camakaris J, Harper JF, Cobbett CS (2004) P-type ATPase heavy metal transporters with roles in essential zinc homeostasis in Arabidopsis. Plant Cell 16:1327–1339

Jean LM, Schikora A, Mari S, Briat JF, Curie C (2005) A loss-of-function mutation in AtYSL1 reveals its role in iron and nicotianamine seed loading. Plant J 44: 769–782

Jovani M, Barbera R, Farre R (2001) Lactoferrin and its possible role in iron enrichment of infant formulas. Food Sci Technol Int 7:97–103

Kanobe MN, Rodermel SR, Bailey T, Scott MP (2013) Changes in endogenous gene transcript and protein levels in maize plants expressing the soybean ferritin transgene. Front Plant Sci. doi:10.3389/fpls.2013.00196

Kapil U, Bhavna A (2008) Adverse effects of poor micronutrient status during childhood and adolescence. Nutr Rev 60:S84–S90

Keurentjes JJ, Fu J, deVos CH, Lommen A, Hall RD, Bino RJ, vanderPlas LH, Jansen RC, Vreugdenhil D, Koornneef M (2006) The genetics of plant metabolism. Nat Genet 38:842–849

Khush GS, Lee S, Cho J, Jeon JS (2012) Biofortification of crops for reducing malnutrition. Plant Biotechnol Rep. doi:10.1007/s11816-012-0216-5

King JC, Shames DM, Woodhouse LR (2000) Zinc homeostasis in humans. J Nutr 130:S 1360–S 1366

Klopfenstein CF, Hoseney RC (1995) Nutritional properties of sorghum and the millets. In: Dendy DAV (ed) Sorghum and millets: chemistry and technology. American Association of Cereal Chemists, St. Paul, pp 125–168

Kovacevic V, Brkic I, Simic D, Bukvic G, Rastija M (2004) The role of genotypes on phosphorus, zinc, manganese and iron status and their relations in leaves of maize on hydromorphic soil. Plant Soil Environ 50: 535–539

Kuwano M, Mimura T, Takaiwa F, Yoshida KT (2009) Generation of stable 'low phytic acid' transgenic rice through antisense repression of the 1D-myo-inositol 3-phosphate synthase gene (RINO1) using the 18-kDa oleosin promoter. Plant Biotechnol J 7:96–105

Larson SR, Young KA, Cook A, Blake TK, Raboy V (1998) Linkage mapping two mutations that reduce phytic acid content of barley grain. Theor Appl Genet 97:141–146

Larson SR, Rutger JN, Young KA, Raboy V (2000) Isolation and genetic mapping of a non-lethal rice low phytic acid mutation. Crop Sci 40:1397–1405

Layrisse M, Garcia-Casal MN, Solano L, Baron MA, Arguello F, Llovera D, Ramirez J, Leets I, Tropper E (2000) Iron bioavailability in humans from breakfasts enriched with iron bis-glycine chelate, phytates and polyphenols. J Nutr 130:2195–2199

Long JK, Banziger M, Smith ME (2004) Diallel analysis of grain iron and zinc density in Southern African-adapted maize inbreds. Crop Sci 44:2019–2026

Lonnerdal B (2002) Phytic acid-trace element (Zn, Cu, Mn) interactions. Int J Food Sci Technol 37:749–758

Lopez HW, Leenhardt F, Coudray C, Remesy C (2002) Minerals and phytic acid interactions: is it a real problem for human nutrition? Int J Food Sci Technol 37: 727–739

Lozoff B, Jimenez E, Xolf AW (1991) Long term development outcome of infants with iron deficiency. N Engl J Med 325:687–694

Lungaho MG, Mwaniki AM, Szalma SJ, Hart J, Rutzke MA, Kochian LV, Glahn R, Hoekenga OA (2011) Genetic and physiological analysis of iron biofortification in maize kernels. PLoS One 6:e20429

Lynch S, Dassenko S, Morck T, Beard J, Cook J (1985) Soy protein and haem iron absorption in humans. Am J Clin Nutr 41:13–20

Marschner H (1995) Mineral nutrition of higher plants. Academic, London

Mascotti DP, Rup D, Thach RE (1995) Regulation of iron metabolism: translational effects medicated by iron, heme and cytokines. Annu Rev Nutr 15:239–261

Maser P, Thomine S, Schroeder JI, Ward JM, Hirschi K, Sze H, Talke IN, Amtmann A, Maathuis FJ, Sanders D, Harper JF, Tchieu J, Gribskov M, Persans MW, Salt DE, Kim SA, Guerinot ML (2001) Phylogenetic relationships within cation transporter families of Arabidopsis. Plant Physiol 126:1646–1667

Maziya-Dixon B, Kling JG, Menkir A, Dixon A (2000) Genetic variation in total carotene, iron, and zinc contents of maize and cassava genotypes. Food Nutr Bull 21:419–422

Menkir A (2008) Genetic variation for grain mineral content in tropical-adapted maize inbred lines. Food Chem 110:454–464

Mertz ET, Bates LS, Nelson OE (1964) Mutant gene that changes protein composition and increases lysine content of maize endosperm. Science 145:279–280

Mi GH, Chen FJ, Liu XS, Chun L, Song JL (2004) Genotype difference in iron content in kernels of maize. J Maize Sci 12:13–15

Morris H (1995) Phytate: a good or a bad food component? Nutr Res 15:754–773

Moussavi-Nik M, Pearson JN, Hollamby GJ, Graham RD (1998) Dynamics of nutrient remobilization during germination and early seedling development in wheat. J Plant Nutr 21:421–434

Mulvihill B, Morrissey PA (1998) An investigation of factors influencing the bioavailability of non-haem iron from meat systems. Ir J Agric Food Res 37:219–226

Naqvi S, Zhu C, Farre G, Ramessar K, Bassie L, Breitenbach J, Conesa D, Ros G, Sandmann G, Capell T, Christou P (2009) Transgenic multivitamin corn through biofortification of endosperm with three vitamins representing three distinct metabolic pathways. Proc Natl Acad Sci U S A 106:7762–7767

Nelson O, Chang MT (1974) Effect of multiple aleurone layers on the protein and amino acid content of maize endosperm. Crop Sci 14:374–376

Nicolas G, Bennoun M, Porteu A, Mativet S, Beaumont C, Grandchamp B, Sirito M, Sawadogo M, Kahn A, Vaulont S (2002) Severe iron deficiency anaemia in transgenic mice expressing liver hepcidin. Proc Natl Acad Sci U S A 99:4596–4601

Norvell WA, Welch RM (1993) Growth and nutrient uptake by barley (*Hordeum vulgare* L. cv Herta). Studies using an N-(2-hydroxyethyl) ethylenedinitrilotriacetic acid-buffered nutrient solution technique. I. Zinc ion requirements. Plant Physiol 101:619–625

Nozoye T, Nakanishi H, Nishizawa NK (2013) Characterizing the crucial components of iron homeostasis in the maize mutants *ys1* and *ys3*. PLoS One 8:e62567

Nunes ACS, Vianna GR, Cuneo F, Amaya-Farfan J, de Capdeville G, Rech EL, Aragao FJL (2006) RNAi-mediated silencing of the myo-inositol-1-phosphate synthase gene (GmMIPS1) in transgenic soybean inhibited seed development and reduced phytate content. Planta 224:125–132

Oikeh SO, Menkir A, Maziya-Dixon B, Welch R, Glahn RP (2003a) Assessment of concentrations of iron and zinc and bioavailable iron in grains of early-maturing tropical maize varieties. J Agric Food Chem 51:3688–3694

Oikeh SO, Menkir A, Maziya-Dixon B, Welch R, Glahn RP (2003b) Genotypic differences in concentration and bioavailability of kernel-iron in tropical maize varieties grown under field conditions. J Plant Nutr 26:2307–2319

Oikeh SO, Menkir A, Maziya-Dixon B, Welch RM, Glahn RP (2004a) Assessment of iron bioavailability from twenty elite late-maturing tropical maize varieties using an in vitro digestion/Caco-2 cell model. J Sci Food Agric 84:1202–1206

Oikeh SO, Menkir A, Maziya-Dixon B, Welch RM, Glahn RP, Gauch G (2004b) Environmental stability of iron and zinc concentrations in grain of elite early-maturing tropical maize genotypes grown under field conditions. J Agric Sci 142:543–551

Ortiz-Monasterio JI, Palacios-Rojas N, Meng E, Pixley K, Trethowan R, Pena RJ (2007) Enhancing the mineral and vitamin content of wheat and maize through plant breeding. J Cereal Sci 46:293–307

Payne KA, Bowen HC, Hammond JP, Hampton CR, Lynn JR, Mead A, Swarup K, Bennett MJ, White PJ, Broadley MR (2004) Natural genetic variation in cesium (Cs) accumulation by *Arabidopsis thaliana*. New Phytol 162:535–548

Pfeiffer WH, McClafferty B (2007) HarvestPlus: breeding crops for better nutrition. Crop Sci 47:S88–S105

Pilu R, Panzeri D, Gavazzi G, Rasmussen S, Consonni G, Nielsen E (2003) Phenotypic, genetic and molecular characterization of a maize low phytic acid mutant (*lpa241*). Theor Appl Genet 107:980–987

Pilu R, Landoni M, Cassani E, Doria E, Nielsen E (2005) The maize *lpa241* cause a remarkable variability of expression and some pleiotropic effects. Crop Sci 45:2096–2105

Pixley KV, Palacios N, Glahn RP (2011) The usefulness of iron bioavailability as a target trait for breeding maize (*Zea mays* L.) with enhanced nutritional value. Field Crop Res 123:153–160

Pollock JD (2002) Gene expression profiling: methodological challenges results and prospects for addiction research. Chem Phys Lipids 121:241–256

Prasad AS (1996) Zinc deficiency in women, infants and children. J Am Coll Nutr 15:113–120

Prasanna BM, Vasal SK, Kassahun B, Singh NN (2001) Quality protein maize. Curr Sci 81:1308–1319

Prasanna BM, Mazumdar S, Chakraborti M, Hossain F, Manjaiah KM, Agrawal PK, Guleria SK, Gupta HS (2011) Genetic variability and genotype × year interactions for kernel iron and zinc concentration in maize (*Zea mays* L.). Indian J Agric Sci 81:704–711

Prioul JL, Pelleschi S, Sene M, Thevenot C, Causse M, deVienne D, Leonardi A (1999) From QTL for enzyme activity to candidate gene. J Exp Bot 50: 1281–1288

Puga P, Kutka F (2012) Identification of multiple aleurone in CG Coroico flour. Maize Genet Coop Newsl 86:1–2

Qin H, Cai Y, Liu Z, Wang G, Wang J, Guo Y, Wang H (2012) Identification of QTL for zinc and iron concentration in maize kernel and cob. Euphytica 187: 345–358

Queiroz VAV, Guimaraes PEO, Queiroz LR, Guedes EO, Vasconcelos VDB, Guimares LJ, Ribeiro PEA, Schaffert RE (2011) Iron and zinc variability in maize lines. Cienc Tecnol Aliment Campinas 31: 577–583

Raboy V (1997) Accumulation and storage of phosphate and minerals. In: Larkins BA, Vasil IK (eds) Cellular and molecular biology of plant seed development. Kluwer Academic Publishers, Dordrecht, pp 441–477

Raboy V (2001) Genetics and breeding of seed phosphorus and phytic acid. J Plant Physiol 158:489–497

Raboy V, Gerbasi PF, Young KA, Stoneberg SD, Pickett SG, Bauman AT, Murthy PPN, Sheridan WF, Ertl DS (2000) Origin and seed phenotype of maize low phytic acid 1–1 and low phytic acid 2–1. Plant Physiol 124:355–368

Revy PS, Jondreville C, Dourmad JY, Nys Y (2006) Assessment of dietary zinc requirement of weaned piglets fed diets with or without microbial phytase. J Anim Physiol Anim Nutr 90:50–59

Riedelsheimer C, Lisec J, Czedik-Eysenberg A, Sulpice R, Flis A, Grieder C, Altmann T, Stitt M, Willmitzer L, Melchinger AE (2012) Genome-wide association mapping of leaf metabolic profiles for dissecting complex traits in maize. Proc Natl Acad Sci 109:8872–8877

Roberts LA, Pierson AJ, Panaviene Z, Walker EL (2004) Yellow Stripe1 expanded roles for the maize iron-phytosiderophore transporter. Plant Physiol 135: 115–120

Robinson NJ, Procter CM, Connolly EL, Guerinot ML (1999) A ferric-chelate reductase for iron uptake from soils. Nature 397:694–697

Roeser HP (1986) Iron. J Food Nutr 42:82–92

Ruel MIT, Bouis HE (1998) Plant breeding: a long-term strategy for the control of zinc deficiency in vulnerable populations. Am J Clin Nutr 68:S488–S494

Salovaara S, Sandberg AS, Andlid T (2003) Combined impact of pH and organic acids on iron uptake by Caco-2 cells. J Agric Food Chem 51:7820–7824

Sandstead H (1995) Is zinc deficiency a public health problem? Nutrition 11:87–92

Sandstorm B (1997) Bioavailability of zinc. Eur J Clin Nutr 51:S17–S19

Savage GP (2002) Oxalates in human foods. Proc Nutr Soc NZ 27:4–24

Schauer N, Semel Y, Roessner U, Gur A, Balbo I, Carrari F, Pleban T, Perez-Melis A, Bruedigam C, Kopka J, Willmitzer L, Zamir D, Fernie AR (2006) Comprehensive metabolic profiling and phenotyping of interspecific introgression lines for tomato improvement. Nat Biotechnol 24:447–454

Scholz-Ahrens KE, Schrezenmeir J (2002) Inulin, oligofructose and mineral metabolism - experimental data and mechanism. Br J Nutr 87:S179–S186

Scrimshaw NS (1984) Functional consequences of iron deficiency in human populations. J Nutr Sci Vitaminol 30:47–63

Shankar AH, Prasad AS (1998) Zinc and immune function: the biological basis of altered resistance to infection. Am J Clin Nutr 68:S447–S463

Sharma A, Chauhan RS (2008) Identification of candidate gene-based markers (SNP and SSRs) in zinc and iron transporter sequences of maize (Zea mays L.). Curr Sci 95:1051–1059

Shen B, Li C, Min Z, Meeley RB, Tarczynski MC, Olsen OA (2003) sal1 determines the number of aleurone cell layers in maize endosperm and encodes a class E vacuolar sorting protein. Proc Natl Acad Sci 100:6552–6557

Shi R, Li H, Tong Y, Jing R, Zhang F, Zou C (2008) Identification of quantitative trait locus of zinc and phosphorus density in wheat (Triticum aestivum L.) grain. Plant Soil 306:95–104

Simic D, Sudar R, Jambrovic A, Ledencan T, Zdunic Z, Kovacevic V, Brkic I (2009) Genetic variation of bioavailable iron and zinc in grain of a maize population. J Cereal Sci 50:392–397

Simic D, Mladenovic DS, Zdunic Z, Jambrovic A, Ledencan T, Brkic J, Brkic A, Brkic I (2011) Quantitative trait Loci for biofortification traits in maize grain. J Hered 103:47–54

Solomons NW (2003) Zinc deficiency. In: Ceballero B (ed) Encyclopedia of food sciences, 2nd edn. Elsevier Science Ltd., Oxford

Soric R, Ledencan T, Zdunic Z, Jambrovic A, Brkic I, Loncaric Z, Kovacevic V, Simic D (2011) Quantitative trait loci for metal accumulation in maize leaf. Maydica 56:323–329

Stein AJ (2010) Global impact of human mineral malnutrition. Plant Soil 335:133–154

Storcksdieck S, Bonsmann G, Hurrell RF (2007) Iron-binding properties, amino acid composition, and structure of muscle tissue peptides from in vitro digestion of different meat sources. J Food Sci 72:S19–S29

Tako E, Rutzke MA, Glahn RP (2010) Using the domestic chicken (Gallus gallus) as an in vivo model for iron bioavailability. Poult Sci 89:514–521

Tako E, Hoekenga OA, Kochian LV, Glahn RP (2013) High bioavailability iron maize (Zea mays L.) developed through molecular breeding provides more absorbable iron in vitro (Caco-2 model) and in vivo (Gallus gallus). Nutr J 12:3–11

Tanksley SD, Ganal MW, Martin GB (1995) Chromosome landing: a paradigm for map-based gene cloning in plants with large genomes. Trends Genet 11:63–68

Terry N, Low G (1982) Leaf chlorophyll content and its relation to the intracellular localization of iron. J Plant Nutr 5:301–310

Teucher B, Olivares M, Cori H (2004) Enhancers of iron absorption: ascorbic acid and other organic acids. Int J Vitam Nutr Res 74:403–419

Thomine S, Lelievre F, Debarbieux E, Schroeder JI, Barbier-Brygoo H (2003) AtNRAMP3, a multispecific vacuolar metal transporter involved in plant responses to iron deficiency. Plant J 34:685–695

Ueno D, Yamaji N, Ma JF (2009) Further characterization of ferric-phytosiderophore transporters ZmYS1 and HvYS1 in maize and barley. J Exp Bot 60:3513–3520

UNSCN (United Nations System Standing Committee on Nutrition) (2004) Fifth report on the world nutrition situation: nutrition for improved development outcomes. Available at http://www.unsystem.org/scn/publications/AnnualMeeting/ SCN31/SCN5report.pdf

Urbany C (2012) Combined omics approaches to unravel the molecular basis of maize iron homeostasis. In: PAG XXII. San Diego. Available via https://pag.confex.com/pag/xx/webprogram/Paper3716.html. Accessed 15 Aug 2013

Vignesh M, Hossain F, Nepolean T, Saha S, Agrawal PK, Guleria SK, Prasanna BM, Gupta HS (2012) Genetic variability for kernel β-carotene and utilization of crtRB1 3'TE gene for biofortification in maize (Zea mays L.). Indian J Genet 72:189–194

Von-Wiren N, Marschner H, Romheld V (1996) Roots of iron-efficient maize also absorb phytosiderophore-chelated zinc. Plant Physiol 111:1119–1125

Von-Wiren N, Klair S, Bansal S, Briat JF, Khodr H, Shioiri T, Leigh RA, Hider RC (1999) Nicotianamine chelates both FeIII and FeII-Implications for metal transport in plants. Plant Physiol 119:1107–1114

Walezyk T, Davidson L, Rossander-Hulthen L, Hallberg L, Hurrell RF (2003) No enhancing effect of vitamin A on iron absorption in humans. Am J Clin Nutr 77:144–149

Walter A, Romheld V, Marschner H, Mori S (1994) Is the release of phytosiderophores in zinc-deficient wheat plants a response to impaired iron utilization? Physiol Plant 92:493–500

Wang T, Sun X, Raddatz J, Chen G (2013) Effects of microfluidization on microstructure and physico-chemical properties of corn bran. J Cereal Sci 58:355–361

Watson SA, Ramstad PT (eds) (1987) Corn: chemistry and technology. American Association of Cereal Chemists, St Paul

Weckwerth W, Loureiro ME, Wenzel K, Fiehn O (2004) Differential metabolic networks unravel the effects of silent plant phenotypes. Proc Natl Acad Sci 101: 7809–7814

Welch RM, Graham RD (2004) Breeding for micronutrients in staple food crops from a human nutrition perspective. J Exp Bot 55:353–364

Welch RM, Graham RD (2005) Agriculture: the real nexus for enhancing bioavailable micronutrients in food crops. J Trace Elem Med Biol 18:299–307

Welch RM, Smith ME, Van-Campen DR, Schaeffer SC (1993) Improving the mineral reserves and protein quality of maize (*Zea mays* L.) kernels unique genes. Plant Soil 155(156):215–218

White PJ, Broadley MR (2009) Biofortification of crops with seven mineral elements often lacking in human diets- iron, zinc, copper, calcium, magnesium, selenium and iodine. New Phytol 182:49–84

Wolf MJ, Cutler HC, Zuber MS, Khoo U (1972) Maize with multilayer aleurone of high protein content. Crop Sci 12:440–442

Wu H, Li L, Du J, Yuan Y, Cheng X, Ling HQ (2005) Molecular and biochemical characterization of the Fe (III) chelate reductase gene family in *Arabidopsis thaliana*. Plant Cell Physiol 46:1505–1514

Xu Y, An D, Li H, Xu H (2011) Review: breeding wheat for enhanced micronutrients. Can J Plant Sci 91: 231–237

Yan J, Kandianis BC, Harjes EC, Bai L, Kim HE, Yang X, Skinner DJ, Fu Z, Mitchell S, Li Q, Fernandez GSM, Zaharoeva M, Babu R, Fu Y, Palacios N, Li J, DellaPenna D, Brutnell T, Buckler SE, Warburton LM, Rocheford T (2010) Rare genetic variation at *Zea mays crtRB1* increases beta carotene in maize grain. Nat Genet 42:322–329

Yi Y, Guerinot ML (1996) Genetic evidence that induction of root Fe (III) chelate reductase activity is necessary for iron uptake under iron deficiency. Plant J 10:835–844

Zhang LH, Byrne PF, Pilon-Smits EAH (2006) Mapping quantitative trait loci associated with selenate tolerance in *Arabidopsis thaliana*. New Phytol 170:33–42

Zhang X, Pfeiffer WH, Palacios-Rojas N, Babu R, Bouis H, Wang J (2012) Probability of success of breeding strategies for improving provitamin A content in maize. Theor Appl Genet 125:235–246

Zheng LQ, Huang FL, Narsai R, He F, Giraud E, Wu JJ, Cheng LJ, Wang F, Wu P, Whelan J, Shou HX (2009) Physiological and transcriptome analysis of iron and phosphorus interaction in rice seedlings. Plant Physiol 151:262–274

Zheng L, Chengz AC, Jiang X, Bei X, Zheng Y, Glahn RP, Welch RP, Miller DD, Lei XG, Shou H (2010) Nicotianamine, a novel enhancer of rice iron bioavailability to humans. PLoS One 5:e10190

Zhu C, Naqvi S, Gomez-Galera S, Pelacho AM, Capell T, Christou P (2007) Transgenic strategies for the nutritional enhancement of plants. Trends Plant Sci 12: 548–555

Zhu C, Naqvi S, Breitenbach J, Sandmann G, Christou P, Capell T (2008) Combinatorial genetic transformation generates a library of metabolic phenotypes for the carotenoid pathway in maize. Proc Natl Acad Sci U S A 105:18232–18237

Part IV
Nutrient Use Efficiency of Crop Species

Nitrogen Uptake and Use Efficiency in Rice

N.K. Fageria, V.C. Baligar, A.B. Heinemann, and M.C.S. Carvalho

Abstract

Rice is a stable food for a large proportion of the world's population. Most of the rice is produced and consumed in Asia. Rice is produced under both upland and lowland rice systems, with about 76 % of the global rice produced from irrigated-lowland rice systems. Nitrogen (N) is one of the most important inputs in the production of rice. Recovery efficiency of N is less than 50 % in both upland and lowland systems. Most of the applied N is lost due to volatilization, leaching, denitrification, and soil erosion. In addition, fertilizers account for almost half of energy used in world agriculture, and the manufacture of N fertilizer is about 10 times more energy intensive than that of P and K fertilizers. Therefore, improving N use efficiency is important not only to improve yield and reduce cost of production but to avoid environmental pollution and to maintain sustainability of the cropping system. Production practices which can improve N use efficiency are liming acid soils, supplying N in adequate rates, use of proper sources, use of suitable methods and time of application, use of crop rotation, use of cover crops, adopting conservation tillage system, planting N efficient genotypes, and control of diseases, insects, and weeds.

Keywords

Rice ecosystems • Nitrogen harvest index • Water use efficiency • Genotypes • Nutrient use efficiency

N.K. Fageria (✉) • A.B. Heinemann • M.C.S. Carvalho
National Rice and Bean Research Center of EMBRAPA,
Caixa Postal 179, Santo Antônio de Goiás, GO, Brazil
e-mail: nand.fageria@embrapa.br

V.C. Baligar
USDA-ARS Beltsville Agricultural Research center,
Beltsville, MD 20705-2350, USA
e-mail: VC.Baligar@ars.usda.gov

1 Introduction

Rice is the staple food for more than 50 % of the world population. Xiong et al. (2013) reported that rice is the staple food for about 60 % of the population in China. Furthermore, rice provides 35–60 % of the dietary calories consumed by

A. Rakshit et al. (eds.), *Nutrient Use Efficiency: from Basics to Advances*,
DOI 10.1007/978-81-322-2169-2_18, © Springer India 2015

nearly 3.5 billion people (Fageria et al. 2003a). It occupies about 23 % of the total area under cereal production in the world (Wassmann et al. 2009; Jagadish et al. 2010). Rice is produced and consumed in all the continents, except Antarctica. However, major part of the rice is produced and consumed in Asia. In Asia, India and China are the major producers as well as consumers of rice. Rice is also consumed in large quantities in North America and Europe by native and immigrants from Asia, Africa, and South America. The population in Africa and South America consume substantial amount of rice. In South America, rice is eaten everyday with dry bean (*Phaseolus vulgarsi* L.) by all section of society (Fageria 2013).

Nitrogen is one of the most yield-limiting nutrients in rice production in all rice-growing regions worldwide. Uptake of N is maximum by rice (sometimes equal to K) compared to other essential nutrients. Recovery efficiency of N is lower than 50 % in most cropping systems (Fageria et al. 2011; Fageria 2013). The major part of N in soil is lost through volatilization, leaching, denitrification, and soil erosion (Fageria and Baligar 2005). Low recovery of N is not only responsible for higher cost of crop production but also for environmental pollution. Hence, improving N use efficiency (NUE) is desirable to improve crop yields, reduce cost of production, and maintain environmental quality. To improve N efficiency in agriculture, integrated N management strategies that take into consideration improved fertilizer along with soil and crop that includes improved fertilizer application timing and methods along with soil and crop management practices are necessary (Fageria and Baligar 2005).

By the year 2025, it is estimated that it will be necessary to produce about 60 % more rice than what is currently produced to meet the food needs of a growing world population (Fageria 2013). Similarly, Normile (2008) reported that increase of 1.2 % per year of rice production will be required to meet the growing demand for food that will result from population growth and economic development in the next decade. Enhancement of rice production and sustainability

are important features of grain production to benefit the world's 3.5 billion people who depend on rice for their livelihood and as their basic food. Adequate amounts of essential nutrients are needed by modern rice cultivars with improved cultural practices to achieve higher yields. In this context, efficient use of inputs is vital to safely produce the additional food from limited resources with minimal adverse impact on the environment. The objective of this chapter is to discuss nitrogen uptake and use efficiency and a summation of best management practices that could help scientists and rice farmers to develop practical, integrated recommendations that improve nitrogen use efficiency in various types of rice production systems.

2   Rice Ecosystems

Ecosystem is defined as the environmental factors around the crop plants. The main environmental factors which influence rice growth are related to climatic and soil conditions. Rice is mainly grown under two ecosystems known as upland and lowland. Upland rice is defined as the rice grown on undulated well-drained soils, without water stagnation in the plots or fields, and totally depends on rainfall for its water requirements. Upland rice is also known as aerobic rice because it is grown on well-drained soils. Lowland rice is defined as the rice grown on flat, saturated soils, with water stagnation in the plots or field for most of the growing season, and generally has controlled irrigation. Lowland rice is also known as irrigated, flooded, or submerged rice. Such systems of productions contributes about 76 % of the total world rice production (Fageria et al. 2003a). The yield of upland rice is relatively low compared to lowland rice, because many abiotic and biotic stresses are associated with low yield. The main abiotic stresses in upland rice are drought, low soil fertility, and use of low technology by the farmers. The main biotic stresses are diseases, insects, and weeds.

For example, in Brazil, yield of upland rice is less than half of lowland rice. Average yields of lowland rice in Brazil is about 5.5 Mg ha^{-1}, whereas average yields of upland rice is about 2.2 Mg ha^{-1}. In upland rice blast disease is a serious constraint in reducing yield in most of the upland rice growing areas. In addition upland rice in Brazil is mainly cultivated in the central part of Brazil, locally known as "Cerrado" region. Most soils of the Cerrado region are acidic and deficient in most of the essential plant nutrients (Fageria and Baligar 2008).

3 Soil Used for Rice Cultivation

The Soil Science Society of America (2008) defined soil as the unconsolidated mineral or organic material on the immediate surface of the earth that serves as a natural medium for the growth of plants. Soil quality is an important factor in crop production. Soil quality is mainly determined by crop yields. When a determined soil produces higher crop yields, it indicates a productive soil. Soil quality is defined in terms of its physical, chemical, and biological properties. Physical properties which determine soil quality are texture, structure, bulk density, and water infiltration rate or porosity, whereas chemical properties of soil are mainly associated with soil fertility. The biomass of a soil is represented by its microbial population. The microbes may be beneficial or may be harmful for the growth of plants. Soils are classified into orders and according to US soil taxonomy, there are 12 soil orders. These soil orders are Alfisols, Andisols, Aridisols, Entisols, Gelisols, Histosols, Inceptisols, Mollisols, Oxisols, Spodosols, Ultisols, and Vertisols. Rice can be grown on all the 12 soil orders. Detailed discussion of characteristics of these soils under rice cultivations are given by Fageria (2014).

4 Functions and Deficiency Symptoms

Nitrogen plays significant role in many physiological and biochemical processes in the plants. It improved tillering in rice and consequently panicle density. Nitrogen also increases panicle length and grain weight and reduces spikelet sterility (Fageria and Baligar 2001a; Fageria 2007). Adequate rate of nitrogen improves root growth (Fig. 1) in rice which is very important for the absorption of water and nutrients (Fageria 2013). In rice N deficiency symptoms are characterized by yellowing of the leaves. Since nitrogen is mobile nutrient in the plants, hence, deficiency symptoms first start in the older leaves. If deficiency persists longer, all the leaves become yellow. Figure 2 shows growth of two lowland rice genotypes at low and high N rates. Nitrogen deficiency symptoms are very clear at low level of N.

5 Nitrogen Uptake and Partitioning

Uptake of N is maximum in rice, except K. It is absorbed in the form of NO_3^- and NH_4^+ by plants. In oxidized soil NO_3^- is the dominant form of N uptake and in reduced soils N is mainly absorbed as NH_4^+. The topic of NO_3^- vs. NH_4^+ uptake of N by rice is discussed in detail by Fageria (2014).

5.1 Nitrogen Concentration

Uptake of N is expressed in concentration which is defined as the N content per unit of dry matter. The unit of N concentration in plants is generally g kg^{-1} or percentage. Generally, concentration values are used to diagnose nutrient sufficiency, deficiency, or excess in plants. Nutrient concentrations can be extrapolated or used for identifying nutritional disorders in the same crop species from different agroecological regions. This is possible because nutrient uptake in plants is an integral part of all factors affecting nutrient availability.

One of the most important considerations in defining adequate concentrations is plant age (Fig. 3). Fageria (2003) determined a relationship between dry matter yield of shoots or grain and N concentration in the shoot or grain of

Fig. 1 Influence of low and high N levels on root growth of two upland rice genotypes

Fig. 2 Growth of two lowland rice cultivars at two N levels

lowland rice at different growth stages. Based on this relationship, optimum N concentrations in shoots at different growth stages and in the grain at harvest were determined in rice. Optimum N concentrations in shoots varied from 43.4 g kg^{-1} at initiation of tillering to 6.5 g kg^{-1} at physiological maturity. The N concentration in the grain at physiological maturity was 11 g kg^{-1}. Hence, optimal N concentration in shoots of rice decreased with

Fig. 3 Nitrogen
concentration in principal
crop species during growth
cycle (Source: Adapted
from Fageria 2004)

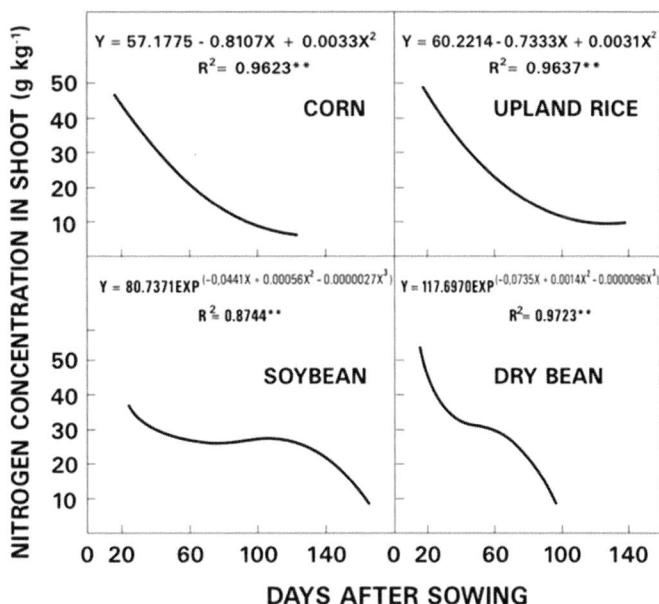

advanced plant age. During grain filling, N content of non-grain tissue generally decreases, while grain N content increases (Fageria and Baligar 2005). However, shoot dry weight increased with age advancement up to the flowering growth stage and then decreased (Fageria 2003). Decreases in shoot dry weight at harvest was related to translocation of assimilate to the panicle from flowering to maturity (Fageria and Baligar 2005). In rice, 60–90 % of the total C accumulated in panicles at the time of harvest was derived from photosynthetic after heading, and the flag leaves are the organs that contribute most to grain filling (Yoshida 1981).

5.2   Nitrogen Accumulation

When dry matter or grain yield is multiplied by concentration, the results are a measure of nutrient uptake and expressed in accumulation or uptake units. Under field conditions, the nutrient uptake or accumulation unit is kg ha^{-1} for macronutrients and g ha^{-1} for micronutrients. Nutrient uptake values are useful indicators of soil fertility depletion and are related to crop yield levels. Nutrient accumulation patterns in

crop plants, including rice, followed dry matter accumulation (Fageria 2004). A study was conducted at the National Rice and Bean Research Center of EMBRAPA, Santo Antônio de Goiás, Brazil, to study the association between dry matter and grain yield of lowland rice and N accumulation during growth cycle (Table 1). The N uptake into shoots as well as into grain of lowland rice was significantly related to shoot dry weight and grain yield (Table 1). Fageria and Baligar (2005) reviewed the literature on this topic and reported that accumulation of N in cereals, including rice, dry matter production, is closely related to N accumulation. Nitrogen uptake as well as shoot dry weight increased up to the flowering stage (Fageria 2003). At harvest, more N was accumulated in grain than in dry matter. Yoshida (1981) reported that during plant ripening, about 70 % of the N absorbed by the straw will be translocated to the grain and maintain N contents of the grain at certain percentages. Nitrogen absorbed by rice during the vegetative growth stage contributes to growth during the reproductive and grain filling growth stages via translocation (Fageria and Baligar 2005).

Accumulation and distribution of N in the vegetative and reproductive organs of rice are

Table 1 Relationship between grain yield (Y) and N uptake in the shoot and grain of lowland rice at different growth stages

Plant growth stage	Regression	R^2	N uptake for maximum shoot or grain yield (kg ha^{-1})
IT (22)	$Y = 166.46 + 9.4552X - 0.1565X^2$	0.61^{NS}	16
AT (35)	$Y = -391.29 + 63.8885X - 0.5898X^2$	0.93**	54
IP (71)	$Y = 40.32 + 101.2576X - 0.3939X^2$	0.97**	129
B (97)	$Y = -2069.44 + 185.7829X - 0.6725X^2$	0.94**	138
F (112)	$Y = -367.39 + 167.8636X - 0.4528X^2$	0.97**	185
PM (140)	$Y = -2330.74 + 335.1191X - 2.3641X^2$	0.99**	71
PM (140)^a	$Y = -3547.09 + 261.4988X - 1.7099X^2$	0.99**	76

Source: Adapted from Fageria (2003)
Where regression equation was nonsignificant, average value across the N rates was considered as quantity of N uptake for maximum yield
Values are averages of 3 years field experimentation
**, NS, Significant at the 1 % probability level and nonsignificant, respectively. *IT* initiation of tillering, *AT* active tillering, *IP* initiation of panicle, *B* booting, *F* flowering, *PM* physiological maturity. Values in the parentheses represent age of the plants in days after sowing
^{a}In this line, values are for grain yield

the important processes in determining grain yield (Fageria 2014). Xiong et al. (2013) reported that super high yielding early cultivars had higher total N content at heading and maturity compared to ordinary early rice cultivars. Xiong et al. (2013) also reported that the differences in N translocation parameters among rice varieties or variety group were associated with the N accumulated in plants before heading. Mae (1997) reported that the amount of N absorbed by the plant during grain-filling period is much smaller than the amount of N accumulated in mature grain, and a large part of grain N is translocated from vegetative organs. Nitrogen distribution studies showed that 30–80 % of the N accumulated in the rice grain originated from translocation from vegetative tissue after heading (Ntanos and Koutroubas 2002).

5.3   Nitrogen Harvest Index

Nitrogen harvest index (NHI) is defined as the portioning of total plant N into grain. It is calculated by N accumulation in grain divided by N accumulation in grain plus straw. It is an important index in defining rice yield. Because it is positively related to grain yield (Fageria and Baligar 2005; Fageria 2007; Fageria et al. 2011). In addition, it is also an important index in

measuring N partitioning in crop plants, which provide an indication of how efficiently the plant utilizes acquired N for grain production (Fageria and Baligar 2005).

The NHI values varied from crop species to crop species and among genotypes of the same species. This trait is important for selecting crop genotypes for higher yield. Fageria (2007) reported that NHI in lowland rice varied from 0.53 to 0.64, with an average value of 0.60.

5.4   Nitrogen Use Efficiency

Efficiency is defined as the output divided by input. The higher the output value, the higher is the efficiency. In case of N use efficiency in crop plants, it can be defined as the maximum economic yield produced per unit of N applied, absorbed, or utilized by the plant to produce grain and straw. However, nutrient use efficiency has been defined in several ways in the literature, although most of them denote the ability of a system to convert inputs into outputs. Definitions of nutrient use efficiencies have been grouped or classified as agronomic efficiency, physiological efficiency, agrophysiological efficiency, apparent recovery efficiency, and utilization efficiency. Fageria and Baligar (2001a) calculated these efficiencies for lowland rice and results are

Table 2 Nitrogen use efficiencies as affected by N fertilizer rate

N rate (kg ha^{-1})	AE (kg ha^{-1})	PE (kg ha^{-1})	APE (kg ha^{-1})	ARE (%)	UE (kg kg^{-1})
30	35	156	72	49	76
60	32	166	73	50	83
90	22	182	75	37	67
120	22	132	66	38	50
150	18	146	57	34	50
180	16	126	51	33	42
210	13	113	46	32	36
Average	23	146	63	39	58
R^2	0.93**	0.62*	0.87**	0.82**	0.90**

Source: Fageria and Baligar (2001a)
AE agronomic efficiency, *PE* physiological efficiency, *APE* agrophysiological efficiency, *ARE* apparent recovery efficiency, *UE* utilization efficiency
*, ** Significant at the 0.05 and 0.01 probability levels, respectively

presented in Table 2. The determination of NUE in crop plants is an important approach to evaluate the fate of applied chemical fertilizers and their role in improving crop yields.

6 Management Practices to Improve Nitrogen Use Efficiency

Recovery efficiency of applied fertilizer N is less than 40 % for lowland rice (Fageria and Baligar 2001a; Fageria 2014). It is reported by Raun and Johnson (1999) that average world N recover efficiency of cereals is about 33 %. Hence, a large part of the N is lost in the soil-plant system by leaching, denitrification, volatilization, and surface runoff. Improving nitrogen use efficiency (NUE) is fundamental to improve crop yield, reducing crop production cost and keeping clean environment. Nitrogen use efficiency of rice can be improved with the adoption of appropriate soil, fertilizer, and plant management practices. These management practices that could improve NUE are discussed in the succeeding sections.

6.1 Liming Acid Soils

Soil acidity is one of the major constraints in crop production throughout the world. Acid soils are found over extensive areas in the tropics, subtropics, and temperate zones. Globally, soil acidity affects land area of about 3.95 billion hectares (Sumner and Noble 2003). This is about 30 % of the world's ice-free land area. Soil acidity is a main constraint in crop production in South America. In South America, 85 % of the soils are acidic, and approximately 850 million ha of such land area is underutilized (Fageria and Baligar 2001b). Liming is the most economical and effective practice to reduce soil acidity. Liming has many beneficial effects in the soils. These includes improvement in soil physical (structure), chemical (Ca, Mg, pH), and biological properties. In addition, liming also neutralize Al, Mn, and H^+ ions toxicity (Fageria 2001). All these practices improve N use efficiency in crop plants.

6.2 Use of Effective Source, Appropriate Method, and Timing of Application

Use of effective source of N is fundamental in improving N use efficiency and consequently achieving higher yields of crops. There are several sources of nitrogen. Urea and ammonium sulfate are the main nitrogen carriers worldwide in annual crop production. However, urea is generally favored by the growers over ammonium sulfate due to lower application cost because urea has a higher N analysis than ammonium sulfate (46 % vs. 21 % N). In developed countries like the USA, anhydrous NH_3 is an important N source for annual crop production. At normal pressures, NH_3 is a gas and is transported and handled as liquid under pressure. It is injected into the soil to prevent loss through volatilization. The NH_3 protonates to form NH_4^+ in the soil and becomes XNH_4^+ which is stable (Foth and Ellis 1988). The major advantages of anhydrous NH_3 are its high N analysis (82 % N) and low cost of transportation and handling. However, specific equipment is required for storage, handling, and application. Hence, NH_3 is not a popular N carrier in developing countries. Stanford (1973) and Campbell et al. (1995) reported that ammonium nitrate is generally superior to urea which may volatilize easily.

Nitrogen is a mobile nutrient in soil-plant system and different from P and K which are immobile in the soil-plant system. Hence, it can be moved from distance to plant roots and can be absorbed. However, if it is broadcast and there is no rainfall or irrigation, it may be lost due to evaporation or volatilization. Hence, application in the band or incorporating into the soil may reduce its loss and improve N use efficiency (Fageria 2009). Campbell et al. (1995) also reported that banding of N fertilizers is superior to broadcasting.

To improve applied fertilizer efficiency, for different cropping systems and environmental conditions, fertilizer industries are manufacturing slow release fertilizers (SRF) and controlled release fertilizers (CRF) either with single nutrient or with multiple nutrients. These fertilizers when added to soil improve recovery efficiency by plants by lowering the rate of release, thereby reducing leaching losses (NO_3) and emission/volatilization (N_2O, NH_3) and providing N during the entire plant growing season (Trenkel 2010). Nitrification inhibitors (N-Serve, Nitropyrin, DCD, DMPP) improve applied fertilizer efficiency by keeping N in ammonia form longer, thereby control loss of nitrate by leaching (Peoples et al. 1995; Prasad and Power 1995; Trenkel 2010). Urease inhibitors (NBPT, PPD/PPDA, hydroquinone) improve urea efficiency

by suppressing the transformation of urea to ammonia and ammonium hydroxide thereby preventing volatile losses of ammonia to air (Hendrickson 1992; Trenkel 2010).

Timing of N application during crop growth is an important strategy in improving N use efficiency. The N application according to plant needs may improve its efficiency and avoid its loss from soil-plant system, In other words, synchronizing of N application with N demand of plants. It has been reported by Matson et al. (1996) and Tilman et al. (2002) that nutrient use efficiency is increased by appropriately applying fertilizers and by better matching temporal and spatial nutrient supplies with plant uptake. Applying fertilizer during periods of highest crop uptake, at or near the point of uptake (roots and leaves), as well as smaller and more frequent applications have the potential to reduce losses while maintaining or improving crop yield quantity and quality (Matson et al. 1996; Cassman et al. 2002). Rose and Bowden (2013) reported that split application of N fertilizer after crop emergence improves N use efficiency. Because plant roots had a chance to penetrate to depth and crop sink sizes are sufficient to take up significant quantities of the soil-mobile nitrate.

6.3   Use of Adequate Rate

Use of adequate rate of N is very important to increase yield and reduce cost of crop production and environmental pollution. Since N is a mobile nutrient in the soil-plant system, field trials are required to determine adequate rate for a given crop under given agroclimatic conditions. Fageria and Baligar (2001a) conducted field experiment involving lowland rice with different N rates in a Brazilian Inceptisol. Based on regression equations, in the first year, maximum grain yield (6,937 kg ha^{-1}) was obtained at 209 kg N ha^{-1}, in the second year maximum grain yield (6,958 kg ha^{-1}) was obtained at 163 kg N ha^{-1}, and in the third year maximum grain yield of 5,682 kg ha^{-1} was obtained at 149 kg N ha^{-1}. The average data of 3 years showed that maximum grain yield of 6,465 kg ha^{-1} was obtained with the application of 171 N ha^{-1}.

Singh et al. (1998) reported that maximum average grain yield of 7,700 kg ha^{-1} of 20 lowland rice genotypes was obtained at 150–200 kg N ha^{-1} at the International Rice Research Institute in the Philippines. Results of this study fall more or less in the same range as reported by Fageria and Baligar (2001a). In our fertilizer experimentations, however, 90 % of maximum rice yield is considered as an economical rate (Fageria and Baligar 2001a); in the first year it was (6,298 kg kg^{-1}) achieved at 120 kg N ha^{-1}. In the second and third years, 90 % of the maximum grain yields (6,345 and 5,203 kg ha^{-1}) were achieved at 90 and 78 kg N ha^{-1}, respectively. The average of 3 years data showed that 90 % of the maximum grain yield (5,731 kg ha^{-1}) was obtained at 84 kg N ha^{-1}. This means, that there was a residual effect of N application in lowland rice grown on an Inceptisol. The increase in grain yield of lowland rice at the economical rate (120 kg N ha^{-1}) in the first year was 76 % as compared to control N treatment. Similarly, the increase in grain yield in the second and third years at the economical N rates (90 and 78 kg ha^{-1}) was 69 and 41 %, respectively. The average increase of grain yield across the 3 years was 56 % at the economical N rate of 84 kg ha^{-1}. At the zero N level, the grain yield was 3,579, 3,754, and 3,702 kg ha^{-1} in the first, second, and third years, respectively. The average value of grain yield across the 3 years was 3,678 kg ha^{-1} at zero N rates. This means rice grain yield under the control treatment (no N application) was quite good during 3 years of experimentation. In control N treatment, rice yields increased during the second and third years of cultivation as compared to the first year of cultivation. Fageria and Baligar (1996) also reported significant increases in grain yields of lowland rice grown on an Inceptisol in the central part of Brazil. These authors reported that an average yield of 3 years (5,523 kg ha^{-1}) of lowland rice was achieved with the application of 100 kg N ha^{-1} and that in grain yields at low soil fertility level increased with succeeding cropping years.

6.4 Use of Crop Rotation

Planting rice in rotation with legumes can improve crop yield, reduce N application rates and nitrate leaching from soil-plant system. In Brazil rice is rotated with dry bean or soybean to get beneficial effects of legume-cereal rotation. Randall et al. (1997) reported that changing from continuous corn to a corn-soybean rotation has been shown to reduce NO_3^- leaching. Crop rotation also controls diseases, insects, and weed infestations which may improve N use efficiency (Fageria 1992). Dinnes et al. (2002) reported that diversifying crop rotation can reduce nitrate leaching and consequently improved nitrogen use efficiency.

6.5 Use of Conservation Tillage

Tillage improved microbial oxidation of organic matter and improved nitrification processes in the soil profile. Nitrogen mineralization increased with tillage. If nitrification or N mineralization exceeds the N demand of the plant, nitrate leaching may occur. On the other hand, conservation tillage may reduce microbial activities and release the nitrate slowly. In addition, conservation tillage reduces soil erosion and conserves more moisture in the soil profile. These favorable effects may improve plant growth and consequently higher N use efficiency.

6.6 Use of Cover Crops

Use of cover crops with main crops or cash crops is an important strategy in reducing nitrate leaching from soil-plant system. Cover crops function by accumulating the inorganic soil N between main crop seasons and holding it in an organic form, thus preventing it from leaching (Magdoff 1991; Dinnes et al. 2002). The N is subsequently released to the next crop as the cover crop residue decomposes. Cover crops also protect against soil erosion and thus preventing N losses from soil-plant systems.

Fig. 4 Response of lowland rice genotypes to N fertilization (Source: Fageria et al 2003b)

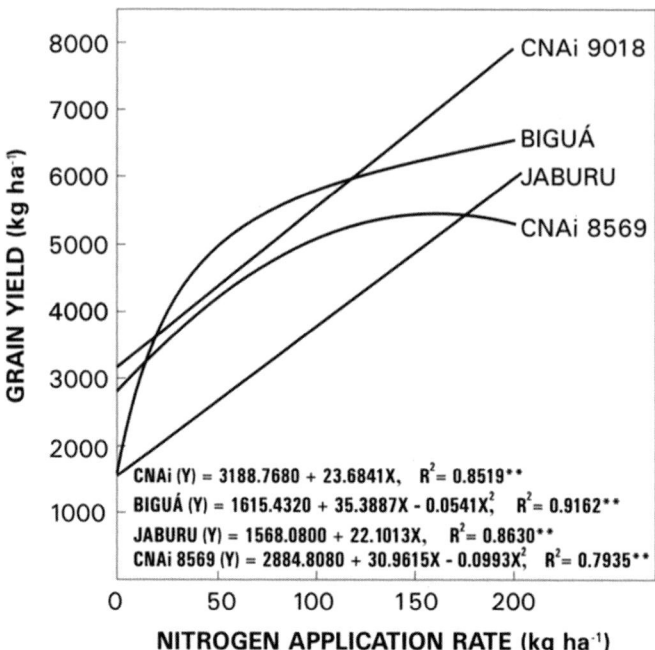

In addition, organic matter of soil can be increased by the decomposed cover crop residues which help in improving physical, chemical, and biological properties of the soil and improve N use efficiency.

6.7 Improve Water Use Efficiency

Soil moisture is one of the most important factors affecting nutrient use efficiency in crop plants (Fageria 2009, 2013). The solubility and transport of nutrient in the rhizosphere is controlled by water availability to plants. Soil moisture at field capacity known to improve nutrient movement and availability to plants. If soils are deficient in water, nutrient use efficiency decreased significantly.

6.8 Use of Efficient Genotypes

Planting N-efficient genotypes is an important strategy. Differences in N uptake and use efficiency of upland and lowland rice genotypes has been reported widely (Fageria et al. 2011;

Fageria 2013, 2014). Figure 4 shows different responses of four lowland rice genotypes. Two were having quadratic responses and two were having linear responses when N was applied in the range of 0–200 kg ha^{-1}. The difference in N uptake and utilization may be associated with better root geometry, ability of plants to take up sufficient nutrients from lower or subsoil concentrations, plants' ability to solubilize nutrients in the rhizosphere, better transport, distribution and utilization within plants, and balanced source-sink relationships (Fageria et al. 2008).

6.9 Control of Diseases, Insects, and Weeds

Diseases, insects, and weeds of agricultural crops are as old as agriculture itself (Fageria 1992). The resultant losses in economic terms are impossible to estimate accurately because the severity of diseases, insects, and weeds varies greatly from place to place, crop to crop, season to season, and year to year owing to changes in environmental factors (Fageria 1992). Kramer (1967) estimated that average worldwide losses for the main agricultural crops were 11.8 % for diseases and 12.2 % for insect pests. The average combined losses caused by diseases, insects, and weeds are put at 33.7 %. Control of diseases, insects, and weeds is an important factor in improving N use efficiency in crop production (Fageria and Gheyi 1999). Crops infested with diseases, insects, and weeds have lower photosynthetic efficiency, lower rate of absorption of water and nutrients, and competition for light, water, and nutrients, consequently reducing yields and resulting in low N use efficiency.

Conclusions

In the twenty-first century, improving nutrient use efficiency, including N, will play a major role in increasing crop yields compared to the twentieth century, mainly due to limited land and water resources available for crop production, higher cost of inorganic fertilizer inputs, declining trends in crop yields globally, and increasing environmental concerns. Furthermore, at least 60 % of the world's arable lands have mineral deficiencies or elemental toxicity problems, and on such soils fertilizers and lime amendments are essential for achieving improved crop yields. Fertilizer inputs are increasing cost of production of farmers, and there is a major concern for environmental pollution due to excess fertilizer inputs. Higher demands for food and fiber by increasing world populations further enhance the importance of improving nitrogen use efficiency. Rice is the staple food for more than 50 % world population, and use of N balanced with other essential nutrients is fundamental to improve rice yield and to maintain sustainability of cropping system. In this chapter N uptake and use efficiency and adopting practices which can improve N use efficiency are discussed.

Yields of modern cultivars is primarily source limited (supply of carbohydrates), and the source capacity should be increased, either genetically or by adopting appropriate cultural practices. More information should be generated about physiological and biochemical mechanisms involved in the efficient use of nutrients by crop plants. The use of biotechnology in identifying and creating nutrient-efficient crop species or genotypes offers exciting potential. However, this needs to be put in appropriate perspective.

References

Campbell CA, Myers RJK, Curtin D (1995) Managing nitrogen for sustainable crop production. Fertil Res 42:277–296

Cassman KG, Dobermann A, Walters D (2002) Agro-ecosystems, nitrogen use efficiency, and nitrogen management. AMBIO 31:132–140

Dinnes DL, Karlen DL, Jaynes DB, Kaspar TC, Hatfield JL, Colvin TS, Cambardella CA (2002) Nitrogen management strategies to reduce nitrate leaching in tile-drained Midwestern soils. Agron J 94:153–171

Fageria NK (1992) Maximizing crop yields. Marcel Dekker, New York

Fageria NK (2001) Effect of liming on upland rice, common bean, corn, and soybean production in cerrado soil. Pesq Agrop Brasileira 36:1419–1424

Fageria NK (2003) Plant tissue test for determination of optimum concentration and uptake of nitrogen at different growth stages in lowland rice. Commun Soil Sci Plant Anal 34:259–270

Fageria NK (2004) Dry matter yield and shoot nutrient concentration of upland rice, common bean, corn, and soybean grown in rotation on an oxisol. Commun Soil Sci Plant Anal 35:961–974

Fageria NK (2007) Yield physiology of rice. J Plant Nutr 30:843–879

Fageria NK (2009) The use of nutrients in crop plants. CRC Press, Boca Raton

Fageria NK (2013) The role of plant roots in crop production. CRC Press, Boca Raton

Fageria NK (2014) Mineral nutrition of rice. CRC Press, Boca Raton

Fageria NK, Baligar VC (1996) Response of lowland rice and common bean grown in rotation to soil fertility levels on a varzea soil. Fertil Res 45:13–20

Fageria NK, Baligar VC (2001a) Lowland rice response to nitrogen fertilization. Commun Soil Sci Plant Anal 32:1405–1429

Fageria NK, Baligar VC (2001b) Improving nutrient use efficiency of annual crops in Brazilian acid soils for sustainable crop production. Commun Soil Sci Plant Anal 32:1303–1319

Fageria NK, Baligar VC (2005) Enhancing nitrogen use efficiency in crop plants. Adv Agron 88:97–185

Fageria NK, Baligar VC (2008) Ameliorating soil acidity of tropical oxisols by liming for sustainable crop production. Adv Agron 99:345–399

Fageria NK, Gheyi HR (1999) Efficient crop production. Federal University of Paraiba, Campina Grande

Fageria NK, Slaton NA, Baligar VC (2003a) Nutrient management for improving lowland rice productivity and sustainability. Adv Agron 80:63–152

Fageria NK, Stone LF, Santos AB (2003b) Soil fertility management of lowland rice. EMBRAPA-Rice and Bean Research Center, Santo Antônio de Goiás

Fageria NK, Baligr VC, Li YC (2008) The role of nutrient efficient plants in improving crop yields in the twenty first century. J Plant Nutr 31:1121–1157

Fageria NK, Baligar VC, Jones CA (2011) Growth and mineral nutrition field crops, 3rd edn. CRC Press, Boca Raton

Foth HD, Ellis BG (1988) Soil fertility. Willey, New York

Hendrickson LL (1992) Corn yield response to the urease inhibitor NBPT: five-year summary. J Prod Agric 5: 131–137

Jagadish SVK, Cairns J, Lafitte R, Wheeler TR, Price AH, Craufurd PQ (2010) Genetic analysis of heat tolerance at anthesis in rice. Crop Sci 50:1633–1641

Kramer HH (1967) Plant protection and world crop production. Bayer AG, Leverkusen, Germany, p 524

Mae T (1997) Physiological nitrogen efficiency in rice: nitrogen utilization, photosynthesis, and yield potential. Plant Soil 196:201–210

Magdoff FR (1991) Managing nitrogen for sustainable corn systems: problems and possibilities. Am J Altern Agric 6:3–8

Matson PA, Billow C, Hall S (1996) Fertilization practices and soil variations control oxide emissions from tropical sugarcane. J Geophys Res 101:18533–18545

Normile D (2008) Reinventing rice to feed the world. Science 321:330–333

Ntanos DA, Koutroubas SD (2002) Dry matter and n accumulation and translocation for Indica and Japonica rice under Mediterranean conditions. Field Crop Res 74:93–101

Peoples MB, Freney JR, Mosier AR (1995) Minimizing gaseous losses of nitrogen. In: Bacon PE (ed) Nitrogen fertilization in the environment. Marcel Dekker, New York, pp 565–602

Prasad R, Power JE (1995) Nitrification inhibitors for agriculture, health and environment. Adv Agron 54: 233–281

Randall GW, Huggisns DR, Russelle MP, Fuchs DJ, Nelson WW, Anderson JL (1997) Nitrate losses through subsurface tile drainage in conservation reserve program, alfalfa, and row crop systems. J Environ Qual 26:1240–1247

Raun WR, Johnson GV (1999) Improving nitrogen use efficiency for cereal production. Agron J 91: 357–363

Rose T, Bowden B (2013) Matching soil nutrient supply and crop demand during the growing season. In: Rengel Z (ed) Improving water and nutrient use efficiency in food production systems. Wiley, Ames, pp 93–103

Singh U, Ladha JK, Castillo EG, Tirol-Padre PA, Duqueza M (1998) Genotypic variation in nitrogen use efficiency in medium and long duration rice. Field Crop Res 58:35–53

Soil Science Society of America (2008) Glossary of soil science terms. SSSA, Madison

Stanford G (1973) Rationale for optimum nitrogen fertilization in corn production. J Environ Qual 2: 159–166

Sumner ME, Noble AD (2003) Soil acidification: the world story. In: Rengel Z (ed) Handbook of soil acidity. Marcel Dekker, New York, pp 1–28

Tilman D, Cassman K, Matson P (2002) Agricultural sustainability and intensive production practices. Nature 418:671–677

Trenkel ME (2010) Slow- and controlled-release and stabilized fertilizers: an option for enhancing nutrient use efficiency in agriculture. International Fertilizer Industry Association (IFA), Paris

Wassmann R, Jagadish SVK, Heuer S, Ismail A, Redona E, Serraj R, Singh RK, Howell G, Pathak H, Sumfleth K (2009) Climate change affecting rice production: the physiological and agronomic basis for possible adaption strategies. Adv Agron 101:59–122

Xiong J, Ding CQ, Wei GB, Ding YF, Wang SH (2013) Characteristic of dry matter accumulation and nitrogen uptake of super-high yielding early rice in China. Agron J 105:1142–1150

Yoshida S (1981) Fundamentals of rice crop science. IRRI, Los Banos

Nutrient-Use Efficiency in Sorghum

J.S. Mishra and J.V. Patil

Abstract

Sorghum [*Sorghum bicolor* (L.) Moench] is an important crop of dryland agriculture. With the threat of climate change looming large on the crop productivity, sorghum being a drought hardy crop will play an important role in food, feed and fodder security in semi-arid tropics. With the development of improved sorghum cultivars, the NPK consumption in sorghum has increased from merely 4 kg/ha during 1974 to 47.5 kg/ha during 2003–2004. The nutrient-use efficiency (NUE) of grain sorghum is quite low (7.06–7.22 kg grain/kg NPK applied). Declining factor productivity, soil health, input-use efficiency and profitability and increasing costs of inputs and their timely availability are the major concerns of resource-poor farmers. Soils of the sorghum-growing regions are deficient in organic carbon, N and Zn, besides shallow in depth, low in water holding capacity, alkaline in reaction and prone to degradation. A system approach that includes sorghum cultivars with high NUE, coupled with best management practices, viz. soil health management, conservation tillage, integrated nutrient management including micronutrients, foliar application of nutrients, inclusion of legumes in sorghum-based cropping systems and efficient weed management, will be required for enhancing the NUE in sorghum.

Keywords

Sorghum • Nutrient-use efficiency • NUE • Nutrient deficiency symptoms • INM

J.S. Mishra (✉) • J.V. Patil
Department of Agronomy, Directorate of Sorghum
Research, Rajendranagar, Hyderabad 500 030, India
e-mail: mishra@sorghum.res.in; jsmishra31@gmail.com;
jvp@sorghum.res.in

1 Introduction

Sorghum, the fifth most important cereal crop on the globe and native to sub-Saharan Africa, is traditionally grown for grain both as food (Africa

A. Rakshit et al. (eds.), *Nutrient Use Efficiency: from Basics to Advances*,
DOI 10.1007/978-81-322-2169-2_19, © Springer India 2015

and India) and as animal feed (developed countries like USA, China, Australia, etc.) and stalks as animal fodder, building material and fuel. Sorghum is called various names in different places in the world. In Western Africa, it is called 'great millet', 'kafir corn' or 'guinea corn', which represents a connection with corn or millet. It is called 'jowar' in India, 'kaolian' in China and 'milo' in Spain. It is the dietary staple of more than 500 million people in 30 countries (Kumar et al. 2011). Sorghum grain is mostly used for food purpose (55 %) followed by feed grain (33 %). Of late, sweet sorghum is emerging as a potential feedstock for biofuels. Because of its drought adaptation capability, sorghum is a preferred crop in tropical, warmer and semi-arid regions of the world with high temperature and water stress (Paterson et al. 2009). With the threat of climate change looming large on the crop productivity, sorghum being a drought hardy crop will play an important role in food, feed and fodder security in dryland economy. Sorghum grain has high nutritive value, with 70–80 % carbohydrate, 11–13 % protein, 2–5 % fat, 1–3 % fiber and 1–2 % ash. Protein in sorghum is gluten free, and thus, it is a specialty food for people who suffer from celiac disease, as well as diabetic patients (Prasad and Staggenborg 2009). Sorghum fibers are used in wallboard, fences, biodegradable packaging materials and solvents. Dried stalks are used for cooking fuel, and dye can be extracted from the plant to colour leather (Maunder 2000).

2   Nutrient Use in Sorghum in India

Adequate supply and balance of mineral elements are required for proper growth and development of sorghum plant. Sorghum is generally grown under less favourable conditions, and meagre amounts of fertilizers are applied. Prior to 1950s relatively very little or no fertilizer was used on sorghum. In a survey during 1968–1971, sorghum accounted for 3.5 % of the total fertilizer used with overall nutrient consumption of 4 kg/ha N + P_2O_5 + K_2O (NCAER

and FAI 1974). However, with the development of improved sorghum cultivars and other improved production practices, the average nutrient consumption reached to 5.5–22.7 kg/ha during 1978–1980 (Tandon and Kanwar 1984) and 47.5 kg/ha (29.2, 14.2 and 4.1 kg N + P_2O_5 + K_2O) in 2003–2004 as against 60.2 kg/in maize, 119.1 kg/ha in paddy and 136.7 kg/ha in wheat (FAO 2005). Development of better adapted, high-yielding sorghum cultivars has increased the yield potential and the amounts of plant nutrients required by the crop. Consequently, the fertilizer application in sorghum has increased substantially. However, increasing fertilizer prices and decreasing purchasing power of the resource-poor sorghum farmers are the major reasons for less fertilizer use in sorghum. Improving plant efficiency for fertilizer use is important to reduce costs of crop production (Bernal et al. 2002). Policy interventions to reduce fertilizer cost and improve grain marketing efficiency will further enable smallholders to increase fertilizer use for substantial increases in sorghum production.

3   Nutrients Removal by Sorghum

Many factors are involved in determining the mineral requirement of sorghum (Maiti 1996).

1. Amount of available and residual mineral elements in soil.
2. Physicochemical properties of soil.
3. Availibility of soil moisture.
4. Yield and end product desired.

Cultivars producing large amounts of biomass remove greater quantities of soil nutrients. Sorghum crop producing 5.5 t/ha grain removes a total of 335 kg nutrients (149 kg N + 61 kg P_2O_5 + 125 kg K_2O)/ha from soil. High-yielding varieties of sorghum removed 22 kg N, 9 kg P_2O_5 and 30 kg K_2O to produce 1.0 t of grain (Tandon and Kanwar 1984). Sorghum crop yielding approximately 8 t of grain/ha removes about 250 kg N, 40 kg P_2O_5, 160 kg K_2O, 45 kg Mg and 40 kg S/ha from soil (Maiti 1996). Nutrients removed by sorghum hybrid

Table 1 Nutrients removal by rainfed hybrid sorghum

Nutrients	Grain yield (t/ha)	Total uptake by grain and stover
N	4.4	78 kg
P_2O_5	4.4	35 kg
K_2O	4.4	117 kg
Ca^a	2.6	28 kg
Mg^a	2.6	17 kg
Fe	4.4	705 g
Mn	4.4	447 g
Zn	4.4	132 g
Cu	4.4	37

[a]Vertisols (CSH 1) (Lakhdive and Gore 1978)

'CSH 5' in Alfisols under rainfed conditions is given in Table 1 (Vijayalakshmi 1979). Further studies revealed that sorghum grown in India removes on an average 22 kg N, 13.3 kg P_2O_5 and 34 kg K_2O to produce one tone of grains (Kaore 2006).

Large quantities of N and P and some potassium are translocated from the other plant parts to the grain as it develops. Unless adequate nutrients are available during grain filling, this translocation may cause deficiencies in leaves and premature leaf loss that reduce leaf area duration and may decrease yields (Roy and Wright 1974). Nitrogen and P accumulation by whole plants increased almost linearly until maturity, but K accumulation was more rapid early in the season. Nitrogen, P and K accumulation rates were higher during the 35th to 42nd day and 70th to 91st day which coincided with the peak vegetative growth period and the grain-filling stage, respectively. In unfertilized plants relatively higher translocation of N and P from the vegetative parts to the developing grain occurred. Little K was translocated. A much smaller percentage of total K was found in the head and more K accumulated in the stem than N and P.

A grain crop of 8.5 t/ha contains (in the total aboveground plant) 207 kg of N, 39 kg of P and 241 kg of K (Vanderlip 1972). Pal et al. (1982) reported that in early stage of crop growth, N and P accumulated slowly compared with the rapid accumulation of K. In later stages, uptake of K decreased relative to that of N and P.

4 Nutrient Deficiency Symptoms in Sorghum

There is a widespread deficiency of nitrogen, phosphorus, iron and zinc under both rainfed and irrigated conditions. Nitrogen, phosphorus, potassium and magnesium are phloem-mobile elements. When a deficiency of these elements occurs, plants tend to withdraw these elements from older leaves and redistribute them to young, actively growing parts of the plant through phloem (Robson and Snowball 1986). Hence, the first and most obvious symptoms of deficiency of these elements occur on lower, older leaves. Elements such as calcium, iron, manganese and boron are phloem-immobile elements and, hence, are not redistributed to any great extent under deficiency conditions (Robson and Snowball 1986). The first and most obvious symptoms of deficiency of these elements occur on young, actively growing parts of the plant, including root tips. The nutrient elements such as sulphur, zinc, copper and molybdenum often have variable mobility in the phloem (Robson and Snowball 1986). Hence, for these elements, symptoms may appear on young or old growth depending on the species, nitrogen supply, etc. However, Grundon et al. (1987) reported that in grain sorghum, only sulphur and zinc exhibited variation in the location of visible symptoms of deficiency and then only when the deficiency was very severe and persisted for some period. The key deficiency symptoms of nutrient elements in sorghum are listed in Table 2.

5 Nutrient-Use Efficiency (NUE)

Nutrient-use efficiency may be defined as 'the mass of nutrient required to produce a given quantity of biomass'. It is estimated that the overall efficiency of applied fertilizer is about or lower than 50 % for N, less than 10 % for P and about 40 % for K (Baligar et al. 2001). The worldwide nitrogen-use efficiency for cereal production including sorghum was approximately 33 %, and the unaccounted 67 % represents a

Table 2 Visible symptoms of nutrient deficiency in sorghum

Nutrients	Deficiency symptoms
Nitrogen	Deficient plants appear pale green to pale yellow in colour, stunted growth and thin and spindly stem and often show delayed flowering and maturity
	Nitrogen is mobile in plants and under conditions of low soil supply; it is easily mobilized from older to younger leaves. Hence, the deficiency symptoms appear first on older leaves and then advance up the stem to younger leaves
	The leaf blades progressively become pale green with pale yellow chlorosis and pale brown necrosis
Phosphorus	Phosphorus deficiency symptoms appear first on the older leaves with purple suffused pigmentation and progress upwards
	Affected plants appear stunted with thin stems and dark green leaves
	Under severe deficiency, plant growth is greatly reduced and dark green older leaves turn purple or purple-red in colour
Potassium	Potassium deficiency symptoms appear first on older leaves with marginal yellow chlorosis and brown necrosis
	Deficient plants lose stalk strength and are prone to lodging
	The internodes are shortened and thin and the older leaves develop a marginal necrosis
Magnesium	Older leaves are pale green to yellow in colour with many brown lesions. The symptoms advance upwards to younger leaves
	In case of severe deficiency, the whole plants appear pale green or pale yellow in colour
Calcium	Young leaves with torn or serrated leaf margins and leaf tips deformed, missing or joined together
	Under severe deficiency, the upper internodes may be very short and the young leaves crowded together to give the appearance of 'rosette'
Sulphur	Sulphur is not easily mobilized from older to younger leaves; the deficiency symptoms appear first on younger leaves
	Young leaves faint yellow interveinal chlorosis and turn pale green in colour, while older leaves remain dark green
Iron	Sorghum is the best indicator plant for iron deficiency
	Prominent pale yellow or white interveinal chlorosis, leaving the veins green and prominent on young leaves
Zinc	Young leaves with broad yellow or white bands between the margins and midvein in lower half leaf. In case of severe deficiency, the chlorosis extends towards the leaf tip and often turns nearly white or pale brown
	Shortening of internodes resulting in stunted plants
	Delayed flowering and maturity
Boron	Young leaves with transparent white interveinal lesions
	Shortening of internodes and stunted plant growth
	Short, erect and dark green leaves
	In case of severe deficiency, apical meristem often dies and tillers develop
Manganese	Plants are pale green to yellow in colour with thin spindly stem
	Young leaves with yellow interveinal chlorosis and red-brown interveinal lesions
Copper	The young leaves and the leaves which are still within the whorl turn pale green in colour
	The whorl of the expanding leaves may remain tightly rolled and become bent to one side
	Young leaves with brown twisted leaf tips
	Stunted plant growth with thin stems and pale green foliage

$15.9 billion annual loss of N fertilizer (Raun and Johnson 1999). The loss of N results from soil denitrification, surface runoff, leaching and volatilization. Continued low NUE in crops could have a drastic impact on land use and food supplies worldwide (Frink et al. 1999). Efficient plants absorb and utilize the nutrients and increase the efficiency of applied fertilizers, reduce cost of inputs and prevent nutrient losses to ecosystems and reduce environmental pollution. The overall NUE of a cropping system can be increased by achieving greater uptake efficiency from applied N inputs, by reducing the amount of N lost from soil inorganic and organic pools or both.

6 Nutrient-Use Efficiency Indices

Nutrient-use efficiency can be expressed in many ways. Prasad (2009) described 4 agronomic indices in relation to nutrient-use efficiency. These are as follows: agronomic efficiency (*AE*), recovery efficiency (*RE*), physiological efficiency (*PE*) and partial factor productivity of fertilizers (*PFPf*). Details of different indices are given in Chap. 1.

7 Factors Affecting NUE in Sorghum

Production practices that lead to affect crop yields will have impact on nutrient-use efficiency. Nutrient requirements and NUE in sorghum vary with soil, climate, cultivar and management practices.

7.1 Soil Factors

Sorghum is grown on diverse range of soils. This range is so wide that some soils are unusually low in certain nutrients or have excessive quantities of certain nutrients. Nutritional stress problems in soils are often related to the type of parent material and the soil-forming processes characteristic of that soil (Dudal 1976; Clark 1982a, b). Acid soils (oxisols, ultisols and some entisols, alfisols and inceptisols) are usually low in exchangeable bases. Acidity increases the solubility of iron, aluminium and manganese, and hence, these elements may reach to the toxic levels. However, acid soils are deficient in phosphorus, calcium, magnesium, molybdenum and

zinc (Clark 1982a). Alkaline soils (mollisols, vertisols and some inceptisols) often contain fairly high concentration of salt in the soil profile. These soils are rich in calcium, magnesium and potassium but deficient in sulphur. The deficiency symptoms of iron, zinc and manganese are the most common in sorghum grown on alkaline soils (Tandon and Kanwar 1984). The nutrient-use efficiency in these soils is greatly influenced by the time and method of fertilizer application. High bulk density, poor soil structure and crust formation, low water holding capacity, water logging and poor soil aeration can also reduce NUE.

Soil management practices like crop rotations and intercropping systems that affect the soil carbon balance will also affect the N balance because the C/N ratio of soil organic matter is relatively constant. In such cropping systems, the overall NUE of the cropping system must include changes in the size of the soil organic and inorganic N pools. When soil-N content is increasing, the amount of sequestered N contributes to a higher NUE of the cropping system, and the amount of sequestered N derived from applied N contributes to a higher NUE. Any decrease in soil-N stocks will reduce the NUE.

7.2 Tillage

After harvest, lots of sorghum stubbles are left in/on the soil. Decomposition of these stubbles prior to planting the next crop is usually desirable. These residues/stalks should be incorporated as soon after harvest as possible. Proper decomposition of residues before planting next crop reduces the problems with tillage and other planting operations. The early decomposition also makes plant nutrients found in residues available for the subsequent crop. Undecomposed stubbles may lead to N immobilization due to high C:N ratio, and N deficiency may occur early in the growth of subsequent crop. A C:N ratio greater than 20 indicates that soil microorganisms feeding on the stubble will require some N from the soil in addition to the

N in stubble for decomposition to occur. The C:N ratio in grain sorghum ranges from 40:1 to 80:1. Nitrogen may be applied to the stubbles to speed up the decomposition and prevent the temporary N deficiency. The amount of N applied usually ranges between 4.5 and 6.8 kg per 450 kg of stubbles produced (Bennett et al. 1990).

Leaving crop residues on the soil surface is the most cost-effective method of reducing soil erosion. Covering 20 % of the surface with crop residues can reduce soil erosion caused by rainfall and runoff water by 50 % compared to residue-free condition (Shelton et al. 1995). Stubbles also control wind erosion and assists in soil moisture conservation. However, N management becomes even more important in high residue (no-till) farming.

7.3   Climate and Weather Factors

Nutrient availability in soil and the ability of plants to absorb and utilize the nutrients and subsequent yields are greatly influenced by temperature, solar radiation, and rainfall during crop growth (Arkin and Taylor 1981; Baligar and Fageria 1997). The rate of nutrient release from organic and inorganic sources and the uptake by roots and subsequent translocation and utilization in plants is influenced by soil temperature (Cooper 1973). Solar radiation directly affects photosynthesis which in turn influences a plants' demand for nutrients (Baligar et al. 2001). Higher rainfall and humid weather during the growing season favours weed growth and more attack of insect pests and diseases in sorghum, which reduces crop yields and nutrient-use efficiency.

7.4   Cultivars

In semi-arid tropics where sorghum is an important crop, inorganic fertilizer use is limited due to high cost and non-availability and limited soil moisture availability. To reduce the impact of nutrient deficiency on sorghum production, the selection of genotypes that are superior in the utilization of available nutrients either due to enhanced uptake capacity or because of more efficient use of the absorbed nutrients in grain production can be a desirable option. Sorghum cultivars differ in growth, rooting pattern, maturity duration, etc., and hence the nutrient uptake pattern and the efficiency are also likely to differ. Exploiting these differences in nutrient demand and efficiency is a possible alternative for reducing the cost and reliance upon fertilizer. Gardner et al. (1994) demonstrated the genetic diversity for N-use efficiency in grain sorghum and concluded that the differences among sorghum cultivars for higher NUE mechanisms were associated with individual morphological, anatomical and biophysical traits, viz. larger canopies comprised of fewer but larger leaves with low N concentration, thicker leaves, larger leaf phloem transactional area, rapid solubilization and remobilization of N from older to younger leaves and lower dark respiration rates. At low N levels (50 kg N/ha), the improved genotype had the highest nitrogen-use efficiency and the commercial hybrid had the lowest. However, at high levels of N (200 kg N/ha), the commercial hybrid showed the highest NUE (Bernal et al. 2002). Landrace cultivars that have adapted to low N environments may possess different stress-coping mechanisms than do domesticated cultivars developed in contemporary breeding programme (Pearson 1985). Indian improved line M35-1 was found superior in NUE among all environments, and the traits related to high NUE included larger canopies comprised of fewer but larger leaves with low N concentration, thicker leaves, larger leaf phloem transactional area, rapid solubilization and remobilization of N from older to younger leaves and lower dark respiration rates (Gardner et al. 1994). The nutrient-use efficiency of rainy season grain sorghum was influenced by nutrient levels. Hybrid sorghum 'CSH 16' had maximum NUE (7.06 kg grain/kg NPK applied) with 150 % RDF (150:60:60 kg NPK/ha) (Fig. 1), but sorghum variety 'SPV 462' recorded maximum NUE (7.22 kg grain/kg NPK applied) at 100 % RDF (80:40:40 kg/ha) (AICSIP 2010–11).

Fig. 1 NUE of grain sorghum cultivars

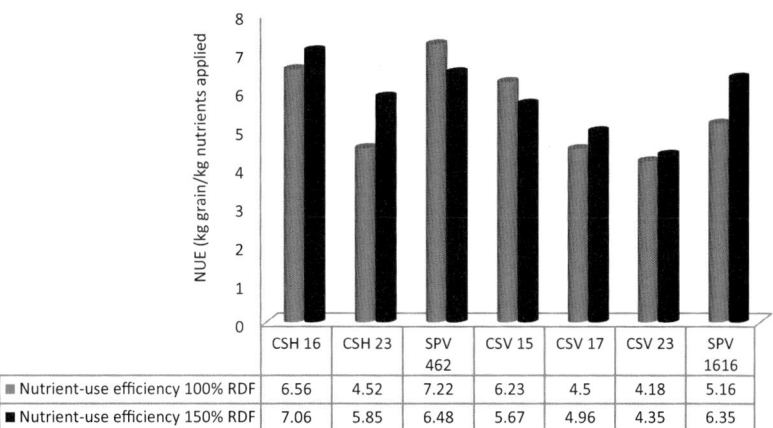

	CSH 16	CSH 23	SPV 462	CSV 15	CSV 17	CSV 23	SPV 1616
■ Nutrient-use efficiency 100% RDF	6.56	4.52	7.22	6.23	4.5	4.18	5.16
■ Nutrient-use efficiency 150% RDF	7.06	5.85	6.48	5.67	4.96	4.35	6.35

7.5 Fertilizer Management

The nutrient-use efficiency is affected by fertilizer dose, sources of nutrients, method and time of application, interaction of different nutrients, soil moisture, mycorrhiza and others. The availability and recovery efficiency are greatly influenced by addition of organic matter, liming, inclusion of legumes in sorghum-based cropping systems and others.

7.5.1 Nitrogen

The availability of nitrogen is a primary factor limiting the growth of sorghum plant. Nitrogen is one of the most abundant mineral nutrients required for sorghum growth. The level of nitrogen fertility has more influence on the growth and yield of grain sorghum than any other single plant nutrient. The amount of fertilizer N required will vary depending on the yield potential of the cultivar and the amount of residual N available in the soil prior to planting. Preplant soil analysis can be very useful in estimating the nitrogen need of the crop. Most of the sorghum-growing soils contain low amount of nitrogen and hence require supplemental nitrogen applied in the form of fertilizer for optimal productivity. The previous studies have shown that most crop plants utilize less than half of nitrogen added to the soil. According to Maiti (1996), only about 50 % of the N applied to soil is taken up and used by plants. The reminder is left for microbial use, leaching, denitrification or incorporated into the

organic fractions through immobilization and many other reactions and processes occurring in the soil. It has been estimated that 1.95–3.2 t soil/ha is lost annually due to wind and water erosion in sorghum belt (ICRISAT 1986). This eroded soil carries away precious nutrients and reduces topsoil depth. Alfisols, vertisols and red lateritic soils where sorghum is prominently grown are prone to soil erosion (Sharda and Singh 2003). Therefore, appropriate nutrient management strategies need to be adopted for improving soil health, N-use efficiency and productivity. Vitousek (1982) reported that the nitrogen-use efficiency decreased with increasing nitrogen availability. In sweet sorghum N-use efficiency defined as theoretical ethanol yield per unit of N taken up decreased with increasing total N uptake (Wiedenfeld 1984). Sawargaonkar et al. (2013) observed that in sweet sorghum NUE increased with N application rate up to 90 kg N/ha and then NUE decreased as N application rate increased.

7.5.1.1 Nitrogen Concentration in Plant Tissues

Nitrogen accumulation in sorghum plants usually continued until maturity of the crop (Srivastava and Singh 1971). The young plants accumulate relatively high concentration of N, and the N content decreases in the various plant parts with age. Most of the plant N is absorbed during the vegetative and by early grain filling stages. Singh and Bains (1973) observed a continuous decline

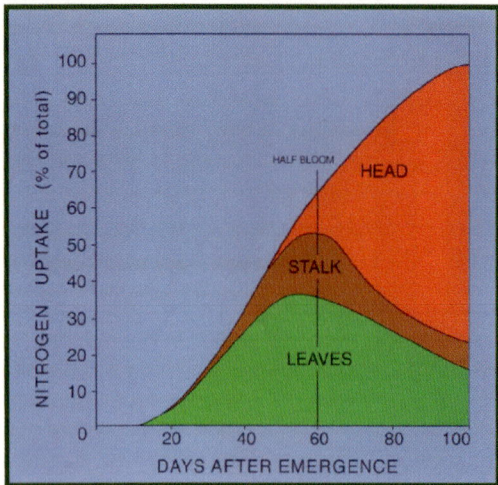
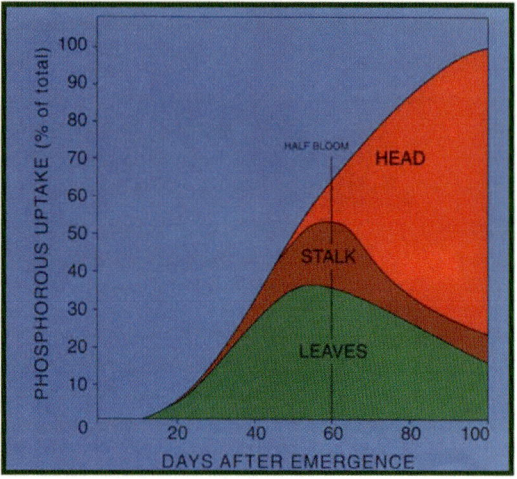
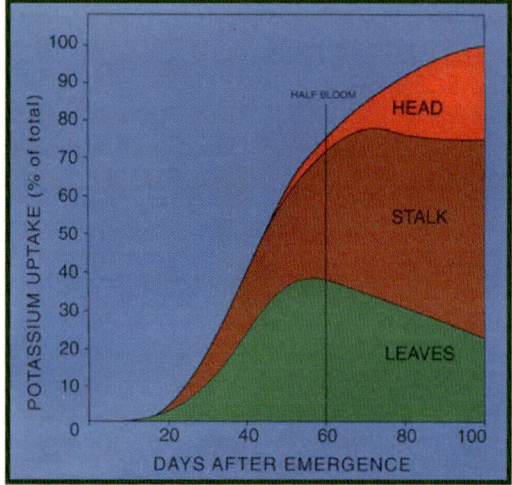

Fig. 2 NPK uptake pattern in grain sorghum (*Source:* R. L. Vanderlip, *How a Sorghum Plant Develops*, Kansas State University, January 1993. (http://www.nasecoseeds.com/products/sorghum/47.html))

in N content in whole plant tissues until 75 days after planting followed by increased N content up to maturity. This late season increase in N content in whole plant after boot leaf stage was the result of the build-up of N in grains. The N content in plant tissues is influenced by the dose and time of fertilizer application, plant population, variety, irrigation and other management practice. There was a strong association between N content in whole plant at 30 and 60 days after sowing and grain yield (Hariprakash 1979). Jones (1983) reported that the N concentration in sorghum grain ranged from 1.02 to 3.20 %

(mean 1.67 %) and in stover 0.36–1.26 % (mean 0.80 %).

7.5.1.2 Nitrogen Uptake

The N-uptake curve is generally similar to sigmoidal growth curve in sorghum (Fig. 2). Nitrogen accumulation rate by the whole plant was usually slower in the early growth stage, became faster in the log phase of crop growth and again slowed down at maturity (Table 3) (Srivastava and Singh 1971). Upon reaching the maximum accumulation in the vegetative plant parts, coinciding mostly with heading stage, nitrogen from vegetative parts starts getting translocated

Table 3 N accumulation rates by sorghum (CSH-1)

Growth stages (days after sowing)	N accumulation rates (mg N/plant/day)
0–30	4.05–5.14
30–45	17.24–21.75
45–60	18.53–23.75
60–75	20.00–21.10
75–90	6.47–6.71
90-full maturity	0.40–2.10

into the panicles (Singh and Bains 1973a, b). More N is translocated from leaves than from stem. Earheads contain the major portion of the N accumulated by the whole plant. Only 12–47 % of applied N is utilized by sorghum (Pal et al. 1982). The recovery of N influenced by the rates and method of N application, soil types, variety, soil moisture and management practices. The results from a study with 15_N labeled urea at ICRISAT, 1982 and 1983, revealed that sorghum recovered 62.5 % of added N in the alfisol and 55.0 % in the vertisol. About 27.1 % of applied N was distributed in the alfisol profile and 38.6 % in the vertisol profile accounting to 89.6 % and 93.6 % N by the soil + crop system. In alfisols crop recovery of N varied from 46.3 to 51.1 % as N levels increased from 40 to 160 kg N/ha. At the highest N level tested, soil + crop system could account for 78.9 % of added N, as compared to 93.2 % at 40 kg N/ha. Although considerable fertilizer N was present in the soil profile after the harvest of rainy season sorghum, this residual N was of limited value either for safflower grown in the post-rainy season or for sorghum grown in the following rainy season (Moraghan et al. 1984). The N response (kg grain/kg nitrogen applied) of rainfed sorghum to optimum or near optimum levels of N during rainy season varied from 21.7 kg in alfisols, 18.32 kg in vertisols 11.9 kg in molisols and 20.15 kg in entisols (Tandon and Kanwar 1984). A significant positive interaction between nitrogen and moisture has been well established in sorghum, and this interaction is stronger in an alfisol than in a vertisol (Kanwar 1978). Nitrogen uptake was also improved by P and Fe fertilization in calcareous soils (Patil 1979).

7.5.1.3 Nitrogen × Moisture Interaction

Moisture availability, moisture use and nutrient supply to the plants are closely interacting factors influencing plant growth and yield production (Viets 1972). There is a significant and positive correlation between fertilizer N and soil moisture for sorghum grain yield. The response was more in alfisols than that of vertisols (Kanwar 1978). Water application in the alfisols probably compensates for it comparatively shallow depth and low moisture storage, as compared with the vertisol. With 58 or 120 kg N/ha, grain yield in nonirrigated vertisol was similar to those in the alfisol irrigated at 50 % moisture depletion (Tandon and Kanwar 1984).

7.5.1.4 Nitrogen × Genotype Interaction

Sorghum genotypes greatly influence the nutrient accumulation in plants due to variation in rate of absorption, translocation and accumulation of nutrients in plant tissues. Genotypic difference in N uptake partitioning and NUE (unit dry matter per unit N in dry matter) has been reported for grain sorghum (Maraanville et al. 2002). The varietal differences for N and P uptake might be due to additive gene action for N and nonadditive for P (Krishna et al. 1985). In general hybrids deplete greater amount of nutrients than that of varieties. Sorghum genotypes vary significantly for various root characteristics which may affect the nutrient uptake (Seetharama et al. 1990). Sorghum genotypes, viz. CSH 1, CSH 5 and CSV 3, were highly responsive to phosphorus (7.8 kg grain/kg P) compared to CSV 5 (3.5 kg grain/kg P) (Krishna 2010). Long-term studies conducted in vertisols indicated that sorghum absorbs mere 5 % of the total N during first 5 weeks followed by rapid N uptake. The crop accumulated 88 kg N/ha in 40–70 days, at the rate of about 3 kg N/ha/day (ICRISAT 1986).

7.5.1.5 N × P Interaction

Long-term studies have indicated positive interaction between N and P in sorghum. The response to N may subside, if sufficient levels of P are not maintained on vertisols and alfisols. The positive N × P interactions have resulted in net advantage of 300–500 kg grains/ha (ICRISAT 1986). N × P interactions may contribute up to 48–50 % of total response of sorghum to nutrient supply (Tiwari 2006).

7.5.2 Phosphorus

Phosphorus is important in plant bioenergetics. As a component of ATP (adenosine triphosphate), phosphorus is needed for the conversion of light energy to chemical energy during photosynthesis. There is widespread deficiency of phosphorus in soils of semi-arid tropics. It is estimated that only about 10 % of the P added to the soil is absorbed by plants and remaining 90 % become unavailable in the soil by adsorption or fixation by various soil fractions. Phosphorus accumulates extensively in the kernels (as phytin). A small fraction (16–22 % of total P uptake) of P is accumulated during early growth of the crop (42 days after sowing) owing to slower accumulation rate, whereas the major portion of the P is accumulated during later stages of crop growth (Roy and Wright 1974).

Phosphorus uptake is enhanced by P fertilization. Excess P can interact with other nutrients (especially, Fe, Zn and Cu) and depress plant growth but causing deficiency of other plant nutrients. The response of phosphorus (kg grain/kg P_2O_5 applied) varies with soil types in order of alfisols (17–32 kg) > entisols (11–34 kg) > vertisols (7–27 kg). In post-rainy crop, a response of 11 kg grain/kg P_2O_5 was obtained in vertisols. Vertisols may require higher P application than other soils because of their high clay content and greater reactive surfaces/components (Rao and Das 1982). On calcareous soils, P uptake by hybrid sorghum was highest when phosphatic fertilizers were applied on the surface, followed by 5 and 10 cm deep placement, but the reverse was the case in non-calcareous soils (Venkatachalam et al. 1969). Apart from soil types, response to P is strongly affected by the yield potential of the cultivars, level of N applied, available soil P and favourable environment. The residual response of P applied to sorghum on succeeding wheat crop is small and not consistent (Tandon and Kanwar 1984). In general, 40–50 kg P_2O_5/ha is recommended for rainfed *kharif* sorghum and 20–30 kg/ha for *rabi* sorghum grown in medium and deep soils. In irrigated *rabi* sorghum, 40–50 kg P_2O_5/ha is recommended.

7.5.3 Potassium

Among the essential plant nutrients, potassium assumes greater significance since it is required in relatively larger quantities by plants. Besides increasing the yield, it largely improves the quality of the crop produce. Potassium regulates the opening and closing of stomata. Since stomata are important in water regulation, adequate potassium content in plants is associated with higher tolerance to drought and higher resistance to frost and salinity damage and resistance to fungal diseases.

Potassium deficiency may not be a serious problem for sorghum in Indian soils. In general, black soils with higher clay and CEC showed high levels of exchangeable K and medium to high non-exchangeable K content; alluvial soils with higher contents of K-rich mica with light texture showed medium in exchangeable K and high in non-exchangeable K content; and red and lateritic soils with kaolinite as a dominant clay mineral and light texture showed low in exchangeable as well as non-exchangeable K content (Srinivasarao et al. 2011). However, recent studies indicated an application of 40–50 kg K_2O/ha in rainfed *kharif* and irrigated *rabi* sorghum.

Similar to N and P, K content in plant tissues also decreases as the crop advances from seedling stage (2.16–2.26 % in the leaves) to maturity (1.33 % in the leaves), and the earheads contain less K than the leaves (Gopalkrishnan 1960). At harvest, the K content in grain declined from 0.41 to 0.39 % and increased in stover from 1.24 to 1.29 % (Venkateswarlu 1973). The potassium accumulated in sorghum plants rapidly during the early growth period and slowly at later stages (Roy and Wright 1974). They further observed that 50–60 % of the total K uptake was completed before heading and around 68–78 % of total K was contained in the vegetative parts and 22–32 % in the heads.

7.5.4 Micronutrients

Among micronutrients, deficiency of zinc is more widespread in sorghum-growing areas. Of the 2,51,660 soil samples analyzed for micronutrients, 49 % were deficient in Zn and 12 % in Fe content (Singh 2001). Most Zn in sorghum is taken by the early grain-fill stage. Next to Zn, iron nutrition to sorghum has importance in some soils. Sorghum is sensitive to iron stress and is less efficient in its absorption and translocation. Since Fe uptake decreases with increased $CaCO_3$ content of the soil, the problem of Fe deficiency is more on calcareous soils. Soils containing more than 1.2 ppm Zn and 3–5 ppm Fe (critical limit for sorghum) did not respond to Zn/Fe application (Tandon and Kanwar 1984). Application of 20 ppm Fe as $FeSO_4$ increased the grain yield by 0.9 t/ha (Babaria and Patel 1981). Koraddi et al. (1969) observed complete recovery from lime-induced chlorosis and obtained higher sorghum yield with spraying of $FeSO_4$. Singh and Vyas (1970) reported 5.1 % and 13.9 % increase in grain yields application of manganese and zinc, respectively, in Jodhpur. Joshi (1956) reported significant increase in yield of sorghum due to $CuSO_4$ application in Maharashtra. Kanwar and Randhawa (1967) observed 35 % and 40 % increase in the yield of sorghum due to application of boron and boron (B) + manganese (Mn). Foliar application of $MnSO_4$ @ 10 kg/ha was found to increase the sorghum grain yield by 24–35 % (Gill and Abichandani 1972). Experiments conducted under All India Coordinated Sorghum Improvement Project (AICSIP) revealed that deficiency of Zn and Fe can be corrected either through soil application of respective sulphate forms or through foliar application (Table 4).

An antagonistic relationship was reported with Fe and Cu, Zn and Mn, whereas Cu showed antagonism with Fe and Zn and synergism with Mn in sorghum shoot (Singh and Yadav 1980)

7.5.5 Biofertilizers

A biofertilizer is a substance which contains living microorganisms which, when applied to seed, plant surfaces or soil, colonizes the rhizosphere or the interior of the plant and promotes growth by increasing the supply or availability of primary nutrients to the host plant (Vessey 2003). Biofertilizers add nutrients through nitrogen fixation, solubilizing phosphorus and stimulating plant growth through the synthesis of growth promoting substances. Biofertilizers can be expected to reduce the use of chemical fertilizers and pesticides. Through the use of biofertilizers, healthy plants can be grown while enhancing the sustainability and health of soil. Biofertilizers form an important component in the integrated nutrient management. A number of biofertilizers, viz. Azotobacter, Azospirillum, vermicompost, etc., are now commercially available for cereals.

Field research on using microorganisms on increasing nutrient-use efficiency was started during 1970s. Various strains of microorganisms like Azotobacter, Azospirillum, Phosphobacterin and Mycorrhiza were found promising. Senthil Kumar and Arockiasami (1995) reported that arbuscular mycorrhizas (AM) inoculated sorghum seedlings contained 11.5 mg Zn/g dry root but non-mycorrhizal seedlings had 7.5 mg Zn/g dry root. Sorghum genotypes also vary with regard to AM mycorrhizal colonization in roots and P uptake (Seetharama et al. 1988). The net advantage from AM symbiosis to sorghum seems to be 10–20 kg P/ha (Krishna et al. 1985). Studies conducted at TNAU, Coimbatore, revealed that fertilizer-N application could be reduced by inoculating with Azospirillum (TNAU 2003). In sorghum-chickpea system, biofertilizer [Azospirillum and phosphate-solubilizing bacteria (PSB)] gave significantly higher grain and fodder yields (Gawai and Pawar 2006).

7.5.6 Method and Time of Fertilizer Application

Nitrogen fertilizers should be applied in a method that ensures a high level of N availability to the crop and high N-use efficiency. It should be placed as close to planting as possible. Fertilizer placement below the soil surface should be more effective than broadcasting or banding on the soil surface, both in ensuring quick

Table 4 Effect of iron and zinc on grain and dry fodder yield of sorghum (AICSIP 2011)

| Treatment | Grain yield (kg/ha) | | | | | |
	Coimbatore	Parbhani	Akola	Dharwad	Surat	Mean
RDF + ZnSO$_4$ 25 Kg (soil application)	1,833	2,737	3,178	2,462	3,045	2,651
RDF + FeSO$_4$ 25 Kg (soil application)	1,502	2,312	3,114	2,862	3,491	2,656
RDF + 0.2 % ZnSO$_4$ foliar spray at 15 and 30 DAS	1,730	2,328	2,609	2,289	2,955	2,382
RDF + 0.5 % FeSO$_4$ foliar spray at 15 and 30 DAS	1,553	2,197	2,525	2,466	3,024	2,353
RDF + ZnSO$_4$ 15 kg (soil application) + 0.20 % as foliar spray at 15 and 30 DAS	1,936	2,662	3,136	2,882	2,826	2,689
RDF + FeSO$_4$ 15 kg (soil application) + 0.50 % as foliar spray at 15 and 30 DAS	1,562	2,009	3,093	2,598	2,971	2,447
RDF + soil application of 15 kg ZnSO$_4$ + 15 kg FeSO$_4$	1,636	2,793	4,798	2,953	3,676	3,171
RDF + foliar application of 0.20 % ZnSO$_4$ + 0.50 % FeSO$_4$	1,698	2,036	2,925	3,184	3,367	2,642
RDF (80:40:40 kg NPK/ha) alone	1,438	1,847	2,883	2,939	2,868	2,395
Mean	1,567	2,252	3,072	2,509	3,028	2,486
C.D. (P = 0.05)	160	437	643	355	228	482
CV%	5.94	11.3	12.2	8.24	4.38	15.1

RDF: Recommended Dose of Fertilizers
Source: (AICSIP 2010–2011)

availability and in enhancing N-use efficiency. In no-till grain sorghum, Lamonds et al. (1991) reported higher yields with knifed UAN (urea-ammonium nitrate) than broadcast. Placement of urea or diammonium phosphate with or near the seed is not recommended due to the risk of seedling injury due to ammonia toxicity. Nitrogen utilization by sorghum plant is quite rapid after the plant reaches to five-leaf stage, with 65–70 % of the total N accumulated by the bloom stage of growth (Cothren et al. 2000). Apparent N recovery was also markedly improves when N is applied in 2 or 3 splits in a high rainfall year (Venkateswarlu et al. 1978). Sorghum yields are adversely affected if the dose of N at planting is either reduced to less than 50 % of the total dose or the top dressing is delayed beyond the flower primordia initiation stage (Tandon and Kanwar 1984). Application of half amount of N at planting and half at 30 days after sowing produced significantly higher yields of hybrid sorghum (Lingegowda et al. 1971; Sharma and Singh 1974; Turkhede and Prasad 1978). However, in light soils and in high rainfall areas, three splits of N fertilizer, 50 % at sowing, 25 % at floral primordial initiation and 25 % at flowering, has

been found beneficial (Choudhary 1978). In heavy black soils of Maharashtra, Bodade (1964, 1966) concluded that the application of 50 % N through foliar application was as effective as the full dose of N through soil application. Choudhary (1978) recommended 2 equal splits of N for foliage application: first at floral primordial initiation and second at mid-bloom stage of crop. Narayana Reddy et al. (1972) reported 6 % concentration of urea solution as the best for foliar spray. It is generally recommended that all phosphatic and potassium fertilizers should be applied as basal and deep placed.

7.5.7 Integrated Nutrients Management (INM)

Continuous application of only mineral fertilizer ultimately results in yield declines. However, with a combination of mineral and organic sources of nutrients yield levels can be maintained (Bationo and Buerkert 2001). It is widely accepted that addition of organic is essential to maintain soil health. Importance of the use of organic sources of nutrients along with chemical fertilizers for maintaining soil health has been emphasized by Katyal (2000). The use of chemical fertilizer or biofertilizer has advantages and disadvantages in the context of nutrient supply, crop growth and environmental quality. The advantages need to be integrated in order to make optimum use of each of the fertilizers to achieve balanced nutrient management for crop growth (Jen-Hshuan 2006). Combined use of inorganic and organic manures improves physical and chemical properties of soils. At a dose equivalent to 40 kg N/ha, crop yield was better secured with organic N than with urea N. Combining organic and mineral sources of nutrients do not have only additive effects but real interaction, which significantly affect crop yield and water-use efficiency (Ouedraogo and Mando 2010). Application of sorghum stubbles, sun hemp and *Gliricidia* has recommended dose of fertilizer resulted in maximum response with only 50 % under rainfed condition. Application of 75 % recommended dose of fertilizer (RDF) + farmyard manure (FYM) + biofertilizer [*Azospirillum* and phosphate-solubilizing bacteria (PSB)] gave significantly higher plant height, dry mater, yield attributes and grain and fodder yields of sorghum and was on a par with application of 100 % RDF through inorganics alone showing 25 % saving of nutrients (Gawai and Pawar 2006; Patil et al. 2008). Incorporation of FYM, wheat straw and *Gliricidia* leaves for 25 or 50 % N substitution in conjunction with balanced dose of NPK fertilizers increased infiltration rate, water stable aggregates and organic matter, the values of which ranged from 0.88 to 0.92 cm h^{-1}, 0.82 to 0.96 mm and 1.10–1.27 %, respectively, whereas bulk density decreased from 1.32 to 1.22 Mg m^{-3}. The soil reaction and electrical conductivity remained unaffected while the organic carbon content increased appreciably and ranged from 0.68 to 0.74 %. The available N, P$_2$O$_5$ and K$_2$O status improved after harvest of both the crops due to integrated nutrient management by the application of 50 % recommended dose of fertilizers and 50 % N equivalent with FYM to sorghum in *kharif* and recommended dose of fertilizers to wheat in *rabi* than the continuous application of recommended dose of

fertilizers to both the crops (Bhonde and Bhakare 2008). Crop residue recycling is a vital aspect of sorghum cultivation as it reduces run-off induced soil and nutrient loss (Dhruvanarayan and Rambabu 1983). Among the residues, prunings from *Leucaena* and *Gliricidia* enhance carbon sequestration better than cereals residue. Integration of vermicompost at 2 t/ha + 50 % RDF was found on a par with RDF in sorghum – chickpea/ field pea/lentil system (AICSIP 2007). Similarly, integration of organic and inorganic sources of N to supplement N requirement of sorghum significantly improved the productivity of succeeding chickpea crop as compared to applying 100 % N through inorganic fertilizer (AICSIP 2014). Minimum tillage with 80:40:40 kg NPK/ha or conventional tillage with 60:30:30 kg NPK/ha, of which 75 % through inorganic + PSB + *Azospirillum* + *dhaincha* incorporation/ mulching at 30 DAS were found promising (Mishra et al. 2012a). In *rabi* sorghum, studies were conducted three consecutive years under All India Coordinated Sorghum Improvement Project (AICSIP) to see the effect of INM practices on N-use efficiency. Results revealed that growing cowpea/green gram/*dhaincha* in preceding *kharif* season (Fig. 3) significantly improved the productivity of succeeding *rabi* sorghum and could save 20–40 kg N/ha as compared to *kharif* fallow. Growing short duration legume crops during *kharif* season significantly improved the NPK content (Fig. 4) and population of soil microflora (Table 5).

7.6   Weed Management

Sorghum is grown on marginal lands with poor fertility. Weeds compete with the sorghum for available nutrients and make the crop deprive of the essential nutrients resulting in poor crop growth and lower NUE. Uncontrolled weeds in sorghum removed 29.94–51.05, 5.03–11.58 and 30.38–74.34 kg/ha NPK, respectively, from soil (Satao and Nalamwar 1993; Mishra et al. 2012b). Effective management of weeds is therefore

Fig. 3 Showing the effect of preceding legume crops on succeeding *rabi* sorghum

Preceding *kharif* crops (Greengram, *Dhaincha*)

Green manuring with *Dhaincha*

Succeeding *Rabi* sorghum

very essential for increasing the NUE. Kondap et al. (1985) reported that increasing levels of nitrogen decreased the population of *Cyperus rotundus* and *Panicum emeciforme* in sorghum. This study revealed the possibility of saving 30–90 kg N/ha by adopting either chemical or manual weed control. Okafor and Zitta (1991) observed that reduction in grain yield due to weed competition by 51.0, 37.8 and 32.2 % at zero, 60 and 120 kg N/ha, respectively, indicating that yield reduction due to weeds decreased at higher N levels.

Fig. 4 Changes in nutrient status of soil after 3 years of INM treatment

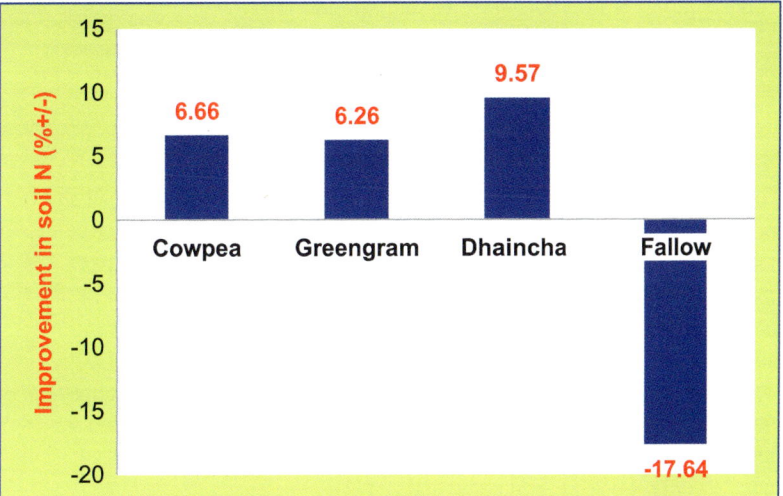

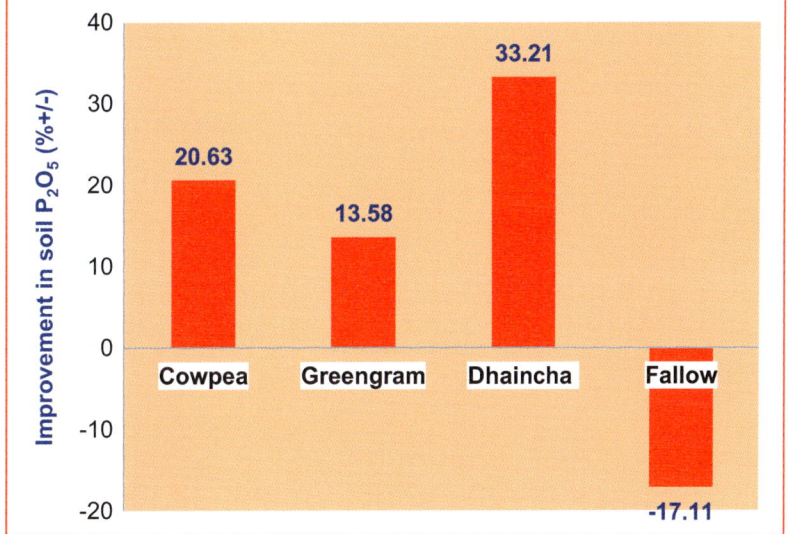

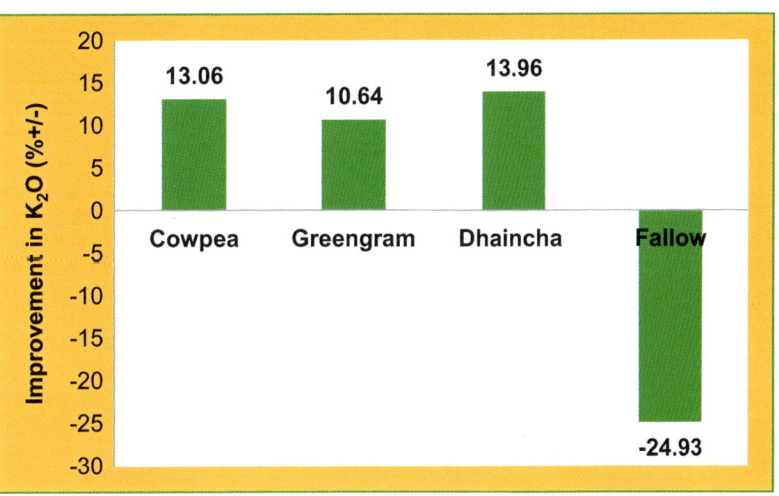

Table 5 Effect of INM treatments on soil microflora after 3 years (2013–2014)

Treatment	Actenomycetes (10^4 CFU g^{-1} soil)	Bacteria (10^6 CFU g^{-1} soil)	Fungi (10^2 CFU g^{-1} soil)
Kharif season (main plots)			
Cowpea fodder-rabi sorghum	216.42	153.17	35.92
Green gram/black gram-rabi sorghum	198.50	121.17	36.75
Dhaincha-rabi sorghum	290.83	164.42	46.25
Fallow-rabi sorghum	126.92	58.33	31.00
LSD (P = 0.05)	7.92	4.45	4.77
Rabi season [N levels (kg/ha)]			
0	249.33	124.25	47.17
20	226.25	132.25	39.75
40	176.42	118.67	34.67
60	144.67	121.92	28.33
LSD (P = 0.05)	7.59	8.20	4.34
Initial value (2011–2012) before start of the experiment	132	72	47

Conclusion

With the increasing cost of chemical fertilizers and decreasing nutrient-use efficiency, it is very important to find out the solutions to address this problem either through developing nutrient efficient genotypes to effectively utilize supplemental nitrogen added to the soil or through developing alternative crop and soil management practices that will minimize the nutrient loss. Genotypes and management conditions can significantly influence nutrient-use efficiency in sorghum. Post flowering drought-tolerant genotypes have been reported to have improved nitrogen-use efficiency over senescent genotypes. Some plant morphological attributes such as leaf thickness and specific leaf weight have been shown to be positively related to nitrogen-use efficiency. While such information are useful for better targeting the problem in future research, most of the results so far generated are based on a small set of entries with relatively narrow genetic backgrounds. Evaluation of larger set of genotypes representing an array of genetic backgrounds having contrasting characteristics for traits assumed to be related to nitrogen-use efficiency may help generate more robust information.

References

AICSIP (2007) All India Coordinated Sorghum Improvement Project. Annual progress report 2006–07. Directorate of Sorghum Research, Hyderabad, AP, India

AICSIP (2011) All India Coordinated Sorghum Improvement Project. Sorghum Agronomy Kharif 2010–11. Directorate of Sorghum Research, Hyderabad, India

AICSIP (2014) All India Coordinated Sorghum Improvement Project. Sorghum Agronomy Kharif 2013–14. Directorate of Sorghum Research, Hyderabad, India

Arkin GF, Taylor HM (eds) (1981) Modifying the root environment to reduce crop stress, ASAE monograph no. 4. American Society of Agricultural Engineers, St. Joseph

Babaria CJ, Patel CL (1981) Response and uptake of Fe by sorghum to application of iron, farm yard manure and sulphur in calcareous soils. GAU Res J 6 (2):121–124

Baligar VC, Fageria NK (1997) Nutrient use efficiency in acid soils: nutrient management and plant use efficiency. In: Monitz AC, Furlani AMC, Fageria NK, Rosolem CA, Cantarells H (eds) Plant-soil interactions at low pH: sustainable agriculture and forestry production. Brazilian Soil Science Society Campinas, Brazil, pp 75–93

Baligar VC, Fageria NK, He ZL (2001) Nutrient use efficiency in plants. Commun Soil Sci Plant Anal 32:921–950

Bationo A, Buerkert A (2001) Soil organic carbon management for sustainable land use in Soudano-Sahelian West Africa. Nutr Cycl Agroecosyst 61:131–142

Bennett WF, Tucker BB, Maunder AB (1990) Modern grain sorghum production. Iowa State University Press, Ames

Bernal JH, Navas GE, Clark RB (2002) Sorghum nitrogen use efficiency in Colombia. Dev Plant Soil Sci 92:66–67

Bhonde MB, Bhakare BD (2008) Influence of integrated nutrient management on soil properties of vertisol under sorghum (*Sorghum bicolor*)-wheat (*Triticum aestivum*) cropping sequence. J Res ANGRAU 36:1–8

Bodade VN (1964) Agronomic trials on jowar (*Sorghum vulgare*). Indian J Agron 9:184–195

Bodade VN (1966) Effect of foliar application of nitrogen and phosphorus on yield of jowar (*Sorghum vulgare*). Indian J Agron 11:267–269

Choudhary SD (1978) Efficiency of nitrogen applied through soil and foliar on grain sorghum. J Maharashtra Agric Univ 3:26–27

Clark RB (1982a) Mineral nutritional factors reducing sorghum yields; micronutrients and acidity. Sorghum in the Eighties. In: Proc int symp sorghum. ICRISAT, Patancheru, pp 179–190, 2–7 November 1981

Clark RB (1982b) Plant response to mineral element toxicity and deficiency. In: Christiansen MN, Lewis CF (eds) Breeding plants for less favourable environments. Wiley, New York, pp 71–142

Cooper AJ (1973) Root temperature and plant growth, Commonwealth Bureau of Horticultural and Plantation Crops, Research review no. 4. East Malling Maidstone, Kent

Cothren JT, Matocha JE, Clark LE (2000) Integrated crop management for sorghum. In: Smith CW, Frederiksen RA (eds) Sorghum: origin, history, technology, and production. Wiley, New York, pp 409–441

Dhruvanarayan VV, Rambabu B (1983) Estimation of soil erosion in India. J Irrig Drain Eng 109:419–433

Dudal R (1976) Inventory of the major soils of the world with special reference to mineral stress hazards. In: Wright MJ (ed) Plant adaptation to mineral stress in problem soils. Cornell University Agricultural Experiment Station, Ithaca, pp 3–13

FAO (2005) Fertilizer use by crops in India. Land and Plant Nutrition Management Service, Land and Water Development Division, Food and Agricultural Organizations of the United Nations, Rome

Frink CR, Waggoner PE, Ausubel JH (1999) Nitrogen fertilizer: retrospect and prospect. Proc Natl Acad Sci U S A 96:1175–1180

Gardner JC, Maranville JW, Paparozzi ET (1994) Nitrogen use efficiency among diverse sorghum cultivars. Crop Sci 34:728–733

Gawai PP, Pawar VS (2006) Integrated nutrient management in sorghum (*Sorghum bicolor*)–chickpea (*Cicer arietinum*) cropping sequence under irrigated conditions. Indian J Agron 51:17–20

Gill AS, Abichandani CT (1972) A note on response of hybrid jowar to micronutrients. Indian J Agron 17:231–232

Gopalkrishnan S (1960) Copper nutrition of millets. Part II. Madras Agric J 47:95–108

Grundon NJ, Edwards DG, Takkar PN, Asher CJ, Clark RB (1987) Nutritional disorders of grain sorghum, ACIAR monograph no. 2. ACIAR, Canberra, p 99

Hariprakash M (1979) Soil testing and plant analysis studies on hybrid sorghum CSH 1. Mysore J Agric Sci 13:178–181

ICRISAT (1986) Annual report. International Crop Research Institute for the Semi-Arid Tropics, Patancheru, p 489

Jen-Hshuan Chen (2006) The combined use of chemical and organic fertilizers and or biofertilizer for crop growth and soil fertility. In: International Workshop on Sustained Management of the Soil-Rhizosphere System for Efficient Crop Production and Fertilizer Use. Land Development Department, Bangkok-10900, Thailand, pp 125–130

Jones CA (1983) A survey of the variability in tissue nitrogen and phosphorus concentration in maize and green sorghum. Field Crop Res 6:133–147

Joshi SG (1956) An examination of the results of the factorial design field experiment for the response of jowar crop to the application of different micronutrients. J Indian Soc Soil Sci 4:147–159

Kanwar JS (1978) Fertilization of sorghum, millets and other food crops for optimum yield under dry farming conditions. Pages AGR II 3/1-16, Proc Annual Seminar, Dec. 1977. Fertilizer Association of India (FAI), New Delhi

Kanwar JS, Randhawa NS (1967) Micronutrient research in soil and plants in India – a review. Indian Council of Agricultural Research, New Delhi

Kaore (2006) An approach for crop wise plant nutrition prescription. Indian J Fertl 1:57–62

Katyal JC (2000) Organic matter maintenance. J Indian Soc Soil Sci 48:704–716

Kondap SM, Ramoji GVNS, Bucha Reddy B, Rao AN (1985) Influence of nitrogen fertilization and weed growth on weed control efficiency of certain herbicides in sorghum crop. In: *Abstracts* annual conf Indian Soc Weed Sci. Gujarat Agricultural University, Anand, pp 31–32, 4–5 April 1985

Koraddi UR, Kulkarni RY, Kajjar NB (1969) Lime induced iron chlorosis in hybrid sorghum. Mysore J Agric Sci 3:116–117

Krishna KR (2010) Agroecosystems of South India: nutrient dynamics, ecology and productivity. BrownWalker Press, Boca Raton

Krishna KR, Dart PJ, Papavinasasundaram KG, Shetty KG (1985) Growth and phosphorus uptake responses of *Sorghum bicolor* to mycorrhyzal inoculations. In: Proc 6th North America conference on Mycorrhiza (NACM), Bend, Oregon, USA, p 404

Kumar AA, Reddy BVS, Sharma HC, Hash TC, Rao PS, Ramaiah B, Reddy PS (2011) Recent advances in sorghum genetic enhancement research at ICRISAT. Am J Plant Sci 2:589–600

Lakhdive BA, Gore NB (1978) Rate and source of phosphorus for hybrid sorghum and blackgram. In: Proceedings symposium on non monetary inputs.

Indian Society of Agronomy, New Delhi, India, pp 213–218

Lamond RE, Whitney DA, Hickman JS, Bonczkowski LC (1991) Nitrogen rate and placement for grain sorghum production in no-tillage systems. J Prod Agric 4:531–535

Lingegowda BK, Inamdar SS, Krishnamoorthy K (1971) Studies on the split application of nitrogen to rainfed hybrid sorghum. Indian J Agron 16:157–158

Maiti RK (1996) Panicle development and productivity. In: Sorghum Science. Oxford and IBH Publishing Co, New Delhi, pp 140–181

Maraanville JW, Pandey RK, Sirifi S (2002) Comparison of nitrogen use efficiency of a newly developed sorghum hybrid and two improved cultivars in the Sahel of West Africa. Commun Soil Sci Plant Anal 33:1519–1536

Maunder AB (2000) History of cultivar development in the United States: from "Memoirs of A. B. Maunder-Sorghum Breeder". In: Smith CW, Frederiksen RA (eds) Sorghum: origin, history, technology, and production. Wiley, New York, pp 191–223

Mishra JS, Talwar HS, Patil, JV (2012a) Conservation tillage and integrated nutrient management in kharif grain sorghum. In: Summaries. National seminar on Indian agriculture: Preparedness for climate change, 24–25 March 2012, New Delhi, pp 53–54

Mishra JS, Rao SS, Dixit A (2012b) Evaluation of new herbicides for weed control and crop safety in rainy season sorghum. Indian J Weed Sci 44(1):71–72

Moraghan JT, Rego TJ, Buresh RJ, Vlek PL, Burfora JR, Singh S, Sahrawat KL (1984) Labeled nitrogen fertilizer research with urea in the semi-arid tropics II. Field studies on a Vertisol. Plant and Soil 80:21–33

Narayana Reddy S, Rangamannan KT, Reddy SR, Shankara Reddy GS (1972) A note on the foliar application of urea to jowar variety 'Swarna'. Indian J Agron 17:363–367

NCAER and FAI (National Council of Applied Economic Research and Fertilizer Association of India) (1974) Fertilizer use on selected crops in India, New Delhi, India, p 52

Okafor LI, Zitta C (1991) The influence of nitrogen on sorghum – weed competition in the tropics. Trop Pest Manag 37:138–143

Ouedraogo E, Mando A (2010) Effect of tillage and organic matter quality on sorghum fertilizer use and water use efficiency in Semi-arid West Africa. In: 19th world congress of soil science, soil solutions for a changing world. Brisbane, 1–6 August 2010

Pal UR, Upadhyay UC, Singh SP, Umrani MK (1982) Mineral nutrition and fertilizer response of grain sorghum in India – a review over the last 25 years. Fertl Res 3:141–159

Paterson AH, Bowers JE, Bruggmann R, Dubchak I, Grimwood J, Gundlach H, Haberer G, Hellsten U, Mitros T, Poliakov A, Schmutz J, Spannagl M, Tang H, Wang X, Wicker T, Bharti AK, Chapman J, Feltus FA, Gowik U, Grigoriev IV, Lyons E, Maher CA, Martis M, Narechania A, Otillar RP, Penning BW, Salamov AA, Wang Y, Zhang L, Carpita NC, Freeling M, Gingle AR, Hash CT, Keller B, Klein P, Kresovich S, McCann MC, Ming R, Peterson DG, Mehboob ur R, Ware D, Westhoff P, Mayer KFX, Messing J, Rokhsar DS (2009) The Sorghum bicolor genome and the diversification of grasses. Nature 457:551–556

Patil MD (1979) Status of iron, zinc and manganese in calcareous soils and response of CSH 1 sorghum to application of phosphorus, iron and zinc in calcareous soils. Ph.D. thesis, University of Agricultural Sciences, Bangalore, India

Patil HM, Tuwar SS, Wani AG (2008) Integrated nutrient management in sorghum (Sorghum bicolor)-chickpea (Cicer arietinum) cropping sequence under irrigated conditions. Int J Agric Sci 4:220–224

Pearson CJ (1985) Editorial: research and development for yield of pearl millet. Field Crop Res 11:113–121

Prasad R (2009) Efficient fertilizer use: the key to food security and better environment. J Trop Agric 47:1–17

Prasad PVV, Staggenborg SA (2009) Growth and production of sorghum and millets. In. soils, plant growth and crop production – volume II. In: Encyclopedia of life support systems. Eolss Publishers, Oxford. http://www.eolss.net

Rao ACS, Das SK (1982) Soil fertility management and fertilizer use in dry land agriculture. In: A decade of dry land agriculture in India 1971–80. AICRPDA, Hyderabad, pp 120–139

Raun WR, Johnson GV (1999) Improving nitrogen use efficiency for cereal production. Agron J 91:357–363

Robson AD, Snowball K (1986) Nutrient deficiency and toxicity symptoms. In: Reuter DJ, Robinson JB (eds) Plant analysis: an interpretation manual. Inkata Press Pvt. Ltd, Melbourne, pp 13–19

Roy RN, Wright BC (1974) Sorghum growth and nutrient uptake in relation to soil fertility, II. N, P and K uptake pattern by various plant parts. Agron J 66:5–10

Satao RN, Nalamwar RV (1993) Studies on uptake of nitrogen, phosphorus and potassium by weeds and sorghum as influenced by integrated weed control. Integrated weed management for sustainable agriculture. In: Proceedings of the Indian Society of Weed Science international symposium Hisar, India, 18–20 November 1993 vol. III, pp 103–107

Sawargaonkar GL, Patil MD, Wani SP, Pavani E, Reddy BVSR, Marimuthu S (2013) Nitrogen response and water use efficiency of sweet sorghum cultivars. Field Crop Res 149:245–251

Seetharama N, Krishna KR, Rego, TJ, Burford J (1988) Prospects for improvement of phosphorus efficiency in sorghum for acid soils. In: Proceedings of the workshop on evaluating sorghum for tolerance to aluminium toxic, tropic soils in Latin America. Centro International Agricultural Tropicale (CIAT), Cali, Columbia, pp 229–249

Seetharama N, Flower DJ, Jayachandran R, Krishna KR, Peacock JM, Singh S, Soman P, Usharani A, Wani SP

(1990) Assessment of genotypic differences in sorghum root characteristics. Proceedings of the international congress on plant physiology, New Delhi, pp 215–219

Senthil Kumar S, Arockiasami DI (1995) VAM fungi in integrated nutrient management. Indian J Microbiol 35:185–188

Sharda VN, Singh R (2003) Erosion control measures for improving productivity of farmer's profitability. Fertl News 48:71–78

Sharma AK, Singh M (1974) A note on the efficiency of nitrogen fertilizers in relation to time and method of application in hybrid sorghum. Indian J Agron 19:158–160

Shelton DP, Smith JA, Jasa PJ, Kanable R (1995) Estimating percent residue covers using calculation method, G 95-1135-A. www.ianr.unl.edu/pubs/fieldcrops/g135.htm

Singh MV (2001) Evaluation of current micronutrient stocks in different agro-ecological zones of India for sustainable crops production. Fertl News 42:25–42

Singh A, Bains SS (1973) Yield, grain quality and nutrient uptake of CSH 1 and Swarna sorghum at different levels of nitrogen and plant population. Indian J Agric Sci 43:408–413

Singh RM, Vyas DL (1970) A note on response of grain sorghum to micronutrients. Indian J Agron 15:309–310

Singh M, Yadav DS (1980) Effect of copper, iron and liming on growth, concentration and uptake of copper, iron, manganese and zinc in sorghum (*Sorghum bicolor*). J Indian Soc Soil Sci 28:113–118

Srinivasarao C, Satyanarayana T, Venkateswarlu B (2011) Potassium mining in Indian agriculture: input and output balance. Karnataka J Agric Sci 24:20–28

Srivastava SP, Singh A (1971) Utilization of nitrogen by dwarf sorghum. Indian J Agric Sci 41:543–546

Tandon HLS, Kanwar JS (1984) A review of fertilizer use research on sorghum in India, Research bulletin no. 8. International Crops Research Institute for the Semi-Arid Tropics. International Crops Research Institute for the Semi-Arid Tropics, Patancheru

Tiwari KN (2006) Future of plant nutrition research in India. Indian J Fertl 2:73–98

TNAU (2003) ^{15}N studies: a report. Department of Soil Science and Agricultural Chemistry. Tamil Nadu Agricultural University, Coimbatore, pp 1–6. http://www.tnau.ac.in/scms/ssac/res/sacn151.htm

Turkhede BB, Prasad R (1978) Effect of rates and timing of nitrogen application on hybrid sorghum. Indian J Agron 23:113–126

Vanderlip RL (1972) How a sorghum plant develops. Cooperative Extension Service, Kansas State University, Manhattan

Venkatachalam S, Premanathan S, Arunachelam G, Vivekanandan SN (1969) Soil fertility studies in Madras State using radio tracer technique. I. Placement of phosphate to hybrid sorghum. Madras Agric J 56:104–109

Venkateswarlu K (1973) Efficiency of nitrogen utilization by hybrid sorghum (CSH 1) and assessment of residual fertility. Ph.D. thesis, G.B. Pant University of Agriculture and Technology, Pantnagar, India

Venkateswarlu K, Sharma KC, Lal B (1978) Recovery of fertilizer nitrogen applied to grain sorghum and assessment of residual effect. Pantnagar J Res 3:36–40

Vessey JK (2003) Plant growth promoting rhizobacteria as biofertilizers. Plant Soil 255:571–586

Viets FG (1972) Water deficits and nutrient availability. In: Kozlowski TT (ed) Water deficit and plant growth, vol 3. Plant responses and control of water balance, vol 3. Academic, New York/London

Vijayalakshmi K (1979) Research achievements with special reference to fertilizer use in drylands. In: Proceedings of the group discussion on fertilizer use in drylands, August 1979. Fertilizer Association of India, New Delhi, pp 31–38

Vitousek PM (1982) Nutrient cycling and nitrogen use efficiency. Am Nat 119:553–572

Wiedenfeld RP (1984) Nutrient requirements and use efficiency by sweet sorghum. Energy Agric 3:49–59

Improving Nutrient Use Efficiency in Oilseeds Brassica

S.S. Rathore, Kapila Shekhawat, B.K. Kandpal, and O.P. Premi

Abstract

The food and edible oil demand is increasing continuously and is expected to be doubled by 2050, while the production system and natural resources are continuously deteriorating. In this context enhancing nutrient use efficiency is the need of the hour for increasing crop productivity and reducing the nutrient waste, which is very high presently. The efficiency means the ability of a system to convert inputs into preferred outputs or to minimize input requirement losses. Enhancement of nutrient use efficiency (NUE) by plants could lessen fertilizer input, reduce the nutrient losses, and boost up the crop productivity. There is scope to increase the mustard productivity up to 2,000.0 kg/ha, from present national average of 1,145 kg/ha enhancing input use efficiency in which fertilizers nutrient sources have great role to play. Nutrient use efficiency enhancement is prerequisite not only for primary nutrients but also for secondary and micronutrients for oilseeds Brassica. Mustard, in general, is very sensitive to micronutrient deficiency, specially zinc and boron. The response of various ideotype to the applied micronutrients varies considerably. The precise information of the bio-physiological mechanism for adaptation to nutrient stress will help in enhancing NUE at plant level. It is important to exploit the potential of organic manures, composts, crop residues, agricultural wastes, bio-fertilizers, and their synergistic effect in combination with chemical fertilizers. This is needed for improving balanced nutrient supply and their use efficiency for increasing productivity, sustainability of agriculture, and soil health. INM improves the nutrient uptake by mustard and hence enhances the use efficiency of various nutrients from the soil.

S.S. Rathore (✉) • K. Shekhawat • B.K. Kandpal •
O.P. Premi
Directorate of Rapeseed-Mustard Research, Sewar,
Bharatpur, Rajasthan 321303, India
e-mail: sanjayrathorears@gmail.com;
drrathorekapila@gmail.com; basantkandpal@rediffmail.
com; oppremidrmr@gmail.com

A. Rakshit et al. (eds.), *Nutrient Use Efficiency: from Basics to Advances*,
DOI 10.1007/978-81-322-2169-2_20, © Springer India 2015

Keywords

Nutrient use efficiency • Integrated nutrient management • Oilseed Brassica • Micro nutrients • Mustard productivity

1   Introduction

The food demand by 2050 is expected to double globally the background of continuous deterioration of natural resources. India needs to double the food grain production from limited arable land. Annually, India is losing nearly 0.8 million tonnes of nitrogen, 1.8 million tonnes of phosphorus and 26.3 million tonnes of potassium thereby deteriorating quality and health of soil which is something to be checked. Problems are further aggravated by imbalanced application of nutrients (especially nitrogen, phosphorus and potash) and excessive mining of micronutrients, leading to deficiency of macro- and micro-nutrients in the soils (Vision 2030 2011). Due to the imbalanced use of plant nutrients, mining of nutrients is considered as the main cause for decline in crop yield and crop response ratio. About 8–10 million tonnes of NPK is mined annually in India. Soils are also being depleted of secondary and micronutrients. About 42 % of the soils are deficient in sulphur, 48.5 % deficient in zinc and 33 % deficient in boron (Gupta et al. 2007).

Nutrient use efficiency has been defined in many ways in diverse contexts (Clark 1990; Blair 1993). Most definitions refer efficiency as the ability of a system to convert inputs into preferred outputs or to minimize input waste. Genetic and physiological components of plants have profound effects on their abilities to absorb and utilize nutrients under various environmental and ecological conditions. In nutrient efficiency, supply or amount of a mineral nutrient is considered as input, while plant growth, physiological activity or yield as typical outputs. Efficiency is the relationship of output to input. This is expressed as a simple ratio, such as kg yield per kg fertilizer or g of plant dry weight per mg of nutrient, but as the amount of input and output varies, the ratio between them is rarely fixed, so efficiency is most comprehensively described by the entire relationship of output as a function of input. Agricultural productivity could only be enhanced through increasing resource use efficiency and soil fertility. Rapeseed-mustard area, production and productivity in India during 2011–2012 was 5.92 M ha, 6.78 mt and 1,145 kg/ha. Enhancement of NUE by plants could lessen fertilizer input costs, reduce nutrient losses and boost up crop productivity. There is scope to increase the mustard productivity up to 2,000 kg/ha by enhancing input use efficiency of the fertilizer nutrients.

2   Nutrient Use Efficiency

It is difficult to explain nutrient use efficiency due to the fact that there is no sole or generally accepted definition of nutrient use efficiency. Different definitions are available in literature to describe the agronomic and physiological range of nutrient use efficiency which refers to external and internal nutrient statuses. However, the evaluation of NUE is useful to differentiate plant species, genotypes and cultivars for their ability to absorb and utilize nutrients for maximum yields. Baligar et al. (2001) reported that NUE is based on (a) uptake efficiency (acquire from soil, influx rate into roots, influx kinetics, radial transport in roots, based on root parameters and uptake related to the amounts of the particular nutrient applied or present in soil), (b) incorporation efficiency (the transportation to shoot and leaves based on shoot parameters) and (c) utilization efficiency (based on remobilization, whole plant, i.e. root and shoot parameters). Most definitions of nutrient use efficiency refer to the external nutrient supply in terms of the agronomic meaning benchmarking seed yield as the essential objective. Nutrient-efficient cultivars are resulting in high seed yield under conditions of limited nutrient supply (Graham 1984).

2.1 Mechanisms for Enhancing Plant Nutrient Use Efficiency

Nutrient-efficient crops/cultivars, uptake efficiency, root morphology, root architecture, root-shoot ratio, root hairs, root radius, cluster root formation, association of roots with arbuscular mycorrhizae, root exudation as organic acids, utilization efficiency and cytoplasmic homeostasis are some of the plant adaptations required to face nutritional stress and increasing nutrient use efficiency. Nutrient-efficient cultivars produce reasonably high yield in low nutrient soils through either ways and thus can reduce mineral nutrient fertilizer input requirement in agricultural production.

2.1.1 Root System and Nutrient Use Efficiency

Among the morphological plant characteristics associated with the adaptation to N-depleted soils, the qualitative and quantitative significance of the root system in taking up N under N-limiting conditions has been reported (Kamara et al. 2003). Root architecture refers to the complexity of root spatial configurations that arise in response to soil conditions (Vance et al. 2003). Some plant species/genotypes alter the architecture of their root systems for efficient nutrient use within them (Richardson et al. 2011). Efficient genotypes develop an architecture that places active roots in regions of the soil more likely to contain available P (Smith 2001). Root morphology parameters such as length, thickness, surface area density, root hairs and root growth rate expressed as dry mass and/or root-shoot ratios are affected by deficiencies and/or essential minerals and/or excess of essential minerals (Baligar et al. 1998; Bennett 1993).

2.1.2 Root-Shoot Ratio

The nutrient-efficient cultivars have the ability to exploit greater soil volume for accessing more

nutrient through larger root system (higher root-shoot ratio), longer root hairs or by forming association with arbuscular mycorrhiza (AM). The ability of a crop/genotype to give higher yield under P-limiting condition may be related to the plant to take up more P from the soil under P-limiting condition (uptake efficiency) or the ability to produce higher dry matter per unit of P in the plant tissue (utilization efficiency) or a combination of both (Gahoonia and Nielsen 1996). Plant species as well as genotypes within the same species may differ in efficiency (Gunes et al. 2006). Difference in P uptake efficiency between crop species (Fohse et al. 1988) and genotypes was noticed, which was accounted to difference in root-shoot ratio.

2.1.3 Optimum Uptake and Nutrient Use Efficiency

Optimum uptake of nitrate is the first step to enhance N use in any plant. It has been established from a number of physiological studies that plants acquire their nitrate from the soil through the combined activities of a set of high- and low-affinity transporter systems, with the influx of NO_3^- being driven by the H^+ gradient across the plasma membrane. For increased P uptake efficiency, plant species/genotypes may use various adaptation mechanisms to gain access to previously unavailable soil P reserves such as through altered root morphology, exudation of chemical compounds into the rhizosphere and association of roots with mycorrhiza (Vance et al. 2003; Lambers et al. 2006). Higher P uptake efficiency is usually related to either larger root system size (usually higher root-shoot ratio) or to higher uptake rate per unit of root length (Fohse et al. 1988). The optimal root-shoot ratio corresponding to the optimal leaf-nitrogen concentration which maximize relative growth rate are reported quantitatively as a function of root-specific activity assumed to be governed by soil nitrogen availability. The plants respond optimally to soil nitrogen with higher root-shoot ratio (David 1989).

Because of low mobility of phosphorus in the soil, some plant species/genotypes develop larger root systems that allow a plant to have access to greater soil volume so that higher quantity of soil P can reach the root surface for being taken up (Jungk 2001). Preferential root growth thus helps the stressed plants to acquire more nutrients from the ambient environment. Root surface area alone may not be adequate to feed plants, especially with nutrient of low mobility like phosphorus. The presence of root hairs is also equally important for the acquisition of poorly mobile nutrients such as P. Root hairs substantially increase the root surface area for ion uptake (Jungk 2001).

2.1.4 Root Hairs

Root hairs form as much as 77 % of the root surface area of field crops (Parker et al. 2000). Some plant species/genotypes are adapted to produce longer and more number of root hairs under nutrient-deficient conditions (Eticha and Schenk 2001). Under nutrient-deficient condition, plant species/genotypes produce fine roots that facilitate a contact of larger soil volume per unit of root surface area, thereby increasing nutrient uptake rates (Fohse et al. 1988). Thus, plant species/genotypes with thinner roots may be more effective in absorbing soil nutrient.

2.1.5 AM Association and Nutrient Use Efficiency

Association of roots with AM of Brassica species results in better nutrient use efficiency. The vast majority (82 %) of higher plant species have the capacity to form a symbiotic association with mycorrhizal fungi. The symbiotic association of plant roots with AM enhances the uptake of nutrients with low mobility like P especially when the species has a root system that is relatively coarse with few root hairs. A significant contribution of AM fungi to plant P uptake has been reported especially for soils with low P content and with high P fixing capacity (Marschner and Dell 1994). Increased P and other nutrient absorption by mycorrhizal hyphae is related to both increased physical exploration of the soil and modification of the root environment (Smith and Read 1997; Tinker and Nye 2000).

2.1.6 Well-Designed Biochemical Mechanisms and Physiological Adaptations

Many plants have developed elegant biochemical mechanisms to solubilize P from insoluble P complexes thereby increasing the pool of P available for uptake (Raghothama and Karthikeyan 2005). Besides increased acquisition of soil nutrients, efficient utilization of acquired nutrient is also considered an important adaptation for plant growth on poor soils. Nutrient utilization efficiency refers to the ability of a plant species/genotype to produce higher dry matter per unit of nutrient absorbed (Blair 1993; Richardson et al. 2011). A portion of the nitrate taken up is utilized/stored in the root cells, while the rest is transported to other parts of the plant. Due to the abundant availability of photosynthetic reductants, leaf mesophyll cells are the main sites of nitrate reduction. Cytoplasmic P homeostasis is the physiological adaptations for P stress to maintain cytoplasmic Pi either through effective buffering with vacuolar Pi (Plaxton and Carswell 1999; Raghothama 1999) or possibly through selective allocation of Pi between cytoplasm and vacuole to constantly keep sufficient Pi in metabolically active compartment (cytoplasm) despite P stress (Lauer et al. 1989). The efficiency of this process is, however, dependent on the relative permeability of the tonoplast to Pi and may vary between different plant species. Thus, the decline in cytoplasmic Pi, due to the absence of effective Pi homeostasis directly affects sugar-phosphate export from the chloroplast (Flügge et al. 1980).

3 Enhancement of NUE in Oilseed Brassica

Nutrient use efficiency in oilseed *Brassica* is greatly influenced by the rate, source and method of fertilizer application. The rate of a particular nutrient to be applied depends upon the initial soil status, climate, topography, cropping system in practice and crop. In plants, N management can be divided into two main phases: the first phase, vegetative, during which sink organs (roots, young leaves) evolve for the assimilation of inorganic N via nitrate assimilatory pathway (Hirel and Lea 2001). In oilseed rape, the requirement for N per kg produce is higher than in cereal crops (Hocking and Strapper 2001). Oilseed rape has a high capacity to take up nitrate from the soil (Lainé et al. 1993), which is accumulated and stored in the vegetative parts at the beginning of flowering. However, in oilseed rape, yield is half of wheat, due to the production of oil. Since N content in the seed of rape is not much high (3 % in oilseed rape and 2 % in wheat on average), a significant portion of the N stored in the vegetative organs is not used. Moreover, a large quantity of N is also lost in early falling of leaves (Malagoli et al. 2005).

3.1 Plant Breeding for Enhancing NUE

There exists considerable genotypic difference in rapeseed-mustard cultivars for nutrient response. The advancement in plant breeding and molecular genetics research improved developments of genotypes with improved characteristics under conditions of high N supply. Genotypic variation in N efficiency is attributed to high N uptake and/or high N utilization. The development of small cultivars (dwarf, semidwarf) is of interest. Actually partitioning of N and carbohydrates into seeds can be improved. But, despite of reductions in stem formation resulting in increased HI, yield formation and N uptake are reduced compared with commercial cultivars. Growing of N-efficient cultivars might contribute to integrated nutrient management strategies in both low-input (improving crop productivity) and high-input (reduction of environmental pollution) agriculture (Wiesler et al. 2001). Habekotte (1997) and Horst et al. (2002) suggested that ideotype characters be combined with the aim of simultaneously increasing sink and source capacity for seed filling, providing best prospects for boosting yield increase. An N-efficient ideotype is characterized by reduced vegetative growth until the beginning of flowering; high N

uptake and less dry mass reduction during the reproductive growth mainly attributed to the interaction between shoot and root during yield formation.

3.2 Improved Agro-Techniques for Enhancement of Nutrient Use Efficiency

Nutrient use efficiency in oilseed Brassica is greatly influenced by the rate, source and method of fertilizer application. Improved agro-techniques also enhance nutrient use efficiency of oilseed Brassica (Table 1).

Conservation tillage is also more productive and enhances nitrogen use efficiency (AICRP-RM 2007). Increase in the nitrogen level up to 60 kg N/ha consistently and significantly increased the number of primary branches, number of seeds per siliqua and 1,000 seed weight (Sharma et al. 2007), while, increasing the nitrogen level up to 90 kg/ha increase the number of secondary branches per plant, seed and straw yields (Sah et al. 2006). Split application of total nitrogen in three equal doses, one each as basal, second at first irrigation and remaining one-third at second irrigation, also leads to maximum increase in yield attributes and yield of Brassica juncea compared to application of total nitrogen in two split doses (Reager et al. 2006). Top-dressing of N fertilizers should be done immediately after first irrigation. Delaying of first irrigation results in yield reduction of mustard crop. The application of nitrogen with pre-sowing irrigation is superior (Sidhu and Sandhu 1995).

The dry matter/plant significantly enhanced with the application of phosphorus up to 60 kg/ha. P application up to 40 kg/ha increased the plant height, branches per plant and leaf chlorophyll content. The uptake of NPK and sulphur by both seed and stover increased significantly with successive increase in nitrogen levels up to 120 kg N/ha, sulphur levels up to 60 kg S/ha and P_2O_5 level up to 60 kg P_2O_5/ha. Seed yield and yield attributes increased, while oil content decreased with increasing level of nitrogen up to

Table 1 Integrated N-management strategies affecting nutrient use efficiency in plant production

S. No.	Techniques	Aim
1	Cropping system	Increased uptake and utilization of soil and fertilizer nutrient by cultivation of nutrient-specific efficient crops, reduction of fallow frequency and rotation of shallow/deep rooting crops. Uptake of soil nutrient and mineralized plant residue nutrient thereby reducing nutrient losses by leaching and increasing supply to succeeding crops
2	Oilseed Brassica cultivar	Increased uptake and utilization of soil and fertilizer nutrient by cultivation of nutrient-efficient cultivars
3	Irrigation and soil management practices	Increased uptake and utilization of soil and fertilizer nutrient by well-grown crops. Timing, intensity and depth of soil cultivation control the soil mineral nutrient release.
4	Precise forecast of fertilizer nutrient requirement (e.g. soil and plant nutrient tests, sensor-controlled fertilization, modelling soil nutrient supply)	Increased uptake and utilization of soil and fertilizer nutrient by considering available soil mineral nutrient at the beginning of the growing season and nutrient mineralization during the growing season
5	Form of fertilizer (e.g. mineral fertilizer vs. organic manure, urea vs. ammonium vs. nitrate, use of urease and nitrification inhibitors)	Avoidance of nutrient losses caused by specific nutrient forms/nutrient transformations in the soil, increased physiological efficiency of nutrient by considering plant specific preferences of certain forms of N (NH_4 + vs. NO_3)
6	Timing of nutrient application	Reduction of nutrient losses (as in the case of NO_2, NO_3, N_2) at the beginning of the growing season, increased physiological efficiency by specific stimulation of harvestable organs
7	Technique of fertilizer nutrient application (e.g. surface vs. incorporation vs. broadcast vs. banded)	Reduction of losses (NH_2), improved spatial availability of nutrient, reduction of N immobilization
8	Management of crop residues	Control of N mineralization during fallow and immobilization of soil mineral nutrients

120 kg/ha. Different levels of phosphorus increased seed yield, maximum being at 80 kg P/ha due to higher number of secondary branches/plant and consequent siliquae/plant. Oil content also increased with increase in levels of N, P_2O_5 and S. Activities of all nitrogen assimilating enzymes, viz. nitrate reductase, nitrite reductase, glutamine synthetase and glutamate synthetase, were found to be maximum at 100 kg N/ha. Nitrogen plays a crucial role in many critical physiology and biochemical mechanism of rapeseed-mustard as components of amino acids and proteins (which form enzymes), genetic material (nucleotides and nucleic acids) and other components found in membranes (such as amines), coenzymes and others. The majority of the N in green plant tissue is present as enzyme protein in chloroplasts where chlorophyll is located. Thus,

plant biomass and its partitioning in seed depend on nitrogen. By harvest, the majority of the N in a canola plant is found as seed protein. The relative N proportions in the plant changes over time and growth stage.

3.3 Sulphur Fertilization and Nutrient Use Efficiency

Among the oilseed crops, rapeseed-mustard has the highest requirement of sulphur which promotes oil synthesis. It is an important constituent of seed protein, amino acid, enzymes and glucosinolate and is needed for chlorophyll formation (Holmes 1980). Sulphur increased the yield of mustard by 12–48 % under irrigated and by 17–124 % under rainfed conditions

(Aulakh and Pasricha 1988). In terms of agronomic efficiency, each kilogram of sulphur increases the yield of mustard by 7.7 kg (Katyal et al. 1997).

Oil content in Canola-4 and Hyola-401 is 3 % higher than the hybrid 'PGSH-51' due to the effect of various doses of nitrogen and sulphur, while the oleic acid content in these hybrids is double than 'PGSH-51' which has erucic acid ranging from 23.2 to 29.4 %. Higher sulphur level causes 2–3 % reduction in erucic acid content, while lower level of nitrogen reduces erucic acid content by 3 % with a concomitant increase in oleic acid (Table 2). Higher doses of sulphur along with low doses of nitrogen affect the chain elongation enzyme system thereby leading to reduction in erucic synthesis.

A significant increase in yield is observed with increase in sulphur levels up to 40 kg S/ha in mustard-based cropping system. At Bawal the highest seed yield of mustard is reported to be in green gram-mustard cropping sequence with the lowest (2,686 kg/ha) being in pearl millet-mustard sequence. In rice-mustard sequence, the optimum seed yield of mustard has been found at 40 kg S/ha at Berhampore and for black gram-mustard at Dholi. Each successive increase in S level increases seed yield up to 20 kg S/ha at Dholi and Ludhiana, 40 kg S/ha at SK Nagar and 60 kg S/ha at Berhampore and Morena conditions (AICRP-RM 2008).

3.4 Role of Micronutrients in Enhancing Nutrient Use Efficiency of Oilseeds Brassica

Mustard, in general, is very sensitive to micronutrient deficiency, specially zinc and boron. The increase in seed yield has been found 8.5 % at 12.5 kg ZnSO4/ha. The harvest index (HI) was significantly affected by Zn application, although seed yield generally showed diminishing return with additional $ZnSO_4$ doses (Table 3).

The response of various ideotypes to the applied micronutrients varies considerably. The response of Indian mustard varieties, viz. 'Pusa Bold' and 'Vardan', to applied zinc has been found higher (AICRP-RM 2000) than Varuna, RH- 30 and Aravali.

The concentration of Zn at flowering, pod formation stage, in straw and grain at maturity of Indian mustard increased significantly with increases in Zn level (Gupta and Kaushik 2006). Similarly, the seed yield increases significantly (16–47 %) with the application of boron. The average response to boron application ranges from 21 to 31 %. The yield increase is due to an increase in seeds/siliqua and 1,000 seed weight, which indicates its role in seed formation (AICRP-RM 2005).

4 Integrated Approaches for Enhancing Nutrient Use Efficiency of Oilseed Brassica

It is important to exploit the potential of organic manures, composts, crop residues, agricultural wastes, fertilizers and their synergistic effect on productivity, sustainability soil health and environmental safety. Balanced fertilization at the right time by proper method increases nutrient use efficiency in mustard. Experiments have been conducted at different AICRP centres with the integrated use of organic manure, green manure, crop residue and bio-fertilizers along with inorganic fertilizers, which show that INM not only reduces the demand of inorganic fertilizers but also increases the efficiency of applied nutrients due to their favourable effect on physical, chemical and biological properties of soil. The introduction of leguminous crops in the rotational and intercropping sequence and use of bacterial and algal cultures play an important role in increasing the nutrient use efficiency (Prasad et al. 1992).

INM improves the nutrient uptake by mustard and hence enhances the use efficiency of various nutrients from the soil. The incorporation of 25 % nitrogen through FYM + 75 % by chemical fertilizer + 100 % sulphur has been found to significantly enhance the uptake use efficiency of nitrogen and sulphur in both seed and stover (Bhat et al. 2005). The highest mustard equivalent yield, which includes converted yield of

Table 2 Effect of N and S levels (kg/ha) application on fatty acid composition and glucosinolate content in *Brassica juncea* cv. Varuna at Ludhiana

N (kg/ha)	S (kg/ha)	Glucosinolate content (μ moles/g in defatted meal)	Palmitic acid	Stearic acid	Oleic acid	Linoleic acid	Linolenic acid	Eicosenoic acid	Erucic acid
75	0	64	2.61	1.17	11.78	14.99	6.48	50.91	11.80
75	20	72	2.88	1.31	10.15	14.53	5.14	52.75	12.28
100	0	52	2.58	1.58	13.16	15.31	7.01	49.55	10.57
100	20	42	2.91	1.65	11.94	15.06	6.13	49.63	12.18
125	0	52	3.01	1.33	12.19	16.17	5.91	47.71	12.26
125	20	42	4.42	1.31	16.12	16.55	6.57	44.77	9.55

Source: AICRP-RM (2007)

Table 3 Effect of Zn on yield and yield attributes of Indian mustard

ZnSO$_4$ (Kg/ha) levels	Seed yield (kg/ha)	Secondary branches/plant	Oil content (%)	Oil yield (kg/ha)	Protein (%)	Protein yield (kg/ha)	Harvest index (%)
0	1,161	6.5	40.2	465.6	22.1	255.2	21.6
12.5	1,260	8.1	39.9	501.1	22.5	281.9	22.4
25.0	1,336	9.6	39.9	532.4	22.6	301.6	22.9
50	1,414	12.4	39.9	570.0	22.5	318.6	22.2
CD at 5 %	33	0.7	NS	22.8	NS	18.8	0.8

Source: AICRP-RM (2000)

other crops into mustard seed yield based on market price of the crops (24.88 q/ha), net monetary returns (Rs. 15,537/ha), B-C ratio (2.07) and agronomic efficiency (16.1), is achieved by the application of 100 % recommended N in the rainy season through FYM and 100 % recommended NP in the winter through inorganic fertilizers (Kumpawat 2004).

4.1 Organic Sources of Nutrients

Bulky organic manures are applied to improve overall soil health and reduce evaporation losses of soil moisture. Depending upon the availability of raw material and land use conditions various organic sources, viz. cluster bean (green manure), Sesbania (green manure), mustard straw at 3 t/ha and vermicompost (5–7.5 t/ha) have been evaluated at Bharatpur. Green manure with *Sesbania* produces significantly higher mustard seed yield at Bharatpur and Bawal and improves soil environment (AICRP-RM 2006).

Many bio-stimulants also encourage higher production. Spray of bioforce (an organic

formulation) at 2 ml/l at the flowering and siliqua formation stage enhances mustard seed yield (2,059 kg/ha) (AICRP-RM 2007).

4.2 Growth Promoter, Bio-fertilizer as a Component of INM

Bio-fertilizers are inoculants or preparations containing microorganism that apply nutrients especially N and P. Two types of N-fixing microorganisms, viz. free living (*Azotobacter*) and associative symbiosis (*Azospirillum*), and two P supplying microorganisms, viz. phosphate solubilizing bacteria and AM, have been extensively tested. Inoculation of mustard seeds with efficient strains of *Azotobacter* and *Azospirillum* enhances the seed yield up to 389 and 305 kg, respectively, with 40 Kg N/ha. The total NPK uptake also increases with Azotobacter inoculation. The combined application of 10 t FYM 90-45-45 NPK kg/ha with Azotobacter inoculation produces the highest B-C ratio of 1.51. At lower N levels, without inoculation, the seed yield declines as compared to seed inoculation.

Table 4 Effect of INM on quality of mustard (Kanti-RK 9807) under maize-mustard sequence

| Treatment | Legends | Oil content (%) | Fatty acid composition (%) | | | | | |
			16:1 Palmitic acid	18:1 Oleic acid	18:2 Linoleic acid	18:3 Linolenic acid	20:1 Eicosenoic acid	22:1 Erucic acid
RDF (120-40-40)	T1	40.4	2.8	18.4	10.1	10.6	4.3	52.7
T$_1$ + 10 t FYM/ha	T2	40.9	2.8	16.3	13.3	10.4	4.1	52.2
T$_2$ + 40 Kg S/ha	T3	40.4	2.9	18.0	14.4	12.2	3.2	48.6
T$_3$ + Zn SO4 25 kg/ha	T4	40.3	2.8	17.8	14.9	10.1	6.1	47.3
T$_4$ + B 1 kg/ha	T5	40.7	2.7	23.0	16.2	9.0	5.2	43.3
T$_1$ + Crop residue (Maize)	T6	40.1	2.7	20.0	14.3	9.2	4.4	48.6
75 % RDF		40.4	2.6	17.8	15.1	7.9	6.3	49.7

Source: Modified from AICRP-RM (2002)

Growth promoter formulations like bioforce and bio-power containing bio-amino acid, plant growth promoting terpenoid, siderophores and attenuated bacteria fortified with BGA help increase water and nutrient absorptions from the soil. Similarly, bioforce having natural free amino acid, phytohormones, macro- and micro-elements and plant growth promoting terpenoid activates the cell division and stimulates plant growth and photosynthate translocation. RDF (80:40:0) along with 25 kg bio-power/ha + spray of bioforce (1 1 in 500 l of water) at 50 % flowering and pod filling stage produces higher yield of mustard than other combinations (AICRP-RM 2005). Premi et al. (2012) reported a synergistic effect of PSB and PSB + VAM and antagonistic effect of VAM in agronomic P efficiency and apparent recovery efficiency.

29.9 % lesser seed yield over RDF at Jobner (AICRP-RM 2008). Amount of available phosphorus increased over initial value when organic manures and crop residues were incorporated. Organic carbon status builds up in organic source-incorporated plots. The application of 10 t FYM/ha in addition to recommended dose of fertilizer (RDF) benefits soil physical condition by improving aggregation, increased saturated hydraulic conductivity, reduced bulk density and penetration resistance of the surface soil (Hati et al. 2006). Nutrient use efficiency of primary, secondary and micronutrient can be enhanced through sound, crop soil-based integrated nutrient management of oilseed Brassica especially rapeseed-mustard crops.

4.3 Effect of INM on Quality of Mustard Oil

At Kanpur, INM studies have been evaluated in maize-mustard, bajra-mustard and fallow mustard sequences. In maize-mustard sequence, 100/75 % of RDF + 2 t FYM have produced the highest seed yield and quality of the oil (Table 4).

At Bharatpur and Jobner, 17.8 and 8.6 % increase in seed yield was recorded with 50 % RDF + 50 % N through FYM and vermicompost. Sole organic-treated plot recorded

Conclusion

Increased NUE of oilseed Brassica is vital to enhance the yield and quality of crops, reduce nutrient input cost and improve soil, water and air quality. Selection of nutrient-efficient genotypes and incorporation of these in breeding programs will result in better NUE. Nonetheless, the poorly developed nutritional genetics of crop plants and its response to external environmental factors and the complexity of identifying nutrient efficiency traits by rapid, reliable techniques have contributed to a lack of progress and success in breeding plant cultivars with high NUE. The different

crop cultivars differ in absorption and utilization of nutrients and such differences are
attributed to morphological, physiological
and biochemical processes in plants and their
interaction with climatic, soil, fertilizer,
biological and agronomic practices. Nutrient
use efficiency of oilseed Brassica can be
enhanced by cautious manipulation of plant,
soil, fertilizer, biological, environmental
factors and best management practices.

References

AICRP-RM (2000) Annual progress report of National
Research Centre on Rapeseed-mustard, pp 15–21

AICRP-RM (2002) Annual progress report of National
Research Centre on Rapeseed-mustard. Directorate
of Rapeseed Mustard research, Sewar, Bharatpur,
Rajasthan, pp 29–34

AICRP-RM (2005) Annual progress report of National
Research Centre on Rapeseed-mustard. Directorate
of Rapeseed Mustard research, Sewar, Bharatpur,
Rajasthan, pp 9–11

AICRP-RM (2006) Annual progress report of National
Research Centre on Rapeseed-mustard, pp 17–22

AICRP-RM (2007) Annual progress report of All India
Coordinated Research Project on Rapeseed-Mustard.
Directorate of Rapeseed Mustard research, Sewar,
Bharatpur, Rajasthan, pp A1–A16

AICRP-RM (2008) Annual progress report of All India
Coordinated Research Project on Rapeseed-Mustard.
Directorate of Rapeseed Mustard research, Sewar,
Bharatpur, Rajasthan, pp A1–A22

Aulakh MS, Pasricha NS (1988) Sulphur fertilization of
oilseeds for yield and quality. In: Sulphur in Indian
agriculture, section II. 3. The Sulphur Institute,
Washington, DC, pp 1–14

Baligar VC, Fageria NK, Elrashidi MA (1998) Toxicity
and nutrient constraints in root growth. Hort Sci 36:
960–965

Baligar VC, Fageria NK, He ZL (2001) Nutrient use
efficiency in plants. Commun Soil Sci Plant Anal
32(7&8):921–950

Bennett WF (1993) Plant nutrient utilization and diagnostic
plant symptoms. In: Bennett WF (ed) Nutrient deficiencies and toxicities in crop plants. The American
Phytopathological Society Press, St. Paul, pp 1–7

Bhat MA, Singh R, Dash D (2005) Effect of INM on
uptake and use efficiency of N and S in Indian mustard
on an inceptisol. Crop Res (Hisar) 30:23–25

Blair G (1993) Nutrient efficiency-what do we really
mean? In: Randall PJ, Delhaize E, Richards RA,
Munns R (eds) Genetic aspects of plant nutrition.
Kluwer Academic Publishers, Dordrecht, pp 204–213

Clark RB (1990) Physiology of cereals for mineral nutrient
uptake use and efficiency. In: Baligar VC, Duncan RR

(eds) Crops as enhancers of nutrient use. Academic
Press, San Diego, pp 131–209

David WH (1989) Optimization of plant root: shoot ratios
and internal nitrogen concentration. Ann Bot 66(1):
91–99

Eticha D, Schenk MK (2001) Phosphorus efficiency of
cabbage varieties. In: Horst WJ et al (eds) Plant
nutrition-food security and sustainability of agroecosystems through basic and applied research.
Kluwer Academic Publisher, Dordrecht, pp 542–543

Flügge UI, Freisl M, Heldt HW (1980) Balance between
metabolite accumulation and transport in relation to
photosynthesis by spinach chloroplasts. Plant Physiol
65:574–577

Fohse D, Claassen N, Jungk A (1988) Phosphorus efficiency of plants. I. External and internal P requirement
and P uptake efficiency of different plant species.
Plant Soil 110:101–109

Gahoonia TS, Nielsen NE (1996) Variation in acquisition
of soil phosphorus among wheat and barley genotypes.
Plant Soil 178:223–230

Graham RD (1984) Breeding for nutritional character-
istics in cereals. In: Tinker PB, Lauchli A (eds)
Advances in plant nutrition, vol 1. Praiger Publisher,
New York, pp 57–102

Gunes A, Inal A, Aplaslan M, Cakmak I (2006) Genotypic
variation in phosphorus efficiency between wheat
cultivars grown under greenhouse and field conditions. Soil Sci Plant Nutr 52:470–478

Gupta M, Kaushik RDE (2006) Effect of saline irrigation
water and Zn on the concentration and uptake of Zn by
mustard. In: Proceedings of the 18th world congress of
soil science, Philadelphia, PA, USA, July 2006

Gupta SP, Singh MV, Dixit ML (2007) Deficiency
and management of micronutrients. Indian J Fertil 3:
57–60

Habekotte B (1997) Option for increasing seed yield of
winter oilseed rape (Brassica napus L.): a simulation
study. Field Crops Res 54:109–126

Hati KM, Mishra AK, Mandal KG, Ghosh PK,
Bandopadhyay KK (2006) Irrigation and nutrient management effect on soil physical properties under
soybean-mustard cropping system. Agric Water Manag
85(3):279–286

Hirel B, Lea PJ (2001) Ammonium assimilation. In:
Lea PJ, Morot-Gaudry JF (eds) Plant nitrogen.
Springer, Berlin, pp 79–99

Hocking PJ, Strapper M (2001) Effect of sowing time and
nitrogen fertiliser on canola and wheat, and nitrogen
fertiliser on Indian mustard. II. Nitrogen concentrations, N accumulation, and N use efficiency.
Aust J Agric Res 52:635–644

Holmes MRJ (1980) Nutrition of the oilseed rape crops.
In: TSI/FAI/IFA symposium. Applied Science Publishers Ltd., Essex, p 158

Horst WJ, Behrens T, Heuberger H, Kamh M,
Reidenbach G, Wiesler F (2002) Genotypic differences
in nitrogen use-efficiency in crop plants. In: Lynch JM,
Schepers JS, Unver I (eds) Innovative soil-plant systems
for sustainable agricultural practices, OECD workshop
2002. OECD Publications, Paris, pp 75–92

Jungk A (2001) Root hairs and acquisition of plant nutrients from soil. J Plant Nutr Soil Sci 164: 121–129

Kamara AY, Kling JG, Menkir A, Ibikunle G (2003) Agronomic performance of maize (Zea mays L.) breeding lines derived from a low nitrogen maize population. J Agric Sci 141:221–230

Katyal JC, Sharma KL, Srinivas K (1997) Proceedings of the TSI/FAI/IFA symposium on sulphur in balanced fertilisation: proceedings of the TSI/FAI/IFA symposium held on 13–14 Feb 1997 at New Delhi

Kumpawat BS (2004) Integrated nutrient management for maize-mustard cropping system. Indian J Agron 49: 4–7

Lainé P, Ourry A, Macduff JH, Boucaud J, Salette J (1993) Kinetic parameters of nitrate uptake by different catch crop species: effect of low temperatures or previous nitrate starvation. Physiol Plant 88:85–92

Lambers H, Shane MW, Cramer MD, Pearse SJ, Veneklaas EJ (2006) Root structure and functioning for efficient acquisition of phosphorus: matching morphological and physiological traits. Ann Bot 98:693–713

Lauer MJ, Blevins DG, Sierputowska-Gracz H (1989) ^{31}P-nuclear magnetic resonance determination of phosphate compartmentation in leaves of reproductive soybeans (Glycine max L.) as affected by phosphate nutrition. Plant Physiol 89:1331–1336

Malagoli P, Laine P, Rossato L, Ourry A (2005) Dynamics of nitrogen uptake and mobilization in field-grown winter oilseed rape (Brassica napus) from stem extension to harvest. Ann Bot 95:853–861

Marschner H, Dell B (1994) Nutrient uptake in mycorrhizal symbiosis. Plant Soil 159:89–102

Parker JS, Cavell AC, Dolan L, Roberts K, Grierson CS (2000) Genetic interactions during root hair morphogenesis in Arabidopsis. Plant Cell 12:1961–1974

Plaxton WC, Carswell MC (1999) Metabolic aspects of the phosphate starvation response in plants. In: Lerner HR (ed) Plant response to environmental stress: from phytohormones to genome reorganization. Marcel-Dekker, New York, pp 350–372

Prasad R, Sharma SN, Singh S, Lakshaman R (1992) Agronomic practices for increasing nutrient use efficiency and sustained crop production. Paper presented in national seminar on resource management for sustainable production, New Delhi

Premi OP, Kandpal BK, Kumar S, Rathore SS, Sekhawat K, Bhogal NS (2012) Phosphorus use efficiency of Indian mustard (Brassica juncea) under

semi arid conditions in relation to phosphorus solubilizing and mobilizing microorganisms and P fertilization. Natl Acad Sci Lett 35(6):547–553

Raghothama KG (1999) Phosphate acquisition. Annu Rev Plant Physiol Plant Mol Biol 50:665–693

Raghothama KG, Karthikeyan AS (2005) Phosphate acquisition. Plant Soil 274:37–49

Reager ML, Sharma SK, Yadav RS (2006) Yield attributes, yield and nutrient uptake of Indian mustard as influenced by N levels and its split application in arid western Rajasthan. Indian J Agron 51(3): 123–126

Richardson AE, Lynch JP, Ryan PR, Delhaize E, Smith FA, Smith SE, Harvey PR, Ryan MH, Veneklaas EJ, Lambers H, Oberson A, Culvenor RA, Simpson RJ (2011) Plant and microbial strategies to improve the phosphorus efficiency of agriculture. Plant Soil 349:121–156

Sah D, Bohra JS, Shukla DN (2006) Effect of N, P, S on growth attributes and nutrient uptake of mustard. Crop Res 31(1):234–236

Sharma R, Thakur KS, Chopra P (2007) Response of N and spacing on production of Ethiopian mustard under mid-hill conditions of Himachal Pradesh. Res Crops 8(1):65–68

Sidhu AS, Sandhu KS (1995) Response of mustard to method of N application and timing of first irrigation. J Indian Soc Soil Sci 43(3):331–334

Smith FW (2001) Plant response to nutritional stresses. In: Hawkesford MJ, Buchner P (eds) Molecular analysis of plant adaptation to the environment. Kluwer Academic Publishers, Dordrecht, pp 249–269

Smith SE, Read DJ (1997) Mycorrhizal symbiosis. Academic, San Diego

Tinker PB, Nye PH (2000) Solute movement in the rhizosphere. Oxford University Press, Inc, New York

Vance CP, Uhde-Stone C, Allan D (2003) Phosphorus acquisition and use: critical adaptation by plants for securing non-renewable resources. New Phytol 15: 423–447

Vision 2030 (2011) Project Director, Directorate of Knowledge Management in Agriculture (formerly DIPA). Indian Council of Agricultural Research, Krishi Anusandhan Bhavan, Pusa, New Delhi, p 38

Wiesler F, Behrens T, Horst WJ (2001) The role of nitrogen-efficient cultivars in sustainable agriculture. In: Optimizing nitrogen management in food and energy production and environmental protection: proceedings of the 2nd international nitrogen conference on science and policy, The Scientific World (1). 14–18 Oct 2001, Washington, DC in Potomac, Maryland, USA

Strategies for Higher Nutrient Use Efficiency and Productivity in Forage Crops

P.K. Ghosh, D.R. Palsaniya, A.K. Rai, and Sunil Kumar

Abstract

Adequate supply of quality forage is essential for sustainable livestock production and productivity. There is a net deficit of 35.6 % green fodder and 10.95 % dry fodder in India at present. This gap in forage supply can effectively be reduced through integrated crop management practices with greater emphasis on nutrient management. However, nutrient management is specific and dynamic in nature in forage crops viz-a-viz grain crops due to factors such as seasonality, perenniality, fodder as end product and multicut behaviour. High-intensity cropping has led to the multi-nutrient deficiencies in forage-based cropping systems. Area-specific nutrient management studies involving NPK along with S, Zn, B and Mo has been attempted to correct the deficiency and balancing the nutrient for quality fodder and livestock health. This chapter thoroughly reviews and discusses the nutrient management strategies in annual and perennial cultivated forage crops, range grasses and legumes, forage-based intercropping and cropping systems and rotations. Integration of secondary and micronutrients has been considered for increased quality biomass production. Application of 40 kg S/ha in sorghum shows significant increase in the Zn, Fe, Cu and cellulose content while decrease in NDF and ADF content of fodder. The efficiency of fertilizer N, P, K, S and micronutrients is reported as 50–60, 15–20, 60–80, 8–12 and 5 %, respectively, in most crops. The nutrient use efficiency can be increased further in forage crop-based cropping system by adopting conservation agriculture, balanced nutrient management, use of biological fertilizers and synergizing the cropping system approaches.

Keywords

Nutrient management • Forage crops • Cropping systems • Nutrient use efficiency • Forage production • Productivity • Soil health

P.K. Ghosh (✉) • D.R. Palsaniya • A.K. Rai • S. Kumar
Indian Grassland and Fodder Research Institute, Jhansi
284003, UP, India
e-mail: ghosh_pk2006@yahoo.com; drpalsaniya@gmail.
com; rai_arvindkumar@rediffmail.com;
sktiwari98@gmail.com

A. Rakshit et al. (eds.), *Nutrient Use Efficiency: from Basics to Advances*,
DOI 10.1007/978-81-322-2169-2_21, © Springer India 2015

1   Introduction

Forages include a variety of crops grown both under irrigated and rainfed conditions. The importance of forages in integrated farming system, crop diversification, watershed management, restoration of degraded lands and climate-resilient agriculture is increasingly being recognized. At present, India faces a net deficit of 35.6 % green fodder, 10.95 % dry crop residues and 44 % concentrate feed ingredients and is likely to encounter a demand of 1,012 and 631 million tonnes of green and dry fodder, respectively, by the year 2050 (IGFRI Vision 2050). Further, there are also seasonal and regional imbalances in the fodder production in the country. The gap in feed and forage supply can effectively be reduced through suitable soil and agronomic management options. Due to multiplicity of region- and season-specific nature, resource management in forages becomes dynamic in nature. The nutrient management practices in fodder crops slightly differ from that of food and other crops. The end product or economic part in forage crops is foliage which largely affects the nutrient management practices for these crops. Further, nutrient management in forage crops is also governed by soil type, cutting regimes, availability of water for irrigation, plant density, etc. The proper nutrient management of forage-based cropping system is one of the most crucial management practices to obtain higher yields and quality of forage species. The nutrient management strategies in forage crops aim at increasing herbage yield per unit area per unit time and also insure improved quality of forages for healthy and productive livestock (Menhi and Tripathi 1987).

2   Nutrient Management in Forages

The nutrient requirement and management strategies in forage crops are influenced by many factors like type of forage crop, variety (single, double or multicut), irrigation water availability, cutting management, soil type, crop rotations and cropping systems followed and other management practices. Indian soils have low total nitrogen; however, uptake of nutrients by forage crops is much higher. Therefore, forages are to be adequately supplemented with nitrogen through available organic and inorganic sources so that higher biomass is obtained from the unit piece of land. The nutrient requirement of forages under multicut system is much higher than under single cut. Several workers have worked out the nutrients schedule, sources of supplementation, application time and methods of application of NPK and sulphur to forage crops. The work on nutrition to forages with reference to secondary nutrients like Ca and Mg is meagre. However, with emerging multi-nutrient deficiencies in specific areas, NPK along with S-, Zn-, B- and Mo-based nutrient management has been attempted to correct the deficiency and balancing the nutrient for quality fodder and livestock health (Kumar and Faruqui 2010).

3   Nutrient Management in Cultivated Forage Crops

3.1   Primary Nutrients

3.1.1 Nitrogen (N)

N is the most important primary nutrient for forage crops, and its management has great significance due to its role in enhancing luxuriant vegetative growth, higher biomass and quick regeneration following cutting or defoliation. Further, optimum N nutrition improves leaf-stem ratio, succulence and palatability of forage crops. Studies conducted at IGFRI, Jhansi revealed that N application increases the crude protein, widens the ratio of true protein to NPN and increases NO_3^{-1} content and metabolizable energy of fodder (Mannikar 1980). Application of 100 kg N/ha in pearl millet increased the green fodder yield (29.2, 19.5 and 10.9 %) and dry matter yield (21.5, 16.0 and 8.7 %) over 25, 50 and 75 kg N/ha, respectively (Puri and Tiwana 2005). Similarly, Hazra and Tripathi (1994) worked out N requirement as 30, 60 and 90 kg/

ha for barley, oat and triticale, respectively. Further, the placement of nitrogen at 10 cm depth or its combination with foliar spraying improved the forage yield. Shukla and Lal (1994) obtained response of applied nitrogen up to 25 kg/ha to oat grown under rainfed condition. Among sources of N, CAN was found better than urea and FYM in acid soils (Tripathi and Mannikar 1985). Neem/Mahua-coated urea gave better response with combined application of urea and FYM on 50 % N basis along with P application (Tripathi et al. 1991). In round the year fodder production, the availability of soil nutrients (N and P) was more after *rabi* legume cultivation than after *kharif* legume system.

3.1.2 Phosphorus (P)

Phosphorous is especially critical at initial crop growth stage. Forage legumes require substantial amount of P for higher biomass yield and persistency. Several workers have reported positive response of P nutrition to cereals, legumes and cereal – legume mixtures in different agroclimatic situations. Doses, sources, time and method of application of P in forages have been thoroughly researched upon. Patel and Kotecha (2008) found that application of 40 and 80 kg P/ha increased dry matter yield of forage sorghum by 8.5 and 12.4 %, respectively, over control at Anand (Gujarat).

Band placement of fertilizer at 5 cm depth was better than traditional fertilizer application practices (surface broadcasting) or placing at 10 cm depth (Minhas and Gill 1984). In normal to alkaline soil, single super phosphate (SSP) and, in acid soil, basic slag and rock phosphate were reported to be superior for forages. In vertisol of Bundelkhand, combined form of organic (FYM) and inorganic P (SSP) source in 1:1 ratio showed better response (103.5 kg/kg P) in giving higher dry fodder yield (Tripathi and Hazra 1986). Application of SSP with rock phosphate + phosphoric acid (partially acidulated) in the ratio of 60:40 was found beneficial for direct effect on berseem and for the residual effect on maize (Marwah et al. 1981). In another study, Shukla et al. (1973) reported that young sorghum crop (30–40 days) developed HCN toxicity to animals. This can be reduced to safer limit by the application of P @ 50 kg P_2O_5/ha and irrigation at 50 % available soil moisture. Efficacy of P application during *rabi + zaid* season was higher than *zaid + kharif* and *kharif + rabi* season (Tripathi and Tripathi 2001).

3.1.3 Potassium (K)

Potassium is required in fairly large quantities by tropical grasses. However, K nutrition to cultivated forages is rarely reported. Perennial grass like napier bajra hybrid and guinea grass in association with *Leucaena leucocephala* under agroforestry system depleted available K content of the soil considerably, and the K fertilization was recommended (Rawat and Hazra 1990). Similarly, Menhi Lal and Tripathi (1987) elucidated the beneficial effect of K in holding higher concentration of nonstructural carbohydrates (soluble carbohydrates) in root, which is essential for regeneration of lucerne crop following cuttings. Tripathi et al. (2004) reported that combined application of 40 kg N/ha through half as urea and the rest half as FYM slurry along with 80 kg K_2O/ha produced significantly higher forage yield of *Cenchrus ciliaris + Stylosanthes hamata* grass (11.40 t/ha) with highest K (116 %) and N use efficiency (260 %). At Anand, Patel and Kotecha (2007) recorded maximum green fodder yield with the application of 300 kg K_2O/ha to the tune of 15.71 and 10.16 % higher than 50 kg and 150 kg K_2O/ha, respectively.

3.1.4 NPK Application

It was observed that combined application of N, P and K is more beneficial to all forages. The combined application of N and P yielded significant results in comparison to the application of N alone in oat with improvement in available N, P and organic carbon in calcareous red soil (Tripathi et al. 1989) and black soil (Tripathi et al. 1991). Nanjundappa et al. (1994) reported that application of 50–75 kg K_2O/ha decreased yield in the absence of N but increased yield when applied together with N. Uptake of K was increased by applied N but slightly decreased by K fertilizer. Sharma and Agrawal (2003a) reported that application of NPK (90, 40,

Table 1 Green, dry fodder and crude protein yield of sorghum as affected by primary nutrients (mean of 2 years)

Treatment	Green fodder yield (t/ha)	Dry fodder yield (t/ha)	Crude protein yield (kg/ha)
$N_0P_0K_0$	30.1	6.51	396.7
$N_{30}P_{20}K_{10}$	40.8	8.77	542.1
$N_{60}P_{30}K_{20}$	54.6	11.67	726.6
$N_{90}P_{40}K_{30}$	62.1	12.74	804.6
SEm	1.3	0.28	18.1
C.D. ($p = 0.05$)	2.6	0.57	36.9

30 kg/ha) under semiarid conditions significantly increased green fodder (62.1 t/ha), dry matter (12.74 t/ha) and crude protein yield (0.8 t/ha) in forage sorghum (Table 1).

3.2 Secondary Nutrients

Ca, Mg and S are essential secondary nutrients required by crops. The work on Ca and Mg nutrition to forages is meagre, though legumes in particular have high demand for calcium and magnesium. Sulphur is essential not only in increasing fodder yields but maintaining desired levels of protein, sugar, amino acid and mineral salts also. Sulphur nutrition is more important for maximum production particularly in crop where sulphur removal is more, viz., pasture legume (6.5–16.9 kg/kg S), than in cultivated legumes (4.6–7.6 kg/kg S). Legume removes 20–24 kg S/ha against 8–10 kg/ha by grasses (Singh et al. 1979). A series of experiments conducted at IGFRI (Hazra and Tripathi 1994; Rawat et al. 1999) on red sandy loam soils indicated higher response to S application for *rabi* fodder crops (15–71 kg dry fodder/kg S) than *kharif* (14–30 kg dry fodder/kg S) and *zaid* fodder crops (14–46 kg dry fodder/kg S).

Among different sources of S, SSP is the most effective source than pyrite, gypsum and elemental sulphur for cultivated rabi fodder crop (Hazra and Tripathi 1994). Effectiveness of sources like pyrite may be enhanced by addition of rock phosphate. The yield of range legumes like *Siratro* and *Stylsanthes* were improved by

34–40 %, respectively, with NPK fertilizer along with 20 kg S/ha over fertilizer NPK without S addition (Gill et al. 1986). In stress moisture condition, sorghum grown on S-deficient soil was found to have excess concentration of HCN. Application of S @ 30 kg/ha is advantageous in preventing and reducing toxic effect of HCN on animal growth (Singh 1992). The N:S ratio in important forage crops and range species has been well documented. The optimum N:S ratio for ruminants is considered to be 10:1 or less. Not much difference was observed among the sources of S in terms of its effect on sugar content, while ammonium sulphate was significantly superior to pyrite and elemental sulphur in methionine content. In general, the NDF and ADF content of fodder sorghum decreased and cellulose content increased with S addition (Tripathi et al. 1992). The application of S helped in the uptake of Zn, Fe and Cu in plants. Although there was a numerical increase from 20 to 80 kg/ha, it was in diminishing pattern after 40 kg/ha with significant increase in the plant contents of Zn, Fe and Cu up to 40 kg S/ha (Tripathi et al. 1992, 1993).

3.3 Micronutrients and Balanced Fertilization

Fe, Mn, Cu, Zn, B, Mo and Cl are micronutrients essential for forage crops. Balanced nutrition to forage crops plays a very important role in increasing fodder production, protein content, mineral contents and their ratios. Most of the range species, cereals straw (major source of forages) and wheat flour are comparatively low in Ca, Mg, Cu and Zn and considered as deficit in these nutrients, when compared to their critical levels for animal nutrition. Forage crops like oat, cowpea and berseem have optimum concentration of Cu and Zn. The Indian soils are becoming deficient of micronutrients like Zn, Mo and B in many pockets where intensive cultivation is practiced with straight fertilizers. The adoption of proper cropping systems (inclusion of legume crop in cropping cycle), use of farm yard manure, green manure crops, vermicomposting, use of

Table 2 Effect of S + micronutrients application on forage yield (q/ha) and some important fodder quality parameters (based on 3 years data)

Treatment	Green (dry) forage yield (q/ha)	Crude protein (%)	NDF (%)	ADF (%)	IVD (%)	N:S
T1 – Control-NPK	325.2 (84.7)	8.8	71.9	44.4	45.4	12.0
T2 – S	345.7 (91.3)	9.3	71.4	44.2	46.9	9.1
T3 – S + Zn	358.8 (95.5)	9.4	70.8	43.6	47.9	8.6
T4 – S + Zn + Mn	375.9 (98.7)	9.8	71.1	43.9	47.4	8.3
T5 – S + Zn + Mn + Cu	379.9 (100.5)	9.8	70.9	43.9	47.8	8.1
T6 – S + Zn + Mn + Cu + Mo	380.1 (99.8)	9.9	70.9	43.7	47.7	8.2
C.D. ($p = 0.05$) at 5 %	15.2 (3.2)	–	–	–	–	–

proper strain of biofertilizer, shifting from straight fertilizers to multiple NPK fertilizers, recycling of crop residues, industry by-products, selection of crops and their varieties according to soil fertility status and soil amendments may help in rectifying micronutrient deficiency in Indian soils.

The reports from various researchers indicated positive response of S, Zn and Mn at many locations. Verma et al. (2005) from Pantnagar reported optimum dose of N and Zn as 108.3 kg and 1.36 kg/ha, respectively, for sorghum; however, maximum yields were with the application of 126.67 and 5.53 kg/ha, respectively. Tripathi et al. (2007) while working at Jhansi (Table 2) reported that application of sulphur (@40 kg/ha), Zn (@20 kg/ha) and Mn (@10 kg/ha) along with recommended NPK to sorghum gave significantly higher yield by 16.5 % (dry fodder) over NPK alone (32.52 t/ha green and 8.48 t/ha dry fodder).

In most of the Zn-deficient situations, the yield increase ranged from 12 to 27 % with the application Zn. However, the combined application of Zn with gypsum + FYM further added 45 % yield over Zn application alone. In sodic and acidic soils, micronutrient-based studies have been conducted by many workers. Proper Cu and Mo ratio is of vital significance in pastures and feed. Application of Mo to first crop of berseem is sufficient for next crop of maize. In pasture grasses, the soil application of Zn, Fe, Cu, Mo and B along with nitrogen proved better in *C. ciliaris* (Hazra 1992). Micronutrients particularly Mo and B are found to influence the seed production positively in forage legumes. Application of B and Mo is beneficial for seed crop of berseem and lucerne in all growing areas.

At Palampur, in soil of pH 5–5.2, application of 5 t/ha of lime with 40 kg S/ha and 1 kg MoO_4/ha helped in seed production of red clover and lucerne (Kumar and Faruqui 2010). In acid soils of Jharkhand, application of lime @ 5 t/ha once in 3 years and Boron @ 3 kg/ha and 1 kg MoO_4/ha annually help in increased lucerne seed production by about 10 times (Prasad 2002).

3.4 Integrated Nutrient Management

Judicious integration of organic and inorganic sources of nutrients as well as bio-fertilizers is essential for optimum growth and quality of forages, sustainability of production system and soil health. The chemical fertilizers supply nutrient for immediate need of plants and give high production during initial years only, but factor productivity declines in subsequent years. Further, continuous application of chemical fertilizers alone in an intensive fodder production system deteriorates soil health and affects crop productivity. At Anand, Yadav et al. (2007) observed that by application of 75 kg N through urea + 25 kg N/ha through farm yard manure (FYM), there was an increase of 11.1–18.6 % in dry matter yield and 19.4–20.0 % in crude protein yield of sorghum over application of 100 kg N/ha through urea. Gangwar and Niranjan (1991) obtained 71.66 and 36.7 % increase in dry matter yield of sorghum with of 60 kg N + 13 kg P_2O_5 and 6 t FYM/ha, respectively, under rainfed condition at Jhansi. Similarly, Sunil Kumar et al. (2005) reported that application of vermicompost and

Table 3 Yield, quality and economics of forage oat as influenced by organic and inorganic sources of nutrient (pooled mean of 3 years)

| Treatments | Yield (q/ha) | | | Cost of cultivation (Rs/ha) | Gross return (Rs/ha) | Net return (Rs/ha) | Benefit/ cost ratio |
	Green fodder	Dry matter	Crude protein				
Control	21.7	4.7	3.8	6,110	10,850	4,740	0.78
Vermicompost 10 t/ha	32.8	7.5	7.1	10,096	16,400	6,304	0.62
FYM 10 t/ha	31.1	7.0	6.4	8,015	15,550	7,535	0.94
100 % NPK	37.6	8.9	8.6	7,225	18,800	11,575	1.60
50 % NPK	33.0	7.4	6.6	6,530	16,650	10,120	1.55
Vermicompost 5 t/ha	29.8	6.7	6.1	8,270	14,900	6,630	0.80
FYM 5 t/ha	26.3	5.8	5.3	7,375	13,150	5,775	0.78
50 % NPK + vermicompost 5 t/ha	35.7	8.8	8.4	9,279	17,850	8,571	0.92
50 % NPK + FYM 5 t/ha	35	8.6	8.5	7,680	17,500	9,820	1.28
50 % NPK + vermicompost 5 t/ha + FYM 5 t/ha	40.6	10.2	10.4	9,488	20,300	10,812	1.14
C.D. ($p = 0.05$)	4.5	1.1	0.6		1,582	560	0.04

FYM @ 5 t/ha to sorghum recorded higher NP uptake than other levels with inorganic sources only except the 100 % recommended dose of NP. At Coimbatore, Jayanthi et al. (2002) found that application of 50 % recommended NPK (40: 20: 0 kg/ha) fertilizer + vermicompost + FYM each at 5 tonnes/ha recorded significantly higher yield of oats. Similarly, application of N @ 150 kg/ha along with 40 kg P_2O_5/ha and dual inoculation of seed with *Azotobacter chroococcum* (N fixer) + *Pseudomonas striata* (phosphate solubilizer) in multicut fodder oat improved the vegetative growth.

Under arid situation of Rajasthan, Singh (2002) found that application of FYM @ 5 tonnes/ha to the preceding crop of cluster bean increased wheat-grain-equivalent yield by 20 % over control. The highest benefit/cost ratio (3.63) and net returns over the control (Rs 5,040) were also accrued with the application of FYM 5 @ tonnes/ha in cluster bean-wheat cropping systems. Pahwa (1995) from Jhansi reported an added benefit of combined inoculation of *Rhizobium trifolii* + *Azospirillum brasilense* as well as *R. trifolii* + *Azotobacter*. The higher nodule number (46/plant), green fodder (724.8 q/ha), dry fodder (93.1 q/ha) and crude protein yield (15.5 q/ha) was obtained with *Rhizobium* + *Azospirillum* inoculation in presence of 20 kg N/ha as compared to uninoculated control. Application of 60 kg P_2O_5/ha, phosphate solubilizing bacteria (*Pseudomonas striata*) and 11 t gypsum or 50 t FYM/ha or gypsum + FYM to berseem in highly sodic soils realized yields of 9.49, 1.74 and 31.89 t/ha, respectively (Sharma and Agrawal 2003b). Application of 20 kg N + 60 kg P + mixture of *Rhizobium trifolii* and phosphate solubilizing bacteria (PSB) recorded highest green fodder (65.45 t/ha), dry matter yield (16.98 t/ha) and protein content (19.71 %) of berseem (Meena and Mann 2006). At Jhansi, considering the yield and economics, application of 50 % N through FYM and rest 50 % NPK through inorganic fertilizer to berseem proved to be economically viable as compared to 100 % NPK through fertilizer (Kumar et al. 2007). Sunil Kumar and Shiva Dhar (2006) reported that application of 50 % recommended dose of NPK, 5 t/ha vermicompost and 5 t/ha FYM may be adopted for getting higher, sustainable and quality fodder from single cut oat under irrigated condition (Table 3).

Effect of inorganic and biofertilizer on napier bajra hybrid grass at Coimbatore revealed that highest green (323.9 t/ha) and dry fodder (79 t/ha) yield could be obtained with the application of biofertilizer mixture (*Azospirillium* + *Phosphobacterium*) along with 100 %

recommended dose of N and P fertilizer together (Chellamuthu et al. 2000). Application of 40 kg N/ha to the freshly introduced velvet bean of a natural grassland of Kangra Valley of Himachal Pradesh recorded 5.8 t/ha green forage herbage (Sood et al. 1994). Application of 50 % recommended N through inorganic sources along with 25 % vermicompost and 25 % sheep manure in cowpea + *Cenchrus* (2:1) ratio and in aonla-based intercropping under semiarid conditions improved the dry matter yield by 122.35 % over 100 % organic-inorganic supplementation (Meena et al. 2011). Fly ash application @ 50 t/ha in sorghum + cowpea (2:2) –oat (red soils) and sorghum + cowpea (2:2) – berseem (black soil) in combination with manure, fertilizer and biofertilizer registered significant increase in yield than no fly ash (Das et al. 2007).

Table 4 Biomass and crude protein yields of grass and legumes as influenced by N and P_2O_5

Treatments	Biomass yields (t/ha)		Crude protein yields (kg/ha)
	Green	Dry	
Crops			
TSH	38.7	9.58	717
Stylosanthes	23.4	6.62	654
TSH + stylosanthes	33.5	8.37	794
C.D. ($p = 0.05$)	1.8	0.31	27
Nitrogen level (kg/ha)			
0	26.4	6.69	602
30	30.3	7.76	683
60	33.4	8.71	561
90	37.5	9.61	842
C.D. ($p = 0.05$)	1.50	0.26	23
Phosphate level (kg/ha)			
0	29.2	7.30	659
30	31.8	8.27	724
60	34.7	9.00	783
C.D. ($p = 0.05$)	1.0	0.28	19

4 Nutrient Management in Range Grasses and Legumes

The application of fertilizer plays an important role in improving herbage productivity of range grasses and legumes. Nitrogen application in *Sehima*, *Heteropogon* and *Iseilema* grasslands significantly increased herbage production. The economically optimum dose was found to be in the range of 40–60 kg N/ha, the lower dose during periods of subnormal rainfall and the higher dose when more soil moisture is available. In northeastern region, Singh (1999) reported that Guatemala grass and guinea grass cv. Hamil, Gatton and Makueni responded up to 200 kg N/ha, while broom grass, guinea grass cv. PGG -1, palm grass and thin napier produced maximum dry matter at 100 kg N/ha. Application of N increased the specific root length and root length density in most of the grasses.

Earlier workers reported that response to applied phosphorus was of lower order while potassium did not play any role in increasing forage production in pastures (Dabadghao et al. 1965; Rai 1990; Kanodia 1995). Rathore et al. (1998) at Barmer (Rajasthan) reported that application of N + P (40,20) kg/ha to *C. ciliaris*

resulted in significantly high forage yield (4.66 t/ha) and crude protein (0.22 t/ha). In a natural grassland at Palampur, Premi and Sood (1999) observed that the application of 80 kg N + 60 Kg P/ha to *Setaria* produced 18.38 t/ha green fodder and 3.32 t/ha dry matter yield, which was 114.9 and 115 % higher over no fertilization. Sunil Kumar et al. (2007) also reported that application of 60 kg N + 40 kg P/ha to *Sehima* resulted in higher green fodder (21.3 t/ha), dry matter (8.2 t/ha) and crude protein yield (0.4 t/ha) registering an increase of 22.25 % over control. Niranjan et al. (2004) while working at Jhansi reported that application of 90 kg N and 60 kg P_2O_5/ha to both TSH (Trishankar hybrid) and *Stylosanthes* recorded maximum green and dry fodder and protein yield (Table 4). In range grasses and legumes, N supplementation plays a significant role in forage production. Split application of N was superior in coarse-textured soil while basal application proved better in fine-textured soil (Verma et al. 2005).

With the intercropping of *Stylosanthes* in Dinanath grass, saving of additional 40 kg N/ha was achieved (Prasad and Mukherjee 1987). In agroforestry system, an additional dose of

50 % N over recommended dose was found beneficial in cereal + legume mixture. Application of nitrogen @ 60 kg/ha improved the dry matter yields of *Cenchrus, Panicum* and *Setaria*. Sulphur-coated urea proved more effective over prilled urea. *Setaria* responded positively to N and P application up to 120 and 60 kg/ha, respectively (Rai and Kanodia 1981), while *Panicum maximum* responded significantly up to 80 kg/ha to biomass production and the CP content increased up to 120 kg/ha (Singh and Rai 1984). The result on method of application indicated the placement of fertilizer in root zone which is beneficial for cereal fodder crops. Similarly, Hazra and Tripathi (1989) noted the beneficial effect of P application to sweet clover on increased yield and soil parameters both under *Albizia lebbeck* tree and open-field situation at 90 kg P_2O_5/ha in red soils. The integration of forage bushes/perennial grasses in a fixed geometry successfully supplied the green fodder round the year from rainfed fields. The cultivation of sorghum (fodder) + pigeon pea (grain) in 3 m wide alleys formed with planting two rows each of subabul + TSH yielded 53.27 tonnes green and 13.28 tonnes dry matter/ha, respectively. The nutrient supplementation to the system as 75 % organic + 25 % inorganic sources was better as compared to other combinations of organics and in organics (Agrawal et al. 2007).

5 Nutrient Use Efficiency in Forages

The nutrient use efficiency of all the major, secondary and micronutrients is low in India and is a major challenge for sustainability and profitability. The efficiency of fertilizer N, P, K, S and micronutrients is reported as 50–60, 15–20, 60–80, 8–12 and 5 %, respectively, in most crops. Large acreage under rainfed cropping, unscientific water application in irrigated agriculture, imbalanced fertilization, tropical climate, cultivation of traditional crops and varieties, poor weed management and investment capacity of farmers result in low nutrient use efficiency in India. Adoption of best forage husbandry practices is a prerequisite for higher input use efficiency and profitability.

6 Strategies for Enhancing Nutrient Use Efficiency

Nutrient use efficiency depends on several agronomic factors including tillage, sowing time, planting or seeding techniques, appropriate crop variety, irrigation management, weed control and balanced and proper nutrient use. These factors largely influence nutrient use efficiency, either individually or collectively. Two pronged strategies can be adopted to improve nutrient use efficiency in forages as well: (a) *product strategy* where focus is on fertilizer product features such as coated fertilizers, slow-release fertilizers, nitrification inhibitors and urease inhibitors and (b) *management strategy* where focus will be on nutrient management practices like split application, balanced application, integrated application, soil or plant test-based application, conservation agricultural practices, variable rate technology or precision farming, etc.

6.1 Slow-Release Fertilizers

Application of slow or controlled release fertilizers (S/CRFs) is a very good approach for minimizing non point contamination in agriculture and achieving higher nutrient use efficiency. They release nutrients slowly as per the crop requirement and thus reduce the losses due to leaching, denitrification and volatilization. A number of fertilizer products supplying plant nutrients have been developed offering a variety of nutrient contents, physical forms and other properties to meet farmer's needs. Much of the work has been done on N and urea as source. The approaches have been to manipulate the granule size, coatings with neem or coal tar or sulphur or modifiers or additives to control the nutrient release rate (Aulakh and Malhi 2005).

A slow-release and superabsorbent N fertilizer (SSNF) was synthesized by Liu et al. (2005) which could improve both fertilizer and water use efficiency simultaneously. The neem-coated urea of National Fertilizers Limited recorded better shelf life, slow dissolution as well as nitrification inhibition property, and with marginal additional cost for coating, it increased NUE in Punjab, Uttar Pradesh and Himachal Pradesh (Mangat 2004). New gel-based CRFs were developed by mixing and processing N, P and K fertilizers with natural and seminatural organic materials and inorganic materials (Hong and Zhang 2006). The gel-based CRFs increased dry biological yield of maize by 26.8–42.3 % and improved N use efficiency by 17.0–31.7 %, P use efficiency by 8.0–16.0 % and K use efficiency by 4.6–18.3 %. Besides, the nutrients (N, P and K) in gel-based fertilizers were leached more slowly into the soil than common fertilizers. The IARI's urea coating technology employing neem oil emulsion needing 0.5–1.0 kg neem oil per tonne of urea was found superior to prilled urea (Prasad et al. 2001).

Large granular forms of N and P fertilizer found better than powdered and prilled forms for increasing nutrient uptake and yield. The nitrogen use efficiency can also be increased through suitable modification of urea by compacting it with acid- and non-acid-producing fertilizers such as NH_4Cl, KCl, $ZnSO_4$ and DAP or industrial by-product like PG. Compacting phosphogypsum (PG), diammonium phosphate (DAP), $ZnSO_4$ and KCl separately with urea slowed down urea hydrolysis and reduced NH_3 volatilization loss (Purakayastha and Katyal 1998). The polymer coating on monoammonium phosphate (MAP) in barley improved plant recovery of fertilizer P and provided modest grain yield advantage compared to uncoated MAP (Malhi et al. 2002).

6.2 Nitrification Inhibitors

The use of nitrification inhibitors contribute to increased NUE or apparent N recovery efficiency. Nitrification inhibitors (ethylene diamine-based chelating agents) like ethylene diamine tetra acetic acid (EDTA), diethylene triamine pentaacetic acid (DTPA) and ethylene diamine (EDA) inhibit ammonium oxidation (Hu et al. 2003). Subbarao (2006) found out that some forage crops, e.g. *Brachiaria* grasses, have the ability to regulate nitrification in soils by releasing inhibitors through root exudates which minimize N losses associated with nitrification and improve NUE. This self-generated inhibitory activity of plants is termed as biological nitrification inhibition (BNI). Maintenance of more NH_4^+ available in the soil might also increase P absorption and therefore increase P use efficiency (Ortega 2006). The use of ammonium N sources with the nitrification inhibitor 3, 4- dimethylpyrazole phosphate (DMPP) shows promise to increase NUE and PUE.

6.3 Nanotechnology for Higher Nutrient Use Efficiency

The nutrient use efficiency in major crops can be improved through nanotechnology. Nano fertilizers (scale below 100 microns) are being developed for slow release and efficient use of plant nutrients. The nanoporous zeolites are being utilized for slow release and efficient dosage of water and fertilizer for plants. Fudao et al. (2006) tried different nanoparticles (clay polyester, humus polyester and plastic starch) in China for slow release of N to wheat and reported 4.5 % increase in yield due to clay and plastic (nanomaterial coating) over chemical fertilizer application. At the same time, it also reduced leaching losses of N fertilizer.

6.4 Crops and Cropping Systems for High Nutrient Use Efficiency

Different crops, varieties and cropping systems differ in their nutrient use efficiency capability. The beneficial effect of important legumes on increasing productivity and nutrient use efficiency in various systems was recently reviewed by Ghosh et al. (2007). Fodder legumes in general are more potent in increasing the productivity of succeeding cereals. Symbiotic N_2 fixation with compatible rhizobial strains resulted in carryover of N for succeeding crops to the tune of 60–120 kg in berseem, 75 kg in Indian clover, 75 kg in clusterbean, 35–60 kg in fodder cowpea, 68 kg in chickpea, 55 kg in urdbean, 54–58 kg in groundnut, 50–51 kg in soybean, 50 kg in *Lathyrus* and 36–42 kg/ha in pigeon pea. Legumes with indeterminate growth habit are more efficient in N_2 fixation than determinate types.

6.5 High Nutrient Use Efficiency Through Organic Farming and Integrated Nutrient Management

Organic sources of plant nutrients maintain buffering capacity, physical and biological properties of soil and also affect soil temperature and moisture and influence nutrient release and adsorption in soil. The build-up or depletion of soil organic matter affects N mineralization and immobilization potential, which in turn controls nutrient availability and use efficiency. It was found that interactions among inputs (manure, cover crops and fertilizer) and soil organic matter influence the rate of soil N mineralization (Horwath et al. 2006). It is also well known that N from many organic fertilizers often shows little effect on crop growth in the year of application because of the slow-release characteristics of organically bound N. Nitrogen immobilization after application can occur, leading to enrichment of the soil N pool. This process increases the long-term efficiency of organic fertilizers.

6.6 Balanced Fertilization for Higher Nutrient Use Efficiency

Balanced fertilizer use is a prerequisite for high nutrient efficiency. With the adoption of exhaustive cropping systems, widespread deficiencies of S, Zn and B and sporadic deficiencies of Fe and Mn have been noticed in intensively cultivated areas. Acharya and Sharma (2008) made a detailed review on integrated input management for improving N use efficiency and crop productivity. Application of S along with N, P and K to pulses and oilseeds showed greater response than cereals. Sulphur not only improved grain yield but also improved the quality of crops (Hegde and Sudhakara Babu 2004). Interactions of S with N and P are positive, while its application decreases the contents of Zn, B and Mo in plant system.

6.7 High Nutrient Use Efficiency Through Conservation Agriculture

Conservation agricultural practices like reduced tillage, crop rotations, residue management, mulching and cover cropping are known for enhancing nutrient and other input use efficiencies. Nitrogen use efficiency was lower on zero till plots when straw was either burned or removed compared with straw incorporation and straw mulch treatments, particularly at low N rates (Singh et al. 2006). A higher N recycling efficiency was observed when residues were incorporated (Sakonnakhon et al. 2005). Alfalfa mulch holds promise for low-input cropping systems, when used on wheat-oat cropping system @ 3.9–5.2 t/ha. It registered higher N uptake and grain yield compared with ammonium nitrate. Nitrogen use efficiency of mulch supplied N by wheat and oats was between 11 and 68 %. The higher nutrient use efficiency achieved through mulching was due to weed suppression and increased soil moisture conservation (Wiens et al. 2006). Nitrogen use efficiency for maize could be improved when *Gliricidia sepium* prunings were incorporated 4 weeks ahead of maize planting. Addition of small doses of inorganic N fertilizer increased N uptake and yield (Makumba et al. 2006).

Reducing tillage and optimizing N fertilization are important strategies for soil and water conservation and N use efficiency for sustainable agriculture. Based on soil type, it is clear that no tillage system does not require increased amounts of N fertilizer for barley production (Angas et al. 2006). Conservation tillage and deep tillage increased N, P and K uptake compared to minimum tillage. Availability of nutrients and water at different growth stages of sorghum was increased by deep tillage (Patil and Sheelavantar 2006). Nitrogen use efficiency and N uptake efficiency were greater with conventional tillage than with no tillage (López-Bellido and López-Bellido 2001). In general, crop yields under no-till practices are more stable than under tilled systems with greater efficiency in the use of nutrients (Martin 2006).

Improper land levelling is the serious cause for losses in water, nutrients and erratic population resulting in low yields and profits. Choudhary et al. (2002) observed higher fertilizer use efficiency in wheat in laser-levelled fields. On-farm investigations in western Uttar Pradesh showed significant improvement in NUE in rice-wheat cropping system from 45.1 to 48.4 and 34.7 to 36.9 kg grain/kg applied N in rice and wheat, respectively. A significant increase in the uptake efficiency as well as apparent recovery fraction of the applied N, P and K in rice was observed due to precision land levelling (Pal et al. 2003).

7 Future Thrust Areas in Nutrient Management in Forage Crops

The relevance of nutrient management in present day forage production systems is increasing fast, and their management for integrated crop-livestock model needs more in-depth system analyses for greater production sustainability and quality forage. From the above discussion, it is imperative to conclude that nutrient management research on forages has made great strides over the last few decades. However, in the changing scenario, following thrust areas are identified for making forage nutrient management research more rewarding and problem solving in nature.

- Agro techniques for higher nutrient use efficiencies for forage and food-fodder cropping systems
- Nutrient management practices especially under tree-forage crop-food crop-livestock-based integrated farming system situations

and understanding the nutrient recycling and balances

- Residual effect of nutrients especially under fodder-based long-term cropping sequences
- Soil microbial consortia and rhizospheric interactions studies for enhancing nutrient acquisitions
- Soil test-based balanced use of nutrients for sustainable herbage production
- Nutrient dynamics studies in silvipasture, hortipasture and perennial fodder production systems
- Nutrient management studies in periurban forage production with special emphasis on use of sewage and sludge and heavy metal toxicities
- Nutrient interaction studies under intensive forage-based cropping systems
- Nutrient management strategies with reference to climate change

References

Acharya CL, Sharma AR (2008) Integrated input management for improving nitrogen-use efficiency and crop productivity. Indian J Fertil 4(2):33–40, 43–50

Agrawal RK, Niranjan KP, Rai SK (2007) Evaluation of perennial grasses based cropping system for prolonging forage productivity under semi-arid rainfed situation. Range Manag Agrofor 28(2B):386–387

Angas P, Lampurlanes J, Cantero MC (2006) Tillage and N fertilization: effects on N dynamics and barley yield under semiarid Mediterranean conditions. Soil Tillage Res 87(1):59–71

Aulakh MS, Malhi SS (2005) Interactions of nitrogen with other nutrients and water: effect on crop yield and quality, nutrient-use efficiency, carbon sequestration and environmental pollution. Adv Agron 86: 341–409

Chellamuthu V, Khan AKF, Malarvizhi P (2000) Studies on the effect of inorganic and biofertilizers on Bajra-Napier hybrid grass. Range Manag Agrofor 21: 135–138

Choudhary MA, Mushtaq A, Gill M, Kahlown A, Hobbs PR (2002) Evaluation of resource conservation technologies in ricewheat system of Pakistan. In: Rice-wheat consortium paper series 14, New Delhi, India. Rice–wheat Consortium for the Indo-Gangetic Plains, p 148

Dabadghao PM, Chaturvedi RB, Das RB, Debroy R, Marwaha SP (1965) Response of some promising desert grasses to fertilizer treatments. Ann Arid Zone 4:120–135

Das SK, Bhatt RK, Yadava RB, Suresh G, Kareemulla K, Rai AK, Mojumder AB, Pathak PS, Singh DK, Singh MK (2007) Effect of fly ash application on forage productivity, nutrient content and physiology of sorghum- cowpea intercrops. Range Manag Agrofor 28(2):406–408

Fudao Z, Rutang W, Qiang X, Yajun W, Jianfeng Z (2006) Effects of slow/controlled release fertilizer cemented by nano materials on biology. II. Effects of slow/controlled release fertilizer cemented and coated by nano materials on plants. Nanoscience 11(1):18–26

Gangwar KS, Niranjan KP (1991) Effect of organic manures and inorganic fertilizers on rainfed fodder sorghum (Sorghum bicolor). Indian J Agric Sci 61 (3):193–194

Ghosh PK, Bandyopadhyay KK, Wanjari RH, Manna MC, Misra AK, Mohanty M, Rao AS (2007) Legume effect for enhancing productivity and nutrient use-efficiency in major cropping systems – an Indian perspective: a review. J Sustain Agric 30(1):61–86

Gill AS, Tripathi SN, Raut MS, Gangwar KS (1986) Sulphur fertilization in forage crops. Paper presented at the National seminar in advances in forage agronomy and future strategy for increasing biomass production held at Jhansi during October 6–7, 1986

Hazra CR (1992) Fertilizer use in forage, pasture and grassland production. In: Tandon HLS (ed) - Non-traditional sectors for fertilizer use. FDCO, New Delhi, pp 30–47

Hazra CR, Tripathi SB (1989) Soil properties and forages yield of sweet clover as influenced by phosphate application under agroforestry. Forage Res 15(I):69–78

Hazra CR, Tripathi SB (1994) Effect of nitrogen fertilization to Anjan grass (Cenchrus ciliaris) with and without pasture legumes in silvipasture. Forage Res 20:287–290

Hegde DM, Sudhakara Babu SN (2004). Role of balanced fertilization in improving crop yield and quality. Fertil News 49(12):103–110, 113–114, 131

Hong D, Zhang Y (2006) Effect of Gel-based controlled release fertilizers on crop yield and nutrient-use efficiency. In: 18th world congress of soil science, held during 9–15 July, Philadelphia, Pennsylvania, USA

Horwath W, Kabir Z, Reed K, Kaffka S, Miyao G, Kent (2006) Long-term assessment of N use and loss in irrigated organic, low-input and conventional cropping systems. In: 18th world congress of soil science, held during 9–15 July, Philadelphia, Pennsylvania, USA

Hu Z, Kartik C, Domenico G, Barth FS (2003) Nitrification inhibition by ethylenediamine- based chelating agents. Environ Eng Sci 20(3):219–228

Jayanthi C, Malarvizhi P, Khan AKF, Chinnusamy C (2002) Integrated nutrient management in forage oat (Avena sativa). Indian J Agron 47(1):130–133

Kanodia KC (1995) Forage production from degraded land in tropical environment. In: Hazra CR,

Misri BK (eds) New vistas in forage production. IGFRI, Jhansi, pp 73–96

Kumar S, Dhar S (2006) Influence of organic and inorganic sources of nutrients on forage productivity and economics of oat (Avena sativa L.). Ann Agric Res 27(3):205–209

Kumar S, Faruqui SA (2010) Forage production technologies for different agro-ecological regions, Tech. Pub. No. 01/2010. AICRP on Forage Crops, IGFRI, Jhansi, p 64

Kumar S, Rawat CR, Dhar S, Rai SK (2005) Dry matter accumulation, nutrient uptake and changes in soil fertility status as influenced by different organic and inorganic sources of nutrients to forage sorghum (Sorghum bicolor). Indian J Agric Sci 75(6): 340–342

Kumar S, Roy AK, Faruqui SA (2007) Performance of Sehima nervosum genotypes under different fertility levels and cutting intervals in semiarid region. Range Manag Agrofor 28(2):375–376

Liu LJ, Xu W, Tang C, Wang ZQ, Yang JC (2005) Effect of indigenous nitrogen supply of soil on the grain yield and fertilizer-N use efficiency in rice. Rice Sci 12: 267–274

López-Bellido RJ, López-Bellido L (2001) Efficiency of nitrogen in wheat under Mediterranean conditions: effect of tillage, crop rotation and N fertilization. Field Crops Res 71(1):31–46

Makumba W, Janssen B, Oenema O, Akinnifesi FK (2006) Influence of application of Gliricidia prunings as a source of N on the performance for maize. Exp Agric 42(1):51–63

Malhi SS, Haserlein DG, Pauly DG, Jhonston AM (2002) Improving fertilizer phosphorus use efficiency. Better Crops 86(4):8–9

Mangat GS (2004) Relative efficiency of NFL [National Fertilizers Limited] – neem coated urea and urea for rice. Fertil News 49(2):63–64

Mannikar ND (1980) Fertilizer use efficiency in relation to fodder crops. Paper presented in seminar "On Maximum Fertilizer Use Efficiency", Ministry of Agriculture, FAO, Norway, New Delhi, 15–19 Sept

Martin DZ (2006) Innovations for Improving productivity and nutrient-use efficiency: no-till grain cropping systems of South America. In: 18th world congress of soil science, held during 9–15 July, Philadelphia, Pennsylvania, USA

Marwah BC, Kanwar BS, Tripathi BR (1981) Direct and residual effect of Mussoorie rock phosphate related to crop science in acid soil. J Indian Soc Soil Sci 29: 349–355

Meena LR, Mann JS (2006) Strategic nutrient supplementation in berseem for higher biomass productivity and economic return under semiarid conditions. Range Manag Agrofor 27(1):40–43

Meena LR, Mann JS, Meena KR (2011) Performance evaluation of cowpea and Cenchrus setigerous intercropping and nitrogen supplementation through organic and inorganic sources in Aonla (Emblica officinalis Gaertn.) based horti – pasture system. Range Manag Agrofor 32(1):33–39

Menhi L, Tripathi SN (1987) Role of fertilizer in production of forage and fodder crops: gaps and future needs. Fertil News 32(12):77–83

Minhas PS, Gill AS (1984) Forage production in sorghum-oats rotation as effected by tillage and methods of fertilizer application under rainfed conditions. Int J Trop Agric 2(3):245–250

Nanjundappa G, Manure GR, Badiger MK (1994) Yield and uptake of fodder maize (Zea mays) as influenced by nitrogen and potassium. Indian J Agron 39(3): 473–475

Niranjan KP, Burman D, Arya AL, Agrawal RK, Dhar S (2004) Influence of nitrogen and phosphate levels on forage production of Pennisetum trispecific hybrid and Stylosanthes hamata. Range Manag Agrofor 25(2): 98–101

Ortega R (2006) Increasing nitrogen and phosphorus fertilizer- use efficiency by using the nitrification inhibitor 3, 4- dimethylpyrazole phosphate (DMPP) in Chile. In: 18th world congress of soil science, held during 9–15 July, Philadelphia, Pennsylvania, USA

Pahwa MR (1995) Biofertiliser for nutrient economy and forage production. In: Singh RP (ed) Forage production and utilization. IGFRI, Jhansi

Pal SS, Jat ML, Subba Rao AVM (2003) Laser land levelling for improving water productivity in rice-wheat system. PDCSR Newsletter, Modipuram

Patel P, Kotecha AV (2007) Effect of phosphorus and potassium on fodder production, quality and nutrient uptake of Lucerne. Range Manag Agrofor 28(2): 411–413

Patel P, Kotecha AV (2008) Influence of P and S nutrition on yield and quality of forage Sorghum. Range Manag Agrofor 29(1):53–57

Patil SL, Sheelavantar MN (2006) Soil water conservation and yield of winter sorghum as influenced by tillage, organic materials and nitrogen fertilizer in semi arid tropical India. Soil Tillage Res 89(2):246–257

Prasad NK (2002) Fifty years of forage & agroforestry research. Directorate of Research, Birsa Agricultural University, Ranchi, p 53

Prasad LK, Mukherjee SK (1987) Effect of nitrogen and phosphorous on the herbage yield of stylo and stylo grass association. Forage Res 13(1):45–48

Prasad R, Sharma SN, Singh S, Saxena VS, Shivay YS (2001) Pusa neem emulsion as an ecofriendly coating agent for urea quality and efficiency. Fertil News 46 (7):73–74

Premi OP, Sood BR (1999) Forage yield and seasonal distribution as influenced by the introduction of improved grass, legume and tree components. Range Manag Agrofor 20(2):115–119

Purakayastha TJ, Katyal JC (1998) Evaluation of compacted urea fertilizers with acid and non-acid producing chemical additives in three soils varying in pH and cation exchange capacity. I. NH3 volatilization. Nutr Cycl Agroecosyst 51:107–115

Puri KP, Tiwana US (2005) Response of pearl millet varieties to nitrogen levels under irrigated conditions. Range Manag Agrofor 26(2):124–126

Rai P (1990) Effect of row spacing on productivity of Anjan grass (Cenchrus ciliaris) with Caribbean stylo under different level of phosphorus. Ann Arid Zone 29 (1):29–33

Rai P, Kanodia KC (1981) Response of Setaria sphacelata nitrogen and phosphorus under rainfed conditions. Indian J Agron 26(2):205–206

Rathore SS, Bohra HC, Bhati TK (1998) Forage and protein yield in Cenchrus ciliaris. Range Manag Agrofor 19:29–31

Rawat CR, Hazra CR (1990) Herbage productivity from subabul (lucerne leucaphala) based agroforestry system under limited irrigation. Forage Res 16(1):31–37

Rawat CR, Kumar S, Arya ON (1999) Effect of sulphur fertilization on productivity of fodder oats. Forage Res 25(2):145–147

Sakonnakhon SPN, Toomsan B, Cadisch G, Baggs EM, Vityakon P, Limpinuntana V, Jogloy S, Patanothai A (2005) Dry season groundnut stover management practices determine nitrogen cycling efficiency and subsequent maize yields. Plant Soil 272(1/2): 183–199

Sharma KC, Agrawal RK (2003a) Effect of chemical fertilizers and vermicompost on the productivity and economics of forage sorghum and their residual effects on oat. Range Manag Agrofor 24 (2):127–131

Sharma KC, Agrawal RK (2003b) Effect of phosphate and phosphate solubilizing bacteria on the productivity and economics of Egyptian clover (Trifolium alexandrinum). Range Manag Agrofor 24(1):49–52

Shukla NP, Lal M (1994) Response of oat (Avena sativa) to nitrogen in relation to moisture conservation techniques under restricted irrigations. Indian J Agron 39 (2):229–232

Shukla NP, Lal M, Lal R (1973) A note on the effect of soil moisture stress on HCN content of MP Chari (Sorghum bicolor). Indian J Agric Sci 43:977–979

Singh KC (1992) Grassland and pasture development in the arid regions of Rajasthan. In: Kaplarkar AS, Joshi DC, Sharma KD (eds) Rehabilitation of degraded arid ecosystems. Scientific Publishers, Jodhpur, pp 121–126

Singh KA (1999) Response of promising forage grasses to nitrogen. Indian J Agron 44(2):419–423

Singh R (2002) Effect of integrated nutrient management in cluster bean (Cyamopsis tetragonoloba)-wheat (Triticum aestivum) cropping system under western Rajasthan condition. Indian J Agron 47(1):41–45

Singh KA, Rai P (1984) Response of forage Panicum species to nitrogen fertilization on semi-arid lands. Indian J Agric Sci 54(5):382–386

Singh D, Mannikar ND, Srivas NC (1979) Comparative performance of indigenous rock phosphates and super phosphate in forage legume cropping pattern. J Indian Soc Soil Sci 27(2):170–173

Singh Y, Singh B, Ladha JK, Gupta R, Pannu R (2006) Tillage and residue management effects on yield and nitrogen-use efficiency in wheat following rice in the Indo-Gangetic plains of India. In: 18th world congress of soil science, held during 9–15 July, Philadelphia, Pennsylvania, USA

Sood BR, Sharma VK, Kumar P (1994) Performance of some improved grass cultivars under tree canopy and open grassland of Kangra valley. In: International conference on sustainable development of degraded land through agroforestry in Asia and Pacific, New Delhi, pp 700–704

Subbarao GV (2006) Biological nitrification inhibition (BNI): is it a widespread phenomenon? Plant Soil 294(1–2):5–18

Tripathi SB, Hazra CR (1986) Effect of P from different sources on the herbage yield of seed clover and soil properties. J Indian Soc Soil Sci 7 (11):809–814

Tripathi SB, Mannikar ND (1985) Forage production of berseem in acid soils in relation to P fertilization. Forage Res 11(2):127–131

Tripathi SB, Tripathi SN (2001) Sulphur in soils and its effect on yield and quality of fodder oats in sole and intercropping system. Range Manag Agrofor 22(1): 20–27

Tripathi SB, Mannikar ND, Hazra CR (1989) Effect of molybdenum and phosphate on berseem and residual effect on fodder maize in acid soil. J Indian Soc Soil Sci 37(1):200–201

Tripathi SB, Hazra CR, Srivas NC (1991) Effect of N sources with and without phosphorus on oats. Indian J Agric Res 25(2):78–84

Tripathi SB, Singh RS, Tripathi RK (1992) Effect of S fertilization on quality constituents of fodder sorghum cv. MP. Chari. Forage Res 18:9–14

Tripathi SB, Singh RS, Tripathi RK (1993) Influence of fertilizer sulphur on fodder quality of summer cowpea (Vigna sinensis L.). Forage Res 19:205–211

Tripathi SB, Pahwa MR, Singh A, Patra AK (2004) Effect of Nitrogen and potassium fertilization on herbage yield of Cenchrus ciliaris + Stylosanthes hamata mixed pasture and soil fertility. Range Manag Agrofor 25(2):92–97

Tripathi SB, Tripathi SN, Singh KK (2007) Influence of sulphur and micronutrients application on yield and quality of fodder sorghum grown with and without FYM. Range Manag Agrofor 28:162–164

Verma SS, Singh N, Joshi YP, Deorari V (2005) Effect of nitrogen and zinc on growth characters, herbage yield, nutrient uptake and quality of fodder sorghum (Sorghum bicolor). Indian J Agron 50(2):167–169

Wiens MJ, Entz MH, Martin RC, Hammermeister AM (2006) Agronomic benefits of alfalfa mulch applied to organically managed spring wheat. Can J Plant Sci 86 (1):121–131

Yadav PC, Sadhu AC, Swarnkar PK (2007) Yield and quality of forage sorghum (Sorghum sudanense) as influenced by the integrated N management. Indian J Agron 52(4):330–334

Integrated Nutrient Management in Potato for Increasing Nutrient-Use Efficiency and Sustainable Productivity

D.C. Ghosh

Abstract

Sustainable agricultural production uses the natural resources to generate increased output and income, especially for low-income groups, without depleting the natural resource base. In this context, integrated nutrient management (INM) integrates the use of all natural and man-made sources of plant nutrients in an efficient and environmentally benign manner to increase crop productivity without sacrificing soil productivity for future generations. Sufficient and balanced application of organic manures and inorganic fertilizers is a major component of INM in potato. The integrated use of organic manures, chemical fertilizers and biofertilizers in potato was found encouraging not only for increasing tuber productivity and nutrient-use efficiency but also for improving soil fertility and sustaining crop productivity at high level.

Keywords

INM • Potato productivity • Nutrient-use efficiency • Sustainability

1 Introduction

The world population is estimated to rise to 9–10 billion by 2050. The proportion of land devoted to agriculture is generally more than one-third of the land in most countries, and there is no scope to further increase the availability of agricultural land. On the other hand, there is an increasing pressure on land to build new homes, public institutions (schools, colleges, health centres, community halls, etc.), roadways, railways and others to accommodate the growing which and that may decrease the availability of agricultural land in the future. The global food security is, thus, heavily dependent upon technological advances in order to avoid Malthusians scenario of poverty and famine due to overpopulation (Hole et al. 2005). The effective maintenance of soil quality, residue and waste management, water infiltration, soil moisture retention, runoff and erosion control and carbon sequestration are all dependent on sustainable agricultural systems. Increased and sustainable food production is a must for food security of the growing

D.C. Ghosh (✉)
Institute of Agriculture, Visva-Bharati, Sriniketan-731236, Birbhum, West Bengal, India
e-mail: dcghosh2011@gmail.com

A. Rakshit et al. (eds.), *Nutrient Use Efficiency: from Basics to Advances*,
DOI 10.1007/978-81-322-2169-2_22, © Springer India 2015

population. We have to search for possible ways of increasing food production in a sustainable manner without deteriorating the soil quality. Modern large-scale conventional agriculture with intensive monoculture often results in unacceptable soil erosion, in runoff and in associated losses of plant nutrients. The highly productive fertilizer and seed technologies introduced over the past four decades are now reaching almost a point of diminishing returns (Cassman et al. 1995; Dawe et al. 2000). Possibilities of converting marginal lands into productive arable land (Crosson and Anderson 1992; Das et al. 2000; Karforma et al. 2012) as an option for yield improvement are now becoming more and more limited. The genetically engineered plants also may not be the major factors in increasing food production in the near future (Peng et al. 1994; Hazell 1995). A sustainable system that can maintain crop productivity at higher level without deteriorating the ecosystem is the need of the hour (Dobermann and White 1999).

Organic farming has potential for reducing some of the negative impacts of conventional agriculture on the environment. The nutrient availability in soil can be increased by improving organic carbon status of the soil through the use of organic manures and green manures (Trehan and Grewal 1983). Organic farming avoids depletion of soil organic matter and plant nutrients (Gaur 2001). It can reduce the need of chemicals for pest and disease control. Govindakrishnan and Kushwah (2003) and Kumar et al. (2005) recorded prolonged effect of organic manures on soil fertility and soil moisture balance. Hole et al. (2005) also identified a wide range of beneficial effects of organic farming on agriculture. But organic farming alone may not meet the increased food requirement and other problems in agriculture.

Soils in India are often not only 'thirsty' but also 'hungry'. What we need is a reduction in the use of costly market-purchased chemical inputs. In this context, low-cost integrated system of nutrient supply suitable for easy adoption like crop rotations, green manure, biofertilizers and biodynamic systems that make significant use of

compost and humus will help to improve soil structure and fertility (Corselius et al. 2001; Anonymous 2003). Highly weathered (*oxisols/ultisols/alfisols*) soils, being inherently low in nutrient reserves, must have a regular nutrient supply to facilitate intensive cultivation for increased food production. Intensive land use and high yields on soils of low inherent fertility can be achieved only by raising the nutrient levels through the use of organic amendments, nutrient recycling, biofertilizers and inorganic fertilizers (Barrios et al. 2006). Many small land holders and resource-poor farmers can afford the expense of this low-cost nutrient management technology. Further, it can avoid the overdependence on synthetic fertilizers and other agricultural chemicals in crop production (Place et al. 2003; Mugwe et al. 2009a). A combination of organic amendments, nutrient recycling, inorganic fertilizers and biofertilizers is, therefore, a useful strategy to minimize dependence on synthetic fertilizers in increasing crop yield and enhancing soil structure, soil fertility and sustainability (Dixit and Gupta 2000; Cobo et al. 2010; Kato and Place 2011).

The accelerated agricultural progress based on the principles of sustainable intensification, value addition and diversification is the best safety against hunger and poverty in rural areas. It is essential to improve soil fertility and soil health through better utilization of natural resources along with balanced use of chemical fertilizer and other inputs (Swaminathan 2004). An appropriate combination of organic and inorganic sources of plant nutrients has been found to be the best option for increasing crop productivity, improving soil quality and maintaining sustainability. Sustainability of agricultural productivity and environmental safety are the priority issues for increasing the food production to fulfil the food requirement of the burgeoning population. There is a need to develop and demonstrate the balanced use of organic manures and chemical fertilizers to avoid wastages of precious natural resources and to minimize the environmental damage. This will not only improve the crop production in sustainable way but also economize the production system. Higher food production needs higher amount of plant nutrients. As no single source is capable of supplying the required amount of plant nutrients, integrated use of all sources of plant nutrients is a must to supply the balanced nutrition to crops (Arora 2008; Ngetich et al. 2011).

Integrated nutrient management (INM) refers to making the best use of inherent soil nutrient stocks, locally available soil amendments (for instance, crop residues, compost, animal manure, green manure) and inorganic fertilizers to increase productivity while maintaining or enhancing the agricultural resource base (Nkonya et al. 2004). It is a holistic approach to soil fertility research that embraces the full range of driving factors and consequences – biological, physical, chemical, social, economic and political – of soil degradation (Barrios et al. 2006). Strategically targeted fertilizer use together with organic nutrient resources to ensure fertilizer use efficiency and crop productivity at farm scale are basic principles of INM (Vanlauwe and Giller 2006). Although INM recognizes the absolute necessity of mineral fertilizer use (Rufino et al. 2006), it advocates the best combination of available nutrient management technologies that are economically profitable and socially acceptable to different categories of farmers (Gentile et al. 2009). It is rapidly becoming more accepted by development and extension programs and, most importantly, by smallholder farmers (Rufino et al. 2006; Mugwe et al. 2009b). Beneficial effects of INM on soil fertility have been shown to increase nutrient-use efficiency associated with combined nutritional and non-nutritional effects of organic and inorganic inputs compared to inorganic fertilizer applied alone (Sanginga and Woomer 2009; Bekunda et al. 2010; Kumar et al. 2012).

Potato (*Solanumtuberosum* L.) gives exceptionally high yield per unit area and time than many other crops. It is cultivated over an area of 18.6 million hectares in 150 countries of the world with a total production of 322 million tonnes (Singh 2008). It is widely cultivated in China, India, Russian Federation, Poland, USA, Germany and Spain. It plays an important role in food security of the world population. In India, potato is cultivated in about 1.86 million hectares

with a total production of about 42.34 million tonnes (Anonymous 2012). Potato is a short duration heavy feeder crop and takes 90–100 days to produce 25–30 t/ha tubers in the subtropical plains. A mature potato crop yielding 25–30 t/ha removes about 165–200 kg N, 14–17 kg P and 185–225 kg K/ha (Baishya et al. 2010b). Therefore, high amount of chemical fertilizers is generally applied in potato. Indiscriminate use of chemical fertilizers and adoption of nutrient-responsive high-yielding varieties though boost up potato production cause degradation of soil and environment (Ali et al. 1997; Mondal et al. 2007; Islam et al. 2009). The introduction of input-intensive agricultural technologies, in which high-yielding varieties, water supply and vast amount of only a few major nutrients were added round the year, led to increase not only the yields of various crops but also mining of all kinds of plant nutrients. This resulted in decrease in soil organic matter, imbalances in plant nutrients in soil and decline in soil fertility. The chain reaction of undesirable consequences of high input agriculture is becoming insurmountable. The soil fertility and environmental quality can be restored by integrated nutrient management (INM). The INM systems envisage the use of different sources of plant nutrients such as organic manures, chemical fertilizers and biofertilizers. A number of diverse organic sources are available for use in agriculture. These sources can reduce the mining of soil nutrient and improve overall soil productivity (Thind et al. 2007; Khan et al. 2008; Kumar et al. 2012). This system improves the physical condition of the soil such as structure, aggregate stability, soil moisture retentivity and hydraulic conductivity that contribute to soil fertility and productivity. In acid soils, biofertilizers (*Azotobacter* and phosphate-solubilizing bacteria, PSB) play an important role in the productivity of potato (Singh 2002). The beneficial effects of organic manuring are manifested through increase in soil organic matter and humus over the period. Soil organic matter and humus acts in several ways; it serves as slow-release source of plant nutrients and increases water-holding capacity to help maintain the water regime of the soil, and it acts as a buffer against change in pH of the soil (Upadhayay and Singh 2003). Its dark colour increases absorption of sun energy and regulates heat of the soil. It acts as 'cement' for water-holding clay and soil particles together; this contributing to the crumb structure of the soil providing resistance against soil erosion binds micronutrient metal ions in the soil to check leaching out of surface soils. Organic constituents in the *humic* substances also act as plant growth stimulants (Singh et al. 1997; Thind et al. 2007). However, organic source alone may not supply the huge amount of N required for potato within short period. Even in N-rich soils of temperate and subtropical high-altitude region, potato requires chemical fertilizers because of slow mineralization rate of organic matter due to low temperature prevailing during its growing period.

Phosphorus (P) is regarded as the most important soil nutrient after nitrogen (N) for plant growth and development. It plays key roles in plant metabolism, structure and energy transformation. The P dynamics in soils and cycling in agro-ecosystems are of increased interest due to its contribution to the current environmental, agronomic and economic issues (Sharpley and Tunney 2000). P involves in a wide range of plant processes like cell division, development of root system, providing resistance to late blight and enhancing crop maturity. Its application increases the number of medium-size tubers which have special significance to the seed crop (Ghosh 1985; Mondal et al. 2005). The P cycle in soil is a cohesive dynamic system under the influence of long-term chemical transformations and short-term changes due to plant uptake or cropping. The leaching of bases, the removal of carbonates and the increasing Fe and Al activity cause a shift from primary to secondary Pi forms and also influence the stabilization of organic matter and its associated Po (Zheng et al. 2002; Zhang et al. 2004). Soil organic carbon contributes energy for microbial activity and promotes processes involved in P transformation. Long-term organic residues or manure application increases microbial activity and mineralization of soil organic matter (Wright 2009). This

mineralization of Po during the growing season is very important for P availability to plants (Zheng et al. 2004; Zhang et al. 2006). The P transformation in soils plays important roles in P bioavailability and mobility. Integrated nutrient supply, tillage and cropping systems affect P transformation (Zheng and Zhang 2011).

Crop response to potassium (K) is highly inconsistent and variable due to inherent soil K status and dynamic K transformations. K extraction from lower depth in the soil profile also influences K balance and its uptake from soil. K management requires long-term strategy for working out the required K inputs as biological and chemical transformations do not add or deplete K from the rhizosphere easily (Praharaj et al. 2007). K increases plant height, crop vigour and impart resistance to drought, frost and diseases in potato (Baishya 2009). It activates a number of enzyme systems involved in photosynthesis, carbohydrates metabolism and protein formation, control of ionic balance, regulation of plant stomata and water use activation of plant enzyme and many other processes (Marschner 1995). Thus, a judicious combination of organic and inorganic sources of plant nutrient helps in obtaining high economic return and improving soil health (Sud and Grewal 1990; Hedge and Dwivedi 1993; Reddy and Reddy 2002). INM is an essential tool for balanced fertilization and sustainability of crop production on long-term basis (Roy et al. 2001; Praharaj et al. 2002; Arora 2008; Baishya et al. 2013).

2   Integrated Nutrient Management on Potato Productivity and Profitability

The INM practices in potato conducted in the hilly region of northeastern India showed that potato plants that received 75 % NPK through chemical fertilizers and 25 % N through FYM resulted in increased tuber yield and net profit (Table 1). Further increase in organic N supply considerably decreased the tuber productivity and profit. Slow mineralization rate of organic matter at low temperature condition prevailing in potato-growing

Table 1 Effect of integrated nutrient management on tuber yield and economics of potato cultivation

	Tuber yield (t/ha)		Net return (INR/ha)		Benefit–cost ratio	
Nutrient management practices	2005	2006	2005	2006	2005	2006
25 % organic N (FYM) + 75 % inorganic fertilizer	24.25	26.59	30,143	33,051	3.83	4.37
50 % organic N (FYM) + 50 % inorganic fertilizer	22.35	24.96	27,781	31,025	3.26	3.16
75 % organic N (FYM) + 25 % inorganic fertilizer	21.15	24.70	26,289	30,702	2.94	2.77
100 % organic N (FYM)[a]	19.27	21.84	23,953	27,147	2.09	1.69
100 % inorganic fertilizer	24.00	26.30	29,832	32,691	3.74	4.11
Control	16.57	17.57	20,597	21,840	–	–
LSD (0.05)	0.83	1.02	3,126	2,827	0.32	0.35

[a]100 % organic N = 120 kg N/ha through farm yard manure (FYM); 100 % inorganic fertilizers = 120 kg N, 120 kg P_2O_5 and 60 kg K_2O/ha (recommended dose) through chemical fertilizers (Source: Baishya et al. 2010a)

season in the northeastern hill region might not meet the nutrient demand of the crop and, thus, decreased the tuber production (Kumar et al. 2005). The tuber yield decreased by 9.1, 12.6 and 25.6 % due to replacement of 50, 75 and 100 % N through FYM, respectively, when compared to that of 25 % N supply through FYM (Baishya et al. 2010a). However, Kumar et al. (2011) noticed that 50 % NPK through chemical fertilizers and 50 % N through organic manures [FYM/poultry manure (PM)/ vermicompost (VC)] enhanced tuber yield and paid greater profit comparable to that of the crop having only chemical fertilizers (Table 2). Seed treatment with biofertilizers (*Azotobacter*/PSB/ *Azotobacter* + PSB) also exerted beneficial effect on tuber yield and profit as compared to the crop without seed treatment. Seed treatment with both *Azotobacter* and PSB produced higher tuber yield and paid greater profit than that of having only *Azotobacter* or PSB.

The benefit of seed treatment with biofertilizer was noticed even in control plots where no plant nutrient was added. Integrating fertility treatments (organic and inorganic) with biofertilizer showed significant interaction effect on potato productivity and profitability (Table 3). Integrated use of 50 % N as organic manures (FYM/PM/VC) and 50 % NPK as chemical fertilizers and seed treatment with biofertilizers (*Azotobacter* + PSB) seemed to be very promising in enhancing tuber productivity and profitability (Kumar et al. 2013). Seed treatment with biofertilizer under integrated organic–inorganic nutrient management might enhance the availability of N from FYM due to higher microbial activities, which led to increase

crop productivity and profitability (Singh 2002; Zaller and Köpke 2004; van Diepeningena et al. 2006; Singh et al. 2007). The tuber yield and net profit of potato increased significantly by integrated use of organic manure and chemical fertilizers (Mondal et al. 2005). The highest yield (25.2 t/ha) and the maximum marginal benefit–cost ratio were achieved with the use of 3 t/ha PM along with reduced rate (70 %) of the recommended dose of chemical fertilizers (Islam et al. 2013). Narayan et al. (2013) also found that integrated use of inorganic and organic sources of nutrients significantly improved the yield of potato and paid high profit. They further reported that application of biofertilizers provided additional benefits and VC proved better than FYM in improving the tuber yield of potato under temperate condition of Kashmir valley.

Integrated use of organics (FYM, PM, poultry manure, green manure and crop residues) with chemical fertilizers in major cropping systems of subtropical northwestern India made better synchrony of crop N needs because of slower mineralization of organics and reduced N losses due to denitrification and nitrate leaching; thus, it produced significantly greater crop yields over that of recommended dose of chemical fertilizers (Aulakh 2010). Similarly, Sharma et al. (2013) revealed that substitution of 25 % NPK through FYM and 75 % NPK through chemical fertilizers along with 5 kg Zn/ha and PSB plus *Azotobacter* increased grain yield of wheat and enhanced profit. These findings suggest that integrated use of organic manure, chemical fertilizers, biofertilizers and micronutrients is necessary for increasing crop yield, improving crop quality and profitability.

Table 2 Effect of nutrient management and biofertilizers on tuber yield and economics of potato

Nutrient management practices	Tuber yield (t/ha)			Net return (INR/ha)		
	2005	2006	2007	2005	2006	2007
Control	10.60	9.64	9.64	23,665	27,880	27,950
100 % organic N (FYM)[a]	15.65	16.25	16.67	36,719	40,104	41,984
100 % organic N (PM)	16.46	17.05	17.08	37,210	41,670	42,045
100 % organic N (VC)	16.09	16.64	16.88	33,480	36,205	37,295
50 % organic N (FYM) + 50 % inorganic fertilizers	21.37	21.85	21.92	64,196	68,871	69,831
50 % organic N (PM) + 50 % inorganic fertilizers	22.18	22.85	23.17	69,724	71,194	71,814
50 % organic N (VC) + 50 % inorganic fertilizers	22.01	22.27	22.21	66,582	67,292	68,457
100 % inorganic fertilizers	21.64	22.38	22.58	67,879	71,559	71,664
LSD (0.05)	1.61	1.62	1.56	4,993	5,309	5,388
Biofertilizer						
Azotobacter	17.69	17.93	18.09	46,695	49,597	50,406
Phosphate-solubilizing bacteria (PSB)	18.05	18.29	18.66	48,688	51,315	53,724
Azotobacter + PSB	19.12	19.62	19.55	54,413	58,380	57,509
LSD (0.05)	0.80	0.88	0.89	4,589	4,715	4,657

[a]100 % organic N = 120 kg N/ha through farm yard manure (FYM)/poultry manure (PM)/vermicompost (VC); 100 % inorganic fertilizers = 120 kg N, 120 kg P_2O_5 and 60 kg K_2O/ha (recommended dose) through chemical fertilizers (Source: Kumar et al. 2011)

Table 3 Interaction effect of nutrient management and biofertilizer on tuber yield and net return from potato

Biofertilizer	Tuber yield (t/ha)			Net return (INR/ha)		
	Azotobacter	PSB	Azo + PSB	Azotobacter	PSB	Azo + PSB
Nutrients management						
Control	9.41	9.77	10.70	25,036	25,984	28,475
100 % organic N (FYM)[a]	15.54	16.03	17.00	38,006	39,222	41,579
100 % organic N (PM)	16.01	16.60	17.98	38,275	39,679	42,971
100 % organic N (VC)	15.86	16.25	17.50	34,193	35,040	37,746
50 % organic N (FYM) + 50 % inorganic fertilizers	21.09	21.49	22.56	65,690	66,944	70,264
50 % organic N (PM) + 50 % inorganic fertilizers	22.09	22.54	23.58	68,894	70,300	73,538
50 % organic N (VC) + 50 % inorganic fertilizers	21.50	21.98	23.01	65,432	66,874	70,025
100 % inorganic fertilizers	21.60	21.97	23.03	68,466	69,638	72,998
LSD between N M at same biofertilizer	1.53			4,854		
LSD between biofertilizer at same/diff. N M	1.65			5,215		

[a]100 % organic N = 120 kg N/ha through farm yard manure (FYM)/poultry manure (PM)/vermicompost (VC); 100 % inorganic fertilizers = 120 kg N, 120 kg P_2O_5 and 60 kg K_2O/ha (recommended dose) through chemical fertilizers Source: Kumar et al. (2011)

3 Integrated Nutrient Management on Nutrient Uptake and Nutrient-Use Efficiency

The nutrient management practices in potato showed significant effect on nutrient uptake and nutrient-use efficiency by the crop. The use of 100 % recommended dose of NPK through chemical fertilizers recorded the highest NPK (193.8, 16.56 and 213.1 kg/ha N, P and K, respectively) uptake and nutrient-use efficiency (27.0 kg/kg) and was comparable to those of the crop having 75 % recommended dose of NPK through chemical fertilizers along with 25 % N through FYM (Table 4). The above treatments significantly outperformed the other nutrient

Table 4 Effect of integrated nutrient management on nutrient uptake by potato (average value of 2 years)

Nutrient management practices	Nutrient uptake by the crop (kg/ha)			Nutrient-use efficiency
	N uptake	P uptake	K uptake	
Control	99.1	8.78	109.2	–
25 % organic N + 75 % inorganic fertilizer	191.0	16.25	210.4	24.2
50 % organic N + 50 % inorganic fertilizer	167.1	14.17	182.2	18.2
75 % organic N + 25 % inorganic fertilizer	149.7	12.50	168.8	14.9
100 % organic N[a]	121.0	10.45	136.7	7.6
100 % inorganic fertilizer	193.8	16.56	213.1	27.0
LSD (0.05)	5.0	0.70	5.1	4.1

[a]100 % organic N = 120 kg N/ha through farm yard manure (FYM); 100 % inorganic fertilizers = 120 kg N, 120 kg P_2O_5 and 60 kg K_2O/ha (recommended dose) through chemical fertilizers (Source: Baishya et al. 2010a)

management treatments with respect to nutrient recovery and nutrient-use efficiency. Nutrient uptake and nutrient-use efficiency decreased steadily due to proportional decrease in nutrient supply through inorganic fertilizers, and the crop receiving 100 % N through FYM (only organic source) recorded the lowest nutrient uptake (121.0, 10.45 and 136.7 kg/ha N, P and K, respectively) and nutrient-use efficiency (7.6 kg/kg) among the other fertility treatments (Baishya et al. 2010a).

High NPK uptake and nutrient-use efficiency in potato were also found by Islam et al. (2013) due to integrated use of 3 t/ha PM along with 70 % recommended dose of chemical fertilizers. Significant effect of nutrient management practices in potato on nutrient uptake and nutrient-use efficiency were noted by Kumar et al. (2012). The highest NPK uptake by the crop and nutrient-use efficiency was obtained with application of 50 % organic N (PM) and 50 % through chemical fertilizers (60 kg N, 60 kg P_2O_5 and 30 kg K_2O/ha) but was statistically similar to those obtained from the crop at 50 % organic N (VC) and 50 % NPK through chemical fertilizers or 100 % NPK through chemical fertilizers (Table 5). Replacement of 50 % N through FYM was found less effective in comparison to the above treatments. Replacement of 100 % N through organic manures (PM, VC or FYM) further reduced NPK uptake by the crop and their use efficiency but increased markedly over that of the control plots. Seed treatment with both biofertilizers (*Azotobacter*, PSB and *Azotobacter* + PSB) exerted beneficial effect on NPK

uptake and their use efficiency in potato and use of both *Azotobacter* and PSB proved superior to either *Azotobacter* or PSB. The results emphasized the need of integrated use of 50 % N through organic manures (PM/VC/FYM) and 50 % NPK through chemical fertilizers in addition to seed treatment with biofertilizers (*Azotobacter* + PSB) for enhancing the NPK uptake and their use efficiency in potato. Favourable effect of integrated nutrient management through organic manures and inorganic fertilizers on increasing NPK uptake and their use efficiency in potato was also noticed by Thind et al. (2007) and Baishya (2009). The study clearly showed that integrated use of organic manure, chemical fertilizers and biofertilizers not only increased tuber productivity but also enhanced nutrient uptake and nutrient-use efficiency in potato.

4   INM on Soil Fertility

The overall fertility status of the soil improved considerably due to integrated use of organic manures and chemical fertilizers. The soil fertility status increased steadily by increasing the rate of organic manuring (Jayaram et al. 1990; Mondal et al. 2007; Mohapatra et al. 2008). Application of 100 % N through organic sources recorded maximum increase in organic carbon and available NPK contents in soil over the initial values (van Diepeningena et al. 2006; Baishya et al. 2010b). They noticed that this benefit was gradually decreasing with

Table 5 Effect of integrated nutrient management on nutrient uptake by potato (average value of 3 years)

Nutrient management practices	Nutrient uptake by the crop (kg/ha)			
	N uptake	P uptake	K uptake	Nutrient-use efficiency
Control	58.1	12.46	74.8	–
100 % organic N (FYM)[a]	84.9	20.50	108.5	20.76
100 % organic N (PM)	89.9	22.65	113.7	23.02
100 % organic N (VC)	88.1	21.10	112.2	21.93
50 % organic N (FYM) + 50 % inorganic fertilizers	109.9	29.64	142.6	39.18
50 % organic N (PM) + 50 % inorganic fertilizers	118.8	33.63	150.4	42.58
50 % organic N (VC) + 50 % inorganic fertilizers	114.9	31.66	148.0	40.68
100 % inorganic fertilizers	116.0	32.89	148.6	40.80
LSD (0.05)	4.6	2.25	6.3	3.56
Biofertilizer				
Azotobacter	95.3	23.67	120.8	26.48
Phosphate-solubilizing bacteria (PSB)	97.7	25.79	125.5	27.91
Azotobacter + PSB	99.6	27.23	128.3	31.57
LSD (0.05)	2.7	1.23	4.5	3.45

[a]RDN (recommend dose of N) = 120 kg N/ha; RD (recommend dose) = 120 kg N/ha, 120 kg P_2O_5/ha and 60 kg K_2O/ha, respectively (Source: Kumar et al. 2012)

Table 6 Effect of integrated nutrient management on fertility status of the soil after end of the experiment (two-crop cycle)

Fertility management practices	Soil properties of the experimental field			
	Organic C (%)	Av. N (kg/ha)	Av. P_2O_5 (kg/ha)	Av. K_2O (kg/ha)
Initial	0.69	163.00	13.05	390.13
Control	0.52	142.67	9.20	361.93
25 % organic N + 75 % inorganic fertilizer	0.83	188.98	20.98	423.70
50 % organic N + 50 % inorganic fertilizer	0.86	200.41	23.01	436.33
75 % organic N + 25 % inorganic fertilizer	0.93	218.93	25.93	453.48
100 % organic N[a]	1.04	235.42	29.02	467.34
100 % inorganic fertilizer	0.75	203.23	19.95	420.30
LSD (0.05)	0.14	38.39	3.10	53.63

[a]100 % organic N = 120 kg N/ha through farm yard manure (FYM); 100 % inorganic fertilizers = 120 kg N, 120 kg P_2O_5 and 60 kg K_2O/ha (recommended dose) through chemical fertilizers (Source: Baishya et al. 2010b)

proportionate reduction in organic manuring. They further observed that nutrient management (100 % NPK) through only chemical fertilizers helped in improving the above soil parameters to a less extent, and the control plot showed considerable negative nutrient balance in soil (Table 6).

In another study, Kumar et al. (2012) noticed that total organic carbon (TOC), soil microbial biomass carbon (SMBC) and ratio of SMBC to TOC varied significantly among the different nutrient management treatments (Table 7). The TOC varied from 9.13 in the control to 14.43 g/ kg of soil in 100 % organic N (FYM) plots, whereas the SMBC ranged from 121.5 in the

control to 197.0 mg/kg of soil in 100 % organic N (VC)-treated plots. Both TOC and SMBC reduced substantially in the control plots from their initial values (12.3 g/kg and 145.0 mg/kg of soil, respectively). The ratio of SMBC to TOC increased over its initial value (11.79 mg/g) due to different fertility management treatments except in plots receiving 100 % NPK through inorganic fertilizers alone which recorded SMBC to TOC ratio much below the control and initial value. Maximum values of TOC and SMBC were recorded with the application of 100 % N through organic manures (PM, VC and FYM), which were markedly higher than

Table 7 Effect of integrated nutrient management on fertility status of the soil after the end of the experiment (three-crop cycle)

Nutrient management	Soil properties of the experimental field						
	pH	TOC (g/kg)	SMBC (mg/kg)	SMBC/TOC (mg/g)	Av. N (kg/ha)	Av. P (kg/ha)	Av. K (kg/ha)
Control	5.40	9.13	121.5	13.31	172.0	11.23	190.2
100 % organic N (FYM)[a]	5.55	14.43	196.4	13.61	244.8	21.91	263.6
100 % organic N (PM)	5.73	14.33	193.8	13.52	246.7	22.70	264.7
100 % organic N (VC)	5.53	14.23	197.0	13.84	244.0	21.60	262.0
50 % organic N (FYM) + 50 % inorganic fertilizers	5.56	13.73	173.2	12.61	252.0	24.85	269.2
50 % organic N (PM) + 50 % inorganic fertilizers	5.61	13.63	171.4	12.58	253.3	26.37	271.1
50 % organic N (VC) + 50 % inorganic fertilizers	5.58	13.63	175.4	12.87	251.6	24.01	268.8
100 % inorganic fertilizers	5.52	13.13	143.9	10.96	236.5	21.63	252.8
LSD (0.05)	0.12	0.37	10.75	1.60	10.8	1.51	14.2
Biofertilizer							
Azotobacter	5.53	13.09	170.4	13.02	237.0	21.32	259.1
Phosphate-solubilizing bacteria (PSB)	5.55	13.25	172.0	12.98	237.5	21.66	260.7
Azotobacter + PSB	5.60	13.51	173.9	12.87	238.3	22.10	261.3
LSD (0.05)	NS	NS	NS	NS	NS	NS	NS
Initial value	**5.30**	**12.30**	**145.0**	**11.79**	**179.5**	**13.35**	**195.1**

[a]RDN (recommend dose of N) = 120 kg N/ha; RD (recommend dose) = 120 kg N/ha, 120 kg P_2O_5/ha and 60 kg K_2O/ha, respectively. TOC = Total Organic Carbon; SMBC = Soil Microbial Biomass Carbon (Source: Kumar et al. 2012)

those of 50 % NPK through inorganic fertilizers and remaining 50 % N through organic manures. Both the treatments again significantly increased TOC and SMBC over their initial values and also over those of the control plots. Kumar et al. (2012) further observed that the available NPK contents in the soil increased due to the application of the recommended dose of nutrients either through inorganic fertilizers or through integrated application of inorganic fertilizers and organic manures. The highest available NPK contents in soil were obtained with the application of 50 % NPK through inorganic fertilizers, and the remaining 50 % N through PM and were comparable to those of 50 % NPK through inorganic fertilizers and remaining 50 % N through VC or FYM. These treatments recorded markedly higher NPK contents in soil than those of 100 % NPK through only inorganic fertilizers (Table 7). Application of 100 % N either through organic manures (PM, VC and FYM) or through inorganic fertilizers also

enhanced the available NPK status of soil over its initial values and those of the control plots. The control plots recorded negative balance of available NPK status in the soil. Biofertilizers treatments did not exert significant effect on TOC, SMBC, ratio of SMBC/TOC and available NPK status of the soil under the study. Similar favourable effects of integrated nutrient management involving inorganic fertilizers and organic manures on improving organic carbon pool and available NPK status of the soil have been noticed by Manna et al. (2006), Kumar et al. (2008), Ghosh et al. (2009) and Zaman et al. (2011).

A combination of organic manures and chemical fertilizers along with biofertilizers is, therefore, a useful strategy to minimize dependence on synthetic fertilizers, enhance soil physico-chemical properties and improve its quality (Iqbal et al. 2002; Jiang et al. 2006; Kumar et al. 2013). The rate of application of inorganic fertilizers can also be reduced by minimizing losses and increasing the recycling of nutrients (Lal 2006; Shafi et al. 2009; Shafi et al. 2012). Losses of nutrients can be controlled through nutrient recycling and organic manuring (Shaaban 2006; Gopinath et al. 2008). There are some advantages in substituting biological nitrogen fixation for inorganic fertilizers (Singh et al. 2007; Kumar et al. 2012). However, the economics of growing nitrogen versus buying nitrogen have to be carefully evaluated in terms of land scarcity and efficiency of nitrogen availability (Aulakh 2010; Baishya et al. 2013; Sharma et al. 2013).

Sustainable agricultural production incorporates the idea that natural resources should be used to generate increased output and income, especially for low-income groups, without depleting the natural resource base. In this context, INM maintains soils as storehouses of plant nutrients that are essential for growth and productivity of crops (Stephen 2001; Prasad et al. 2002; Premi 2003).

Conclusion

INM's goal is to integrate the use of all natural and man-made sources of plant nutrients, so that crop productivity increases in an efficient and environmentally benign manner, without sacrificing soil productivity for future generations (Mondal et al. 2007; Zaman et al. 2011; Kumar et al. 2012). INM relies on a number of factors, including appropriate nutrient application and conservation and the transfer of knowledge about INM practices to farmers and researchers. Sufficient and balanced application of organic and inorganic fertilizers is a major component of INM. In potato, the beneficial effect of the integrated use of organic manures and chemical fertilizers on improving soil fertility was noticed by Mondal et al. (2007), Kumar et al. (2008), Baishya (2009) and Zaman et al. (2011) and on soil quality (TOC, SMBC and SMBC/TOC ratio) by Manna et al. (2006), Ghosh et al. (2009) and Kumar et al. (2012) and was found to be responsible for sustaining crop productivity.

References

Ali MM, Shaheed SM, Kubota D (1997) Soil degradation during the period 1967–1995 in Bangladesh. II. Selected chemical characters. Soil Sci Plant Nutr 43:879–890

Anonymous (2003) Report of the inter ministerial task force on integrated plant nutrient management using city compost, pp 5–10

Anonymous (2012) Agricultural statistics at a glance, pp 10–45

Arora S (2008) Balanced nutrition for sustainable crop production. Krishi World (Pulse of Indian Agriculture), pp 1–5

Aulakh MS (2010) Integrated nutrient management for sustainable crop production, improving crop quality and soil health, and minimizing environmental pollution. In: Proceedings of the 19th world congress of soil science, soil solutions for a changing world held at Brisbane, Australia, pp 70–82

Baishya LK (2009) Response of potato varieties to organic and inorganic sources of nutrients. Ph.D. thesis submitted to Visva-Bharati (Central University), West Bengal, pp 99–102

Baishya LK, Kumar M, Ghosh DC (2010a) Effect of different proportion of organic and inorganic nutrients on productivity and profitability of potato (Solanum tuberosum) varieties in Meghalaya hills. Indian J Agron 55(3):230–234

Baishya LK, Kumar M, Ghosh DC, Ghosh M, Dubey SK (2010b) Effect of organic and inorganic nutrient management in potato varieties on nutrient content and uptake, nutrient use efficiency and soil fertility status in Meghalaya hill. Environ Ecol 28(3):1745–1751

Baishya LK, Kumar M, Ghosh M, Ghosh DC (2013) Effect of integrated nutrient management on growth, productivity and economics of rainfed potato in Meghalaya hills. Int J Agric Environ Biotechnol 6 (1):1–9

Barrios E, Delve RJ, Bekunda M, Mowo J, Agunda J, Ramisch J, Trejo MT, Thomas RJ (2006) Indicators of soil quality: a South-South development of a methodological guide for linking local and technical knowledge. Geoderma 135:248–259

Bekunda B, Sanginga N, Woomer PL (2010) Restoring soil fertility in Sub-Sahara Africa. Adv Agron 108:184–236

Cassman KG, De Datta SK, Olk DC, Alcantara J, Samson M, Descalsota J, Dizon M (1995) Yield decline and the nitrogen economy of long term experiments on continuous, irrigated rice systems in the tropics. In: Lal R, Stewart B (eds) Soil management: experimental basis for sustainability and environmental quality. CRC Press, Boca Raton

Cobo JG, Dercon G, Cadisch G (2010) Nutrient balances in African land use systems across different spatial scales: a review of approaches, challenges and progress. Agric Ecosyst Environ 136:1–15

Corselius K, Wisniewski S, Ritchie M (2001) Sustainable agriculture: making money, making sense. The Institute for Agriculture and Trade Policy, Washington, DC, pp 1–15

Crosson P, Anderson JR (1992) Resources and global food prospects: supply and demand for cereals to 2030, World Bank technical paper 184. World Bank, Washington, DC

Das AK, Mookherjee S, Ghosh DC (2000) Productivity, economics and soil fertility status as influenced by integrated nutrient management in wheat. J Interacad 4(1):39–43

Dawe D, Dobermann A, Moya P, Abdulracman S, Singh B, Lal P, Li SY, Lin B, Panaullah G, Sariam O, Singh Y, Swarup A, Tan PS, Zhen QX (2000) How widespread are yield declines in long-term rice experiments in Asia? Field Crops Res 66:175–193

Dixit KG, Gupta BR (2000) Effect of FYM, chemical and biofertilizer on yield and quality of rice and soil properties. J Indian Soc Soil Sci 48:773–780

Dobermann A, White PF (1999) Strategies for nutrient management in irrigated and rainfed lowland rice systems. Nutr Cycl Agroecosyst 53:1–18

Gaur AC (2001) Organic manure – a basic input of organic farming. Indian Fmg 51(3):3–5

Gentile R, Vanlauwe B, van Kessel C, Six J (2009) Managing N availability and losses by combining fertilizer-N with different quality residues in Kenya. Agric Ecosyst Environ 131:308–314

Ghosh DC (1985) Studies on potato agronomy I. Effect of plant density and time of fertilizer application on growth and development of potato varieties. Madras Agric J 72(1):39–46

Ghosh PK, Saha R, Gupta JJ, Ramesh T, Das A, Lama TD, Munda GC, Bordoloi JS, Verma MR, Ngachan SV (2009) Long term effect of pastures on soil quality in acid soil of North-East India. Aust J Soil Res 47:372–379. http://dx.doi.org/10.1071/SR08169

Gopinath KA, Saha S, Mina BL (2008) Influence of organic amendments on growth, yield and quality of wheat and on soil properties during transition to organic production. Nutr Cycl Agroecosyst 82:51–60

Govindakrishnan PM, Kushwah VS (2003) Low input technology for potato production. In: The potato. Mehta Publishers, New Delhi, pp 130–135

Hazell P (1995) Technology's contribution to feeding the world in 2020. In: Speeches made at an international conference, International Food Policy Research Institute, Washington, DC

Hedge DM, Dwivedi BS (1993) Integrated nutrient supply and management as a strategy meet nutrient demand. Fert News 38:49–50

Hole DG, Perkins AJ, Wilson JD, Alexander IH, Grice F, Evans AD (2005) Does organic farming benefit biodiversity? Biol Conserv 122(1):113–130

Iqbal A, Abbasi MK, Rasool G (2002) Integrated plant nutrition system (IPNS) in wheat under rainfed conditions. Pak J Soil Sci 21:41–50

Islam MZ, Zamam MM, Hossain MM, Hossain A (2009) Integrated nutrient management with liming for potato production in North-West region of Bangladesh. Annual report 2008–2009, Tuber Crops Research Centre, Bangladesh Agricultural Research Institute, Gazipur, Bangladesh

Islam MM, Akhter S, Majid NM, Ferdous J, Alam MS (2013) Integrated nutrient management for potato (Solanum tuberosum) in grey terrace soil (AricAlbaquipt). Aust J Crop Sci 7(9):1235–1241

Jayaram D, Chatterjee BN, Mondal SS (1990) Effect of FYM, crop residues and fertilizers management in sustaining productivity under intensive cropping. J Potassium Res 6(4):172–179

Jiang D, Ting-Bo HHD, Boer WD, Qi J, Xing CW (2006) Long-term effects of manure and inorganic fertilizers on yield and soil fertility for a winter wheat-maize system in Jiangsu, China. Pedosphere 16:25–32

Karforma J, Ghosh M, Ghosh DC, Mandal S (2012) Effect of integrated nutrient management on growth, productivity, quality and economics of fodder maize in rainfed upland of Terai Region of West Bengal. Int J Agric Environ Biotechnol 5(4):419–427

Kato E, Place FM (2011) Heterogeneous treatment effects of integrated soil fertility management on crop productivity: evidence from Nigeria. IFPRI discussion paper 01089 pp 20

Khan MS, Shil NC, Noor S (2008) Integrated nutrient management for sustainable yield of major vegetable crops in Bangladesh. Bangladesh J Agric Environ 4:81–94

Kumar M, Gupta VK, Gogoi MB, Kumar S, Lal SS, Baishya LK (2005) Effect of poultry manure in potato production under rainfed condition of Meghalaya. Potato J (Special issue) 32(3–4):242

Kumar M, Jadav MK, Trehan SP (2008) Contributing of organic sources to potato nutrition at varying nitrogen levels. Global potato conference held at New Delhi

Kumar M, Baishya LK, Ghosh DC, Gupta VK (2011) Yield and quality of potato (Solanum tuberosum) tubers as influenced by nutrients sources under rained conditions of Meghalaya. Indian J Agron 56(3):260–266

Kumar M, Baishaya LK, Ghosh DC, Gupta VK, Dubey SK, Das A, Patel DP (2012) Productivity and soil health of potato (Solanumtuberosum L.) field as influenced by organic manures, inorganic fertilizers and biofertilizers under high altitudes of eastern Himalayas. J Agric Sci 4:223–234

Kumar M, Baishya LK, Ghosh DC, Ghosh M, Gupta VK, Verma MR (2013) Effect of organic manures, chemical fertilizers and biofertilizers on growth and productivity of rainfed potato in the eastern Himalayas. J Plant Nutr 36(7):1065–1082

Lal R (2006) Enhancing crop yield in developing countries through restoration of soil organic pool in agricultural lands. Land Degrad Dev 17:197–206

Manna MC, Swarup A, Wanjari RH, Singh YV, Ghosh PK, Singh KN, Tripathi AK, Saha MN (2006) Soil organic matter in West Bengal inceptisol after 30 years of multiple cropping and fertilization. Soil Sci Soc Am J 70:121–129. http://dx.doi.org/10.2136/sssaj2005.0180

Marschner H (1995) Mineral nutrition of higher plants, 2nd edn. Academic, London

Mohapatra BK, Maiti S, Satapathy MR (2008) Integrated nutrient management in potato (*Solanum tuberosum*)–jute (*Corchorus olitorius*) sequence. Indian J Agron 53(3):205–209

Mondal SS, Acharya D, Ghosh A, Bug A (2005) Integrated nutrient management on the growth, productivity and quality of potato in Indo-Gangetic plains of West Bengal. Potato J 32:75–78

Mondal SS, Saha M, Acharya D, Putra D, Chatterjee S (2007) Integrated effect of nitrogen and potassium with or without sulphur and farm yard manure on potato tuber yield, storage quality and soil fertility status. Potato J 34:97–98

Mugwe J, Mugendi DN, Mucheru-Muna M, Odee D, Mairura F (2009a) Effect of selected organic materials and inorganic fertilizer on the soil fertility of a Humic Nitisol in the central highlands of Kenya. Soil Use Manag 25:434–440

Mugwe J, Mugendi D, Mucheru-Muna M, Merckx R, Chianu J, Vanlauwe B (2009b) Determinants of the decision to adopt integrated soil fertility management practices by smallholder farmers in the central highlands of Kenya. Exp Agric 45:61–75

Narayan S, Kanth RH, Narayan R, Khan FA, Singh P, Rehman SU (2013) Effect of integrated nutrient management practices on yield of potato. Potato J 40(1):84–86

Ngetich FK, Shisanya CA, Mugwe J, Mucheru-Muna M, Mugendi D (2011) The potential of organic and inorganic nutrient sources in Sub-Saharan African crop farming systems. In: Whalen JK (ed) Soil fertility improvement and integrated nutrient management – a global perspective. In Tech Publisher, Rijeka, pp 135–156. ISBN 978-953-307-945-5

Nkonya E, Pender J, Jagger P, Serunkuuma D, Kaizzi C, Sali H (2004) Strategies for sustainable land management and poverty reduction in Uganda. International Food Policy Research Institute (IFPRI) Research Report 133, Washington, DC

Peng S, Khush GS, Cassman KG (1994) Evolution of the new plant ideotype for increased yield potential. In: Cassman KG (ed) Breaking the yield barrier: proceedings of a workshop on rice yield potential in favourable environments, International Rice Research Institute, Manila

Place F, Barrett CB, Freeman HA, Ramisch JJ, Vanlauwe B (2003) Prospects for integrated soil fertility management using organic and inorganic inputs: evidence from smallholder African agricultural systems. Food Policy 28:365–378

Praharaj CS, Kumar D, Sharma RC (2002) In extended summaries of the 2nd international congress on balancing food and environmental security – a continuing challenge, held at New Delhi, IARI, pp 232–233

Praharaj CS, Bandyopadhyay KK, Sankaranarayanan K (2007) Integrated nutrient management for increasing cotton productivity. In: Model training course on 'cultivation on long staple cotton' Central Institute for Cotton Research, Regional Station, Coimbatore, pp 158–170

Prasad PVV, Satyanarayana V, Murthy VRK, Boote KJ (2002) Maximizing yields in rice–groundnut cropping sequence through integrated nutrient management. Field Crops Res 75(1):9–21

Premi OP (2003) Integrated nutrient supply for sustainable rice production in an acid *alfisol*. Indian J Agric Res 37(2):132–135

Reddy TY, Reddy GHS (2002) Mineral nutrition, manures and fertilizers. In: Principles of agronomy, 3rd edn. Kalyani Publishers, Ludhiana, pp 204–256

Roy SK, Sharma RC, Trehan SP (2001) Integrated nutrient management by using farmyard manure and fertilizers in potato–sunflower–paddy rice rotation in the Punjab. J Agric Sci Cambridge 137:271–278

Rufino MC, Rowea EC, Delve RJ, Giller KE (2006) Nitrogen cycling efficiencies through resource-poor African crop – livestock systems. Agric Ecosyst Environ 112:261–282

Sanginga N, Woomer PL (eds) (2009) Integrated soil fertility management in Africa: principles, practice and developmental process. Tropical Soil Biology and Fertility, Institute of International Centre of Tropical Agriculture, Nairobi, 266

Shaaban M (2006) Effect of organic and inorganic nitrogen fertilizer on wheat plants under varying water regime. J Appl Sci Res 2:650–656

Shafi J, Shafi M, Jan MT, Shah Z (2009) Influence of crop residue management, cropping system and N fertilizer on soil N and C dynamics and sustainable wheat (*Triticum aestivum* L.) production. Soil Tillage Res 104:233–240

Shafi M, Shah A, Bakht J, Shah M, Wisal M (2012) Integrated effect of inorganic and organic nitrogen sources on soil fertility and productivity of maize. J Plant Nutr 35(4):524–537

Sharma GD, Thakur R, Raj S, Kauraw DL, Kulkhare PS (2013) Impact of integrated nutrient management on yield, nutrient uptake, protein content of wheat (*Triticum aestivum*) and soil fertility in a typichaplustert. The Bioscan 8(4):1159–1164

Sharpley AN, Tunney H (2000) Phosphorus research strategies to meet agricultural and environmental challenges of 21st century. J Environ Qual 29:176–181

Singh K (2002) Role of biofertilizers in increasing the efficiency of nitrogen to potato crop under north eastern hill conditions. J Indian Potato Assoc 2:904–990

Singh HP (2008) Policies and strategies conducive to potato development in Asia and the pacific region. In: Proceedings of the workshop to commemorate the international year of the potato-2008 held at Bangkok, Thailand, pp 18–29

Singh JP, Trehan SP, Sharma RC (1997) Crop residue management for sustaining soil fertility and productivity of potato based cropping systems in Punjab. J Indian Potato Assoc 24:85–99

Singh SN, Singh BP, Singh OP, Singh R, Singh RK (2007) Effect of nitrogen application in conjunction with bio-inoculants on the growth, yield and quality of potato under Indo-Gangetic plain region. Potato J 34:103–104

Stephen MN (2001) Nutrient cycling in agroecosystems – soil organic carbon (SOC) management for sustainable productivity of cropping and agro-forestry systems in Eastern and Southern Africa. Earth Environ Sci 61(1–2):143–158

Sud KC, Grewal JS (1990) Integrated use of farmyard manure and potassium in potato production in acidic hill soils of Shimla. J Potassium Res 6:83–95

Swaminathan MS (2004) Extending the "Feel Good Factor" to rural and farming families. In: Proceedings of the international conference on organic food, pp 3–5

Thind SS, Sidhu AS, Sekhon NK, Hira GS (2007) Integrated nutrient management for sustainable crop production in potato-sunflower sequence. J Sustain Agric 29:173–188

Trehan SP, Grewal JS (1983) Available micronutrient status of alluvial and hill soil as influenced by intensive cropping and manuring. J Indian Soc Soil Sci 31:343–361

Upadhayay NC, Singh JP (2003) The potato – production and utilization in sub-tropics. In: Paul Khurana SM, Minhas JS, Pandey SK (eds), Mehta Publishers, A-16 (East), Naraina II, New Delhi-110028, India pp 25–145

van Diepeningena AD, de Vosa OJ, Korthalsb GW, van Bruggena AHC (2006) Effects of organic versus conventional management on chemical and biological parameters in agricultural soils. Appl Soil Ecol 31:120–135

Vanlauwe B, Giller KE (2006) Popular myths around soil fertility management in sub-Saharan Africa. Agric Ecosyst Environ 116:34–46

Wright AL (2009) Phosphorus sequestration in soil aggregates after long-term tillage and cropping. Soil Tillage Res 103:406–411

Zaller JG, Köpke U (2004) Effects of traditional and biodynamic farmyard manure amendment on yields, soil chemical, biochemical and biological properties in a long-term field experiment. Biol Fertil Soils 40:222–229

Zaman A, Sarkar A, Sarkar S, Devi WP (2011) Effect of organic and inorganic sources of nutrients on productivity, specific gravity and processing quality of potato (*Solanum tuberosum*). Indian J Agric Sci 81 (12):1137–1142

Zhang TQ, Mackenzie AF, Liang BC, Drury CF (2004) Soil test phosphorus and phosphorus fractions with long-term phosphorus addition and depletion. Soil Sci Soc Am J 68:519–528

Zhang TQ, Tan CS, Drury CF, Reynolds WD (2006) Long-term ($\geq$43 years) fate of soil phosphorus as related to cropping systems and fertilization. World congress of soil science. Philadelphia, USA, 9–16 July

Zheng M, Zhang TQ (2011) Soil phosphorus tests and transformation analysis to quantify plant availability: a review. In: Whalen JK (ed) Soil fertility improvement and integrated nutrient management – a global perspective, pp 19–36

Zheng Z, Simard RR, Lafond J, Parent LE (2002) Pathways of soil phosphorus transformations after 8 years of cultivation under contrasting cropping practices. Soil Sci Soc Am J 66:999–1007

Zheng Z, MacLoed JA, Sanderson JB, Lafond J (2004) Soil phosphorus dynamics during 10 annual applications of manure and mineral fertilizers: fractionation and path analyses. Soil Sci 169:449–456

Part V

Specialised Case Studies

Enhancing Nutrient Use Efficiencies in Rainfed Systems

Suhas P. Wani, Girish Chander, and Rajneet K. Uppal

Abstract

Successful and sustained crop production to feed burgeoning population in rainfed areas, facing soil fertility-related degradation through low and imbalanced amounts of nutrients, requires regular nutrient inputs through biological, organic or inorganic sources of fertilizers. Intensification of fertilizer (all forms) use has given rise to concerns about efficiency of nutrient use, primarily driven by economic and environmental considerations. Inefficient nutrient use is a key factor pushing up the cost of cultivation and pulling down the profitability in farming while putting at stake the sustainability of rainfed farming systems. Nutrient use efficiency implies more produce per unit of nutrient applied; therefore, any soil-water-crop management practices that promote crop productivity at same level of fertilizer use are expected to enhance nutrient use efficiency. Pervasive nutrient depletion and imbalances in rainfed soils are primarily responsible for decreasing yields and declining response to applied macronutrient fertilizers. Studies have indicated soil test-based balanced fertilization an important driver for enhancing yields and improving nutrient use efficiency in terms of uptake, utilization and use efficiency for grain yield and harvest index indicating improved grain nutritional quality. Recycling of on-farm wastes is a big opportunity to cut use and cost of chemical fertilizers while getting higher yield levels at same macronutrient levels. Best management practices like adoption of high-yielding and nutrient-efficient cultivars, landform management for soil structure and health, checking pathways of nutrient losses or reversing nutrient losses through management at watershed scale and other holistic crop management practices have great scope to result in enhancing nutrient and resource use efficiency through higher yields. The best practices have been found to promote soil organic

S.P. Wani (✉) • G. Chander • R.K. Uppal
Resilient Dryland Systems, International Crops Research
Institute for the Semi-Arid Tropics (ICRISAT),
Patancheru 502324, Andhra Pradesh, India
e-mail: s.wani@cgiar.org; g.chander@cgiar.org;
rajwaraich@gmail.com

A. Rakshit et al. (eds.), *Nutrient Use Efficiency: from Basics to Advances*,
DOI 10.1007/978-81-322-2169-2_23, © Springer India 2015

carbon storage that is critical for optimum soil processes and improve soil health and enhance nutrient use efficiency for sustainable intensification in the rainfed systems.

Keywords

N use efficiency • Nutrient efficient genotypes • P use efficiency • Rainfed agriculture • Soil health • Sustainable intensification

1 Introduction

Awareness of and interest in enhancing nutrient use efficiency have never been greater than as of today mainly due to the need to produce more food from limited land and to protect the environment through sustainable intensification. Regular nutrient inputs through chemical fertilizers have become an integral component of the production systems as the systems have become open to exporting of nutrients through food production areas (rural farming areas) to urban areas as well as to outside countries as against the traditional closed systems wherein nutrients were recycled. It is essential to recognize that in rainfed production systems, even with relatively low productivity level, the quantity of nutrient removal is quite substantial over the years, as these soils did not receive balanced nutrient applications. Furthermore, the quantum of nutrients available for recycling via crop residues and animal manures is grossly inadequate to compensate for the amounts removed in crop production. Thus, mineral fertilizers have come to play a key role where increased agricultural production is required to meet growing food demand and particularly in soils having low fertility. Though the consumption of chemical fertilizers has increased steadily over the years, the use efficiency of nutrients applied as fertilizers continues to remain awfully low. A review of best available information suggests that the average N recovery efficiency for fields managed by farmers ranges from about 20 to 30 % under rainfed conditions and 30 to 40 % under irrigated conditions (Roberts 2008).

Improving nutrient efficiency is a worthy goal and fundamental challenge. The opportunities are there, and tools are available to accomplish the task of improving the efficiency of applied nutrients. However, we must be cautious that improvements in efficiency do not come at the expense of the farmers' economic viability or the environment. Judicious application of nutrients targeting both high yields and nutrient efficiency will benefit farmers, society and the environment alike.

2 Importance of Rainfed Agricultural Systems

Addressing rainfed agricultural systems is very important as 80 % of the cultivated area worldwide is rainfed and contributes to about 60 % of the world's food (Wani et al. 2012a). Rainfed regions are the homes to the world's poor and malnourished people, and maximum population growth (95 %) is taking place here (Wani et al. 2012a). In India also, the rainfed-cropped areas comprise about 60 % (89 million ha) of the net-cultivated area (Wani et al. 2008). Irrigated regions in India have reached a productivity plateau, and today there is a big issue of concern to feed the burgeoning population. In spite of best efforts to increase irrigation, around 45 % of cultivated will still continue to remain rainfed by the year 2050 (Bhatia et al. 2006; Amarasinghe et al. 2007). There is no option of increasing arable land, and with burgeoning population, per capita arable land availability in India has decreased from 0.39 ha in 1951 to 0.12 ha in 2011 and is expected to be 0.09 ha by the year 2050 (Ministry of Agriculture, Government of India 2013; FAOSTAT 2013). Within existing land and water constraints, India must sustainably increase the productivity levels of the major rainfed crops to meet the ever-increasing demand of food to around 380 million tonnes in 2050 (Amarasinghe et al. 2007). Moreover, due to the role of agriculture in economic development and poverty reduction (Irz and Roe 2000; Thirtle et al. 2002; World Bank 2005), the upgradation of rainfed agriculture is priority of the government. So, in current context of suboptimal input use in rainfed systems, a regular use of nutrient inputs through chemical fertilizers is going to be increased with needs and opportunities for enhancing nutrient use efficiencies.

3 Large Yield Gaps and Untapped Potential

Yield gap analyses for major rainfed crops in semi-arid tropics (SAT) in Asia (Fig. 1) and Africa reveal large yield gaps, with farmers' yields being a factor of two- to fourfold lower than achievable yields for major rainfed crops grown in Asia and Africa (Rockström et al. 2007). At the same time, the dry subhumid and semi-arid regions experience the lowest yields and the lowest productivity improvements. Here, yields oscillate between 0.5 and 2 t ha^{-1}, with an average of 1 t ha^{-1}, in sub-Saharan Africa and 1–1.5 t ha^{-1} in SAT Asia (Rockström and Falkenmark 2000; Wani et al. 2003a, b; Rockström et al. 2007). Farmers' yields continue to be very low compared with the experimental yields (attainable yields) as well as simulated crop yields (potential yields), resulting in a very significant yield gap between actual and attainable rainfed yields. The difference is largely explained by inappropriate soil, water and crop management options used at the farm level, combined with persistent land degradation and inappropriate institutional and policy mechanisms. The vast potential of rainfed agriculture needs to be unlocked through knowledge-based management of soil, water and crop resources for increasing productivity and nutrient use efficiency through sustainable intensification.

4 Intensification to Bridge Yield Gaps and Environmental Implications

The intensive use of chemical fertilizers during the past four to five decades undoubtedly quadrupled global food grain production but has created implications for the environmental safety (Tilman et al. 2001, 2002; Hungate et al. 2003; Sutton et al. 2011). Worldwide, chemical fertilizer consumption has increased fourfold during the last 50 years (FAO 2011). As regards to N fertilizers, the increase in agricultural food production worldwide over the past four decades has been associated with a sevenfold increase in the use of N fertilizers (Rahimizadeh et al. 2010), with 33 % nitrogen use efficiency (Raun and Johnson 1999). Similarly, an overview of agriculture in India indicates that since the late 1960s (1966–1971), the period that coincides with the launch of green revolution, the food grain production is more than doubled during 2006–2009 with almost no change in area but accompanied by more than 12 times increase in nitrogenous fertilizer consumption (Ministry of Agriculture, Government of India 2013). High nitrifying nature of intensive production systems results in loss of nearly 70 % of the overall N-fertilizer inputs (Peterjohn and Schlesinger 1990; Raun and Johnson 1999). Rapid and unregulated nitrification from agricultural systems results in increased N leakage to the environment (Schlesinger 2009). Nitrogen-fertilizer-based pollution is also becoming a serious issue for many agricultural regions (Garnett et al. 2009). Inefficient use of N fertilizer is causing serious environmental problems associated with the emission of NH_3, N_2 and N_2O (the last being an important greenhouse gas implicated both in the global warming and ozone layer depletion in the stratosphere) to the atmosphere. N_2O is a powerful greenhouse gas having a global warming potential (GWP) 300 times greater than that of CO_2 (Kroeze 1994; IPCC 2007), while the earth's protective ozone layer is damaged by NOs that reach the stratosphere (Crutzen and Ehhalt 1977). The loss of NO_3 from the root

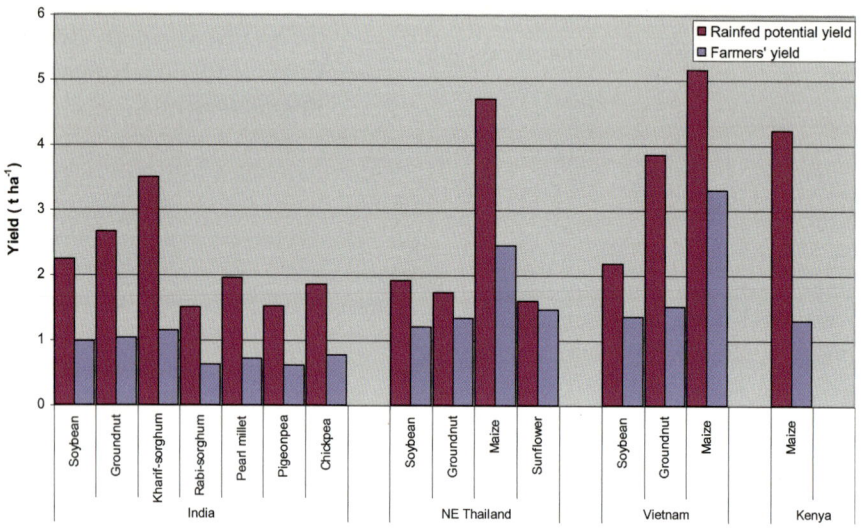

Fig. 1 Yield gap of important rainfed crops in different countries (Source: Rockström et al. 2007)

zone and NO_3 contamination of ground and surface water via nitrate leaching or run-off are major environmental concerns (Singh and Verma 2007; Tilman et al. 2001; Galloway et al. 2008; Schlesinger 2009). Current estimates indicate that N lost by NO_3 leaching from agricultural systems could reach 61.5 Tg N year^{-1} by 2050 (Schlesinger 2009). Excessive fertilizer run-off in water bodies results in growth of algal blooms leading to eutrophication, shifting the state of lake systems from clear to turbid water (Carpenter 2003). It was recently documented by Rockstorm et al. (2009) that planetary boundaries for nitrogen cycle have already crossed the biophysical thresholds. Similarly excessive phosphate fertilizer can be a significant contributor of potentially hazardous trace elements such as arsenic, cadmium and lead in croplands. These trace elements have the potential to accumulate in soils and be transferred through the food chain (Jiao et al. 2012). In response to continually increasing economic and environmental pressures, there is an urgent need to enhance efficient use of nitrogenous fertilizers and increase profitability by developing sustainable farming systems (Mahler et al. 1994).

5   Potential for Sustainable Intensification

Evidence from a long-term experiment at the International Crops Research Institute for the Semi-Arid Tropics (ICRISAT), Patancheru, India, since 1976 demonstrated the virtuous cycle of persistent yield increase through improved land, water and nutrient management in rainfed agriculture. Improved systems of sorghum + pigeon pea intercrops produced higher mean grain yields (5.1 t ha^{-1}) through increased rainwater use efficiency compared with 1.1 t ha^{-1}, the average yield of sole sorghum in the traditional (farmers') post-rainy system, where crops are grown on stored soil moisture (Figs. 2 and 3). The annual gain in grain yield in the improved system was 70 kg ha^{-1} year^{-1} compared with 20 kg ha^{-1} year^{-1} in the traditional system. The large yield gap between attainable yield and farmers' practice as well as between the attainable yield of 5.1 t ha^{-1} and potential yield of 7 t ha^{-1} shows that a large potential of rainfed agriculture remains to be tapped. Moreover, the improved management system is still continuing to provide an increase

Fig. 2 Effects of improved management and farmers' management systems on crop yields during 1976–2012 at ICRISAT, Patancheru, India (Source: Wani et al. 2012a)

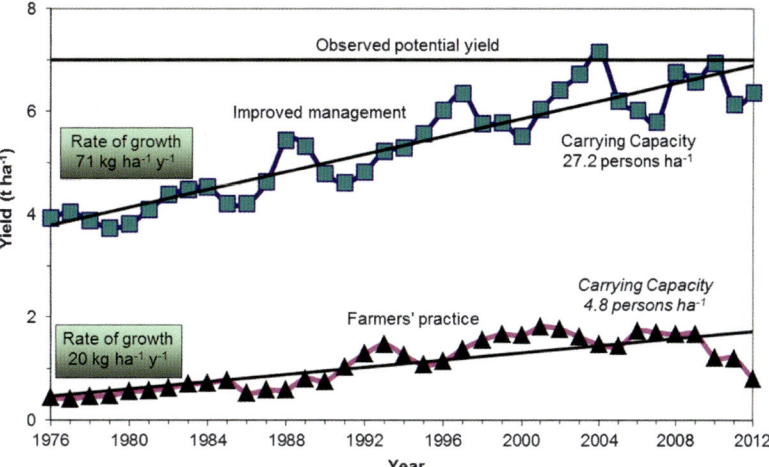

Fig. 3 Effects of improved management and farmers' management systems on rainfall use efficiency during 1976–2012 at ICRISAT, Patancheru, India

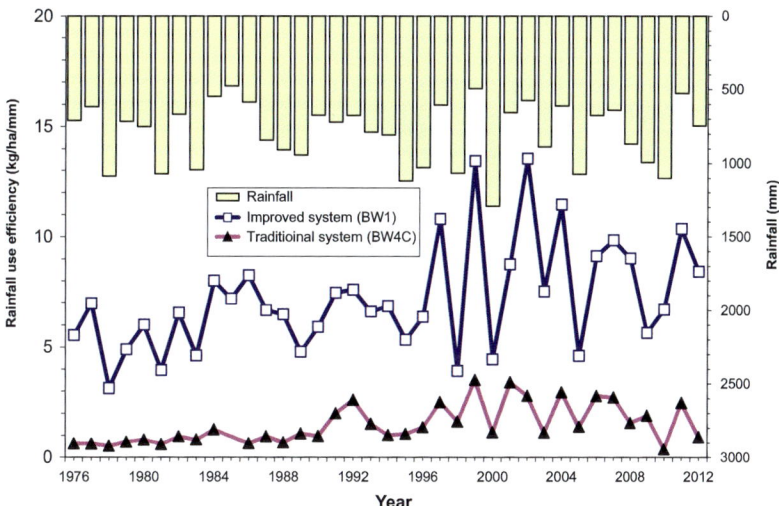

in productivity as well as improving soil quality (physical, chemical and biological parameters) along with increased carbon sequestration which is very much required to promote soil organic carbon storage critical for optimum soil processes to enhance nutrient use efficiency.

Long-term studies at ICRISAT showed that an improved system having balanced fertilization not only increased crop productivity but also increased soil organic C and nutrients like total and available N and Olsen P (Wani et al 2003a) in the system. This study showed that an additional quantity of 7.3 t C ha^{-1} (335 kg C ha^{-1} year^{-1}) was sequestered in soil under the improved system compared with the traditional system over the

24-year period. With an increase in biomass C (89 %), there was 83 % increase in mineral N, 105 % increase in microbial biomass N and about 18 % increase in total N in the improved system compared with the traditional system. Microbial biomass is one of the most labile pools of organic matter and serves as an important reservoir of plant nutrients such as N and P (Jenkinson and Ladd 1981). Biomass C, as a proportion of total soil C, serves as a surrogate for soil quality (Jenkinson and Ladd 1981). ICRISAT long-term study showed that under improved management practices, biomass C constituted a higher proportion of soil organic C up to 10.3 % as compared with 6.4 % under farmers' practice. Biomass N is comprised of about 2.6 % of total soil N in the improved system, whereas in the traditional system, it constituted only 1.6 %.

6   What Does Increased Nutrient Use Efficiency Imply?

Nutrient use efficiency can be defined in many ways and is easily misunderstood and misrepresented. Definitions differ, depending on the perspective. Increased nutrient use efficiency implies the following:

- Lesser nutrient need for obtaining a given level of production or more produce per unit of nutrient applied
- Lower cost of production per unit of produce
- Higher returns per $ invested on nutrient use
- Reduced risk of environmental pollution

Over- or under-application of needed nutrients will result in reduced nutrient use efficiency or losses in yield and crop quality. Improving nutrient efficiency is an appropriate goal for all involved in agriculture. However, maximizing efficiency may not always be advisable or effective, and effectiveness cannot be sacrificed for the sake of efficiency. Much higher nutrient efficiencies could be achieved simply by sacrificing yield, but that would not be economically effective or viable for the farmer or the environment. For a typical yield response curve, nutrient use efficiency is high at a low yield level, because any small amount of nutrient applied could give a large yield response. If nutrient use efficiency were the only goal, it would be achieved here in the lower part of the yield curve. As we move up the response curve, yields continue to increase, albeit at a slower rate, and nutrient use efficiency typically declines. However, the extent of the decline is dictated by the best management practices (BMPs) employed (i.e. right rate, right time, right place, improved balance in nutrient inputs, etc.) as well as soil and climatic conditions and is the target area of researchers to enhance the nutrient use efficiency through optimization of BMPs.

6.1   Measures of Nutrient Use Efficiency

The nutrient use efficiency is measured in different ways depending upon the perspective in which it is computed and considered. The agronomists, soil scientists, plant physiologists and agricultural economists use different expressions/measures for nutrient use efficiency. Taking nitrogen (N) as an example of plant nutrients, different measures of nutrient use efficiency can be defined as follows (Delogu et al. 1998; Lopez-Bellido and Lopez-Bellido 2001):

Nitrogen uptake efficiency (NUpE) is worked out by dividing total plant N uptake with N supply (Eq. 1).

$$NUpE(kg\ kg^{-1}) = Nt/N\ supply \qquad (1)$$

where Nt is the total plant N uptake and is determined by multiplying dry weight of plant parts by N concentration and summing over parts for total plant uptake. N supply is the sum of soil N content at sowing, mineralized N and N fertilizer. N supply is defined (Limon-Ortega et al. 2000) as the sum of (i) N applied as fertilizer and (ii) total N uptake in control (0 N applied).

Nitrogen utilization efficiency (NUtE) is worked out by dividing grain yield with total plant N uptake (Eq. 2).

$$NUtE(kg\ kg^{-1}) = Y/Nt \qquad (2)$$

where Y is grain yield.

Nitrogen use efficiency (NUE) is estimated by dividing grain yield with N supply (Eq. 3).

$$NUE(kg\ kg^{-1}) = Y/N\ supply \qquad (3)$$

The nitrogen harvest index (NHI) is determined by dividing total grain N uptake with total plant N uptake and multiplying by 100 (Eq. 4).

$$NHI(\%) = (Ng/Nt) \times 100 \qquad (4)$$

where Ng is the total grain N uptake. Ng is determined by multiplying dry weight of grain by N concentration.

There are some incremental efficiency measures under Reddy (2013).

Agronomic efficiency of N (AEN) is the increase in crop yield per unit of N applied, i.e. ratio of the increase in yield to the amount of N applied (Eq. 5).

$$AEN(kg\ kg^{-1}) = (Y_N - Y_0)/N\ applied \quad (5)$$

where Y_N (kg ha^{-1}) is the economic yield with N application, Y_0 (kg ha^{-1}) is the economic yield without N application and N applied (kg ha^{-1}) is the amount of N applied.

Recovery efficiency of N (REN) refers to the increase in N uptake by plant (aboveground parts) per unit of N applied (Eq. 6).

$$REN(\%) = (NnNo)/N\ applied \times 100 \quad (6)$$

where Nn (kg ha^{-1}) is the N uptake by crop with N application and No (kg ha^{-1}) is the N uptake by crop without N application.

Physiological efficiency of N (PEN) indicates the efficiency with which the plant utilizes the absorbed N to produce economic yield (Eq. 7).

$$PEP(kg\ kg^{-1}) = (Y_N - Y_0)/(NnNo) \quad (7)$$

Economic efficiency of N (EEN) refers to agronomic efficiency (AEP) expressed in monetary terms (Eq. 8). It can be equated with most popularly used benefit to cost ratio.

$$\begin{aligned} EEP = {} & (Y_N - Y_0)/N\ applied \\ & \times Value\ of\ the\ produce(Rs) \\ & /Cost\ of\ the\ nutrient(Rs) \quad (8) \end{aligned}$$

Partial factor productivity for N (PFPN) from applied N is the ratio of grain yield to amount of N applied (Eq. 9).

$$PFPN(kg\ kg^{-1}) = Y/N\ applied \quad (9)$$

7 Enhancing Nutrient Use Efficiency Through Bridging Yield Gaps

Crop yield directly or indirectly is the numerator in different terms of nutrient use efficiency, and the practices that increase crop yield may therefore increase nutrient use efficiency. The soil-water-crop management practices that promote crop productivity at the same level of fertilizer

use are expected to enhance nutrient use efficiency. Similarly, all the management practices that minimize nutrient requirement while achieving desired productivity targets would also lead to increased nutrient use efficiency.

7.1 Integrated Watershed Management

In rainfed areas, watershed management is the approach used for conservation of water and other natural resources as well as for sustainable management of natural resources while enhancing ecosystem services such as provisioning production (food, fodder and fuel), erosion control, groundwater recharge, transportation of nutrients, recreation, etc. Watershed management is the process of organizing land use and use of other resources in a watershed to provide desired goods and services to people while enhancing the resource base without adversely affecting natural resources and the environment (Wani et al. 2001). The soil and water management measures in the treated watershed include field bunding, gully plugging and check dams across the main watercourse, along with improved soil, water, nutrient and crop management technologies.

In Adarsha watershed in Kothapally, Andhra Pradesh, India, there was a significant reduction in run-off from the treated watershed compared to the untreated area in 2000 and 2001 (Table 1). In high rainfall year (2000), run-off from the treated watershed was 45 % less than the untreated area. During a subnormal rainfall year (2001), run-off from the treated watershed was 29 % less than the untreated area. Of the 3 years during 1999–2001, 2 years (1999 and 2001) were low rainfall years. Besides low rainfall, most of the rainfall events were of low intensity. This resulted in very low seasonal run-off during 1999 and 2001. Generally, during the low run-off years, the differences between the treated and untreated watersheds are very small. During good rainfall, i.e. 2000, a significant difference in

Table 1 Seasonal rainfall, run-off and soil loss from the Adarsha watershed in Kothapally, Andhra Pradesh, India, 1999–2001

	Run-off (mm)			Soil loss (t ha^{-1})	
Year	Rainfall	Untreated	Treated	Untreated	Treated
1999	584	16	NR	–	–
2000	1,161	118	65	1.04	–
2001	612	31	22	1.48	0.51

Source: Sreedevi et al. (2004)

Untreated = control with no development work; treated = with improved soil, water and crop management technologies; NR = not recorded

the run-off was seen between treated and untreated watersheds (Table 1). The soil loss was measured both from treated and untreated watersheds during 2001. There was a significant reduction in soil loss from treated watershed (only 1/3 soil loss) compared to untreated watershed in 2001. Thus, integrated watershed management is an important vehicle of technologies to check nutrient losses or reversing nutrient losses through run-off water or along with soil lost. Thus, management at watershed scale is another important aspect that needs urgent attention to enhance efficiency of inherent nutrients in soil and added through fertilizers and manures.

More infiltrations through reduced run-off under watersheds (Wani et al. 2012b) also strengthen the green-water sources to create synergy with nutrients to get higher yields and nutrient use efficiency. For food production worldwide, the consumption of green water is almost threefold more than blue water (5,000 vs. 1,800 km^3 year^{-1}) (Karlberg et al. 2009) and thereby changes in it can result large impact on yields and also nutrient use efficiencies. Evidences from different watersheds (Table 2) have shown substantial productivity improvement as compared to non-watershed regions leading to efficient nutrient and resource use efficiency. As a result of watershed interventions, the rainwater use efficiency by different crops increased by 15–29 % at Xiaoxincun (China), 13–160 % at Lucheba (China) and 32–37 % at Tad Fa (Thailand), which brought in substantial productivity improvement (Table 2). The

watershed interventions which improve substantially the green-water resources apparently led to better utilization of available water resources in productive transpiration and resulted in more food per drop of water. The run-off water harvested in tanks facilitated supplementary irrigation at critical stages and brought a change in production scenario. The results proved that integrated soil, crop and water management with the objective of increasing the proportion of the water balance as productive transpiration, which constitutes one of the most important rainwater management strategies to improve yields and water productivity, is effectively addressed through participatory watershed interventions. In addition to long-term sustainable benefits, crop production with watershed intervention is also a profitable option in terms of benefit: cost ratio.

7.2   Soil Health Management and Nutrient Use Efficiency

7.2.1 Widespread Soil Fertility Degradation Resulting Low Crop Yields and Nutrient Use Efficiency

Land degradation represents a diminished ability of ecosystems or landscapes to support the functions or services required for sustainable intensification. Agricultural production over a period of time particularly in marginal and fragile lands has resulted in degradation of the natural resource base, with increasing impact on productivity and nutrient use efficiency. Pervasive nutrient depletion and nutrient imbalances in agricultural soils are primary causes of decreasing yields and declining response to applied fertilizers. This depletion of selected soil nutrients often leads to fertility levels that limit production and severely affect nutrient use efficiency. Shorter fallow periods do not compensate for losses in soil organic matter and nutrients, leading to the mining of soil nutrients. In many African, Asian and Latin American countries, the nutrient depletion of agricultural soils is so high that current agricultural land use is not sustainable.

Table 2 Crop yield and rainwater use efficiency during pre- and post-watershed interventions in watersheds in China and Thailand

Crop	Pre-project period			Post-project period		
	Crop yield (kg ha^{-1})	RWUE (kg mm^{-1} ha^{-1})	B:C ratio	Crop yield (kg ha^{-1})	RWUE (kg mm^{-1} ha^{-1})	B:C ratio
Xiaoxincun, China						
Rice	5,800	9.5	1.9	6,300	11.2	2
Maize	4,500	7	1.9	5,200	8.1	2.2
Groundnut	1,400	2.2	1.8	1,800	2.8	2.2
Watermelon	10,500	16.4	3.4	12,500	19.5	3.9
Sweet potato	19,500	30.4	2.5	22,500	35.1	3
Lucheba, China						
Vegetables	36,900	28.8	1.4	41,900	32.6	1.8
Watermelon	11,300	8.8	1.5	29,300	22.8	1.6
Tad Fa, Thailand						
Maize	3,218	2.7	2.3	4,500	3.7	2.7
Cabbage	36,343	29.8	3.9	49,063	40.2	4.3
Chillies	2,406	2	4	3,188	2.6	4.6

Source: Wani et al. (2012a)

Table 3 Soil fertility status of farmers' fields in rainfed semi-arid tropics of India

State	No. of farmers	% deficiency (range of available nutrients)					
		Org-C	Av P	Av K	Av S	Av B	Av Zn
[a]Andhra Pradesh	3,650	76 (0.08–3.00)	38 (0.0–248)	12 (0–1,263)	79 (0.0–801)	85 (0.02–4.58)	69 (0.08–35.6)
[b]Gujarat	82	12 (0.21–1.90)	60 (0.4–42.0)	10 (30–635)	46 (1.1–150)	100 (0.06–0.49)	85 (0.18–2.45)
[c]Karnataka	92,904	52 (0.01–9.58)	41 (traces-544)	23 (traces-3,750)	52 (0.9–237)	62 (0.02–4.60)	55 (traces-235)
[a]Madhya Pradesh	341	22 (0.28–2.19)	74 (0.1–68)	1 (46–716)	74 (1.8–134)	79 (0.06–2.20)	66 (0.10–3.82)
[a]Rajasthan	421	38 (0.09–2.37)	45 (0.2–44)	15 (14–1,358)	71 (1.9–274)	56 (0.08–2.46)	46 (0.06–28.6)
[b]Tamil Nadu	119	57 (0.14–1.37)	51 (0.2–67.2)	24 (13–690)	71 (1.0–93.6)	89 (0.06–2.18)	61 (0.18–5.12)

Source: [a]Wani et al. (2012b), [b]Sahrawat et al. (2007), [c]Wani et al. (2011)
The figures in the parentheses indicate the range of nutrients % for Org-C and mg kg^{-1} for P, K, S, B and Zn

Nutrient depletion is now considered the chief biophysical factor limiting small-scale production in Africa (Drechsel et al. 2004). Recent characterization of farmers' fields in different states across India revealed a widespread deficiency of zinc (Zn), boron (B) and sulphur (S) in addition to known deficiencies of macronutrients such as nitrogen (N) and phosphorus (P) (Table 3). New widespread deficiencies of secondary and micronutrients are apparently the reason for holding back the productivity potential (Sahrawat et al. 2007, 2011; Wani et al. 2012b; Chander et al. 2013a, b, 2014a, b) and declining response to macronutrients and so decreasing nutrient use efficiency. In view of observed deficiencies, the application of major nutrients N, P and K as currently practiced is important for the SAT soils (El-Swaify et al. 1985; Rego et al. 2003), but very little attention has been paid to diagnose and take corrective measures for deficiencies of secondary nutrients and micronutrients in various crop production systems (Rego et al. 2005; Sahrawat et al. 2007, 2011; Wani et al. 2012b) followed in millions of small and marginal farmers' fields in the rainfed SAT. The role of soil organic carbon (C) in maintaining soil health is also well documented (Wani et al. 2012c). However, low soil organic C in SAT soils is another factor contributing to poor crop productivity (Lee and Wani 1989; Edmeades 2003; Ghosh et al. 2009; Materechera 2010; Chander et al. 2013a). Soil organic matter, an important driving force for supporting biological activity in soil, is very much in short supply, particularly in tropical countries. Management practices that augment soil organic matter and maintain it at a threshold level are needed (Chander et al. 2013a). Therefore, there is need to identify and promote management interventions with high carbon sequestration potential to promote soil organic carbon storage which is very critical for optimum soil processes to enhance nutrient use efficiency.

7.2.2 Soil Health Management: An Important Driver for Enhancing Nutrient Use Efficiency

Often, soil fertility is the limiting factor to increased yields in rainfed agriculture. With experiences of green revolution and in a quest to get higher yields, farmers have started adding macronutrients in quantities higher than required and getting declining response to nutrient inputs. Based on soil analysis results, ICRISAT-led consortium has designed and is promoting balanced nutrient management practices which also include deficient secondary nutrients and micronutrients. Soil test-based fertilizer recommendations are designed at cluster of villages called block, a lower administrative unit in a district, by considering practical aspects like available infrastructure, human power and economics in research for impact for smallholders in the Indian SAT. Fertilizer recommendations at block level cater well to soil fertility needs in contrast to current blanket recommendations at state level. We recommend to apply full dose of a particular nutrient if its deficiency was on >50 % farms in a block and half dose of a nutrient if its deficiency was on <50 % farms. This way of nutrient recommendation was adopted to manage existing risks in rainfed agriculture in the SAT while targeting optimum yields to improve livelihoods of poor SAT farmers. Scaling up of such soil test-based balanced fertilization through farmer participatory trials in rainfed systems in India and particularly in Karnataka through extensive government support has shown substantial increase (~20–70 %) in crop yields after micro- and secondary nutrient amendments and at same levels of primary macronutrients indicating enhanced use efficiency of macronutrients (Fig. 4).

Based on diagnosed deficiencies and using soil test-based nutrient management, on-farm trial results indicated improvements in soil fertility parameters in spite of getting higher yields (Fig. 5). In simple terms soil test-based balanced fertilization not only enhances nutrient use efficiencies of macronutrients through increased yields under same levels of macronutrients but also captured more nutrients in the soil system. On-farm studies have shown residual benefits of soil test-based applied secondary nutrients and micronutrients as increased yields over farmers' practice plots up to three succeeding seasons (Chander et al. 2013a, 2014a), and thereby enhancing use efficiencies of macronutrients on a sustainable basis.

7.2.3 Nitrogen and Phosphorus Use Efficiency Under Balanced Nutrition

Nitrogen is often the most limiting nutrient for crop yield in many regions of the world and, in a quest to achieve high yields, is applied in large quantity from external sources resulting in low-N use efficiency. Along with N, the deficiencies of P are common in SAT soils (Sahrawat et al. 2007, 2010), and P is the next nutrient added in large quantities. On these soils, it can be necessary to apply up to fivefold more P as fertilizer than is exported in products (Simpson et al. 2011) due to extensive fixation in the soil.

Fig. 4 Maize grain yield response to improved management and farmers' management practices in various districts of Karnataka during 2011 rainy season

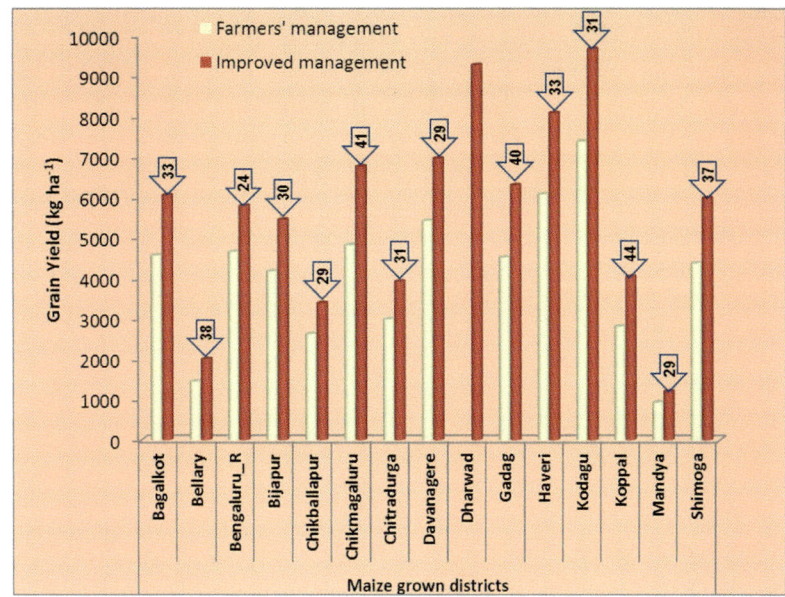

Fig. 5 Postharvest soil fertility status after 2010 rainy season groundnut in Nalgonda (Source: Chander et al. 2014a)

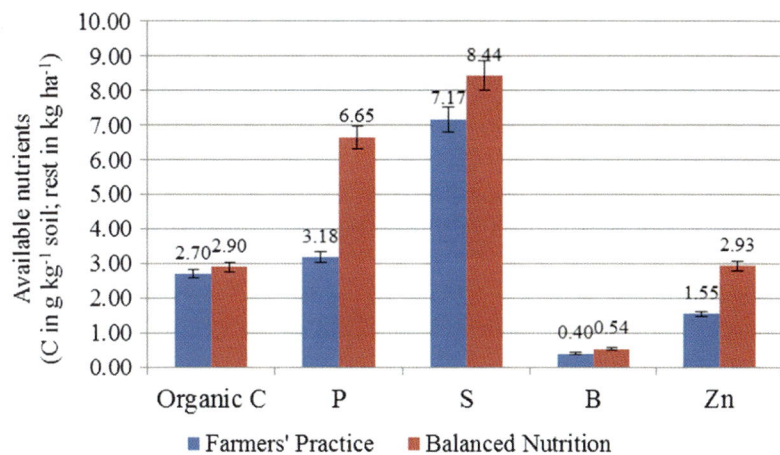

Phosphorus fertilizer is expensive for smallholder farmers, and given the finite nature of global P sources, it is important that such inefficiencies be addressed. Plant nutrients rarely work in isolation. Interactions among nutrients are important because a deficiency of one restricts the uptake and use of another. We hypothesized that multiple nutrient deficiencies could result into low-nutrient use efficiency in N and P and therefore studied different aspects of it.

Nutrient uptake efficiency (NUpE/PUpE) reflects the efficiency of the crop in obtaining it

from the soil (Rahimizadeh et al. 2010). Uptake of supplied nutrient is the first crucial step and an issue of concern worldwide, and hence, increased nutrient uptake efficiency has been proposed as a strategy to increase nutrient use efficiency by Raun and Johnson (1999). Nutrient utilization efficiency (NUtE/PUtE) reflects the ability of the plant to transport the nutrient uptakes into grain (Delogu et al. 1998). The nutrient harvest index (NHI/PHI), defined as nutrient in grain to total nutrient uptake, is an important consideration in cereals. The NHI/PHI reflects the grain

Table 4 Effects of balanced nutrient management strategies on nitrogen efficiency indices in maize at ICRISAT, Patancheru, India, 2010 rainy season

Treatment	NUpE	NUtE	NUE	NHI
Control	1.00	60.2	60.2	46.8
NP	0.37	80.7	30.1	67.3
NP + SBZn (every year)	0.46	78.5	36.0	60.5
NP + 50 %SBZn (every year)	0.51	92.5	47.3	65.8
NP + SBZn (alternate year)	0.47	84.4	39.7	69.3
NP + 50 %SBZn (alternate year)	0.42	80.8	34.1	67.0
LSD (5 %)	0.11	17.4	8.85	11.3

Source: Chander et al. (2014b)

Table 5 Effects of balanced nutrient management strategies on phosphorus efficiency indices in maize at ICRISAT, Patancheru, India, 2010 rainy season

Treatment	PUpE	PUtE	PUE	PHI
Control	1.00	172	172	60.4
NP	0.49	228	111	83.5
NP + SBZn (every year)	0.41	328	134	83.9
NP + 50 %SBZn (every year)	0.51	343	176	87.9
NP + SBZn (alternate year)	0.53	281	146	90.1
NP + 50 %SBZn (alternate year)	0.44	299	125	84.9
LSD (5 %)	0.15	83.7	38.6	9.40

Source: Chander et al. (2014b)

nutritional quality (Hirel et al. 2007). The results showed that the addition of deficient S, B and Zn recorded the highest uptake efficiency, utilization efficiency, use efficiency and harvest index in N and P in maize (Tables 4 and 5). The treatment N, P plus 50 % S, B and Zn added every year proved best over generally followed 100 % S, B and Zn addition once in 2 years. The nutrient uptake efficiency is positively correlated with plant dry matter and grain yield (Lee et al. 2004), which were favourably affected under S, B and Zn addition and explain the increase in NUpE. The findings showed that the balanced nutrition is the best strategy to increase cereal nitrogen uptake efficiency and thereby minimize N loss and environmental damage. Similar findings were also recorded in case of P. The study proved here that balancing N and P with deficient nutrients (Potarzycki 2010), which

in current context are S, B and Zn in the SAT soils, is an important strategy to improve utilization efficiency, use efficiency and harvest index in both N and P.

7.2.4 Recycling Nutrients in On-farm Wastes

In view of widespread low levels of soil organic carbon in rainfed soils, additions through organic sources of nutrients are very important to maintain optimum soil processes and enhance nutrient use efficiencies. Presently in India, about 960 million tonnes of solid wastes are being generated annually as by-products during municipal, agricultural, industrial, mining and other processes, and solely 350 million tonnes are organic wastes from agricultural sources (Pappu et al. 2007). Such large quantities of organic wastes can be converted through simple vermicomposting technique into valuable manure called vermicompost (VC) (Wani et al. 2002; Nagavallemma et al. 2004). Vermicomposting is faster than other composting processes due to biomass breakdown while passing through the earthworm gut and enhanced microbial activity in earthworm castings. Some earlier studies showed that vermicompost is an enriched source of nutrients with additional plant growth promoting properties and vermicompost application can improve nutrient availability, crop growth, yield and nutrient uptake (Nagavallemma et al. 2004). So, the on-farm produced vermicompost can enhance soil health and save costs of chemical fertilizers leading to nutrient use efficiency and economic productivity improvement.

Enriched vermicompost may be prepared from on-farm organic wastes and cow dung. Rock phosphate being a cheap source of P is added at 3 % of composting biomass to improve P content in vermicompost due to solubilization action of humic acids and phosphate solubilizing bacteria (Hameeda et al. 2006) during the vermicomposting process. *Eudrilus eugeniae* and *Eisenia foetida* species of earthworms are used for vermicomposting. The mature vermicompost is contained on an average of 1.0 % N, 0.8 % P, 0.7 % K, 0.26 % S, 110 mg

Table 6 Effects of nutrient managements on soybean (Glycine max) grain yield, benefit/cost ratio under rainfed conditions in Madhya Pradesh, India, during 2010 rainy season

| District | Grain yield (kg ha^{-1}) | | | LSD (5 %) | Benefit/cost ratio | |
	FP	BN	INM		BN	INM
Guna	1,270	1,440	1,580	34	1.31	4.58
Raisen	1,360	1,600	1,600	115	1.85	3.55
Shajapur	1,900	2,120	2,410	69	2.99	10.2
Vidisha	1,130	1,410	1,700	640	2.16	8.43

Source: Chander et al. (2013a)
Note: *FP* farmers' practice (application of N, P, K only), *BN* balanced nutrition (FP inputs plus S + B + Zn), *INM* integrated nutrient management (50 % BN inputs + vermicompost)

Table 7 Effects of nutrient managements on soybean (Glycine max) grain nutrient contents and total nutrient uptake in Raisen district, Madhya Pradesh, India, during 2010 rainy season

| Treatment | Total nutrient uptake | | | | | |
| | N | P | K | S | B | Zn |
	kg ha^{-1}			g ha^{-1}		
FP	98	9.71	53.5	5.78	88	101
BN	134	12.5	61.8	8.20	103	156
INM	138	13.8	65.1	9.29	108	179
LSD (5 %)	26	2.96	8.53	1.71	20	30

Source: Chander et al. (2013a)
Note: *FP* farmers' practice (application of N, P, K only), *BN* balanced nutrition (FP inputs plus S + B + Zn), *INM* integrated nutrient management (50 % BN inputs + vermicompost)

B kg^{-1}, 60 mg Zn kg^{-1} and 14 % organic C (Chander et al. 2013a).

On-farm results showed that with the use of vermicompost, the use and cost of chemical fertilizers can be reduced up to 50 % while getting higher productivity as compared to balanced nutrition solely through chemical fertilizers (Table 6), thereby enhancing nutrient use efficiency. More nutrients are captured as plant uptake under BN and INM practices due to enhanced contents and yields (Table 7). This is expected due to synergy created through nutrient balancing and specific roles of roles of nutrients like B which is necessary to maintain membrane integrity (Cakmak et al. 1995) and hence can enhance the ability of membranes to transport available nutrients. The INM practice results in economic benefits and efficient resource utilization including on-farm wastes and so is a sound-scalable technology.

7.3 Landform Management

Through efficient in situ water management using landform management like broad bed and furrow (BBF) or conservation furrow (CF) in poorly drained Vertisols, nutrient and other inputs can be efficiently utilized to get higher crop yields (Dwivedi et al. 2001; Sreedevi et al. 2004; Wani et al. 2003a). Rainwater management practices in rainfed agriculture are very critical particularly when most rainfall occurs in a limited period of the year. Initial downpours distort soil structure and also adversely affect water infiltration into soil and thereby ultimately negatively affect crop productivity and thereby resource use efficiency. Participatory evaluation clearly showed that landform management like BBF and CF keeps soil surface intact for more effective infiltration and safely allows excess run-off through furrows. The landform management practices in Sujala watersheds in Karnataka, India, increased crop yields over the farmers' practice of cultivating on flatbed by 12–20 % with CF and 30 % with BBF (Table 8).

7.4 Supplemental Irrigation

Water scarcity is a major limiting factor under rainfed agriculture, and thus the role of lifesaving one or two irrigations through harvested water in enhancing crop productivity and nutrient use efficiency is well understood and documented. However, studies have indicated micro-irrigation practices more effective than traditional flood irrigation practices in enhancing yields, nutrient and water use efficiency. On-station experiments at ICRISAT headquarter at Patancheru recorded significantly higher yields under drip irrigation as compared to flood irrigation (Table 9). The drip irrigation practice proved economically more remunerative while saving water resources also.

Table 8 Effects of land form management practices on crop yield in Sujala watersheds, Karnataka, India, 2006–2007

| District/watershed | Crop | Crop yields (kg ha^{-1}) | | |
		Farmers practice	Cultivation across slope with conservation furrow	Broad bed and furrow
Haveri				
Aremallapur	Maize	3,110	3,610 (16)*	–
Hedigonda	Maize	4,030	4,560 (13)	
Dharwad				
Parsapur	Soybean	1,500	1,800 (20)	
Kolar				
Diggur	Groundnut	1,010	1,200 (19)	–
Venkatesh Halli	Groundnut	950	1,070 (12)	–
Chitradurga				
Toparamalige	Maize	3,530	–	4,560 (30)

Source: ICRISAT (2007)
*Note: Figures in () indicate per cent increase over the farmers' practice

Table 9 Pooled data on yield of maize-chickpea cropping system (2009–2011) at ICRISAT, Patancheru

Treatment	Maize (t ha^{-1})	Chickpea (t ha^{-1})	Maize equivalent yield (t ha^{-1})	B:C
Flood irrigation	3.87	1.99	9.15	2.97
Drip irrigation	3.97	2.24	9.91	3.26
LSD (5 %)	NS	0.14	0.33	

Source: Sawargaonkar et al. (2012)

7.5 Integrated Genetic and Natural Resource Management

Cultivation of low-yielding cultivars in rainfed semi-arid tropics is one of the major factors for low yields leading to inefficient use of nutrient resources. This is a big opportunity to enhance nutrient use efficiencies through replacing low-yielding cultivars with high-yielding ones. On-farm research showed enhanced nutrient use efficiencies with high-yielding cultivars (Table 10). However, nutrient imbalances do not allow the high-yielding varieties to show potential, and participatory trials showed the highest yields and use efficiency of nutrients under integrated approach of improved variety and balanced nutrition.

Table 10 Integrated improved crop cultivar and balanced nutrient management enhance maize grain yield and RWUE in different districts of Rajasthan during 2009 rainy season

| District | Yield (kg ha^{-1}) | | | LSD (5 %) | B:C ratio |
	FP	IC	IC + BN		
Tonk	1,150	1,930	3,160	280	4.26
Sawai Madhopur	1,430	2,030	3,000	420	3.33
Bundi	1,380	2,180	4,240	714	6.05
Bhilwara	2,990	4,340	6,510	860	7.45
Jhalawar	2,550	3,520	4,960	316	5.11
Udaipur	2,530	3,090	6,320	509	8.03

Source: Chander et al. (2013b)

7.6 Improved Genotypes and Nutrient Use Efficiency

7.6.1 Need for Exploring Genotypic Diversity

Nitrogen use efficiency is a fundamental issue when discussing crucial topics related to yield improvements with fertilizer nitrogen application in an eco-friendly manner. The efficient use of nitrogen is important for the economic and environmental sustainability of production systems. Improving nitrogen uptake and partitioning to grain reduces the amount of nitrogen at risk of loss to the environment (Raun and

Johnson 1999). Enhanced grain N recovery is important for maintaining protein concentrations in high-yielding crops (Cox et al. 1986). In cereal cropping systems, nutrient use efficiency can be improved through two main strategies: by adopting more efficient farming techniques and by breeding more nutrient use-efficient cultivars (Ortiz-Monasterio et al. 1997). The efficient crop management practices have been discussed. Breeding strategies include identification and selection of desirable traits which increase the uptake and/or utilization efficiency of the crop (Foulkes et al. 2009) and identifying quantitative trait loci for NUE (Hirel et al. 2007). Therefore, development of N-efficient cultivars is needed to sustain or increase yield and quality while reducing the negative impacts of crop and fertilizer production on the environment (Hirel et al. 2007).

7.6.2 Genotypic Diversity for NUE Components

Genotypic diversity for NUE is well documented in wheat (Cox et al. 1985; Gooding et al. 2012), corn (Chevalier and Schrader 1977), sorghum (Maranville et al. 1980) and pearl millet (Wani et al. 1992; Uppal et al. 2014). As discussed earlier NUE can be expressed by two components NUpE and NUtE which express differently at various N input conditions. Various studies worldwide have identified genetic association between cereal grain yield and NUE components under contrasting conditions of high and low-N input supply. Some studies indicate that NUpE accounts for more genetic variations in NUE under low-N supply (Ortiz-Monasterio et al. 1997; Le Gouis et al. 2000), some indicate NUtE accounts for NUE in low-N supply (Wani et al. 1992; Alagarswamy and Bidinger 1982), whereas some studies conclude that both NUpE and NUtE contribute equally to NUE at all levels (Dhugga and Waines 1989). For NUE, genetic variability and genotype × nitrogen interactions reflecting differences in responsiveness have been observed in several studies on maize (Moll et al. 1982; Bertin and Gallais 2000), pearl millet (Wani et al. 1992) and sorghum. In addition, it has been found that correlations among various

agronomic traits such as grain protein yield and its components are different according to the level of nitrogen fertilization. At high N input, genetic variation in NUE was explained by variation in N uptake, whereas at low-N input, NUE variability was mainly due to differences in nitrogen utilization efficiency. This suggests that the limiting steps in N assimilation may be different when plants are grown under high or low levels of nitrogen fertilization.

Millets are staple food for millions of people in semi-arid tropics of Asia and sub-Saharan Africa which are generally grown on poor soils and low rainfall conditions with low fertilizer inputs. Genotype screening and selection for tolerance to low N and low P is an important strategy to increase productivity in nutrient-stressed environment. Various experiments on fertility management in pearl millet indicate that response of pearl millet varies widely among N studies with optimum rates from 0 to greater than 150 kg ha^{-1} N (Gascho et al. 1995). Most of the studies concluded that genotype × fertility interaction for grain yield and N utilization efficiency depends on grain production efficiency, i.e. cultivars yielding ability at a given level of fertilizer. A study conducted at two sites in ICRISAT with 12 genotypes and two N and P levels reported that millet hybrids have higher N, P and K use efficiency than composites and landraces which are conferred by higher harvest index and translocation of nutrients to developing grain in hybrids (Wani et al. 1992). The correlation between grain yield and NUtE suggests that direct selection for NUE may have value in improvement of yielding ability under low-fertility conditions (Alagarswamy and Bidinger 1982). A recent attempt to resynthesize earlier data sets from strategic research experiments on pearl millet reveals that NUtE is a more important contributor to NUE than NUpE under low to medium N supply (Uppal et al. 2014) (Fig. 6).

Similarly in a study at different agroecological systems, 15 genotypes of sorghum were evaluated for N and P concentrations at different growth stages in low-N or low-P Alfisols. Hybrids and improved varieties produced higher

Fig. 6 Linear regression of N uptake efficiency (NUpE) (y = 3.39 + 9.189; R^2 = 0.016) and N utilization efficiency (NUtE) (y = 0.788 + 1.93; R^2 = 0.72) on nitrogen use efficiency among four pearl millet cultivars. Symbols represent cultivar means over N rates (♦) = 700256, (■) = BJ 104, (▲) = Ex-Bornu and (●) = GAM 73. Regression was significant for **b**

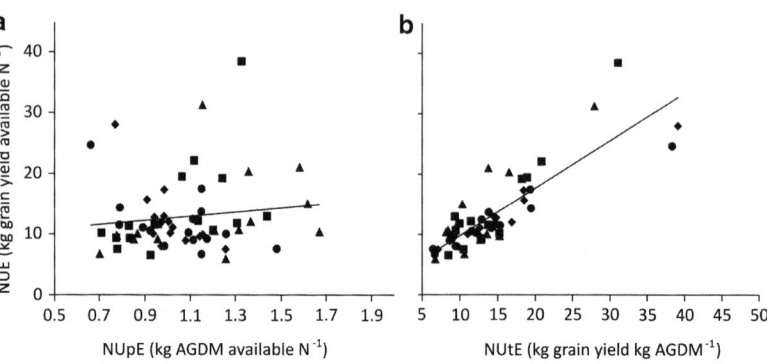

Table 11 Sorghum grain yield (GY, kg ha^{-1}), above-ground dry matter (AGDM, kg ha^{-1}), harvest index (HI), N uptake efficiency (NUpE = kg aboveground dry matter kg soil available N^{-1}), N utilization efficiency (NUtE = kg grain yield kg aboveground dry matter^{-1}) and nitrogen use efficiency (NUE = NUpE × NUtE = kg grain yield soil available N^{-1}) in a long-term trial (1978–1986)

Cultivar	GY	AGDM	HI	NUpE	NUtE	NUE
FLR101	1,899	3,913	0.33	1.03	46.06	47.48
CSV5	1,017	4,690	0.18	0.94	26.95	25.43
CSH5	2,173	5,037	0.30	1.11	48.97	54.33
IS889	1,405	2,203	0.39	0.84	41.84	35.13
DIALL	1,666	4,101	0.29	0.98	42.39	41.65

biomass and grain yield. In P-stressed situations, P from leaves and stem reserves is rapidly and efficiently translocated to support grain filling (Adu-Gyamfi et al. 2002). A P32 study revealed that in low-P conditions, P-efficient genotype translocates more P from roots to flag leaves (Adu-Gyamfi et al. 2002). In a study three maize genotypes that were grown in two sites with different soil types revealed that N-efficient trait of genotype is closely related to its adaptability to soil characteristics and water availability. ICRISAT's long-term experiments on sorghum reveal that genotypic diversity for NUE and its components exist among sorghum genotypes and genotypes with higher yield potential have higher NUE in Alfisols which are low in N and P (Table 11).

There is a lot of controversy about the performance of landraces, and farmers preferred varieties compared to hybrids and improved varieties in a low-nutrient environment. Various studies have showed that hybrids and new cultivars have more yield potential than landraces and old cultivars due to improved efficiency to fertilizer application (Wani et al. 1992; Adu-Gyamfi et al. 2002). On the contrary, some studies (Bationo et al. 1989; Payne et al. 1995) reported that local landraces or farmer-selected local lines of sorghum and pearl millet are better adapted to low-fertility regimes. There are various biotic and abiotic factors that influence the adaptation of crop plants to low-nutrient environments. Also crop response to nutrients depends on agronomic traits of the cultivar which contribute to grain yield and nitrogen use. Improvement in grain yield is more closely associated with grain N uptake in pearl millet (Fig. 7) leading to higher NHI (Uppal et al. 2014). Wani et al. (1992) found that selection for improved HI in modern pearl millet cultivars has inadvertently improved traits for NUE resulting in improved nutrient use efficiencies and nutrient translocation indices (Fig. 8).

Selection for nutrient-efficient cultivars is typically conducted under favourable field conditions with only the difference in soil nutrient availability. However, in practical field conditions, variation in soil types and/or seasonal weather conditions may have a strong influence on soil nutrient dynamics and plant growth and, therefore, nutrient uptake and its subsequent utilization in plants. Screening should take into

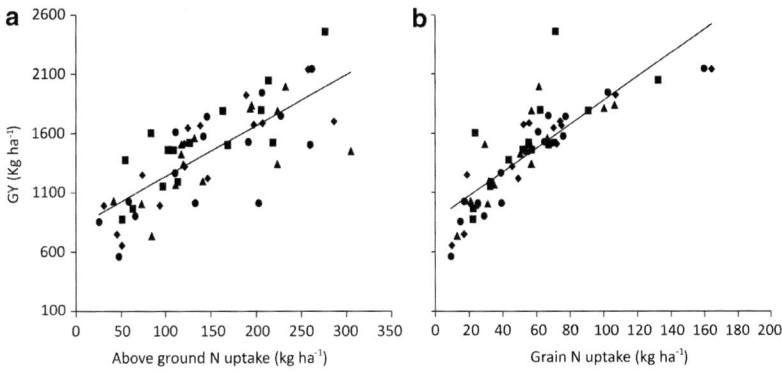

Fig. 7 Linear regression of aboveground N uptake (y = 4.28x + 806.79; R^2 = 0.58) and grain N uptake (y = 10.06 + 869.8; R^2 = 0.70) on grain yield among four pearl millet cultivars. Symbols represent cultivar means over N rates (♦) = 700256, (■) = BJ104, (▲) = Ex-Bornu and (●) = GAM 73

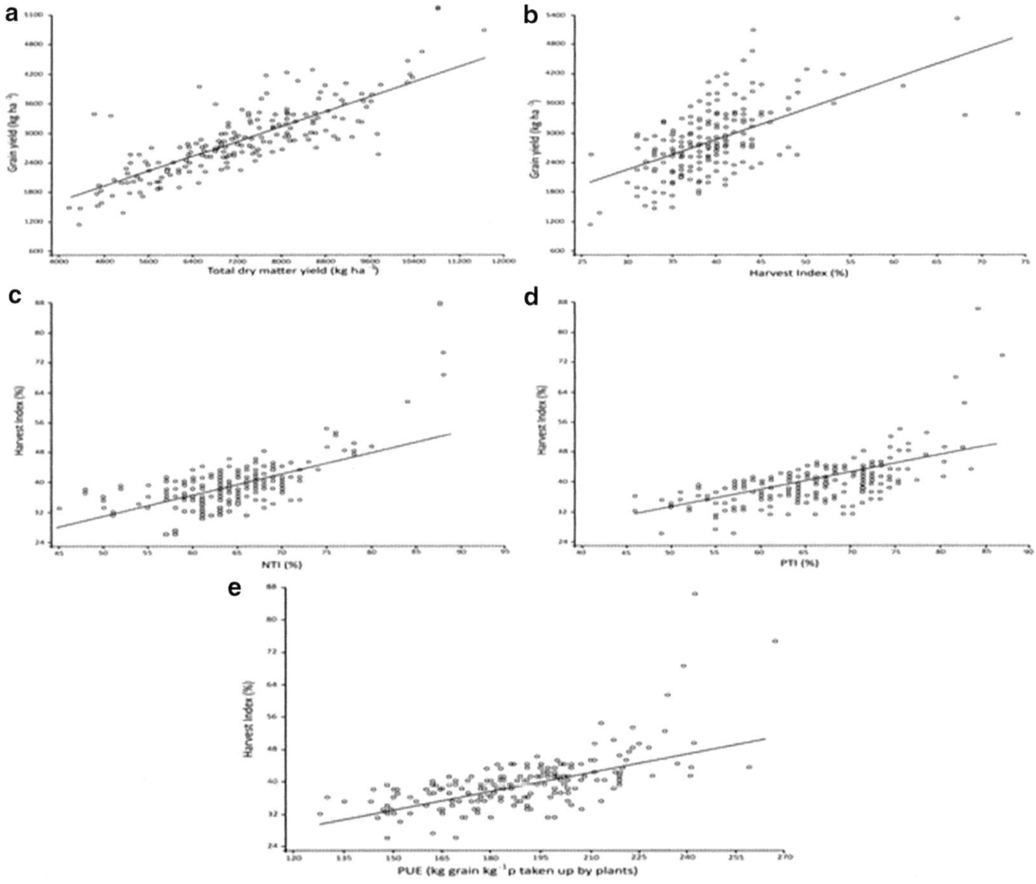

Fig. 8 Relationship between (**a**) grain yield and total dry matter (y = 84 + 0.380x; R^2 = 0.67), (**b**) grain yield and harvest index (y = 472 + 60.10x; R^2 = 0.28), (**c**) harvest index and nitrogen translocation index (NTI) (y = 1.41 + 0.589x; R^2 = 0.44), (**d**) harvest index and phosphorous translocation index (PTI) (y = 7.86 + 0.478x; R^2 = 0.38) and (**e**) harvest index and phosphorous use efficiency (y = 8.64 + 0.162x; R^2 = 0.48) of pearl millet genotypes

Table 12 Severity of foliar diseases, pod and haulm yields of IDM and non-IDM plots in a watershed in Dharwad District, Karnataka, 2006 rainy season

	FD score 1–9 scale		Pod yield (kg ha^{-1})		Haulm yield (kg ha^{-1})	
District	IDM	Non-IDM	IDM	Non-IDM	IDM	Non-IDM
Dharwad	5.5	8.3	860	660	1,530	1,140

Source: ICRISAT (2007)

Note: FD = foliar diseases; IDM = improved dual purpose cultivar ICGV 91114; seed treatment with bavistin + thirum (1:1) @ 2.5 g kg^{-1} seed; foliar application of fungicide kavach/bavistin at 60–65 DAS; Non-IDM = farmers' practice

consideration the interaction of nutrients, water, soil type, climatic variables and cropping system.

7.6.3 Candidate Traits for High NUE and Mechanism

Promising traits for selection by breeders to increase NUE have been identified which include increased root length density, higher N uptake, low-leaf lamina N concentration, more efficient post-anthesis N remobilization to developing grain and reduced N concentration in feed crops may be of particular value for increasing NUE. We will be discussing N remobilization in detail as it affects the nitrogen harvest index of the crop.

During leaf senescence NH3 is liable to be lost from plants by volatilization. This loss can be reduced by high glutamine synthetase (GS1) activity (Mattsson et al. 1998). A positive relationship between GS1 activity and NUtE and grain yield has been reported in maize grown under low-N conditions (Masclaux et al. 2001), and QTLs for NUE and a structural gene for GS1 are co-localized (Hirel et al. 2007). Over 80 % of the aboveground N at harvest can be present in the aboveground crop at flowering and can account for 50–80 % of the nitrogen accumulated in the grains at maturity depending on crop species (Hirel et al. 2001). N remobilization is an important trait affecting the utilization of canopy N, and the efficiency of the N remobilization from aboveground parts to the grain can be measured by the nitrogen harvest index (NHI). The NHI is a heritable characteristic (Cox et al. 1985). The nitrogen harvest index has a positive association with N uptake by grain and a negative trend with straw N concentration and quantity (Tripathi et al. 2004).

7.7 Integrated Pest Management

Crop diseases, insects, weeds are one of the major constraints to increase food production and higher resource use efficiency. Though reliable estimates on crop losses are limited, Oerke et al. (1995) brought out about 42 % loss in global output due to insect pests, diseases and weeds despite the use of plant protection options. In India, the pre-harvest loss was up to 30 % in cereals and pulses, and it can be up to 50 % in cotton and oilseed crops (Dhaliwal and Arora 1993).

In rainfed systems, unawareness about and lack of good agronomic practices is leading to low yields resulting in poor nutrient use efficiency. Participatory trials in Dharwad District of Karnataka, India, showed that foliar disease severity was low in holistic integrated disease management (IDM) plots of groundnut variety ICGV-91114 than non-IDM plots of local cultivar. Its mean severity was 5.5 on a 1–9 rating scale in IDM plots compared to an 8.3 rating in non-IDM plots (Table 12). Under IDM plots, pod yield was significantly higher as compared to non-IDM plot under the same level of nutrient use.

The agricultural sector in India or elsewhere has long been recognized for its dependence on chemical control for the management of biotic stresses (insects, diseases and weeds). The excessive dependence on chemical pesticides led to the development of resistance in pests to pesticides, outbreaks of secondary pests and pathogens/biotypes and occurrence of residues in the food chain (Ranga Rao et al. 2009). To overcome such situations and minimize damage to human and animal health, several organizations have started advocating the concept of IPM with better profits. Studies have indicated that crop- and need-based IPM technologies which are very effective tools to reduce chemical use, also result into better pest control (Ranga Rao et al. 2009; Chuachin et al. 2012) to get higher productivity and nutrient use efficiency.

8 Conclusions and Way Forward

The rising use of nutrient inputs to meet future food security is unavoidable. However, in current scenario as discussed in this chapter, there is lot of scope to improve nutrient use efficiency through optimizing crop-growing environment and other inputs to get the maximum productivity. Scientific awareness and solutions to most problems are available and, however, have not reached on farmers' fields particularly in rainfed systems. Ensuring implementation of holistic solutions at farm level through consortium of technical institutions should be the priority of all stakeholders. Strengthening of on-farm research for impact and innovative extension systems is a very important aspect that needs immediate attention to see changes on ground.

References

Adu-Gyamfi JJ, Ishikawa S, Nakamura T, Nakano H (2002) Genotypic variability and physiological characteristics of crop plants adapted to low-nutrient environments. Food security in nutrient-stressed environments: exploring plants' genetic capabilities. Kluwer Academic Publishers, Dordrecht. pp 67–79

Alagarswamy G, Bidinger FR (1982) Nitrogen uptake and utilization by pearl millet (*Pennisetum americanum* (L.) Leeke). In: 9th international plant nutrition congress, Warwick University, UK, 22–27 Aug 1982

Amarasinghe UA, Shah T, Turral H, Anand BK (2007) India's water future to 2025–2050: business-as-usual scenario and deviations, IWMI research report 123. International Water Management Institute (IWMI), Colombo

Bationo A, Christianson CB, Baethgen WE (1989) Plant density and fertilizer effects on pearl millet production in Niger. Agron J 82:290–295

Bertin P, Gallais A (2000) Physiological and genetic basis of nitrogen use efficiency in maize. I Agrophysiological results. Maydica 45:53–66

Bhatia VS, Singh P, Wani SP, Kesava Rao AVR, Srinivas K (2006) Yield gap analysis of soybean, groundnut, pigeonpea and chickpea in India using simulation modeling, Global Theme on Agroecosystems Report No. 31. International Crops Research Institute for the Semi-Arid Tropics (ICRISAT), Patancheru, Andhra Pradesh

Cakmak I, Kurtz H, Marschner H (1995) Short term effects of boron, germanium and high light intensity on membrane permeability in boron deficient leaves of sunflower. Physiol Plant 95:11–18

Carpenter SR (2003) Regime shifts in lake ecosystems: pattern and variation, Excellence in ecology series, 15. Ecology Institute, Oldendorf

Chander G, Wani SP, Sahrawat KL, Kamdi PJ, Pal CK, Pal DK, Mathur TP (2013a) Balanced and integrated nutrient management for enhanced and economic food production: case study from rainfed semi-arid tropics in India. Arch Agron Soil Sci 59:1643–1658

Chander G, Wani SP, Sahrawat KL, Pal CK, Mathur TP (2013b) Integrated plant genetic and balanced nutrient management enhances crop and water productivity of rainfed production systems in Rajasthan, India. Commun Soil Sci Plant Anal 44:3456–3464

Chander G, Wani SP, Sahrawat KL, Dixit S, Venkateswarlu B, Rajesh C, Rao PN, Pardhasaradhi G (2014a) Soil test based nutrient balancing improved crop productivity and rural livelihoods: case study from rainfed semi-arid tropics in Andhra Pradesh, India. Arch Agron Soil Sci 60(8):1051–1066

Chander G, Wani SP, Sahrawat KL, Rajesh C (2014b) Enhanced nutrient and rainwater use efficiency in maize and soybean with secondary and micro nutrient amendments in the rainfed semi-arid tropics. Arch Agron Soil Sci. doi:10.1080/03650340.2014.928928

Chevalier P, Schrader LE (1977) Genotypic differences in nitrate absorption and partitioning of N among plant parts in maize. Crop Sci 17:897–901

Chuachin S, Wangkahart T, Wani SP, Rego TJ, Pathak P (2012) Simple and effective integrated pest management technique for vegetables in northeast Thailand. In: Wani SP, Pathak P, Sahrawat KL (eds) Community watershed management for sustainable intensification in Northeast Thailand. International Crops Research Institute for the Semi-arid Tropics, Patancheru

Cox MC, Qualset CO, William RD (1985) Genetic variation for nitrogen assimilation and translocation in wheat. I. Dry matter and nitrogen accumulation. Crop Sci 25:430–435

Cox MC, Qualset CO, Rains DW (1986) Genetic variation for nitrogen assimilation and translocation in wheat. III. Nitrogen translocation in relations to grain yield and protein. Crop Sci 26:737–740

Crutzen PJ, Ehhalt DH (1977) Effects of nitrogen fertilizers and combustion on the stratospheric ozone layer. Ambio 6:112–116

Delogu G, Cattivelli L, Pecchioni N, Defalcis D, Maggiore T, Stanca AM (1998) Uptake and agronomic efficiency of nitrogen in winter barley and winter wheat. Eur J Agron 9:11–20

Dhaliwal GS, Arora R (1993) Changing status of insect pests and their management strategies. In: Gill KS, Dhaliwal GS, Hansara BS (eds) Changing scenario of Indian agriculture. Commonwealth Publishers, New Delhi

Dhugga KS, Waines JG (1989) Analysis of nitrogen accumulation and use in bread and durum wheat. Crop Sci 29:1232–1239

Drechsel P, Giordano M, Gyiele L (2004) Valuing nutrition in soil and water: concepts and techniques with examples from IWMI studies in developing world, IWMI research paper 82. International Water Management Institute (IWMI), Colombo

Dwivedi RS, Ramana KV, Wani SP, Pathak P (2001) Use of satellite data for watershed management and impact assessment, In: Wani SP, Maglinao AR, Ramakrishna A, Rego TJ (ed) Integrated watershed management for land and water conservation and sustainable agricultural production in Asia: proceedings of the ADB-ICRISAT-IWMI project review and planning meeting during 10–14 December 2001, Hanoi, Vietnam. International Crops Research Institute for the Semi-Arid Tropics (ICRISAT), Patancheru, Andhra Pradesh, pp 149–157

Edmeades DC (2003) The long-term effects of manures and fertilisers on soil productivity and quality: a review. Nutr Cycl Agroecosyst 66:165–180

El-Swaify SA, Pathak P, Rego TJ, Singh S (1985) Soil management for optimized productivity under rainfed conditions in the semi-arid tropics. Adv Soil Sci 1:1–64

FAO (2011) FAOSTAT database – agriculture production. Food and Agriculture Organization of the United Nations, Rome

FAOSTAT (2013) FAOSTAT database - total population in India by 2050. [online] Available from: http://faostat.fao.org/site/550/default.aspx#ancor

Foulkes MJ, Hawkesford MJ, Barraclough PB, Holdsworth MJ, Kerr S, Kightley S, Shewry PR (2009) Identifying traits to improve the nitrogen economy of wheat: recent advances and future prospects. Field Crops Res 114:329–342

Galloway JN, Townsend AR, Erisman JW et al (2008) Transformation of the nitrogen cycle: recent trends, questions and potential solutions. Science 320:889–892

Garnett T, Conn V, Kaiser BN (2009) Root based approaches to improving nitrogen use efficiency in plants. Plant Cell Environ 32:1272–1283

Gascho GJ, Menezes RSC, Hanna WW, Hubbard RK, Wilson JP (1995) Nutrient requirements of pearl millet. In: Proceedings First National Grain Pearl Millet Symposium, 17–18 January, Georgia

Ghosh K, Nayak DC, Ahmed N (2009) Soil organic matter. J Indian Soc Soil Sci 57:494–501

Gooding MJ, Addisu M, Uppal RK, Snape JW, Jones HE (2012) Effect of wheat dwarfing genes on nitrogen-use efficiency. J Agric Sci 150:3–22

Hameeda B, Reddy Y, Rupela OP, Kumar GN, Reddy G (2006) Effect of carbon substrates on rock phosphate solubilization by bacteria from composts and macrofauna. Curr Microbiol 53:298–302

Hirel B, Bertin I, Quilleré W, Bourdoncle C, Attagnant C, Dellay A, Gouy S, Cadiou C, Retailliau M, Falque A, Gallais A (2001) Towards a better understanding of the genetic and physiological basis for nitrogen use efficiency in maize. Plant Physiol 125:1258–1270

Hirel B, Gouis JL, Ney B, Gallais A (2007) The challenge of improving nitrogen use efficiency in crop plants: toward a more central role for genetic variability and quantitative genetics within integrated approaches. J Exp Bot 58:2369–2387

Hungate BA, Dukes JS, Shaw MR, Luo Y, Field CB (2003) Nitrogen and climate change. Science 302:1512–1513

ICRISAT (International Crops Research Institute for the Semi-Arid Tropics) (2007) Establishing participatory research-cum-demonstrations for enhancing productivity with sustainable use of natural resources in Sujala watersheds of Karnataka. Annual report 2006 to 2007. Submitted to the Commissioner, Watershed Development Department, Sujala Watershed Project, Government of Karnataka, pp 40

IPCC (2007) Climate change: the physical science basis – summary for policy makers. World Meteorological Organization/United Nations Environmental Program, Paris

Irz X, Roe T (2000) Can the world feed itself? Some insights from growth theory. Agrekon 39:513–528

Jenkinson DS, Ladd JN (1981) Microbial biomass in soil: measurement and turnover. In: Paul EA, Ladd JN (eds) Soil biochemistry, vol 5. Dekker, New York, pp 415–471

Jiao W, Chen W, Chang AC, Page AL (2012) Environmental risk of trace elements associated with long-term phosphate fertilizers applications: a review. Environ Pollut 168:44–53

Karlberg L, Rockstrom J, Falkenmark M (2009) Water resource implications of upgrading rainfed agriculture – focus on green and blue water trade-offs. In: Wani SP, Rockstrom J, Oweis T (eds) Rainfed agriculture: unlocking the potential. CAB International, Wallingford, pp 44–53

Kroeze C (1994) Nitrous oxide and global warming. Sci Total Environ 143:193–209

Le Gouis J, Beghin D, Heumez E, Pluchard P (2000) Genetic differences for nitrogen uptake and nitrogen utilization efficiencies in winter wheat. Eur J Agron 12:163–173

Lee KK, Wani SP (1989) Significance of biological nitrogen fixation and organic manures in soil fertility management. In: Christianson CB (ed) Soil fertility and fertility management in semi-arid tropical India. IFDC, Muscle Shoals, pp 89–108

Lee HJ, Lee SH, Chung JH (2004) Variation of nitrogen use efficiency and its relationships with growth characteristics in Korean rice cultivars. In: Fischer T, Turner N, Angus J, McIntyre L, Robertson M, Borrell A, Lloyd D (eds) Proceedings of the 4th international crop science congress, Brisbane, Australia, 2004. International Crop Science, Brisbane

Limon-Ortega A, Sayre KD, Francis CA (2000) Wheat nitrogen use efficiency in a bed planting system in Northwest Mexico. Agron J 92:303–308

Lopez-Bellido RJ, Lopez-Bellido L (2001) Efficiency of nitrogen in wheat under Mediterranean condition: effect of tillage, crop rotation and N fertilization. Field Crops Res 71:31–64

Mahler RL, Koehler FE, Lutcher LK (1994) Nitrogen source, timing of application and placement: effects on winter wheat production. Agron J 86:637–642

Maranville JW, Clark RB, Ross WM (1980) Nitrogen efficiency in grain sorghum. J Plant Nutr 2:577–589

Masclaux C, Quilleré I, Gallais A, Hirel B (2001) The challenge of remobilization in plant nitrogen economy. A survey of physioagronomic and molecular approaches. Ann Appl Biol 138:69–81

Materechera SA (2010) Utilization and management practices of animal manure for replenishing soil fertility among smallscale crop farmers in semi-arid farming districts of the North West Province, South Africa. Nutr Cycl Agroecosyst 87:415–428

Mattsson M, Husted S, Schjoerring JK (1998) Influence of nitrogen nutrition and metabolism on ammonia volatilization in plants. Nutr Cycl Agroecosyst 51:35–40

Ministry of Agriculture, Government of India, Agricultural Statistics at a Glance (2013) [Internet] Directorate of Economics and Statistics, Department of Agriculture and Cooperation, Ministry of Agriculture, Government of India, New Delhi, India. Available from: http://eands. dacnet.nic.in/latest_2006.htm. (Accessed April 2013)

Moll RH, Kamprath EJ, Jackson WA (1982) Analysis and interpretation of factors which contribute to efficiency to nitrogen utilization. Agron J 74:562–564

Nagavallemma KP, Wani SP, Stephane Lacroix, Padmaja VV, Vineela C, Babu Rao M, Sahrawat KL (2004) Vermicomposting: recycling wastes into valuable organic fertilizer. Global theme on agroecosystems report no. 8. International Crops Research Institute for the Semi-Arid Tropics (ICRISAT), Patancheru, Andhra Pradesh

Oerke EC, Delhne HW, Schohnbeck F, Weber A (1995) Crop production and crop protection: estimated losses in major food and cash crops. Elsevier, Amsterdam

Ortiz-Monasterio JI, Sayre KD, Rajaram S, McMahom M (1997) Genetic progress in wheat yield and nitrogen use efficiency under four nitrogen rates. Crop Sci 37:898–904

Pappu A, Saxena M, Asolekar SR (2007) Solid wastes generation in India and their recycling potential in building materials. Build Environ 42:2311–2320

Payne WA, Hossner LR, Onken AB, Wendt CW (1995) Nitrogen and phosphorous uptake in pearl millet and its relation to nutrient and transpiration efficiency. Agron J 87:425–431

Peterjohn WT, Schlesinger WH (1990) Nitrogen loss from deserts in the South Western United States. Biogeochemistry 10:67–79

Potarzycki J (2010) Improving nitrogen use efficiency of maize by better fertilizing practices: review. Fertilizers Fertilization 39:5–24

Rahimizadeh M, Kashani A, Zare-Feizabadi A, Koocheki AR, Nassiri-Mahallati M (2010) Nitrogen use efficiency of wheat as affected by preceding crop, application rate of nitrogen and crop residues. Aust J Crop Sci 4:363–368

Ranga Rao GV, Desai S, Rupela OP, Krishnappa K, Wani SP (2009) Integrated pest management options for better crop production. In: Best-bet options for integrated watershed management – proceedings of the comprehensive assessment of watershed programs in India, 25–27 July 2007. ICRISAT, Patancheru, Andhra Pradesh

Raun WR, Johnson GV (1999) Improving nitrogen use efficiency for cereal production. Agron J 91:357–363

Reddy DD (2013) Nutrient use efficiency in rainfed agroecosystems: concepts, computations and improvement interventions. Available online at: http://www.crida. in/DRM2-Winter%20School/DDR.pdf

Rego TJ, Rao VN, Seeling B, Pardhasaradhi G, Kumar Rao JVDK (2003) Nutrient balances – a guide to improving sorghum and groundnut-based dryland cropping systems in semi-arid tropical India. Field Crops Res 81:53–68

Rego TJ, Wani SP, Sahrawat KL, Pardhasaradhi G (2005) Macro-benefits from boron, zinc and sulfur application in Indian SAT: a step for grey to green revolution in agriculture. Global theme on agroecosystems report no. 16. International Crops Research Institute for the Semi-Arid Tropics (ICRISAT), Patancheru, Andhra Pradesh

Roberts TL (2008) Improving nutrient use efficiency. Turk J Agric For 32:177–182

Rockström J, Falkenmark M (2000) Semiarid crop production from a hydrological perspective: gap between potential and actual yields. Crit Rev Plant Sci 19:319–346

Rockström J, Hatibu N, Oweis T, Wani SP (2007) Managing water in rain-fed agriculture. In: Molden D (ed) Water for food, water for life: a comprehensive assessment of water management in agriculture. Earthscan/International Water Management Institute (IWMI), London/Colombo, pp 315–348

Rockstorm J, Steffen W, Noone K, Persson A, Chapin AS III et al (2009) A safe operating space for humanity. Science 461:472–475

Sahrawat KL, Wani SP, Rego TJ, Pardhasaradhi G, Murthy KVS (2007) Widespread deficiencies of sulphur, boron and zinc in dryland soils of the Indian semi-arid tropics. Curr Sci 93:1428–1432

Sahrawat KL, Wani SP, Parthasaradhi G, Murthy KVS (2010) Diagnosis of secondary and micronutrient deficiencies and their management in rainfed agroecosystems: case study from Indian semi-arid tropics. Commun Soil Sci Plant Anal 41:346–360

Sahrawat KL, Wani SP, Subba Rao A, Pardhasaradhi G (2011) Management of emerging multinutrient deficiencies: a prerequisite for sustainable enhancement of rainfed agricultural productivity. In: Wani SP, Rockstrom J, Sahrawat KL (eds) Integrated watershed management. CRC Press, The Netherlands, pp 281–314

Sawargaonkar GL, Wani SP, Patil MD (2012) Enhancing water use efficiency of maize-chickpea sequence under semi-arid conditions of southern India. In: Extended summaries vol 2: 3rd international agronomy congress, November 26–30, 2012, Indian Society of Agronomy, ICAR, New Delhi, India, pp 576–578

Schlesinger WH (2009) On the fate of anthropogenic nitrogen. Proc Natl Acad Sci U S A 106:203–208

Simpson RJ, Oberson A, Culvenor RA, Ryan MH, Veneklaas EJ, Lambers H, Lynch JP, Ryan PR, Delhaize E, Smith FA, Smith SE, Harvey PR, Richardson AE (2011) Strategies and agronomic interventions to improve the phosphorus-use efficiency of farming systems. Plant Soil 349:89–120

Singh SN, Verma A (2007) The potential of nitrification inhibitors to manage the pollution effect of nitrogen fertilizers in agricultural and other soils: a review. Environ Pract 9:266–279

Sreedevi TK, Shiferaw B, Wani SP (2004) Adarsha watershed in Kothapally: understanding the drivers of higher impact, Global theme on agroecosystems report no. 10. International Crops Research Institute for the Semi-Arid Tropics, Patancheru, Andhra Pradesh

Sutton MA, Oenema O, Erisman JW, Leip A, van Grinsven H, Winiwarter W (2011) Too much of a good thing. Nature 472:159–161

Thirtle C, Beyers L, Lin L, McKenzie-Hill V, Irz X, Wiggins S, Piesse J (2002) The impacts of changes in agricultural productivity on the incidence of poverty in developing countries. DFID report no. 7946.

Tripathi SC, Sayre KD, Kaul JN (2004) Genotypic effects on yield, N uptake, NUTE and NHI of spring wheat. New directions for a diverse planet. In: Proceedings of the 4th international crop science congress, Brisbane, Australia, 26 September–1 October 2004

Uppal RK, Wani SP, Garg KK, Alagarswamy G (2014) Validating sustainability and resilience of pearl millet for the impacts of climate change. In: van Beuischem ML (ed) Plant nutrition – physiology and applications. Kluwer Academic Publishers, Dordrecht

Wani SP, Zambre MA, Lee KK (1992) Genotypic diversity in pearl millet (*Pennisetum glaucum*) for nitrogen, phosphorous and potassium use efficiencies. In: van Beuischem ML (eds) Plant nutrition - physiology and applications. Kluwer Academic Publishers, Dordrecht, pp 595–601

Wani SP, Maglinao AR, Ramakrishna A, Rego RJ (2001) Integrated watershed management for land and water conservation and sustainable agricultural production in Asia. In: Proceedings of the ADB-ICRISAT-IWMI project review and planning meeting during December 10–14 2001, Hanoi, Vietnam. International Crops Research Institute for the Semi-Arid Tropics (ICRISAT), Patancheru, pp 259

Wani SP, Rego TJ, Pathak P (2002) Improving management of natural resources for sustainable rainfed agriculture. In: Proceedings of the training workshop on on-farm participatory research methodology during July 26–31 2001, Khon Kaen, Bangkok, Thailand. International Crops Research Institute for the Semi-Arid Tropics (ICRISAT), Patancheru, pp 1–68

Wani SP, Pathak P, Jangawad LS, Eswaran H, Singh P (2003a) Improved management of vertisols in the semi-arid tropics for increased productivity and soil carbon sequestration. Soil Use Manag 19:217–222

Wani SP, Pathak P, Sreedevi TK, Singh HP, Singh P (2003b) Efficient management of rainwater for increased crop productivity and groundwater recharge in Asia. In: Kijney JW, Barker R, Molden D (eds) Water productivity in agriculture: limits and opportunities for improvement. CAB International, Wallingford, pp 199–215

Wani SP, Sreedevi TK, Sahrawat KL, Ramakrishna YS (2008) Integrated watershed management – a food security approach for SAT rainfed areas. J Agrometeorol 10:18–30

Wani SP, Sahrawat KL, Sarvesh KV, Baburao M, Krishnappa K (2011) Soil fertility atlas for Karnataka, India. International Crops Research Institute for the Semi-Arid Tropics, Patancheru, Andhra Pradesh, India and Department of Agriculture, Government of Karnataka... knowledge and aspirations of the farmers. In: Sarode SV, Deshmukh JP, Kharche VK, Sable YR (ed) Proceedings of national seminar on "Soil security for sustainable agriculture" during February 27–28, 2010, Dr. Panjabrao Deshmukh Krishi Vidyapeet, Akola, Maharashtra, India, pp 1–9

Wani SP, Chander G, Sahrawat KL, Srinivasa Rao C, Raghvendra G, Susanna P, Pavani M (2012c) Carbon sequestration and land rehabilitation through Jatropha curcas (L.) plantation in degraded lands. Agric Ecosyst Environ 161:112–120

World Bank (2005) Agricultural growth for the poor: an agenda for development. The International Bank for Reconstruction and Development, Washington, DC

Dynamics of Plant Nutrients, Utilization and Uptake, and Soil Microbial Community in Crops Under Ambient and Elevated Carbon Dioxide

Shardendu K. Singh, Vangimalla R. Reddy, Mahaveer P. Sharma, and Richa Agnihotri

Abstract

In natural settings such as under field conditions, the plant-available soil nutrients in conjunction with other environmental factors such as solar radiation, temperature, precipitation, and atmospheric carbon dioxide (CO_2) concentration determine crop adaptation and productivity. Therefore, crop success depends on the intricate balance among these multiple environmental factors. Plant nutrients are the major constraint for crop productivity worldwide because it must be supplied externally to achieve maximum production. The depleting natural resources of mineral nutrients in addition to the global changes in climate caused by the emission of green house gases including CO_2 are among the major concerns of crop production and food security. Moreover, crop demand for nutrients has been increased due to use of modern cultivars and improved irrigation facilities and is expected to be even higher under elevated CO_2. Soil microorganisms including arbuscular mycorrhizal (AM) fungi partly enhance crop nutrient availability and acquisition in many soil types through symbiotic or non-symbiotic relationships. Atmospheric CO_2 concentration is expected to be doubled from its current level

S.K. Singh (✉)
Crop Systems and Global Change Laboratory, United States Department of Agriculture, Agricultural Research Service (USDA-ARS), Bldg 001, Rm 342, BARC-West, 10300 Baltimore Ave., Beltsville, MD 20705, USA

Wye Research and Education Center, University of Maryland, Queenstown, MD, USA
e-mail: shardendu.singh@ars.usda.gov;
singh.shardendu@gmail.com

V.R. Reddy
Crop Systems and Global Change Laboratory, United States Department of Agriculture, Agricultural Research Service (USDA-ARS), Bldg 001, Rm 342, BARC-West, 10300 Baltimore Ave., Beltsville, MD 20705, USA

M.P. Sharma • R. Agnihotri
Microbiology Section, ICAR-Directorate of Soybean Research, DARE, Ministry of Agriculture, Indore, India

A. Rakshit et al. (eds.), *Nutrient Use Efficiency: from Basics to Advances,*
DOI 10.1007/978-81-322-2169-2_24, © Springer India 2015

of 400 μmol mol^{-1} at the end of this twenty-first century. Elevated CO_2 increases growth and yield of many crops upon which humans depend for food and clothing. However, plant nutrient availability exerts major control on the degree of stimulation by elevated CO_2 on crop growth and yield. One of the objectives of this chapter is to provide a summary of crop responses to plant nutrients mainly nitrogen, phosphorus, and potassium and underline in part the dynamics of soil microorganisms including AM fungi in the nutrient accessibility under current and elevated CO_2 concentrations. Regardless of the CO_2 levels, nutrient deficiencies negatively affect crop photosynthesis, growth and biomass production, yield, and yield quality. Elevated CO_2 tends to compensate, at least partly, for the losses caused by nutrient deficiency especially by increasing plant growth due to improved efficiency of nutrient acquisition and utilization. However, crop species, deficiency of the specific nutrient, and its severity greatly influence the nutrient efficiency in crop plants. The critical tissue nutrient concentration required to achieve 90 % of maximum productivity of some plant nutrients is likely to be higher at elevated CO_2. Another objective of this chapter is to discuss the influence of crop species, soil nutrient status, and elevated CO_2 on the dynamics of nutrient uptake and utilization efficiency and resultant tissue nutrient concentration. Future research methods utilizing the combined effect of plant nutrient status and elevated CO_2 on crops will improve our understanding of the complex relationships among various plant processes leading to efficient use of nutrient under field conditions.

Keywords

Arbuscular mycorrhizal fungi • Climate change • Critical nutrient concentration • Crop growth and productivity • Nutrient deficiency • Nutrient use efficiency

1   Introduction

Plant primary nutrients such as nitrogen (N), phosphorus (P), and potassium (K) are needed in the large quantities. These nutrients are essential for plant metabolism and components such as chlorophyll pigments, enzymes, proteins, and nucleic acid; are required for energy transfer and enzyme activation; and act as anion and cation to maintain the osmotic balance in plant cells. Their deficiency causes stunted growth, decreased leaf greenness and canopy area, reduced photosynthesis, and thus reduces crop yield.

The natural soil reserves of nutrients are limited and their release into the rhizosphere results in the improved growth and high yield. For nutrients such as N, almost all the agricultural land is deficient and extra N supply is needed to obtain optimum crop yield. P deficiency in soils is a limiting growth factor in over 30 % of crop lands, and a major production constraint in acidic soils comprising up to 70 % worldwide (Vance et al. 2003; Cordell et al. 2009; Lenka and Lal 2012). In the recent past, the nutrient requirement of agricultural crops has been increased due to the use of modern high-yielding cultivars and improved irrigation facilities. Moreover, the nutrient demand of crops is expected to be even higher under rising atmospheric carbon dioxide (CO_2) concentration due to increased plant growth (Rogers et al. 1993; Lewis et al. 1994). The current atmospheric CO_2 of approximately 400 μmol mol^{-1} is projected to be doubled by the end of twenty-first century (IPCC 2007). Crops grown under nutrient deficiency lead to decreased biomass and crop yield, whereas opposite is the case for plants grown under elevated CO_2. However, plant nutrition status exerts a major control in the degree of growth stimulation by elevated CO_2. Therefore, the degree of crop growth enhancement under CO_2-enriched environment is expected to be greatly influenced by nutrient availability (Cure et al. 1988; Campbell and Sage 2006; Lenka and Lal 2012; Singh et al. 2013a, b). Thus, the deficiency of mineral nutrients such as N, P, and K in the soil and elevated CO_2 often has opposite effect on crop growth, and their coexistence under natural conditions is inevitable. This leads to the need for developing suitable crop cultivars that can efficiently utilize plant-available nutrients by enhancing nutrient uptake and utilization efficiency under current and projected atmospheric CO_2 concentration.

Nutrient deficiency often reduces the degree of crop response to elevated CO_2. Increased growth at elevated CO_2 is often associated with increased plant size, leaf area, photosynthesis at the leaf and canopy levels, and higher efficiency for nutrient utilization. However, the stimulatory effect of elevated CO_2 on crop growth is highly impacted under nutrient deficient conditions. For instance, over 80 % increase in soybean dry mass observed under elevated CO_2 was not observed when grown under P-deficient environment (Sa and Israel 1998). Similar results have also been observed in other legumes (Lam et al. 2012). However, the degree of crop response to elevated CO_2 depends on the severity of P deficiency (Cure et al. 1988). The decreased growth response at elevated CO_2 under nutrient deficiency may also include accumulation of nonstructural carbohydrates and reduced utilization of the starch in the leaves (Sa and Israel 1998).

Enhanced stress tolerance potential for increasing crop productivity has become prerequisite for fulfilling the food needs of the growing population. The global climate is said to be influenced by the changes in soil carbon (C) pools which act as carbon sink and affect the CO_2 concentration in the atmosphere (Drigo et al. 2009). The composition and functioning of soil microbial communities in rhizosphere is affected by temperature and atmospheric CO_2 concentration. Plant-associated microorganisms including arbuscular mycorrhizal fungi (AM) influence plant responses to various soil and environmental factors by helping plants capture nutrients from the soil (Compant et al. 2010). Terrestrial ecosystems are intimately associated with atmospheric CO_2 levels through photosynthetic fixation of CO_2, sequestration of C into the organic carbon in soil and plants, and subsequent release of CO_2 through respiration and decomposition of organic matter (Deng et al. 2012). Relationships exist between heterotrophic soil respiration and soil moisture, CO_2 and elevated temperature as with the rise in temperature the wetter soils may emit more CO_2 into the atmosphere (Moyano et al. 2013). The AM fungi are one of the most important consumers of plant-derived carbon in plant-soil systems (Staddon 2005). The symbiotic relationship between AM fungi and roots of higher plants contributes significantly to plant nutrition and growth (Sharma and Adholeya 2004). The role of AM fungi in stress mitigation has been suggested via several mechanisms such as altering root morphology and physiology exhibiting efficient nutrient uptake and higher enzymatic activities (Ahanager et al. 2014). Changes in microbial biomass and community structure (fungi to bacteria ratio) were found to relate with moisture-induced changes and increase in soil respiration rate by 10 °C rise in temperature (Zhou et al. 2014).

Increased efficiency of nutrient acquisition and utilization by plants under both the nutrient-rich and nutrient-poor soils are of an urgent global need. In the following sections the crop response to N, P, and K nutrition, their uptake and utilization efficiency, and the influence of soil microbial community on nutrient dynamics under ambient and elevated CO_2 concentration are discussed. The possible effect of nutrient deficiency and elevated CO_2 in the quality of edible part of the crop production and future perspective of the crop improvement in relation to the nutrient acquisition and utilization has also been highlighted.

2 Crop Response to N, P, K Under Ambient and Elevated CO_2

Plant growth response to mineral nutrition should be viewed by taking into account both the availability and the tissue concentration. Large uncertainty exists about the changes in the mineral availability and the response of the tissue nutrient concentration under elevated CO_2 conditions (Sinclair 1992). Nutrient acquisition and assimilation in plants are strongly influenced by elevated CO_2. Elevated CO_2-mediated decreases in the tissue nutrient concentration have commonly been observed in many crops while grown under similar nutrient supply as of ambient CO_2 (Cure et al. 1988; Conroy 1992; Singh et al. 2013a). However, the decreases in the tissue nutrient concentration under elevated CO_2 were not associated with the limitation to growth or photosynthetic processes albeit increased the nutrient utilization efficiency especially for nitrogen (N) (Barrett and Gifford 1995; Prior et al. 1998; Singh et al. 2013b). The beneficial effect of elevated CO_2 on plant growth often declines under nutrient-deficient conditions. The dynamics of the mineral nutrition of nitrogen and nitrogenous fertilizers in the crops has frequently been studied; however, nutrients such as P and K have received lesser attention. This is partly due to the nitrogen-driven increase in the crop productivity that has been main focus in the fertilizer industry and for farmers as well. However, the contribution of other nutrients such as P and K and micronutrients for maximizing crop yield are also important under current and future climatic

conditions (Sinclair 1992). For example, compared to N nutrition, the interactions between P and CO_2 have been largely unexplored in agronomic crops. The limited studies indicate that higher foliar concentration of phosphorus in the CO_2-enriched environment may be needed for maximum growth in many agronomic crops and trees (Conroy et al. 1990; Rogers et al. 1993).

2.1 Physiological and Biochemical Processes

Photosynthetic processes in crop plants are one of the primary metabolic processes influenced by either nutrient deficiency or CO_2 enrichment. This leads to changes in the assimilation and utilization of organic and inorganic compounds and the biochemical alterations inside the plant including modification and adjustment in cellular components such as chlorophyll and mineral nutrient concentration and the amount and functionality of protein and enzymes. The rate of photosynthesis in many C_3 crop species is limited by the current atmospheric CO_2 concentration. Moreover, growth stimulation by elevated CO_2 is expected to increase the overall plant demand for mineral nutrients such as nitrogen and phosphorus. Elevated CO_2 tends to increase total nutrient accumulation in plants but decreases their tissue concentration leading to alterations in nutrient uptake and use efficiency (Rogers et al. 1993; Lenka and Lal 2012; Singh et al. 2013a).

The acclimation/downregulation of photosynthetic processes to long-term CO_2 enrichment has been observed in many crops and may be partly caused by imbalance between CO_2 fixation and assimilate utilization (Barrett and Gifford 1995; Singh et al. 2013b). The elevated CO_2 mediated alteration in the photosynthetic processes may also be adjusted by distribution of absorbed light energy between photochemical and non-photochemical processes in the chloroplast. Gas exchange and chlorophyll fluorescence phenomena are integral parts of photosynthetic processes in leaves which can be divided into light-dependent and carbon-fixation reactions.

The absorbed light energy that exceeds the photochemical processes of CO_2 fixation can be either dissipated as heat or reemitted as chlorophyll *a* fluorescence (Maxwell and Johnson 2000). The extent of changes in the chlorophyll *a* fluorescence parameters has been shown to be influenced more by nutrients than CO_2 (Betsche 1994; Singh and Reddy 2014). One of the important photosynthetic responses of crop plants to rising CO_2 is the reduction in photorespiration. This can partly lead to the lower energy demand at elevated CO_2 versus ambient CO_2 and excess radiant energy may also lead to thylakoid energization causing reduction in photochemical quenching without having any detrimental effect on the CO_2 assimilation (Edwards and Baker 1993; Betsche 1994; Singh et al. 2013b). The reduction in the carboxylation efficiency of photosynthesis has often been observed in many C_3 crops due to photosynthetic acclimation and downregulation. However, at elevated CO_2 concentration, the decrease in photosynthetic capacity due to nutrient deficiency is partly offset by the compensatory growth processes such as rapid growth rate, increased leaf area, and rate of net canopy photosynthesis leading to stimulation of plant growth.

The deficiency of one nutrient may also alter the dynamics of other nutrients. Previous studies suggested that nutrient deficiency and elevated CO_2 may also alter biomass partitioning and allocation of nutrients in the plant organs affecting the assimilation of other nutrients such as nitrogen (Israel et al. 1990; Rufty et al. 1991; Reddy and Zhao 2005; Fleisher et al. 2012; Singh et al. 2013a). Using isotopic N sources (^{15}N), Rufty et al. (1993) reported that P deficiency increased the root dry weight and decreased the rate of nitrate uptake while accumulating more absorbed ^{15}N in the roots. Additionally, increased tissue N concentration at the severe P deficiency has been previously reported suggesting decrease of the N mobility and utilization inside the plant (Singh et al. 2013a, 2014). P deficiency may cause an increase in soluble-reduced N in leaves which fails to incorporate into cellular components such as chlorophyll, proteins (thus, enzymes), and nucleic acids

(Israel and Rufty 1988). Nitrogen is considered as one of the most mobile element in plants (Marschner 1986). Generally, at the early stage, plant leaves act as a sink for mineral nutrients, while later they becomes nutrient sources for the new growth and development of seeds (Himelblau and Amasino 2001). Hanway and Weber (1971) reported mobilization of approximately half the nutrients from vegetative parts to seeds. Lauer et al. (1989) found that P concentration of the vacuole in the leaf cell depleted first followed by P concentration in the cytoplasm during seed development. The increased tissue N in P-deficient plants may also increase the tissue N:P ratio leading to adjustment in the N assimilation and reduced mobility inside the plant (Rufty et al. 1993). Singh et al. (2014) reported a N:P ratios between 11 and 16 for optimal growth in soybean regardless of ambient or elevated CO_2 concentration. A higher than this N:P ratio exhibits P limitations in soybean and other crops (Koerselman and Meuleman 1996; Güsewell 2004; Singh et al. 2014).

2.2 Morphological Processes

The stunted plants with lower number of main stem nodes and leaf area under nutrient deficiency are commonly observed morphological changes in the crop plant (Reddy and Zhao 2005; Fleisher et al. 2012; Singh et al. 2013a). Soybean grown under low nutrient conditions produced fewer flowers, pods, and seeds and caused flower abscission and pod senescence (Sionit 1983). Deficiencies of N severely inhibit synthesis of cellular components such as chlorophyll, proteins, and enzymes that need to sustain plant growth. Radin and Eidenbock (1984) reported that P deficiency limits cell expansion by reducing hydraulic conductance inside plants which may lead to reduced plant size and leaf area expansion. The smaller plant stature under nutrient deficiency is mainly associated with smaller leaf area and decreased canopy photosynthesis which may result in lower biomass and reduced yield. Compared to N and P, plant response to K nutrition is highly dependent on the growth processes. Often, the photosynthesis and vegetative growth are not influenced by a medium K deficiency but suddenly deceases to the minimum at very low level of tissue K concentration (Reddy and Zhao 2005).

The specific leaf weight (leaf dry weight per unit of leaf area) can be used as an indication for leaf thickness and can also have major influence on the tissue nutrient concentration (Singh et al. 2013a). Nutrient deficiency such as P and elevated CO_2 can also lead to increased leaf thickness in soybean (Sionit 1983; Prior and Rogers 1995). However, cotton has not shown a consistent increase in leaf thickness under P deficiency (Longstreth and Nobel 1980; Singh et al. 2013a). The expression of leaf tissue nutrient concentration either based on leaf mass or leaf area may influence its relationship with other parameters such as photosynthesis. Singh et al. (2013a) found no significant difference in the leaf P concentration across CO_2 when expressed on leaf area basis in cotton leaves indicating the dilution of the nutrients due to increased leaf thickness. Specific leaf weight has also been reported to increase in response to elevated CO_2 in a wide range of plant species (Gifford et al. 2000).

The beneficial effect of CO_2 enrichment on crop plants is anticipated even under stress conditions such as nutrient deficiency (Norby et al. 1986; Fleisher et al. 2012; Pérez-López et al. 2012), because elevated CO_2 increases plant carbohydrate supply by stimulating CO_2 assimilation, thus reducing the pressure on energy (carbohydrate) demand of plants coping with stresses (Ahmed et al. 1993; Pérez-López et al. 2010). For instance, elevated CO_2 reduced the diffusive and photo-biochemical limitations to photosynthesis caused by P deficiency in cotton (Singh et al. 2013b) and salinity stress in barley (*Hordeum vulgare* L.) (Pérez-López et al. 2012).

Elevated CO_2 tends to compensate, at least partly, the decrease in the growth and yield caused by nutrient deficiency. However, the degree of compensation by elevated CO_2 mainly depends on the severity of the nutrient deficiency. Under moderate nutrient deficiency,

stimulation of growth, photosynthesis, and yield at elevated CO_2 can still be observed, but at a lesser extent as compared to the stimulation observed under optimum nutrient conditions. However, under severe nutrient-deficient conditions the growth stimulation by elevated CO_2 is none causing similar or sometimes even more decrease in growth and crop yield as observed under ambient CO_2. The increased growth at elevated CO_2 is mainly associated with increased leaf area, photosynthesis, and nutrient utilization efficiency. The increased canopy photosynthesis due to increased leaf area and the rate of photosynthesis to the individual leaves contributed most to the increased growth under P nutrition (Cure et al. 1988; Singh et al. 2013a). Total biomass accumulation for all plant organs under elevated CO_2 leads to stimulation of growth and higher nutrient utilization efficiencies. The biomass partitioning to seeds reflects the harvest index and has not been found to be affected by elevated CO_2 across P nutrition in soybean (Cure et al. 1988). Thus, the increase in the number of pods and seeds rather than the seed size contributes to higher seed yield under elevated CO_2 conditions. Prior and Rogers (1995) also reported that the number of seed per plant increased at elevated CO_2 regardless of different water regimes. The seed size appeared to be insensitive to moderate nutrient deficiency due to the plant capacity to support the fewer number of seeds formed. The seed size tends to be fairly stable across tissue P concentrations except at very low P indicating plants' ability to support more seed if they had been set, as suggested in previous studies (Cassman et al. 1981; Cure et al. 1988).

2.3   Biomass Partitioning and Nutrient Allocation Among the Plant Parts

One of the major impacts of plant nutrition and growth CO_2 on plants is the alteration in the biomass partitioning and nutrient allocation among plant parts such as leaves, stems, roots, pods, and seeds (Prior et al. 1998; Reddy and

Zhao 2005; Fleisher et al. 2012). Moreover, the composition of other nutrients such as N in plant tissues may be altered due the deficiency of other nutrients leading to more complex interactions with the elevated CO_2 (Prior et al. 1998; Fleisher et al. 2012, 2013; Lenka and Lal 2012). Thus, the deficiency of one nutrient may also alter the metabolism of another. For instance, increased tissue N concentrations have been reported under P deficiency in many agronomic crops indicating relatively greater N uptake and accumulation in plant organs (Almeida et al. 2000; Fleisher et al. 2012; Singh et al. 2013a).

The highest biomass partitioning is generally towards the fruiting structures and seed followed by leaves and then stems (Mullins and Burmester 1990). However, an increased root to shoot ratio has commonly been observed under nutrient deficiency such as P, and elevated CO_2 (Radin and Eidenbock 1984; Fleisher et al. 2012; Singh et al. 2013a). Under stress conditions (e.g. low water or nutrient), plants allocate biomass towards the organ (e.g. roots) associated with acquiring the limited resources (Bazzaz 1997). Moreover, due to the nutrient deficiency, the growth of aboveground organs are generally affected sooner than the roots because roots are closer to the sources of the deficient nutrients (Brouwer 1962). Increased biomass partitioning to roots under stress conditions such as nutrient deficiency signifies the allocation of more dry matter to the roots and may be a mechanism to exploit below-ground resources in an effort to supply plant demand for nutrients. Similarly, under elevated CO_2, increased above-ground growth is often associated with increased root biomass, thus indicating adjustment between above- and below-ground plant growths (Lenka and Lal 2012). Nutrient concentrations are often reported to be lower in the stems compared to other plant organs which might be explained by the high mobility of N, P, and K in the stem as it largely serves as a medium for the translocation between root, leaf, and reproductive parts (Prior et al. 1998). Mullins and Burmester (1990) also reported lower shoot N and P concentration as compared to leaves and fruiting structures of cotton.

The reductions in plant tissue nutrient under elevated CO_2 across P treatments have been commonly reported; however, large variability also exists (Prior et al. 1998; Gifford et al. 2000; Prior et al. 2003). Factors such as dilution of tissue nutrients due to increased carbon assimilation and leaf thickness, decreased nutrient demand such as N, and restricted uptake and lower transpiration may contribute to the lower tissue nutrient at elevated CO_2 (Gifford et al. 2000; Taub and Wang 2008; Singh et al. 2013a, 2014). This clearly demonstrates that alteration in the uptake and utilization of nutrients under nutrient deficient condition and under elevated CO_2 (Bloom et al. 2010; Kawakami et al. 2013).

3   Nutrient Uptake and Utilization Efficiency

In general, nutrient efficiency reflects the nutrient acquisition and utilization by plants (Marschner 1986). The genotypic differences for nutrient efficiency offer an opportunity for crop improvement. The nutrient acquisition may be defined by the rate of uptake per unit of root area, length, or per unit of root mass. The nutrient uptake efficiency (NUpE; mg nutrient g^{-1} root mass) can be estimated using the equation

$$NUpE = \text{Total amount of nutrient in plant}$$
$$\left(\text{mg plant}^{-1}\right)/\text{Root dry mass}\left(\text{g plant}^{-1}\right) \tag{1}$$

Similarly, the nutrient utilization efficiency (NUE, g^2 dry mass mg^{-1} nutrient) refers to the total biomass production per unit of the nutrient present in the biomass and can be estimated as per Siddiqi and Glass (1981)

$$NUE = \text{Total plant dry mass}\left(\text{g plant}^{-1}\right)/$$
$$\text{Plant nutrient concentration}\left(\text{mg g}^{-1}\text{ dry mass}\right) \tag{2}$$

Often, the nutrient concentration is determined separately for various plant parts. Therefore, considering the relative dry mass and nutrient concentration for each plant part is crucial in the estimation of total amount of nutrients in the plant (Eq. 1) and plant nutrient concentration (Eq. 2). For example, plant nutrient concentration is not exactly the mean of the nutrient concentration in the leaf, stem, and root. Therefore, the plant nutrient concentration should be estimated as the weighted sum of the products of dry mass of each plant parts and their nutrient concentration divided by total biomass. When estimation from the plant parts (e.g. leaf, stem, and root) is required, the total amount of the nutrient in plant (as in the Eq. 1) can be calculated as

$$
\begin{aligned}
\text{Total amount of nutrient in plant } &\left(\text{mg plant}^{-1}\right) \\
= (\text{LDM} \times \text{LNC}) &+ (\text{SDM} \times \text{SNC}) \\
+ (\text{RDM} &\times \text{RNC})
\end{aligned}
\tag{3}
$$

where LDM, SDM, and RDM are the dry mass (g) of the leaves, stems, and roots, respectively. The LNC, SNC, and RNC are the nutrient concentration (mg g^{-1}) of the leaves, stems, and roots, respectively. Then the plant nutrient concentration (as in the Eq. 2) can be calculated as

$$
\begin{aligned}
\text{Plant nutrient concentration} \\
= \text{Total amount of nutrient in plant}/ \\
\text{Total plant dry mass}
\end{aligned}
\tag{4}
$$

The nutrient utilization efficiency of photosynthesis [NUE$_{Pnet}$, g (μmol CO$_2$ m^{-2} s^{-1}) fixed mg^{-1} nutrient] is a good metric to study the differences in the effectiveness of nutrients on photosynthesis (Singh and Reddy 2014). The estimation of the NUE$_{Pnet}$ requires measurement of photosynthesis and tissue nutrient concentration from the same leaf or plant canopy and can be estimated as

$$
\begin{aligned}
\text{NUE}_{Pnet} = \text{Photosynthetic rate} \\
\left(\mu\text{mol CO}_2 \, \text{m}^{-2} \, \text{s}^{-1}\right)/\text{Nutrient concentration} \\
\left(\text{mg g}^{-1}\right)
\end{aligned}
\tag{5}
$$

In a dynamic system like plants, a minimum concentration of nutrient element is required for metabolic reactions to occur. Therefore, the NUE takes the tissue concentration into account because plant growth actively depends on the tissue nutrient concentration rather than the absolute amount (Siddiqi and Glass 1981). However, in agronomic point of view, the nutrient efficiency is often defined by the gain in the harvestable yield per unit of nutrient applied especially when grown under nutrient-deficient soil. The NUE of N, P, and K have been reported to increase when grown under deficient condition in several crops. However, severe nutrient deficiency may also lead to decrease in NUE (Cure et al. 1988). An increased NUE of N in cotton (Prior et al. 1998), P in soybean (Cure et al. 1988), and K in cotton (Reddy and Zhao 2005) have been reported. While studying the effect of salinity which also causes unavailability of macro nutrients, Pérez-López et al. (2014) reported increased N and K use efficiency but decreased P use efficiency in barley across CO$_2$ levels. Singh et al. (2014) found decreased P utilization efficiency across CO$_2$ levels in cotton as external phosphorus supply was reduced.

Thus, when grown under nutrient-deprived conditions, the NUE varies and depends on the crop species and the severity of the deficiency. The biomass accumulation and tissue nutrient concentration are the two main factors which determine NUE as shown in the Eq. 2. Increase in biomass or decrease in tissue nutrient will lead to the higher NUE. However, under deficient conditions, both the biomass and the tissue nutrient will decrease, and it is their relative rate of reduction that determines the absolute changes in NUE. Therefore, plant growth and nutrient acquisition response to a particular nutrient-deficient situation is critical to improve NUE. This can be clearly illustrated by simulating two scenarios where in the first scenario Species-A showed lesser decrease in biomass versus tissue nutrient concentration, while in the second scenario Species-B showed higher decrease in the biomass than tissue nutrient concentration in response to a given nutrient deficiency (Fig. 1a, b). This clearly resulted into an increased NUE for Species-A while decreased NUE for Species-B in response to nutrient deficiency (Fig. 1c).

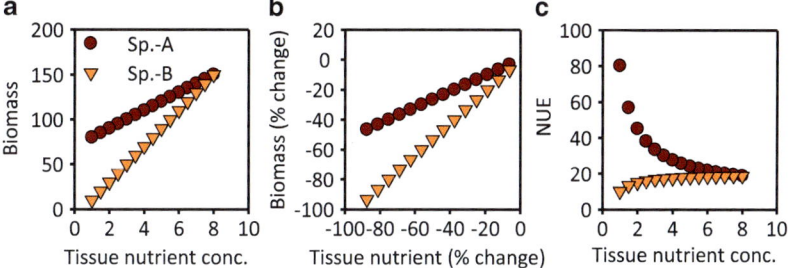

Fig. 1 Simulation of the two contrasting species (Sp.-A and Sp.-B) differing in their nutrient utilization efficiency (NUE, g^2 dry mass mg^{-1} nutrient) when grown into deficient condition. For comparison, here, both the species are assumed to have 8 mg nutrient g^{-1} dry mass when grown under sufficient condition and have a similar biomass. Both Sp.-A and Sp.-B responded to the nutrient deficiency by decreasing the total biomass and the deficient nutrient simultaneously (**a**). The percentage of decrease (% change) in the biomass versus tissue nutrient was lower in Sp.-A, but higher in Sp.-B (**b**). This resulted an increase in NUE of Sp.-A whereas a decrease in the NUE for Sp.-B under deficient condition (**c**)

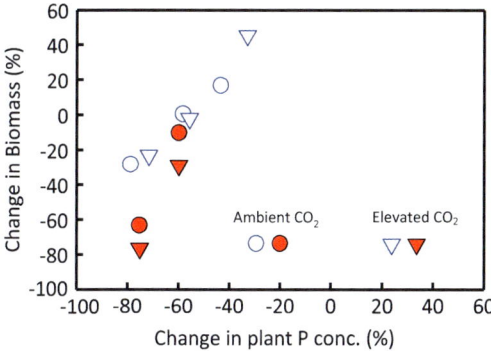

Fig. 2 The percentage changes in total biomass versus plant P concentration in soybean at 44 days after planting (Cure et al. 1988, *unfilled symbols*) and at maturity (Singh et al. 2014, *filled symbols*). Both studies were conducted under controlled environment growth chambers with a range (sufficient to deficient) of external P supply each under 350 and 700 (*unfilled symbols*) or 400 and 800 (*filled symbols*) μmol mol^{-1} CO$_2$ representing ambient (*circles*) and elevated (*triangles*) growth CO$_2$. For the unfilled symbols, the percentage changes were estimated as in Cure et al. (1988) after excluding the lowest external P supply of 0.005 mM

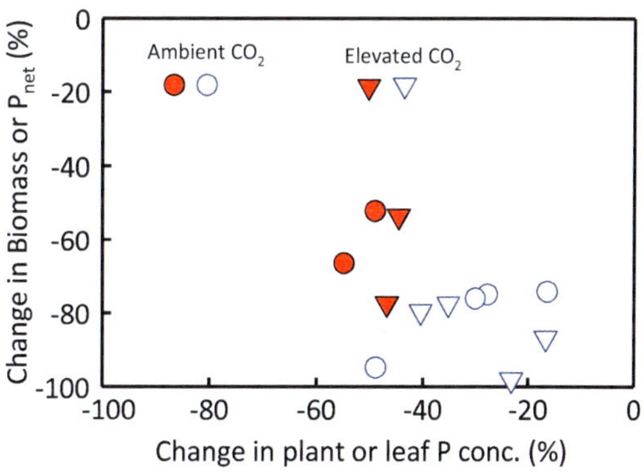

Fig. 3 The percentage changes in cotton total biomass versus plant P concentration (*unfilled symbol*) and rate of leaf photosynthesis (P$_{net}$) versus leaf P concentration (*filled symbol*). The data are 91–112 (*unfilled symbols*) and mean value of 57–112 (*filled symbols*) days after planting estimated from Singh et al. (2013a). The study was conducted in controlled environment growth chambers with a range of external P supply each under 400 (ambient, *circles*) and 800 (elevated, *triangles*) μmol mol^{-1} CO$_2$

When grown under P deficient condition, studies have found NUE of P either increased in soybean (Cure et al. 1988; Singh et al. 2014) or decreased in cotton (Singh and Reddy 2014). This is further illustrated by calculating the percentages change in plant biomass and tissue P concentration obtained under a range of P deficient conditions as compared to the P sufficient condition from previous studies utilizing soybean (Cure et al. 1988; Singh et al. 2014) and cotton (Singh et al. 2013a; Singh and Reddy 2014). When percentage change in biomass was plotted against the percentage change in the tissue nutrient concentration, soybean showed lesser decrease in biomass versus tissue P concentration (Fig. 2). However, cotton showed greater decrease in biomass than in the tissue P concentration (Fig. 3). In this example, aforementioned Species-A and Species-B could represent the soybean and the cotton, respectively. The contrasting results obtained between these two

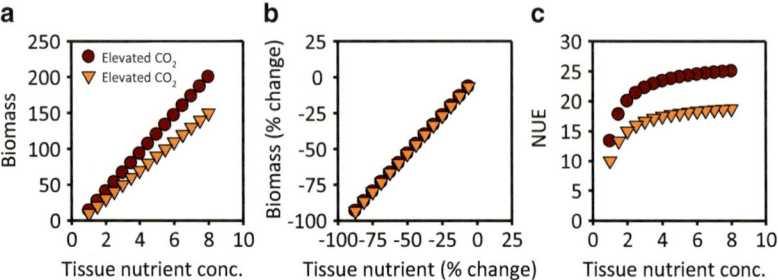

Fig. 4 Simulation of a species exhibiting different nutrient utilization efficiency (NUE, g^2 dry mass mg^{-1} nutrient) due to growth differences under ambient versus elevated CO_2 concentration when grown into deficient condition. For comparison, here, the species are assumed to have 8 mg nutrient g^{-1} dry mass at both CO_2 levels when grown under sufficient condition and the biomass was higher under elevated CO_2 versus ambient CO_2. The species responded to the nutrient deficiency by decreasing the total biomass and the deficient nutrient simultaneously at both CO_2 levels (**a**). The percentage of decrease (% change) as compared to the plant grown at the sufficient nutrient supply in the biomass versus tissue nutrient was also similar at both CO_2 levels (**b**). However, the NUE was higher at elevated versus ambient CO_2 concentration

species might be explained by their growth habit. Soybean is an annual crop while cotton is genetically a perennial which might exhibit a slower growth rate than soybean especially under nutrient deficiency. Therefore, under P deficiency the pace of cotton growth versus nutrient acquisition might have been much slower than it was in soybean resulting in decreased NUE of P in cotton. Another explanation for the higher NUE of P under P deficiency or lower NUE in the soybean at the higher tissue P concentration (or higher external P supply) is the fact that the tissue P continued to increase without any additional growth or biomass accumulation (Cure et al. 1988; Singh et al. 2013a). This might also occur in the case of Species-A. Similar to the total biomass or yield as an assessment of NUE, the rate of photosynthesis (P_{net}) can also be used to estimate the NUE of photosynthesis using Eq. 5 as also shown in the Fig. 3. Cotton P_{net} also decreased more than the leaf tissue P concentration. Therefore, the NUE of P_{net} also followed a similar tend as of biomass production and decreased under P deficiency in cotton across both CO_2 levels (Singh et al. 2013a; Singh and Reddy 2014).

Plants may be more efficient in the nutrient acquisition and utilization under elevated CO_2 due to increased growth and development and decrease in the tissue nutrient concentration (Rogers et al. 1993; Prior et al. 1998; Pérez-López et al. 2014). Moreover, the increased root proliferation and mass under elevated CO_2 also support the stimulated growth by increasing nutrient acquisition. Previous studies reported an increased N, P, and K utilization efficiency in agronomic crops such as soybean and cotton under elevated CO_2 (Cure et al. 1988; Prior et al. 1998; Singh and Reddy 2014). However, under nutrient limited supply this response of elevated CO_2 may be limited (Cure et al. 1988; Pérez-López et al. 2014; Singh and Reddy 2014). Pérez-López et al. (2014) did not find a consistant evidence that nutrient use efficiency was affected by CO_2 concentration in barley (*Hordeum vulgare* L.) grown in either normal or saline conditions. Therefore, the degree of influence by elevated CO_2 on NUE might be determined by crops species and the severity of nutrient deficiency (Cure et al. 1988; Singh and Reddy 2014). Under a given condition, the stimulated growth and decreased tissue nutrient concentration at elevated versus ambient CO_2 should result in higher NUE. However, an increase in NUE can also be achieved under elevated CO_2-mediated higher biomass production, without reduction in the tissue nutrient concentration at elevated versus ambient CO_2. This has been demonstrated in Fig. 4 using a simulated data set across ambient and elevated

CO_2. Plants grown at elevated CO_2 tended to have higher biomass with similar tissue nutrient concentration as compared to the ambient CO_2 (Fig. 4a). In addition, the biomass and tissue nutrient concentration declined similarly at both CO_2 to a given nutrient deficiency (Fig. 4b). However, due to growth stimulation, elevated CO_2 showed an increased NUE (Fig. 4c).

The resulting overall nutrient efficiency of a species or a cultivar in an agronomic point of view is essentially determined by the combination of both the nutrient acquisition or uptake and the nutrient utilization efficiency. The former relates to the nutrient absorption via roots from soil and translocation into the shoots, whereas the latter signifies the assimilation and utilization of the absorbed nutrient in the plants to support the plant metabolism in the synthesis of the tissues and organs. An increased root to shoot ratio in plants grown under nutrient deficient or elevated CO_2 as observed in many of the previous studies might increase the nutrient acquisition, thus providing more root surface for nutrient absorption and enhanced nutrient uptake to support the aboveground biomass (Prior et al. 1998; Singh et al. 2013a). Most importantly, the NUE increases even under declined nutrient uptake efficiency due to favored root growth and overall increase in the total amount of nutrient uptake (Cure et al. 1988; Rogers et al. 1994; Singh et al. 2013a).

4 Influence of Elevated CO_2 on the Critical Tissue Nutrient Concentration

The critical tissue nutrient concentration (CTNC) is dependent upon the crop species and physiological, growth, and development processes. The critical tissue concentration may be defined as the concentration to achieve 90 % of maximum productivity (Conroy 1992; Rogers et al. 1993; Reddy and Zhao 2005). This is generally derived by plotting measured growth processes (such as biomass accumulation, rates of photosynthesis, leaf area expansion, node addition and stem elongation, and yield) versus tissue nutrient concentration (usually leaf or whole plant). The CTNC for N, P, and K for biomass accumulation in cotton may vary as 51, 4.1, and 12 g kg^{-1} dry weight, respectively, near ambient CO_2 concentration (Rogers et al. 1993; Reddy and Zhao 2005). Similarly, the CTNC of N and P for biomass production in 36-days-old wheat is reported to be 45 and 3.9 g kg^{-1} dry weight under ambient CO_2 (Rogers et al. 1993). In addition, the CTNC of K for the growth process such leaf area expansion is higher (17 g kg^{-1} dry weight) than photosynthesis and dry matter accumulation (Reddy and Zhao 2005). However, variability for CTNC also exists for a given growth process and may be attributed to the stage of growth measurements and experimental conditions (Reddy and Zhao 2005).

Previous studies indicate that elevated CO_2 increases the overall nutrient demand and a higher concentration of P and K may be required for maximum plant productivity (Conroy 1992; Rogers et al. 1993; Reddy and Zhao 2005; Singh et al. 2014). As a result, the critical leaf P and K concentration is likely to be higher in plants grown under elevated CO_2. The critical leaf tissue concentration of N has often been reported to decrease under elevated CO_2; however, it is not always true for P and K (Conroy et al. 1990; Conroy 1992; Rogers et al. 1993). Rogers et al. (1993) reported that for biomass production critical N concentration decreased from 51 to 32 g kg^{-1} in cotton and 45 to 38 g kg^{-1} in wheat grown between 350 and 900 μmol mol^{-1} CO_2, whereas the critical P concentration increased from 4.1 to 7.8 g kg^{-1} in cotton and 3.9 to 5.3 g kg^{-1} in wheat from ambient to elevated CO_2 concentration in the same study. Similarly, the critical leaf K concentration increased from 12 g kg^{-1} at ambient (360 μmol mol^{-1}) CO_2 to 19 g kg^{-1} dry weight at elevated (720 μmol mol^{-1}) CO_2 for photosynthesis and biomass production in cotton (Reddy and Zhao 2005). This clearly suggested that it is not only the nutrient demand but also the sensitivity of crop that may be increased especially for P and K nutrition under a CO_2-enriched atmosphere of the future. The lower demand of N but higher requirement of P or K under elevated CO_2

might be one of the major causes for the observed differences of the critical nutrient concentrations among N, P, and K. Therefore, the current understanding of tissue nutrient concentration to achieve optimum crop growth and yield under ambient CO_2 is likely to change into the CO_2-enriched atmosphere of the future. Thus a reassessment will be imperative. Overall this suggests that a lower N but higher foliar concentration of P and K may be needed to attain maximum growth in many agronomic crops and trees (Conroy et al. 1990; Rogers et al. 1993; Reddy and Zhao 2005; Singh et al. 2014).

5 Effect of Elevated Temperature and CO_2 on Soil Microbial Communities

The effects of increased atmosphere CO_2 concentration on soil microbial habitat are mainly a result of alteration in the responses of plant communities to elevated CO_2. In higher plants, elevated CO_2 primarily increases biomass and residues and altered tissue compositions (e.g. minerals, nutrients, proteins) and root exudation upon which soil microbes depend for their food and energy. Therefore, elevated CO_2 indirectly affects various processes in soil including nitrification, denitrification, emission of trace gases (methane and nitrous oxides), decomposition, rhizoremediation, and rhizodeposition processes depending upon the moisture and temperature of the soil (Freeman et al. 2004; Sadowsky and Schortemeyer 1997). Elevated CO_2 is known to stimulate soil respiration across various plant communities (King et al. 2004). The plant derived carbon in rhizosphere is primarily used by microorganism including those associated with the root systems. Drigo et al. (2008) pointed out that soil microbial communities, especially in the vicinity of roots, including mycorrhiza, bacteria, and fungi are altered due to plant metabolisms and root secretion driven by elevated CO_2. Greater plant debris or organic matter production under elevated CO_2 has also been attributed to alteration of soil microbial community composition (Lesaulnier

et al. 2008). Under elevated CO_2, the rate of plant residue decomposition may become limited by lower N concentration which may reduce the release of nitrogen from decomposing organic material.

Runion et al. (1994) found increased total microbial activity under free-air-enriched CO_2 in a selected group of rhizosphere and phyllosphere of cotton. In a Mojave Desert ecosystem, under elevated CO_2 there was a significant decrease in the operational taxonomic units for *Firmicutes* (bacteria) and *Basidiomycota* (fungi), and qRT-PCR (quantitative real time polymerase chain reaction) analysis revealed that under elevated CO_2 there was a 43 % decrease in the population of gram-positive microorganisms (Nguyen et al. 2011). Drigo et al. (2009) observed that under elevated CO_2 *Bacillus* and slow-growing microorganisms like actinomycetes remain unaffected, but the genera of *Pseudomonas* and *Burkholderia* were highly affected depending up on soil types and plant species.

The elevated levels of CO_2 which results in higher carbon assimilation by plants may support higher abundance of fungi and higher activities of soil carbon degrading enzymes by the additional carbon accumulation and increased microbial utilization of soil organic matter (Carney et al. 2007). The CO_2 enrichment significantly increases above- and below-ground plant biomass which as a soil residue may result in an additional organic matter and mineral elements to the existing C pool of the soil. Previous results have suggested that with conservation management in the CO_2 enriched environment, greater residue amounts could increase soil C storage as well as increased ground cover (Prior et al. 2005). The increases in air and soil temperature are most likely to affect soil microbial community by its direct influence on the metabolic processes. Higher temperature leads to higher rate of changes in phospholipid fatty acids (PLFA) pattern of bacteria and activity (Pettersson and Baath 2003). Sugawara and Sadowsky (2013) studied the influence of elevated atmospheric CO_2 on *Bradyrhizobium japonicum* in soybean rhizosphere and presented transcriptome data which suggested that an influence on the gene expression resulting into alteration of carbon/nitrogen metabolism, respiration, and nodulation efficiency. Nelson et al. (2010) reported a shift in soil biochemical processes affecting archaeal community composition in soybean rhizosphere under elevated CO_2 condition. He et al. (2013) studied the distinct response of soil microbial communities to elevated CO_2 and O_3 (ozone) in soybean agro- ecosystem and found that it affects their functional composition, structure, and metabolic potential.

6 Effect of Elevated Temperature and CO₂ on Mycorrhizal Communities

The fungi that are most predominant in agricultural soils are arbuscular mycorrhizal (AM) fungi (phylum *Glomeromycota*). They account for up to 50 % of the biomass of soil microbes (Olsson et al. 1999) and almost all crops are mycorrhizal, and many, if not most, are strongly responsive to AM fungi (Cardoso and Kuyper 2006). However, only a few families and genera of plants do not generally colonized by AM fungi; these include *Brassicaceae* (their root exudates are possibly even toxic to AM fungi), *Caryophyllaceae*, *Cyperaceae*, *Juncaceae*, *Chenopodiaceae*, and *Amaranthaceae* (although each of these families has some representatives that are usually colonized by AM fungi). The AM association has received attention as essential component of soil biological community for increasing the sustainability of agricultural systems. The ability of AM fungi to enhance host–plant uptake of relatively immobile nutrients, in particular P, and several micronutrients, has been highly recognized. AM colonization also protects plants against biotic and abiotic stresses (Charest et al. 1993). However, the research on the influence of elevated temperature and CO_2 on AM fungi and its functions in agro-ecosystems are limited.

Drigo et al. (2008) in their review concluded that the main effects of elevated atmospheric CO_2 occur via plant metabolism and root secretion especially in C_3 plants thereby directly affecting the mycorrhizal, bacterial, and fungal communities in the close vicinity of the roots. Moreover, under CO_2 enrichment, fungal food chain is more strongly stimulated than bacterial food chain, and AM fungi root infection increases at community level (Rillig et al. 1999). In fact, due to elevated CO_2 and nitrogen deposition, mycorrhizal tissues have been demonstrated to have a significant fraction of soil organic matter and may act as a carbon sink (Treseder and Allen 2000). As a consequence of elevated CO_2 there is an increase in labile and stable soil C pool, an increased efficiency in the degradation of organic pollutants by rhizoremediation processes. There is also an enhancement in C inputs and mycorrhizal colonization which stimulates both microbial and plant N acquisition (Formánek et al. 2014). However when a mycocentric model was used to study the mycorrhizal fungal and plant responses, it was observed that ectomycorrhizal systems respond more strongly than AM systems to elevated CO_2 (Alberton et al. 2005).

From the long-term effect of elevated atmospheric CO_2 study where a nonmycorrhizal plant *Carex arenaria* and a mycorrhizal plant *Festuca rubra* were grown, Drigo et al. (2013) reported that mycorrhizal plant exert a greater influence on bacterial and fungal communities. Moreover, the fatty acid biomarker data confirmed that the rhizodeposited carbon is first processed by AM fungi and subsequently transferred to bacterial and fungal communities in rhizospheric soil. Over the course of three years there was a delay in transfer of carbon to bacterial communities. The ^{13}C isotope pulse chase experiment revealed that an increase in ^{13}C enrichment in AM fungi. However, in a Chaparral ecosystem (shrubland or heathland plant community) with CO_2 enrichment, there were no changes in mycorrhizal colonization and hyphal length except an increase in the amount of carbon, glomalin, and hyphae of *Scutellospora* and *Acaulospora* within macro aggregates (Allen et al. 2005). In the experiments of Bunn et al. (2009), AM fungi were found to ameliorate temperature stress in variety of plants tested and the AM colonization levels and length of extraradical hyphae increased with soil temperature. Under the future climatic scenarios simulating high CO_2 and temperature, a stimulation in the mycorrhizal colonization is expected (Büscher et al. 2012). Therefore, mycorrhizal communities tend to support plant growth under the conditions of elevated temperature and carbon dioxide.

7   Influence of Nutrient Stress and CO_2 Enrichment on the Quality of Harvestable Crop Yield

There has been strong evidence of lower mineral nutrients in the seeds of many crops when grown under CO_2-enriched environment (Loladze 2014; Myers et al. 2014). The reduced nutrients in seeds impact not only the human or animal food nutrition but also the early crop growth when seeded for the next crop. For instance, seeds with high P content might help to produce greater proliferated roots when planted under field condition, thus increasing the nutrient uptake (Riley et al. 1993). The quality of seed yield, fruits, and vegetables is also affected by plant nutrition and elevated CO_2 concentration. Grain and seed quality of several crops such as corn, soybean, rice, and wheat is reduced due to their small size and wrinkled seed with altered nutrient composition especially when grown under nutrient-poor soils. Although the genetic character of a crop species determines the seed composition, factors such as plant nutrition and growth CO_2 can drastically alter the seed composition (Conner et al. 2004).

The well-known growth- and yield-stimulating effect (quantity) of rising atmospheric CO_2 also appears to cause reduction in the quality of the crop harvestable products. Elevated CO_2 decreases protein content due to reduction in total nitrogen and storage protein and mineral nutrients such as P, zinc (Zn), and iron (Fe) in wheat, rice, peas, and soybean affecting the overall taste and quality of many food products (Loladze 2014; Myers et al. 2014). These effects have mostly observed in the C_3

species which are the most responsive to the increased CO_2 as comparison to the C_4 species such as corn (*Zea Maize* L.). The C_4 species has already developed CO_2-concentrating mechanism inside the mesophyll and therefore shows minor or insignificant response to increased atmospheric CO_2. Myers et al. (2014) reported roughly up to 12.7 % lower Zn and Fe in almost all the C_3 grasses and legumes when grown under elevated versus ambient CO_2 concentration. Reduction of these mineral elements in the edible portion of the food crops will have profound effect for human nutrition and animal feed. Population living in the countries where staple food from these crops are the major component of their everyday diet and main source of the mineral nutrition are and will be affected the most. As noted above, decline in the nutrients such as N and P in the plant biomass may also lead to low protein content and reduce the quality of feed, affecting animal health. Thus, elevated CO_2-mediated changes in the dynamics of the mineral nutrition in the edible plant products are of a global concern that requires an immediate attention.

Future Perspective and Concluding Remark

It is established that the nutrient deficiency and elevated CO_2 have inverse effect on crop productivity. However, increase in nutrient efficiency is commonly observed under elevated CO_2 and also in nutrient-deficient condition in several crops such as soybean. This provides an opportunity for crop improvement by enhancing plant nutrient metabolism in the CO_2-enriched environment of the future.

Crop productivity depends on the plant-available nutrients in the rhizosphere. There are two primary ways to increase the quantity and quality of crop production on a given piece of land. Firstly, this can be achieved by the increase of the nutrient supply. This is not a sustainable option albeit an expensive one in the terms of both farmer's economics and economy of the environment. The second option is to enhance the efficiency of nutrient acquisition and utilization. This option is sustainable, plausible, and inexpensive but not an easy solution to the problem as it involves identifying existing cultivars or developing new ones with the desirable traits that are related to efficient uptake and use of the nutrients. Even using both aforementioned options is far better than just pertaining to the first option.

Since crops grown under elevated CO_2 show an increased total amount of nutrient uptake and utilization efficiency, better preparedness is required to reap the complete benefit of rising atmospheric CO_2 concentration. For instance, a near optimum soil fertility will be prerequisite to harness the full potential and to avoid the loss in the degree of growth stimulation caused by elevated CO_2. In addition, active breeding programs are needed that involves the identification of desirable cultivars or traits and incorporate these traits via traditional breeding and/or genetic engineering into crop improvements aimed to nutrient deficient and CO_2-enriched environments. The existing cultivar differences in the nutrient uptake and utilization provides opportunity for breeding even under nutrient-deficient condition. This can be achieved by screening a large number of genotypes for the desirable traits in a crop species grown under a specific nutrient-deficient condition. These traits may include nutrient uptake and utilization efficiency, enhanced root system, increased photosynthesis, biomass, or yield. Then a breeding program can be designed that incorporates these traits in the improved cultivars suitable for a wide range of environments. This is a challenging task, which requires several years and state-of-the-art facilities especially when elevated CO_2 is considered as component in the genotype screening and other resources.

The existing crop production, thus the productivity on per unit of land must increase to meet the food demand for the growing world's human population. A lot of existing resources such as water, nutrients, and energy needed for the human recreation and consumption must be diverted toward improving agricultural production. In the past, extensive uses of fertilizers have contributed immensely to the increased crop productivity. The production costs of these fertilizers are enormous and challenge the sustainability of agriculture. The fertilizer application in the crops also leads to the air pollution and contamination of water. Other harmful environmental effects include the release of gases such as nitrous oxide, CO_2, methane, and ammonia that cause greenhouse effect and involve in the depletion of ultraviolet-B protective gases such as stratospheric ozone and causes eutrophication of the aquatic system due to the release of nitrates and phosphates in the water body. In addition, the natural reserves for the phosphate rocks that are being used to produce phosphate fertilizers are depleting and may be exhausted within next 25–100 years. Therefore, elucidating the dynamics of nutrients as affected by soil, plant, and the environment are a daunting challenge.

Climate change is already affecting the natural resources such as terrestrial vegetation, animal husbandry, and fisheries that societies depend on to provide food, fiber, fuel, several industrial products, and recreational services (World-Bank 2010). Therefore, among all, world's food supply for the growing population, which is primarily depending on the agriculture, has become a major concern. The current world's human population of approximately 7.2 billion is expected to increase to 8.0 billion by 2025, about 9.6 billion by 2050, and about 10.6 billion by 2100 (United Nations 2013). About 86 % of this growth is expected to occur in less developed countries where a quarter of population is still living in extreme poverty. This increase in population requires almost double the current agricultural production to meet the future food demand while minimizing the associated environmental damage. Moreover, this vast increase in the productivity must be achieved in the face of changing climate, diminishing natural resources such as plant nutrients and global conflict. The pressure for increased food production will be more in the developing countries. Since there is no new arable land available to bring into agriculture, an intensive cultivation of the exiting arable land will be the primary option by using modern agricultural methodologies. The world will face intensified competition among the human needs for usage of land, water, energy, and other natural resources to produce more food in a sustainable manner. Water has to be used efficiently among the competing demands for human personal usage such as energy and urban consumption, recreation purposes, agriculture, fisheries, and healthy ecosystem. Furthermore, in the face of competing demand for the use of natural resources by growing human population, the vulnerability of the global climates will be critical due to depleting natural reserves needed for fertilizer production, soil erosion, desertification, emission of greenhouse gases, destruction of natural habitat, and pollution to soil, air, and water.

References

Ahanager MA, Hashem A, Abd-Allah EF, Ahmad P (2014) Arbuscular mycorrhiza in crop improvement under environmental stress. In: Ahmad P, Rasool S (eds) Emerging technologies and management of crop stress tolerance. Academic Press, London

Ahmed FE, Hall AE, Madore MA (1993) Interactive effects of high temperature and elevated carbon dioxide concentration on cowpea (*Vigna unguiculata* (L.) Walp.). Plant Cell Environ 16:835–842

Alberton O, Kuyper TW, Gorissen A (2005) Taking mycocentrism seriously: mycorrhizal fungal and plant responses to elevated CO_2. New Phytol 167:859–868

Allen MF, Klironomos JN, Treseder KK, Walter OC (2005) Responses of soil biota to elevated CO_2 in a chaparral ecosystem. Ecol Appl 15:1701–1711

Almeida JF, Hartwig UA, Frehner M, Nösberger J, Lüscher A (2000) Evidence that P deficiency induces N feedback regulation of symbiotic N_2 fixation in white clover (*Trifolium repens* L.). J Exp Bot 51:1289–1297

Barrett DJ, Gifford RM (1995) Acclimation of photosynthesis and growth by cotton to elevated CO_2: interactions with severe phosphate deficiency and restricted rooting volume. Aust J Plant Physiol 22:955–963

Bazzaz FA (1997) Allocation of resources in plants: state of the science and critical questions. In: Bazzaz FA,

Grace J (eds) Plant resource allocation. Academic Press, San Diego, pp 1–38

Betsche T (1994) Atmospheric CO_2 enrichment: kinetics of chlorophyll a fluorescence and photosynthetic CO_2 uptake in individual, attached cotton leaves. Environ Exp Bot 34:75–86

Bloom AJ, Burger M, Asensio JSR, Cousins AB (2010) Carbon dioxide enrichment inhibits nitrate assimilation in wheat and *Arabidopsis*. Science 328:899–903

Brouwer R (1962) Nutritive influences on the distribution of dry matter in the plant. Neth J Agric Sci 10:399–408

Bunn R, Lekberg Y, Zabinski C (2009) Arbuscular mycorrhizal fungi ameliorate temperature stress in thermophilic plants. Ecology 90:1378–1388

Büscher M, Zavalloni C, de Boulois HD, Vicca S, Van den Berge J, Declerck S, Ceulemans R, Janssens IA, Nijs I (2012) Effects of arbuscular mycorrhizal fungi on grassland productivity are altered by future climate and below-ground resource availability. Environ Exp Bot 81:62–71

Campbell CD, Sage RF (2006) Interactions between the effects of atmospheric CO_2 content and P nutrition on photosynthesis in white lupin (*Lupinus albus* L.). Plant Cell Environ 29:844–853

Cardoso IM, Kuyper TW (2006) Mycorrhizas and tropical soil fertility. Agric Ecosyst Environ 116:72–84

Carney KM, Hungate BA, Drake BG, Megonigal JP (2007) Altered soil microbial community at elevated CO_2 leads to loss of soil carbon. Ecology 104:4990–4995

Cassman KG, Whitney AS, Fox RL (1981) Phosphorus requirements of soybean and cowpea as affected by mode of N nutrition. Agron J 73:17–22

Charest C, Dalpé Y, Brown A (1993) The effect of vesicular–arbuscular mycorrhizae and chilling on two hybrids of *Zea mays* L. Mycorrhiza 4:89–92

Compant S, van der Heijden MG, Sessitsch A (2010) Climate change effects on beneficial plant- microorganism interactions. FEMS Microbiol Ecol 73:197–214

Conner T, Paschal EH, Barbero A, Johnson E (2004) The challenges and potential for future agronomic traits in soybeans. AgBioforum 7:47–50

Conroy J (1992) Influence of elevated atmospheric CO_2 concentrations on plant nutrition. Aust J Bot 40:445–456

Conroy JP, Milham PJ, Reed ML, Barlow EW (1990) Increases in phosphorus requirements for CO_2-enriched pine species. Plant Physiol 92:977–982

Cordell D, Drangert JO, White S (2009) The story of phosphorus: global food security and food for thought. Glob Environ Change 19:292–305

Cure JD, Rufty TW, Israel DW (1988) Phosphorus stress effects on growth and seed yield responses of nonnodulated soybean to elevated carbon dioxide. Agron J 80:897–902

Deng Y, He Z, Xu M, Qui Y, Van Nostrand JD, Wu L, Roe BA, Wiley G, Hobbie SE, Reich PB, Zhou J

(2012) Elevated carbon dioxide alters the structure of soil microbial communities. Appl Environ Microbiol 78:2991–2995

Drigo B, Kowalchuk GA, Veen JA (2008) Climate change goes underground: effects of elevated atmospheric CO_2 on microbial community structure and activities in the rhizosphere. Biol Fertil Soils 44:667–679

Drigo B, van Veen JA, Kowalchuk GA (2009) Specific rhizosphere bacterial and fungal groups respond differently to elevated atmospheric CO_2. ISME J 3:1204–1217

Drigo B, Kowalchuk GA, Knapp BA, Pijl AS, Boschker HTS, van Veen J (2013) Impacts of 3 years of elevated atmospheric CO_2 on rhizosphere carbon flow and microbial community dynamics. Glob Change Biol 19:621–636

Edwards GE, Baker NR (1993) Can CO_2 assimilation in maize leaves be predicted accurately from chlorophyll fluorescence analysis? Photosynth Res 37:89–102

Fleisher DH, Wang Q, Timlin DJ, Chun JA, Reddy VR (2012) Response of potato gas exchange and productivity to phosphorus deficiency and carbon dioxide enrichment. Crop Sci 52:1803–1815

Fleisher DH, Wang Q, Timlin DJ, Chun JA, Reddy VR (2013) Effect of carbon dioxide and phosphorus supply on potato dry matter allocation and canopy morphology. J Plant Nutr 36:566–586

Formánek P, Rejšek K, Vranová V (2014) Effect of elevated CO_2, O_3, and UV radiation on soils. Sci World J 2014:8

Freeman C, Kim SY, Lee SH, Kang H (2004) Effects of elevated atmospheric CO_2 concentrations on soil microorganisms. J Microbiol 42:267–277

Gifford R, Barrett D, Lutze J (2000) The effects of elevated [CO_2] on the C:N and C:P mass ratios of plant tissues. Plant Soil 224:1–14

Güsewell S (2004) N : P ratios in terrestrial plants: variation and functional significance. New Phytol 164:243–266

Hanway JJ, Weber CR (1971) Accumulation of N, P, and K by soybean (Glycine max (L.) Merrill) plants. Agron J 63:406–408

He Z, Xiong J, Kent AD, Deng Y, Xue K, Wang G, Wu L, Van Nostrand JD, Zhou J (2013) Distinct responses of soil microbial communities to elevated CO_2 and O_3 in a soybean agro- ecosystem. ISME J 8:714–726

Himelblau E, Amasino RM (2001) Nutrients mobilized from leaves of Arabidopsis thaliana during leaf senescence. J Plant Physiol 158:1317–1323

IPCC (2007) Climate change 2007: the physical science basis. Contribution of working group I to the fourth assessment report of the Intergovernmental Panel on Climate Change. In Solomon S et al (eds). Cambridge University Press, Cambridge/New York

Israel DW, Rufty TW (1988) Influence of phosphorus nutrition on phosporus and nitrogen utilization effeciencies and associated physiological responses in soybean. Crop Sci 28:954–960

Israel DW, Rufty TW, Cure JD (1990) Nitrogen and phosphorus nutritional interactions in a CO_2 enriched environment. J Plant Nutr 13:1419–1433

Kawakami EM, Oosterhuis DM, Snider JL (2013) Nitrogen assimilation and growth of cotton seedlings under NaCl salinity and in response to urea application with NBPT and DCD. J Agron Crop Sci 199:106–117

King JS, Hanson PJ, Bernhardt E, DeAngelis P, Norby RJ, Pregitzer KS (2004) A multiyear synthesis of soil respiration responses to elevated atmospheric CO_2 from four forest FACE experiments Spruce and Peatland responses under climatic and environmental change. Glob Change Biol 10:1027–1042

Koerselman W, Meuleman AFM (1996) The vegetation N:P ratio: a new tool to detect the nature of nutrient limitation. J Appl Ecol 33:1441–1450

Lam SK, Chen D, Norton R, Armstrong R (2012) Does phosphorus stimulate the effect of elevated [CO_2] on growth and symbiotic nitrogen fixation of grain and pasture legumes? Crop Pasture Sci 63:53–62

Lauer MJ, Blevins DG, Sierzputowska-Gracz H (1989) ^{31}P-affected by phosphate nutrition-nuclear magnetic resonance determination of phosphate compartmentation in leaves of reproductive soybeans (Glycine max l.) as affected by phosphate nutrition. Plant Physiol 89:1331–1336

Lenka NK, Lal R (2012) Soil-related constraints to the carbon dioxide fertilization effect. Crit Rev Plant Sci 31:342–357

Lesaulnier C, Papamichail D, McCorkle S, Ollivier B, Skkiena S, Taghavi S, Zak D, van der Lelie D (2008) Elevated atmospheric CO_2 affects soil microbial diversity associated with trembling aspen. Environ Microbiol 10:926–941

Lewis JD, Griffin KL, Thomas RB, Strain BR (1994) Phosphorus supply affects the photosynthetic capacity of loblolly pine grown in elevated carbon dioxide. Tree Physiol 14:1229–1244

Loladze I (2014) Hidden shift of the ionome of plants exposed to elevated CO_2 depletes minerals at the base of human nutrition. eLife 3:e02245

Longstreth DJ, Nobel PS (1980) Nutrient influences on leaf photosynthesis: effects of nitrogen, phosphorus, and potassium for Gossypium hirsutum L. Plant Physiol 65:541–543

Marschner H (1986) Mineral nutrition of higher plants. Academic Press, Orlando

Maxwell K, Johnson GN (2000) Chlorophyll fluorescence: a practical guide. J Exp Bot 51:659–668

Moyano FE, Manzoni S, Chenu C (2013) Responses of soil heterotrophic respiration to moisture availability: an exploration of processes and models. Soil Biol Biochem 59:72–85

Mullins GL, Burmester CH (1990) Dry matter, nitrogen, phosphorus, and potassium accumulation by four cotton varieties. Agron J 82:729–736

Myers SS, Zanobetti A, Kloog I, Huybers P, Leakey ADB, Bloom AJ, Carlisle E, Dietterich LH, Fitzgerald G, Hasegawa T, Holbrook NM, Nelson RL, Ottman MJ, Raboy V, Sakai H, Sartor KA, Schwartz J, Seneweera S, Tausz M, Usui Y (2014) Increasing CO_2 threatens human nutrition. Nature 510:139–142

Nelson DM, Cann Issac KO, Mackie RI (2010) Response of archaeal communities in the rhizosphere of maize and soybean to elevated atmospheric CO_2. PLoS One 5:e15897

Nguyen LM, Buttner MP, Cruz P, Smith SD, Robleto EA (2011) Effect of elevated CO_2 on rhizosphere soil microbial communities in a Mojave Desert ecosystem. J Arid Environ 75:917–925

Norby RJ, O'Neill EG, Luxmoore RJ (1986) Effects of atmospheric CO_2 enrichment on the growth and mineral nutrition of *Quercus alba* seedlings in nutrient-poor soil. Plant Physiol 82:83–89

Olsson PA, Thingstrup I, Jakobsen I, Baath F (1999) Estimation of the biomass of arbuscular mycorrhizal fungi in a linseed field. Soil Biol Biochem 31:1879–1887

Pérez-López U, Robredo A, Lacuesta M, Muñoz-Rueda A, Mena-Petite A (2010) Atmospheric CO_2 concentration influences the contributions of osmolyte accumulation and cell wall elasticity to salt tolerance in barley cultivars. J Plant Physiol 167:15–22

Pérez-López U, Robredo A, Lacuesta M, Mena-Petite A, Muñoz-Rueda A (2012) Elevated CO_2 reduces stomatal and metabolic limitations on photosynthesis caused by salinity in *Hordeum vulgare*. Photosynth Res 111:269–283

Pérez-López U, Miranda-Apodaca J, Mena-Petite A, Muñoz-Rueda A (2014) Responses of nutrient dynamics in barley seedlings to the interaction of salinity and carbon dioxide enrichment. Environ Exp Bot 99:86–99

Pettersson M, Baath E (2003) Temperature-dependent changes in the soil bacterial community in limed and unlimed soil. FEMS Microbiol Ecol 45:13–21

Prior SA, Rogers HH (1995) Soybean growth response to water supply and atmospheric carbon dioxide enrichment. J Plant Nutr 18:617–636

Prior SA, Torbert HA, Runion GB, Mullins GL, Rogers HH, Mauney JR (1998) Effects of carbon dioxide enrichment on cotton nutrient dynamics. J Plant Nutr 21:1407–1426

Prior SA, Rogers HH, Mullins GL, Runion GB (2003) The effects of elevated atmospheric CO_2 and soil P placement on cotton root deployment. Plant Soil 255:179–187

Prior SA, Runion BG, Rogers HH, Torbert HA, Reeves DW (2005) Elevated atmospheric CO_2 effects on biomass production and soil carbon in conventional and conservation cropping systems. Glob Chang Biol 11:657–665

Radin JW, Eidenbock MP (1984) Hydraulic conductance as a factor limiting leaf expansion of phosphorus-deficient cotton plants. Plant Physiol 75:372–377

Reddy KR, Zhao DL (2005) Interactive effects of elevated CO_2 and potassium deficiency on photosynthesis, growth, and biomass partitioning of cotton. Field Crops Res 94:201–213

Riley MM, Adcock KG, Bolland MDA (1993) A small increase in the concentration of phosphorus in the sown seed increased the early growth of wheat. J Plant Nutr 16:851–864

Rillig MC, Field CB, Allen MF (1999) Soil biota responses to long-term atmospheric CO_2 enrichment in two California annual grasslands. Oecologia 119:572–577

Rogers GS, Payne L, Milham P, Conroy J (1993) Nitrogen and phosphorus requirements of cotton and wheat under changing atmospheric CO_2 concentrations. Plant Soil 155–156:231–234

Rogers HH, Runion GB, Krupa SV (1994) Plant responses to atmospheric CO_2 enrichment with emphasis on roots and the rhizosphere. Environ Pollut 83:155–189

Rufty TW, Siddiqi MY, Glass ADM, Ruth TJ (1991) Altered $^{13}NO_3^-$ influx in phosphorus limited plants. Plant Sci 76:43–48

Rufty TW, Israel DW, Volk RJ, Qiu J, Sa T (1993) Phosphate regulation of nitrate assimilation in soybean. J Exp Bot 44:879–891

Runion GB, Curl EA, Rogers HH, Backman PA, Rodriguez-Kabana R, Helms BE (1994) Effects of free-air CO_2 enrichment on microbial populations in the rhizosphere and phyllosphere of cotton. Agric For Meteorol 70:117–130

Sa T, Israel DW (1998) Phosphorus-deficiency effects on response of symbiotic N_2 fixation and carbohydrate status in soybean to atmospheric CO_2 enrichment. J Plant Nutr 21:2207–2218

Sadowsky MJ, Schortemeyer M (1997) Soil microbial responses to increased concentrations of atmospheric CO_2. Glob Change Biol 3:217–224

Sharma MP, Adholeya A (2004) Influence of arbuscular mycorrhizal fungi and phosphorus fertilization on the *post-vitro* growth and yield of micropropagated strawberry in an alfisol. Can J Bot 82(3):322–328

Siddiqi MY, Glass ADM (1981) Utilization index: a modified approach to the estimation and comparison of nutrient utilization efficiency in plants. J Plant Nutr 4:289–302

Sinclair TR (1992) Mineral nutrition and plant growth response to climate change. J Exp Bot 43:1141–1146

Singh SK, Reddy VR (2014) Combined effects of phosphorus nutrition and elevated carbon dioxide concentration on chlorophyll fluorescence, photosynthesis and nutrient efficiency of cotton. J Plant Nutr Soil Sci, in press. doi:http://dx.doi.org/10.1002/jpln.201400117

Singh SK, Badgujar GB, Reddy VR, Fleisher DH, Timlin DJ (2013a) Effect of phosphorus nutrition on growth and physiology of cotton under ambient and elevated carbon dioxide. J Agron Crop Sci 199:436–448

Singh SK, Badgujar G, Reddy VR, Fleisher DH, Bunce JA (2013b) Carbon dioxide diffusion across stomata and mesophyll and photo-biochemical processes as affected by growth CO_2 and phosphorus nutrition in cotton. J Plant Physiol 170:801–813

Singh SK, Reddy VR, Fleisher DH, Timlin DJ (2014) Grow.., nutrient dynamics, and efficiency responses to carbon dioxide and phosphorus nutrition in soybean. J Plant Interact 9:838–849

Sionit N (1983) Response of soybean to two levels of mineral nutrition in CO_2-enriched atmosphere. Crop Sci 23:329–333

Staddon PL (2005) Mycorrhizal fungi and environmental change: the need for a mycocentric approach. New Phytol 167:635–637

Sugawara M, Sadowsky MJ (2013) Influence of elevated atmospheric carbon dioxide on transcriptional responses of Bradyrhizobium japonicum in the soybean rhizoplane. Microbes Environ 28:217–227

Taub DR, Wang X (2008) Why are nitrogen concentrations in plant tissues lower under elevated CO_2? A critical examination of the hypotheses. J Integr Plant Biol 50:1365–1374

Treseder K, Allen MF (2000) Mycorrhizal fungi have a potential role in soil carbon storage under elevated CO_2 and nitrogen deposition. New Phytol 147:189–200

United Nations, Department of Economic and Social Affairs, Population Division (2013) World population prospects: the 2012 revision, volume I: comprehensive tables. United Nations, New York

Vance CP, Uhde-Stone C, Allan DL (2003) Phosphorus acquisition and use: critical adaptations by plants for securing a nonrenewable resource. New Phytol 157:423–447

World-Bank (2010) World development report 2010: development and climate change. The World Bank, Washington, DC

Zhou W, Hui D, Shen W (2014) Effects of soil moisture on the temperature sensitivity of soil heterotrophic respiration: a laboratory incubation study. PLoS One 9:e92531

Phytometallophore-Mediated Nutrient Acquisition by Plants

Tapan Adhikari

Abstract

Soils in many agricultural areas are alkaline and have high amounts of calcium carbonate, resulting in low availability of micronutrients. Plant nutrient uptake is influenced by root architecture, the presence of mycorrhizal fungi, proton exudation from roots, and release of phytometallophores. Some genotypes are able to release phytometallophores from roots to the surrounding rhizosphere soil, which increases the solubility of nutrients and as a result, its availability for plant uptake. Phytometallophore (PM) release occurs under nutrient deficiencies in representative Poaceae and has been speculated to be a general adaptive response to enhance the acquisition of micronutrient metals. Many vascular plant species are unable to colonize calcareous sites. The inability of calcifuge plants to establish in limestone sites could be related to a low capacity of such plants to solubilize and absorb Fe from these soils. Under Fe deficiency, species of Poaceae enhance their Fe uptake by releasing non-proteinogenic amino acids, phytometallophores (PM), from their roots which mobilize Fe from the soil by forming a chelate that is then taken up by the root. Phytometallophore release and uptake is thought to be specific for Fe deficiency. However, a universal role of phytometallophores in the acquisition of micronutrient metals is established. Since PM form stable chelates with Zn, Mn, and Cu, they extract considerable amounts of Zn, Mn, and Cu from calcareous soils and deficiencies of Zn, Mn, and Cu are quite common on calcareous and non-calcareous soils. Fe deficiency in the shoot triggers the production of phytometallophores. Although PM release under nutrient deficiency is not a specific response, it could still have ecological significance. Studying metal extraction by PM from a wide range of calcareous and non-calcareous soils indicated that PM preferentially mobilize Fe but also significant quantities of Zn and

T. Adhikari (✉)
Indian Institute of Soil Science, Berasia Road, Bhopal
462038, MP, India
e-mail: tapan_12000@yahoo.co.uk

A. Rakshit et al. (eds.), *Nutrient Use Efficiency: from Basics to Advances*,
DOI 10.1007/978-81-322-2169-2_25, © Springer India 2015

Cu from soils. Considering that plant Cu demand is much lower than Fe demand, the amounts of Cu mobilized appeared sufficient to meet plant requirements for this metal. This suggests that PM release would be an advantage for plants growing on soils low in available Cu. PM release in response to deficiencies of micronutrient metals is not restricted to Fe and Zn, but might be more widespread than previously thought. The mechanism seems to be specific for Fe and Cu deficiencies, but not for Zn or Mn deficiency in crop. The release of PM under nutrient deficiency could have an ecological significance, regardless of whether it is indirectly caused by impaired metabolism or as a specific response mechanism. Also, it remains to be examined whether nutrient-deficiency-induced PM release is a general phenomenon in crop species.

Keywords

Phytometallophore • Zinc • Copper • Iron • Manganeese • Plants • Soil

1   Introduction

During the past five decades, the role of fertilizer in enhancing food grain production has been widely recognized in both the developed and developing countries. It has been highlighted that fertilizer is the kingpin of the green revolution and the best hope for meeting the food security challenges in future. Despite considerable progress made in the field of fertilizer use research, the recovery efficiency of applied fertilizer nutrients hardly exceeds 50 % and considerable amount is lost from the soil system. Though the consumption of chemical fertilizers in India increased steadily over the years, the use efficiency of nutrients applied through fertilizers continues to remain low (in the range of 30–50 % for N, 20 % for P, 55 % for K, and 2–5 % for Zn, Fe, and Cu) owing to nutrient losses from soils or conversions of nutrients into slowly cycling/recalcitrant pools within soil. Improvement in fertilizer use efficiency, therefore, is necessary to increase crop productivity and reduce environmental pollution. The nutrient use efficiency may be improved by regulating the supply from the fertilizer material and enhancing the uptake and utilization efficiencies by the plant. Efficient use of nutrients in agriculture may be defined differently when viewed from agronomic, economic, or environmental perspectives. Proper definition for the intended use is essential to understand published values and have meaningful discussion. It is a fallacy that the highest possible nutrient efficiencies should be the ultimate goal of fertilizer users. The highest "efficiency" occurs when small amounts of nutrients are applied in deficient soils. While efficiency may be very high in this condition, crop growth in this region is generally stunted and profitability is low, compared with the situation where balanced and appropriate nutrition is provided. Another example of inadequate understanding of "efficiency" is when an insufficient quantity of nutrients is regularly added to meet crop needs. In this condition, soil productivity will gradually decline as crop production continues to be increasingly reliant on nutrient stocks from soil reserves. Nominal nutrient use efficiency may be very high under these circumstances, but it is clearly a non-sustainable scenario.

2   Nutrient Use Efficiency

Nutrient efficiency can be classified into three groups, viz., agronomic efficiency, physiological efficiency, and apparent recovery efficiency. The agronomic efficiency is defined as this economic production obtained per unit of nutrient applied. It can be calculated with the help of following equation:

$$Agronomic\,efficiency = \frac{Grain\,yield\,of\,fertilized\,crop\,in\,kg - Grain\,yield\,of\,unfertilized\,crop\,in\,kg}{Quantity\,of\,fertilizer\,applied\,in\,kg}$$

$$= kg\,kg^{-1}$$

newly emerging concepts, mechanisms, and techniques in cellular and molecular biology need to be explored for better understanding of

Sometime, agronomic efficiency is also called economic efficiency. If the efficiency is determined under greenhouse conditions, the agronomic efficiency may be expressed in $g\,g^{-1}$ or $mg\,mg^{-1}$.

Physiological efficiency is defined as the biological production obtained per unit of nutrient observed. Sometimes, it is also known as biological efficiency or efficiency ratio. It can be calculated with the help of the following equation:

tolerance mechanisms to stress so that appropriate strategies could be developed for identification of crop genotypes with superior resource use efficiency. Integration of such crops into crop rotations in conjunction with improved management practices will play an increasingly important role in enhancing crop production especially under conditions of low availability of nutrients and other stresses. Significant progress has been made in understanding the physiological traits of several crop species responsible for tolerance to a

Physiological efficiency

$$= \frac{Total\,dry\,matter\,yield\,of\,fertilized\,crop\,in\,kg - Total\,dry\,matter\,yield\,of\,unfertilized\,crop\,in\,kg}{Nutrient\,uptake\,by\,fertilized\,crop\,in\,kg \qquad Nutrient\,uptake\,by\,unfertilized\,crop\,in\,kg}$$

$$= kg\,kg^{-1}$$

The apparent recovery efficiency is defined as the quantity of nutrient absorbed per unit of nutrient applied. It can be calculated with the help of the following expression:

wide range of nutrient deficiencies in soils. Preliminary studies have shown that inclusion of such promising crop species into crop rotations or to intercropping would be a useful strategy to

Apparent recovery efficiency

$$= \frac{Nutrient\,uptake\,by\,fertilized\,crop - Nutrient\,uptake\,by\,unfertilized\,crop}{Quantity\,of\,fertilizer\,applied} \times 100\%$$

Physiological and recovery efficiency can be combined to obtain the nutrient use efficiency:
Nutrient use efficiency = physiological efficiency × recovery efficiency

In addition to agricultural intensification on the best arable land, management practices need to be identified for rational utilization of marginal lands for agriculture for enhancing sustainable crop production in developing countries (Lal 2000). In these context possibilities of harnessing

improve nutrient use efficiency, crop nutrition, and yields (Hocking 2001).

2.1 What Is Phytometallophore?

In response to iron-deficiency stress conditions, roots of cereals (Poaceae family) release nonproteinogenic amino acids, called phytometallophores or phytosiderophores that solubilize and chelate inorganic Fe. In cereal, Fe (III) is taken up in toto as the Fe (III)-phytometallophore complex. Apparently, there are specific recognition sites on the plasma membrane of the root cells which allow the binding and transport of the metal-phytometallophore (PM) complex across the plasma membrane and into the cystosol. Absorption across the plasma membrane is thought to be via an amino acid cotransport system. In addition to their role in Fe acquisition, it has been hypothesized that PM has a universal role for acquisition of other trace metals such as Zn, Mn, and Cu that also have low solubilities in alkaline soils. Recent study reported that Zn (II)-PM complexes (i.e., 2-deoxymugineic acid, epi-hydroxymugineic acid, and mugineic acid) were readily absorbed by corn roots. These results suggest that corn root-cell plasma membrane binding sites are not highly specific for Fe (III)-phytometallophore, allowing the transport of other transition metals into cell. PM release may vary with the physiological status and age of the plant and with severity of the deficiency. PM release rates generally increase at first and afterwards decrease with plant age. Photoperiod and temperature also directly influenced by diurnal release patterns. PM release rates and their recovery may be influenced by nonspecific root-microbial interaction and by the physical environment of the roots such as aeration or root contact that may influence root morphology and exudation rates.

2.2   Biosynthesis of PM

In graminaceous species (strategy II), these iron-deficiency-induced morphological and physiological changes are absent. Instead, roots release phytosiderophores (PS) are chelators for Fe (III). The pathway of PS biosynthesis is understood reasonably well. L-Methionine is the dominant precursor, and three molecules of it are used to form one molecule of nicotianamine which, after deamination and hydroxylation, is converted into 2-deoxymugineic acid and further to PS, depending upon plant species. Nicotianamine (NA) is not only a precursor of PS biosynthesis

but is also a strong chelator of Fe (II), but not of Fe (III). It is also essential for the proper functioning of Fe (II)-dependent processes. Nicotianamine might be the link between the two strategies of iron-deficiency-induced root responses, perhaps reflecting differences in codon usage in genes of dicots in comparison with monocots. From DMA, the biosynthesis pathway may diverge in different plant species, resulting in different PS being exuded into the rhizosphore of different species. The PS is synthesized mainly in root tips, even though biosynthesis may occur in the meristematic tissue of the shoot as well. In either case, PS are synthesized continuously and are stored in roots for release into the rhizosphere during a defined period of the day. An increased exudation of PS under Fe deficiency occurs in a distinct diurnal rhythm, with a peak exudation after the onset of illumination, the light ensuring the continuous supply of assimilates from photosynthetically active plant parts. More detailed studies on the diurnal rhythm revealed that it is an increase in temperature during the light period, rather than the onset of light itself, which causes an increase in PS exudation.

2.3   Strategy I and Strategy II Plants

Classical studies of Fe nutrition in plants have resulted in the division of plants into various strategies.

Strategy I, dicotyledonous plants respond to iron stress by inducing a cell surface reduction system and in some instances can be accompanied by the release of protons and reductants that alter chemical conditions in the rhizosphere to increase inorganic Fe solubility. Strategy I is typically for dicots and non-graminaceous monocots and characterized by at least two distinct components of iron-deficiency response, increased reducing capacity, and enhanced net excretion of protons. In many instances also the release is enhanced of reducing and/or chelating compounds, mainly phenolics. These root responses are often related to changes in root morphology and anatomy, particularly in the formation of transfer cell-like structures in rhizodermal cells. The most sensitive and typical response is the increase in activity of a plasma membrane-bound reductase in the rhizodermal cells. The supposed existence of two reductases, a constitutive (basic) reductase with low capacity, and an iron-deficiency-induced high capacity reductase increases the activity Marschner et al. (1989). Although the transmembrane redox pump may contribute to the net excretion of protons, the strongly enhanced net excretion of protons under iron deficiency is most probably the result of higher activity of the plasma membrane proton efflux pump and not of the reductase. The activity of the reductase (R) which is strongly stimulated by low pH, i.e., enhanced proton excretion by the ATPase, is important for the efficiency in Fe (III) reduction. Accordingly, high concentrations of HCO^{-3} counteract this response system in strategy I plants.

Strategy II, monocotyledonous grasses, responds to Fe stress by production of Ps that are secreted into the rhizosphere and which are subsequently transported by a specific uptake system on the root surface. Strategy II is confined to graminaceous plant species and characterized by an iron-deficiency-induced enhanced release of non-proteinogenic amino acids, so-called phytosiderophores. The release follows a distinct diurnal rhythm and is rapidly depressed by resupply of iron. The diurnal rhythm in release of PS in iron-deficient plants is inversely related with the volume of a particular type of vesicles in the cytoplasm of root cortical cells. PS such as mugineic acid form highly stable complexes with Fe (III); the stability constant in water is in the order of 10^{23}. As a second component of strategy II, a highly specific constitutive transport system translocator is present in the plasma membrane of root cells of grasses which transfers the Fe (III) PS into the cytoplasm. In plant species with strategy I, this transport system is also lacking. Although PS form complexes also with other heavy metals such as Zn, Cu, and Mn, the translocator in the plasma membrane has only a low affinity to the corresponding complexes. Nevertheless, release of PS may indirectly enhance the uptake rate of these other metals by

increasing their mobility in the rhizosphere and in the root apoplasm. Under iron deficiency not only the release of phytosiderophores is increased but also the uptake rate of the Fe (III) PS complexes indicating a higher transport capacity due either to an increase in number of the turnover rate of the translocator. Although this PS system resembles features of the siderophore system in microorganism, its affinity to PS is two to three orders of magnitude higher than for siderophores such as ferrioxamine B or for synthetic iron chelates such as FEEDDHA.

3   Methodology

A rapid, simple, and accurate method of determining concentrations of Fe-chelating agents in solution was developed Shenker et al. (1995). The assay employs Cu-CAS (chrome azurol S) complex as a testing agent and measures the equilibrium concentration of this complex in the presence of other Fe chelators. This method is of particular importance for colorless chelates, which are very difficult to determine otherwise.

The Cu-CAS assay—a proposed protocol to quantify unknown chelate concentration.

1. Cu-CAS reagent preparation:

Prepare a stock solution of the Cu-CAS reagent consisting of: 200 µM $CuCl_2$, 210 µM CAS, 40 mM MES, pH adjusted to 5.7 with NaOH. This solution is stable for a long period (months).

2. Sampling of unknown chelate solution:

Prepare a series of six 1.5 mL microtubes for each chelate solution to be assayed. Add an accurate volume containing approximately 140 nmol of the unknown chelate to the first microtube and add distilled water to a final volume of 1.4 mL.

3. Preparation for dilution step:

Add 700 µL of distilled water to each of the other microtubes.

4. Constructing a 1:1 serial dilution of the tested chelate:

Prepare 1:1 serial dilution by transferring 700 µL from each microtube to the next. Ensure good mixing after each dilution step.

5. Ligand exchange color reaction and absorbance reading:

Add 700 µL of Cu-CAS assay solution to each microtube, including reference microtubes containing 700 µL of water, and read absorbance at 582 nm.

6. Plotting of results:

Plot the ratio of $Abs_{(sample)}/AbS_{(ref)}$ vs. the actual chelate solution volume in each microtube (similar to Fig. 7, where x axis is sample volume rather than concentration). Note that one-half of the initial volume added in step 2 is taken for the next microtube in the dilution step (4).

7. Calculations and interpretation:

Calculate chelate concentration according to sample volume at the intercept with the x axis at $y = 0$, the Cu-known concentration (100 µM), and the Cu/ligand ratio of the complex.

3.1   Diurnal Rhythm of Release of PS

The rate of PS release in the Fe-deficient plants showed a distinct diurnal rhythm with a maximum value about 4 h after the onset of the light period. In contract to the Fe-deficient plants, in the Fe-sufficient plants the release of PS was very low and nearly constant throughout the daytime. Among graminaceous plants examined, there are two patterns for the secretion of phytosiderophores. A distinct diurnal rhythm in secretion has been reported in barley (Takagi et al. 1984), wheat (Zhang et al. 1991), *Hordelymus europaeus* (Gries and Runge 1992), and *Festuca rubra* (Ma et al. 2003). In other species, such as maize (Yehuda et al. 1996) and rice (Inoue et al. 2009), there is no diurnal rhythm in the secretion. Species with diurnal secretion patterns have different times of peak secretion rate. For example, in barley and wheat, maximum secretion rates occurs 4 h after the onset of light period (Takagi et al. 1984; Zhang et al. 1991). On the other hand, the secretion peak occurred at 5.5 h in *Hordelymus europaeus* (Gries and Runge 1992) and between 2 and 5 h in *Festuca rubra*

(Ma et al. 2003). Since the growth conditions are different in these previous studies, it is difficult to conclude whether the differences in the secretion time results from the growth conditions or from species itself. In the present study, the secretion pattern between *P. pratensis* and *L. perenne* under the same growth conditions was compared. *P. pratensis* and *L. perenne* secrete different kinds of phytosiderophores in response to Fe deficiency; *P. pratensis* secretes DMA, AVA, and HAVA, while *L. perenne* secretes DMA, HDMA, and epiHDMA (Ueno et al. 2007). The amount of these phytosiderophore secreted is higher in *P. pratensis* than in *L. perenne*, but the secretion amount increased with the progression of Fe deficiency in both species (Ueno et al. 2007). Since the growth rate was different between the two species tested and the experiments were conducted in different seasons, we had to use plants of different ages and with different lengths of Fe-deficiency duration in order to obtain sufficient amounts of phytosiderophores for quantitative determination. However, *P. pratensis* and *L. perenne* showed distinct diurnal rhythms in the secretion of phytosiderophores irrespective of plant age and the duration of Fe deficiency, with a difference of 2–3 h in the secretion peak between the two species. Experimental results revealed that comparing ten cultivars of perennial grasses but with fewer collection times, the secretion pattern differed between the species, but not cultivars (Ueno et al. 2007). These results indicated that the secretion time differ consistently between *P. pratensis* and *L. perenne*. The diurnal rhythm in the secretion of phytosiderophore was suggested to be affected by both temperature and light (Ma et al. 2003; Reichman and Parker 2007). Earlier secretion was correlated with increased temperature; however, shading experiments revealed that phytosiderophore secretion is not triggered by light in both perennial grass species. Furthermore, the temperature of the rooting zone, but not the air temperature, controls secretion time. These results support the idea that the initiation of phytosiderophore secretion is triggered by the temperature around the roots. By using a square-wave light regime, Reichman and Parker (2007) concluded that the secretion of phytosiderophores in wheat is mainly mediated by changes in light rather than temperature. However, in their study, plants were shaded for a longer time, which may have affected phytosiderophore biosynthesis. Therefore, the lack of secretion of phytosiderophores under darkness may be the result of decreased synthesis. Another possibility is that phytosiderophore secretion time is controlled differently between wheat and perennial grasses. The mechanism responsible for diurnal rhythm of phytosiderophore secretion is still unknown. Recently, diurnal changes were reported in the expression of some genes involved in biosynthesis of phytosiderophores in rice (Nozoye et al. 2004) and uptake of Fe (III)-phytosiderophore complex in rice and barley (Inoue et al. 2009). Some elements associated to the diurnal change have been proposed to be present in the promoter region of these genes. Since the gene responsible for the secretion of phytosiderophore has not been cloned yet, it remains to be examined whether similar elements are involved in the diurnal rhythm of phytosiderophore secretion or the secretion is regulated independent of biosynthesis and uptake. Phytosiderophore secretion by grasses may increase Fe availability for coexisting species. Recently, it was reported that citrus can utilize Fe effectively from Fe(III)-phytosiderophore complex secreted from Poa (Cesco et al. 2006). Moreover, the combination of three perennial grasses, *F. rubra*, *L. perenne* and P. pratensis, has been shown to prevent Fe deficiency more effectively compared to single species. The secretion peak time of *F. rubra* is between those of *L. perenne* and *P. pratensis* (Ma et al. 2003). When these perennial grass species are grown together in an orchard, the combined effect of different secretion peak times may maintain Fe availability for orchard trees for longer period of time relative to single grass species.

4 Role of the PM in Acquisition of Different Nutrients in Plants

In addition to their role in Fe acquisition, it has been hypothesized that PS have a universal role for acquisition of other trace metals such as Zn, Mn, and Cu that also have low solubilities in alkaline soils. In support of this hypothesis, it has been shown that PS form stable chelates with Zn, Mn, and Cu and are effective in extracting these elements from calcareous soils. Production of PS has also been shown to be induced by Zn deficiency in wheat. However, the importance of PS as a general response to trace metal deficiencies remains uncertain. Presently there are no data on root chelator exudation for Poaceae species subjected to Cu and Mn deficiencies, and PS release rates under Zn deficiency reported by Zhang and coworkers were considerably lower than those typically observed with Fe-stressed wheat. MA enhanced the solubility of Fe (III) between pH 4 and 9, when added to nutrient solution MA strongly stimulated the uptake of Fe by "Fe-inefficient" rice seedlings. Commonly used chelating agents such as EDTA, EDDHA, and citrate, etc. had no stimulative effects. The MA-mediated Fe uptake proved to be dependent on metabolic energy. These results suggest the possibility of MA as a phytosiderophore for graminaceous plants.

4.1 Iron

Graminaceous species acquire Fe by releasing PS with a high binding affinity for Fe and by taking up ferreted PS through a specific transmembrane uptake system. An increased mobilization of Fe from a calcareous soil, even as far away from the root surface as 4 mm, demonstrated a high capacity of PS to mobilize Fe. The rate of PS exudation from roots (an average exudation rate of $2.9 \ nmol \ cm^{-1} \ root \ h^{-1}$) is possibility related to tolerance of different species and genotypes to Fe deficiency, Zuo et al. (2000) which formed a basis for screening plants for their relative Fe uptake efficiencies. In general, the broad trend seems to be that plants which release low levels of PS in response to Fe deficiency are adapted poorly to Fe-limiting soils. It has been well established that the undissociated Fe (III)-PS complex is taken up by corn and rice roots. The uptake of Fe (III)-PS was inhibited by metabolic inhibitors (DCCD or CCCP) and chilling, indicating that the transport of the Fe (III)-PS complex across the plasma membrane is an energy-dependent process. Aciksoz (2011) reported that root release of phytosiderophores (PSs) is an important step in iron (Fe) acquisition of grasses, and this adaptive reaction of plants is affected by various plant and environmental factors. The results show that the root release of PS, mobilization of Fe from $^{59}Fe(OH)_3$, and root uptake and shoot translocation of Fe(III)-PS by durum wheat are markedly affected by N nutritional status of plants. The complex formation properties of mugineic acid, which is a biologically important molecule for iron uptake, were studied using the density-functional methods combined with the IEF-PCM continuum solvation model. In particular, it has been found that the inclusion of explicit water molecules interacting with mugineic acid is a key factor for obtaining reliable computational results. The present computational results show that the metal coordination structure is somewhat different between the Fe II-mugineic acid and Fe III-mugineic acid complexes; the former has a five-coordinated structure while the latter has a nearly octahedral binding structure. Sugiura et al. have suggested that the reduction of the Fe III complex into the Fe II complex is an important first process in the iron release mechanism in organism's cell since the iron ion can be easily released from the weakly bonded ferrous complex [Fe II(HMA)] formed in the reduction process. The structural difference theoretically predicted in this work may play a role in this iron release mechanism although the detailed mechanism has not yet been understood at a molecular level. The characterization of phytosiderophore secretion patterns of perennial grasses provides important information for designing an optimal combination of species that

can be used for effective correction of Fe-induced chlorosis of crops grown on calcareous soils.

4.2 Zinc

Exudation of PM from roots increases under Zn deficiency in a range of plant species. However, an equivocal experimental proof that PM play a role in mobilization and uptake of Zn from Zn-deficient soils has yet to be reported. This is especially important because PM has a greater affinity for Fe than for Zn (e.g., DMA has a twofold higher stability constant for Fe than for Zn). Either Zn or Fe deficiency may stimulate production of PM in wheat. However, an increased release of PM under Zn deficiency might be due to an indirect effect, for example, as a response to impaired translocation of Fe from roots to shoots under Zn deficiency. Such an imbalance in Fe circulation in plants might cause hidden physiological Fe deficiency, resulting in the increased PM release. For wheat, recent cause study suggested a model in which PM, released across plasma membrane, mobilize Zn in the apoplasm of root cells, But dissociation of the Zn (II)-PM complex occurs at the plasma membrane, and only Zn is taken up into the cytoplasm. However, more recent research has indicated that not only splitting of Zn (II)-PM complex as the plasma membrane and uptake of ionic Zn (II) occur but that the Zn (II)-PM complex can also be taken up undissociated, at least by corn roots. In addition to exudation of PM, Zn deficiency increases root exudation of amino acids, sugars, and phenolics in a range of plant species, including wheat. The importance of this exudation has yet not been assessed in terms of increasing plant capacity to acquire Zn from soils with low Zn availability. Sorghum and wheat plants increased the release of phytosiderophore in response to Zn deficiency but corn did not. The total amount of phytosiderophore released by the roots was in the order wheat > sorghum > corn. The total Zn uptake by the species in this study decreased in the order corn > sorghum > wheat, which is inversely related to their tolerance to Zn deficiency. The absence of a "phytosiderophore" response to Zn deficiency in corn, coupled with the evidence that this species accumulates more Zn than wheat or sorghum, provides an explanation for why Zn deficiencies are more prevalent for corn than wheat or sorghum. Soils in many agricultural areas are alkaline and have high amounts of calcium carbonate, resulting in low availability of Zn (Welch et al. 1991). Wheat grown on such soils suffers from Zn deficiency, although tolerance to Zn deficiency largely varies among wheat genotypes (Khoshgoftar et al. 2006). Wheat genotypes differ in their mechanisms for improved root Zn uptake under Zn-deficient conditions (Hacisalihoglu and Kochian 2003). These differences in the ability of genotypes for root uptake means the available pools of Zn in soil may vary for different genotypes (Marschner 1995). Plant Zn uptake is influenced by root architecture, the presence of mycorrhizal fungi, proton exudation from roots, and release of phytosiderophores (Hacisalihoglu and Kochian 2003). Some genotypes are able to release phytosiderophores from roots to the surrounding rhizosphere soil, which increases the solubility of Zn and, as a result, its availability for plant uptake (Hacisalihoglu and Kochian 2003). To date, nine mugineic acids have been identified in different graminaceous plants (Ueno et al. 2007). Under Zn-deficiency stress, the rate of phytosiderophore release differs among and within cereal species (Cakmak et al. 1996). The well-known differences among durum and bread wheat genotypes in Zn efficiency are closely related to the differences in the rate of phytosiderophore release from roots (Cakmak et al. 1996). Soil salinity is frequently associated with alkaline soils, which are commonly deficient in available Zn (Khoshgoftar et al. 2006). Salinity may reduce Zn uptake by plant roots due to competition of other cations, e.g., Ca and Na, at the root surface (Marschner 1995). For all three of the wheat genotypes studied, salinity stress resulted in greater amounts of phytosiderophores exuded by the roots. In general, for Kavir, the greatest amount of phytosiderophores was exuded from the roots at the highest salinity level (120 mM NaCl). Greater phytosiderophore exudation under Zn-deficiency conditions was accompanied by greater Fe transport from root to shoot. The relationship between Fe transport to shoots and differential exudation of phytosiderophores by wheat genotypes has been proposed as a physiological mechanism behind differential genotypic tolerance to zinc deficiency (Rengel and Graham 1995). Under such circumstances, decreased transport of Fe toward leaves under Zn deficiency would result in physiological Fe deficiency in leaves that would trigger increased exudation of phytosiderophores into the rooting medium by genotypes tolerant to Zn deficiency. In contrast, genotypes sensitive to Zn deficiency would transport relatively large amounts of Fe to leaves, thus avoiding physiological Fe deficiency and lacking a trigger for increasing root exudation of phytosiderophores. Many vascular plant species are unable to colonize calcareous sites. Thus, the floristic composition of adjacent limestone and acid silicate soils varies greatly. The inability of calcifuge plants to establish in limestone sites could be related to a low capacity of such plants to solubilize and absorb Fe from these soils. Under Fe deficiency, species of Poaceae enhance their Fe uptake by releasing non-proteinogenic amino acids, phytosiderophores (PS), from their roots which mobilize Fe from the soil by forming a chelate that is then taken up by the root (Römheld and Marschner 1986). In previous research it was shown that calcicole grasses are better adapted to low Fe availability on calcareous sites, as a consequence of higher PS release rates and lower tissue Fe demand. Phytosiderophore release and uptake is thought to be specific for Fe deficiency. However, a universal role of phytosiderophores in the acquisition of micronutrient metals has been proposed (Crowley et al. 1987), since PS form stable chelates with Zn, Mn, and Cu (Nomoto et al. 1987; Murakami et al. 1989); they extract considerable amounts of Zn, Mn, and Cu from calcareous soils (e.g., Treeby et al. 1989), and deficiencies of Zn, Mn, and Cu are quite common on calcareous and non-calcareous soils. Enhanced PS release in Zn deficient wheat has been reported by Zhang et al. (1989, 1991). The effect of zinc nutritional status of the plant on the release of zinc mobilizing root exudates was studied in various dicotyledonous (apple, bean, cotton, sunflower, tomato) and graminaceous (barley, wheat) plant species grown in nutrient solutions. In all species, zinc deficiency increased root exudation of amino acids, sugars, and phenolics. However, the root exudates of zinc-deficient dicotyledonous species did not enhance zinc mobilization from a synthetic resin (Zn chelate), or a calcareous soil, although mobilization of iron from Fe^{III} hydroxide was increased. By contrast in the graminaceous species, root exudates from zinc-deficient plants greatly increased mobilization of both zinc and iron from the various sources. These differences in capability of mobilization of zinc and iron between the plant species are the result of an enhanced release of phytosiderophores with zinc deficiency in the graminaceous species.

Ptashnyk et al. (2011) reported that rice (*Oryza sativa* L.) secretes far smaller amounts of metals complexing phytosiderophores (PS) than other grasses. But there is increasing evidence that it relies on PS secretion for its zinc (Zn) uptake. After nitrogen, Zn deficiency is the most common nutrient disorder in rice, affecting up to 50 % of lowland rice soils globally. A mathematical model was developed of PS secretion from roots and resulting solubilization and uptake of Zn, allowing for root growth, diurnal variation in secretion, decomposition of the PS in the soil, and the transport and interaction of the PS and Zn in the soil.

4.3   Manganese

Environmentally controlled changes in redox potential occur when oxygen is depleted; NO_3^-, Mn, and Fe then serve as alternative electron acceptors for microbial respiration and are transformed into reduced ionic species. This process greatly increases the solubility and availability of Mn and Fe but is not under direct control of the plant. In some circumstances, such as in poorly aerated soils, this results in Mn and Fe toxicities to plants. Manganese availability may be further influenced by the activity of Mn-oxidizing and Mn-reducing bacteria that colonize plant roots. Since differential Mn efficiency can only be demonstrated for plants growing in soil, but not for those growing in the nutrient solution, it appears obvious that a change in the biology and/or chemistry of the rhizosphere precedes an increase in Mn availability to plants. However, the nature and activity of root exudates components that might be involved in mobilization on Mn is still unclear. Further research on root exudates effective in mobilizing Mn from the high-pH substrates for uptake by plants roots is warranted.

4.4   Copper

Phytosiderophore (PS) release in *H. europaeus* was rapidly induced in response to both Fe and Cu deficiencies. This is the first reported case of Cu-deficiency-induced PS release in grasses. Fe- and Cu-deficient plants were able to maintain release rates well above background levels even when growth was reduced to 6 or 31 % of the control, respectively. However, the plants in the metal-deficiency treatments progressed from normal growth to severely deficient in 30 day. For the induction of Cu deficiency, this is in agreement with Gries et al. (1995) but contrary to the findings of Bell et al. (1991), who suggested that Cu deficiency could only be obtained using a combined BPDS-HEDTA chelator-buffered system. Theoretically, other nonspecific chelators such as citric acid, which has some affinity for Cu and is present in root exudates of Fe- and Cu-deficient *H. europaeus*, could contribute to the Fe-mobilizing capacity of root exudates under Cu deficiency. The fact that almost the same constant ratio between the two assay methods was found for root exudates from both Fe-deficient and Cu-deficient plants suggests that the same chelators are produced under Cu deficiency as under Fe deficiency.

Several studies using wheat (*Triticum aestivum* and *Triticum durum*) have demonstrated that PS release is enhanced under Zn deficiency (Zhang et al. 1989, 1991) and can reach levels comparable to those of PS release by Fe-deficient barley (Cakmak et al. 1994). Theoretically, the same could be true for the response to Cu deficiency that was observed. The hypothesis that PS release is a physiological response to Cu deficiency is further supported by the fact that Cu-deficient plants were able to maintain high rates of PS exudation throughout the experiment even when growth was reduced to less than one-third of the control. Also, the diurnal pattern of PS release under Cu deficiency was identical to that known from Fe-deficient *H. europaeus* plants (Gries and Runge 1992). In combination, these findings suggest that PS release in response to Cu deficiency is a well-regulated mechanism. In barley, uptake rates of PS-complexed Cu are tenfold lower than those of the PS-Fe complex (Ma et al. 1993), suggesting that the PS system functions primarily for Fe transport. This preferential recognition of Fe-PS complexes remains to be examined for *H. europaeus*. Nonetheless, even tenfold lower uptake rates of Cu-PS could still be sufficient to meet plant Cu demand. Based on calculations of plant yield and tissue Cu concentration, the quantities of chelators released under Cu deficiency greatly exceed the Cu uptake rate required for normal growth. Studying metal extraction by PS from a wide range of calcareous and non-calcareous soils revealed that PS preferentially mobilized Fe but also significant quantities of Zn and Cu from soils. Considering that plant Cu demand is much lower than Fe demand, the amounts of Cu mobilized appeared sufficient to meet plant requirements for this metal. This suggests that PS release would be an advantage for plants growing on soils low in available Cu. Release of PS under Cu deficiency could have an ecological significance, regardless of whether it is indirectly caused by impaired metabolism or as a specific response mechanism. This question needs further study. Also, it remains to be examined whether Cu-deficiency-induced PS release is a general phenomenon in native grass species.

5 Intercropping and Phytosiderophore Release

Micronutrient deficiency in plants is becoming an increasingly important global problem. Proper metal transport and homeostasis are critical for the growth and development of plants and in order to potentially fortify plants preharvest. Also, improvement in Fe and Zn content in the edible portions of the plant will be helpful for alleviating human nutritional disorders worldwide (Welch and Graham 2002; Grotz and Guerinot 2006). reported that peanut intercropping with different gramineous species not only improved the iron nutrition of the peanut but also enhanced the Cu and Zinc content in the peanut shoot in the greenhouse experiment. Although, this was not statistically significant difference in the field experiment, the Cu and Zinc content in peanut shoot of intercropping showed a general increasing trend, which means that agronomic intercropping helps mobilize and uptake limiting nutrient elements as well as provides benefits through effects on plant growth, development, and adaptability to adverse environments. The possible reason for such differential effects on Cu and Zn concentrations of peanut plants caused by intercropping could be root exudates from gramineous species. Specifically, production and release from phytosiderophores of gramineous species may improve solubility of Fe, Zn, and Cu by chelation, which helps plants obtain those essential elements from the soil (Inal et al. 2007). The release of phytosiderophore by strategy II plants also improves Zn nutrition (Khalil et al. 1997). In the current study, Zn and Cu nutrition of peanut was improved by the associated maize, barley, oats, and wheat. Enhanced production of phytosiderophore by maize might be responsible for the increases in Zn and Cu concentration of peanut. On the other hand, one metal deficiency might cause an excess of another metal to be absorbed. We have learned from studies to date that the transporters involved in Fe uptake can transport a variety of divalent cations such as Zn and Cu. However, Mn concentration of the peanut shoot is significantly enhanced by serious iron deficiency of peanut in monocropping. The higher Mn concentration may be caused by the enhanced Mn uptake by the peanut roots due to increased reducing capacity of peanut roots in monoculture in conjunction with the appearance of Fe-deficiency chlorosis symptoms in young leaves. Peanut plants have similar uptake mechanisms of Fe and Mn that require the reducing capacity of the root to accumulate Mn^{2+} and Fe^{2+} ions in the rhizosphere soil. The peanut intercropped with strategy II plants could not only improve iron nutrition of peanut but also enhance other critical micronutrients, such as Zn and Cu, which are critical metals for the growth and development of plants. However, systemic mechanisms might be involved in adaptation to these nutrient stresses at the whole plant level. Reasonable intercropping system of nutrient-efficient species should be considered to prevent or mitigate iron and zinc deficiency of plants in agricultural practice.

6   Ecological and Soil Chemical Factors Affecting the Efficacy of PS

The mechanism by which plants acquire Fe from siderophores in soils is not yet known. Many indirect processes such as extracellular reduction, chelate degradation, or passive diffusion may also contribute to root uptake of Fe mobilized by PS. Soil chemical and ecological factors can also affect the efficacy of PS in soils. Among the most obvious soil factors are Fe mineral dissolution rates, exchange kinetics for Fe complexed by organic matter, and solution pH of Fe redox potential. Physical structure and soil moisture content influence the diffusion path of the Fe and are important in soil aeration. Soil clay and organic matter strongly adsorb PS. Important biological factors include plant growth rates which affect Fe demand and the consequent induction of responses to Fe stress.

7   Summary

The availability of nutrients in plants is determined by the type of soil, climate conditions, and crop species, and the cultivars within the species determine the availability of nutrients in plants. Those crop species or cultivars that have the ability to absorb large amounts of nutrients and convert them into useful dry matter on highly enriched soils, in which a less-efficient species of cultivars reaches a yield plateau, have been described as the nutrient-efficient species or cultivars. However, with recent economic developments and the large potential of infertile soils that are located in developing regions of the world, it has been realized that the most significant contribution to world food production must come from crops grown on soils with relatively low fertility. More emphasis is now being given to that plant species or cultivars that should produce more on soils having a low fertility. Plant species or cultivars that produce higher yields under low nutrient supply have evolved one or more of the characteristics like an efficient internal economy, which may result from efficient redistribution within the plant, or lower requirements at functional sites. Due to escalating cost of chemical fertilizers, the nutrient uptake and utilization in crop plants should be most efficient to cause reduction in the cost of production and in achieving a higher profit for the farmers,. To arrive at these objectives, it is important to understand nutrient use efficiency, the factor effecting it, and ways of enhancing it in modern crop production system without reducing the crop/yields. To improve the nutrient use efficiency by crop plants, one of the strategies is to screen out the plants' (high nutrient efficiency) group which can secret phytometallophores. Phytometallophores help the plants in greater absorption of the nutrients under sub-optional supply conditions. Phytometallophores (PM) are released in graminaceous species (Gramineae) under iron (Fe)- and zinc (Zn)-deficiency stress and are of great ecological significance for acquisition of Fe and presumably also of Zn. The potential for release of PS is much higher than reported up to now. Rapid microbial degradation during PM collection from nutrient solution-grown plants is the main cause of this underestimation. Due to spatial separation of PM release and microbial activity in the rhizosphere, a much slower degradation of PM can be assumed in soil-grown plants. Concentrations of PM up to molar levels have been calculated under non-sterile conditions in the rhizosphere of Fe-deficient barley plants. Besides Fe, PM mobilize also Zn, Mn, and Cu. Despite this unspecific mobilization, PS mobilizes appreciable amounts of Fe in calcareous soils and arc of significance for chlorosis resistance of graminaceous species. In most species the rate of PS release is high enough to satisfy the Fe demand for optimal growth on calcareous soils. In contrast to the chelates, ZnPM and MnPM, FePM are preferentially taken up in comparison with other soluble Fe compounds. In addition, the specific uptake system for FePM (translocator) is regulated exclusively by the Fe nutritional status.

References

Aciksoz SB, Ozturk L, Gokmen OO, Römheld V, Cakmak I (2011) Effect of nitrogen on root release of phytosiderophores and root uptake of Fe(III)-phytosiderophore in Fe-deficient wheat plants. Physiol Plant 142:287–296

Bell PB, Angle JS, Chaney RL (1991) Free metal activity and total metal concentration as indices of micronutrient availability to barley [Hordeum vulgare (L.) cv. 'Klages']. Plant Soil 130:51–62

Cakmak I, Gulut K, Marschner H, Graham RD (1994) Effect of zinc and iron deficiency on phytosiderophore release in wheat genotypes differing in zinc efficiency. J Plant Nutr 17:1–17

Cakmak I, Sari N, Marschner H, Ekiz H, Kalayci M, Yilmaz A et al (1996) Release of phytosiderophores in bread and durum wheat genotypes differing in zinc efficiency. Plant Soil 180:183–189

Cesco S, Rombola AD, Tagliavini M, Varanini Z, Pinton R (2006) Phytosiderophores released by graminaceous species promote ^{59}Fe-uptake in citrus. Plant Soil 287:223–233

Crowley DE, Reid CPP, Szaniszlo PT (1987) Microbial siderophores as iron sources for plants. In: Winkelmann G, van der Helm D, Neilands JB (eds)

Iron transport in microbes, plants, and animals. VCH, Weinheim, pp 375–386

Gries D, Runge M (1992) The ecological significance of iron mobilization in wild grasses. J Plant Nutr 15:1727–1737

Gries D, Runge M (1995) Responses of calcicole and calcifuge Poaceae species to iron-limiting conditions. Bot Acta 108:482–489

Grotz N, Guerinot ML (2006) Molecular aspects of Cu, Fe and Zn homeostasis in plants. Biochim Biophys Acta 1763:595–608

Hacisalihoglu G, Kochian LV (2003) How do some plants tolerate low levels of soil zinc? Mechanisms of zinc efficiency in crop plants. New Phytol 159:341–350

Hocking P (2001) Organic acids exuded from roots in phosphorus uptake and aluminium tolerance of plants in acid soils. Adv Agron 74:63–97

Inoue H, Kobayashi T, Nozoye T, Takahashi M, Kakei Y, Suzuki K, Nakazono M, Nakanishi H, Mori S, Nishizawa NK (2009) Rice OsYSL15 is an iron-regulated iron(III)-deoxymugineic acid transporter expressed in the roots and is essential for iron uptake in early growth of the seedlings. J Biol Chem 284:3470–3479

Khalil KW, Hagagg LF, Awad F (1997) Response of guava seedlings to phytosiderophores from cereal species and availability of iron and zinc in calcareous soil. In: The 9th international symposium on iron nutrition and interactions in plants, Stuttgart, p 117

Khoshgoftar AH, Shariatmadari H, Karimian N (2006) Responses of wheat genotypes to zinc fertilization under saline soil conditions. J Plant Nutr 27:1–14

Lal R (2000) Soil management in the developing countries. Soil Sci 165:57–72

Ma JF, Kusano G, Kimura S, Nomoto K (1993) Specific recognition of mugineic acid-ferric complex by barley roots. Phytochemistry 34:599–603

Ma JF, Ueno H, Ueno D, Rombolà AD, Iwashita T (2003) Characterization of phytosiderophore secretion under Fe deficiency stress in Festuca rubra. Plant Soil 256:131–137. doi:10.1023/A:1026285813248

Marschner H (1995) Mineral nutrition of higher plants. Academic, London

Marschner H, Treeby M, Römheld V (1989) Role of root-induced changes in the rhizosphere for iron acquisition in higher plants. Z Pflanzenernaehr Bodenkd 152:197–204

Murakami T, Ise K, Hayakawa M, Kamei S, Takagi S (1989) Stabilities of metal complexes of mugineic acids and their specific affinities for iron (III). Chem Lett 12:2137–2140

Nomoto K, Sugiura Y, Takagi S (1987) Mugineic acids, studies on phytosiderophores. In: Winkelmann G, van der Helm D, Neilands JB (eds) Iron transport in microbes, plants and animals. VCH, Weinheim, pp 401–425

Nozoye T, Itai RN, Nagasaka S, Takahashi M, Nakanishi H, Mori S, Nishizawa NK (2004) Diurnal changes in the expression of genes that participate in phytosiderophore synthesis in rice. Soil Sci Plant Nutr 50:1125–1131

Ptashnyk M, Roose T, Jones DL, Kirk GJD (2011) Enhanced zinc uptake by rice through phytosiderophore secretion: a modelling study. Plant Cell Environ 34(12):2038–2046

Reichman SM, Parker DR (2007) Probing the effects of light and temperature on diurnal rhythms of phytosiderophore release in wheat. New Phytol 174:101–108

Rengel Z, Graham RD (1995) Importance of seed Zn content for wheat growth on Zndeficient soil. 2. Grain yield. Plant Soil 173:267–274

Römheld V, Marschner H (1986) Evidence for a specific uptake system for iron phytosiderophores in roots of grasses. Plant Physiol 80:175–180

Shenker M, Hadar Y, Chen Y (1995) Rapid method for accurate determination of colorless siderophores and synthetic chelates. Soil Sci Soc Am J 59:1612–1618

Takagi S, Nomoto K, Takemoto T (1984) Physiological aspect of mugineic acid, a possible phytosiderophore of gramineous plants. J Plant Nutr 7:469–477

Treeby M, Marschner H, Romheld V (1989) Mobilization of iron and other micronutrient cations from a calcareous soil by plant-borne, microbial, and synthetic chelators. Plant Soil 114:312–315

Ueno D, Rombolá A, Iwashita T, Nomoto K, Ma JF (2007) Identification of two novel phytosiderophores secreted from perennial grasses. New Phytol 174:304–310. doi:10.1111/j.1469-8137.2007.02056.x

Welch RM, Webb MJ, Loneragan JF (1982) Zinc in membrane function and its role in phosphorus toxicity. In: Scaife A (ed) Plant nutrition 1982. Proceedings of the ninth international plant nutrition colloquium, Warwick University, England, 22–27 August 1982. Commonwealth Agricultural Bureaux, Farnham House, Farnham Royal, pp 710–715

Welch RM, Graham RD (2002) Breeding crops for enhanced micronutrient. Plant Soil 245:205–214

Yehuda Z, Shenker M, Römheld V, Marschner H, Hadar Y, Chen Y (1996) The role of ligand exchange in the uptake of iron from microbial siderophores by gramineous plants. Plant Physiol 112:1273–1280

Zhang F.S., Römheld V, Marschner H (1989) Effect of zinc deficiency in wheat on the release of zinc and iron mobilizing root exudates. Zeitschrift fu X r Pflanzenerna X hrung und Bodenkunde 152:205–210

Zhang F.S., Römheld V, Marschner H (1991) Release of zinc mobilizing root exudates in different plant species as affected by zinc nutritional status. J Plant Nutr 14:675–686

Zuo YM, Zhang FS, Li XL, Cao YP (2000) Studies on the improvement in iron nutrition of peanut by intercropping maize on a careous soil. Plant Soil 220:13–25

Index

A

Abiotic stress, 2, 10, 128, 131, 141, 142, 165, 175, 176, 184, 186, 189, 194–196, 198–201, 225, 275, 286, 393
ACC. *See* 1-Aminocyclopropane-1-carboxylate (ACC)
Acinetobacter, 95, 96
Aerobacter aerogenes, 105
Agrobacterium rubi, 186
Agronomic efficiency (AE), 4, 7, 17, 18, 23, 50–52, 54, 55, 200, 290, 291, 301, 323–325, 402, 403
Agronomic fortification, 241
Agrophysiological efficiency (APE), 4, 7, 290, 291
Alum, 31, 36–37
Amendment, 2, 8, 29–41, 62, 77, 88–90, 109, 175, 295, 333, 344, 345, 368
AMF. *See* Arabuscular mycorrhizal fungi (AMF)
2-Amino-4-chloro 6 methyl pyrimidine (AM), 9, 111, 112
1-Aminocyclopropane-1-carboxylate (ACC), 143, 165, 186, 199, 200
3-Amino-1,2,4 triazole (ATC), 111, 113
Anabaena, 147–149, 167, 168
Antinutritional, 267–269, 274–276
APE. *See* Agrophysiological efficiency (APE)
Apparent recovery efficiency (ARE), 4, 18, 325, 402, 403
Arabidopsis, 67, 127, 146, 198, 199, 223, 224, 226–230, 258
Arabuscular mycorrhizal fungi (AMF), 11, 96, 128, 165, 166
Arbuscular mycorrhizal (AM), 64, 128–133, 135
ARE. *See* Apparent recovery efficiency (ARE)
Arthrobacter sp., 105, 143
Aspergillus
    A. awarmori, 37, 145
    A. flavus, 105
ATC. *See* 3-Amino-1,2,4 triazole (ATC)
Aulosira, 147, 149, 168
Azospirillum lipoferum, 186
Azotobacter chroococcum, 37, 145, 186, 334
Azotobacter sp., 6, 105, 143, 146, 149, 168, 175, 200, 307, 324, 345, 347–351

B

Bacillus
    B. cereus, 185
    B. megaterium, 33, 105, 186
    B. polymixa, 33, 131
    B. subtilis, 34, 185
Balanced fertilization, 21, 22, 323, 332–333, 336, 338, 346, 363, 368
Best management practice (BMP), 8, 20, 24–26, 286, 364
Bioavailability, 8, 9, 75, 81, 82, 88, 90, 145, 216, 238–241, 261, 267–271, 275–276, 346
Biofertilizer, 78, 167–169, 176, 177, 183, 184, 194, 307, 309, 323–325, 333–335, 344, 345, 347–351
Biofortification, 95, 237–248, 263, 276
Biomass partitioning, 385–387
Biopriming, 181–189, 193–202
Biotic stress, 2, 10–11, 165, 175, 184, 186, 189, 195, 196, 225, 286, 376, 393
BMP. *See* Best management practice (BMP)
Brachiaria humidicola, 118
Bradyrhizobium japonicum, 392
Brevundimonas sp., 167
Burkholderia, 95, 132, 143, 144, 186, 392

C

Calcite (CaCO$_3$), 31, 34–37, 39, 40, 169, 307
Calothrix, 147, 149, 167, 168
1-Carbamoyl-3-methylpyrazole (CMP), 112
Clinoptilolite, 31, 34
Clonostachys rosea, 186
Conservation furrow (CF), 371, 372
Conservation tillage (CT), 49, 56, 66, 293, 321, 338
Controlled release fertilizer (CRF), 9, 21, 292, 336, 337
Cover crops, 9, 23, 65, 167, 293–294, 338
COX1, 218
CRF. *See* Controlled release fertilizer (CRF)
Critical tissue nutrient concentration, 382, 391–392
Crop rotation, 9, 11, 22, 54, 65, 293, 301, 323, 330, 338, 341, 403
Cyanobacteria, 127, 143, 147–149, 163–169

D

Denitrification, 9, 19, 25, 46, 47, 49, 57, 104, 105, 107, 108, 111, 120, 143, 212, 222, 286, 291, 300, 303, 336, 347, 392
Dicyandiamide cyanoguanidine (DCD), 9, 21, 111, 112, 114, 117, 292
Dolomite, 31, 34–35

A. Rakshit et al. (eds.), *Nutrient Use Efficiency: from Basics to Advances*,   415
DOI 10.1007/978-81-322-2169-2, © Springer India 2015

Printed by Printforce, the Netherlands